VISUALIZING
ANATOMY & PHYSIOLOGY

THE WORLD . . .

The National Geographic Society (NGS) has been inspiring people to care about the planet since 1888. NGS photographers and cartographers study the world and record it *visually,* making their images and maps an ideal resource to help immerse students in the world of Human Geography.

IN YOUR HANDS,

Developed in partnership with the National Geographic Society, *Visualizing Anatomy & Physiology* integrates photos, maps, illustrations, and video with clear and concise text, to deliver an engaging learning experience. NGS verifies every fact in the book with two outside sources, ensuring accuracy, currency, and effective learning.

TODAY!

A portion of the proceeds of *Visualizing Anatomy & Physiology* help further the mission of National Geographic: to increase global understanding through education, exploration, research, and conservation.

VISUALIZING
ANATOMY &
PHYSIOLOGY

CRAIG C. FREUDENRICH

GERARD J. TORTORA

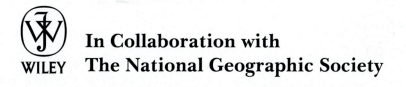

WILEY

In Collaboration with
The National Geographic Society

VICE PRESIDENT AND EXECUTIVE PUBLISHER Kaye Pace
EXECUTIVE EDITOR Bonnie Roesch
DIRECTOR OF DEVELOPMENT Barbara Heaney
MANAGER, PRODUCT DEVELOPMENT Nancy Perry
SENIOR DEVELOPMENT EDITOR Mary O'Sullivan
FREELANCE DEVELOPMENTAL EDITOR Karen Trost
PROJECT EDITOR Lorraina Raccuia
WILEY VISUALIZING PROJECT EDITOR Beth Tripmacher
WILEY VISUALIZING SENIOR EDITORIAL ASSISTANT Tiara Kelly
EDITORIAL ASSISTANTS Marcus Van Harpen, Darnell Sessoms, Brittany Cheetham,
    Lauren Morris
ASSOCIATE DIRECTOR, MARKETING Jeffrey Rucker
SENIOR MARKETING MANAGER Clay Stone
SENIOR PRODUCTION MANAGER Micheline Frederick
SENIOR MEDIA EDITOR Linda Muriello
MEDIA PRODUCTION SPECIALIST Svetlana Barskaya
CREATIVE DIRECTOR Harry Nolan
COVER DESIGN Harry Nolan
INTERIOR DESIGN Jim O'Shea
PHOTO MANAGER Hilary Newman
PHOTO RESEARCHERS Stacy Gold/National Geographic Society
SENIOR ILLUSTRATION EDITOR Sandra Rigby
PRODUCTION SERVICES Furino Production

COVER CREDITS
(Center image): John Burcham/NG Image Collection
(Bottom inset photos, left to right): David Evans/NG Image Collection; MedicalRF.com/
Getty Images, Inc.; Amy White and Al Petteway/NG Image Collection; Anne Keiser/NG Image
Collection; Robert Clark/NG Image Collection
(Back cover inset photo): Anne Keiser/NG Image Collection

This book was set in New Baskerville by Precision Graphics, printed and bound
by Quad/Graphics. The cover was printed by Quad/Graphics.

ISBN-13: 9780470491249
BRV ISBN: 9780470917763

Printed in the United States of America
10 9 8 7 6 5 4 3 2 1

Preface

How is Wiley *Visualizing Anatomy and Physiology* different?

Wiley *Visualizing Anatomy and Physiology* differs from competing textbooks by uniquely combining three powerful elements: a visual pedagogy integrated with comprehensive text, the use of authentic situations and issues from the National Geographic Society collections, and the inclusion of interactive multimedia in the *WileyPLUS* learning environment. Together these elements deliver a level of rigor, as each key concept and its supporting details have been analyzed and carefully crafted to maximize and enhance student learning and engagement.

(1) Visual Pedagogy. Wiley Visualizing is based on decades of research on the use of visuals in learning (Mayer, 2005). Using the Cognitive Theory of Multimedia Learning, which is backed up by hundreds of empirical research studies, Wiley's authors select visualizations for their texts that specifically support students' thinking and learning—for example, the selection of relevant materials, the organization of the new information, or the integration of the new knowledge with prior knowledge. Visuals and text are conceived and planned together in ways that clarify and reinforce major concepts while allowing students to understand the details. This commitment to distinctive and consistent visual pedagogy sets Wiley Visualizing apart from other textbooks.

(2) Authentic Situations and Problems. Through Wiley's exclusive publishing partnership with National Geographic, Visualizing has benefited from National Geographic's more than century-long recording of the world. Accompanying this text is a great selection of videos from the National Geographic Society collections. These authentic materials, which immerse the student in real-life issues in human anatomy and physiology, enhance motivation, learning, and retention (Donovan & Bransford, 2005). These high-quality videos from the National Geographic Society collections are unique to Wiley *Visualizing Anatomy and Physiology*.

(3) Interactive Multimedia. Texts in Wiley Visualizing are based on the understanding that learning is an active process of knowledge construction. *Visualizing Anatomy and Physiology* is therefore tightly integrated with *WileyPLUS,* an online learning environment that provides interactive multimedia activities in which learners can actively engage with the materials. The combination of textbook and *WileyPLUS* provides learners with multiple entry points to the content, giving them greater opportunity to explore concepts, interact with the material, and assess their understanding as they progress through the course. Wiley Visualizing makes this online *WileyPLUS* component a key element of the learning and problem-solving experience.

Wiley Visualizing and the *WileyPLUS* Learning Environment are designed as natural extensions of how we learn

Visuals, comprehensive text coverage, and learning aids are integrated to display facts, concepts, processes, and principles more effectively than words alone. To understand the effectiveness of the visualizing approach, it is first helpful to understand how we learn.

1. Our brains process information using two main channels: visual and verbal. Our *working memory* holds information that our minds process as we learn. Using working memory, we begin to make sense of words and pictures and build verbal and visual models of the information.

2. When the verbal and visual models of corresponding information are integrated in working memory, we form more comprehensive, lasting mental models.

3. When we link these integrated mental models to our prior knowledge, which is stored in our *long-term memory,* we build even stronger mental models. When an integrated (visual plus verbal) mental model is formed and stored in long-term memory, real learning begins.

The effort our brains put forth to make sense of instructional information is called *cognitive load*. There are two kinds of cognitive load: productive cognitive load, when we're engaged in learning or exert positive effort to create mental models; and unproductive cognitive load, which occurs when the brain is trying to make sense of needlessly complex content or when information is not presented well. The learning process can be impaired when the amount of information to be processed exceeds the capacity of working memory. Well-designed visuals and text, along with effective pedagogical guidance, can reduce the unproductive cognitive load in working memory.

Wiley Visuals and Interactive Media are designed for engaging and effective learning

The figures in *Visualizing Anatomy and Physiology* are specifically designed to accomplish three tasks: present complex processes in clear steps and with clear representations, organize related pieces of information, and integrate related information with one another. This helps solve the problem of cognitive overload. This approach, along with the use of interactive multimedia, also provides the level of rigor needed for the course and helps students engage with the content.

Research shows that well-designed visuals can improve the efficacy with which a learner processes information. SEG Research, an independent research firm, conducted a national, multisite study evaluating the effectiveness of the Wiley Visualizing series. Its findings indicate that students using Wiley Visualizing products (both print and multimedia) were more engaged in the course, exhibited greater retention throughout the course, and made significantly greater gains in content area knowledge and skills, as compared to students in similar classes that did not use Wiley Visualizing products.[1]

The use of *WileyPLUS* can also enhance learning. According to a white paper entitled "Leveraging Blended Learning for More Effective Course Management and Enhanced Student Outcomes" by Peggy Wyllie of Evince Market Research & Communications,[2] studies show that effective use of online resources can increase learning outcomes. Pairing supportive online resources with face-to-face instruction can help students to learn and reflect on material, and deploying multimodal learning methods can help students to engage with the material and retain their acquired knowledge. *WileyPLUS* provides students with an environment that stimulates active learning and enables them to optimize the time they spend on their coursework. Continual assessment/remediation is also key to helping students stay on track. The *WileyPLUS* system facilitates instructors' course planning, organization, and delivery and provides a range of flexible tools for easy design and deployment of activities and tracking of student progress for each learning objective.

[1]SEG Research (2009). Improving Student-Learning with Graphically-Enhanced Textbooks: A Study of the Effectiveness of the Wiley Visualizing Series. Available online at www.segmeasurement.com/

[2]Peggy Wyllie (2009). Leveraging Blended Learning for More Effective Course Management and Enhanced Student Outcomes. Available online at http://catalog.WileyPLUS.com/about/instructors/whitepaper.html

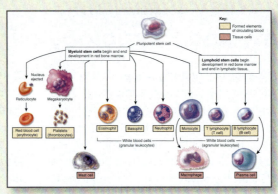

Figure 1: Origin and development of blood cells (Fig. 10.4) Graphic features such as arrows and labels, connect visual information and so direct learner's attention to the underlying concept of the developmental process blood cells undergo.

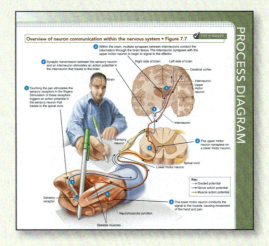

Figure 2: Overview of neuron communication in the nervous system (Fig. 7.7) This Process Diagram visualizes for students the complex processes involved in nerve communication, clearly identifying the sequence steps that enable us to perform routine tasks, such as using a pen to write.

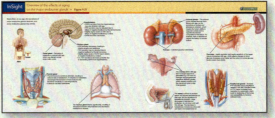

Figure 3: Overview of the effects of aging on the endocrine glands (Fig. 9.21) InSight features explore a major concept or topic in detail through illustrations and photos with accompanying labels and captions that aid student understanding.

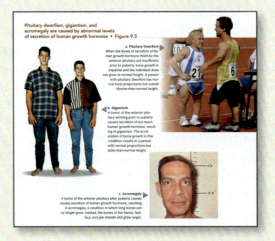

Figure 4: Pituitary dwarfism, gigantism, and acromegaly (Fig. 9.5) Textual and visual elements are physically integrated. This eliminates split attention (when we must divide our attention between several sources of different information).

How is each chapter of Wiley *Visualizing Anatomy and Physiology* organized?

Student engagement requires more than just providing visuals, text, and interactivity—it entails motivating students to learn. Student engagement can be behavioral, cognitive, social, and/or emotional. It is easy for a student to get bored or lose focus when presented with large amounts of information, and to lose motivation when the relevance of the information is unclear. Wiley Visualizing and *WileyPLUS* work together to organize course content into manageable learning objectives and relate it to everyday life. The design of *WileyPLUS* is based on cognitive science, instructional design, and extensive research into user experience. It transforms learning into an interactive, engaging, and outcomes-oriented experience for students.

The content in Wiley *Visualizing Anatomy and Physiology* and *WileyPLUS* is organized in learning modules. Each module has a clear instructional objective, one or more examples, and an opportunity for assessment. These modules are the building blocks of Wiley *Visualizing Anatomy and Physiology*.

Each chapter engages students from the start

Chapter opening text and visuals introduce the subject and connect the student with the material that follows.

Chapter Introductions illustrate key concepts in the chapter with intriguing stories and striking photographs.

Chapter outlines anticipate the content.

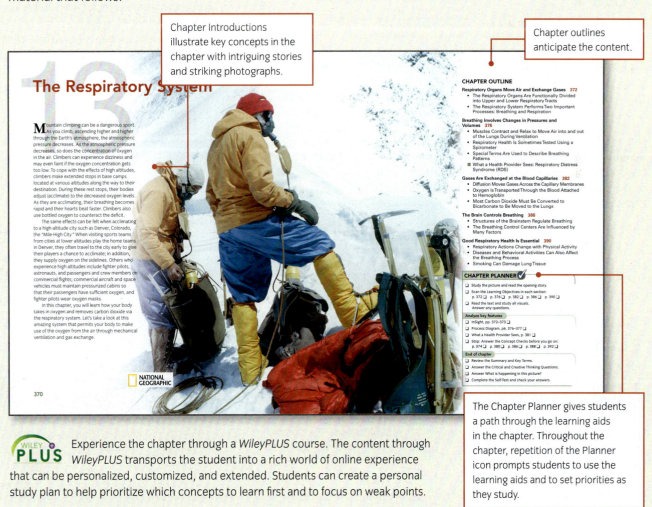

The Chapter Planner gives students a path through the learning aids in the chapter. Throughout the chapter, repetition of the Planner icon prompts students to use the learning aids and to set priorities as they study.

Experience the chapter through a *WileyPLUS* course. The content through *WileyPLUS* transports the student into a rich world of online experience that can be personalized, customized, and extended. Students can create a personal study plan to help prioritize which concepts to learn first and to focus on weak points.

Wiley Visualizing media guides students through the chapter

The content of Wiley *Visualizing Anatomy and Physiology* in *WileyPLUS* gives students a variety of approaches—visuals, words, illustrations, interactions, and assessments—that work together to provide students with a guided path through the content. But this path isn't static: It can be personalized, customized, and extended to suit individual needs, and so it offers students flexibility as to how they want to study and learn the content.

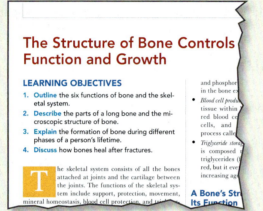

PROCESS DIAGRAM

Impulse transmission at the neuromuscular junction • Figure 6.5

Learning Objectives at the start of each section outline the concepts that students are expected to master while reading the section.

WILEY PLUS Every content resource is related to a specific learning objective so that students will easily discover relevant content organized in a more meaningful way.

The Structure of Bone Controls Function and Growth

LEARNING OBJECTIVES

1. **Outline** the six functions of bone and the skeletal system.
2. **Describe** the parts of a long bone and the microscopic structure of bone.
3. **Explain** the formation of bone during different phases of a person's lifetime.
4. **Discuss** how bones heal after fractures.

The skeletal system consists of all the bones attached at joints and the cartilage between the joints. The functions of the skeletal system include support, protection, movement, mineral homeostasis, blood cell protection, and tri...

Process Diagrams provide in-depth coverage of processes correlated with clear, step-by-step narrative, enabling students to grasp important topics with less effort.

WILEY PLUS There are many visual resources within *WileyPLUS* that provide additional examples of difficult concepts or processes. These can help students master the topic being studied and fully engage with the content. Examples closely integrated with the reading material include narrated animations for all major physiological topics and engaging video clips. Accompanying assignments for this rich media can be graded online and added to the instructor gradebook.

NATIONAL GEOGRAPHIC Video

InSights are multipart visual features that focus on a key concept or topic in the chapter, exploring it in detail or in broader context using a combination of photos, photomicrographs, diagrams, and data.

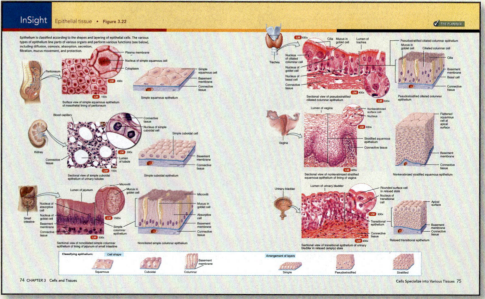

InSight Epithelial tissue • Figure 3.22

74 CHAPTER 3 Cells and Tissues   Cells Specialize into Various Tissues 75

WILEY PLUS In *WileyPLUS*, students can strengthen their grasp of the many anatomical structures highlighted in Insight figures by engaging with *Anatomy Drill and Practice* exercises. These simple-to-use drag-and-drop or fill-in-the-blank labeling exercises let students review illustrations from the text, cadaver photographs, histology micrographs, or lab models.

WHAT A HEALTH PROVIDER SEES ✓ THE PLANNER

Pain Management

Health providers vary their treatment of pain, depending on the type of pain being experienced. Most pain can be managed with medications (analgesics), which act at different places in the somatosensory pathway:

- When an injury causes pain, damaged cells or immune cells release certain products (such as bradykinin, prostaglandins, and histamines) that activate nearby nociceptors. *Anti-inflammatory analgesics*, such as aspirin, ibuprofen, acetaminophen, or naproxen, interfere with the enzymes that make these products.

- A class of neurotransmitters called opioids (such as endorphin, dynorphin, and enkephalin) is responsible for synaptic transmission in CNS pain pathways. At the synapse, *opioid analgesics* (such as morphine, meperidine, oxycodone, and codeine) bind to the opioid receptors but do not activate them. Health providers use opioid analgesics to treat high levels of pain. Opioid analgesics can become addictive and are easily overdosed; therefore, health providers must closely monitor their use.

- Pain relievers called *adjuvant analgesics (co-analgesics)* are used to treat other conditions and also help relieve pain. For example, *anti-epileptic drugs* (such as phenytoin) reduce the ability of CNS neurons to conduct action potentials. *Tricyclic antidepressants* (such as amitriptyline) block synaptic transmission involving the neurotransmitter serotonin. *Anesthetics* (such as lidocaine and benzocaine) block sodium and potassium channels, thereby preventing propagation of action potentials; most are applied topically or injected locally to relieve pain.

Other therapies can also be used to manage chronic pain without the use of medications or in conjunction with smaller doses of pain medications:

- Physical therapy involves a series of exercises, massage, thermal stimulation, and electrotherapy to release muscle tension and build muscle strength that can help to relieve pain caused by pressure on peripheral nerves.

- Transcutaneous electric nerve stimulator (TENS) units produce electrical pulses via the skin to stimulate the release of endorphins and enkephalins within the CNS. These neurotransmitters help to block pain signals from reaching the brain.

WILEY PLUS Video

NATIONAL GEOGRAPHIC

- Acupuncture and acupressure techniques involve the insertion of needles or application of pressure to strategic parts of the body, which helps to relieve pain in other parts of the body. The areas stimulated by the techniques share the spinal segments with the area that will ultimately be "treated" by the procedure.

- Biofeedback can sometimes be used to help patients manage chronic pain by helping individuals learn to control the impact of the pain on their daily lives.

Think Critically **1.** Migraine headaches are thought to involve areas of the brain where the primary neurotransmitter is serotonin. Which pain medication might work best for this condition?
2. Jenny is 10 years old and has a cold. She fell from a step and twisted her ankle, and her ankle is beginning to swell. Pediatricians do not recommend giving children aspirin as it has been implicated in a deadly condition called Reye's syndrome. What medication might be best to give Jenny for the pain?

What a Health Provider Sees highlights a concept or phenomenon and examines it from a clinical point of view. Photos and figures are used to compare how a nonscientist and a scientist see the issues, and students apply their observational and critical thinking skills to answer questions.

WILEY PLUS Numerous additional clinical applications are presented as examples in *WileyPLUS* content modules for each chapter. These engaging discussions of a wide variety of clinical scenarios from disease coverage to tests and procedures fully engage students in the material and help them comprehend the relevance of understanding normal anatomy and physiology.

Think Critically questions encourage students to analyze the material and develop insights into essential concepts.

Coordinated with the section-opening **Learning Objectives**, at the end of each section **Concept Check** questions allow students to test their comprehension of the learning objectives.

CONCEPT CHECK STOP

1. **What** are the functions of bone tissue and the skeletal system?

2. **How** does compact bone differ from spongy bone?

3. **What** are the stages of intramembranous ossification?

4. **How** does an open, complete fracture differ from a closed, comminuted one?

WILEY PLUS At the end of each learning objective module, students can assess their progress with independent practice opportunities and quizzes. This feature gives them the ability to gauge their comprehension and grasp of the material. Practice tests and quizzes help students self-monitor and prepare for graded course assessments.

Student understanding is assessed at different levels

Wiley *Visualizing Anatomy and Physiology* with *WileyPLUS* offers students lots of practice material for assessing their understanding of each study objective. Students know exactly what they are getting out of each study session through immediate feedback and coaching.

The illustrated **Summary** revisits each learning objective, with informative images taken from each module in the chapter. These visual clues reinforce important concepts.

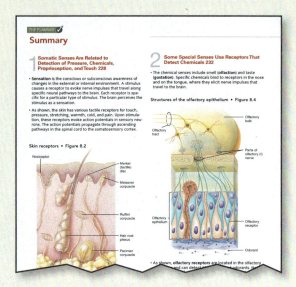

THE PLANNER ✓
Summary

1 Somatic Senses Are Related to Detection of Pressure, Chemicals, Proprioception, and Touch 228

- **Sensation** is the conscious or subconscious awareness of changes in the external or internal environment. A stimulus causes a receptor to evoke nerve impulses that travel along specific neural pathways to the brain. Each receptor is specific for a particular type of stimulus. The brain perceives the stimulus as a sensation.

- As shown, the skin has various tactile receptors for touch, pressure, stretching, warmth, cold, and pain. Upon stimulation, these receptors evoke action potentials in sensory neurons. The action potentials propagate through ascending pathways in the spinal cord to the somatosensory cortex.

Skin receptors • Figure 8.2

2 Some Special Senses Use Receptors That Detect Chemicals 232

- The chemical senses include smell (**olfaction**) and taste (**gustation**). Specific chemicals bind to receptors in the nose and on the tongue, where they elicit nerve impulses that travel to the brain.

Structures of the olfactory epithelium • Figure 8.4

- As shown, **olfactory receptors** are located in the olfactory ... and can detect ... odorants. N...

Critical and Creative Thinking Questions

1. Two teenage boys, Bob and Bill, are diagnosed with diabetes. Bob is slightly underweight for his age, while Bill is overweight, possibly obese. Bob must take daily insulin injections to control his diabetes, while Bill must watch his diet and take oral medications. Identify the reasons for the different treatments and explain what is going on in each of their bodies.

2. Joe's wife of 20 years has noticed changes in his appearance. When they were married at age 18, Joe was a tall, handsome teenager. As Joe has aged, his hands, feet, and jaws have thickened, and his head has become elongated. What might be happening to Joe? What tests might you conduct to support your hypothesis?

3. A scientist has two cultures of the same type of cell. She administers Hormone A to one culture and Hormone B to the other. She notices that Hormone A has an immediate effect on the culture, while the culture with Hormone B takes about 90 minutes to show any activity. When the scientist examines protein synthesis in both cultures, she notices that new proteins are made only in the culture exposed to Hormone B. Explain the differences between the two hormones and indicate what types they might be.

4. Laura recently suffered a head injury in a car accident. As she is recovering, she notices that she is always thirsty, urinates frequently, and passes large volumes of urine. If she does not drink often, she feels weak, faint, and nauseated. What might have happened to Laura, and what is going on in her body that would explain these signs and symptoms?

5. Michael is a 45-year-old man who was recently diagnosed with high blood pressure. His physician prescribed a drug called an angiotensin-converting enzyme (ACE) inhibitor. How will the drug affect Michael's ability to withstand dehydration? Explain.

Critical and Creative Thinking Questions challenge students to think more broadly about chapter concepts. The level of these questions ranges from simple to advanced; they encourage students to think critically and develop an analytical understanding of the ideas discussed in the chapter.

What is happening in this picture? presents an uncaptioned photograph that is relevant to a chapter topic and illustrates a situation students are not likely to have encountered previously. The photograph is paired with questions that ask the student to describe and explain what they can observe in the photo based on what they have learned.

What is happening in this picture?

It is essential that a donor and recipient be closely matched in the millions of blood transfusions that occur each year. The blood being administered here is type A.

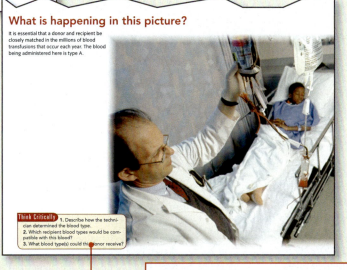

Think Critically 1. Describe how the technician determined the blood type.
2. Which recipient blood types would be compatible with this blood?
3. What blood type(s) could this donor receive?

Think Critically questions ask students to apply what they have learned in order to interpret and explain what they observe in the image.

Self-Test

(Check your answers in Appendix C.)

1. Astronauts spend vast amounts of time in weightlessness, where the effect of gravity is minimal. Which sense would most likely be affected by this environment?
   a. hearing
   b. smell
   c. touch
   d. equilibrium

Use this figure for questions 2–4.

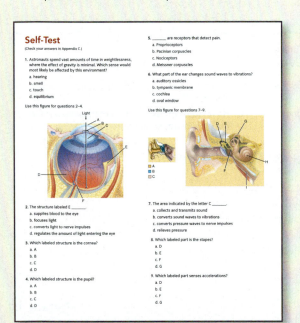

Light

2. The structure labeled E _____.
   a. supplies blood to the eye
   b. focuses light
   c. converts light to nerve impulses
   d. regulates the amount of light entering the eye

3. Which labeled structure is the cornea?
   a. A
   b. B
   c. C
   d. D

4. Which labeled structure is the pupil?
   a. A
   b. B
   c. C
   d. D

5. _____ are receptors that detect pain.
   a. Proprioceptors
   b. Pacinian corpuscles
   c. Nociceptors
   d. Meissner corpuscles

6. What part of the ear changes sound waves to vibrations?
   a. auditory ossicles
   b. tympanic membrane
   c. cochlea
   d. oval window

Use this figure for questions 7–9.

☑ A
■ B
☑ C

7. The area indicated by the letter C _____.
   a. collects and transmits sound
   b. converts sound waves to vibrations
   c. converts pressure waves to nerve impulses
   d. relieves pressure

8. Which labeled part is the stapes?
   a. D
   b. E
   c. F
   d. G

9. Which labeled part senses accelerations?
   a. D
   b. E
   c. F
   d. G

Visual end-of-chapter **Self-Tests** pose review questions that ask students to demonstrate their understanding of key concepts.

Why *Visualizing Anatomy and Physiology?*

Teaching and learning anatomy and physiology in a one-semester course poses particular challenges for both students and professors. The time constraints of one semester make it difficult to cover the essential material effectively. This is compounded by the fact that, although students enter the course with enthusiasm and a will to succeed in order to move forward in their chosen career path, many struggle with the complexities of the course content and the challenges these present.

Visualizing Anatomy and Physiology is designed to provide a fresh, visual presentation of the subject, telling the story of the human body, the interrelatedness of its structures and functions, and the relevance of this knowledge to health and disease. Integrated with *WileyPLUS*, it is a combination of a visually powerful textbook and interactive media that is inviting—not intimidating—to students who are overwhelmed by other scientific texts. The pedagogically innovative illustration program supported by a clear and condensed narrative highlights the relevance of anatomy and physiology to a health-related career and engages and motivates students to study and learn. *WileyPLUS* effectively integrates the teaching process with the learning environment, seamlessly linking classroom presentation, student activity, and assessment.

Visualizing Anatomy and Physiology assumes that students have not previously studied the human body. Our goal is to provide a basic understanding of the structure and functions of the human body with an emphasis on homeostasis. Health and disease topics are integrated throughout the book and media, and topics such as wellness, aging, and personal health choices are discussed within each chapter. Tested and proven visuals and pedagogy that help students learn more effectively are used to teach and explain, not just illustrate. These include A & P *InSight* figures, *Process Diagrams*, and images and videos from trusted sources such as the National Geographic Society. *WileyPLUS 5.0* is an innovative, research-based online environment that integrates relevant resources, including the entire digital textbook, in an easy-to-navigate framework. *WileyPLUS* engages and excites students about the content and builds their confidence. Tools for promoting individual initiative, ways to increase effectiveness, and help in achieving personal goals are all linked to measurable outcomes, ensuring that students always know how their efforts are benefiting them.

Organization

Visualizing Anatomy and Physiology is organized into 16 chapters, designed to fit more easily into the shorter one-semester course. As each chapter unfolds, the core concepts needed to fully comprehend the subject matter—homeostasis and the relevance and results of disruptions to homeostasis, the interrelatedness of structure and function, and the development of critical thinking skills—are highlighted and emphasized. This emphasis begins with the unique illustrated chapter openers designed for this book. Each opening spread eases the student into the chapter's content with an engaging exploration of a contemporary and recognizable example that illustrates the main concepts explored in the chapter. As the chapter material progresses, students will encounter another unique feature of this text—*What a Health Provider Sees*. Each of these puts the chapter's topic into the context of real-life situations that the students moving into allied health careers are likely to encounter. Partnered with a series of critical thinking questions, this exciting feature helps students build the necessary skills to apply their knowledge of anatomy and physiology to their future careers.

Chapters 1 through 3 establish the groundwork for understanding the structure and function of each body system. **Chapter 1** sets the stage with an **overview of the organization of the human body** and the introduction of the major concept of homeostasis. **Chapter 2** provides the **basics of introductory chemistry** in focused modules that minimize the anxiety many students experience with this foundational content. **Cells and tissues** are covered in **Chapter 3**. The outstanding histological micrographs help students grasp the often challenging material and recognition of different tissue types and their roles in different organ systems.

The remaining chapters of the text each focus on a major organ system of the human body. **Chapter 4** introduces the **integumentary system** and its important role in protection and regulation of body temperature. **Chapters 5 and 6,** the **skeletal and muscular systems**, introduce students to the many structures and important functions that enable the support and movement of the human body. The important regulatory systems of the body are introduced in Chapters 7 through 9. Two **chapters, 7 and 8**, are devoted to the complexities of the **nervous system**. **Chapter 7** presents the major structures and functions of both the **central and peripheral nervous systems** while **Chapter 8** focuses on **somatic senses** and special senses. The **endocrine system** is introduced in **Chapter 9** and effectively organizes this complex system by the major functions of each endocrine gland. **Chapters 10 and 11** are devoted to the **cardiovascular system**. The components of blood, the importance and process of clotting, and the difference in blood types and their importance in health care are covered in chapter 10. Chapter 11 features the dynamic functions of the heart and the structures and processes, such as capillary exchange, of the circulatory system. **Chapter 12**, the **lymphatic system and immunity**, overviews the crucial functions of this system and highlights its connection to major health issues such as allergies, autoimmune disease, and HIV. **Chapter 13** explores the **respiratory system** and its important role in the exchange of gases. **Chapter 14** introduces the many structures of the **digestive system**, the breakdown and absorption of nutrients through metabolism and a basic understanding of the role of nutrition in maintaining a healthy body. **Chapter 15** covers the **urinary system** and its important function of removing wastes from the body. In addition, fluid, electrolyte, and acid-base balance is discussed in this chapter. The final **Chapter 16** overviews both the male and female **reproductive systems** and their role in the continuation of human life.

How does Wiley Visualizing support instructors?

Wiley Visualizing Site   WILEY VISUALIZING™ | NATIONAL GEOGRAPHIC

The Wiley Visualizing site hosts a wealth of information for instructors using Wiley Visualizing, including ways to maximize the visual approach in the classroom and a white paper titled "How Visuals Can Help Students Learn," by Matt Leavitt, instructional design consultant. You can also find information about our relationship with the National Geographic Society and other texts published in our program. Visit Wiley Visualizing at www.wiley.com/college/ visualizing.

Wiley Custom Select   WILEY Custom LEARNING SOLUTIONS

Wiley Custom Select gives you the freedom to build your course materials exactly the way you want them. Offer your students a cost-efficient alternative to traditional texts. In a simple three-step process create a solution containing the content you want, in the sequence you want, delivered how you want. Visit Wiley Custom Select at http:// customselect.wiley.com.

Videos From the National Geographic Society, the BBC, and The New York Times

National Geographic Digital Media TV enabled the use of National Geographic videos to accompany *Visualizing Anatomy and Physiology* and enrich the text. Some of our videos were selected from the libraries of the BBC and The New York Times. Researched by Janis Thompson, Lorain Community College; Michael Harman, Lonestar College; and Charles Benton, Madison Area Technical College; the videos presented in each chapter of the textbook, provide visual context for key concepts, ideas, and terms addressed in the chapters. Streaming videos are available to students in the context of *WileyPLUS*, and accompanying assignments can be graded online and added to the instructor gradebook.

Videos for Visualizing Anatomy and Physiology

Chapter 1 NGS Video: The Incredible Human Machine

Chapter 2 BBC Video: Study on Supplements

Chapter 3 NGS Video: Stem Cell Research

Chapter 4 NGS Video: The Incredible Human Machine: Skin

Chapter 5 BBC Video: Osteoporosis

Chapter 6 New York Times Video: Increasing Knee Stability

Chapter 7 BBC Videos: Psychoactive Drugs, Brain Bank

Chapter 8 New York Times Video: Coping with Chronic Pain

Chapter 9 NGS Video: Iodine Deficiency in Ethiopia

Chapter 10 BBC Video: Sickle Cell Anemia

Chapter 11 NGS Video: The Incredible Human Machine: Circulation

New York Times Video: Looking at the Heart

Chapter 12 NGS Video: AIDS

Chapter 13 BBC Video: Asthma

Chapter 14 NGS Video: The Incredible Human Machine: Digestion

Chapter 15 New York Times Video: The Not-So-Strong Kidney

Chapter 16 NGS Video: The Incredible Human Machine: The Miracle of Conception

BBC Video: Teratogens

Book Companion Site   www.wiley.com/college/freudenrich

All instructor resources (the Test Bank, Instructor's Manual, PowerPoint presentations, and all textbook illustrations and photos in jpeg format) are housed on the book companion site (www.wiley.com/college/freudenrich). Student resources include self quizzes and flashcards.

PowerPoint Presentations   (available in *WileyPLUS* and on the book companion site)

A complete set of highly visual PowerPoint presentations—one per chapter—developed by Sandra Hutchinson, Sinclair Community College is available online and in *WileyPLUS* to enhance classroom presentations. Tailored to the text's topical coverage and learning objectives, these presentations are designed to convey key text concepts, illustrated by embedded text art and animations. Video PowerPoints also offer embedded links to videos to help introduce classroom discussions with short, engaging video clips.

Test Bank   (available in *WileyPLUS* and on the book companion site)

The visuals from the textbook are also included in the Test Bank developed by Charles Benton, Madison Area Technical College, Evelyn Biluk, Chippewa Valley Technical College, and Cammie Emory, Bossier Parish Community College. The Test Bank has a diverse selection of test items including multiple-choice and essay questions, with at least 20 percent of them incorporating visuals from the book. The Test Bank is available online in MS Word files, as a Computerized Test Bank, and within *WileyPLUS*. The easy-to-use test-generation program fully supports graphics, print tests, student answer sheets, and answer keys. The software's advanced features allow you to produce an exam to your exact specifications.

Instructor's Resources   (available in *WileyPLUS* and on the book companion site)

The Instructor's Resources includes creative ideas for in-class activities, and suggested lecture outlines by Scott Rahschulte, Ivy Tech Community College. The Critical and Creative Thinking questions are included in the Assignment section of *WileyPLUS* and the Concept Check questions are included in the student Practice section of *WileyPLUS*. Guidance is also provided on how to maximize the effectiveness of visuals in the classroom.

1. **Use visuals during class discussions or presentations.** Point out important information as the students look at the visuals, to help them integrate separate visual and verbal mental models.

2. **Use visuals for assignments and to assess learning.** For example, learners could be asked to identify samples of concepts portrayed in visuals.

3. **Use visuals to encourage group activities.** Students can study together, make sense of, discuss, hypothesize, or make decisions about the content. Students can work together to interpret and describe the diagrams, or use the diagrams to solve problems, conduct related research, or work through a case study activity.

4. **Use visuals during reviews.** Students can review key vocabulary, concepts, principles, processes, and relationships displayed visually. This recall helps link prior knowledge to new information in working memory, building integrated mental models.

5. **Use visuals for assignments and to assess learning.** For example, learners could be asked to identify samples of concepts portrayed in visuals.

6. **Use visuals to apply facts or concepts to realistic situations or examples.** For example, a familiar photograph, such as one of mountain climbers, can illustrate key information about breathing, linking this new concept to prior knowledge.

Anatomy and Physiology Visual Library

All photographs, figures, and other visuals from the text are online and in *WileyPLUS* and can be used as you wish in the classroom. The images are included in a variety of formats, such as labeled, unlabeled, and unlabeled with leader lines. In addition, many illustrations or photos not included in this text can be found in the Visual Library. These online electronic files allow you to easily incorporate images into your PowerPoint presentations as you choose, or to create your own materials.

Wiley Faculty Network

The Wiley Faculty Network (WFN) is a global community of faculty, connected by a passion for teaching and a drive to learn, share, and collaborate. Their mission is to promote the effective use of technology and enrich the teaching experience. Connect with the Wiley Faculty Network to collaborate with your colleagues, find a mentor, attend virtual and live events, and view a wealth of resources all designed to help you grow as an educator. Visit the Wiley Faculty Network at www.wherefacultyconnect.com.

How has Wiley Visualizing been shaped by contributors?

Wiley Visualizing and the *WileyPLUS* learning environment would not have come about without a team of people, each of whom played a part in sharing their research and contributing to this new approach. First and foremost, we begin with the National Geographic Society.

National Geographic Society

Visualizing Anatomy and Physiology offers an array of remarkable photographs, illustrations, multimedia, and video from the National Geographic Society collections. Students using the book benefit from the rich, fascinating resources of National Geographic.

National Geographic School Publishing performed an invaluable service in fact-checking *Visualizing Anatomy and Physiology*. They have verified every fact in the book with two outside sources, to ensure that the text is accurate and up-to-date. This kind of fact-checking is rare in textbooks and unheard-of in most online media.

National Geographic Image Collection provided access to National Geographic's award-winning image and illustrations collection to identify the most appropriate and effective images and illustrations to accompany the content. Each image and illustration has been chosen to be instructive, supporting the processes of selecting, organizing, and integrating information, rather than being merely decorative.

Academic Research Consultants

Richard Mayer, Professor of Psychology, UC Santa Barbara. Mayer's *Cognitive Theory of Multimedia Learning* provided the basis on which we designed our program. He continues to provide guidance to our author and editorial teams on how to develop and implement strong, pedagogically effective visuals and use them in the classroom.

Jan L. Plass, Professor of Educational Communication and Technology in the Steinhardt School of Culture, Education, and Human Development at New York University. Plass co-directs the NYU Games for Learning Institute and is the founding director of the CREATE Consortium for Research and Evaluation of Advanced Technology in Education.

Matthew Leavitt, Instructional Design Consultant, advises the Visualizing team on the effective design and use of visuals in instruction and has made virtual and live presentations to university faculty around the country regarding effective design and use of instructional visuals.

Independent Research Studies

SEG Research, an independent research and assessment firm, conducted a national, multisite effectiveness study of students enrolled in entry-level college Psychology and Geology courses. The study was designed to evaluate the effectiveness of Wiley Visualizing. You can view the full research paper at www.wiley.com/college/visualizing/freudenrich/efficacy.html.

Instructor and Student Contributions

Throughout the process of developing the concept of guided visual pedagogy for Wiley Visualizing, we benefited from the comments and constructive criticism provided by the instructors and colleagues listed below. We offer our sincere appreciation to these individuals for their helpful reviews and general feedback:

Visualizing Reviewers, Focus Group Participants, and Survey Respondents

James Abbott, Temple University
Melissa Acevedo, Westchester Community College
Shiva Achet, Roosevelt University
Denise Addorisio, Westchester Community College
Dave Alan, University of Phoenix
Sue Allen-Long, Indiana University–Purdue
Robert Amey, Bridgewater State College
Nancy Bain, Ohio University
Corinne Balducci, Westchester Community College
Steve Barnhart, Middlesex County Community College
Stefan Becker, University of Washington–Oshkosh
Callan Bentley, NVCC Annandale
Valerie Bergeron, Delaware Technical & Community College
Andrew Berns, Milwaukee Area Technical College
Gregory Bishop, Orange Coast College
Rebecca Boger, Brooklyn College
Scott Brame, Clemson University
Joan Brandt, Central Piedmont Community College
Richard Brinn, Florida International University
Jim Bruno, University of Phoenix
William Chamberlin, Fullerton College
Oiyin Pauline Chow, Harrisburg Area Community College
Laurie Corey, Westchester Community College
Ozeas Costas, Ohio State University at Mansfield
Christopher Di Leonardo, Foothill College
Dani Ducharme, Waubonsee Community College
Mark Eastman, Diablo Valley College
Ben Elman, Baruch College
Staussa Ervin, Tarrant County College
Michael Farabee, Estrella Mountain Community College
Laurie Flaherty, Eastern Washington University
Susan Fuhr, Maryville College
Peter Galvin, Indiana University at Southeast
Andrew Getzfeld, New Jersey City University
Janet Gingold, Prince George's Community College
Donald Glassman, Des Moines Area Community College
Richard Goode, Porterville College
Peggy Green, Broward Community College
Stelian Grigoras, Northwood University
Paul Grogger, University of Colorado

Michael Hackett, Westchester Community College
Duane Hampton, Western Michigan University
Thomas Hancock, Eastern Washington University
Gregory Harris, Polk State College
John Haworth, Chattanooga State Technical Community College
James Hayes-Bohanan, Bridgewater State College
Peter Ingmire, San Francisco State University
Mark Jackson, Central Connecticut State University
Heather Jennings, Mercer County Community College
Eric Jerde, Morehead State University
Jennifer Johnson, Ferris State University
Richard Kandus, Mt. San Jacinto College District
Christopher Kent, Spokane Community College
Gerald Ketterling, North Dakota State University
Lynnel Kiely, Harold Washington College
Eryn Klosko, Westchester Community College
Cary T. Komoto, University of Wisconsin–Barron County
John Kupfer, University of South Carolina
Nicole Lafleur, University of Phoenix
Arthur Lee, Roane State Community College
Mary Lynam, Margrove College
Heidi Marcum, Baylor University
Beth Marshall, Washington State University
Dr. Theresa Martin, Eastern Washington University
Charles Mason, Morehead State University
Susan Massey, Art Institute of Philadelphia
Linda McCollum, Eastern Washington University
Mary L. Meiners, San Diego Miramar College
Shawn Mikulay, Elgin Community College
Cassandra Moe, Century Community College
Lynn Hanson Mooney, Art Institute of Charlotte
Kristy Moreno, University of Phoenix
Jacob Napieralski, University of Michigan–Dearborn
Gisele Nasar, Brevard Community College, Cocoa Campus
Daria Nikitina, West Chester University
Robin O'Quinn, Eastern Washington University
Richard Orndorff, Eastern Washington University
Sharen Orndorff, Eastern Washington University
Clair Ossian, Tarrant County College
Debra Parish, North Harris Montgomery Community College District

Visualizing Reviewers, Focus Group Participants, and Survey Respondents, *continued*

Linda Peters, Holyoke Community College
Robin Popp, Chattanooga State Technical Community College
Michael Priano, Westchester Community College
Alan "Paul" Price, University of Wisconsin–Washington County
Max Reams, Olivet Nazarene University
Mary Celeste Reese, Mississippi State University
Bruce Rengers, Metropolitan State College of Denver
Guillermo Rocha, Brooklyn College
Penny Sadler, College of William and Mary
Shamili Sandiford, College of DuPage
Thomas Sasek, University of Louisiana at Monroe
Donna Seagle, Chattanooga State Technical Community College
Diane Shakes, College of William and Mary
Jennie Silva, Louisiana State University
Michael Siola, Chicago State University
Morgan Slusher, Community College of Baltimore County
Julia Smith, Eastern Washington University
Darlene Smucny, University of Maryland University College
Jeff Snyder, Bowling Green State University
Alice Stefaniak, St. Xavier University

Alicia Steinhardt, Hartnell Community College
Kurt Stellwagen, Eastern Washington University
Charlotte Stromfors, University of Phoenix
Shane Strup, University of Phoenix
Donald Thieme, Georgia Perimeter College
Pamela Thinesen, Century Community College
Chad Thompson, SUNY Westchester Community College
Lensyl Urbano, University of Memphis
Gopal Venugopal, Roosevelt University
Daniel Vogt, University of Washington–College of Forest Resources
Dr. Laura J. Vosejpka, Northwood University
Brenda L. Walker, Kirkwood Community College
Stephen Wareham, Cal State Fullerton
Fred William Whitford, Montana State University
Katie Wiedman, University of St. Francis
Harry Williams, University of North Texas
Emily Williamson, Mississippi State University
Bridget Wyatt, San Francisco State University
Van Youngman, Art Institute of Philadelphia
Alexander Zemcov, Westchester Community College

Student Participants

Karl Beall, Eastern Washington University
Jessica Bryant, Eastern Washington University
Pia Chawla, Westchester Community College
Channel DeWitt, Eastern Washington University
Lucy DiAroscia, Westchester Community College
Heather Gregg, Eastern Washington University
Lindsey Harris, Eastern Washington University
Brenden Hayden, Eastern Washington University
Patty Hosner, Eastern Washington University

Tonya Karunartue, Eastern Washington University
Sydney Lindgren, Eastern Washington University
Michael Maczuga, Westchester Community College
Melissa Michael, Eastern Washington University
Estelle Rizzin, Westchester Community College
Andrew Rowley, Eastern Washington University
Eric Torres, Westchester Community College
Joshua Watson, Eastern Washington University

Reviewers and Survey Respondents of *Visualizing Anatomy and Physiology*

Bryce Abbey, University of Nebraska–Kearney
Dorene Adams, Napa Valley College
Amir Afshar, Academy of the New Church
Caryn Babaian, Bucks County Community College
Cherryl Baker, Orange Coast College
Mary Bath-Balogh, Pierce College
Shawn Bearden, Idaho State University
Charles Benton, Madison Area Technical College
Heather Billings, West Virginia University
Margaret Bolton, College of Southern Maryland
Felicia Brenoe, Glendale Community College
Doug Bruce, Laney College
Carolyn Bunde, Idaho State University
Dave Burns, Carl Sandburg College
Mahlon Cannon, J. Sargeant Reynolds Community College
Daisy Carr, California State University–Dominguez Hills

Tom Carson, Bossier Parish Community College
Craig Castaneda, Illinois Valley Community College
Maryanne Cattieu, Erie Community College
Lorna Chacha, Centennial College
Ed Chang, Imperial Valley College
Irena Ciftja, Lonestar College
LuAnne Clark, Lansing Community College
Pamela Cole, Shelton State Community College
Jean Cremins, Middlesex Community College
Angela Crocker, Erie Community College
David Crow, Augsburg College
Laura Cucci, Community College of Baltimore County
Jean Cuppett, Pennsylvania Highlands Community College
John Danley, Albuquerque Technical Vocational Institute
Thomas Delany, Kilgore College
Robert DeLorme, Gwinnett Technical Institute

Reviewers and Survey Respondents of *Visualizing Anatomy and Physiology,* continued

Stacy Deputy, Tidewater Community College
Chris Donnelly, College of DuPage
Steve Edinger, Ohio University
Cammie Emory, Bossier Parish Community College
Colin Everhart, Wake Technical Community College
James Ezell, J. Sargeant Reynolds Community College
Abdulmunam Fellah, Pima Community College
Amy Fenech, Columbus Technical Institute
Patricia Finkenstadt, Phoenix College
Sally Flesch, Black Hawk College
Maria Florez, Lone Star College–CyFair
Richard Foreman, Darton College
Teri Foster, Graceland University
Karen Fowler, Metropolitan Community College
Deborah Furbish, Wake Technical Community College
Bonnie Futrell, Sinclair Community College
Christina Gan, Highline Community College
Anne Geller, San Diego Mesa College
Cynthia Gerstner, Columbia College
Sharale Golding, Manchester Community College
Pamela Gregory, Tyler Junior College
Ron Groleau, Illinois Valley Community College
Bentley Gubar, Stockton State College
Terrance Hackett, J. Sargeant Reynolds Community College
Michael Harman, North Harris Montgomery Community College
Mary Beth Hawkins, North Carolina State University
Kristen Herman, San Diego Mesa College
Michael Highers, Darton College
Jacki Houghton, Moorpark College
Erik Hoyer, Lone Star College–CyFair
Sandra Hutchinson, Sinclair Community College
Donna Jennings, Sinclair Community College
Rosalba Jepson, Imperial Valley College
Eddie Johnson, Central Oregon Community College
Tom Johnson, Tidewater Community College
Walter Johnson, Merritt College
Tom Jordan, Pima County Community College
Paul Kaseloo, Virginia State University
Daniel Kifle, Community College of Baltimore County
Brian Kipp, Grand Valley State University
Jody Klann, Gateway Community College
Joseph LaFazia, Bristol Community College
Jeffrey Levin, Stockton State College
May Liu, Darton College
James Ludden, College of DuPage
Linda Maluf, Pima Community College
Rich Marsillo, State University of New York at Farmingdale
George Mateja, Community College of Baltimore County
Bhavya Mathur, Chattahoochee Technical Institute
Kate Mathis, San Antonio College
Cheryl Matthias, Chicago State University

Ron May, Pierce College
Elysia Mbuja, Pierce College
Helen Mergenthal, Germanna Community College
Scott Meyer, Pensacola State College
Christopher Migliore, Loyola University Chicago
Milton Miller, Macon Technical Institute
Robert Moats, Ohio University–Chillicothe
Susan Moss, Imperial Valley College
William Mott, J. Sargeant Reynolds Community College
Kelly Neary, Mission College
Kimberly O'Brien, Cornell University
Sima Otsuka, Northern Virginia Community College
TC Parker, Gwinnett Technical College
Lisa Parks, North Carolina State University
Dale Pederson, Augsburg College
Ellengene Peterson, Ohio University
Jessica Peterson, Pensacola State College
Russell Peterson, Indiana University of Pennsylvania
Anne Pinkerton, Tidewater Community College
John Polos, Laney College
Dorothy Puckett, Kilgore College
James Reed, Grand Valley State University
Tricia Richard, Middlesex Community College
Loretta Ritu, Canadian Memorial Chiropractic College
Susan Rohde, Triton College
John Rogers, Albuquerque Technical Vocational Institute
Jimmy Rozell, Tyler Junior College
Jodie Rubens, Ohio University
Marigrace Ryan, Sinclair Community College
Adrienne Sainten, Merritt College
Holly Sanders, Gwinnett Technical Institute
Nancy Schellinger, Indiana University of Pennsylvania
Gaynelle Schmieder, Pennsylvania Highlands Community College
Eric Shargo, Moorpark College
Richard Sikon, John Tyler Community College
Jennifer Sink, Davidson County Community College
Judith Slon, Hilbert College
Tamara Smith, Tidewater Community College
Sharon Smith-Douglas, College of Southern Maryland
James Sosebee, Sinclair Community College
George Spiegel, College of Southern Maryland
Sherry Stewart, Navarro College
Jan Stone, Bristol Community College
Anthony Tesch, College of the Desert
Janis Thompson, Lorain County Community College
James Valente, Glendale Community College
Cecilia Vigil, Arizona Western College
Kathy Webb, Bucks County Community College
Sheila Williams, Houston Community College

Special Thanks

Special thanks to my wife, Theresa, for her patience with me during this project.

Craig

To my children—Lynne, Gerard Jr., Kenneth, Anthony, and Andrew—the wind beneath my wings.

Jerry

About the Authors

Craig C. Freudenrich is a science writer in Durham, NC where he writes various science articles, science educational materials, and science education courses. He earned a bachelor's degree in biology from West Virginia University and a Ph.D. in physiology from the School of Medicine at the University of Pittsburgh. He completed eight years of postdoctoral training in the Physiology Division within the Department of Cell Biology at Duke University Medical Center. For more than 20 years, he has conducted biomedical research in various fields including cell physiology, general physiology, neuroscience, and plant physiology.

Craig completed teacher training at the University of North Carolina at Chapel Hill and taught high school sciences for Durham Public Schools, Auldern Academy, and the Talent Identification Program at Duke University. He has experience teaching biology, anatomy and physiology, physics, physical sciences, astrobiology, science of science fiction, and algebra.

Before becoming a freelance science writer, Craig was a science editor for the popular website *HowStuffWorks*. There, he wrote numerous articles in biology, chemistry, physics, geology, astronomy, and space exploration. Craig is a member of the National Association of Science Writers (NASW) and the National Science Teachers Association (NSTA). Beyond science, Craig has interests in medieval history, European martial arts, and fencing.

Gerard J. Tortora is Professor of Biology and former Biology Coordinator at Bergen Community College in Paramus, New Jersey, where he teaches human anatomy and physiology as well as microbiology. He received his bachelor's degree in biology from Fairleigh Dickinson University and his master's degree in science education from Montclair State College. He is a member of many professional organizations, including the Human Anatomy and Physiology Society (HAPS), the American Society of Microbiology (ASM), American Association for the Advancement of Science (AAAS), National Education Association (NEA), and the Metropolitan Association of College and University Biologists (MACUB).

Above all, Jerry is devoted to his students and their aspirations. In recognition of this commitment, Jerry was the recipient of MACUB's 1992 President's Memorial Award. In 1996, he received a National Institute for Staff and Organizational Development (NISOD) excellence award from the University of Texas and was selected to represent Bergen Community College in a campaign to increase awareness of the contributions of community colleges to higher education.

Jerry is the author of several best-selling science textbooks and laboratory manuals, a calling that often requires an additional 40 hours per week beyond his teaching responsibilities. Nevertheless, he still makes time for four or five weekly aerobic workouts that include biking and running. He also enjoys attending college basketball and professional hockey games and performances at the Metropolitan Opera House.

Contents in Brief

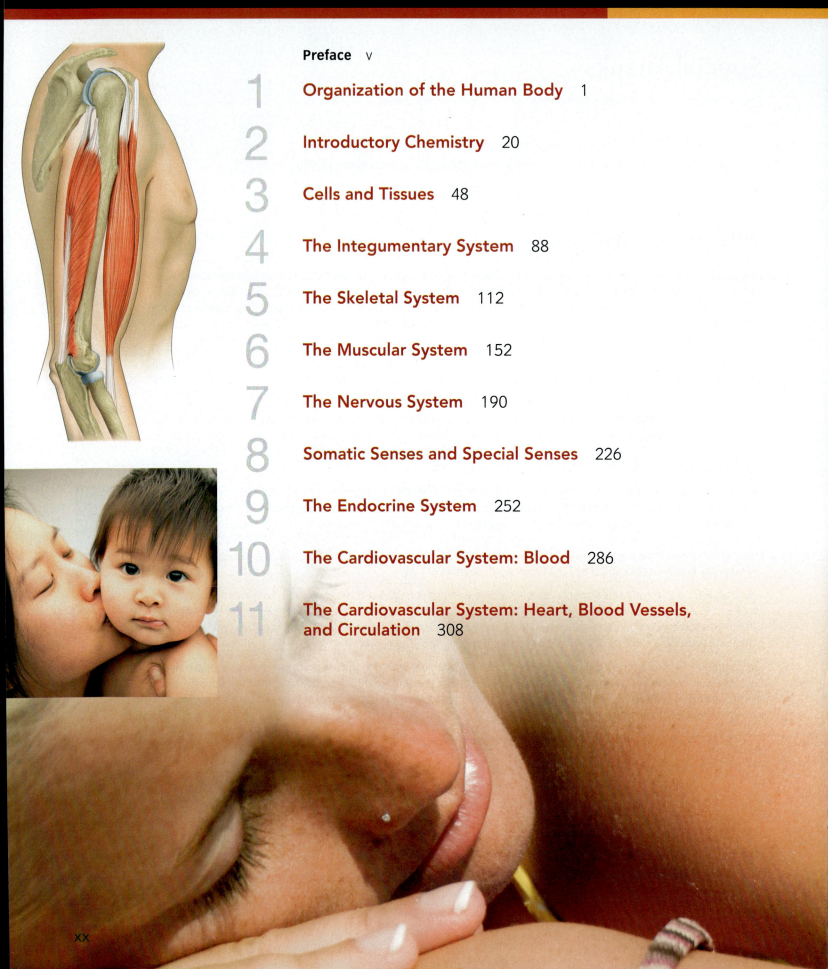

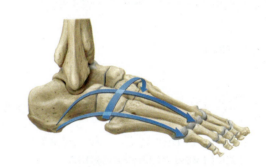

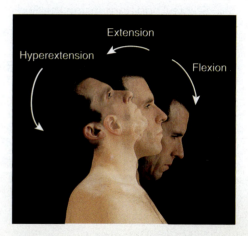

Contents

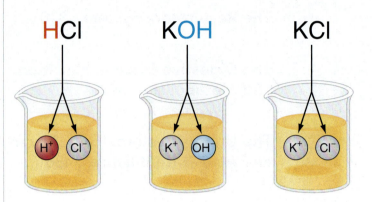

3 Cells and Tissues

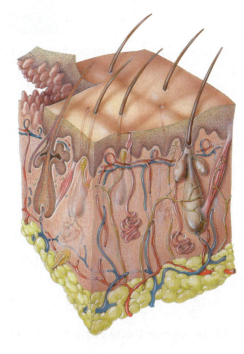

4 The Integumentary System

5 The Skeletal System

6 The Muscular System

7 The Nervous System

8 Somatic Senses and Special Senses

9 The Endocrine System

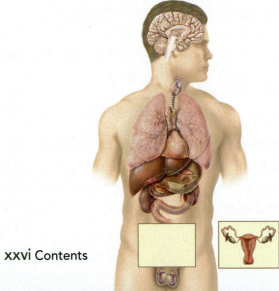

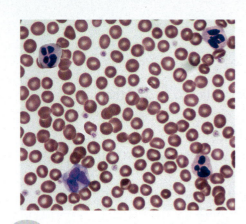

10 The Cardiovascular System: Blood

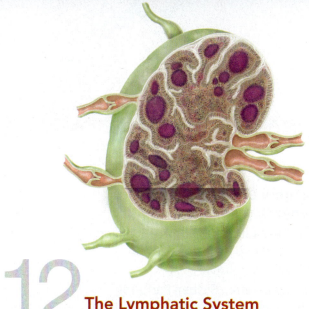

11

The Cardiovascular System: Heart, Blood Vessels, and Circulation

12

The Lymphatic System and Immunity

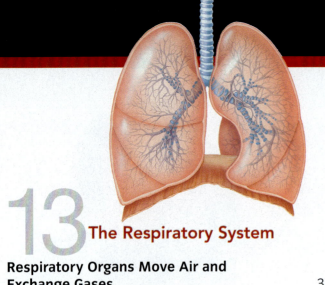

13 The Respiratory System

14 The Digestive System, Nutrition, and Metabolism

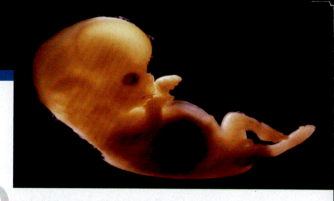

Visual presentations that focus on key anatomical structures in the chapter.

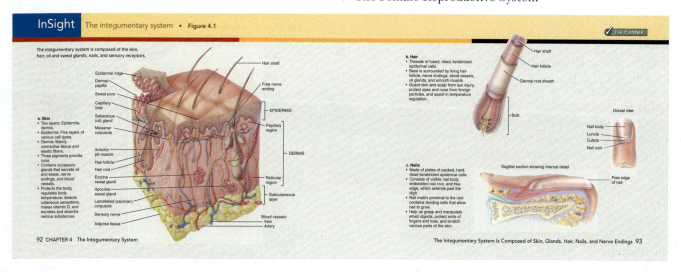

InSight   The integumentary system • Figure 4.1

The integumentary system is composed of the skin, hair, oil and sweat glands, nails, and sensory receptors.

a. Skin
- Two layers: Epidermis, dermis.
- Epidermis: Five layers of various cell types.
- Dermis: Mainly connective tissue and elastic fibers.
- Three pigments provide color.
- Contains accessory glands that secrete oil and sweat, nerve endings, and blood vessels.
- Protects the body, regulates body temperature, detects cutaneous sensations, makes vitamin D, and excretes and absorbs various substances.

b. Hair
- Threads of fused, dead, keratinized epidermal cells.
- Base is surrounded by living hair follicle, nerve endings, blood vessels, oil glands, and smooth muscle.
- Guard skin and scalp from sun injury, protect eyes and nose from foreign particles, and assist in temperature regulation.

c. Nails
- Made of plates of packed, hard, dead keratinized epidermal cells.
- Consists of visible nail body, embedded nail root, and free edge, which extends past the digit.
- Nail matrix proximal to the root contains dividing cells that allow nail to grow.
- Help us grasp and manipulate small objects, protect ends of fingers and toes, and scratch various parts of the skin.

Epidermal ridge · Dermal papilla · Sweat pore · Capillary loop · Sebaceous (oil) gland · Meissner corpuscle · Arrector pili muscle · Hair follicle · Hair root · Eccrine sweat gland · Apocrine sweat gland · Lamellated (pacinian) corpuscle · Sensory nerve · Adipose tissue

Hair shaft · Free nerve ending · EPIDERMIS · Papillary region · DERMIS · Reticular region · Subcutaneous layer · Blood vessels: Vein, Artery

Hair shaft · Hair follicle · Dermal root sheath · Bulb

Dorsal view · Nail body · Lunula · Cuticle · Nail root

Sagittal section showing internal detail · Free edge of nail

PROCESS DIAGRAMS

A series or combination of figures and photos that describe and depict a complex physiological process.

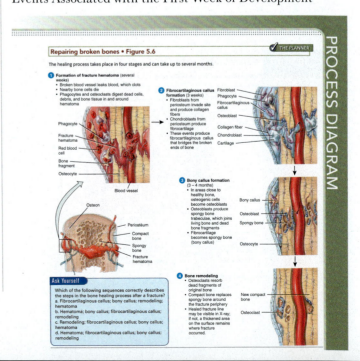

VISUALIZING
ANATOMY & PHYSIOLOGY

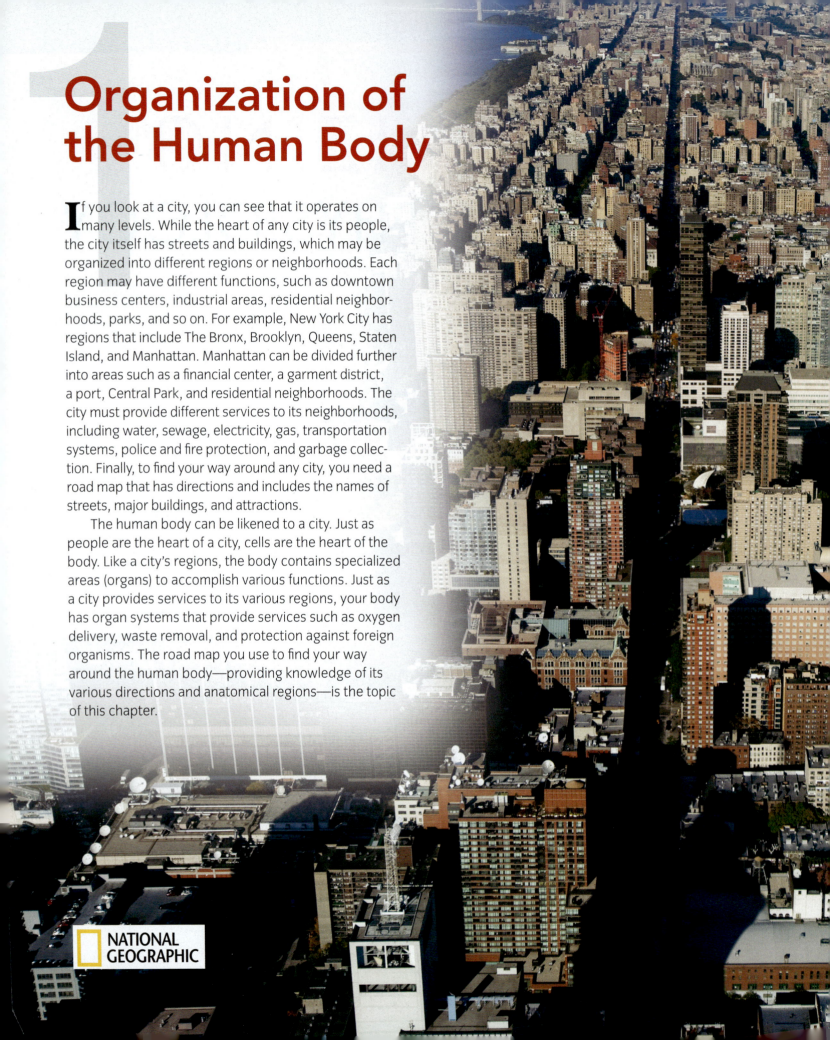

1 Organization of the Human Body

If you look at a city, you can see that it operates on many levels. While the heart of any city is its people, the city itself has streets and buildings, which may be organized into different regions or neighborhoods. Each region may have different functions, such as downtown business centers, industrial areas, residential neighborhoods, parks, and so on. For example, New York City has regions that include The Bronx, Brooklyn, Queens, Staten Island, and Manhattan. Manhattan can be divided further into areas such as a financial center, a garment district, a port, Central Park, and residential neighborhoods. The city must provide different services to its neighborhoods, including water, sewage, electricity, gas, transportation systems, police and fire protection, and garbage collection. Finally, to find your way around any city, you need a road map that has directions and includes the names of streets, major buildings, and attractions.

The human body can be likened to a city. Just as people are the heart of a city, cells are the heart of the body. Like a city's regions, the body contains specialized areas (organs) to accomplish various functions. Just as a city provides services to its various regions, your body has organ systems that provide services such as oxygen delivery, waste removal, and protection against foreign organisms. The road map you use to find your way around the human body—providing knowledge of its various directions and anatomical regions—is the topic of this chapter.

CHAPTER OUTLINE

CHAPTER PLANNER ✓

- ❑ Study the picture and read the opening story.
- ❑ Scan the Learning Objectives in each section:
  p. 2 ❑   p. 6 ❑   p. 10 ❑
- ❑ Read the text and study all visuals.
  Answer any questions.

Analyze key features

- ❑ InSight, pp. 2–3 ❑   p. 7 ❑
- ❑ Process Diagram, pp. 8–9 ❑
- ❑ What a Health Provider Sees, p. 15 ❑
- ❑ Stop: Answer the Concept Checks before you go on:
  p. 6 ❑   p. 10 ❑   p. 14 ❑

End of chapter

- ❑ Review the Summary and Key Terms.
- ❑ Answer the Critical and Creative Thinking Questions.
- ❑ Answer What is happening in this picture?
- ❑ Complete the Self-Test and check your answers.

Body Structure Is Closely Linked to Function at All Levels of Organization

LEARNING OBJECTIVES

1. **Compare** anatomy and physiology.
2. **Outline** the levels of organization of the human body.
3. **Describe** the function of each major organ system in the body.

A s a city has structures and each structure has a function, so does the human body. We refer to the sciences of the structure and function of the human body as anatomy and physiology. **Anatomy** (a-NAT-ō-mē; *ana-* = up; *-tomy* = process of cutting) is the science of *structure* and the relationships among structures. **Physiology** (fiz'-ē-OL-ō-jē; *physio-* = nature, *-logy* = study of) is the science of body *functions*— that is, how the body parts work. Because function can never be separated completely from structure, the human body is easiest to understand by studying anatomy and physiology together. We will look at how each structure of the body is designed to carry out a particular function and how the structure of a part often determines the functions it can perform. For example, the bones of the skull are rigidly joined to protect the brain, whereas

---

InSight | Levels of structural organization of the human body • Figure 1.1

Level 1. Chemical
The chemical level contains atoms, the smallest units of matter. Common atoms in living things include carbon (C), hydrogen (H), oxygen (O), nitrogen (N), and phosphorus (P). Atoms combine to form molecules such as deoxyribonucleic acid (DNA), the cell's genetic material.

Level 2. Cellular
Cells are the basic structural and functional unit of any organism. Your body has about 1 trillion cells. Although each cell must carry out the basic functions of life, most cells in your body are specialized to do one particular function. For example, the smooth muscle cells of the stomach can be stretched to accommodate a large meal.

Level 3. Tissue
Tissues are composed of groups of cells and their surroundings (intercellular substances) that work together to perform a particular function. There are four types of tissues: epithelial, muscle, connective, and nervous.

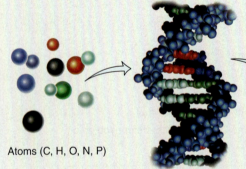

Atoms (C, H, O, N, P)

Molecule (DNA)

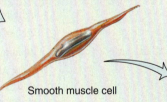

Smooth muscle cell

Smooth muscle tissue

WILEY PLUS | NATIONAL GEOGRAPHIC **Video**

the bones of the finger are loosely joined for freedom of movement, but both places where the bones come together are considered joints.

Let's look at how the body is organized.

Levels of Organization Extend from Atoms to the Human Organism

Your body has many organizational levels (**Figure 1.1**). Like all other things, your body is made of matter; the basic building blocks of matter are **atoms**, such as hydrogen, oxygen, carbon, and nitrogen. Atoms combine to form **molecules**, which perform various **biochemical** functions. For example, carbohydrates provide energy, fats store energy, proteins carry out many functions, and DNA codes information. Molecules compose **cells**, which distinguish living things from non-living chemicals. Examples of cells include neurons, epithelial cells, and myocytes. Cells that have similar functions are organized into **tissues**, such as muscle tissue, connective tissue, and nervous tissue. Different types of tissue join together to form structures called **organs**, including the heart, kidney, and brain. Related organs with a common function join to form a **system** (organ-system); body systems include, among others, the digestive system, cardiovascular system, nervous system, and respiratory system. Finally, all the systems combine to form an **organism** such as a human. In anatomy and physiology, the organism is considered the highest level of organization.

Let's look at the various systems of the human body.

Every Body System Performs Vital Functions

If you think about all the things that your body must do to survive, you will see that you have a system that covers each one (see **Figure 1.2** on the following page).

THE PLANNER

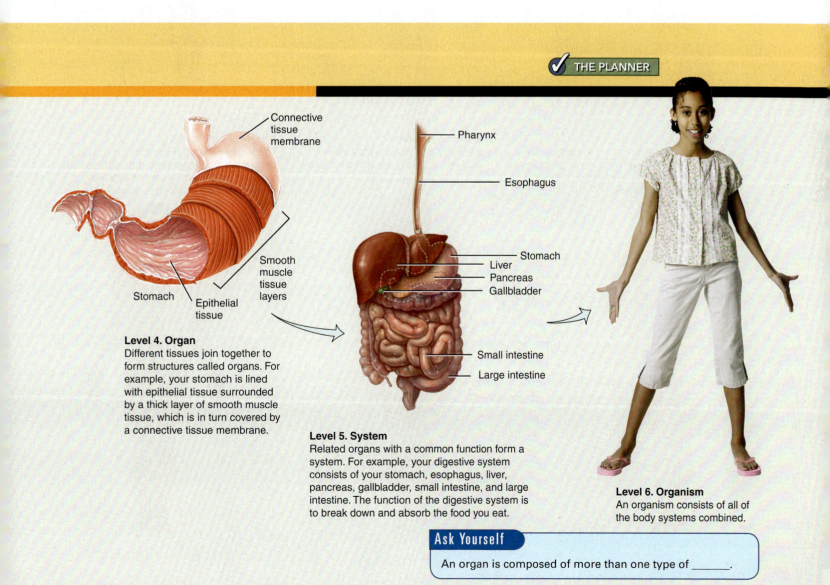

Connective tissue membrane

Pharynx

Esophagus

Smooth muscle tissue layers

Stomach
Liver
Pancreas
Gallbladder

Stomach

Epithelial tissue

Small intestine

Large intestine

Level 4. Organ
Different tissues join together to form structures called organs. For example, your stomach is lined with epithelial tissue surrounded by a thick layer of smooth muscle tissue, which is in turn covered by a connective tissue membrane.

Level 5. System
Related organs with a common function form a system. For example, your digestive system consists of your stomach, esophagus, liver, pancreas, gallbladder, small intestine, and large intestine. The function of the digestive system is to break down and absorb the food you eat.

Level 6. Organism
An organism consists of all of the body systems combined.

Ask Yourself

An organ is composed of more than one type of _____.

Components and functions of the body's organ systems • Figure 1.2

Integumentary system
Skin, hair, and nails provide an outer covering that protects the internal systems, senses the outside environment, regulates body temperature, and eliminates some wastes.

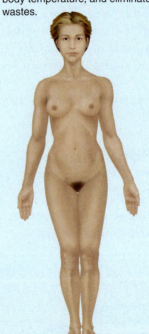

Skeletal system
Bones and joints provide a framework for all the organs and systems of the body. Bones also produce blood cells and store minerals.

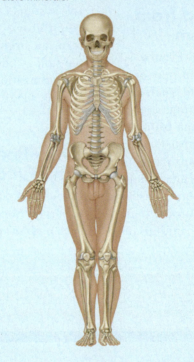

Muscular system
Muscles move the parts of the framework and provide force for various functions.

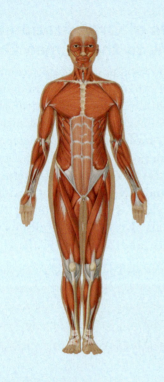

Nervous system
The brain, spinal cord, and nerves rapidly sense the internal and external environments, process information, and enable the various systems to communicate and coordinate.

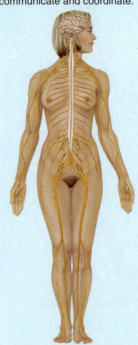

Endocrine system
Various glands communicate chemically with target organs and coordinate numerous body functions.

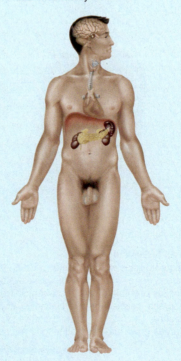

Cardiovascular system
The heart and blood vessels form a pumping system that delivers oxygen and nutrients to all cells and tissues, while removing carbon dioxide and wastes. Blood components also help fight disease and regulate acidity of body fluids.

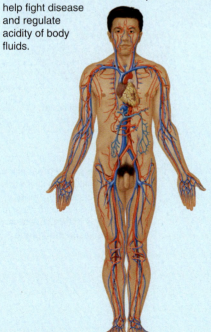

Lymphatic system
Lymph vessels and nodes filter fluid from spaces between cells and tissues into the blood and produce cells that fight diseases and foreign organisms.

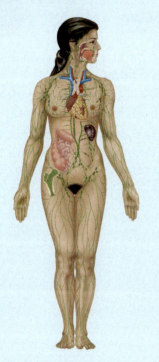

Respiratory system
The pharynx, trachea, bronchial passageways, and lungs form a ventilation system that brings in oxygen and removes carbon dioxide from the blood, while helping to maintain the acidity of the blood. The larynx produces sounds.

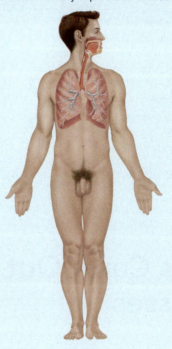

Digestive system
The mouth, esophagus, stomach, intestines, liver, gallbladder, and pancreas form a system that mechanically and chemically breaks down food, absorbs the molecules from it, and eliminates solid wastes.

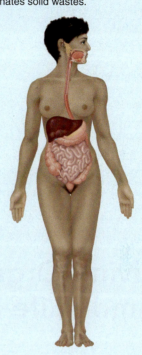

Urinary system
The kidneys, ureters, bladder, and urethra form a filtration system that regulates the ionic composition of the blood and body fluids and eliminates wastes.

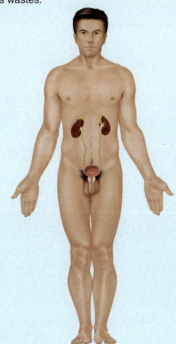

Reproductive system
The male's testes make sperm cells, and the seminal vesicles, prostate, and penis provide a delivery system for the sperm. The female's ovaries make the eggs, and the uterine (*Fallopian*) tubes, uterus, and vagina provide places to receive the sperm, fertilize the egg, incubate the developing offspring, and deliver the baby.

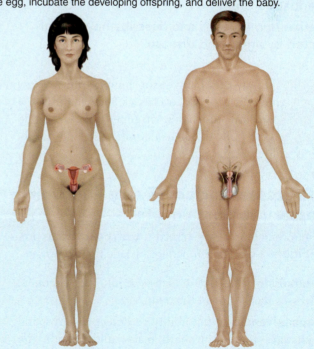

Table 1.1

System(s)	Function
Integumentary and lymphatic systems	Protect your body from disease
Integumentary and nervous systems	Sense the outside and process information
Skeletal and muscular systems	Move
Digestive system	Eat and get nutrients from food
Respiratory and cardiovascular systems	Exchange gases with the air and body tissues
Cardiovascular system	Distribute nutrients from food and gases throughout your body
Digestive and urinary systems	Eliminate wastes
Integumentary, cardiovascular, and endocrine systems	Regulate body temperature
Nervous and endocrine systems	Coordinate body functions
Reproductive and endocrine systems	Reproduce

As you proceed through this book, you will learn the anatomy and function of each system separately, but keep in mind that they all work together in a normally functioning individual (**Table 1.1**).

CONCEPT CHECK 🛑 STOP

1. **What** is the difference between anatomy and physiology?
2. **What** are the levels of organization of the body, from lowest to highest?
3. **Which** body system helps distribute oxygen and nutrients to cells and tissues, while removing carbon dioxide and wastes?

All Living Organisms Carry Out Common Life Processes

LEARNING OBJECTIVES

1. **Outline** the important life processes of humans.
2. **Distinguish** between negative and positive feedback systems.
3. **Define** *homeostasis* and describe how it is affected by aging and disease.

Like any other organism, humans must carry out certain life processes and maintain a constant internal environment. Failure to do these things can lead to disease and death. Let's look at life processes.

Life Processes Include Every Function Necessary to Sustain Life

Your organ systems enable your body to carry out various processes (**Figure 1.3**). Their functions include the following:

1. **Metabolism** (me-TAB-ō-lizm) is all the chemical processes that occur in the body.
2. **Responsiveness** is the ability to detect and respond to changes that occur both inside and outside the body. Different cells in the body detect different types of changes and respond in characteristic ways. If you touch a hot pot, nerve cells sense the change or possible damage, make electrical signals called nerve impulses, and send these impulses to muscle cells, which contract to move your hand away, sometimes even before you are aware of the heat.

3. **Movement** includes motion of the whole body, individual organs, single cells, and even tiny structures inside cells.

4. **Growth** is an increase in body size due to an increase in the size of existing cells, the number of cells, or the amount of material surrounding cells.

5. **Differentiation** (dif'-er-en-shē-Ā-shun) is the process that unspecialized cells undergo to become specialized cells. Specialized cells differ in structure and function from the unspecialized cells that gave rise to them.

6. **Reproduction** (rē-prō-DUK-shun) refers to either the production of a new individual or the formation of new cells for growth, repair, or replacement.

Not all of these processes continually occur in cells throughout the body. However, when life processes cease to occur properly, cell death may occur. When cell death is extensive and leads to organ failure, the organism usually dies.

Your body must carry out several processes.

Responsiveness
You respond to a hot temperature by sweating.

Movement
Running is an example of large-scale movement.

Metabolism
You break down complex molecules in food into simpler ones for energy and for building new, complex molecules.

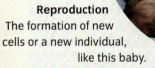

Reproduction
The formation of new cells or a new individual, like this baby.

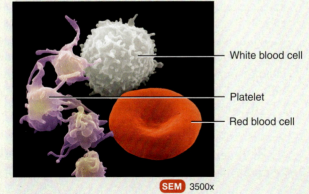

— White blood cell

— Platelet

— Red blood cell

SEM 3500x

Differentiation
A single type of cell differentiated into these very different blood cells.

Growth
Increase in body size (4–6 fold).

Homeostatic Balance Maintains Relatively Stable Conditions Inside the Body

You would not feel well if your body temperature changed from normal one moment to high fever in the next followed by chills in the moment after that. Likewise, you would be in serious danger if your heart rate changed erratically and spontaneously every minute. Your body and its cells require relatively stable conditions; the maintenance of these stable conditions is called **homeostasis**. Homeostasis protects the internal environment against changes both inside and outside the body.

> **homeostasis**
> (hō-mē-ō-STĀ-sis) The condition in which the body's internal environment remains relatively constant, within physiological limits.

Every body structure, from cells to systems, has one or more homeostatic mechanisms, which are mainly under the control of two systems:

- The nervous system detects changes from the balanced state and sends messages in the form of nerve impulses to organs that can counteract them. For example, when body temperature rises, nerve impulses cause sweat glands to secrete more sweat, which evaporates and cools the body.

> **hormone** (HOR-mōn)
> A secretion of endocrine cells that alters the physiological activity of target cells of the body.
>
> **feedback system**
> A sequence of events in which information about the status of a situation is continually reported (fed back) to a control center.

- The endocrine system corrects changes by secreting chemicals called **hormones** into the blood. Hormones affect specific body cells where they cause responses that restore homeostasis. For example, the hormone insulin reduces the blood glucose level when it is too high.

Nerve impulses typically cause rapid corrections, while hormones usually work more slowly.

Homeostasis is maintained by many **feedback systems** (**Figure 1.4**). Each monitored condition in a feedback system, or *feedback loop*, is termed a **controlled condition**. Any disruption that causes a change in a controlled condition is called a *stimulus*. Some stimuli come from outside the body, while others come from within.

In addition to the controlled condition and the stimulus, feedback systems have three other components:

1. A **receptor** monitors the controlled condition and sends information (input) to a control center.

2. A **control center** receives the input, compares it to a set of values that the controlled condition should have (set point) and sends output commands (nerve impulses or chemical signals) to an effector.

3. An **effector** receives output commands and produces a response that changes the controlled condition.

Feedback systems • Figure 1.4

Characteristics of a feedback system **a.**, how a negative feedback system controls blood pressure **b.**, and how a positive feedback system operates in labor **c.**

1 Some stimulus alters homeostasis by altering a controlled condition. *Negative feedback:* the stimulus increases blood pressure. *Positive feedback:* Contractions of the uterus stretch the cervix.

2 A receptor senses a change in the controlled condition and provides input to a control center. *Negative feedback:* Pressure receptors in certain blood vessels send nerve impulses to the brain. *Positive feedback:* Stretch receptors in the cervix send signals to the brain.

3 The control center receives the input, compares it to some set point, and sends output to an effector. *Negative feedback:* The brain compares the increase in blood pressure to the normal value and sends nerve impulses to the heart. *Positive feedback:* The brain (hypothalamus) interprets the impulses and releases oxytocin.

4 The effector receives the output and produces a response. *Negative feedback:* The heart receives nerve impulses to decrease its rate of beating. *Positive feedback:* Uterine muscles contract.

If a feedback system reverses the change in the controlled condition to restore it to the set point, this is a **negative feedback system**; for example, a negative feedback system controls blood pressure. However, if a feedback system further strengthens a change in the controlled condition, this is a **positive feedback system**. For example, childbirth is an example of positive feedback. During labor, uterine contractions force the baby's head into the cervix, which stretches. The stretching causes the hypothalamus to secrete a hormone called oxytocin, which induces more uterine contractions. Negative

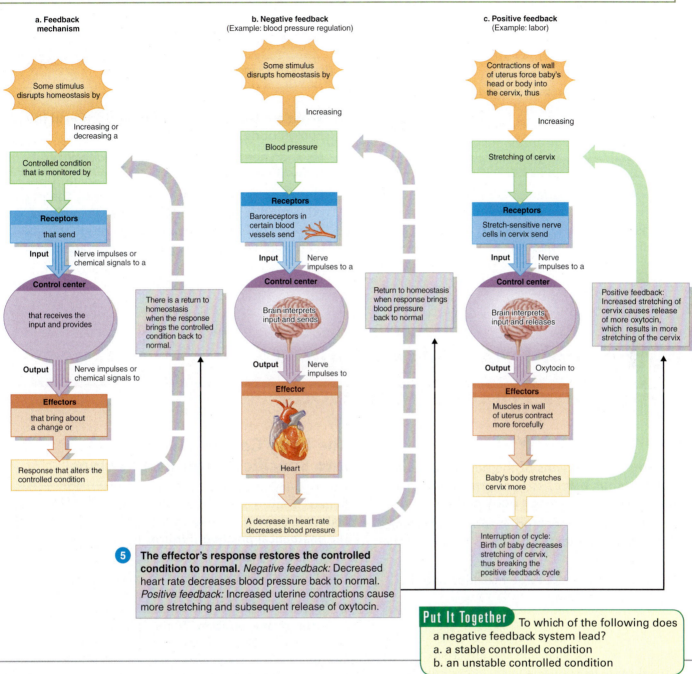

a. Feedback mechanism

Some stimulus disrupts homeostasis by

Increasing or decreasing a

Controlled condition that is monitored by

Receptors
that send

Input Nerve impulses or chemical signals to a

Control center
that receives the input and provides

There is a return to homeostasis when the response brings the controlled condition back to normal.

Output Nerve impulses or chemical signals to

Effectors
that bring about a change or

Response that alters the controlled condition

b. Negative feedback
(Example: blood pressure regulation)

Some stimulus disrupts homeostasis by

Increasing

Blood pressure

Receptors
Baroreceptors in certain blood vessels send

Input Nerve impulses to a

Control center
Brain interprets input and sends

Return to homeostasis when response brings blood pressure back to normal

Output Nerve impulses to

Effector
Heart

A decrease in heart rate decreases blood pressure

5 **The effector's response restores the controlled condition to normal.** *Negative feedback:* Decreased heart rate decreases blood pressure back to normal. *Positive feedback:* Increased uterine contractions cause more stretching and subsequent release of oxytocin.

c. Positive feedback
(Example: labor)

Contractions of wall of uterus force baby's head or body into the cervix, thus

Increasing

Stretching of cervix

Receptors
Stretch-sensitive nerve cells in cervix send

Input Nerve impulses to a

Control center
Brain interprets input and releases

Positive feedback: Increased stretching of cervix causes release of more oxytocin, which results in more stretching of the cervix

Output Oxytocin to

Effectors
Muscles in wall of uterus contract more forcefully

Baby's body stretches cervix more

Interruption of cycle: Birth of baby decreases stretching of cervix, thus breaking the positive feedback cycle

Put It Together To which of the following does a negative feedback system lead?
a. a stable controlled condition
b. an unstable controlled condition

feedback systems tend to maintain stable conditions, whereas positive feedback systems tend to be unstable and must be shut off by some event that is outside the feedback loop, such as the delivery of the child.

Homeostasis is *dynamic*; that is, it can change over a narrow range that is compatible with maintaining cellular life processes. Your physiological state can move from one **steady state** to another, depending upon various stimuli. For example, when you are resting, your body is in one steady state of conditions, but when you are exercising, your homeostatic mechanisms adapt to create another steady state. As a result of the many feedback mechanisms, most disruptions of homeostasis are mild and temporary; the responses of body cells quickly restore balance in the internal environment. For example, when you stand, your blood pressure momentarily drops, but it is restored to normal within a heartbeat or two. (If this did not happen, you would faint.) In other cases, the disruptions of homeostasis may be intense and prolonged, such as with severe infection or poisoning.

steady state A set of conditions that remains constant over some period of time.

All Living Organisms Carry Out Common Life Processes **9**

Aging and Disease Upset Homeostasis

Aging represents a progressive decline in the body's homeostatic mechanisms. Some components of feedback mechanisms may weaken or fail. For example, as we age, arterial walls become stiffer. As a result, the baroreceptors (see Figure 1.4) in certain arteries become less able to sense increases or decreases in blood pressure. Furthermore, the heart muscle becomes stiffer and weaker, and it becomes less able to increase the force of contraction to increase blood pressure.

Similarly, disease may alter the body's ability to respond to various stimuli, affecting components of feedback systems. For example, ovarian cancer may cause an ovary to secrete too much of the female sex hormone estrogen. The increased estrogen in the blood is communicated to the pituitary gland and hypothalamus, which depress the secretions of three other hormones that help maintain normal functioning of the ovaries and signal ovulation. Because their secretions are depressed, ovulation does not occur, and the woman becomes infertile.

1. **What** are two examples of reproduction in the human body?

2. **How** do negative and positive feedback systems differ from one another?

3. **How** would age affect the response by the negative feedback system to an increase in blood pressure? (Keep in mind that age alters the ability of the heart to change its rate of beating.)

Anatomical Road Maps Guide Navigation Through the Body

LEARNING OBJECTIVES

1. **Identify** major regions of the body and relate their common names to the corresponding anatomical terms for various body parts.

2. **Define** the directional terms and the anatomical planes and sections used to locate parts of the human body.

3. **Describe** the principal body cavities and the organs they contain.

If you ask someone for directions to some place in a city, you might get a response like, "Go two blocks north to Main Street, take a right, and proceed 0.5 miles." Scientists need to describe locations and directions on the human body in a similarly clinical fashion to avoid misinterpretation. For example, how would you describe a cut on your forearm? You might say that it is 2 cm above the right elbow. But does "above the elbow" mean toward the wrist or toward the shoulder? Anatomists have devised a system of terms and directions for naming various body parts and locating them precisely. How does this system work?

The Body Can Be Divided into Specific Anatomical Regions

Descriptions of any part of the human body assume that the body is in a specific stance called the **anatomical position** (**Figure 1.5**). In the anatomical position, the body is upright; however, two terms describe a reclining body. If the body is lying face down, it is in the **prone** position. If the body is lying face up, it is in the **supine** position. The face view of the body is called the **anterior** side, or **ventral** side, and the back view is called the **posterior** side, or **dorsal** side.

> **anatomical position** (an-a-TOM-i-kal) The subject stands erect with the head level, eyes facing forward, feet flat on the floor, and directed forward, and arms at the sides, with the palms turned forward.

The body is further divided into the following major regions:

- *Head (cephalic)*—Skull and face
- *Neck (cervical)*—Supports head and attaches it to trunk
- *Trunk (thoracic, abdominal, pelvic)*—Chest, back, abdomen, pelvis, and buttock
- *Upper limb*—Shoulder, armpit, arm, forearm, wrist, and hand
- *Lower limb*—Buttock, thigh, leg, ankle, and foot

This figure shows the anatomical positions, with directions and names of the major body regions (anatomical terms in parentheses).

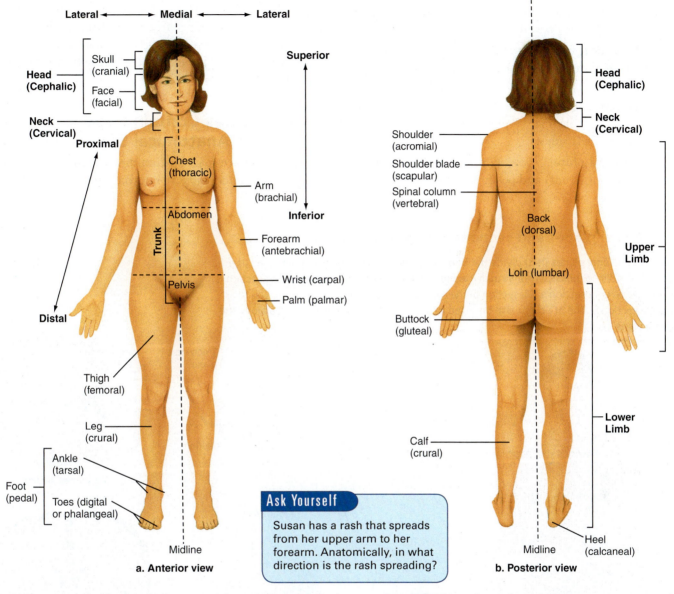

a. Anterior view

b. Posterior view

Ask Yourself

Susan has a rash that spreads from her upper arm to her forearm. Anatomically, in what direction is the rash spreading?

Now that you have learned about positions and regions, let's look at how a clinician would use directions to describe the location of that cut on your arm.

Directional Terms Describe the Location of Body Parts Relative to Each Other

Your body is a three-dimensional object, so anatomical locations must be specific in all three dimensions. In addition, a reference line called the **midline** is used to divide the body vertically into halves. With this in mind, let's look at the various directions (see Figure 1.5):

1. Trunk:
   • *Medial*—Toward the midline
   • *Lateral*—Away from the midline.
   • *Superior*—Toward the head
   • *Inferior*—Away from the head
2. Limbs:
   • *Proximal*—Toward the point of attachment
   • *Distal*—Away from the point of attachment
3. Trunk and limbs:
   • *Superficial*—Toward the surface of the body
   • *Deep*—Away from the surface

For example, you might describe that cut below your right elbow more precisely as a superficial cut on the anterior side of the right antebrachial region, 2 cm distal to the elbow.

To orient a viewer, anatomists use various **planes** (imaginary flat surfaces) to divide the body, body cavities, and organs into slices or flat surfaces called **sections** (**Figure 1.6**):

1. The **sagittal plane** (SAJ-i-tal; *sagitt-* = arrow) divides the object into right and left sides:
    *   The **midsagittal plane** passes through the midline of the body and divides the object into *equal* right and left sides.
    *   The **parasagittal plane** does not pass through the midline but instead divides the object into *unequal* right and left sides.
2. The **frontal plane**, or *coronal plane*, divides the object into anterior (front) and posterior (back) portions.

3. The **transverse plane** divides the object into superior (upper) and inferior (lower) portions. A transverse plane may also be called a *cross-sectional plane*, or *horizontal plane*.

Sagittal, frontal, and transverse planes are all at right angles to one another. By contrast, an **oblique plane** passes through the object at an angle between the transverse plane and a sagittal plane or between the transverse plane and the frontal plane.

Body Cavities Contain Organs and Other Anatomical Structures

Body cavities are spaces within the body that contain, protect, separate, and support internal organs (**Figure 1.7**).

Anatomical planes and sections • Figure 1.6

Three-dimensional planes (left) divide the body, its organs, and cavities into sections (right). If the brain is sliced (physically or virtually) along the transverse plane, the resulting section is called a transverse section **a**. Dissection of the brain along the frontal plane is referred to as a frontal section **b**. The brain can also be sliced into sagittal or midsagittal sections along the sagittal plane **c**.

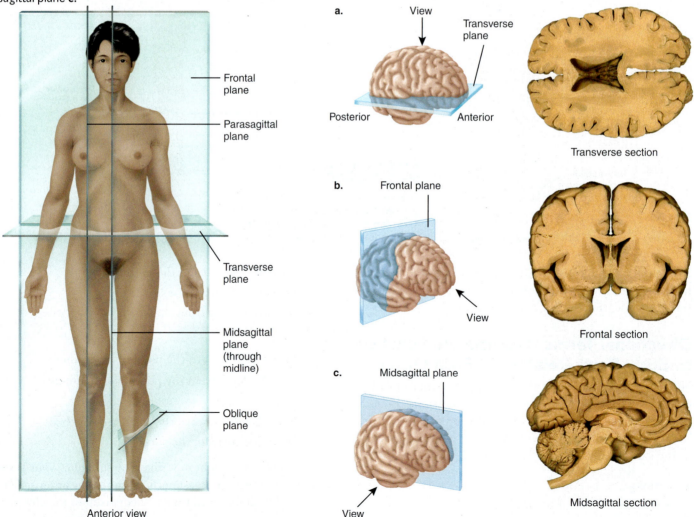

Frontal plane

Parasagittal plane

Transverse plane

Midsagittal plane (through midline)

Oblique plane

Anterior view

a. View

Transverse plane

Posterior · Anterior

Transverse section

b. Frontal plane

View

Frontal section

c. Midsagittal plane

View

Midsagittal section

Body cavities • Figure 1.7

The properties of each cavity are summarized in the table.

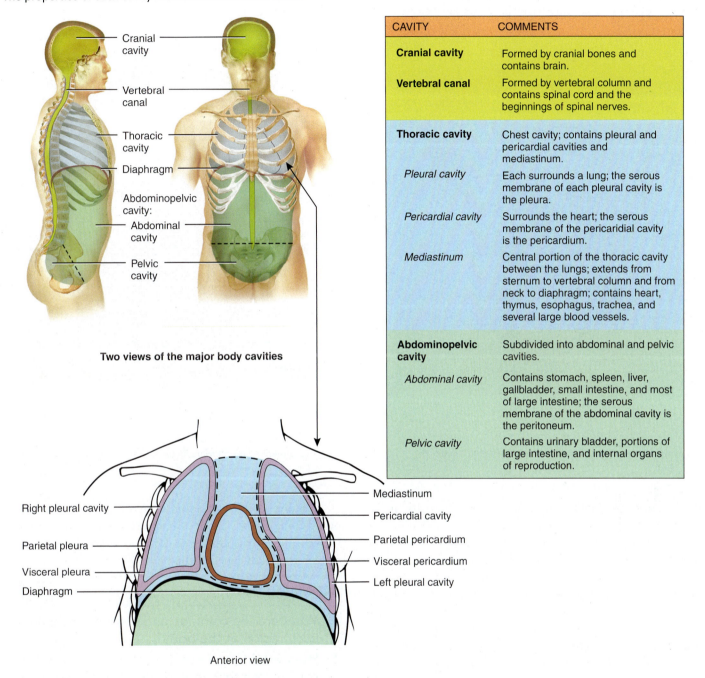

Two views of the major body cavities

CAVITY	COMMENTS
Cranial cavity	Formed by cranial bones and contains brain.
Vertebral canal	Formed by vertebral column and contains spinal cord and the beginnings of spinal nerves.
Thoracic cavity	Chest cavity; contains pleural and pericardial cavities and mediastinum.
Pleural cavity	Each surrounds a lung; the serous membrane of each pleural cavity is the pleura.
Pericardial cavity	Surrounds the heart; the serous membrane of the pericaridial cavity is the pericardium.
Mediastinum	Central portion of the thoracic cavity between the lungs; extends from sternum to vertebral column and from neck to diaphragm; contains heart, thymus, esophagus, trachea, and several large blood vessels.
Abdominopelvic cavity	Subdivided into abdominal and pelvic cavities.
Abdominal cavity	Contains stomach, spleen, liver, gallbladder, small intestine, and most of large intestine; the serous membrane of the abdominal cavity is the peritoneum.
Pelvic cavity	Contains urinary bladder, portions of large intestine, and internal organs of reproduction.

Anterior view

The cranial cavity contains the brain, and the vertebral cavity contains the spinal cord. The thoracic cavity is divided into two pleural cavities (left and right), containing the lungs, one pericardial cavity surrounding the heart, and the mediastinum, containing the heart, esophagus, trachea, and several large blood vessels. The diaphragm, a muscle that powers breathing, separates the thoracic cavity from the abdominopelvic cavity. The abdominopelvic cavity is further divided into the abdominal cavity and the pelvic cavity. The contents of the thoracic and abdominopelvic cavities are called **viscera**. The viscera of the thoracic and abdominopelvic cavities are covered by a thin, slippery double-layered membrane called a **serous membrane**. The serous membrane covering the lungs is called the **pleura**, the membrane covering the heart is called the **pericardium**, and the membrane covering the abdominopelvic cavity is called the **peritoneum**.

viscera (VIS-e-ra) The organs inside the thoracic and abdominopelvic cavities.

Divisions of the abdominopelvic cavity • Figure 1.8

To describe the precise location of organs within the abdominopelvic cavity, anatomists use a nine-region system **a.**, while health providers find it easier to use a quadrant system **b.**

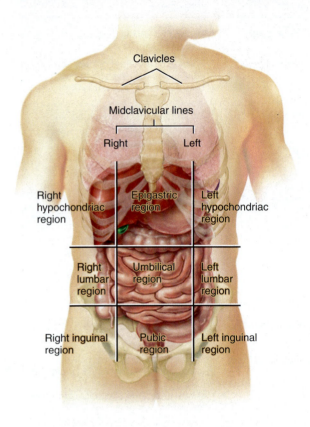

a. Anterior view showing location of abdominopelvic regions

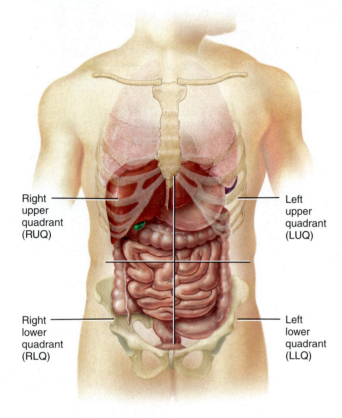

b. Anterior view showing location of abdominopelvic quadrants

To describe the location of many abdominal and pelvic organs more precisely, scientists divide the abdominopelvic cavity into smaller compartments. Anatomical scientists use a grid of two horizontal and vertical lines (like a tic-tac-toe grid) to separate the cavity into nine abdominopelvic regions (**Figure 1.8a**). Each region has a distinct name: the *right hypochondriac* (hī-pō-KON-drē-ak), *epigastric* (ep-i-GAS-trik), *left hypochondriac*, *right lumbar*, *umbilical* (um-BIL-i-kul), *left lumbar*, *right inguinal (iliac)* (IL-ē-ak), *hypogastric* (hī-pō-GAS-trik), and *left inguinal (iliac)*. However, because it is the patient who often describes the location of abdominopelvic pain, health providers find it easier to divide the abdominopelvic region into quadrants, using horizontal and vertical lines through the belly button (navel) (**Figure 1.8b**).

Finally, modern imaging technologies, such as computed tomography (CT), magnetic resonance imaging (MRI), and ultrasound (see *What a Health Provider Sees*), can show virtual slices of the body.

CONCEPT CHECK

1. **What** are the common and anatomical names of each region of the body?

2. **How** would you describe the location of a poison ivy rash that spreads up your left leg from your ankle toward your knee, using anatomical terminology?

3. **Why** do health providers use a quadrant system for describing the abdominopelvic cavity?

WHAT A HEALTH PROVIDER SEES

Medical Imaging

Medical images give health providers a look inside the patient for diagnostic purposes. There are many types of images, and we will look at a few common ones:

- *Radiographs (X-rays)*
  A single burst of X-rays passes through an area of the body and gets captured on sensitive film or digital imaging device. Dense structures (for example, bones) scatter X-rays and appear white. X-rays pass through less dense areas, which appear dark. Chemicals that create contrast can be injected to image tubular structures (for example, GI tract, blood vessels, ureters).

- *Magnetic resonance imaging (MRI)*
  A highly magnetic field aligns the protons in body fluids of the exposed area, a pulse of radio waves reads the pattern, and a computer constructs the image. MRI is good for imaging soft tissues but not bones.

- *Computed tomography (CT)*
  An X-ray beam traces an arc around a section of the body. Detectors read the scattering and feed the information to a computer, which constructs a sectional image of the body, where differing tissue densities show as shades of gray. Sections can be assembled to reveal three-dimensional images of soft tissues.

- *Ultrasound*
  High-frequency sound waves are passed through and reflected by tissues within the exposed area. A detector reads the reflection, and a computer constructs an image (a sonogram). This is a safe, non-invasive imaging method used most commonly to show fetal development, growths (for example, tumors, cysts, stones), and organ blood flow (Doppler ultrasound).

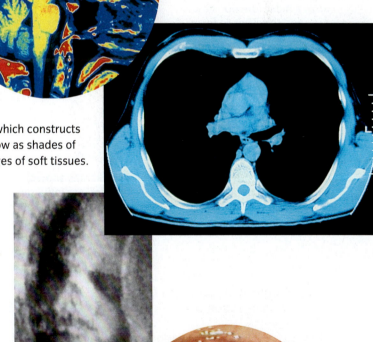

- *Direct visualization*
  Small cameras with fiber-optic cables can be inserted into the airways (bronchoscopy), abdominal cavity (laparoscopy), joints (arthroscopy), stomach (endoscopy), and large intestine (colonoscopy—image shown). The images are fed to a computer or TV screen.

Images, depending upon the type, can be interpreted by many health professionals.

Think Critically

1. A football player is injured, cannot stand on his right leg when weight is applied, and complains of pain. Which imaging technique(s) would you use to determine whether he broke his leg or ankle?

2. You suspect that a woman has a cyst (a sac with a distinct connective tissue wall, containing fluid or some other material) on her right ovary. Which imaging technique(s) would you use to confirm your diagnosis? Why?

Summary

1 Body Structure Is Closely Linked to Function at All Levels of Organization 2

- **Anatomy** is the science of structure and the relationships among structures; **physiology** is the science of how body structures function.

The cardiovascular system • Figure 1.2

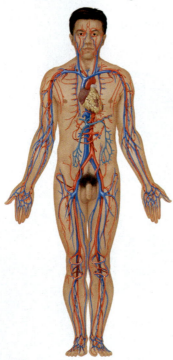

- The human body consists of six levels of organization: chemical, cellular, tissue, organ, system, and organism. Cells are the basic structural and functional unit of any organism. Cells that perform similar functions are grouped together to form tissues. Different tissues join together to form organs. Related organs with common functions form systems. Systems combine to form an organism.

- There are the 11 systems of the human body: integumentary, skeletal, muscular, nervous, endocrine, cardiovascular (as shown), lymphatic, respiratory, digestive, urinary, and reproductive.

2 All Living Organisms Carry Out Common Life Processes 6

- All living organisms have certain characteristics that set them apart from non-living things. Among the life processes in humans are **metabolism**, **responsiveness**, **movement**, **growth**, **differentiation**, and **reproduction**.

- Living organisms require a stable internal environment. **Homeostasis** is the maintenance of that stable environment within certain limits. Homeostasis is regulated by the nervous and endocrine systems acting together or separately. Homeostasis is maintained by **feedback systems** (as shown), which consist of receptors, a control center, and effectors. **Receptors** monitor changes in a controlled condition and send input to a **control center** that sets the value at which a controlled condition should be maintained, evaluates the input it receives, and generates output commands when they are needed. **Effectors** receive output from the control center and produce a response (effect) that alters the controlled condition.

- **Negative feedback systems** reverse the change in the controlled condition and are stable, while **positive feedback systems** strengthen the change and are unstable. Aging and disease alter feedback systems and their ability to maintain homeostasis.

Feedback systems • Figure 1.4

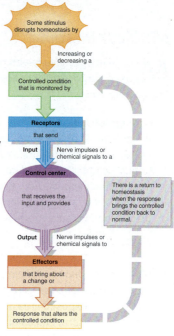

3 Anatomical Road Maps Guide Navigation Through the Body 10

- Descriptions of any region of the body assume the body is in the **anatomical position**. The human body is divided into several major regions: the **head**, **neck**, **trunk**, **upper limbs**, and **lower limbs**. Within body regions, specific body parts have common names and corresponding anatomical names.

Anatomical planes • Figure 1.6

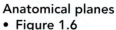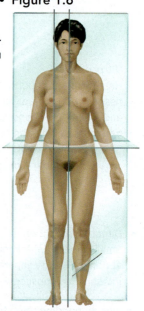

- Directional terms indicate the relationship of one part of the body to another and include **medial**, **lateral**, **superior**, **inferior**, **proximal**, **distal**, **superficial**, and **deep**.

- **Planes** (as shown) are imaginary flat surfaces, as shown, that divide the body or organs into parts or sections, which are named according to the plane on which the cut is made: **sagittal**, **frontal**, or **transverse**.

- **Body cavities** are spaces in the body that contain, protect, separate, and support internal organs. They include the **cranial**, **cervical**, **vertebral**, **thoracic**, and **abdominopelvic** cavities.

- To describe the location of organs easily, the abdominopelvic cavity may be divided into nine regions or four quadrants.

Key Terms

- anatomical position 10
- anatomy 2
- anterior 10
- atom 3
- biochemical 3
- body cavities 12
- cell 3
- control center 8
- controlled condition 8
- deep 11
- differentiation 6
- distal 11
- dorsal 10
- effector 8

- feedback system 8
- frontal plane 12
- growth 6
- homeostasis 8
- hormone 8
- inferior 11
- lateral 11
- medial 11
- metabolism 6
- midline 11
- midsagittal plane 12
- molecule 3
- movement 6
- negative feedback system 8

- oblique plane 12
- organ 3
- organism 3
- parasagittal plane 12
- pericardium 13
- peritoneum 13
- physiology 2
- planes 12
- pleura 13
- positive feedback system 8
- posterior 10
- prone 10
- proximal 11
- receptor 8

- reproduction 6
- responsiveness 6
- sagittal plane 12
- sections 12
- serous membrane 13
- steady state 9
- superficial 11
- superior 11
- supine 10
- system 3
- tissue 3
- transverse plane 12
- ventral 10
- viscera 13

Critical and Creative Thinking Questions

1. Nerve cells in the hypothalamus sense the level of the adrenal hormone cortisol in the blood. When cortisol levels are low, the hypothalamus secretes a hormone, called corticotropin-releasing factor (CRF), which stimulates the anterior pituitary gland to secrete adrenocortical hormone (ACTH). ACTH acts on the adrenal cortex, which secretes cortisol. The increase in blood cortisol shuts down CRF and ACTH secretions by the hypothalamus and pituitary gland. What type of feedback system is this? What is the controlled condition? Identify the receptors, control center, and effectors in this system.

2. A severely burned patient has tissue damage deep into his arm down to the bone and is bleeding. His blood pressure is falling, and he passes out. What body systems do his burn and its consequences affect?

3. A patient with a knife wound appears in the emergency room. The knife entered the victim just below the diaphragm at the midline and was thrust upward into the chest, past the lower tip of the heart and into the right lung. What is the correct anatomical description of the wound's entry and direction? Which body cavities are affected?

4. You dissect an object that has an inner lining of epithelium, enclosed by a layer of smooth muscle and an outer tough connective tissue covering. Based on the hierarchy of body organization, what have you dissected? Explain.

5. A patient complains of shooting pain from his left groin to his stomach. How would you describe the location of the pain for a medical report?

What is happening in this picture?

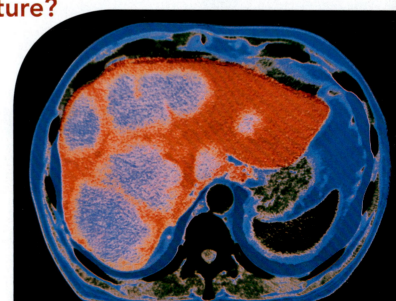

Think Critically This X-ray image with false colors shows tumors (pink/blue) inside the liver (red).
1. What is the anatomical orientation of this picture?
2. In what direction are you looking?
What type of section is this? In what cavity are you looking?

Self-Test

(Check your answers in Appendix C.)

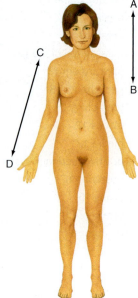

1. The term _____ correctly describes the relationship indicated as B on this figure.

   a. inferior

   b. superior

   c. proximal

   d. distal

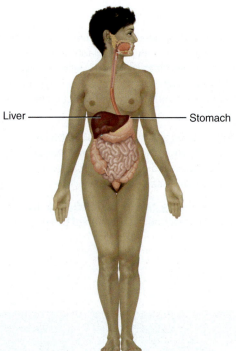

Liver ———————   ——————— Stomach

Use this figure for questions 2–3.

2. In this figure, the position of the stomach is _____ relative to the liver.

   a. proximal

   b. distal

   c. superior

   d. inferior

3. The system shown in the figure _____.

   a. distributes oxygen and nutrients to body tissues and removes carbon dioxide and wastes

   b. chemically coordinates various body functions

   c. breaks down food, absorbs nutrients, and excretes solid waste

   d. controls the ionic composition of the blood and removes wastes

4. The _____ is the part of a feedback system that receives output.

   a. receptor

   b. effector

   c. control center

   d. controlled condition

5. Aerobic respiration is a physiological process that breaks down glucose into carbon dioxide and water, while capturing chemical energy into ATP. The life process that best describes this pathway is _____.

   a. movement

   b. reproduction

   c. responsiveness

   d. metabolism

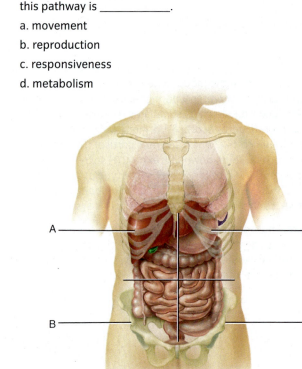

6. Which quadrant would best pinpoint the location of a patient's stomach pain?

   a. A   b. B   c. C   d. D

7. A patient has injured his scapula. The common name for this body part is the _____. (*Hint*: See Figure 1.5b.)

   a. hip

   b. shoulder

   c. shoulder blade

   d. arm

8. A patient has carpal tunnel syndrome. This patient is hurting in the _____.

  a. ankle

  b. wrist

  c. knee

  d. elbow

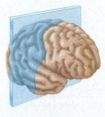

9. The system shown in the figure _____.

  a. distributes oxygen and nutrients to body tissues and removes carbon dioxide and wastes

  b. chemically coordinates various body functions

  c. breaks down food, absorbs nutrients, and excretes solid waste

  d. controls the ionic composition of the blood and removes wastes

10. The _____ plane is indicated in the figure at right.

  a. parasagittal

  b. transverse

  c. frontal

  d. midsagittal

11. In the temperature control feedback system, the skin's sweat glands act as the _____.

  a. receptor

  b. control center

  c. controlled condition

  d. effector

12. The _____ is responsible for rapidly sensing and processing information, communicating, and coordinating various systems.

  a. nervous system

  b. endocrine system

  c. lymphatic system

  d. integumentary system

13. _____ is not a negative feedback system.

  a. Blood glucose regulation

  b. Childbirth

  c. Temperature regulation

  d. Blood pressure regulation

14. The _____ would be found in the abdominopelvic cavity.

  a. eye

  b. lung

  c. heart

  d. bladder

15. The _____ system is affected by sunburn.

  a. respiratory

  b. endocrine

  c. integumentary

  d. lymphatic

THE PLANNER ✓

Review your Chapter Planner on the chapter opener and check off your completed work.

Introductory Chemistry

Many children play with building blocks such as Legos. They can combine blocks of different shapes and colors to make almost any structure they can imagine. For example, a child may build an airplane and then may decide to take it apart and make a car, a dinosaur, or even a model of the human body. An almost infinite number of toys can be made using these basic "building blocks."

Nature also uses building blocks—they're called *atoms*. Atoms vary in size and properties. They can combine to make more complex structures called *molecules*, which have different properties than their atomic building blocks. Like a an airplane made out of blocks, molecules can also be broken down. For example, your body can break down the complex molecules in the fast food meal you had for lunch into simpler ones that it absorbs. From these simpler molecules, it can build more complex ones that it uses to make various structures, such as parts of cells, tissues, and organs. It can also string together many simple molecules to make more complex ones that it will use to store *energy*, the ability to do work. (If you eat too much fast food, the molecules may be stored as fat.) Finally, your body can break down some simple molecules to extract the stored energy. This is the chemistry of your body.

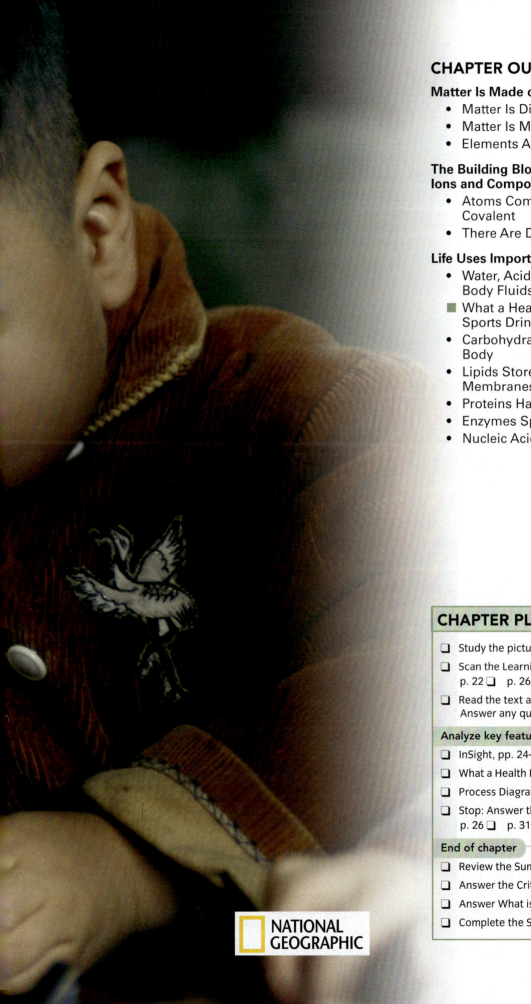

CHAPTER OUTLINE

CHAPTER PLANNER ✓

- ❑ Study the picture and read the opening story.
- ❑ Scan the Learning Objectives in each section:
  p. 22 ❑ p. 26 ❑ p. 31 ❑
- ❑ Read the text and study all visuals. Answer any questions.

Analyze key features

- ❑ InSight, pp. 24–25 ❑ p. 30 ❑ p. 33 ❑ pp. 40–41 ❑
- ❑ What a Health Provider Sees, p. 34 ❑
- ❑ Process Diagram, p. 38 ❑ p. 39 ❑
- ❑ Stop: Answer the Concept Checks before you go on:
  p. 26 ❑ p. 31 ❑ p. 42 ❑

End of chapter

- ❑ Review the Summary and Key Terms.
- ❑ Answer the Critical and Creative Thinking Questions.
- ❑ Answer What is happening in this picture?
- ❑ Complete the Self-Test and check your answers.

NATIONAL GEOGRAPHIC

Matter Is Made of Elements and Atoms

LEARNING OBJECTIVES

1. **Distinguish** between matter and energy.
2. **List** the major chemical elements in your body.
3. **Describe** the parts of an atom.
4. **Define** *isotope*.
5. **Define** *radioisotope* and describe the uses of radioisotopes.

Chemistry (KEM-is-trē) is the science of the structure and interactions of matter and energy. Let's take a closer look at the different states of matter and energy and examine the changes and interactions that each undergo.

Matter Is Different from Energy

Matter is anything that occupies space and has mass. Matter comes in three forms, or states:

- *Solid*—A **solid** has definite shape and **volume**. Examples of solids are bones and muscles.

- *Liquid*—A **liquid** has definite volume but not shape. Examples of liquids are blood plasma and urine.

- *Gas*—A **gas** has no definite shape or volume. An example of a gas is air.

> **volume** The amount of space taken up by matter.

Matter can undergo **physical changes**, such as the change of state of water from a liquid to a gas when it boils. As another example, ice melts into liquid water, but both forms are water; the basic substance does not change. Matter can also undergo **chemical changes**. For example, when a nail rusts, the iron combines with oxygen to form iron oxide (rust). When wood burns in a fire (**Figure 2.1**), the carbon-containing molecules in the wood combine with oxygen in the air to form gaseous carbon dioxide and water vapor.

> **physical change** A change in form that matter can undergo without altering its basic nature.
>
> **chemical change** A change in matter that alters its basic nature.

Unlike matter, energy has no mass and does not occupy space. **Energy** is defined as the ability to do work. Energy is usually classified as either kinetic energy or potential energy. **Kinetic energy** is the energy of motion, such as moving your body or arms or moving smaller pieces of matter, such as atoms and molecules. **Potential energy** is stored energy. For example, a book on a high shelf has gravitational potential energy. If the book fell off the shelf because of the constant vibration from your roommate hitting the wall with a basketball, the book's downward motion would be kinetic energy. So, poten-

Matter and energy in a campfire • Figure 2.1

Cooking over a campfire shows many states of matter, energy, physical changes, and chemical changes.

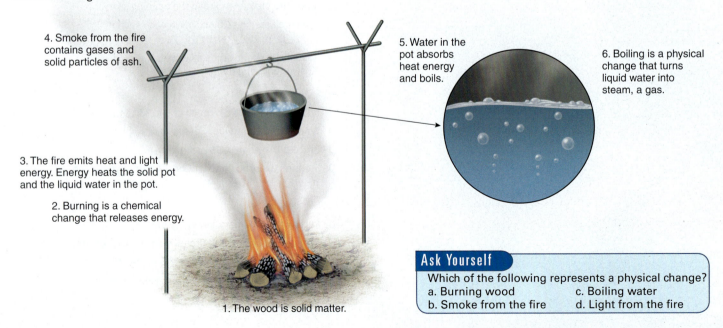

4. Smoke from the fire contains gases and solid particles of ash.

5. Water in the pot absorbs heat energy and boils.

6. Boiling is a physical change that turns liquid water into steam, a gas.

3. The fire emits heat and light energy. Energy heats the solid pot and the liquid water in the pot.

2. Burning is a chemical change that releases energy.

1. The wood is solid matter.

Ask Yourself

Which of the following represents a physical change?
a. Burning wood   c. Boiling water
b. Smoke from the fire   d. Light from the fire

Chemical elements in the body • Figure 2.2

The 26 elements in the human body fall into three categories: major elements, lesser elements, and trace elements.

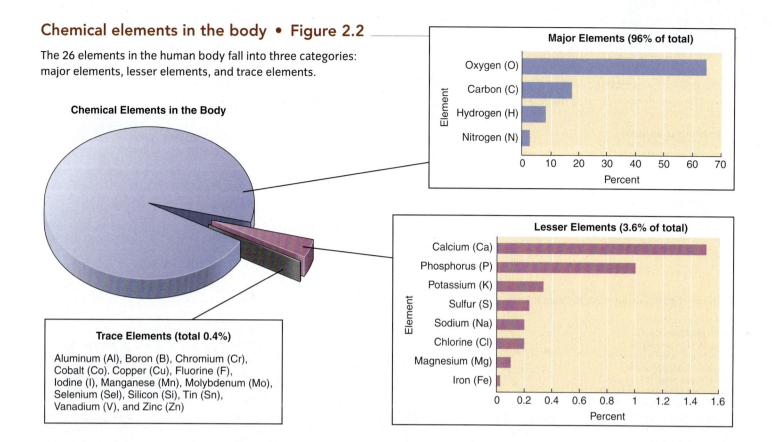

Chemical Elements in the Body

Major Elements (96% of total)

(bar chart: Element vs. Percent, 0 to 70)
- Oxygen (O)
- Carbon (C)
- Hydrogen (H)
- Nitrogen (N)

Lesser Elements (3.6% of total)

(bar chart: Element vs. Percent, 0 to 1.6)
- Calcium (Ca)
- Phosphorus (P)
- Potassium (K)
- Sulfur (S)
- Sodium (Na)
- Chlorine (Cl)
- Magnesium (Mg)
- Iron (Fe)

Trace Elements (total 0.4%)

Aluminum (Al), Boron (B), Chromium (Cr), Cobalt (Co). Copper (Cu), Fluorine (F), Iodine (I), Manganese (Mn), Molybdenum (Mo), Selenium (Sel), Silicon (Si), Tin (Sn), Vanadium (V), and Zinc (Zn)

tial energy and kinetic energy can be exchanged. Both kinetic energy and potential energy can be classified as other forms of energy, including the following:

- **Mechanical energy** is energy directly involved in moving matter. For example, when you lift a box, you use mechanical energy. Specifically, your muscles generate a force to lift the box against the force of gravity.

- **Chemical energy** is energy that is stored in the chemical bonds of matter. For example, the wood for a campfire stores energy in the bonds of its carbon-containing molecules.

- **Electrical energy** comes from moving positively or negatively charged particles. For example, when you plug a lamp into a wall outlet, negatively charged electrons move through the lamp's lightbulb. The energy driving the flow of electrons is electrical energy.

- **Radiant energy** is energy from the **electromagnetic spectrum**. Electromagnetic radiation travels in waves.

The most common type of energy conversion that you will see as you study anatomy and physiology is the shift between chemical potential energy and kinetic energy or heat energy. In your body, you break down the chemical energy in food and

electromagnetic spectrum The distribution of frequencies of light, including gamma rays, X-rays, ultraviolet light, visible light, microwaves, and radio waves.

store it in other substances, such as adenosine triphosphate (ATP), carbohydrates, and fats. The processes that convert energy from one form to another are not 100% efficient, so some energy is lost as heat energy. It is the heat energy lost from the chemical reactions in your body that keeps your body temperature above that of its surroundings.

Matter Is Made of Chemical Elements

Matter is made of building blocks called **chemical elements**. Chemical elements are substances that cannot be broken down into any simpler units by chemical means and still retain the same chemical properties. There are 117 different chemical elements; 94 occur in nature, and 23 are synthetic. Only 26 of the chemical elements are normally present in the human body (**Figure 2.2**). Each element is represented by a one- or two-letter symbol representing the element's name, usually an English or Latin name or a name representing a famous scientist. For example, C is carbon, H is hydrogen, O is oxygen, N is nitrogen, P is phosphorus, Ca is calcium, and Na, which comes from the Latin word *natrium*, is sodium. Figure 2.2 shows the organization of the elements in the human body into three categories: major (96%), lesser (3.6%), and trace (0.4%). The chemical symbol of each element is also given.

Atoms and isotopes make up all matter. They are made of basic particles: protons, neutrons, and electrons. The arrangement of these particles in the atom's structure dictates its chemical properties.

a. Structure of an atom

- An atom is the smallest form of an element that retains the properties of that element.
- An atom is electrically neutral, so the number of protons = number of electrons.
- Atoms of different elements have different numbers of protons and electrons.

Nucleus
- The nucleus contains most of the mass of the atom.

Electrons
- Electrons orbit the nucleus in a cloud. The electrons do not have distinct locations, but they fill discrete energy levels or "shells."

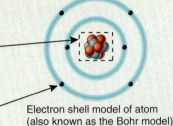

Electron shell model of atom (also known as the Bohr model)

Electron cloud model of atom

● Protons (p^+)   ● Neutrons (n^0)   • Electrons (e^-)

b. Atoms of common elements in living things

- Each electron shell can accommodate a maximum number of electrons (level 1: up to 2; level 2: up to 8, and so on). Each shell above the first has sub-levels that fill in a specific order.
- The number of electrons in the outer (valence) shell determines the element's chemical properties and reactivity.

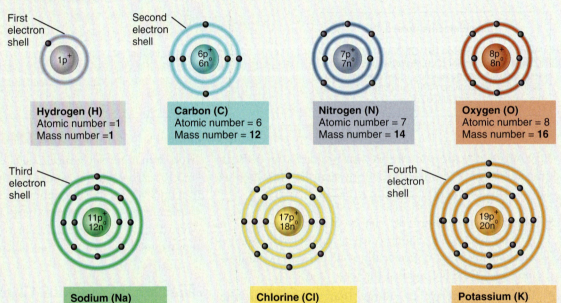

First electron shell

Second electron shell

Third electron shell

Fourth electron shell

Hydrogen (H)
Atomic number = 1
Mass number = 1

Carbon (C)
Atomic number = 6
Mass number = **12**

Nitrogen (N)
Atomic number = 7
Mass number = **14**

Oxygen (O)
Atomic number = 8
Mass number = **16**

Sodium (Na)
Atomic number = 11
Mass number = **23**

Chlorine (Cl)
Atomic number = 17
Mass number = **35**

Potassium (K)
Atomic number = 19
Mass number = **39**

Atomic number = number of protons in an atom
Mass number = number of protons + number of neutrons in an atom (boldface indicates most common isotope)

Elements Are Made of Atoms

Each element is made of **atoms**, the smallest units of matter that retain the properties and characteristics of the element. For example, pure coal contains only carbon atoms and oxygen gas contains only oxygen atoms.

An atom consists of two basic parts: a nucleus and one or more electrons (**Figure 2.3a**). The centrally located **nucleus** contains positively charged **protons (p^+)** and uncharged (neutral) **neutrons (n^0)**. Because each proton has one positive charge, the nucleus is positively charged. **Electrons (e^-)** are tiny, negatively charged particles that move about in a large space surrounding the nucleus.

They do not follow a fixed path or orbit but instead form a negatively charged "cloud" that surrounds the nucleus. The number of electrons in an atom equals the number of protons. Because each electron carries one negative charge, the negatively charged electrons and the positively charged protons balance each other. As a result, each atom is electrically neutral, meaning its total charge is zero.

The number of protons in the nucleus of an atom is called the atom's **atomic number**. The atoms of each different element have a different number of protons in the nucleus and, hence, a different atomic number (**Figure 2.3b**). The total number of protons plus neutrons in an atom is its **mass number**. For instance, an atom of so-

c. Isotopes

- **Isotopes** are atoms of the same element that have the same number of protons and electrons, but different numbers of neutrons.
- Isotopes of an element have the same chemical properties because the number of electrons in the outermost shell is the same.
- Some isotopes have stable nuclei, while others have unstable nuclei. Unstable nuclei change or decay into stable ones by emitting radioactive particles and energy. These unstable isotopes are called **radioactive isotopes**, **radioisotopes**, or **radionuclides**.
- Isotopes are usually named by giving the element number and the mass number. Hydrogen-2 and hydrogen-3 are isotopes that have other names as well.

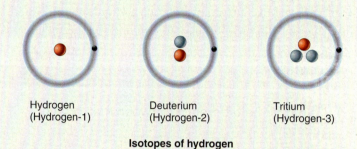

Hydrogen
(Hydrogen-1)

Deuterium
(Hydrogen-2)

Tritium
(Hydrogen-3)

Isotopes of hydrogen

d. Radioisotopes in medicine

Examples of radioisotopes used in medicine

Name	Use
Technetium-99	Imaging
Chromium-51	Labeling red blood cells
Iodine-125	Treating cancer, assessing kidney function
Iodine-131	Imaging thyroid, treating thyroid cancer
Iridium-92	Treating cancer
Strontium-89	Treating bone cancer pain
Xenon-133	Assessing lung ventilation
Carbon-11	Imaging
Oxygen-15	Imaging
Nitrogen-13	Imaging
Fluorine-18	Imaging
Thallium-201	Diagnosing heart conditions

- Radioisotopes are produced in nuclear reactors.
- Radioisotopes are ingested or injected into patients.

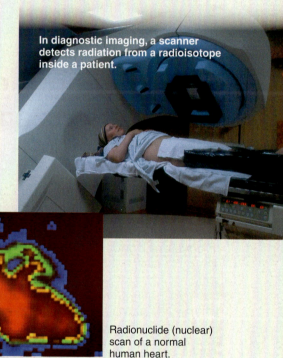

In diagnostic imaging, a scanner detects radiation from a radioisotope inside a patient.

Radionuclide (nuclear) scan of a normal human heart.

dium, with 11 protons and 12 neutrons in its nucleus, has an atomic number of 11 and a mass number of 23.

Even though their exact positions cannot be predicted, specific groups of electrons are most likely to move about within certain regions around the nucleus. These regions are called **electron shells**, or **electron levels**, and they are depicted as circles in Figure 2.3a, b. The electron shells can hold different numbers of electrons. The outermost shell of an atom is called the **valence shell**. The number of electrons in the valence shell determines the atom's chemical reactivity; the maximum number of electrons that the valence shell can hold is eight.

Atoms of almost all elements have some variations in which the number of protons in the nucleus is the same but the number of neutrons within the nucleus is different. These variations are called **isotopes** (**Figure 2.3c**). Isotopes have the same atomic number but different atomic masses. Many isotopes have more neutrons than the nucleus can hold and remain stable. The nucleus of such unstable isotopes changes or decays to a stable nucleus by releasing radioactive particles and energy; such unstable isotopes are called **radioisotopes**. Radioisotopes are used in nuclear medicine for diagnostic imaging and cancer treatment (**Figure 2.3d**).

Matter Is Made of Elements and Atoms **25**

Information about atoms and elements can be found in the **periodic table of the elements**, or simply the *periodic table* (see Appendix). Here, the elements are arranged in rows called **periods** and columns, in order of increasing atomic number. In each element's square are its name, symbol (one- or two-letter abbreviation), atomic number, and mass number. (The atomic mass number in the table is a weighted average of the atomic masses of the element's isotopes. This is why you see decimal numbers in the mass numbers of the table.) The period number indicates the number of energy levels in the element's atom, and the column number indicates the number of valence electrons. Chemists use the information in the periodic table to predict how atoms of different elements will react chemically.

CONCEPT CHECK	STOP

1. **What** is the difference between matter and energy?
2. **What** are the major chemical elements in the human body?
3. **What** particles are found in the nucleus of an atom?
4. **What** is an isotope of an element?
5. **How** are radioisotopes used in nuclear medicine?

The Building Blocks of Matter Fit Together to Make Ions and Compounds

LEARNING OBJECTIVES

1. **Distinguish** between an atom, an ion, a molecule, and a compound.
2. **Compare** and contrast an ionic bond with a covalent bond.
3. **Distinguish** between a polar covalent bond and a non-polar covalent bond.
4. **Describe** the various types of chemical reactions.

hen atoms interact by exchanging electrons, they form **chemical bonds**, which bind them together in molecules and compounds and resist separation. Let's take a look at two types of important bonds: ionic and covalent.

Atoms Combine to Form Compounds: Ionic and Covalent

Atoms of various elements can interact with each other by exchanging electrons of their valence shells in one of three ways:

- Give up one or more valence electrons to another atom
- Take one or more valence electrons from another atom
- Share one or more valence electrons with one or more atoms

The chance that an atom will form a chemical bond with another atom depends on the number of electrons in its valence shell. An atom with an outer shell holding eight electrons is *chemically stable*, which means it is unlikely to form chemical bonds with other atoms. Neon, for example, has eight electrons in its outer shell, and for this reason it rarely forms bonds with other atoms. In contrast, the atoms of most biologically important elements do not have eight valence electrons. However, atoms tend to interact in such a way as to produce a chemically stable arrangement of eight electrons in the outer shell of each atom. (Note that hydrogen can hold only two electrons in its valence shell.) This tendency toward achieving a full valence shell is called the *octet rule*.

Let's look at the various ways that atoms can interact.

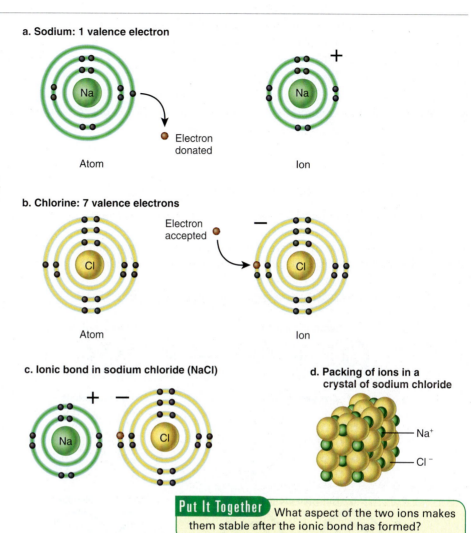

a. Sodium: 1 valence electron

Electron donated

Atom

Ion

When sodium and chlorine atoms interact, sodium gives up a valence electron to become a stable sodium ion with 8 electrons in its outer shell. Chlorine takes an electron to become a chloride ion with 8 electrons in its outer shell.

b. Chlorine: 7 valence electrons

Electron accepted

Atom

Ion

c. Ionic bond in sodium chloride (NaCl)

d. Packing of ions in a crystal of sodium chloride

Both ions are electrically attracted by an ionic bond and become a new ionic compound called sodium chloride.

Na^+

Cl^-

Put It Together What aspect of the two ions makes them stable after the ionic bond has formed?

Ions and ionic bonds If an atom either *gives up* or *gains* electrons to conform to the octet rule, it becomes an **ion**, an atom that has a positive or negative charge due to having unequal numbers of protons and electrons. An ion of an atom is symbolized by writing its chemical symbol followed by the number of positive (+) or negative (–) charges. For example, Ca^{2+} stands for a calcium ion that has two positive charges because it has given up two electrons.

Let's look at the ionic reaction between sodium, a highly reactive metal, and chlorine, a poisonous gas (**Figure 2.4a**). When sodium gives up its valence electron to chlorine, elemental sodium becomes a sodium ion (Na^+) that has eight electrons in its outer shell. When chlorine accepts the electron from sodium, it becomes a chloride ion (Cl^-), also with eight electrons in its outer shell.

These two stable ions are electrically attracted (due to the attraction between positively and negatively charged ions). This electrical attraction holds the ions together and is called an **ionic bond**. The new ionic compound, sodium chloride, is chemically different from either sodium or chlorine; it is what we know as common table salt (**Figure 2.4b**). Typically, ionic compounds dissolve in water. Both positively and negatively charged ions remain as ions, but become attracted to water molecules in solution.

Positively charged ions are called *cations* because they are attracted to the negatively charged electrodes of electrical sources called **cathodes**. In contrast, negatively charged ions are called *anions* because they are attracted to positively charged electrodes of electrical sources called **anodes**. So, Na^+ is a cation, and Cl^- is an anion.

Covalent bonds and molecules When two or more atoms *share* electrons, they form a **molecule** (MOL-e-kūl). A molecule may consist of two or more atoms of the same element (**Figure 2.5a-c**) or two or more atoms of different elements (**Figure 2.5d,e**). When *different* elements combine to form a molecule, we call that substance a *compound*. Like ionic compounds, covalent compounds have chemical properties that are different from those of the individual elements.

A **molecular formula** indicates the number and type of atoms that make up a molecule. In the molecular formula O_2, the subscript 2 indicates there are two atoms of oxygen. In the water molecule, H_2O, one atom of oxygen shares electrons with two atoms of hydrogen. (Note that O_2 is a molecule but not a compound because it is composed of identical elements.)

The shared electron pair between two atoms is called a **covalent bond**. The greater the number of covalent

Covalent bond formation and molecules • Figure 2.5 _____

When two or more atoms interact by sharing valence electrons, they form molecules. The shared electron pairs are called covalent bonds. In some molecules—for example, oxygen **b.** and nitrogen **c.**—two atoms may share two or three pairs of electrons and form double or triple covalent bonds. Lines in structural formulas represent covalent bonds.

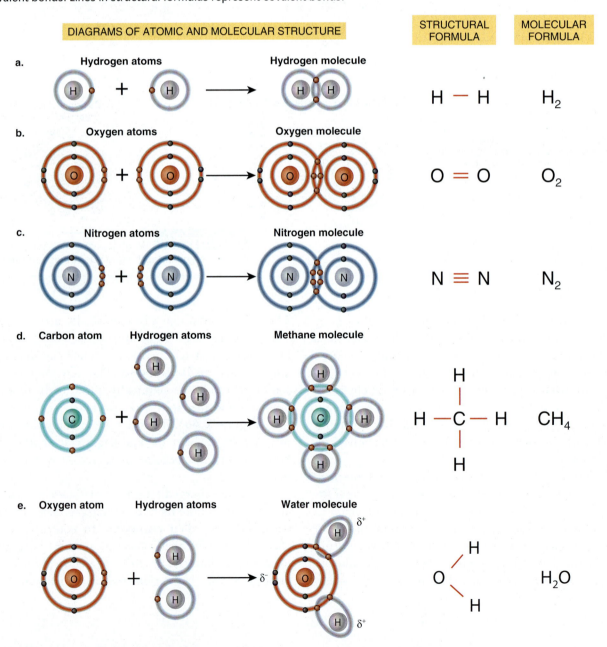

bonds, the more strongly the atoms are bound together. In some molecules, two atoms can form single, double, or triple covalent bonds (Figure 2.5). Unlike ionic bonds, covalent bonds do not break apart when the molecule is dissolved in water. Covalent bonds are the most common chemical bonds in the body.

Polar molecules and hydrogen bonds

In some covalent compounds, the electrons are shared unevenly between the two atoms. One atom attracts electrons more strongly than the other, so the electrons spend more time around the stronger atom than around the weaker one. This results in one end of the molecule having a partial negative charge (denoted by the symbol δ^-), while the other has a partial positive charge (δ^+). These types of molecules are called **polar molecules**, and the bonds between them are referred to as **polar covalent bonds**. Polar molecules usually occur when a weak atom such as hydrogen forms covalent bonds with strong atoms such as oxygen or nitrogen. Water is an example of a polar molecule (Figure 2.5e).

In contrast, other covalent compounds share valence electrons equally between the atoms. The molecules in these compounds are **non-polar molecules**, and the bonds between them are referred to as **non-polar covalent bonds**. For example, hydrogen molecules (H_2) and methane molecules (CH_4) are non-polar molecules (Figure 2.5a, d).

You can think of polar molecules as small bar magnets. The partially positive end of one polar molecule will attract the partially negative end of another, and vice versa. When a partially positive end (δ^+) of one polar molecule is attracted to the partially negative end (δ^-) of another polar molecule, a weak bond is formed. The partially positive end usually involves a hydrogen atom, so this type of bond is called a **hydrogen bond**. The hydrogen bonds between water molecules (**Figure 2.6**) give water some unique properties, such as a relatively high boiling point and low freezing point compared to other liquids. In addition, the hydrogen bonding between water molecules affects their formation of crystals when water freezes. Because the crystals take up more space than the number of water molecules that make it up, water expands when it freezes.

Hydrogen bonds are weak, so they are easily formed and broken. They play important roles in the shapes of biologically important molecules such as proteins and deoxyribonucleic acid (DNA) (see Figures 2.12 and 2.14).

Free radicals

Most of an atom's electrons associate in pairs. However, when an ion or a molecule has an unpaired electron in its outer shell, it is called a **free radical**. A common example of a free radical is *superoxide*, which is formed by the addition of an electron to an oxygen molecule. Having an unpaired electron makes a free radical unstable and destructive to nearby molecules. Free radicals break apart important body molecules by giving an unpaired electron to another molecule or accepting an electron from another molecule.

Hydrogen bonds in water • Figure 2.6

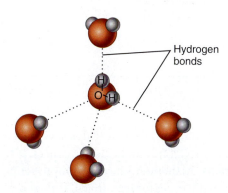

a. Weak hydrogen bonds form between the partially positive hydrogen atoms of one water molecule and the partially negative oxygen atoms of others.

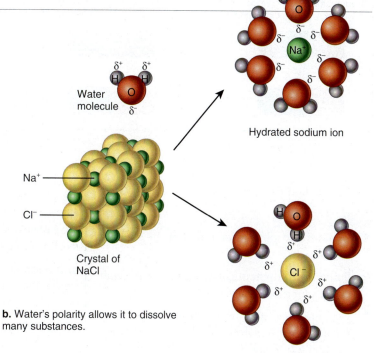

b. Water's polarity allows it to dissolve many substances.

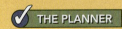
All chemical reactions change reactants into products, but do so in different ways.

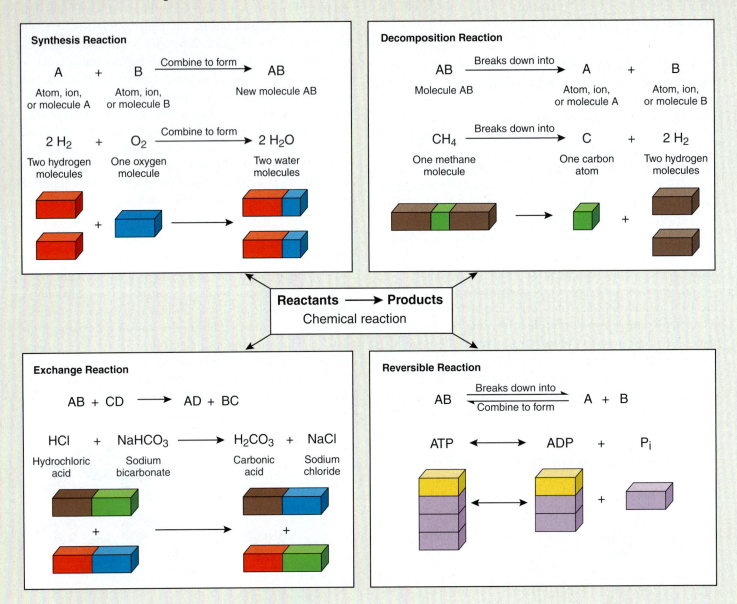

Synthesis Reaction

A + B —Combine to form→ AB

Atom, ion, or molecule A   Atom, ion, or molecule B   New molecule AB

$2 H_2$ + O_2 —Combine to form→ $2 H_2O$

Two hydrogen molecules   One oxygen molecule   Two water molecules

Decomposition Reaction

AB —Breaks down into→ A + B

Molecule AB   Atom, ion, or molecule A   Atom, ion, or molecule B

CH_4 —Breaks down into→ C + $2 H_2$

One methane molecule   One carbon atom   Two hydrogen molecules

Reactants ——→ Products
Chemical reaction

Exchange Reaction

AB + CD ——→ AD + BC

HCl + $NaHCO_3$ ——→ H_2CO_3 + NaCl

Hydrochloric acid   Sodium bicarbonate   Carbonic acid   Sodium chloride

Reversible Reaction

AB ⇌ A + B   (Breaks down into / Combine to form)

ATP ⟷ ADP + P_i

In our bodies, several processes can generate free radicals:

- Exposure to ultraviolet radiation in sunlight or to X-rays.
- Some normal chemical reactions that occur during metabolic processes.
- Certain harmful substances, such as carbon tetrachloride (a solvent used in dry cleaning). Carbon tetrachloride can strip electrons away from other molecules and produce free radicals.

atherosclerosis (ath'-e-rō-skle-RŌ-sis) The buildup of fatty materials in blood vessels.

antioxidants Substances that inactivate oxygen-derived free radicals.

Among the many disorders and diseases linked to oxygen-derived free radicals are cancer, **atherosclerosis**, Alzheimer's disease, emphysema, diabetes mellitus, cataracts, macular degeneration, rheumatoid arthritis, and deterioration associated with aging. Consuming more **antioxidants** is thought to slow the pace of damage caused by free radicals. Important dietary antioxidants include selenium, zinc, beta-carotene, and vitamins C and E. Red, blue, and purple fruits and vegetables contain high levels of antioxidants.

A **chemical reaction** occurs when new bonds form or old bonds break between atoms. Chemical reactions involve transfers of energy. Through chemical reactions, body structures are built and body functions are carried out. Let's take a look at four types of chemical reactions.

There Are Different Types of Chemical Reactions

Chemical reactions can be classified into four types (**Figure 2.7**):

- *Synthesis*—In **synthesis**, one or more atoms, ions, or molecules combine to form new and larger molecules.
- *Decomposition*—In **decomposition**, a molecule is split apart.
- *Exchange*—An **exchange** consists of both synthesis and decomposition reactions simultaneously.
- *Reversible*—**Reversible** reactions can go in either direction under different conditions. A double-ended arrow or two half arrows pointing in opposite directions indicates a reversible reaction.

All the synthesis reactions that occur in the human body are collectively referred to as **anabolism** (a-NAB-ō-lizm). An example of anabolism is combining simple amino acids to form the large proteins that form structures within cells, speed up chemical reactions, transport substances, and so on. (Glycogen synthesis is an example of anabolism that will be discussed later in this chapter.) In contrast, all the decomposition reactions that occur in the body are collectively referred to as **catabolism** (ka-TAB-ō-lizm). The breakdown of large starch molecules into many small glucose molecules during digestion is an example of catabolism.

CONCEPT CHECK	

1. **What** is the significance of the valence shell electrons of an atom?
2. **How** are ionic and covalent bonds different?"
3. **Why** do water molecules and methane molecules have different properties?"
4. **What** type of reaction is this: $2H_2O \rightarrow 2H_2 + O_2$?

Life Uses Important Chemicals

LEARNING OBJECTIVES

Video   See this in your WileyPLUS course.

1. **Describe** the properties of water.
2. **Define** *pH* and briefly explain how the body regulates it.
3. **Discuss** the structure and functions of carbohydrates.
4. **Explain** the structure and functions of lipids.
5. **Identify** the structure and function of proteins.
6. **Describe** how enzymes work.
7. **Distinguish** between DNA, RNA, and ATP.

rganic compounds are found in all living systems. Let's take a look at the roles these compounds play in maintaining homeostasis in the human body.

Water, Acids, and Bases Make Up a Major Part of Body Fluids

Water is the most important and most abundant inorganic compound in all living systems, making up 55% to 60% of body mass in lean adults. With few exceptions, most of the volume of cells and body fluids is water. Several of its properties explain why water is such a vital compound for life:

1. **Water is an excellent solvent**. Water is the solvent that carries nutrients, oxygen,

and wastes throughout the body. The versatility of water as a solvent is due to its polar covalent bonds (see Figure 2.6), and water is often referred to as the *universal solvent*. Polar substances dissolve easily in water and are called **hydrophilic** (*hydro-* = water; *-philic* = loving). Common examples of hydrophilic **solutes** are sugar and salt. The combination of solvent plus solute is called a **solution**. Salt water is considered a solution. In contrast, molecules that contain mainly non-polar covalent bonds are not very water soluble and are called **hydrophobic** (*-phobic* = fearing). Examples of hydrophobic compounds include animal fats and vegetable oils.

> **solvent** A liquid or gas in which some other material, a solute, has been dissolved.

> **solute** A substance or material that has been dissolved in a solvent.

2. **Water participates in chemical reactions.** Because water can dissolve many different substances, it is an ideal medium for chemical reactions. Water also actively participates in some decomposition and synthesis reactions. For example, decomposition reactions during digestion break down large nutrient molecules into smaller molecules through the addition of water molecules. This type of reaction is called **hydrolysis** (hī-DROL-i-sis; *-lysis* = to loosen or break apart). As you will see later in this chapter, the double sugar sucrose gets broken down into two single sugars, glucose and fructose, by the addition of a water molecule. Water also participates in synthesis reactions. In one type of synthesis reaction, called **dehydration synthesis** (dē-hī-DRĀ-shun), a water molecule is removed from the reactants. For example, sucrose is formed from glucose and galactose by the removal of a water molecule.

3. **Water absorbs and releases heat very slowly.** In comparison to most other substances, water can absorb or release a relatively large amount of heat with only a slight change in its own temperature. The large amount of water in the human body thus moderates the effect of changes in the environmental temperature, thereby helping maintain the homeostasis of body temperature.

4. **Water requires a large amount of heat to change from a liquid to a gas.** When the water in sweat evaporates from the skin surface, it takes with it large quantities of heat and with it provides an excellent cooling mechanism.

5. **Water serves as a lubricant.** Water is a major part of saliva, mucus, and other lubricating fluids. Lubrication is especially necessary in the thoracic and abdominal cavities, where internal organs touch and slide over one another.

Many **inorganic compounds** can be classified as acids, bases, or salts. An **acid** is a substance

> **inorganic compound** A substance that does not contain carbon.

that breaks apart or *dissociates* (dis-SŌ-sē-āts′) into one or more *hydrogen ions* (H⁺) when it dissolves in water (**Figure 2.8a**); an alternative definition of an acid is a substance that releases one or more H⁺ when dissolved in water. In contrast, a **base** usually dissociates into one or more *hydroxide ions* (OH⁻) when it dissolves in water (**Figure 2.8b**); an alternative definition of a base is a substance that absorbs one or more H⁺ when dissolved in water. A **salt**, when dissolved in water, dissociates into **cations** and **anions**, neither of which is H⁺ or OH⁻ (**Figure 2.8c**). Acids and bases react with one another to form salts. For example, the reaction of hydrochloric acid (HCl) and the base potassium hydroxide (KOH) produces the salt potassium chloride (KCl), along with water (H_2O). This exchange reaction is written as follows:

> **cation** A positively charged ion that is attracted to a negatively charged electrode called a cathode.
>
> **anion** A negatively charged ion that is attracted to a positively charged electrode called an anode.

$$HCl + KOH \rightarrow KCl + H_2O$$
Acid   Base   Salt   Water

To ensure homeostasis, body fluids must contain almost balanced quantities of acids and bases. The more H⁺ dissolved in a solution, the more acidic it is; conversely, the more OH⁻, the more basic (alkaline) it is. The chemical reactions that take place in the body are very sensitive to even small changes in the acidity or alkalinity of body fluids. Any departure from the narrow limits of normal H⁺ and OH⁻ concentrations greatly disrupts body functions. Acidity or alkalinity is expressed on the **pH scale**, which ranges from 0 to 14 (**Figure 2.9**). This scale is based on the concentration of hydrogen ions in a solution. The midpoint of the pH scale, where the concentrations of H⁺ and OH⁻ are equal, is 7. A solution with a pH of 7,

Acids, bases, and salts • Figure 2.8

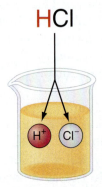

HCl

a. Acids dissociate into hydrogen ions HH⁺).

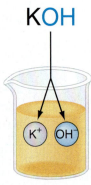

KOH

b. Bases dissociate into hydroxide ions (OH⁻).

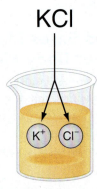

KCl

c. Salts dissociate into anions and cations that are neither H⁺ nor OH⁻.

The strengths of acid and bases are measured on the pH scale. Some common examples are shown.

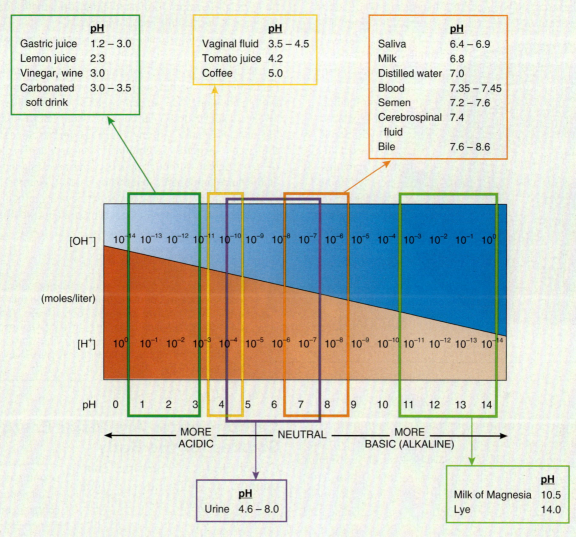

	pH
Gastric juice	1.2 – 3.0
Lemon juice	2.3
Vinegar, wine	3.0
Carbonated soft drink	3.0 – 3.5

	pH
Vaginal fluid	3.5 – 4.5
Tomato juice	4.2
Coffee	5.0

	pH
Saliva	6.4 – 6.9
Milk	6.8
Distilled water	7.0
Blood	7.35 – 7.45
Semen	7.2 – 7.6
Cerebrospinal fluid	7.4
Bile	7.6 – 8.6

	pH
Urine	4.6 – 8.0

	pH
Milk of Magnesia	10.5
Lye	14.0

$[OH^-]$ 10^{-14} 10^{-13} 10^{-12} 10^{-11} 10^{-10} 10^{-9} 10^{-8} 10^{-7} 10^{-6} 10^{-5} 10^{-4} 10^{-3} 10^{-2} 10^{-1} 10^{0}

(moles/liter)

$[H^+]$ 10^{0} 10^{-1} 10^{-2} 10^{-3} 10^{-4} 10^{-5} 10^{-6} 10^{-7} 10^{-8} 10^{-9} 10^{-10} 10^{-11} 10^{-12} 10^{-13} 10^{-14}

pH 0 1 2 3 4 5 6 7 8 9 10 11 12 13 14

◄—— MORE ACIDIC —— NEUTRAL —— MORE BASIC (ALKALINE) ——►

such as pure water, is neutral—neither acidic nor alkaline. An acidic solution has a pH below 7, while a basic solution has a pH above 7. A change of one whole number on the pH scale represents a *10-fold* change in H^+ concentration. Thus, a solution with a pH of 6 is 10 times more acidic than one with a pH of 7 and 100 times more acidic than one with a pH of 8.

Although the pH of various body fluids may differ, the normal limits for each are quite narrow. Figure 2.9 shows the pH values for certain body fluids compared with those of common household substances. Homeostatic mechanisms maintain the pH of blood between 7.35 and 7.45, so that it is slightly more basic than pure water.

Even though strong acids and bases may be taken into the body or be formed by body cells, the pH of fluids inside and outside cells remains almost constant, in part because of the presence of **buffers**. Buffers are chemical compounds that act quickly to temporarily bind H^+, removing the highly reactive, excess H^+ from solution (but not from the body). Buffers can also release H^+ that they have stored into the body to maintain pH. Buffers prevent rapid, drastic changes in the pH of a body fluid by converting strong acids and strong bases into weak acids and weak bases. Strong acids release H^+ more readily than weak acids and thus contribute more free H^+. Similarly, strong bases raise pH more than weak ones.

WHAT A HEALTH PROVIDER SEES

Dehydration and Sports Drinks

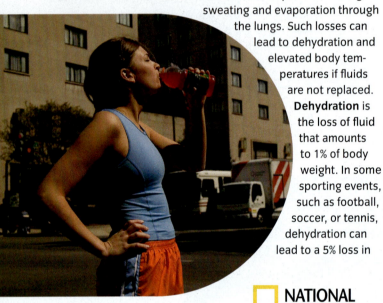

Strenuous exercise in hot weather can cause a loss of up to 2 liters or about 68 fluid ounces of water per hour through sweating and evaporation through the lungs. Such losses can lead to dehydration and elevated body temperatures if fluids are not replaced. **Dehydration** is the loss of fluid that amounts to 1% of body weight. In some sporting events, such as football, soccer, or tennis, dehydration can lead to a 5% loss in body weight. Symptoms of simple dehydration include irritability, fatigue, headache, and loss of appetite.

To prevent dehydration under such conditions, you should drink plenty of water. But what about sports drinks? Are they any better than plain water? Sports drinks have three basic components that help combat dehydration. Water is the most important. In addition, you lose some salts or electrolytes when you sweat, so sports drinks contain potassium salts. Finally, exercising muscles require carbohydrates as energy sources. Sports drinks contain generous amounts of sugars such as glucose, fructose, or sucrose. Although sports drinks are designed to replace all the fluid components lost through sweat, water is the most essential component. So, if you don't have a sports drink handy, drink plenty of water when exercising or in hot weather.

NATIONAL GEOGRAPHIC

Think Critically 1. When young children get sick and vomit frequently, pediatricians recommend that parents administer Pedialyte to prevent dehydration. The composition of Pedialyte is similar to that of a sports drink. What substances are children losing when they vomit? 2. What steps might you take prior to intensive exercise to prevent dehydration?

One important example of a buffer system in the human body is the *carbonic acid–bicarbonate buffer*. It is based on the *bicarbonate ion* (HCO_3^-), a weak base, a significant anion in both intracellular and extracellular fluids, and *carbonic acid* (H_2CO_3), a weak acid. If there is an excess of strong acid (H^+), HCO_3^- can function as a weak base and remove it as follows:

$$H^+ \quad + \quad HCO_3^- \quad \rightarrow \quad H_2CO_3$$

Hydrogen ion   Bicarbonate ion   Carbonic acid
(weak base)

Conversely, if there is an excess of strong base (a shortage of H^+), carbonic acid can function as a weak acid and increase the amount of H^+:

$$H_2CO_3 \quad \rightarrow \quad H^+ \quad + \quad HCO_3^-$$

Carbonic acid   Hydrogen ion   Bicarbonate ion
(weak acid)

You will learn more about the carbonic acid–bicarbonate buffer system, which helps regulate the pH of the blood and other body fluids, in Chapter 15. For more information on dehydration and how to prevent it see *What A Health Provider Sees*.

Carbohydrates Are Major Energy Sources for the Body

Carbohydrates are **organic compounds**. Carbohydrates include sugars, starches, and cellulose. Carbohydrates contain carbon, hydrogen, and oxygen in a ratio of 1:2:1. For example, the molecular formula of the simple carbohydrate glucose is $C_6H_{12}O_6$. Carbohydrates are divided into three categories, based on size:

> **organic compound**
> A compound that contains carbon.

1. **Monosaccharides** (mon'-ō-SAK-a-rīds; *mono-* = one; *sacchar-* = sugar) or simple sugars are the building blocks of carbohydrates (**Figure 2.10**). They include glucose, galactose, fructose, ribose, and deoxyribose. In your body, glucose is the main source of chemical energy. As you will see later, chemical energy is captured in a molecule called *adenosine triphosphate (ATP)* that fuels metabolic reactions. Two monosaccharide sugars, ribose and deoxyribose, make up ribonucleic acid (RNA) and deoxyribonucleic acid (DNA), two nucleic acids important to human body functions that are described later in the chapter.

Carbohydrates • Figure 2.10

Carbohydrate monosaccharides, like glucose, galactose, and fructose, combine through dehydration reactions to form disaccharides, like sucrose and lactose. Many monosaccharides can combine to form polysaccharides like glycogen or cellulose.

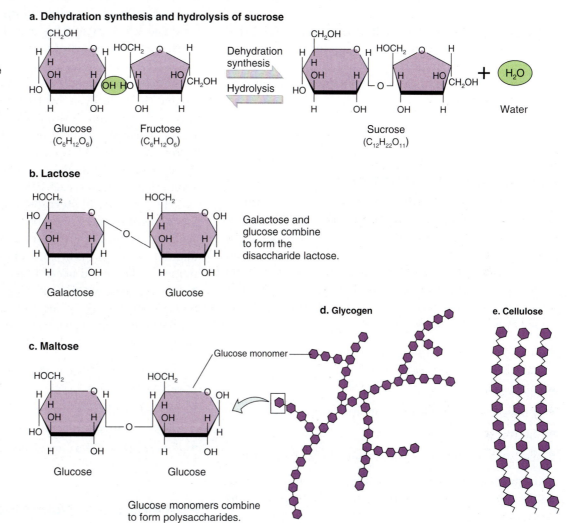

a. Dehydration synthesis and hydrolysis of sucrose

In a dehydration synthesis reaction, glucose and fructose (both monosaccharides) combine to form sucrose (disaccharide) plus water. In the opposite reaction (hydrolysis), sucrose combines with water and splits into glucose and fructose.

Glucose
($C_6H_{12}O_6$)

Fructose
($C_6H_{12}O_6$)

Dehydration synthesis

Hydrolysis

Sucrose
($C_{12}H_{22}O_{11}$)

H_2O

Water

b. Lactose

Galactose and glucose combine to form the disaccharide lactose.

Galactose

Glucose

c. Maltose

If many glucose molecules are combined in long chains, you get polysaccharides like glycogen or cellulose.

Glucose

Glucose

Glucose monomer

Glucose monomers combine to form polysaccharides.

d. Glycogen

e. Cellulose

---

2. **Disaccharides** (dī-SAK-a-rīds; *di-* = two), or double sugars, consist of two monosaccharides joined by a covalent bond. When two monosaccharides combine to form a disaccharide, a molecule of water is formed and removed in a reaction called dehydration synthesis (*de-* = from, down, or out; *hydra-* = water). Such reactions are common during synthesis of large molecules. For example, glucose and fructose combine via dehydration synthesis to form sucrose (table sugar) (Figure 2.10a). Conversely, disaccharides can be split into monosaccharides by adding a molecule of water; you learned about this type of reaction, called a hydrolysis reaction, earlier in the chapter. For example, sucrose can be hydrolyzed into glucose and fructose (Figure 2.10). Other disaccharides include maltose, or malt sugar (glucose + glucose), and lactose, or milk sugar (glucose + galactose).

3. **Polysaccharides** (pol′-ē-SAK-a-rīds; *poly-* = many) are large, complex carbohydrates that contain tens or hundreds of monosaccharides joined through dehydration synthesis reactions.

   Like disaccharides, polysaccharides can be broken down into monosaccharides through hydrolysis reactions. The main polysaccharide in the human body is *glycogen* (Figure 2.10d), which is made entirely of glucose units joined together in branching chains. Glycogen is stored in liver cells and skeletal muscle cells. Glycogen is broken down when glucose is needed and built up when glucose is abundant. *Starches* are polysaccharides of glucose that are found mostly in plants such as corn, wheat, potatoes, and barley. The body digests starches to glucose as another energy source. *Cellulose* is a polysaccharide found in plant cell walls (Figure 2.10e). Although humans cannot digest cellulose, it does provide bulk (roughage, or fiber) that helps move feces through the large intestine. Unlike simple sugars, polysaccharides usually are not soluble in water and do not taste sweet. Examples include corn starch and potato starch.

Lipids Store Energy and Comprise Cell Membranes and Hormones

Like carbohydrates, **lipids** (LIP-ids; *lip-* = "fat") contain carbon, hydrogen, and oxygen. Unlike carbohydrates, they do not have a 2:1 ratio of hydrogen to oxygen. The proportion of oxygen atoms in lipids is usually smaller than in carbohydrates, so there are fewer polar covalent bonds. As a result, most lipids are hydrophobic (insoluble in water), but they are *fat soluble* (dissolve in fat). The diverse lipid family includes:

- Triglycerides (trī-GLI-cer-īdes; *tri-* = "three"), fats and oils that store chemical energy
- Fatty acids
- Phospholipids, lipids that contain phosphorus
- Steroids
- Fat-soluble vitamins, such as vitamins A, D, E, and K

A **triglyceride** consists of two types of building blocks, a single glycerol molecule that forms the backbone and three fatty acid molecules, one attached to each carbon of glycerol by dehydration reactions (**Figure 2.11a**). The fatty acid chains of a triglyceride may be saturated (only single covalent bonds), monounsaturated (one double covalent bond), or polyunsaturated (more than one double covalent bond). Triglycerides that consist mainly of saturated fatty acids are solid at room temperature and occur mostly in meats (especially red meats) and non-skim dairy products (whole milk, cheese, and butter). They also occur in a few tropical plants, such as cocoa, palm, and coconut. Diets that contain large amounts of saturated fats are associated with disorders such as heart disease and colorectal cancer. Monounsaturated fats are found in olive oil, peanut oil, canola oil, and most nuts; they are thought to decrease the risk of heart disease. Polyunsaturated fats are found in corn oil, safflower oil, sunflower oil, soybean oil, and fatty fish (salmon, tuna, and mackerel); they are also believed to decrease the

risk of heart disease. However, when products such as margarine and vegetable shortening are made from poly-unsaturated fats, compounds called *trans* fatty acids are produced. Like saturated fats, *trans* fatty acids increase the risk of cardiovascular disease.

Like triglycerides, **phospholipids** have a glycerol backbone and two fatty acids attached to the first two carbons (**Figure 2.11b**). Attached to the third carbon is a phosphate group (PO_4^{3-}) that links a small negatively charged group to the glycerol backbone. If you look at Figure 2.11b, you can see that the molecule is divided into two parts: the non-polar fatty acids form the hydrophobic "tails" of a phospholipid, and the polar phosphate group and charged group form the hydrophilic "head." Phospholipids line up tail-to-tail in a double row called a lipid bilayer to make up much of the membranes that surrounds each cell or internal parts of cells.

The structure of **steroids**, with their four rings of carbon atoms, differs considerably from triglycerides and phospholipids. Cholesterol (**Figure 2.11c**), which has received a lot of negative press because of its role in heart disease, is an important steroid needed for membrane structure; it is the steroid from which body cells make other steroids. For example, cells in the ovaries of females make the female sex hormones estrogens, which regulate sexual functions. Other steroids include the following:

- *Testosterone* (the main male sex hormone), which regulates sexual functions
- *Cortisol*, which is necessary for maintaining normal blood sugar levels
- *Vitamin D*, which is necessary for bone growth

Proteins Have Many Functions

Proteins are large molecules that contain carbon, hydrogen, oxygen, nitrogen, and sometimes sulfur. Proteins are classified as **fibrous proteins** (structural proteins) or **globular proteins** (mobile, spherical proteins with many functions). Proteins have a more complex structure than carbohydrates or lipids. Among their many roles in the body, proteins are largely responsible for the structure of body cells. Some proteins, called *enzymes*, speed up particular chemical reactions, while others are responsible for contraction of muscles. Proteins called *antibodies* help defend the body against invading microbes, and hormones are protein messengers.

Proteins are made of building blocks called **amino acids** (a-MĒ-nō). All amino acids have an *amino group* (—NH_2) at one end and a *carboxyl group* (—COOH) at the other end. Each of the 20 different amino acids has a different *side chain* called an R group. The amino group from one amino

Lipids • Figure 2.11

There are three classes of lipids: **a.** triglycerides, **b.** phospholipids, and **c.** steroids.

a. Triglycerides

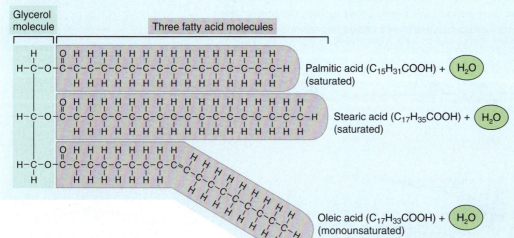

Glycerol molecule

Three fatty acid molecules

Palmitic acid ($C_{15}H_{31}COOH$) + H_2O (saturated)

Stearic acid ($C_{17}H_{35}COOH$) + H_2O (saturated)

Oleic acid ($C_{17}H_{33}COOH$) + H_2O (monounsaturated)

- Triglycerides consist of a glycerol backbone and 3 fatty acids.
- The fatty acids may be saturated (as in butter) or unsaturated (as in cooking oil).

b. Phospholipids

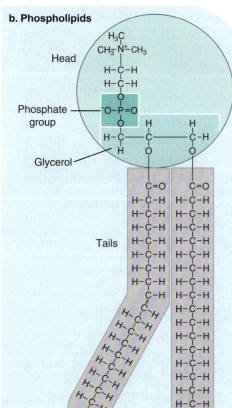

Head

Phosphate group

Glycerol

Tails

Simplified way to draw a phospholipid

Polar head

Nonpolar tails

- Phospholipids have a glycerol backbone, a phosphate head, and two fatty acids.
- Phospholipids have a polar head that is hydrophilic and a non-polar tail that is hydrophobic.
- Phospholipids are the major constituents of membranes found in cells.

Arrangement of phospholipids in a portion of a cell membrane

Polar heads

Non-polar tails

Polar heads

Cell membrane

c. Steroids

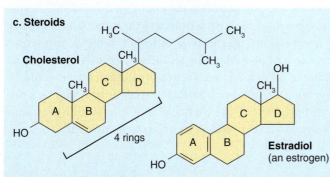

Cholesterol

4 rings

Estradiol (an estrogen)

- Steroids comprise the steroid hormones, such as estrogens, progesterone, testosterone, and cortisol.
- Estrogens and progesterone are found in birth control pills.
- Cortisol and its derivatives, like prednisone, are widely used as anti-inflammatory agents.

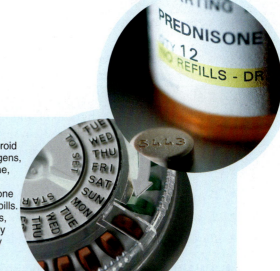

Amino acids bind together to form proteins • Figure 2.12

Amino acids combine to form peptides and proteins. The amino acid sequence, called the *primary structure*, determines the higher-order twists and folds, called the secondary and **tertiary structures**. When two or more polypeptide chains bind together, the protein has quaternary structure. The protein's function depends on its three-dimensional structure (both secondary and tertiary).

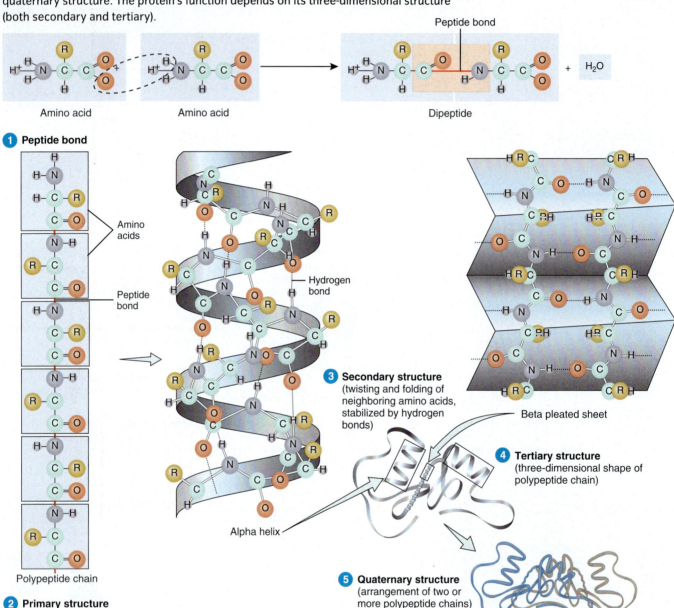

① Peptide bond

② Primary structure (amino acid sequence)

③ Secondary structure (twisting and folding of neighboring amino acids, stabilized by hydrogen bonds)

④ Tertiary structure (three-dimensional shape of polypeptide chain)

⑤ Quaternary structure (arrangement of two or more polypeptide chains)

---

acid links to the carboxyl group of another by a dehydration synthesis reaction to form a covalent bond called a **peptide bond** (**Figure 2.12**). When amino acids link together, they form *peptides* (PEP-tīdz):

- *Dipeptide*—Two amino acids linked together
- *Tripeptide*—Three amino acids linked together
- *Polypeptide*—Four or more amino acids linked together

Proteins are polypeptides that contain 50 to 2,000 amino acids. Because each variation in the number and sequence of amino acids produces a different protein, numerous proteins are possible. The situation is similar to using an alphabet of 20 letters to form words. Each letter would be equivalent to an amino acid, and each word would be a different protein. The sequence of amino acids in a protein is referred to as the *primary structure* of the protein.

A protein may consist of a single polypeptide or several intertwined polypeptides. A given type of protein has a unique three-dimensional shape because of the ways that each polypeptide twists and folds as associated polypeptides come together (Figure 2.12). The twists and folds, which form the secondary and tertiary structures of a protein, are due to hydrogen bonds between nearby amino acids within the polypeptide chain. In addition to hydrogen bonds, **disulfide bonds**, which are bonds between sulfur-containing groups of nearby amino acids, also contribute to secondary and tertiary structures. For example, a hair stylist uses chemicals to give a client a "permanent wave." These chemicals cause the proteins in hair to form disulfide bonds, which cause the hair to curl. Other chemicals break these disulfide bonds to reverse a permanent wave.

The sequence of amino acids in a protein is critical to its function. A mistake of even one amino acid can have serious consequences. For example, a single substitution of an amino acid in hemoglobin, a blood protein, can result in a deformed molecule that produces sickle-cell disease. The defective protein causes the cells to become sickle-shaped, which interferes with oxygen transport and blood flow.

The shape of a protein is also critical to its function. Changes in temperature, radiation, pH, and ion concentrations can break the hydrogen and disulfide bonds within a protein, thereby making it dysfunctional. This nonreversible process is called **denaturation** (dē-nā′-chur-Ā-shun). A common example of denaturation is frying an egg. In a raw egg, the egg-white protein (albumin) is soluble, and the egg white appears as a clear, viscous fluid. When heat is applied to the egg, the albumin denatures; it changes shape, becomes insoluble, and turns white.

Enzymes Speed Up Chemical Reactions

When atoms and molecules collide with sufficient energy, chemical bonds are made and broken. However, at room temperature, this does not happen rapidly enough to maintain life. So, all cells have a class of proteins called **enzymes** (EN-zīms) that speed up chemical reactions by bringing together and properly orienting the reacting molecules (called **substrates**) to form or break chemical bonds. Enzymes are biological **catalysts** (KAT-a-lists); they speed up a chemical reaction without being consumed.

The names of enzymes usually end in -ase. All enzymes can be grouped according to the types of chemical reactions they catalyze. For example, *oxidases* add oxygen, *kinases* add phosphate, and *dehydrogenases* remove hydrogen.

Enzymes have three important properties:

- *Specificity*—Each enzyme catalyzes a particular chemical reaction with a specific reactant. Substrates come together on the enzyme's surface, where they react. The enzyme fits the substrate like a lock accepts a key, so the shape of the enzyme is critical to its function.

- *Efficiency*—Enzymes catalyze reactions at high rates, millions to billions of times faster than the reaction would occur alone. One enzyme molecule can catalyze a reaction up to 600,000 times per second.

- *Control*—Enzymes are controlled by many factors. For example, genes control the rates at which enzymes are made. Some substances can bind to an enzyme and inhibit or enhance its activity. Some enzymes require non-protein substances called **cofactors** or **coenzymes** for activity. Cofactors can be minerals or vitamins, while coenzymes are vitamins.

During an enzymatic reaction (**Figure 2.13**), substrates bind to the surface of the enzyme at particular sites called **active sites**. Once bound, the compound is called the **enzyme–substrate complex**.

How an enzyme works • Figure 2.13

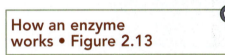
THE PLANNER

Enzymes speed up chemical reactions by bringing together and properly orienting the reacting molecules to form or break chemical bonds without being consumed in the process.

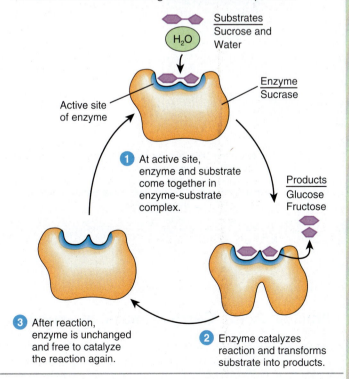

Substrates
Sucrose and Water

H_2O

Enzyme
Sucrase

Active site of enzyme

1 At active site, enzyme and substrate come together in enzyme-substrate complex.

Products
Glucose
Fructose

3 After reaction, enzyme is unchanged and free to catalyze the reaction again.

2 Enzyme catalyzes reaction and transforms substrate into products.

DNA is located in the cell's nucleus **a**. The structure of DNA is essential to its function and replication **b**.

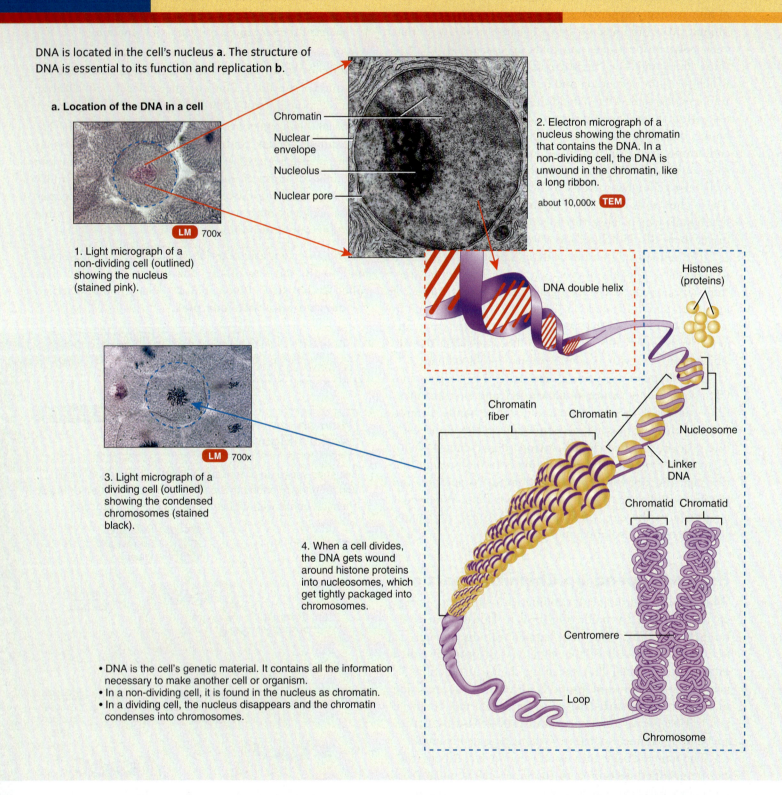

a. Location of the DNA in a cell

LM 700x

1. Light micrograph of a non-dividing cell (outlined) showing the nucleus (stained pink).

Chromatin

Nuclear envelope

Nucleolus

Nuclear pore

2. Electron micrograph of a nucleus showing the chromatin that contains the DNA. In a non-dividing cell, the DNA is unwound in the chromatin, like a long ribbon.

about 10,000x TEM

DNA double helix

Histones (proteins)

Chromatin fiber

Chromatin

Nucleosome

Linker DNA

Chromatid Chromatid

Centromere

Loop

Chromosome

LM 700x

3. Light micrograph of a dividing cell (outlined) showing the condensed chromosomes (stained black).

4. When a cell divides, the DNA gets wound around histone proteins into nucleosomes, which get tightly packaged into chromosomes.

- DNA is the cell's genetic material. It contains all the information necessary to make another cell or organism.
- In a non-dividing cell, it is found in the nucleus as chromatin.
- In a dividing cell, the nucleus disappears and the chromatin condenses into chromosomes.

Nucleic Acids Carry Genetic Instructions

Nucleic acids (noo-KLĒ-ic) were first discovered in the nuclei of cells (**Figure 2.14a**). They are huge organic molecules that contain carbon, hydrogen, oxygen, nitrogen, and phosphorus. A nucleic acid molecule is composed of repeating building blocks called **nucleotides** (**Figure 2.14b**). Each nucleotide consists of three parts:

b. Structure and replication of DNA

- DNA is made of two strands twisted in a spiral staircase-like structure called a double helix.
- Each strand consists of nucleotides bound together.
- Each nucleotide consists of a deoxyribose sugar bound to a phosphate group and one of 4 nitrogenous bases [adenine (A), thymine (T), guanine (G), cytosine (C)].
- The nitrogenous bases pair together through hydrogen bonding to form the "steps" of the double helix.
- Adenine pairs with thymine and guanine pairs with cytosine.

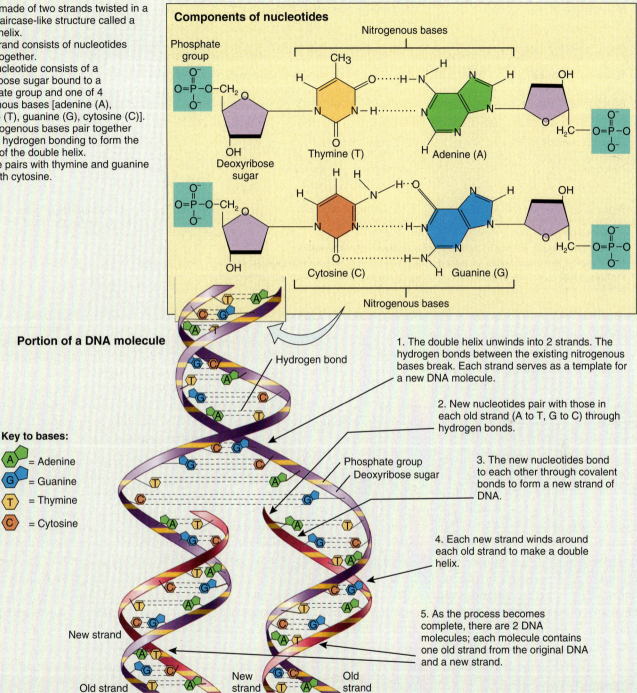

Components of nucleotides

Nitrogenous bases

Phosphate group

Deoxyribose sugar

Thymine (T)   Adenine (A)

Cytosine (C)   Guanine (G)

Nitrogenous bases

Portion of a DNA molecule

Hydrogen bond

Key to bases:

A = Adenine
G = Guanine
T = Thymine
C = Cytosine

New strand

Old strand

New strand

Old strand

Phosphate group
Deoxyribose sugar

1. The double helix unwinds into 2 strands. The hydrogen bonds between the existing nitrogenous bases break. Each strand serves as a template for a new DNA molecule.

2. New nucleotides pair with those in each old strand (A to T, G to C) through hydrogen bonds.

3. The new nucleotides bond to each other through covalent bonds to form a new strand of DNA.

4. Each new strand winds around each old strand to make a double helix.

5. As the process becomes complete, there are 2 DNA molecules; each molecule contains one old strand from the original DNA and a new strand.

- One of four different **nitrogenous bases**, ring-shaped molecules that contain C, H, O, and N
- A five-carbon monosaccharide—deoxyribose or ribose
- A phosphate group (PO_4^{3-})

The two kinds of nucleic acids are **deoxyribonucleic acid (DNA)** (dē-ok′-sē-rī′-bō-noo-KLĒ-ik), which forms the genetic material inside each human cell, and **ribonucleic acid (RNA)**, which relays instructions from genes to guide each

Life Uses Important Chemicals   **41**

Comparing DNA and RNA   Table 2.1

Feature	DNA	RNA
Nitrogen bases	Adenine (A), Cytosine (C), Guanine (G), Thymine (T)*	Adenine (A), Cytosine (C), Guanine (G), Uracil (U)
Sugar in nucleotides	Deoxyribose	Ribose
Number of strands	Two (double-helix, like a twisted ladder)	One
Nitrogen base pairing (number of hydrogen bonds)	A with T (2), G with C (3)	A with U (2), G with C (3)
How is it copied?	Self-replicating	Made by using DNA as a blueprint
Function	Encodes information for making proteins	Carries the genetic code and assists in making proteins
Types	Nuclear, mitochondrial**	Messenger RNA (mRNA), transfer RNA (tRNA), ribosomal RNA (rRNA)***

* Letters and words in red emphasize the differences between DNA and RNA.

**The nucleus and mitochondria are cell organelles, which will be discussed in Chapter 3.

***These RNAs participate in the process of protein synthesis, which will be discussed in Chapter 3.

cell's synthesis of proteins from amino acids. **Table 2.1** shows the characteristics of the nucleic acids.

DNA has a double-helix structure. Prior to cell division, the DNA copies itself as shown in Figure 2.14b. The double-helix structure allows DNA to copy itself easily prior to cell division. We will discuss cell division in more detail in Chapter 3.

In its sequence of nucleotides, DNA encodes information on how to make proteins. (See Chapter 3 for more details.) It takes about 1,000 nucleotides to make a gene, which codes for one protein. In 2003, the Human Genome Project originally estimated that humans have about 30,000 genes but reduced this estimate to 20,000 to 25,000 in 2008. Genes determine which traits each individual inherits (your mother's brown eyes or your father's blue ones, for example) and control all the activities that take place in a person's cells throughout a lifetime. Any change that occurs in the sequence of nitrogenous bases of a gene is called a *mutation*, which can result in the death of a cell, cause cancer, or produce genetic defects in future generations.

One other important nucleotide compound vital to cell functions is **adenosine triphosphate (ATP)**. ATP is the chemical that stores and releases energy for chemical reactions within cells. It consists of the nitrogenous base adenine bound to the sugar ribose and three phosphate groups (**Figure 2.15a**). When the cell needs energy for a chemical reaction or some other type of work, ATP gets split into ADP and a phosphate group by an enzyme called ATPase via a hydrolysis reaction (**Figure 2.15b**). When the cell absorbs energy, ADP combines with a phosphate to make ATP by an enzyme called ATP synthetase in a dehydration synthesis reaction.

> **adenosine triphosphate (ATP)**
> (a-DEN-ō-sēn trī-FOS-fāt) The main energy currency in living cells; used to transfer the chemical energy needed for metabolic reactions.

CONCEPT CHECK STOP

1. **Why** is water referred to as the universal solvent?
2. **What** is the difference in hydrogen ion concentration between a solution of pH 8.5 and one of pH 5.5?
3. **How** do sucrose and glucose differ from one another?
4. **What** do phospholipids do?
5. **What** types of bonds are important for the folding and twisting of polypeptides?
6. **What** is the importance of an enzyme?
7. **What** is the structure of DNA?

ATP and ADP: Structure and function • Figure 2.15

a. ATP structure and hydrolysis

When ATP is required for transferring energy, the terminal phosphate is hydrolyzed to produce ADP and a free phosphate group. The reverse reaction occurs when energy is transferred to ADP to make ATP.

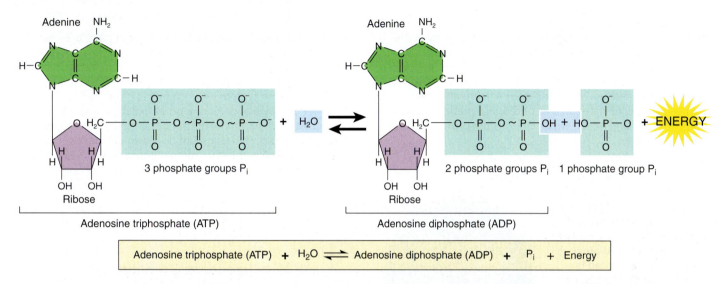

Adenosine triphosphate (ATP) + H_2O ⇌ Adenosine diphosphate (ADP) + P_i + Energy

b. ATP provides energy for work done by cells such as membrane transport, muscle contraction and chemical reactions.

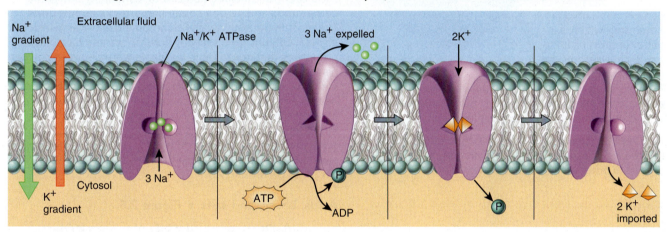

Transporting ions across membranes like the sodium-potassium pump (transport work)

Muscle contraction (mechanical work)

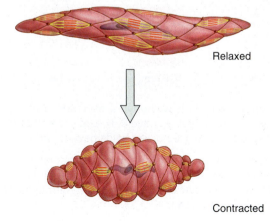

Relaxed

Contracted

Chemical reactions (chemical work)

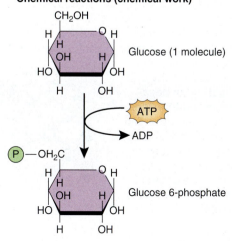

Glucose (1 molecule)

Glucose 6-phosphate

Summary

1 Matter Is Made of Elements and Atoms 22

- **Chemistry** deals with the interactions of matter and energy. **Matter** has mass and occupies space, while **energy** is the ability to do work. Matter is made of up to 117 chemical **elements**; only 26 of them are found naturally in the body. Elements such as oxygen (O), carbon (C), hydrogen (H), and nitrogen (N) make up 96% of the body's elements. Energy comes in many forms—such as potential, kinetic, mechanical, chemical, and heat—which can be converted from one form to another.

- Chemical elements are made of **atoms**. An atom consists of a nucleus, which contains positively charged **protons** and neutral **neutrons**, and is surrounded by a cloud of negatively charged **electrons** that are located in various electron levels, or **electron shells**.

Visualizing atoms • Figure 2.3

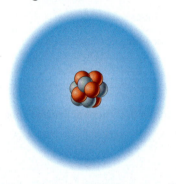

- Each element has a unique **atomic number**, which is the number of protons it contains. Because atoms are electrically neutral, the number of protons in an atom equals the number of electrons. Most of an atom's mass is in its nucleus; the atomic **mass number** is the total number of protons plus neutrons.

Atoms interact by exchanging electrons in their outer shells, the **valence shells**. Atoms can give up valence electrons, take them from other atoms, or share them.

2 The Building Blocks of Matter Fit Together to Make Ions and Compounds 26

- When an atom gives up or accepts one or more valence electrons, it becomes an ion. Ions can be positively charged (**cation**) or negatively charged (**anion**). Ions remain electrically attracted to **ionic bonds** and form **ionic compounds**, such as sodium chloride. Ionic compounds have different properties than the constituent elements.

Ionic bonds • Figure 2.4

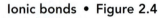
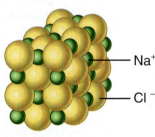

Na⁺

Cl⁻

- When one or more atoms share one or more valence electrons, they become covalent molecules. Each shared electron pair forms a **covalent bond**. Atoms of the same element can form covalent bonds to become molecules; when atoms of different elements form covalent bonds, they are called covalent compounds. Two atoms may form up to four covalent bonds; however, single, double, or triple bonds are most common.

- In some covalent compounds, the electron pairs are not shared equally. One end of the molecule becomes partially positive, while the other becomes partially negative (like a bar magnet). These **polar molecules** attract each other electrically. One type of weak chemical bond that forms between the hydrogen of one polar molecule and a strongly attractive atom (O, N) of another polar molecule is called a **hydrogen bond**. Hydrogen bonds are easily made and broken; they stabilize the structures of many molecules (such as DNA) and provide water with unique chemical properties.

3 Life Uses Important Chemicals 31

- Life's chemicals include non-carbon-containing **inorganic compounds** and carbon-containing **organic compounds**. The most important inorganic compound is water, which is an ideal **solvent** for dissolving substances and has unique properties that are suitable for life. In water, inorganic **acids** dissociate into hydrogen ions (H^+), **bases** dissociate into hydroxyl ions (OH^-), and **salts** dissociate into ions that are neither H^+ nor OH^-. The concentration of H^+ in a substance is measured by the **pH scale**, which ranges from 0 to 14. Acids have high H^+ concentrations and pH less than 7, bases have low H^+ concentrations and pH greater than 7, and neutral substances have equal concentrations greater than H^+ and OH^-, with pH around 7.

Acids, bases, and salts • Figure 2.8

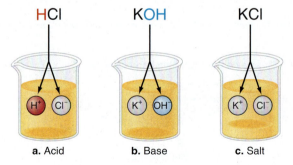

HCl   KOH   KCl

a. Acid   b. Base   c. Salt

- Organic compounds include carbohydrates, lipids, proteins, nucleic acids, and ATP. They all have C, H, and O. In addition, proteins have N and S, while nucleic acids have N and P.

- Carbohydrates are primary energy sources that consist of simple sugars (**monosaccharides**, **disaccharides**) and complex starches (**polysaccharides**). Starches are composed of simple sugars linked together.

- **Lipids**, specifically **triglycerides** (such as fats and oils), are another energy source. Other lipids include structural **phospholipids** and multi-ringed **steroids** derived from cholesterol. Steroids have structural functions and serve as chemical messengers (hormones).

- **Proteins** are large molecules consisting of long chains of amino acids bound together by **peptide bonds**. Amino acid **polypeptide** chains twist and fold together to give a protein its three-dimensional structure, which is essential to its function. Proteins have many functions.

- **Enzymes**, a class of proteins, are biological **catalysts** that speed up chemical reactions.

- **Nucleic acids** are proteins composed of long chains of **nucleotides**. Nucleic acids, including **DNA** and **RNA**, contain information for building and regulating proteins. One nucleotide, **ATP**, is important for storing and transferring chemical energy in biochemical reactions.

Key Terms

- acid 32
- active site 39
- adenosine triphosphate (ATP) 42
- amino acid 36
- anabolism 31
- anion 32
- anode 27
- antioxidant 30
- atherosclerosis 30
- atom 24
- atomic number 24
- base 32
- buffer 33
- catabolism 31
- catalyst 39
- cathode 27
- cation 32
- chemical bond 26
- chemical change 22
- chemical element 23
- chemical energy 23
- chemical reaction 31
- chemistry 22
- coenzyme 39
- cofactor 39

- covalent bond 28
- decomposition 31
- dehydration synthesis 32
- denaturation 39
- deoxyribonucleic acid (DNA) 41
- disaccharide 35
- disulfide bond 39
- electrical energy 23
- electromagnetic spectrum 23
- electron (e⁻) 24
- electron level 25
- electron shell 25
- energy 22
- enzyme 39
- enzyme–substrate complex 39
- exchange 31
- fibrous protein 36
- free radical 29
- gas 22
- globular protein 36
- hydrogen bond 29
- hydrolysis 32
- hydrophilic 31
- hydrophobic 31
- inorganic compound 32

- ion 27
- ionic bond 27
- isotope 25
- kinetic energy 22
- lipid 36
- liquid 22
- mass number 24
- matter 22
- mechanical energy 23
- molecular formula 28
- molecule 28
- monosaccharide 34
- neutron (n⁰) 24
- nitrogenous base 41
- non-polar covalent bond 29
- non-polar molecule 29
- nucleic acid 40
- nucleus 24
- nucleotide 40
- organic compound 34
- peptide bond 38
- period 26
- periodic table of the elements 26
- pH scale 32

- phospholipid 36
- physical change 22
- polar covalent bond 29
- polar molecule 29
- polysaccharide 35
- potential energy 22
- protein 36
- proton (p⁺) 24
- radiant energy 23
- radioisotope 25
- reversible 31
- ribonucleic acid (RNA) 41
- salt 32
- solid 22
- solute 31
- solution 31
- solvent 31
- steroid 36
- substrate 39
- synthesis 31
- triglyceride 36
- valence shell 25
- volume 22
- water 31

Critical and Creative Thinking Questions

1. Some chefs "cook" raw fish filets by soaking them in lemon juice or marinades containing lemon juice, rice vinegar, and wine. How do these techniques produce "cooked" fish?

2. After cooking bacon in a frying pan, Susan notices that the grease solidifies upon cooling. What can you deduce about the triglycerides in the bacon?

3. Jim has heartburn, which is caused by excess stomach acid. He takes a bicarbonate antacid drink to make him feel better. Using what you have learned in this chapter about acids and bases, explain how the antacid works.

4. In a chemistry lab, Jill stirs sucrose into water, and the sucrose disappears. Bill heats a tablespoon of sucrose in a flame, and it turns black, with droplets of water around the edges. Who has actually changed the sucrose, and why?

5. Jane makes a salad dressing by mixing vegetable oil, water, and salt. She shakes it up and lets it sit on the table. After a while, the oil is floating on top of the water. Why do the ingredients separate?

What is happening in this picture?

Pizza is a meal enjoyed by many children and adults. This one has bacon, olives, and extra cheese.

Self-Test

(Check your answers in Appendix C.)

1. _____ is the most common compound in the body.
   a. Protein
   b. Fat
   c. Carbohydrate
   d. Water

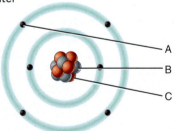

Use this figure for questions 2–4.

2. In the figure, _____ contributes to atomic mass number.
   a. A   b. B   c. C   d. B + C

3. The particle labeled A in the figure is _____.
   a. a proton
   b. an electron
   c. a neutron
   d. a valence electron

4. What is the atomic number of the atom shown in the figure?
   a. 2   b. 4   c. 6   d. 12

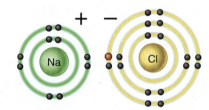

Use this figure for questions 5–6.

5. The chemical bond between the two atoms shown in the figure is _____ bond.
   a. a covalent
   b. a hydrogen
   c. an ionic
   d. a polar covalent

6. When dissolved in water, the sodium and chlorine shown in the figure separate as ions. Which of the following terms best describes the solution that is formed?
   a. acid   b. salt   c. base   d. carbohydrate

7. The biological chemical _____ does not contain nitrogen.

   a. glucose

   b. hemoglobin

   c. RNA

   d. ATP

8. Which of the following describes this reaction: amino acid A + amino acid B → dipeptide AB?

   a. synthesis

   b. decomposition

   c. exchange

   d. reversible

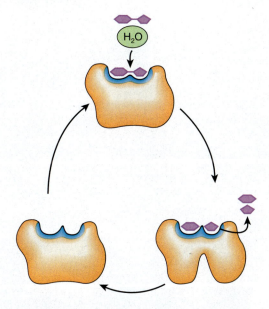

Use this figure for questions 9–10.

9. The figure shows an _____ reaction.

   a. enzyme-catalyzed synthesis

   b. enzyme-catalyzed exchange

   c. enzyme-catalyzed hydrolysis

   d. enzyme-catalyzed reversible

10. The figure shows a _____ molecule.

   a. carbohydrate

   b. nucleic acid

   c. lipid

   d. protein

11. The pH values of substances A, B, C, and D are 5.7, 3.2, 8.7, and 7.4, respectively. Which of the following is the correct order from most acidic to most basic?

   a. A, B, C, D

   b. C, D, A, B

   c. B, A, D, C

   d. D, C, B, A

12. For the substances A and C in question 11, which of the following statements is true?

   a. A is an acid and has a hydrogen ion concentration that is 3 times higher than that of C.

   b. C is a base and has a hydrogen ion concentration that is 3 times lower than that of A.

   c. A is an acid and has a hydrogen ion concentration that is 1000 times higher than that of C.

   d. C is a base and has a hydrogen ion concentration that is 1000 times higher than that of A.

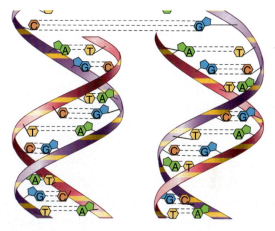

Use this figure for questions 13–15.

13. In the figure shown, _____ bonds hold the nucleotides together.

   a. covalent

   b. hydrogen

   c. ionic

   d. polar covalent

14. The figure shows _____ molecules.

   a. RNA

   b. ATP

   c. DNA

   d. protein

15. The molecules shown _____.

   a. store energy

   b. code information for proteins

   c. provide structural support for cells and tissues

   d. code information for carbohydrates

THE PLANNER ✓

Review your Chapter Planner on the chapter opener and check off your completed work.

3 Cells and Tissues

In medieval times, the city of Carcassonne in southern France was an entirely self-contained community. As you can see from this aerial view, the city was protected by two walls and each was equipped with guarded gates to control who entered or exited the city. Craftspeople, merchants, and residents lived and worked in the buildings inside the city walls. They interacted with each other, with inhabitants who dwelled outside the city walls, and with travelers. Residents and visitors traveled along city streets. At the heart of the city was the castle, where the nobility and the army lived. The nobles governed, and the army defended the city from attack. Every part of the city and each individual within it had distinct functions. Beyond the walls, many medieval cities and their surroundings formed countries.

Each cell in your body is like the medieval city of Carcassonne. The cell membrane serves as a wall that separates the inside from the outside and has "gates" that control which substances enter and leave the cell. Just as the buildings within the city house shops and workplaces, organelles carry out different functions. Like traffic moving along the streets, materials move within cells. The nucleus can be likened to the castle, governing the cell and coordinating its functions. Finally, many cells combine to form tissues, similar to the grouping of cities and their surroundings into countries. This chapter examines the structures and functions of various cells and how they are assembled into tissues.

NATIONAL GEOGRAPHIC

CHAPTER OUTLINE

CHAPTER PLANNER ✔

- ❑ Study the picture and read the opening story.
- ❑ Scan the Learning Objectives in each section:
  p. 50 ❑   p. 57 ❑   p. 71 ❑   p. 84 ❑
- ❑ Read the text and study all visuals. Answer any questions.

Analyze key features

- ❑ InSight, pp. 50–51 ❑   pp. 74–75 ❑
- ❑ Process Diagram, p. 58 ❑   p. 62 ❑   p. 65 ❑
  p. 66 ❑   pp. 68–69 ❑   p. 70 ❑
- ❑ What a Health Provider Sees, p. 83 ❑
- ❑ Stop: Answer the Concept Checks before you go on:
  p. 57 ❑   p. 71 ❑   p. 83 ❑   p. 84 ❑

End of chapter

- ❑ Review the Summary and Key Terms.
- ❑ Answer the Critical and Creative Thinking Questions.
- ❑ Answer What is happening in this picture?
- ❑ Complete the Self-Test and check your answers.

Cells Have Distinct Parts

LEARNING OBJECTIVES

1. **Describe** the structure and function of the plasma membrane.
2. **Identify** the organelles that make up the cytoplasm of the cell and define the function of each.
3. **Explain** the role of the nucleus.

There are over 100 trillion cells in your body, each one a distinct structural and functional unit. There are about 200 different cell types, in many sizes and shapes. All cells come from other cells through a process called **cell division**. With the exception of red blood cells, all the cells in your body consist of three major portions (**Figure 3.1**):

- The **plasma membrane**, or cell membrane, separates the inside from the outside.
- The **cytoplasm** contains **organelles**.
- The **nucleus** controls the cell's activities.

Let's take a closer look at the structures within the cell and their functions.

plasma membrane
The outer, limiting membrane that separates a cell's internal parts from extracellular fluid or the external environment.

cytoplasm (SĪ-tō-plazm) The fluid inside a cell and various internal membrane-bound compartments.

organelle (or-gan-EL) A membrane-bound structure within a cell that carries out specific functions.

nucleus (NOO-klē-us) The largest organelle, which houses the genetic material and controls the cell's activities.

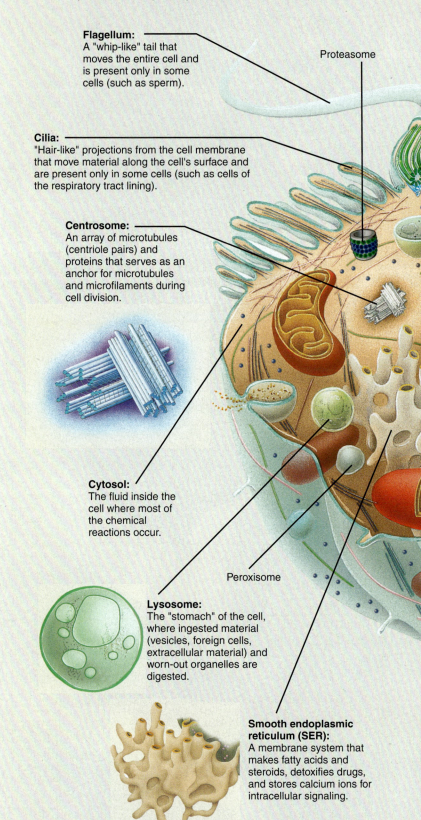

InSight
The animal cell
- **Figure 3.1**

The cell is a complex organization with specific parts that carry out specialized functions.

Flagellum:
A "whip-like" tail that moves the entire cell and is present only in some cells (such as sperm).

Proteasome

Cilia:
"Hair-like" projections from the cell membrane that move material along the cell's surface and are present only in some cells (such as cells of the respiratory tract lining).

Centrosome:
An array of microtubules (centriole pairs) and proteins that serves as an anchor for microtubules and microfilaments during cell division.

Cytosol:
The fluid inside the cell where most of the chemical reactions occur.

Peroxisome

Lysosome:
The "stomach" of the cell, where ingested material (vesicles, foreign cells, extracellular material) and worn-out organelles are digested.

Smooth endoplasmic reticulum (SER):
A membrane system that makes fatty acids and steroids, detoxifies drugs, and stores calcium ions for intracellular signaling.

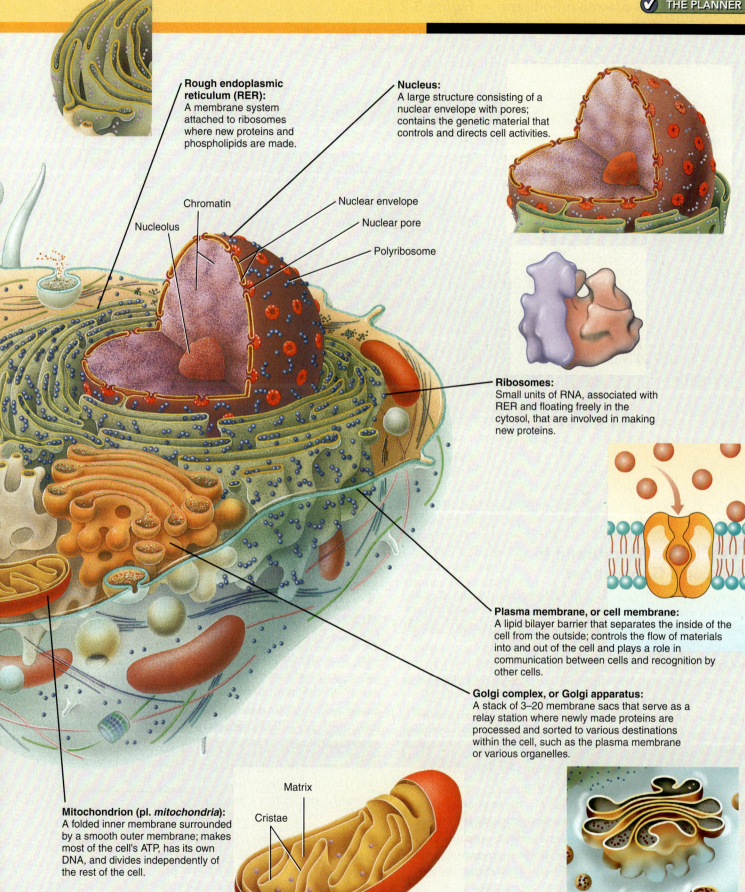

Rough endoplasmic reticulum (RER):
A membrane system attached to ribosomes where new proteins and phospholipids are made.

Nucleus:
A large structure consisting of a nuclear envelope with pores; contains the genetic material that controls and directs cell activities.

Chromatin

Nuclear envelope

Nuclear pore

Nucleolus

Polyribosome

Ribosomes:
Small units of RNA, associated with RER and floating freely in the cytosol, that are involved in making new proteins.

Plasma membrane, or cell membrane:
A lipid bilayer barrier that separates the inside of the cell from the outside; controls the flow of materials into and out of the cell and plays a role in communication between cells and recognition by other cells.

Golgi complex, or Golgi apparatus:
A stack of 3–20 membrane sacs that serve as a relay station where newly made proteins are processed and sorted to various destinations within the cell, such as the plasma membrane or various organelles.

Mitochondrion (pl. *mitochondria*):
A folded inner membrane surrounded by a smooth outer membrane; makes most of the cell's ATP, has its own DNA, and divides independently of the rest of the cell.

Matrix

Cristae

Cells Have Distinct Parts **51**

The plasma membrane's phospholipid bilayer separates the inside from the outside. The proteins embedded within the membrane help transport material across the membrane, distinguish the cell's identity, and participate in intercellular communication.

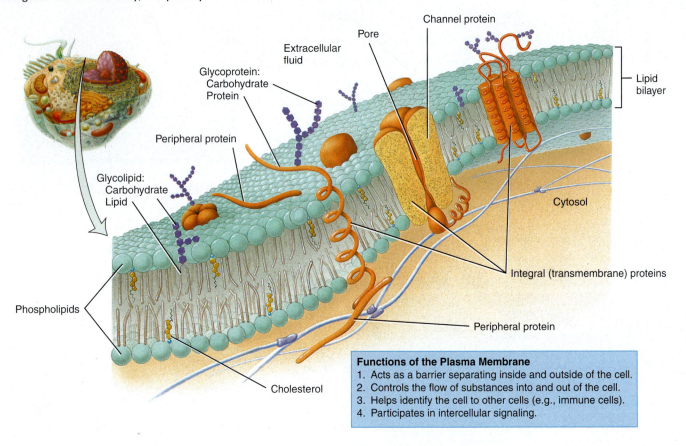

Functions of the Plasma Membrane
1. Acts as a barrier separating inside and outside of the cell.
2. Controls the flow of substances into and out of the cell.
3. Helps identify the cell to other cells (e.g., immune cells).
4. Participates in intercellular signaling.

The Plasma Membrane Is the Cell's Gatekeeper

The plasma membrane is a flexible barrier made of two layers of phospholipids that is referred to as a **lipid bilayer** (**Figure 3.2**; see also Chapter 2). There are other lipids within the bilayer, mostly cholesterol and some **glycolipids** (lipids with sugar attachments).

Associated with the lipid bilayer are membrane proteins. Some proteins are firmly embedded in the bilayer and either extend into the bilayer or span the entire bilayer. These proteins are called **integral proteins**. Other proteins are not firmly embedded but are rather loosely associated with the bilayer. These proteins are called **peripheral proteins**. Many membrane proteins (integral or peripheral) have sugar attachments and are called **glycoproteins**. Membrane proteins have various functions, including transporting substances across the membrane, binding substances to either side of the membrane, catalyzing chemical reactions (enzymes), and enabling the cell to be recognized by immune cells.

Think of the plasma membrane as a sandwich with two layers of raisin bread and a filling held together by toothpicks. The plasma membrane has four basic functions:

- It acts as a barrier that separates the inside of the cell from its surroundings.

- Its transport proteins (for water-soluble material) and the bilayer itself (for lipid-soluble material) control the flow of material into and out of the cell via a property called **selective permeability**.

- Certain external glycoproteins on the plasma membrane help identify the cell to immune cells.

- Its receptor proteins and some enzymes participate in intercellular signaling

Let's look inside the cell.

selective permeability The property of a membrane that allows some substances to pass through but not others.

Cytoskeleton • Figure 3.3

Filaments and tubules of the cytoskeleton give the cell shape and generate movement.

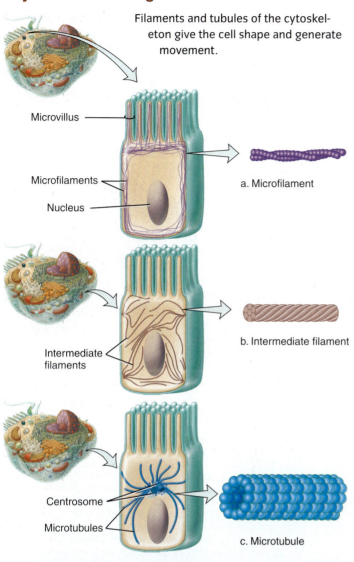

Microvillus

Microfilaments

Nucleus

a. Microfilament

Intermediate filaments

b. Intermediate filament

Centrosome

Microtubules

c. Microtubule

Functions of the Cytoskeleton
1. Serves as a scaffold that helps to determine a cell's shape and to organize the cellular contents.
2. Aids movement of organelles within the cell, of chromosomes during cell division, and of whole cells such as phagocytes.

The Cytoplasm Contains Many Organelles

The cytoplasm is the region between the plasma membrane and the nucleus. It consists of the intracellular fluid called cytosol and organelles. **Cytosol** consists of water plus dissolved ions, proteins, amino acids, fatty acids, ATP, and gases. The organelles are membrane-bound compartments where specialized functions happen. An overview of the cell's organelles is provided in Figure 3.1, and here we will take a closer look at many of them.

For a long time, biologists thought that cells were like bags containing water and organelles. However, cells actually have an internal "skeleton" that gives them specific shapes.

Cytoskeleton The **cytoskeleton** is made of networks of the following protein elements (**Figure 3.3**):

- **Microfilaments** (mī-krō-FIL-a-mentz)—Small protein strands that provide mechanical support and generate force for movement. They are analogous to muscles in your body. They also anchor proteins within the plasma membrane and provide support for **microvilli**.

  > **microvilli** Microscopic finger-like projections of the plasma membrane in some cells that increase the surface area of the cell for absorption.

- **Intermediate filaments**—Protein strands that are larger than microfilaments but smaller than microtubules. They hold organelles in place and attach cells to one another.

- **Microtubules** (mī-krō-TOO-būl′z)—Long, hollow protein tubes that determine shape and movement similar to the way bones shape your body. They are also the stiff components of cilia and flagella.

Centrosomes **Centrosomes** (SEN-trō-sōmz) are located near the nucleus. They consist of a pair of centrioles and pericentriolar material. **Centrioles** (SEN-trē-ōlz) are hollow cylinders, each made of nine sets of three microtubules. They are surrounded by proteins called tubulins, which make up the pericentriolar material (**Figure 3.4**). As you will see later in the chapter, centrosomes play a role in cell division.

Centrosomes • Figure 3.4

Centrosomes consist of a pair of centrioles and pericentriolar material and play a role in cell division.

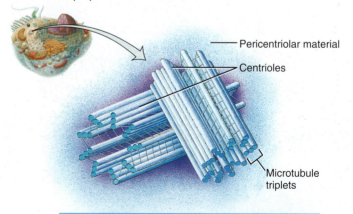

Pericentriolar material

Centrioles

Microtubule triplets

Function of the Centrosome
The pericentriolar material of the centrosome contains tubulins that build microtubules in nondividing cells and form the mitotic spindle during cell division.

Cilia and flagella Microtubules form the main parts of cilia and flagella. **Cilia** (SIL-ē-a; singular is *cilium*) are short, hair-like projections of the plasma membrane that sweep material across the surface of the cell (See Figure 3.1). Cells may have hundreds of cilia with coordinated movements. For example, cells that line the respiratory passages have cilia that help sweep mucus containing foreign particles out of the lungs to keep the airways clear.

Flagella (fla-JEL-a; singular is *flagellum*) have longer whip-like structures than cilia (see Figure 3.1). In contrast to cilia, which move material along the cell surface, a flagellum moves the entire cell. In humans, the sperm cell is the only type of cell with a flagellum, which allows it to swim through the uterine tube to fertilize an egg cell during the process of sexual reproduction.

Ribosomes Ribosomes (RĪ-bō-sōmz) are made of RNA and proteins. They consist of a large subunit and a small subunit, both of which are made in the nucleolus inside the cell's nucleus (**Figure 3.5**). The subunits leave the nucleus and are assembled in the cytosol. Some ribosomes are associated with the rough endoplasmic reticulum, some are located within mitochondria, and some are free-

Ribosomes • Figure 3.5 _____

Ribosome subunits combine to make proteins in the cytosol, rough endoplasmic reticulum, and mitochondria.

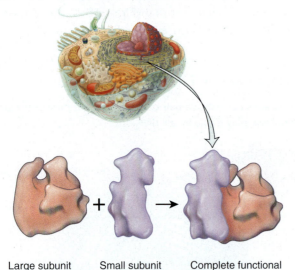

Large subunit   Small subunit   Complete functional ribosome

Functions of the Ribosomes
1. Ribosomes associated with endoplasmic reticulum synthesize proteins destined for insertion in the plasma membrane or secretion from the cell.
2. Free ribosomes synthesize proteins used in the cytosol.

Endoplasmic reticulum (ER) • Figure 3.6 _____

ER is an extensive membrane system that participates in protein synthesis (RER), detoxifying substances (SER), making fatty acids and steroids (SER), and storing calcium for intracellular signaling (SER).

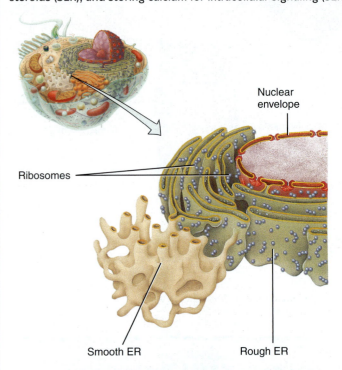

Nuclear envelope

Ribosomes

Smooth ER   Rough ER

Functions of the Endoplasmic Reticulum
1. Rough ER synthesizes glycoproteins and phospholipids that are transferred into the plasma membrane, or secreted during exocytosis.
2. Smooth ER synthesizes fatty acids and steroids, such as estrogens and testosterone; inactivates or detoxifies drugs and other potentially harmful substances; removes the phosphate group from glucose-6-phosphate; and stores and releases calcium ions that trigger contraction in muscle cells.

floating. Whether free-floating or associated with organelles, ribosomes are involved in making new proteins.

Endoplasmic reticulum The **endoplasmic reticulum** (en′-dō-PLAS-mik re-TIK-ū-lum) **(ER)** is a large membrane system that extends outward from the outer nuclear membrane throughout the cytoplasm (**Figure 3.6**). The ER makes up half of the membranes within the cytoplasm. ER comes in two forms:

• **Rough ER (RER)**—RER is a portion of the ER that extends immediately from the nuclear envelope and is studded with ribosomes. Proteins made within RER and inserted into the RER membrane are destined for the plasma membrane or the membranes of other organelles. Proteins inserted through the RER membrane into its lumen (space inside) will be secreted from the cell.

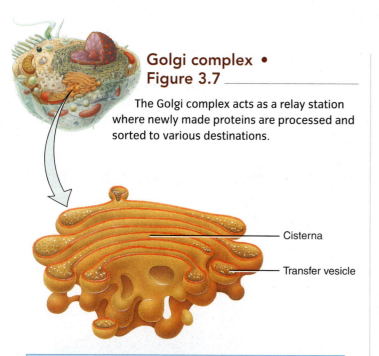

Golgi complex • Figure 3.7

The Golgi complex acts as a relay station where newly made proteins are processed and sorted to various destinations.

Cisterna

Transfer vesicle

Functions of the Golgi Complex
1. Modifies, sorts, packages, and transports proteins received from the rough ER.
2. Forms secretory vesicles that discharge processed proteins via exocytosis into extracellular fluid; forms membrane vesicles that ferry new molecules to the plasma membrane; forms transport vesicles that carry molecules to other organelles, such as lysosomes.

• **Smooth ER (SER)**—SER is a portion of the ER that extends from the RER outward and ends as a series of microtubules. Fatty acids and steroids are made in the SER. SER also detoxifies harmful substances and serves as a storage site for ionized calcium, which gets released as part of an intracellular signal for muscle contraction and the actions of some chemical messengers called *hormones*. SER in skeletal muscle cells and heart muscle cells has a special name, reflecting its specialized function: **sarcoplasmic reticulum** (sar′-kō-PLAZ-mik). You will learn more about the role of the sarcoplasmic reticulum in Chapter 6.

Golgi complex The **Golgi complex** (GOL-jē) is an array of 3 to 20 flat membrane sacs or cisternal (**Figure 3.7**). Most cells have several Golgi complexes within them. The Golgi complex is like a train yard, where trains arrive and are sorted onto the appropriate tracks for their destinations. In cells, the Golgi complex processes newly made proteins and sorts them to one of two destinations within the cell: the plasma membrane or other organelles. Transfer vesicles containing the proteins bud off the ends of the cisterns and proceed to their destinations.

Lysosomes Lysosomes (LĪ-sō-sōmz) are membrane-enclosed vesicles that contain digestive enzymes (**Figure 3.8**). Their job is to break down material ingested by the cell from the extracellular environment, such as proteins and bacteria, as well as worn-out organelles from inside the cell. Lysosomes fuse with the vesicle or organelle to be digested. Once the material has been digested, the lysosome releases it into the cytosol, and it gets recycled.

Besides lysosomes, there are two other digestive organelles, peroxisomes and proteasomes (see Figure 3.1). The **peroxisome** (per-OK-si-sōm) contains enzymes called **oxidases** that remove hydrogen atoms from various molecules, such as amino acids and fatty acids. Peroxisomes in liver cells detoxify alcohol. Hydrogen peroxide (H_2O_2) is a byproduct of the chemical reactions within peroxisomes. Left to itself, H_2O_2 would be harmful to the cells, so peroxisomes also contain an enzyme called **catalase**, which breaks down H_2O_2 into water and oxygen.

Proteasomes (PRŌ-tē-a-sōmz) are tiny, barrel-shaped organelles. Each cell contains thousands of proteasomes in the cytosol and nucleus. Proteasomes contain enzymes called **proteases**, which break down faulty, damaged, and unneeded proteins into smaller peptides. Other enzymes then break the peptides into amino acids, which are recycled (that is, reused within the cell).

Lysosome • Figure 3.8

Lysosomes have a number of digestive enzymes that break down ingested material and worn-out organelles and release their components into the cytosol.

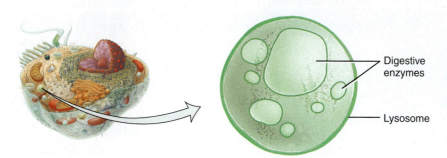

Digestive enzymes

Lysosome

Functions of the Lysosome
1. Digest substances that enter the cell.
2. Digest worn-out organelles.
3. Digest entire cells (autolysis).
4. Carry out extracellular digestion.

Mitochondria Depending on how active it is, a cell may have anywhere from 100 to thousands of **mitochondria** (mī′-tō-KON-drē-a; singular is *mitochondrion*). Each of these small, kidney bean–shaped organelles consists of a smooth outer membrane and an inner folded membrane (**Figure 3.9**). Each fold in the inner membrane is called a **mitochondrial crista** (KRIS-ta; plural is *cristae*). The fluid compartment inside the inner membrane is called the **mitochondrial matrix**. The inner membrane is studded with numerous enzymes that participate in a process called **oxidative phosphorylation**. (You will learn more about oxidative phosphorylation in Chapter 14.)

> **oxidative phosphorylation** The process by which the movement of electrons through the inner membrane of the mitochondria is coupled in making ATP.

Although their primary function is to make ATP, mitochondria also participate in the regulation of intracellular ionized calcium. They contain their own DNA and ribosomes, both of which make them capable of reproducing themselves and making new proteins, but the main source of the cell's DNA is contained within its largest organelle, the nucleus.

The Nucleus Controls the Cell's Activities

The largest organelle in the cell is the nucleus (**Figure 3.10**). The nucleus is usually round or oval shaped. Most cells have only one, but some cells, such as skeletal muscle

Mitochondria • Figure 3.9

Each mitochondrion is composed of a smooth outer membrane and a folded inner membrane, which contain numerous enzymes that are involved in making ATP.

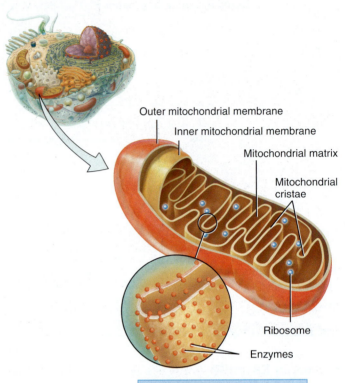

Outer mitochondrial membrane
Inner mitochondrial membrane
Mitochondrial matrix
Mitochondrial cristae
Ribosome
Enzymes

Function of the Mitochondrion
Generates ATP through reactions of aerobic cellular respiration.

The nucleus • Figure 3.10

The nucleus consists of three parts: the nuclear envelope, the nucleolus, and the chromatin. The nuclear envelope is studded with nuclear pores that control the flow of material into and out of the nucleus. The nucleolus makes ribosomes, and the chromatin is the cell's genetic material.

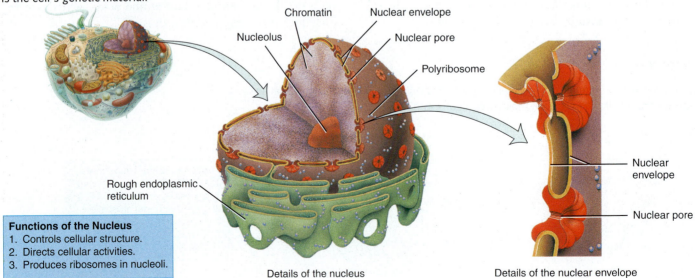

Chromatin
Nucleolus
Nuclear envelope
Nuclear pore
Polyribosome
Rough endoplasmic reticulum
Nuclear envelope
Nuclear pore

Functions of the Nucleus
1. Controls cellular structure.
2. Directs cellular activities.
3. Produces ribosomes in nucleoli.

Details of the nucleus

Details of the nuclear envelope

cells, have many nuclei. The nucleus is enclosed by a double membrane called the **nuclear envelope**. The nuclear envelope has numerous openings called **nuclear pores**, which control the movement of substances into and out of the nucleus. Inside the nucleus is a large, round structure called the **nucleolus** (noo-KLĒ-ō-lus), which is made of DNA, RNA, and proteins. The nucleolus makes ribosomes. The nucleus also houses genetic material, which contains all the information necessary to control cell activities and make new cells. In a non-dividing cell, the genetic material is spread out in the form of **chromatin** (KRŌ-ma-tin). In a dividing cell, the genetic material is condensed into structures called **chromosomes** (KRŌ-mō-sōmz), which we will talk about when we discuss cell division later in this chapter.

CONCEPT CHECK STOP

1. **What** are the functions of the plasma membrane?
2. **What** organelle consists of a smooth outer membrane and an inner folded membrane?
3. **What** part of the nucleus controls the flow of substances into and out of the nuclear membrane?

Cells Carry Out Many Processes

LEARNING OBJECTIVES

1. **Outline** the various processes of membrane transport.
2. **Explain** the steps of protein synthesis.
3. **Compare** and contrast the processes of mitosis and meiosis.

 ells must carry out several processes to function properly. Materials must be able to move across cell and organelle membranes in a controlled manner. The various proteins that carry out many cell functions must be made, using instructions encoded in the cell's genes. Finally, cells must be able to make copies of themselves, whether to grow, repair, or reproduce. Here we will discuss all these processes. Let's begin with membrane transport.

Membranes Transport Substances

Movement of materials across membranes is essential to the life of a cell. There are essentially two fluid compartments important to cellular function: intracellular fluid (inside the cell) and extracellular fluid (outside the cell). The name of the extracellular fluid is based on its location:

- **Interstitial fluid**—Fluid between cells within a tissue
- **Plasma**—Fluid within a blood vessel
- **Lymph**—Fluid within a lymphatic vessel
- **Cerebrospinal fluid**—Fluid surrounding the brain and spinal cord

As they move across and within the cells, substances—including gases, nutrients, and ions—are dissolved in the various fluids. As you learned in Chapter 2, the following terms apply to solutions:

- **Solute**—A substance that is dissolved within a fluid
- **Solvent**—A fluid (or gas) in which a solute is dissolved
- **Concentration**—The amount of solute dissolved in a given volume of solvent
- **Concentration gradient**—The difference in concentration of a substance between two areas

Let's begin with the simplest process by which substances move from one place to another, a process called *diffusion*.

PROCESS DIAGRAM

Diffusion • Figure 3.11

In response to concentration gradients, molecules move by diffusion. Diffusion occurs freely in solution, as shown in the graduated cylinders on the left, or across semi-permeable membranes, as in the beakers on the right.

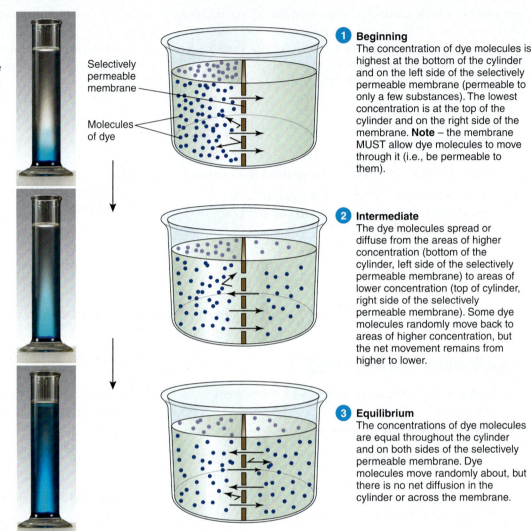

Selectively permeable membrane

Molecules of dye

1 Beginning
The concentration of dye molecules is highest at the bottom of the cylinder and on the left side of the selectively permeable membrane (permeable to only a few substances). The lowest concentration is at the top of the cylinder and on the right side of the membrane. **Note** – the membrane MUST allow dye molecules to move through it (i.e., be permeable to them).

2 Intermediate
The dye molecules spread or diffuse from the areas of higher concentration (bottom of the cylinder, left side of the selectively permeable membrane) to areas of lower concentration (top of cylinder, right side of the selectively permeable membrane). Some dye molecules randomly move back to areas of higher concentration, but the net movement remains from higher to lower.

3 Equilibrium
The concentrations of dye molecules are equal throughout the cylinder and on both sides of the selectively permeable membrane. Dye molecules move randomly about, but there is no net diffusion in the cylinder or across the membrane.

Diffusion Diffusion (di-FŪ-zhun) is the process by which solutes move from an area of high concentration to areas of low concentration (**Figure 3.11**). You encounter diffusion when you walk into a house and smell something cooking in the kitchen, such as a pot of soup. The soup molecules are at the highest concentration in the pot in the kitchen and the lowest concentration in the room where you are. The soup molecules (solute) diffuse or spread out through the air (solvent) from the area of high concentration to areas of low concentration, which is why you can smell them upon entering the house. The soup molecules in the air will continue to diffuse until the concentrations are equal (equilibrium)—that is, until the smell is equally strong everywhere in the house.

Substances always diffuse from an area of high concentration to areas of low concentration; this is often referred to as movement down a concentration gradient. Diffusion itself requires no added energy. The solute molecules are in constant motion. They have internal kinetic energy (energy of motion). This internal kinetic energy drives diffusion. Because diffusion requires no added energy, it is often referred to as a *passive process*.

Diffusion occurs across membranes such as the plasma membrane under two conditions:

1. The membrane must allow the particular substance to move across it (that is, it must be permeable to that substance).
2. There must be a concentration gradient of the particular substance across the membrane. Physiologists refer to the concentration gradient as the "driving force" for diffusion.

If these conditions are met, the substance will diffuse from the side of the membrane with the higher concentration to the side with the lower concentration until equilibrium is reached (see Figure 3.11).

What makes a biological membrane permeable to a substance? The answer depends on both the nature of the substance being transported and characteristics of proteins in the membrane that move substances across it— that is, **transport proteins**:

- If the substance can dissolve in the lipids of the membrane, it will pass directly through it (**Figure 3.12a**).

- If the substance can be dissolved in water and is larger than the space between the lipids, it must pass through a transport protein within the membrane to get to the other side.

- Transport proteins can be **channels** or **pores**, which are tunnels with an opening large enough to allow the substance to pass through. The channel may have a gate to control the movement of the substance (**Figure 3.12b**).

- Transport proteins may be **carriers**. The substance binds to the outside surface of the carrier, the carrier changes shape, the substance is released on the inside, and the carrier returns to its original shape so the process can occur again (**Figure 3.12c**).

Diffusion across a membrane with the aid of a transport protein is called **facilitated diffusion** or **facilitated transport**. Like diffusion, facilitated transport requires the presence of a concentration gradient across the membrane for any net movement of the substance to occur.

Diffusion and facilitated diffusion • Figure 3.12

There are three types of diffusion: simple diffusion (**a**), facilitated diffusion through a channel (**b**), and diffusion facilitated by a carrier (**c**).

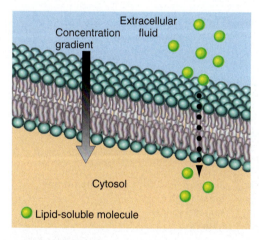

a. Simple Diffusion
Lipid-soluble molecules diffuse across the membrane down their concentration gradient.

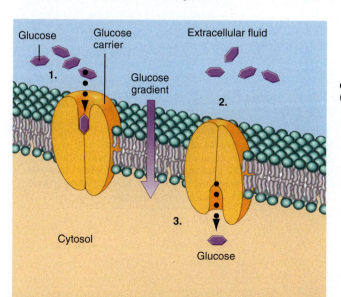

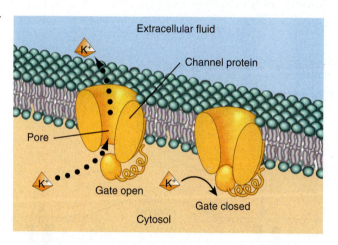

Details of the K⁺ channel

b. Facilitated Diffusion Through a Channel
(e.g., K⁺ channel)
- K^+ concentration is higher inside the cell than outside.
- K^+ diffuse through the opening of the channel to the outside.
- The channel's gate controls the movement of K^+ ions through the channel.

c. Facilitated Diffusion Through a Carrier
(e.g., glucose transport)

1. Glucose binds to the outside surface of the glucose carrier.

2. The carrier undergoes a change in shape that moves glucose through the membrane.

3. The carrier releases glucose inside the cell. Once glucose is released, the carrier returns to its original shape and the process repeats itself as glucose continues to move down its gradient.

Osmosis Typically, the term *diffusion* refers to molecules or substances such as ions, sugars, fatty acids, gases, and other substances. The diffusion of one substance—water—has a special name. The diffusion of water through a selectively permeable membrane is called **osmosis** (oz-MŌ-sis). Osmosis follows the same rules as other types of diffusion; it requires a membrane permeable to water and a concentration gradient.

What exactly is a concentration gradient for water? Think of it this way: Pure water is the most concentrated that you can get. When you dissolve a substance in water, you are actually diluting the concentration of water in the solution. So a concentrated salt solution has a lower water concentration than a dilute salt solution. Here's an example of an osmosis problem: You have a dialysis bag, which is a selectively permeable membrane tied at both ends. Inside the bag is a concentrated sugar solution. You place the bag in a beaker with a dilute sugar solution. To understand what will happen, ask yourself these questions:

- Is the membrane permeable to water? If the answer is no, the question of osmosis is irrelevant. In this case, the dialysis tubing is permeable to water.

- Which side of the membrane has the higher water concentration? The side with the lower solute (sugar) concentration, which is outside the bag.

- Which side of the membrane has the lower water concentration? The side with the higher solute (sugar) concentration, which is inside the bag.

- Which way will water flow? Water will always move from the area of higher water concentration (lower solute concentration) to lower water concentration (higher solute concentration). So water will move from the outside of the bag (lower sugar concentration) to the inside of the bag (higher sugar concentration) (**Figure 3.13**). The bag will swell as water flows into it. However, as noted in Figure 3.13, other forces are at work to counter the effects of osmosis, including hydrostatic pressure (the pressure of the water on the bag).

Osmosis • Figure 3.13

Water moves through the selectively permeable membrane by osmosis until equilibrium is reached.

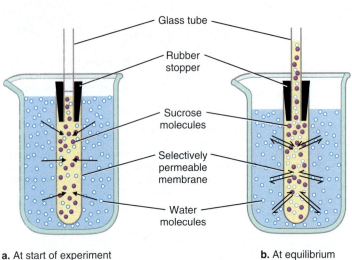

Glass tube

Rubber stopper

Sucrose molecules

Selectively permeable membrane

Water molecules

a. At start of experiment

b. At equilibrium

Question	a. At start	b. At equilibrium
Is the membrane permeable to water?	Yes	Yes
Where is the water concentration higher?	Outside membrane (No sucrose)	Outside membrane
Where is the water concentration lower?	Inside membrane (high sucrose concentration)	Inside membrane
Which way will there be net movement of water?	Outside to inside (higher water concentration to lower water concentration)	None. Water flowing into the membrane by osmosis will build up an outward pressure (hydrostatic pressure). This pressure forces water molecules back out of the membrane despite the concentration gradient.

From the start of the experiment, water will flow through the membrane and the level of water in the tube will rise. At equilibrium, water pressure will drive some water molecules and there will be a balance between the force of osmosis driving water into the membrane and the force of water pressure inside, which drives water out of the membrane.

Principles of osmosis applied to red blood cells • Figure 3.14

The arrows indicate the direction and degree of water movement into and out of cells.
The scanning electron micrographs of the cells have magnifications of 15,000X.

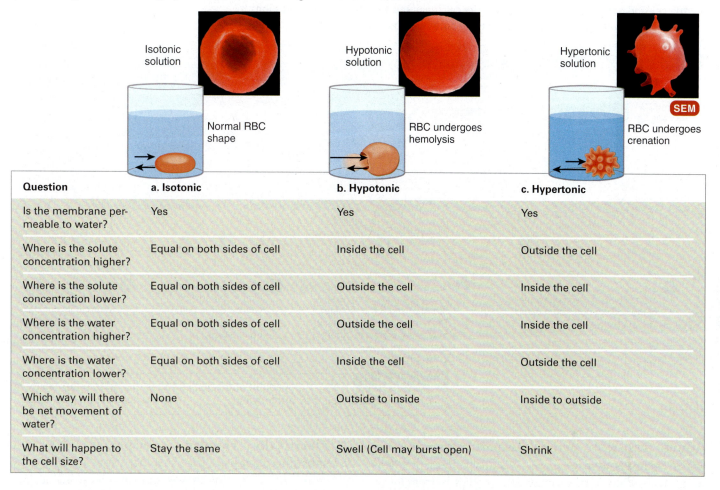

Question	a. Isotonic	b. Hypotonic	c. Hypertonic
Is the membrane permeable to water?	Yes	Yes	Yes
Where is the solute concentration higher?	Equal on both sides of cell	Inside the cell	Outside the cell
Where is the solute concentration lower?	Equal on both sides of cell	Outside the cell	Inside the cell
Where is the water concentration higher?	Equal on both sides of cell	Outside the cell	Inside the cell
Where is the water concentration lower?	Equal on both sides of cell	Inside the cell	Outside the cell
Which way will there be net movement of water?	None	Outside to inside	Inside to outside
What will happen to the cell size?	Stay the same	Swell (Cell may burst open)	Shrink

Why is osmosis so important to your health? As a result of fluid intake, transfusions, injuries, and diseases, the salt and water concentrations of various fluid compartments within your body change. So, cells within those compartments, like red blood cells traveling through blood vessels, may find themselves in environments with different solute and water concentrations (**Figure 3.14**). There are specific names for such environments; it is important to note that these terms refer to the *concentrations of solutes,* not the concentration of water:

- **Isotonic** (ī′-sō-TON-ik)—The solute concentration outside the cell is the same as that inside the cell. Therefore, water concentration is also the same on both sides of the cell, and the net movement of water is zero.
- **Hypotonic** (hī′-pō-TON-ik; *hypo-* = less)—The solute concentration outside the cell is less than the concentration inside the cell. Therefore, the water

concentration outside is greater than that inside, and water flows into the cell.

- **Hypertonic** (hī′-per-TON-ik; *hyper-* = more)—The solute concentration outside the cell is greater than the concentration inside the cell. Therefore, the water concentration is greater inside the cell than outside, and water flows out of the cell.

The balance of water concentration across the cell membrane will change under these various conditions, and the cell shape and volume may remain the same, swell, or shrink (Figure 3.14). Dramatic changes in cell shape and volume may lead to cell death.

So far, we have only talked about substances moving down a concentration gradient. However, it is possible for substances to be transported against a concentration gradient via a process called *active transport.*

How the sodium-potassium pump works • Figure 3.15

Because the cell membrane is leaky to sodium and potassium, sodium ions diffuse into the cell and potassium ions diffuse out. The cell maintains the gradients of these ions by pumping sodium out of the cell and bringing potassium back in.

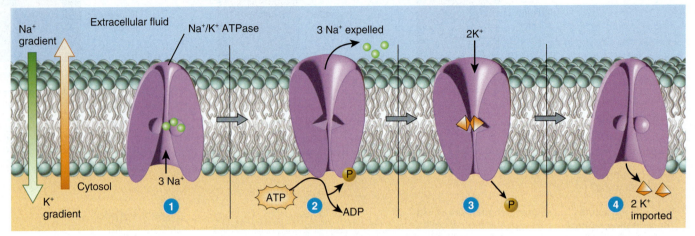

Na⁺ gradient • Extracellular fluid • Na⁺/K⁺ ATPase • 3 Na⁺ expelled • 2K⁺ • Cytosol • K⁺ gradient • 3 Na⁺ • ATP • ADP • P • P • 2 K⁺ imported

1 3 sodium ions (Na⁺) from the cytosol bind to the inside surface of the sodium-potassium pump.

2 Na⁺ binding triggers ATP to bind to the pump and be split into ADP and P (phosphate). The energy from ATP splitting causes the protein to change shape, which moves the Na⁺ to the outside.

3 2 potassium ions (K⁺) bind to the outside surface of the pump and cause the P to be released.

4 The release of the P causes the pump to return to its original shape, which moves the K⁺ into the cell.

Put It Together If Na⁺ is moved out of the cell and K⁺ is moved into the cell by active transport, then they are moving _____ their concentration gradients.

Active transport

Active transport is the process in which energy is used to move substances across a membrane against a concentration gradient (that is, from lower concentration to higher concentration). The source of energy depends on which of three transport mechanisms is used:

- **Pumps** use energy from splitting ATP to power the movement of substances. The most common example is the **sodium-potassium pump (Na⁺/K⁺ pump**, or **Na⁺/K⁺-ATPase)**, which is found in all cells (**Figure 3.15**). The sodium-potassium pump transports sodium out of the cell and transports potassium into the cell.

- **Exchangers** move a substance against its concentration gradient by combining it with the movement of a second substance down its concentration gradient. For example, many cells, such as heart and muscle cells, have a sodium-calcium exchanger. The sodium-calcium exchanger moves calcium out of the cell against its concentration gradient by coupling this movement to the movement of sodium from outside the cell to inside the cell (down the sodium concentration gradient).

- **Electrically coupled transporters** move a substance against its concentration gradient by coupling it to the movement of electrons across some membranes. As you will see in Chapter 14, the movement of electrons from one protein to another in the inner mitochondrial membrane provides the energy to transport proteins out of the mitochondrial matrix.

The active and passive transport mechanisms just described exist throughout the membranes of organelles and the cell itself. They work very well for moving small substances—such as ions, water molecules, glucose molecules, and amino acids—across membranes. However, large proteins and even invading bacteria require other bulk transport processes.

Endocytosis and exocytosis

To transport large materials, a membrane engulfs the material into the inside of a small round sac, or **vesicle** (VES-i-kul). These sacs move about the cell interior with energy supplied by ATP and contractions of microfilaments, which pull them along. There are two major types of vesicle transport, endocytosis and exocytosis. **Endocy-**

tosis (en'-dō-sī-TŌ-sis; *endo-* = into) involves ingesting material by forming a vesicle from the plasma membrane. The sac of ingested material buds off inside the cell and usually fuses with lysosomes. There are three types of endocytosis:

- **Phagocytosis** (fag'-ō-sī-TŌ-sis; *phago-* = eat). The cell "eats" large particles such as bacteria, viruses, and dead cells. The plasma membrane forms projections called **pseudopods** (SOO-dō-pods) that enclose the material to form a vesicle called a **phagosome** (**Figure 3.16**). The phagosome later fuses with lysosomes. White blood cells kill invading bacteria by phagocytosis.

- **Pinocytosis** (pin'-ō-sī-TŌ-sis; *pino-* = drink). The cell periodically "drinks" by forming small vesicles around droplets of extracellular fluid; these droplets may have small particles dissolved in them as well. These vesicles fuse with lysosomes and release their contents.

- **Receptor-mediated endocytosis**. When **hormones** bind to **receptors** on the plasma membrane, the hormone–receptor complex is often ingested by endocytosis after the hormone has produced its effect. The patch of membrane containing the hormone–receptor complex pinches off to form a vesicle, which then merges with one or more lysosomes. The complex gets broken down; new receptors are made and packaged into vesicles, which later merge again with the plasma membrane. When cells are exposed to constant levels of hormones or drugs, they may become less sensitive to them. This desensitization can be explained by receptor-mediated endocytosis. The hormone–drug receptor complexes are taken inside the cell, thereby leaving fewer receptors available to bind to the hormone or drug.

> **hormone** (HOR-mōn) A chemical message by which one cell produces an effect in another cell.
>
> **receptor** A specific molecule or cluster of molecules that recognize and bind a hormone or drug.

Phagocytosis • Figure 3.16

Phagocytosis is a type of endocytosis in which large particles, such as large proteins, bacteria, viruses, and dead cells, are brought inside a cell and digested.

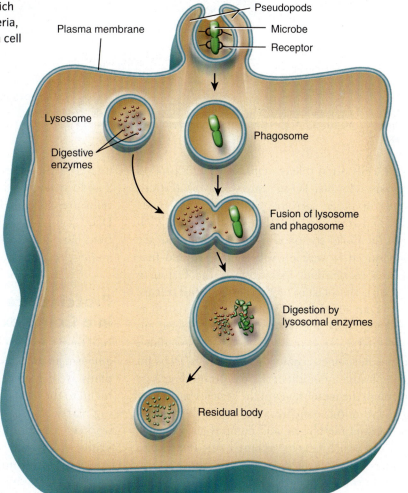

Endocytosis and exocytosis • Figure 3.17

Endocytosis brings material inside the cell, while exocytosis secretes material from the cell. Both types of transport use vesicles that travel between the plasma membrane and organelles, such as lysosomes, ER, and the Golgi complex. These processes recycle segments of the plasma membrane.

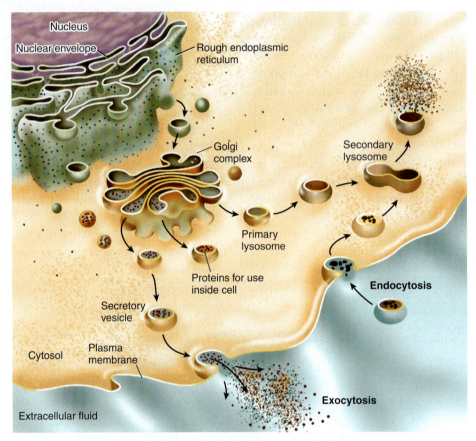

In contrast, cells often release substances by a process called **exocytosis** (eks'-ō-sī-TŌ-sis; *exo-* = out of) (**Figure 3.17**). Proteins or other secretory substances are made within the ER and transported within vesicles to the Golgi complex. In the Golgi complex, the vesicle contents are processed and sorted to vesicles, which move to and fuse with the plasma membrane. Exocytosis is the fusion of these vesicles with the plasma membrane and the release of their contents outside the cell. Many substances (such as neurotransmitters, endocrine hormones, and digestive enzymes) are secreted via exocytosis. The same process allows proteins, such as hormone receptors and enzymes, to be inserted into the plasma membrane; these proteins do not go all the way through the membranes of the vesicles but rather are embedded within those membranes. When the vesicles fuse with the plasma membrane and dump their contents, the membrane of a vesicle gets incorporated into the plasma membrane, where the embedded proteins remain.

In addition to their roles in membrane transport and the structure of the plasma membrane, proteins carry out numerous functions in the cell. As mentioned in Chapter 2 and earlier in this chapter, the instructions to build proteins are contained in the DNA located in the nucleus on DNA segments called **genes**. But how are these instructions carried out? The following section provides a simple analogy to help you understand protein synthesis.

Proteins Are Made in a Complex Process

Building a protein is much like building a house. To build a house, an architectural firm creates a set of plans. Those plans are copied or transcribed into blueprints, which are taken to the home site. Workers at the site bring the building materials to a scaffold at the home's foundation. The workers assemble the materials according to the blueprints, essentially translating the information contained in the blueprint into the final product, the house.

Now let's apply this analogy to protein synthesis. When building a protein, the following events occur:

Transcription • Figure 3.18

THE PLANNER ✓

During transcription, an enzyme called RNA polymerase binds to the DNA and makes a copy (mRNA) of the information encoded in DNA.

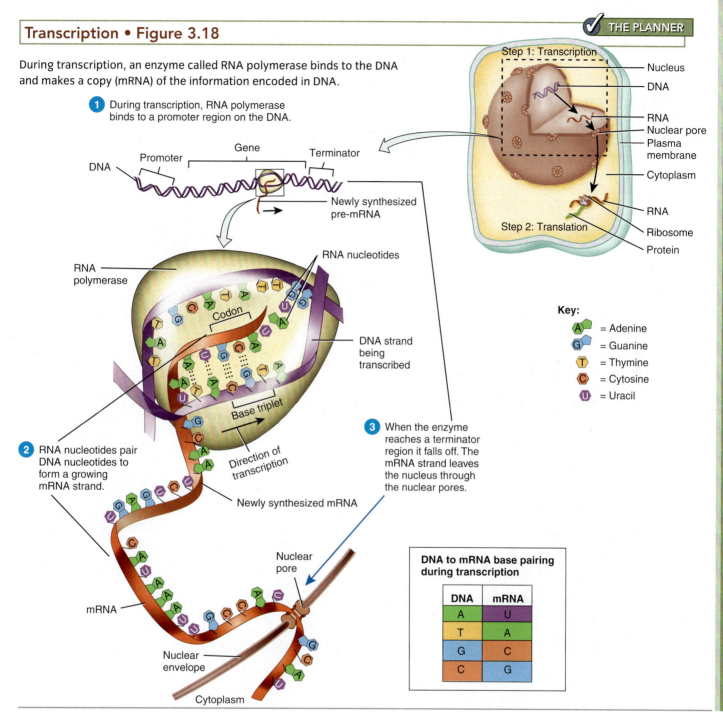

1 During transcription, RNA polymerase binds to a promoter region on the DNA.

Promoter · Gene · Terminator · DNA · Newly synthesized pre-mRNA

Step 1: Transcription · Nucleus · DNA · RNA · Nuclear pore · Plasma membrane · Cytoplasm · Step 2: Translation · RNA · Ribosome · Protein

RNA polymerase · RNA nucleotides · Codon · DNA strand being transcribed · Base triplet · Direction of transcription · Newly synthesized mRNA

2 RNA nucleotides pair DNA nucleotides to form a growing mRNA strand.

3 When the enzyme reaches a terminator region it falls off. The mRNA strand leaves the nucleus through the nuclear pores.

mRNA · Nuclear pore · Nuclear envelope · Cytoplasm

Key:
- A = Adenine
- G = Guanine
- T = Thymine
- C = Cytosine
- U = Uracil

DNA to mRNA base pairing during transcription

DNA	mRNA
A	U
T	A
G	C
C	G

1. The DNA that encodes (puts into code) for the protein (plans) is located in the nucleus (architectural firm). The sequence of nucleotides in the DNA specifies the order of amino acids in the protein. It takes one sequence of 3 nucleotides (codon) to code for one specific amino acid. There are at least 20 codons to account for the various amino acids, plus some codons for start and stop signals.

2. A copy of the DNA is made in the nucleus. This copy is **messenger RNA (mRNA)** (the blueprint). The copying process is called **transcription** (**Figure 3.18**):

    • During transcription, an enzyme called **RNA polymerase** binds to the DNA at a sequence called a **promoter**.

    • RNA polymerase moves along the DNA strand, making a copy of the DNA as it goes. The nucleotides of mRNA pair up with the complementary nucleotides of DNA.

    • When the enzyme reaches a sequence of DNA called a **terminator**, it stops copying, falls off the DNA, and releases the mRNA. The newly made mRNA has a nucleotide sequence complementary to that of the DNA so it can specify the amino acid order in the protein.

Translation • Figure 3.19

Ribosomes, tRNA, and mRNA work together to assemble the appropriate amino acids into a protein.

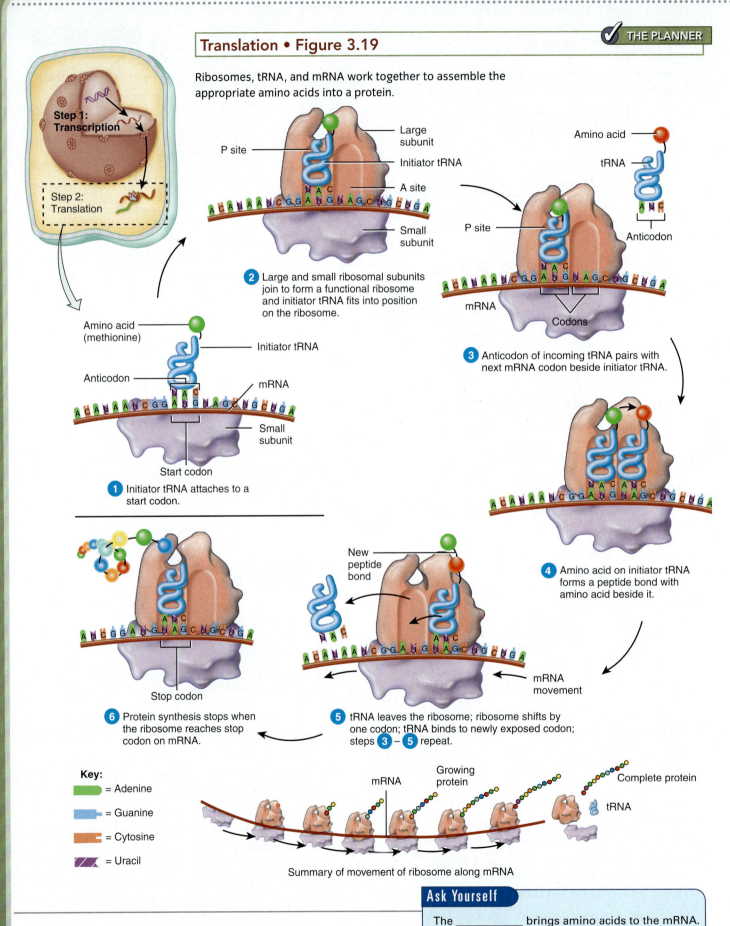

Step 1: Transcription

Step 2: Translation

2 Large and small ribosomal subunits join to form a functional ribosome and initiator tRNA fits into position on the ribosome.

P site
Large subunit
Initiator tRNA
A site
Small subunit

Amino acid
tRNA
P site
Anticodon
mRNA
Codons

3 Anticodon of incoming tRNA pairs with next mRNA codon beside initiator tRNA.

Amino acid (methionine)
Initiator tRNA
Anticodon
mRNA
Small subunit
Start codon

1 Initiator tRNA attaches to a start codon.

4 Amino acid on initiator tRNA forms a peptide bond with amino acid beside it.

New peptide bond

mRNA movement

Stop codon

6 Protein synthesis stops when the ribosome reaches stop codon on mRNA.

5 tRNA leaves the ribosome; ribosome shifts by one codon; tRNA binds to newly exposed codon; steps **3** – **5** repeat.

Key:
= Adenine
= Guanine
= Cytosine
= Uracil

mRNA
Growing protein
Complete protein
tRNA

Summary of movement of ribosome along mRNA

Ask Yourself

The _____ brings amino acids to the mRNA.
a. DNA   c. ribosome
b. tRNA   d. stop codon

3. The mRNA (blueprint) leaves the nucleus through the nuclear pores and goes to the ER (home site).

4. Ribosomes (the scaffolding) attach to the mRNA at a specific start sequence (AUG) and move along the mRNA strand.

5. Sets of RNA molecules called **transfer RNA (tRNA)** (the workers) bring specific amino acids (building materials) to the ribosomes:

   • There are 20 different types of tRNA molecules, one type for each amino acid.

   • As a project manager may tell workers what to do, the sequence of codons in mRNA helps direct the tRNA. Each type of tRNA has a three-nucleotide sequence on one end (anticodon) that is complementary to the codon on mRNA for that specific amino acid. Only the tRNA with the complementary anticodon can bind to the codon in mRNA.

6. The tRNA molecules (workers) assemble the protein (house) according to the instructions in the mRNA (blueprint). The process of assembling the protein from the instructions in mRNA (steps 2–5) is called **translation** (**Figure 3.19**):

   • tRNA molecules bound to specific amino acids enter the ribosome.

   • The anticodons of the appropriate tRNA pair up with the appropriate codons in the mRNA.

   • A peptide bond forms between amino acids within the ribosome.

   • The empty tRNA leaves and the process repeats until the protein (house) gets assembled completely.

   • Eventually, the ribosome reaches a stop sequence in the mRNA (the last part of the blueprint). This stop sequence is called a stop codon (UGA, UAG, UAA). Upon reaching the stop codon, the ribosome falls apart and releases the newly made protein (house).

Approximately 15 amino acids are translated every second, and proteins can have hundreds to thousands of amino acids. It can take as long as 90 minutes to make new proteins from the start of transcription to the end of translation. Therefore, many ribosomes can be involved in making multiple copies of a protein. As one ribosome moves along the mRNA, a second one can attach and begin to make another protein. This process repeats, producing a *polyribosome* (one mRNA with many ribosomes and growing protein chains). The polyribosome activity allows many proteins to be made from a single molecule of mRNA.

Cells Divide by Mitosis or by Meiosis

Cells transport materials and make proteins as part of their normal functions. They grow and, at some point, divide to produce new cells. Cell division is the way your body grows and how it replaces worn-out cells and cells damaged by disease or injury. Most of the cells in your body are **somatic cells** (sō-MAT-ik) and divide through a process called **mitosis** (mī-TŌ-sis). Somatic cells are cells other than sex cells (sperm and egg). During mitosis, one starting cell divides into two identical cells. Each cell has exactly the same genetic makeup as the parent cell (two sets of chromosomes).

Specialized cells called **gametes** undergo a different process of cell division called meiosis. During **meiosis** (mē-Ō-sis), a starting cell undergoes two rounds of cell division to produce four cells. Each cell has one-half the genetic material of the starting cell (only one set of chromosomes). Let's look at mitosis first.

> **gamete** (GAM-ēt) A sex cell, such as an egg or a sperm.

Mitosis is one part of the cell's normal life cycle, called the **cell cycle**. The cell is continually changing from the time it forms until it divides. Although the cell cycle is continuous, it is commonly divided into **interphase** (IN-ter-fāz) and mitosis. During interphase, the cells goes through three stages: G1—a growth phase in which proteins are synthesized, S—when DNA is replicated, and G2—another growth phase in which proteins are made. Interphase, which may take 20 to 22 hours, is followed by mitosis. Mitosis itself consists of four phases: **prophase** (PRŌ-fāz), **metaphase** (MET-a-phāz), **anaphase** (AN-a-fāz), and **telophase** (TEL-ō-fāz) (**Figure 3.20** on the next page). As a result of mitosis, each cell contains 23 pairs of chromosomes. This makeup, which is identical to that of the starting cell, is called **diploid**.

> **diploid** Referring to a cell or an organism that has two sets of chromosomes.

Mitosis and cytokinesis • Figure 3.20

Drawings and light micrographs (700X) show cells in each stage of mitosis.

LM all at 700x

Centrosome:
Centrioles
Pericentriolar material
Nucleolus
Nuclear envelope
Chromatin
Plasma membrane
Cytosol

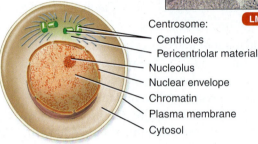

1 Interphase:
Cell carries out its normal functions. As mitosis approaches, the cell prepares for division by making new proteins, copying the entire DNA, and making new organelles. Cells spend most of their time in interphase. Some cells become arrested in interphase and do not divide unless stimulated (for example, muscle cells, liver cells), while other cells do not divide at all once arrested (such as nerve cells). In contrast, some cells are constantly dividing (such as skin cells, cells lining the intestine).

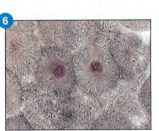

6 Interphase for cells:
Identical cells carry out normal functions and grow until the next division. They are exactly the same as the original starting cell.

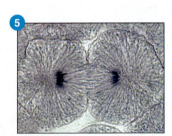

Cleavage furrow

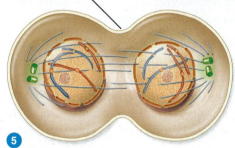

5 Telophase:
Chromatids at opposite ends decondense into chromatin, a nuclear envelope forms around them, and new nucleoli appear. The cytoplasm pinches in the middle (cleavage furrow) and divides (cytokinesis) to form two separate but identical cells.

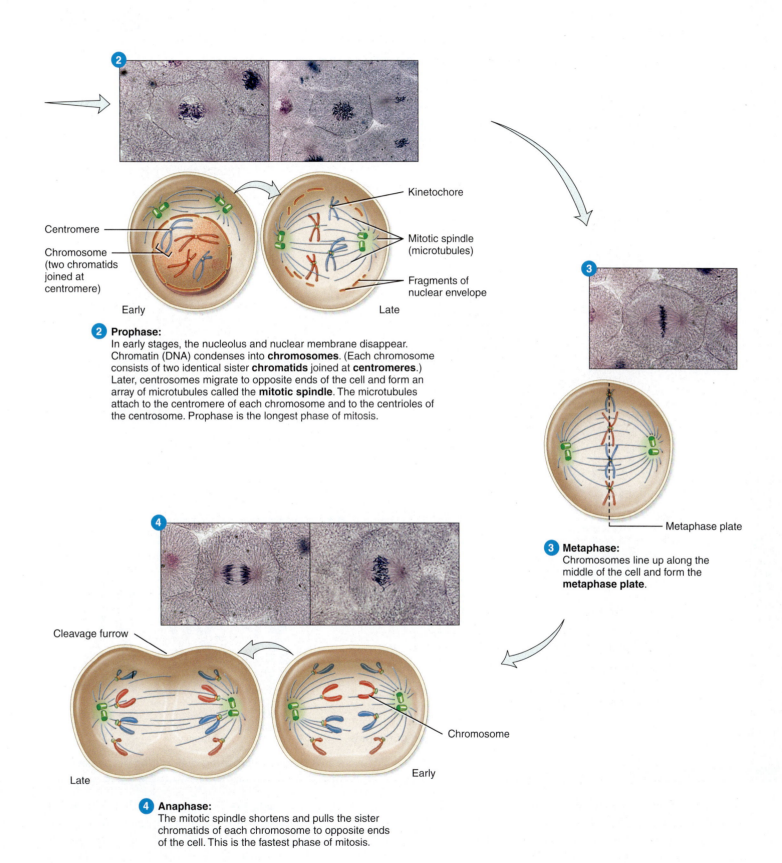

2 **Prophase:**
In early stages, the nucleolus and nuclear membrane disappear. Chromatin (DNA) condenses into **chromosomes**. (Each chromosome consists of two identical sister **chromatids** joined at **centromeres**.) Later, centrosomes migrate to opposite ends of the cell and form an array of microtubules called the **mitotic spindle**. The microtubules attach to the centromere of each chromosome and to the centrioles of the centrosome. Prophase is the longest phase of mitosis.

Centromere

Chromosome (two chromatids joined at centromere)

Early

Late

Kinetochore

Mitotic spindle (microtubules)

Fragments of nuclear envelope

3 **Metaphase:**
Chromosomes line up along the middle of the cell and form the **metaphase plate**.

Metaphase plate

4 **Anaphase:**
The mitotic spindle shortens and pulls the sister chromatids of each chromosome to opposite ends of the cell. This is the fastest phase of mitosis.

Cleavage furrow

Late

Early

Chromosome

Meiosis • Figure 3.21

Meiosis is the orderly distribution of genetic material to newly formed haploid cells. It includes steps very similar to those of mitosis, the main differencve being the formation of tetrads in prophase I. Crossing over at this stage offers even more genetic variation, as the ends of these chromosomes are close enough to exchange genetic material. Telophase I then forms two "cells" that immediately go into prophase II, metaphase II, anaphase II, and telophase II. These phases result in four haploid cells.

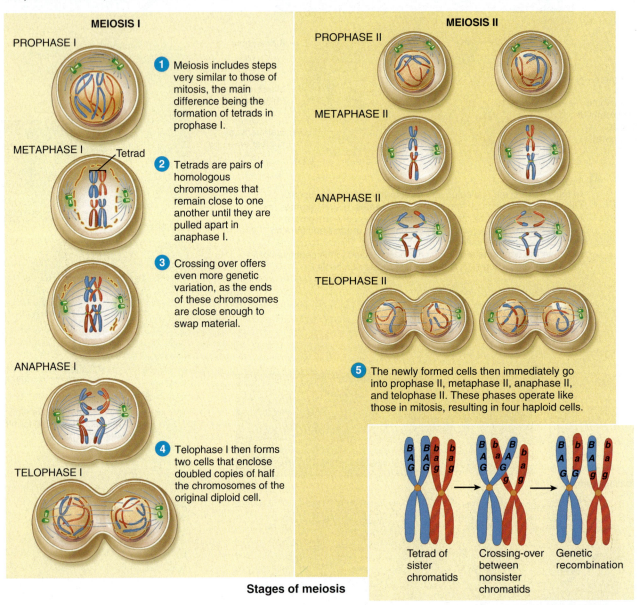

MEIOSIS I

PROPHASE I

METAPHASE I — Tetrad

ANAPHASE I

TELOPHASE I

1 Meiosis includes steps very similar to those of mitosis, the main difference being the formation of tetrads in prophase I.

2 Tetrads are pairs of homologous chromosomes that remain close to one another until they are pulled apart in anaphase I.

3 Crossing over offers even more genetic variation, as the ends of these chromosomes are close enough to swap material.

4 Telophase I then forms two cells that enclose doubled copies of half the chromosomes of the original diploid cell.

Stages of meiosis

MEIOSIS II

PROPHASE II

METAPHASE II

ANAPHASE II

TELOPHASE II

5 The newly formed cells then immediately go into prophase II, metaphase II, anaphase II, and telophase II. These phases operate like those in mitosis, resulting in four haploid cells.

Tetrad of sister chromatids

Crossing-over between nonsister chromatids

Genetic recombination

Details of crossing-over during prophase I

As mentioned previously, gametes undergo meiosis (**Figure 3.21**). Meiosis is very similar to mitosis. In fact, the two are so similar that their stages are rather confusingly referred to by the same names. One difference is that, in the first prophase of meiosis (prophase I), pairs

homologous (huh-MAHL-uh-gus) Similar in structure, function, or sequence of genetic information.

of **homologous** chromosomes (that is, pairs of chromosome #1, pairs of chromosome #2, and so on) remain close together in tight groups called **tetrads**. During this phase, the chromosomes may exchange pieces of DNA in a process called **crossing over**. Crossing over "shuffles"

Comparing mitosis and meiosis   Table 3.1

Point of comparison	Mitosis	Meiosis
Cell type	Somatic	Gamete
Number of divisions	1	2
Stages	Interphase	Interphase I only
	Prophase	Prophase I and II
	Metaphase	Metaphase I and II
	Anaphase	Anaphase I and II
	Telophase	Telophase I and II
Copy DNA?	Yes, interphase	Yes, interphase I; No, interphase II
Tetrads?	No	Yes
Number of cells	2	4
Number of chromosomes per cell	46, or two sets of 23; this makeup, called diploid, is identical to the chromosomes in the starting cell	One set of 23; this makeup, called haploid, represents half of the chromosomes in the starting cell

the genetic material, which allows genetic variation from one generation to the next. In the first anaphase of meiosis (anaphase I), the tetrads get pulled apart. Ultimately, the two stages of meiosis divide the chromosome complement of the parent in half, a status called **haploid**. **Table 3.1** compares mitosis and meiosis.

haploid Referring to a cell or an organism that has only one (unpaired) set of chromosomes.

CONCEPT CHECK   STOP

1. **How** does the sodium-potassium pump work?
2. **What** happens during transcription and translation?
3. **How** is mitosis different from meiosis?

Cells Specialize into Various Tissues

LEARNING OBJECTIVES

1. **Identify** the structures and functions of the various epithelial tissues.
2. **Describe** the types of connective tissues.
3. **Explain** the locations and functions of each type of muscle tissue.
4. **Discuss** the classifications and functions of membranes.

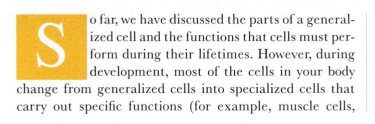 o far, we have discussed the parts of a generalized cell and the functions that cells must perform during their lifetimes. However, during development, most of the cells in your body change from generalized cells into specialized cells that carry out specific functions (for example, muscle cells,

nerve cells, epithelial cells). Groups of cells and surrounding material that work together to perform a specific function are referred to as **tissues**. The science that deals with the study of tissues is called **histology** (hiss-TOL-ō-jē). A **pathologist** (pa-THOL-ō-jist) is a physician who examines tissues for changes that indicate damage or disease.

There are four basic types of tissues in your body:

- **Epithelial** (ep′-i-THĒ-lē-al) **tissue** covers body surfaces, forms glands, and lines body cavities, hollow organs, and ducts.
- **Connective tissue** protects and supports the body and its organs, binds organs together, stores energy reserves as fat, and provides immunity.
- **Muscular tissue** generates the physical force needed to make body structures move.
- **Nervous tissue** detects changes inside and outside the body and generates transmits nerve impulses that coordinate body activities to help maintain homeostasis.

Most epithelial cells and some muscle and nerve cells are tightly joined into functional units by points of contact between their plasma membranes called **cell junctions**. Cell junctions perform different functions in different tissues:

1. **Tight junctions** fuse cells together tightly to prevent substances from passing between the cells. In tissues that line the stomach, intestines, and urinary bladder, tight junctions prevent the contents of these organs from leaking out.

2. Some cell junctions hold cells together so that they don't separate while performing their functions:
   - **Adherens** (ad-HER-ens) **junctions** have a dense layer of proteins just inside the plasma membrane called a plaque that runs along microfilaments to form a belt or strap-like structure called an adhesion belt. Two adjacent cells are joined by transmembrane glycoproteins that insert into the corresponding adhesion belts. This arrangement resists separation even when stretched.
   - **Desmosomes** (DEZ-mō-sōms) are like adherens junctions, but the plaque binds to intermediate filaments and does not form a belt. Instead of two cells adhering along a belt, they adhere at specific spots.
   - **Hemidesmosomes** resemble half of a desmosome. They do not adhere adjacent cells but rather attach cells to membranes.

3. **Gap junctions** form channels that allow ions and molecules to pass between cells. This permits cells in a tissue to communicate and enables nerve or muscle impulses to spread rapidly among cells.

Let's take a closer look at tissues, first examining epithelial tissues.

Epithelial Tissue Covers Body Surfaces

There are two types of epithelial tissue or epithelium: covering and lining epithelium and glandular epithelium. **Covering and lining epithelium** forms the outer coverings of skin and internal organs, lines body cavities, and makes up the sense organs, along with nerve tissue. **Glandular epithelium** forms secretory glands, such as sweat glands. Both types of epithelial tissues have the following five general features:

1. Consist of continuous sheets (single or multiple layers) of closely packed cells with little extracellular material between them
2. Have three types of surfaces:
   - *Apical (free surface)*. Exposed to body cavity, lining of an internal organ, or exterior of the body
   - *Lateral surface*. Faces adjacent surrounding cells
   - *Basal surface*. Attached to a *basement membrane*, which is a thin, extracellular structure made of protein fibers and which lies between the epithelial cells and the underlying connective tissue

   In stratified epithelia, the uppermost layer is the apical layer, while the deepest layer is the basal layer.
3. Lack blood vessels and get their nutrients from the connective tissue below by diffusion
4. Have a nerve supply
5. Divide rapidly and continuously to replace worn-out and injured cells

   Let's look at how epithelial tissue is classified.

Covering and lining epithelium
Covering and lining epithelium is classified according to cell shape and number of layers. Epithelial cells have different shapes that affect their functions (**Table 3.2**):

- **Squamous** (SKWĀ-mus)—Thin and flat cells that allow substances to pass rapidly through them
- **Cuboidal**—Cube-shaped cells that may have microvilli at their apical surface for secretion or absorption

- **Columnar**—Tall and thin cells that may have microvilli or cilia at their apical surface for secretion and absorption

- **Transitional**—Change shape from flat to cuboidal and back, These cells are found in organs that can stretch (for example, the urinary bladder).

Epithelial tissue is also classified according to its cell layers. **Simple epithelium** has one layer of cells. **Pseudostratified epithelium** is a single layer of cells that looks like it has many layers because the cells' nuclei are at many levels. Finally, **stratified epithelium** consists of multiple layers of cells. Cell shape and layering characteristics are combined in the classification of covering and lining epithelium; examples include simple squamous epithelium, simple cuboidal epithelium, pseudostratified columnar epithelium, and stratified columnar epithelium (**Figure 3.22** on next page).

Glandular epithelium Glandular epithelial cells work together to secrete various substances (enzymes, hormones, perspiration, milk, saliva, and so on). Glandular epithelium forms both exocrine glands and endocrine glands. **Exocrine glands** (EK-sō-krin) secrete substances through tubes or ducts; examples include sweat glands, salivary glands, and mammary glands. In contrast, **endocrine glands** (EN-dō-krin) secrete substances (mainly hormones) into interstitial fluid and then the blood; examples include the thyroid gland, the pituitary gland, and the adrenal gland. Some glands contain both endocrine and exocrine glandular epithelium; examples include the pancreas, ovaries, and testes.

The next major type of tissue we will discuss, connective tissue, protects and holds together tissues and organs such as the pancreas.

Classifications of covering and lining epithelium   Table 3.2

Classification	Function	Locations
Simple epithelium		
Simple squamous	Diffusion, osmosis, filtration, secretion	Blood vessels, lining of the heart, lymph vessels, lungs (air sacs), glomerulus of the kidneys, serous membranes (SIR-us), peritoneum
Simple cuboidal	Secretion, absorption	Kidney tubules, small ducts of many glands (for example, thyroid, pancreas), ovary linings, eye surfaces
Simple columnar		
Nonciliated	Secretion, absorption	Lining of digestive tract, gallbladder, and ducts of many glands
Ciliated	Moves mucus and other substances	Lining of upper respiratory tract, uterine tubes, uterus, nasal sinuses, central canal of spinal cord
Pseudostratified columnar		
Nonciliated	Absorption, protection	Lining of larger ducts of many glands, epididymis, male urethra
Ciliated	Secretion, moves mucus	Upper respiratory tract
Stratified epithelium		
Stratified squamous		
Keratinized	Protection	Skin (upper layers)
Nonkeratinized	Protection	Lining of wet surfaces (for example, mouth, esophagus, pharynx, vagina, tongue)
Stratified cuboidal	Protection, secretion, absorption	Lining of ducts of sweat glands, esophageal glands, part of male urethra
Stratified columnar	Protection, secretion	Lining of urethra, large excretory ducts of some glands (for example, esophageal), anus, eye, conjunctiva
Transitional	Stretching	Lining of urinary bladder, parts of ureters, and urethra

Epithelium is classified according to the shapes and layering of epithelial cells. The various types of epithelium line parts of various organs and perform various functions (see below), including diffusion, osmosis, absorption, secretion, filtration, mucus movement, and protection.

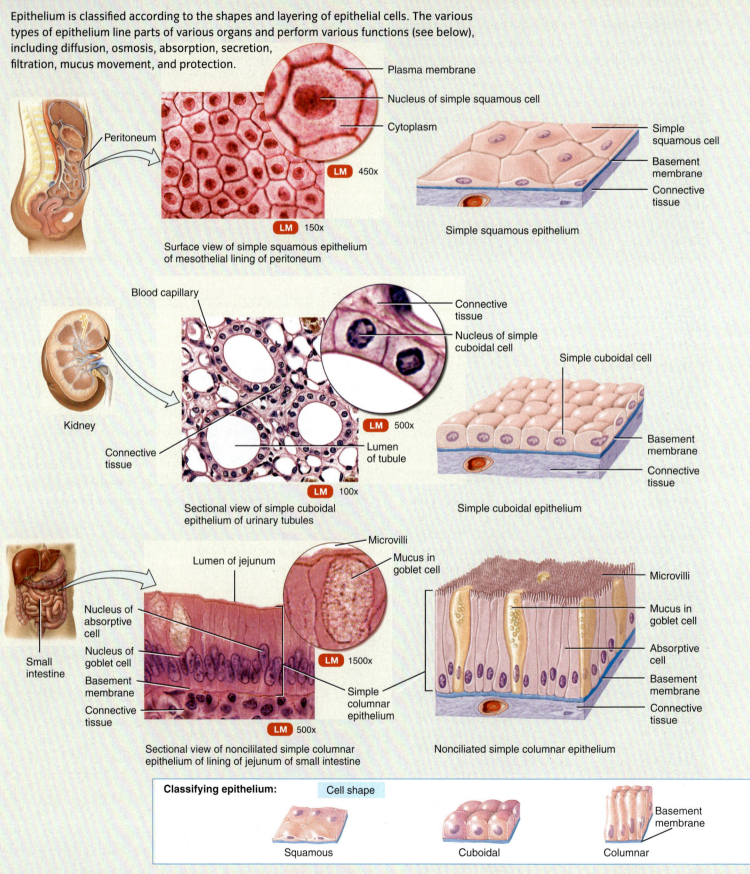

Peritoneum

Plasma membrane

Nucleus of simple squamous cell

Cytoplasm

LM 450x

LM 150x

Surface view of simple squamous epithelium of mesothelial lining of peritoneum

Simple squamous cell

Basement membrane

Connective tissue

Simple squamous epithelium

Blood capillary

Kidney

Connective tissue

Connective tissue

Nucleus of simple cuboidal cell

LM 500x

Lumen of tubule

LM 100x

Sectional view of simple cuboidal epithelium of urinary tubules

Simple cuboidal cell

Basement membrane

Connective tissue

Simple cuboidal epithelium

Microvilli

Mucus in goblet cell

Lumen of jejunum

Nucleus of absorptive cell

Nucleus of goblet cell

Basement membrane

Connective tissue

Small intestine

LM 1500x

Simple columnar epithelium

LM 500x

Sectional view of nonciliated simple columnar epithelium of lining of jejunum of small intestine

Microvilli

Mucus in goblet cell

Absorptive cell

Basement membrane

Connective tissue

Nonciliated simple columnar epithelium

Classifying epithelium: Cell shape

Basement membrane

Squamous

Cuboidal

Columnar

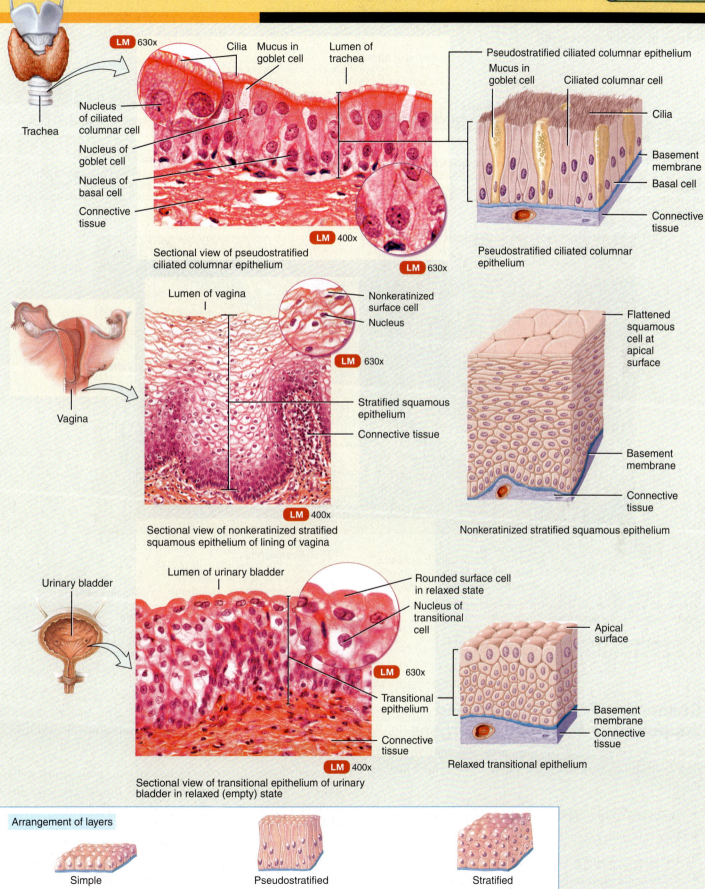

LM 630x

Cilia Mucus in goblet cell Lumen of trachea

Pseudostratified ciliated columnar epithelium

Mucus in goblet cell Ciliated columnar cell

Trachea

Nucleus of ciliated columnar cell

Nucleus of goblet cell

Nucleus of basal cell

Connective tissue

Cilia

Basement membrane

Basal cell

Connective tissue

LM 400x

Sectional view of pseudostratified ciliated columnar epithelium

LM 630x

Pseudostratified ciliated columnar epithelium

Lumen of vagina

Nonkeratinized surface cell

Nucleus

LM 630x

Flattened squamous cell at apical surface

Vagina

Stratified squamous epithelium

Connective tissue

Basement membrane

Connective tissue

LM 400x

Sectional view of nonkeratinized stratified squamous epithelium of lining of vagina

Nonkeratinized stratified squamous epithelium

Lumen of urinary bladder

Rounded surface cell in relaxed state

Nucleus of transitional cell

Apical surface

Urinary bladder

LM 630x

Transitional epithelium

Basement membrane

Connective tissue

Connective tissue

LM 400x

Sectional view of transitional epithelium of urinary bladder in relaxed (empty) state

Relaxed transitional epithelium

Arrangement of layers

Simple   Pseudostratified   Stratified

Cells Specialize into Various Tissues 75

Components of connective tissue • Figure 3.23

Connective tissue consists of various cells, fibers, and other substances (ground substance).

Reticular fibers
(re-TIK-ū-lar) are made of collagen and glycoproteins. They provide support in blood vessel walls and form branching networks around various cells (fat, smooth muscle, nerve).

Fibroblasts
(FI-brō-blasts) are large flat cells that move and secrete fibers and ground substance.

Collagen fibers
(KOL-a-jen) are strong, flexible bundles of the protein collagen, the most abundant protein in your body.

Macrophages
(MAK-rō-fāj-es) develop from white blood cells. They eat bacteria and cell debris by phagocytosis.

Mast cells
are abundant along blood vessels. They produce histamine, which dilates small blood vessels during inflammation and kill bacteria.

Elastic fibers
are stretchable but strong fibers made of proteins, elastin, and fibrillin. They are found in skin, blood vessels, and lung tissue.

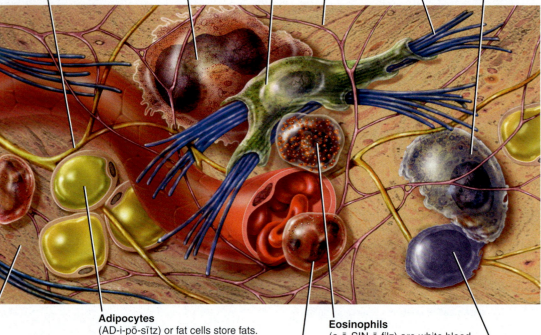

Adipocytes
(AD-i-pō-sītz) or fat cells store fats. They are found below the skin and around organs (heart, kidney).

Eosinophils
(e-ō-SIN-ō-filz) are white blood cells that migrate to sites of parasitic infection and allergic responses.

Plasma cells
are small cells that develop from a white blood cell. They secrete antibodies that attack and neutralize foreign substances.

Ground substance
is the stuff between cells and fibers. It is made of water and organic molecules (hyaluronic acid, chondroitin sulfate, glucosamine). It supports cells and fibers, binds them together, and provides a medium for exchanging substances between blood and cells.

Neutrophils
(NOO-trō-filz) are white blood cells that migrate to sites of infection.

Connective Tissue Protects and Supports the Body and Its Organs

Connective tissue is one of the most abundant tissues in your body. It has various forms and performs the following functions:

- Binds together, supports, and strengthens other tissues
- Protects and insulates internal organs
- Divides structures into compartments
- Transports materials throughout the body

- Stores energy (in fat tissues and cells)
- Protects the body by destroying invading microorganisms and eliminating cellular debris

Connective tissue consists of two major components, cells and the extracellular matrix (**Figure 3.23**). The cells include fibroblasts, macrophages, plasma cells, mast cells, and adipocytes, as well as some white blood cells (neutrophils, eosinophils) (**Figure 3.24**). The extracellular matrix includes various fibers—for example, collagen, elastic fibers, and ground substance made up of water and various organic

Types of connective tissues • Figure 3.24

There are three types of connective tissue: loose (areolar), dense regular, and elastic.

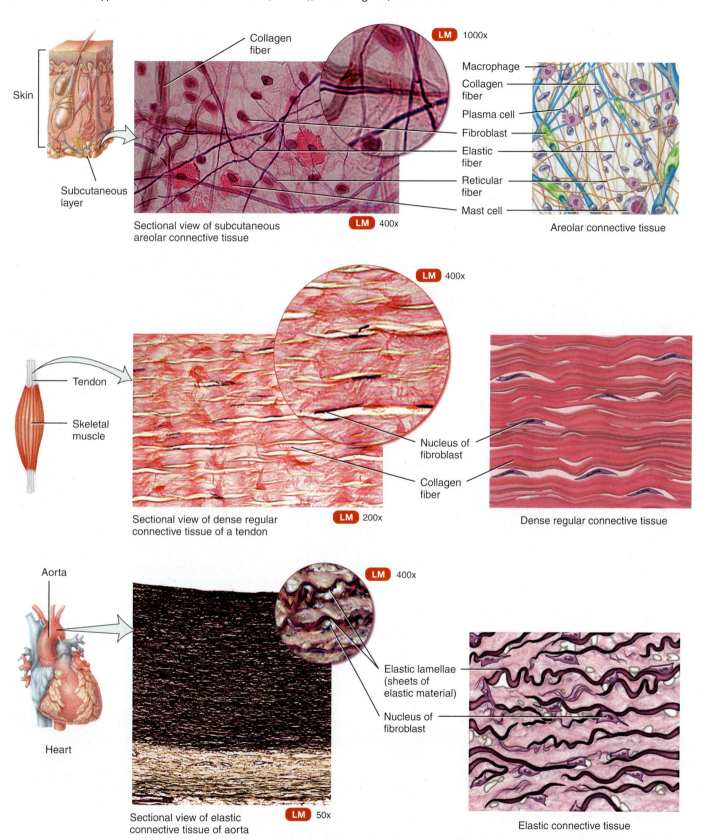

LM 1000x

Collagen fiber

Macrophage
Collagen fiber
Plasma cell
Fibroblast
Elastic fiber
Reticular fiber
Mast cell

Skin

Subcutaneous layer

Sectional view of subcutaneous areolar connective tissue

LM 400x

Areolar connective tissue

LM 400x

Tendon

Skeletal muscle

Nucleus of fibroblast

Collagen fiber

Sectional view of dense regular connective tissue of a tendon

LM 200x

Dense regular connective tissue

Aorta

LM 400x

Elastic lamellae (sheets of elastic material)

Nucleus of fibroblast

Heart

Sectional view of elastic connective tissue of aorta

LM 50x

Elastic connective tissue

molecules. Like epithelium, connective tissue is classified according to its cells, extracellular matrix, and appearance under the microscope (**Table 3.3**). Unlike epithelium, most connective tissue has a rich supply of blood vessels.

A class of connective tissue called *cartilage* is unique because it has no blood vessels (**Figure 3.25**). Specialized cells called **chondrocytes** (KON-drō-sītz) secrete a gel extracellular matrix that surrounds them. A chondrocyte or groups of chondrocytes form an "island" within the surrounding space, which is called a *lacuna* (la-KOO-

na). Because there are no blood vessels within cartilage, substances must diffuse through the extracellular matrix into the chondrocytes. This diffusion is a relatively slow process, which is why cartilage injuries take a long time to heal. There are three types of cartilage that perform three different functions: cushion joints, join structures, and provide flexible shapes (see Table 3.3 and Figure 3.25). Embryonic and fetal skeletons are initially made of cartilage that eventually develops into bone; some portions develop in the womb, and the remainder develops after birth during early childhood.

Classifications of connective tissues   Table 3.3

Classification	Components	Function	Locations
Loose connective tissue			
Areolar	Cells: fibroblasts, macrophages, plasma cells, adipocytes, mast cells Fibers: collagen, elastic, and reticular fibers	Provides strength, elasticity, and support	Skin (subcutaneous layer); mucous membranes; around blood vessels, nerves, and body organs
Adipose tissue	Cells: adipocytes	Reduces heat loss and provide energy reserve, support, and protection	Skin (subcutaneous layer); mucous membranes; around heart, kidneys, yellow bone marrow, and joints; behind eye socket
Reticular	Cells: specialized fibroblasts called reticular cells Fibers: reticular fibers	Forms supporting framework of organs, filters worn-out blood cells and microbes, binds smooth muscle cells together	Spleen; lymph nodes; liver; red bone marrow; basement membranes; around blood vessels and muscles
Dense connective tissue			
Dense regular	Cells: fibroblasts Fibers: collagen fibers	Provides strong attachments	Tendons; ligaments
Dense irregular	Cells: fibroblasts Fibers: collagen fibers	Provides strength	Fascia; deep skin; bone; cartilage; joint capsules; membrane capsules around organs (kidney, liver, testes); pericardium; heart valves
Elastic	Cells: fibroblasts Fibers: elastic fibers	Allows stretching of various organs	Lungs; arterial walls; trachea; bronchial tubes; vocal cords; ligaments
Cartilage			
Hyaline	Cells: differentiated fibroblasts called chondrocytes Fibers: collagen fibers	Provides smooth surface for movement, flexibility, and support	Joints; ends of long bones; ribs; nose; larynx; trachea; bronchial tubes; bronchi; embryonic/fetal skeleton
Fibrocartilage	Cells: differentiated fibroblasts called chondrocytes Fibers: collagen fibers	Provides support by joining structures together	Pelvis; intervertebral discs; knee
Elastic	Cells: differentiated fibroblasts called chondrocytes Fibers: elastic fibers	Provides support by maintaining shape	Epiglottis; external ear; auditory tubes
Bone tissue	Several connective tissues (detailed in Chapter 5)	Provides mechanical support and enables calcium and phosphate storage, blood cell production, fat storage	Bones of the skeletal system
Blood and lymph	Several cell types detailed in Chapters 10 and 12	Enables transport of substances and gases	Blood and lymph vessels

Types of cartilage • Figure 3.25

There are three types of cartilage: hyaline, elastic, and fibrocartilage.

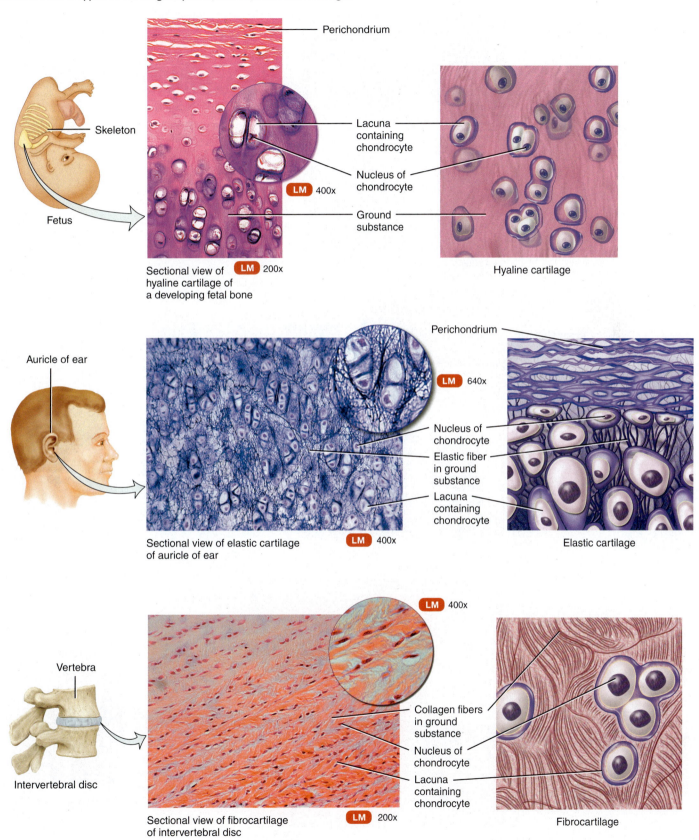

Skeleton

Fetus

Perichondrium

Lacuna containing chondrocyte

Nucleus of chondrocyte

Ground substance

LM 400x

LM 200x

Sectional view of hyaline cartilage of a developing fetal bone

Hyaline cartilage

Auricle of ear

Perichondrium

LM 640x

Nucleus of chondrocyte

Elastic fiber in ground substance

Lacuna containing chondrocyte

LM 400x

Sectional view of elastic cartilage of auricle of ear

Elastic cartilage

Vertebra

Intervertebral disc

LM 400x

Collagen fibers in ground substance

Nucleus of chondrocyte

Lacuna containing chondrocyte

LM 200x

Sectional view of fibrocartilage of intervertebral disc

Fibrocartilage

There are three types of muscular tissue: skeletal, smooth, and cardiac.

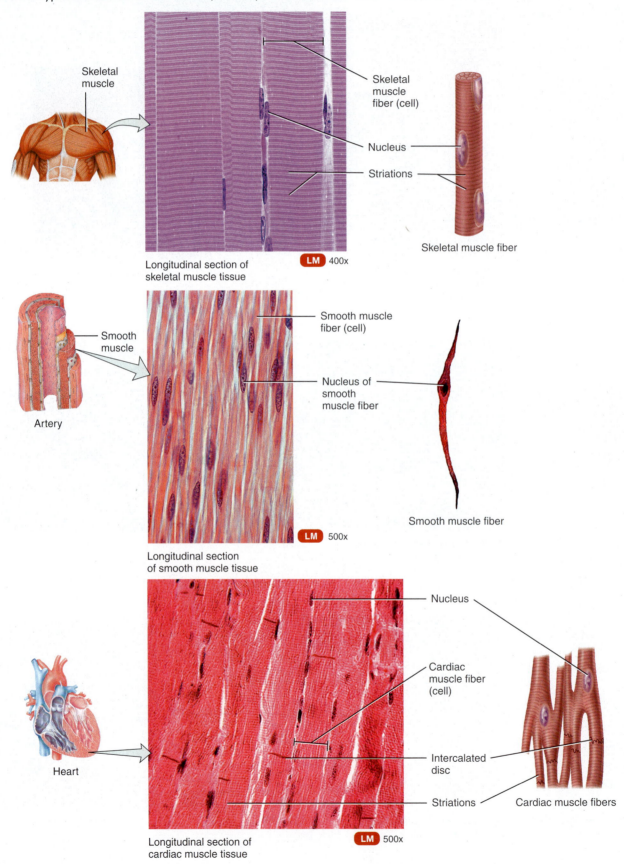

Skeletal muscle

Skeletal muscle fiber (cell)

Nucleus

Striations

Skeletal muscle fiber

LM 400x

Longitudinal section of skeletal muscle tissue

Artery

Smooth muscle

Smooth muscle fiber (cell)

Nucleus of smooth muscle fiber

Smooth muscle fiber

LM 500x

Longitudinal section of smooth muscle tissue

Heart

Nucleus

Cardiac muscle fiber (cell)

Intercalated disc

Striations

Cardiac muscle fibers

LM 500x

Longitudinal section of cardiac muscle tissue

Neurons and neuroglia • Figure 3.27

Numerous projections of the cell body of the neuron called dendrites receive nerve impulses from other cells. A long projection of the neuron cell body called an axon conducts each impulse away from the cell body to other cells. The two types of neuroglial cells—astrocytes and microglia—support the nerve cells.

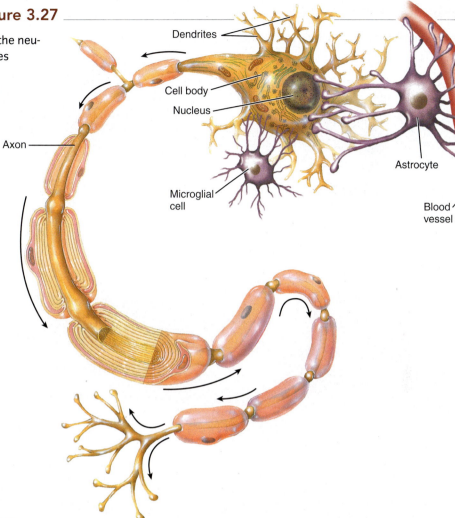

Dendrites

Cell body

Nucleus

Axon

Astrocyte

Microglial cell

Blood vessel

Muscular Tissue Generates Force for Movement

Muscular tissue is composed of elongated muscle cells called **muscle fibers**. The job of muscular tissue is to generate force, which produces motion, maintains posture, and generates heat. There are three types of muscular tissue (**Figure 3.26**):

- **Skeletal muscle**. These groups of long, multinucleated cells with regular bands or striations generate force upon voluntary commands. This muscle is usually attached to the skeleton and contracts voluntarily.

- **Smooth muscle**. These groups of small cells with one nucleus each are capable of stretching and are part of blood vessels, the stomach, intestines, uterus, and bladder. Smooth muscle tissue has no striations and contracts involuntarily.

- **Cardiac muscle**. The intermediate-sized cells that make up this tissue are connected to one another by cell junctions called *intercalated discs*. Cardiac muscle has striations and contracts involuntarily.

You will learn the details of these muscle tissues in Chapters 6, 11, and 14.

Nervous Tissue Transmits Impulses to Coordinate Activities

Despite the complexity of nervous system functions, nervous tissue consists of only two types of cells: neurons and neuroglia (**Figure 3.27**). **Neurons** (NOO-ronz), or **nerve cells**, are specialized cells that are sensitive to various stimuli. They convert stimuli into nerve impulses and conduct these impulses to other neurons, to muscle fibers, or to glands. Neurons can be small or large. For example, some neurons within the brain may extend only a few inches, while a spinal neuron from the spinal cord to a muscle in your foot may extend well over a foot. **Neuroglia** (noo-RŌG-lē-a) are supporting cells that do not generate or conduct nerve impulses but have many other important supportive functions. The detailed structure and function of neurons and neuroglia are considered in Chapter 7.

There are three major types of membranes: mucous, serous, and synovial.

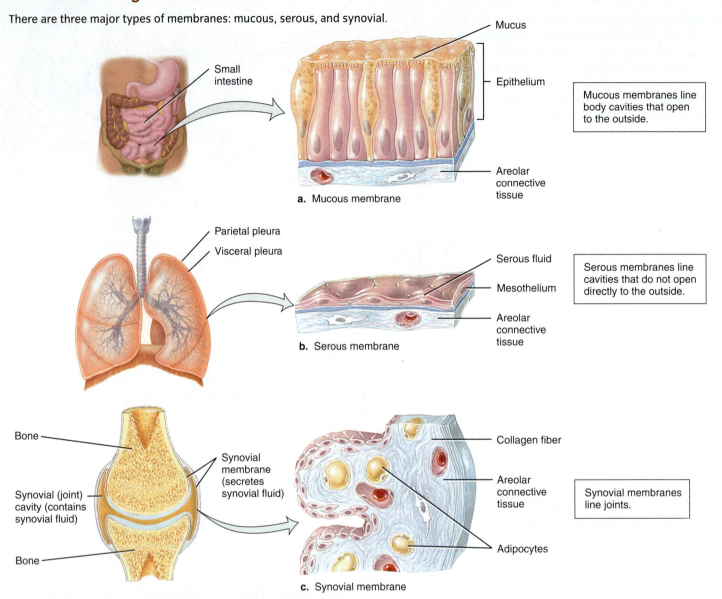

Small intestine

Mucus

Epithelium

Mucous membranes line body cavities that open to the outside.

Areolar connective tissue

a. Mucous membrane

Parietal pleura

Visceral pleura

Serous fluid

Mesothelium

Serous membranes line cavities that do not open directly to the outside.

Areolar connective tissue

b. Serous membrane

Bone

Synovial membrane (secretes synovial fluid)

Synovial (joint) cavity (contains synovial fluid)

Bone

Collagen fiber

Areolar connective tissue

Synovial membranes line joints.

Adipocytes

c. Synovial membrane

Membranes Cover or Line Parts of the Body

Membranes are flat sheets of pliable tissue that cover or line a part of the body. An **epithelial membrane** is composed of an epithelial layer and an underlying connective tissue layer. The principal epithelial membranes of the body are mucous membranes, serous membranes, and the cutaneous membrane or skin. The skin will be discussed in detail in Chapter 4. Another type of membrane, a *synovial membrane*, lines joints and contains connective tissue but no epithelium. Let's look at the various types of membranes (**Figure 3.28**).

A **mucous membrane** (MŪ-kus), or **mucosa** (mū-KŌ-sa), lines a body cavity that opens directly to the outside. Mucous membranes line the entire digestive system,

respiratory system, and reproductive system and much of the urinary system. The epithelial layer of a mucous membrane secretes mucus, which has the following functions:

- Prevents cavities from drying out
- Traps particles in the respiratory passageways
- Lubricates and absorbs food as it moves through the gastrointestinal tract
- Secretes digestive enzymes

The connective tissue layer of the mucous membrane (areolar connective tissue) helps bind the epithelium to the underlying structures, provides it with oxygen and nutrients, and removes wastes via its blood vessels.

A **serous membrane**, or **serosa** (se-RŌ-sa), lines a body cavity that does not open directly to the outside,

WHAT A HEALTH PROVIDER SEES

✓ THE PLANNER

Tissue Culture and Engineering

Scientists have long been able to grow cells in glass or plastic vessels outside the body, using a technology called **cell culture**, or **tissue culture**. Tissue culture requires that cells be supplied with a balanced salt solution, similar to that found in the body (that is, tissue culture medium). The tissue culture medium must contain essential nutrients and growth factors as well. Tissue culture has become important for studying cells and cell processes as well as developing new diagnostic tests. Tissue culture has led to a relatively new technology called **tissue engineering**.

Tissue engineering allows scientists to grow new tissues in a laboratory to replace damaged tissues in a body. Tissue engineers have already developed laboratory-grown versions of skin and cartilage, using the following procedure: Scaffolding beds of biodegradable synthetic materials or collagen are used as substrates that permit body cells such as skin cells or cartilage cells to be cultured. As the cells divide and assemble, the scaffolding degrades, leaving the new, permanent tissue, which is then implanted in the patient.

Other structures that tissue engineers are developing include bones, tendons, heart valves, bone marrow, and intestines. Work is also under way to develop insulin-producing cells for patients with diabetes, dopamine-producing cells for patients with Parkinson disease, and even entire livers and kidneys.

NATIONAL GEOGRAPHIC

Think Critically 1. What type of patient would benefit from tissue-engineered skin?
. If you developed an artificial bone, what tissues would it contain?

---

such as the peritoneum, which covers abdominal organs and lines the abdominal cavity. It also covers the organs that lie within the cavity. Serous membranes consist of two parts: a parietal layer and a visceral layer. The **parietal layer** (pa-RĪ-e-tal) is attached to the cavity wall; the **visceral layer** (VIS-er-al) covers and attaches to the organs inside these cavities. Each layer consists of areolar connective tissue covered by a simple squamous epithelium called **mesothelium** (mez′-ō-THĒ-lē-um). Mesothelium secretes **serous fluid**, a watery lubricating fluid that allows organs to glide easily over one another or to slide against the walls of cavities. Serous membranes are associated with the thoracic cavity (**pleura**), the heart (**pericardium**), and the abdominal cavity (**peritoneum**).

Synovial membranes (si-NŌ-vē-al) line the cavities of some joints. These membranes are composed of areolar connective tissue and adipose tissue with collagen fibers; they do not have an epithelial layer. Synovial membranes contain cells called synoviocytes, which

secrete **synovial fluid**. This fluid lubricates the ends of bones as they move at joints, nourishes the cartilage covering the bones, and removes microbes and debris from the joint cavity.

Despite their complexity, many tissues can be created in the laboratory using either traditional tissue culture techniques or a relatively new approach called tissue engineering (see *What a Health Provider Sees*).

CONCEPT CHECK 🛑 STOP

1. **What** is the difference between a simple squamous epithelium and a stratified columnar epithelium?

2. **What** is the function of fibroblasts in connective tissue?

3. **How** are skeletal, cardiac, and smooth muscle similar? How do they differ from one another?

4. **What** are the differences between a synovial membrane and an epithelial membrane?

Cells Specialize into Various Tissues **83**

Aging Affects Cells and Tissues

LEARNING OBJECTIVE

Describe the cellular and tissue changes that occur with aging.

A s we get older, our wounds take longer to heal, skin becomes less pliable, joints stiffen, and arteries become less elastic. Many of these changes can be explained by the effects of increasing age on cells:

- Cellular metabolism slows, so cells do not make new materials as fast as they once did, and healing occurs more slowly.

- Because cells do not divide as quickly as they once did and some may stop dividing altogether, worn-out cells do not get replaced.

- Cellular DNA erodes after many cell divisions. The protective DNA sequences at the ends of chromosomes, called **telomeres** (TEL-ō-mērz) (**Figure 3.29**), shorten and disappear after many divisions, leading to loss of functional DNA and cell death.

- Increasingly, glucose gets added to proteins both inside and outside the cell. The glucose units cross-link to each other, binding adjacent proteins and stiffening cell membranes and tissues such as skin and connective tissues. Loss of elasticity in connective tissue fibers such as collagen and elastin can also lead to wrinkles, stiffness in blood vessels,

Telomeres • Figure 3.29

Telomeres are sequences of repeating DNA bases located at the tips of chromosomes that protect the chromosomes.

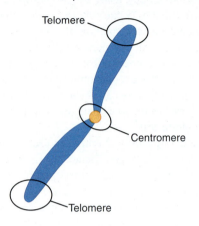

high blood pressure, atherosclerosis, cardiovascular disease, and stroke.

- Chemical reactions within the cell produce highly reactive oxygen free radicals, which damage lipids, proteins, and nucleic acids.

- Cells lose identity markers on the plasma membrane and become susceptible to attack by the body's immune system.

CONCEPT CHECK   **STOP**

Why does connective tissue stiffen with age?

Summary

1 Cells Have Distinct Parts 50

- A cell has a plasma membrane that separates the inside from the outside and controls the movement of substances across it. Within the membrane, the cytoplasm contains fluid and numerous organelles that carry out various cell functions. As shown, the largest organelle is the nucleus, which houses the cell's DNA. DNA controls and coordinates cell functions and reproduction.

- The cell's organelles include a plasma membrane, cytoskeleton, centrosomes, ribosomes, rough ER, smooth ER, Golgi complex, lysomomes, mitochondria, and a nucleus. Each organelle carries out specific functions.

The nucleus • Figure 3.10

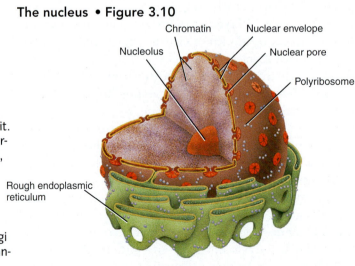

2 Cells Carry Out Many Processes 57

- Several processes allow substances to cross membranes. Diffusion and osmosis involve the movement of substances from areas of higher concentration to areas of lower concentration. Facilitated diffusion involves the movement of substances through membrane channels and sometimes involves carriers like the one illustrated in the figure.

Faciliated diffusion • Figure 3.12c

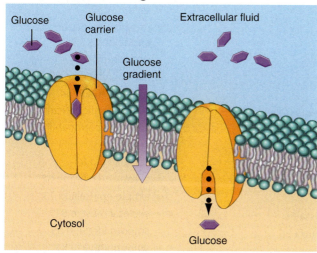

- Active transport, which moves substances against concentration gradients with energy from various sources, utilizes pumps and exchangers. Substances that are too large to take advantage of other transport mechanisms move across membranes via endocytosis, exocytosis, phagocytosis, and pinocytosis.

- Information for making proteins lies within a cell's DNA. This information gets transcribed by RNA polymerase into mRNA; the mRNA is translated into protein sequences through the actions of ribosomes and tRNA.

- Cells divide through the processes of mitosis and meiosis. Somatic cells undergo mitosis, which produces two identical cells with a genetic makeup identical to that of the starting cell. In contrast, gametes undergo meiosis, which produce four cells; each cell has half the amount of genetic material present in the starting cell. The mechanics of mitosis and meiosis are similar, but there are also crucial differences that permit genetic variation from generation to generation.

3 Cells Specialize into Various Tissues 71

- Epithelium covers and lines body cavities. The types of epithelium are based on cell shapes (squamous, cuboidal,

columnar) and arrangements (simple, stratified, pseudostratified); simple cuboidal epithelium is shown here. Various types are located throughout the body and serve specific functions, based on their structures.

Epithelial tissue • Figure 3.22

- Connective tissue binds together, supports, protects, and insulates body tissues. Its extracellular matrix contains many types of fibers and proteins, and the makeup of this matrix varies with the type of connective tissue.

- Muscular tissues generate force for movement, maintain posture, and generate heat. There are three types of muscular tissue: skeletal, cardiac, and smooth.

- Nerve cells conduct nerve impulses, which transmit information to and from various other cells. Nervous tissue consists of neurons, which transmit the nerve impulses, and neuroglia, which support the neurons.

- Membranes cover organs and line various body cavities. Epithelial membranes (mucous, serous) consist of epithelium overlying connective tissue. Synovial membranes line joint cavities and have no epithelial layer.

4 Aging Affects Cells and Tissues 84

- Aging slows and may even stop cell division, and it induces loss of genetic material through shortening of telomeres (shown here), cross-linking of glycoproteins, buildup of free radicals, and development of autoimmunity.

Telomeres • Figure 3.29

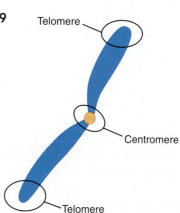

- Cell aging leads to slower wound healing, organ failure, and stiffness of connective tissues and blood vessels. Aging of blood vessels can lead to hypertension, atherosclerosis, cardiovascular disease, and stroke.

Key Terms

- active transport 62
- adherens junction 72
- anaphase 67
- cardiac muscle 81
- carrier 59
- catalase 55
- cell culture 83
- cell cycle 67
- cell division 50
- cell junction 72
- centriole 53
- centrosome 53
- cerebrospinal fluid 57
- channel 59
- chondrocyte 78
- chromatin 57
- chromosome 57
- cilia 54
- columnar epithelium 72
- concentration 57
- concentration gradient 57
- connective tissue 72
- covering and lining epithelium 72
- crossing over 70
- cuboidal epithelium 72
- cytoplasm 50
- cytoskeleton 53
- cytosol 53
- desmosome 72
- diffusion 58
- diploid 67
- electrically coupled transporter 62
- endocrine gland 73
- endocytosis 62
- endoplasmic reticulum (ER) 54
- epithelial membrane 82
- epithelial tissue 72
- exchanger 62
- exocrine gland 73
- exocytosis 64
- facilitated diffusion 59
- facilitated transport 59
- flagellum 54
- gamete 67
- gap junction 72
- genes 64
- glandular epithelium 72
- glycolipid 52
- glycoprotein 52
- Golgi complex 55
- haploid 71

- hemidesmosome 72
- histology 71
- homologous 70
- hormone 63
- hypertonic 61
- hypotonic 61
- integral protein 52
- intermediate filament 53
- interphase 67
- interstitial fluid 57
- isotonic 61
- lipid bilayer 52
- lymph 57
- lysosome 55
- meiosis 67
- membrane 82
- mesothelium 83
- messenger RNA (mRNA) 65
- metaphase 67
- microfilament 53
- microtubule 53
- microvilli 53
- mitochondria 56
- mitochondrial crista 56
- mitochondrial matrix 56
- mitosis 67
- mucosa 82
- mucous membrane 82
- muscle fiber 81
- muscular tissue 72
- nerve cell 81
- nervous tissue 72
- neuroglia 81
- neuron 81
- nuclear envelope 57
- nuclear pore 57
- nucleolus 57
- nucleus 50
- organelle 50
- osmosis 60
- oxidase 55
- oxidative phosphorylation 56
- parietal layer 83
- pathologist 71
- pericardium 83
- peripheral protein 52
- peritoneum 83
- peroxisome 55
- phagocytosis 63
- phagosome 63
- pinocytosis 63

- plasma 57
- plasma membrane 50
- pleura 83
- pore 59
- promoter 65
- prophase 67
- protease 55
- proteasome 55
- pseudopod 63
- pseudostratified epithelium 73
- pump 62
- receptor 63
- receptor-mediated endocytosis 63
- ribosome 54
- RNA polymerase 65
- rough ER (RER) 54
- sarcoplasmic reticulum 55
- selective permeability 52
- serosa 82
- serous fluid 83
- serous membrane 82
- simple epithelium 73
- skeletal muscle 81
- smooth ER (SER) 55
- smooth muscle 81
- sodium-potassium pump (Na+/K+ pump, or Na+/K+-ATPase) 62
- solute 57
- solvent 57
- somatic cell 67
- squamous epithelium 72
- stratified epithelium 73
- synovial fluid 83
- synovial membrane 83
- telomere 84
- telophase 67
- terminator 65
- tetrad 70
- tight junction 72
- tissue 71
- tissue culture 83
- tissue engineering 83
- transcription 65
- transfer RNA (tRNA) 67
- transitional epithelium 73
- translation 67
- transport protein 59
- vesicle 62
- visceral layer 83

Critical and Creative Thinking Questions

1. As you examine various cells under a microscope, you observe one type of cell that has large amounts of rough endoplasmic reticulum, vesicles merging with the plasma membrane, Golgi complexes, and mitochondria. What do you think is the function of this cell? Explain your answer, based on your observations.

2. Cells within a culture may grow and divide, but they do not do so all at the same time (that is, their growth is not synchronized). You want to study the cell cycle and require a synchronous culture. A drug called colchicine blocks microtubule formation, and its effects are reversible. How might you use this drug to synchronize your cell culture?

3. Janet's house is being infested by soft-bodied slugs. Rather than spray pesticide, an exterminator places rock salt around the perimeter of the foundation. Janet notices that upon encountering the salt, the slugs appear to shrink in size before dying. What might the salt do to a slug's cells and tissues? Explain your answer.

4. Chronic smokers often cough and have difficulty clearing mucus from their lungs. What type of epithelium is present in the airways, and what cell parts might be damaged, thus preventing smokers from clearing away foreign particles and mucus?

5. Manufacturers claim that a popular nutritional supplement keeps joints healthy. The supplement contains high concentrations of chondroitin sulfate and glucosamine. How do you suppose these substances might work?

What is happening in this picture?

Snake venom includes the enzymes phospholipase and hyaluronidase. Major effects include red blood cell and muscle cell damage and disruption of blood vessel walls, which leads to bleeding, swelling, cell death, and cardiovascular effects.

NATIONAL GEOGRAPHIC

Think Critically 1. How would phospholipase damage cells?
2. How would hyaluronidase disrupt blood vessels and cause swelling?

Self-Test

(Check your answers in Appendix C.)

1. _____ is made during transcription.

   a. Protein

   b. tRNA

   c. mRNA

   d. DNA

Use this figure for questions 2–3.

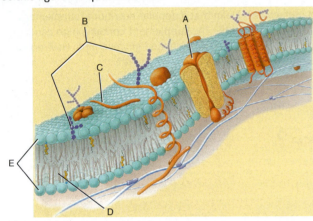

2. What structure is depicted in this figure?

   a. nucleus

   b. mitochondrion

   c. endoplasmic reticulum

   d. plasma membrane

3. Which of the structures labeled in the figure is associated with facilitated diffusion?

   a. A   b. B   c. D   d. E

4. Which of the following cells would *not* undergo mitosis in its lifetime?

   a. intestinal cell

   b. sperm

   c. fibroblast

   d. hair cell

5. The major function of the organelle in this figure is to _____.

   a. make proteins

   b. control flow of substances into and out of cell

   c. control and coordinate cell functions

   d. make ATP

6. _____ is the movement of substances against a concentration gradient using energy.

   a. Facilitated diffusion

   b. Passive transport

   c. Active transport

   d. Osmosis

7. The cell in this photo is in the _____ phase of mitosis.

   a. prophase

   b. metaphase

   c. anaphase

   d. telophase

8. The _____ is the part of the chromosome that shortens and disappears with age.

   a. centromere

   b. telomere

   c. kinetochore

   d. promoter

9. _____ is a tissue that lines organs and is important in absorption.

a. Nervous tissue

b. Muscular tissue

c. Connective tissue

d. Epithelium

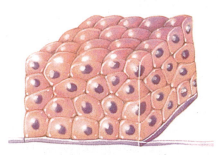

10. The tissue pictured in this figure is _____ epithelium.

a. simple squamous

b. stratified columnar

c. simple cuboidal

d. stratified cuboidal

11. A macrophage accomplishes its primary function through _____.

a. exocytosis

b. phagocytosis

c. pinocytosis

d. endocytosis

12. _____ membranes line joint cavities.

a. Synovial

b. Serous

c. Epithelial

d. Mucous

13. In what type of tissue would you find the cell shown in this figure?

a. heart muscle

b. smooth muscle

c. nerve

d. skeletal muscle

14. Adipocytes are found in _____.

a. epithelium

b. nervous tissue

c. musclar tissue

d. connective tissue

15. _____ have support functions in nerve tissue.

a. Neurons

b. Fibroblasts

c. Neutrophils

d. Neuroglia

THE PLANNER ✔

Review your Chapter Planner on the chapter opener and check off your completed work.

4 The Integumentary System

Every summer, millions of Americans flock to beaches with the dual goals of having fun in the sun and getting that ideal symbol of beauty, the perfect suntan. The quest does not end at the beach. Many people, especially teens, use indoor tanning devices. The Centers for Disease Control and Prevention (CDC) estimates that in 2005, 8.7% of teens aged 14 to 17 used indoor tanning devices. According to that report, girls are seven times more likely than boys to use tanning devices. However, exposure to ultraviolet rays, whether from the sun or a tanning bed, causes both tanning and sunburn. In the same report, the CDC noted that the prevalence of sunburn among adults is on the rise. From 1999 to 2004, the percentage of sunburned adults rose from 31.8% to 33.7%.

Sun exposure causes skin damage and is an important risk factor in skin cancer, most commonly basal cell carcinoma and melanoma. These types of skin cancer are also on the rise. In 2009 (the most recent year for which statistics are available), 68,720 people in the United States were diagnosed with melanomas, and 8,650 people died from this deadly type of skin cancer.

Let's look at the structure and function of skin and see how environmental factors affect it.

CHAPTER OUTLINE

CHAPTER PLANNER ✓

- ❏ Study the picture and read the opening story.
- ❏ Scan the Learning Objectives in each section:
  p. 92 ❏   p. 96 ❏   p. 100 ❏   p. 105 ❏
- ❏ Read the text and study all visuals. Answer any questions.

Analyze key features

- ❏ InSight, pp. 92–93 ❏   p. 103 ❏
- ❏ Process Diagram, p. 100 ❏   p. 101 ❏
- ❏ What a Health Provider Sees, p. 104 ❏
- ❏ Stop: Answer the Concept Checks before you go on:
  p. 96 ❏   p. 99 ❏   p. 104 ❏   p. 106 ❏

End of chapter

- ❏ Review the Summary and Key Terms.
- ❏ Answer the Critical and Creative Thinking Questions.
- ❏ Answer What is happening in this picture?
- ❏ Complete the Self-Test and check your answers.

NATIONAL GEOGRAPHIC

The Integumentary System Is Composed of Skin, Glands, Hair, Nails, and Nerve Endings

LEARNING OBJECTIVES

1. **Identify** various parts of the integumentary system.
2. **Describe** the structure of the skin.
3. **Explain** the basis for skin colors.

Your skin is the most visible organ of your body. Its appearance and health is affected by nutrition, hygiene, blood circulation, age, immunity, genetics, psychological state, and drugs. Because the appearance of the skin contributes so much to a person's self-image, people spend vast amounts of money and time on hygiene, beauty products, and medical treatments to take care of it. The branch of medicine that specializes in diagnosing and treating skin disorders is called **dermatology** (der-ma-TOL-ō-jē; *dermato-* =skin; *-logy* = study of). Let's take a closer look at skin and the other structures that make up the integumentary system.

There Are Many Components of the Integumentary System

The **integumentary system** (in-tēg′-ū-MEN-tar-ē; *in-* = inward; *tegere* = to cover) is composed of the skin, hair, oil and sweat glands, nails, and sensory receptors (**Figure 4.1**).

InSight   The integumentary system   •   Figure 4.1

The integumentary system is composed of the skin, hair, oil and sweat glands, nails, and sensory receptors.

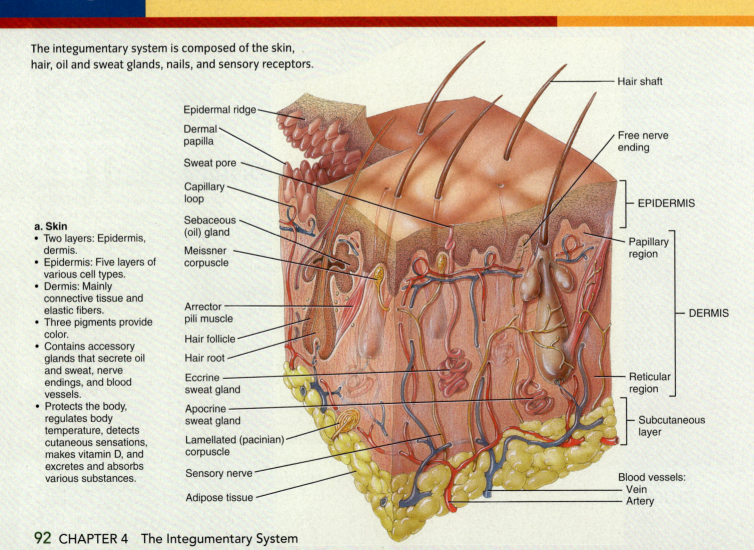

a. Skin
- Two layers: Epidermis, dermis.
- Epidermis: Five layers of various cell types.
- Dermis: Mainly connective tissue and elastic fibers.
- Three pigments provide color.
- Contains accessory glands that secrete oil and sweat, nerve endings, and blood vessels.
- Protects the body, regulates body temperature, detects cutaneous sensations, makes vitamin D, and excretes and absorbs various substances.

Labels:
- Epidermal ridge
- Dermal papilla
- Sweat pore
- Capillary loop
- Sebaceous (oil) gland
- Meissner corpuscle
- Arrector pili muscle
- Hair follicle
- Hair root
- Eccrine sweat gland
- Apocrine sweat gland
- Lamellated (pacinian) corpuscle
- Sensory nerve
- Adipose tissue
- Hair shaft
- Free nerve ending
- EPIDERMIS
- Papillary region
- DERMIS
- Reticular region
- Subcutaneous layer
- Blood vessels: Vein, Artery

cutaneous membrane (kū-TĀ-nē-us) Pertaining to the skin.

epidermis (ep-i-DERM-is; *epi-* = "above") The superficial, thinner layer of the skin, which is composed of epithelial tissue.

The **skin**, or **cutaneous membrane**, covers the external surface of the body. The skin is the largest organ of the body in terms of both surface area and weight. In adults, the skin covers an area of about 2 square meters (22 square feet) and weighs 4.5–5 kg (10–11 lb), about 16% of total body weight.

Let's take a closer look at the structure of the skin.

Skin Is a Multilayered Organ

The skin consists of two main parts (see Figure 4.1a):

- **Epidermis** is the superficial, thinner portion, which is composed of *epithelial tissue*.

- **Dermis** is the deeper, thicker *connective tissue* portion.

Deep to the dermis, but not part of the skin, is the **subcutaneous (subQ) layer**. This layer consists of **areolar connective tissue** and adipose tissue. Fibers that extend from the dermis anchor the skin to the subcutaneous layer, which, in turn, attaches to underlying tissues and organs. The subcutaneous layer serves as a storage depot for fat and contains large blood vessels that supply the skin.

The *epidermis* is composed of keratinized stratified squamous epithelium. It contains four principal types of cells: keratinocytes,

dermis A layer of dense irregular connective tissue lying deep to the epidermis.

subcutaneous (subQ) layer (sub´-kū-TĀ-nē-us) The layer beneath the skin. Also called **hypodermis** (hī-pō-DER-mis).

areolar connective tissue (a-RĒ-ō-lar) A type of fibrous connective tissue in which the fibers are arranged in a loose net or mesh.

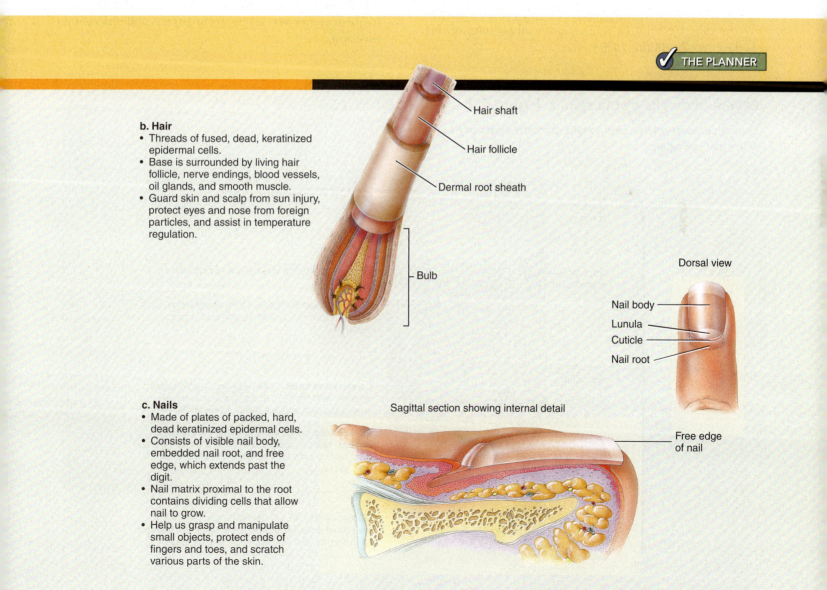

b. Hair
- Threads of fused, dead, keratinized epidermal cells.
- Base is surrounded by living hair follicle, nerve endings, blood vessels, oil glands, and smooth muscle.
- Guard skin and scalp from sun injury, protect eyes and nose from foreign particles, and assist in temperature regulation.

Hair shaft

Hair follicle

Dermal root sheath

Bulb

Dorsal view

Nail body
Lunula
Cuticle
Nail root

c. Nails
- Made of plates of packed, hard, dead keratinized epidermal cells.
- Consists of visible nail body, embedded nail root, and free edge, which extends past the digit.
- Nail matrix proximal to the root contains dividing cells that allow nail to grow.
- Help us grasp and manipulate small objects, protect ends of fingers and toes, and scratch various parts of the skin.

Sagittal section showing internal detail

Free edge of nail

The Integumentary System Is Composed of Skin, Glands, Hair, Nails, and Nerve Endings 93

melanocytes, Langerhans cells, and Merkel cells (**Figure 4.2**). Several distinct layers of keratinocytes in various stages of development form the bulk of the epidermis. This is called **thin skin**. The epidermis has five layers—stratum basale, stratum spinosum, stratum granulosum, stratum lucidum, and a thick stratum corneum, it is called **thick skin**.

The layers of the epidermis are as follows (from deepest to most superficial):

> **stem cell** An unspecialized cell that has the ability to divide for indefinite periods and give rise to specialized cells.

- **Stratum basale** (ba-SA-lē; *basal* = base)—A single row of cuboidal or columnar keratinocytes. Some cells in this layer are **stem cells** that undergo cell division to continually produce new keratinocytes.

- **Stratum spinosum** (spi-NŌ-sum; *spinos* = thornlike)—8 to 10 layers of many-sided keratinocytes that fit closely together. This layer provides strength and flexibility to the skin. Cells in the more superficial portions of this layer are beginning to flatten.

- **Stratum granulosum** (gran-ū-LŌ-sum; *granulos* = little grains)—3 to 5 layers of flattened keratinocytes that are undergoing **apoptosis**. A distinctive feature of cells in this layer is the presence of the protein **keratin** (KER-a-tin) and membrane-enclosed *lamellar granules*.

> **apoptosis** (ap-ō-TŌ-sis *or* ap'-op-TŌ-sis) Programmed cell death in which a cell goes through a limited number of cell divisions and then dies.

- **Stratum lucidum** (LOO-si-dum; *lucid* = clear)—4–6 layers of flattened clear, dead keratinocytes that contain large amounts of keratin. This layer is found only in the areas of the body (thick skin) where exposure to friction is greatest (for example, fingertips, palms, soles).

- **Stratum corneum** (COR-nē-um; *corne-* = horn or horny)—25 to 30 layers of flattened dead cells from the deeper strata. The interior of the cells contains mostly keratin. Its multiple layers of dead cells help to protect deeper layers from injury and microbial invasion.

Layers and cells of the skin • Figure 4.2

The various layers of skin cells consist of mostly keratinocytes intermixed with other cell types.

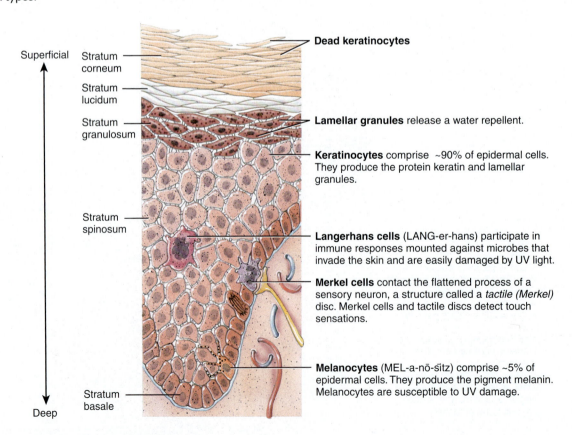

Superficial

Deep

Stratum corneum

Stratum lucidum

Stratum granulosum

Stratum spinosum

Stratum basale

Dead keratinocytes

Lamellar granules release a water repellent.

Keratinocytes comprise ~90% of epidermal cells. They produce the protein keratin and lamellar granules.

Langerhans cells (LANG-er-hans) participate in immune responses mounted against microbes that invade the skin and are easily damaged by UV light.

Merkel cells contact the flattened process of a sensory neuron, a structure called a *tactile (Merkel) disc*. Merkel cells and tactile discs detect touch sensations.

Melanocytes (MEL-a-nō-sītz) comprise ~5% of epidermal cells. They produce the pigment melanin. Melanocytes are susceptible to UV damage.

New skin cells form in the stratum basale and are slowly pushed to the surface. As the cells move from one epidermal layer to the next, they accumulate more and more keratin through a process called **keratinization** (ker'-a-tin-i-ZĀ-shun). Eventually, the keratinized cells slough off and are replaced by underlying cells. The cycle from cell division to sloughing off at the surface takes about 4 weeks in an average epidermis of 0.1 mm (0.004 in.) thickness. When an excessive amount of keratinized cells shed from the skin of the scalp, it is called *dandruff.*

The *dermis* is the second, deeper part of the skin. It is composed mainly of dense irregular connective tissue containing collagen and elastic fibers. The superficial part of the dermis, also known as the *papillary region*, makes up about one-fifth of the thickness of the total layer (see Figure 4.1b). It contains fine elastic fibers. Its surface area is greatly increased by small, fingerlike projections called **dermal papillae** (pa-PIL-ē; *papula* = nipples; singular is papilla). These nipple-shaped structures project into the undersurface of the epidermis and can contain the following:

- *Blood capillaries* (capillary loops)
- *Nerve endings* (sensory receptors):
  - *Corpuscles of touch* or *Meissner corpuscles*—Associated with touch
  - *Free nerve endings*—Associated with sensations of warmth, coolness, pain, tickling, and itching

The deeper part of the dermis, also known as the *reticular region*, which is attached to the subcutaneous layer, contains bundles of collagen and some coarse elastic fibers interspersed with adipose cells, hair follicles, nerves, oil glands, and sweat glands. This combination of collagen and elastic fibers provides the skin with strength, *extensibility* (ability to stretch), and *elasticity* (ability to return to original shape after stretching). The extensibility of skin can readily be seen in pregnancy and obesity. The skin has its limits, however; extreme stretching may produce small tears in the dermis, forming *striae* (STRĪ-ē; *stria* = streaks), or stretch marks, that are visible as red or silvery white streaks on the skin surface.

You can see the difference between the epidermis and dermis when you get a blister. The top part of the blister is the epidermis, while the bottom part beneath the fluid is the papillary region of the dermis.

Skin Color Is Caused by Pigments

A wide variety of skin colors are produced by three pigments: melanin, hemoglobin, and carotene. Differing amounts of **melanin** (MEL-a-nin) and the shade of melanin cause the skin to present in a wide range of colors from pale yellow to reddish brown to black. Melanin absorbs damaging ultraviolet (UV) light to protect the underlying tissues of the skin. Melanin-producing cells, called melanocytes, are present in the skin (stratum basale of the epidermis) and mucous membranes all over the body. However, they are most plentiful in the epidermis of the penis, nipples of the breasts, areas just around the nipples (areolae), face, and limbs. Because the number of melanocytes is about the same in all people, differences in skin color are due mainly to the amount and shade of pigment that the melanocytes produce and transfer to keratinocytes. Consequently, the epidermis has a variation of pigmentation with skin color ranges from yellow to red to tan to black:

- Dark-skinned individuals have large amounts of melanin in the epidermis. The more melanin that is present, the darker the skin.
- Light-skinned individuals have little melanin in the epidermis. Thus, the epidermis appears translucent, and skin color ranges from pink to red, depending on the oxygen content of the blood moving through capillaries in the dermis. The red color is due to **hemoglobin**, the oxygen-carrying pigment in red blood cells.
- **Albinism** (AL-bin-izm; *albin-* = white) is an inherited trait that causes individuals to not produce melanin. People affected by albinism are called **albinos** (al-BĪ-nōs). Because most albinos do not have melanin in their hair, eyes, and skin, they need to take extra precautions when exposed to the sun.

Melanocytes may not be evenly scattered throughout the skin, causing uneven melanin distribution:

- *Freckles*—With freckles, melanin accumulates in patches.
- *Age (liver) spots*—With age spots, melanin accumulates with age, forming flat blemishes that look like freckles but range in color from light brown to black.
- *Nevus or mole*—A **nevus** (NĒ-vus) is a round, flat, or raised area that represents a benign localized overgrowth of melanocytes and usually develops in childhood or adolescence.
- *Vitiligo*—**Vitiligo** (vit-i-LĪ-gō) is a condition in which melanocytes are partially or completely lost from areas of the skin, producing irregular white spots. One possible cause is a malfunction of the immune system in which antibodies attack and destroy the melanocytes.

Melanin sits superficial to the nucleus in the keratinocyte, blocking ionizing radiation from reaching and damaging the DNA. Melanin levels in the skin may change with exposure to sunlight. UV light stimulates melanin production, which increases the amount of melanin that is deposited into the keratinocytes. Melanin gives the skin a tanned appearance and improves protection of the body against UV radiation. A tan "fades," or is lost, when the melanin-containing keratinocytes are shed from the stratum corneum. Despite the protection provided by melanin, repeated or excessive exposure to UV radiation can lead to skin cancer.

Carotene (KAR-ō-tēn; *carot* = carrot) is a yellow-orange pigment that gives egg yolk and carrots their color. This precursor of vitamin A accumulates in the stratum corneum, fatty areas of the dermis, and subcutaneous layer in response to excessive dietary intake. In fact, so much carotene may be deposited in the skin after eating large amounts of carotene-rich foods that the skin color may actually turn orange, which is especially apparent in light-skinned individuals. Decreasing carotene intake eliminates the problem.

CONCEPT CHECK

1. **What** are the main components of the integumentary system?
2. **How** is thick skin different from thin skin?
3. **What** is the main pigment that determines skin color?

Accessory Structures Provide Protection and Help Regulate Body Temperature

LEARNING OBJECTIVES

1. **Describe** the structure and function of hair.
2. **Identify** the glands found in skin and describe what they do.
3. **Describe** the structure and function of nails.

Hair, glands, and nails are accessory structures of the integumentary system that develop from the epidermis of the embryo. Each of these accessory structures performs important functions in the body. For example, hair and nails protect the body, and sweat glands help regulate body temperature. Let's take a look at hair first.

Hair Protects the Skin and Other Structures of the Body

Hairs, or **pili** (PĪ-lī), are present on most skin surfaces except the palms, palmar surfaces of the fingers, soles, and plantar surfaces of the toes. In adults, hair usually is most heavily distributed across the scalp, over the brows of the eyes, and around the external genitalia. The thickness and pattern of distribution of hair is largely determined by genetic and hormonal influences. Hair on the head guards the scalp from injury and the sun's rays; eyebrows and eyelashes protect the eyes from foreign particles; and hair in the nostrils filters insects and foreign particles to protect the tissues of the respiratory system.

Each hair is a thread of fused, dead, keratinized epidermal cells that consists of the following (**Figure 4.3**):

- The **shaft** is the superficial portion that projects above the surface of the skin.
- The **root** is the portion below the surface that penetrates into the dermis and sometimes into the subcutaneous layer.
- The **hair follicle** surrounds the root and is composed of epidermal cells.
- **Hair root plexuses** are nerve endings that surround each hair follicle. They are sensitive to touch and are stimulated if a hair shaft is moved.

The base of each follicle is enlarged into an onion-shaped structure called the **bulb**. In the bulb is a nipple-shaped indentation referred to as the **papilla of the hair**. The papilla of the hair contains many blood vessels and provides nourishment for the growing hair. The bulb also contains a region of cells called the **hair matrix**, which produces new hairs by cell division, thus lengthening the hair.

Associated with each hair are sebaceous (oil) glands and a bundle of smooth muscle cells called **arrector pili** (a-REK-tor; *arrect* = to raise), which extends from the upper dermis to the side of the hair follicle. In its normal position, hair emerges at an angle to the surface of the skin. Under stress, such as cold or fright, nerves stimulate the arrector pili muscles to contract, which pulls the hair shafts perpendicular to the skin surface. This action

Hair anatomy • Figure 4.3

Hairs are composed of dead, keratinized cells.

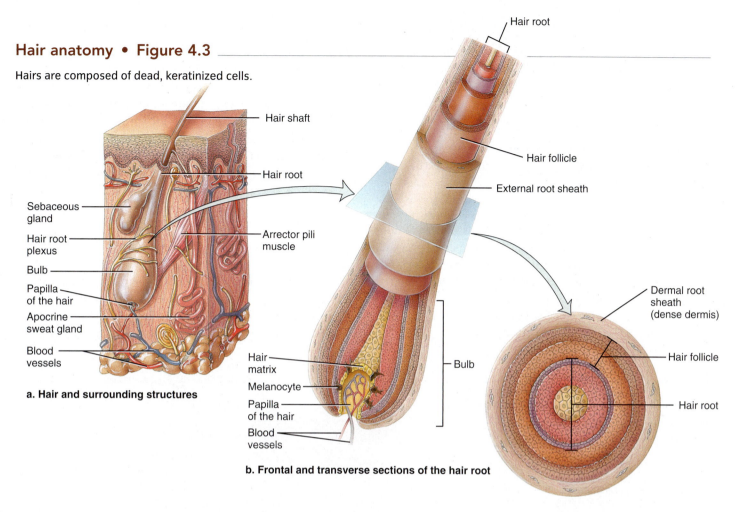

a. Hair and surrounding structures

Labels (a): Hair shaft, Hair root, Sebaceous gland, Hair root plexus, Bulb, Papilla of the hair, Apocrine sweat gland, Blood vessels, Arrector pili muscle

b. Frontal and transverse sections of the hair root

Labels (b): Hair root, Hair follicle, External root sheath, Bulb, Hair matrix, Melanocyte, Papilla of the hair, Blood vessels

c. Transverse section of hair root

Labels (c): Dermal root sheath (dense dermis), Hair follicle, Hair root

causes "goose bumps" because the skin around the shaft forms slight elevations.

The color of hair is due to melanin being synthesized by melanocytes in the matrix of the bulb and passing into cells of the root and shaft. Dark-colored hair contains mostly brown to black melanin. Blond and red hairs contain variants of yellow to red melanin in which there is iron and more sulfur. Gray hair is the result of sporadic production of melanin as the melanocytes begin to die off, usually with aging. White hair is generally due to an absence of melanin in the hair.

Scalp and body hair grow continuously throughout most of your life. However, at puberty, hormones stimulate a different type of hair growth, which varies with gender:

- *Males*—When the testes begin secreting significant quantities of **androgens**, males develop the typical male pattern of hair growth, including facial, chest, pubic, and **axillary** hair.
- *Females*—The ovaries and the adrenal glands produce small quantities of androgens, which promote hair growth in the axillary and pubic regions.

Occasionally, a tumor of the adrenal glands, testes, or ovaries produces an excessive amount of androgens. This results in a condition of excessive body hair called **hirsutism** (HER-soo-tizm; *hirsut-* = shaggy), which is most common in females or prepubertal males.

In genetically predisposed adults, androgens inhibit hair growth, thereby producing the most common form of baldness, **androgenic alopecia**, or **male-pattern baldness**. Men tend to lose hair at the temples and crown, while women are more likely to have thinning of hair on top of the head. To treat baldness and enhance hair growth, the Food and Drug Administration approved the use of minoxidil (sold under the brand name Rogaine). This drug causes **vasodilation**, thus increasing circulation that stimulates scalp follicles to enlarge and lengthens the growth cycle. While minoxidil improves hair growth in about

androgen (AN-drō-jen) A masculinizing sex hormone produced by the testes in males and the adrenal cortex in both genders.

axillary (ak-SIL-ary) Of the axilla (plural is axillae), the small hollow beneath the arm where it joins the body at the shoulders.

vasodilation (vāz′-ō-Dī-lā-shun) An increase in the size of the lumen of a blood vessel caused by relaxation of the smooth muscle in the wall of the vessel.

one-third of the people who try it, many find the hair growth to be meager. This drug will not help individuals who are already bald.

Glands Produce Secretions That Perform a Variety of Functions

Three types of glands are associated with the skin (**Figure 4.4**):

- **Sebaceous glands** (se-BĀ-shus) secrete *sebum*, an oily substance containing lipids and cellular debris. Sebum softens skin, prevents hair from drying out, and prevents water loss from the body surface. When sebaceous glands of the face become enlarged because of accumulated sebum, *blackheads* develop. Bacteria metabolize the sebum and form pimples or boils (especially in the skin of the forehead, palms, and soles).

- **Sudoriferous glands** (soo'-dor-IF-er-us) are sweat glands. Sudoriferous glands come in two varieties: *eccrine* and *apocrine*. The eccrine sweat glands are distributed all over the body and produce a watery sweat that helps regulate body temperature. The apocrine variety is limited to the axillary and pubic regions and produces a thicker sweat often released during emotional stress. This sweat is composed mostly of water but also contains electrolytes (sodium, potassium, chloride, calcium, magnesium, trace elements) and certain waste materials (lactate, urea).

- **Ceruminous glands** (se-ROO-mi-nus) are modified sweat glands found in the ear canal and outer ear. They secrete *earwax*, or *cerumen* (se-ROO-men). Cerumen is composed of lipids and cellular debris, forming a sticky barrier that both waterproofs and inhibits debris, foreign objects, insects, and microbial organisms from entering the ear canal.

Glands of the skin • Figure 4.4

a. The major glands of the skin include the oil glands

b., c. Two types of sweat glands

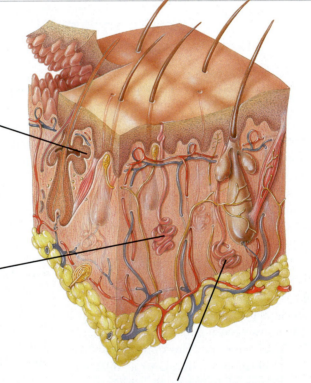

a. Sebaceous oil glands lie in the dermis and open into hair follicles or directly onto the skin. They secrete oily sebum, which does the following:
- Keeps hair from drying out
- Prevents excessive water evaporation from the skin
- Keeps skin soft
- Inhibits growth of certain bacteria

b. Eccrine sudoriferous sweat glands are found throughout the body.
- Secretory portion in the deep dermis
- Open directly to the outside through pores
- Secrete sweat in response to rise in temperature and emotions
- Sweat is mostly H_2O, Na^+, Cl^-, urea, ammonia, amino acids, glucose, and lactic acid
- Remove heat by evaporation, thereby cooling the skin

c. Apocrine sudoriferous sweat glands are found in axillae, groin, areolae, and beards.
- Secretory portion in subcutaneous layer
- Open into hair follicle
- Secrete viscous milky or yellow sweat in response to emotions (for example, fear, embarrassment, sexual activity)
- Sweat is similar to eccrine sweat but also contains lipids and proteins
- Sweat interacts with bacteria and produces body odor

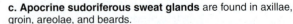

WILEY PLUS | NATIONAL GEOGRAPHIC | Video

Sagittal plane

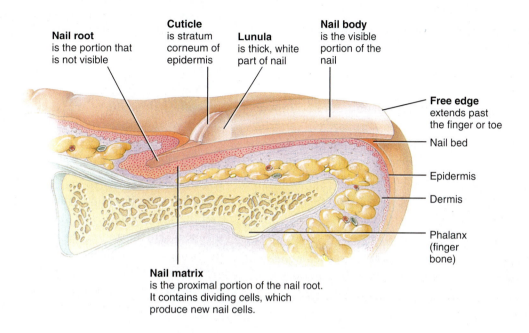

Nail root is the portion that is not visible

Cuticle is stratum corneum of epidermis

Lunula is thick, white part of nail

Nail body is the visible portion of the nail

Free edge extends past the finger or toe

Nail bed

Epidermis

Dermis

Phalanx (finger bone)

Nail matrix is the proximal portion of the nail root. It contains dividing cells, which produce new nail cells.

Sagittal section showing internal detail

Nails Are Composed of Keratinized Dead Cells

Nails are plates of tightly packed, hard, dead, keratinized cells of the epidermis. Functionally, nails help us grasp and manipulate small objects, provide protection to the ends of the fingers and toes, and allow us to scratch various parts of the body. Each nail consists of several parts, or regions: *nail body*, *nail root*, and *nail matrix* (**Figure 4.5**). Dividing cells within the nail matrix become keratinized as the nail grows outward. The average growth of fingernails is about 1 mm (0.04 inch) per week.

Various portions of nails have different colors. For example, the nail body is pink because the blood vessels of the underlying skin partially show through. The free edge is white because it extends past the tip of the finger or toe, and there is no underlying tissue. Finally, the **lunula** (LOO-nyū-la) is white because the nail is too thick in this region for any blood vessels to show through.

Artificial nails have become popular in recent years. Although these acrylic nails can enhance the appearance, they are not without hazards. The adhesives used to attach these nails are toxic and can cause inflammation of the nail bed. Fungal infections are also common. If the chemical adhesives damage the nail root, ridges and peeling of the nail layers may be seen on the nail surface. Occasionally, the nail may even separate from the nail bed, leading to a loss of the nail.

CONCEPT CHECK   STOP

1. **What** causes hair color?
2. **Where** are eccrine sweat glands located, and what do they do?
3. **What** are the parts of the nail?

The Skin Plays a Number of Roles in the Body

LEARNING OBJECTIVES

1. **Describe** how the skin contributes to body temperature regulation.
2. **Outline** the processes of epidermal and deep wound healing.
3. **Explain** the role of the skin in cutaneous sensation.
4. **Identify** types of burns and their consequences.

Y ou may not think much about what the skin does, but it performs essential functions for your health and well being. Let's look at these functions. Your skin helps regulate your body temperature in many ways:

- Because the skin's surface area is so large (2 m², or 22 ft²), your body is able to radiate to the outside air vast amounts of heat that your body has produced.

- The eccrine glands produce sweat to help eliminate heat via evaporation.

- The skin acts as a large reservoir of blood (8%–10% total blood flow). When you are hot, increased blood flow to the skin delivers more heat to be radiated away and contributes to higher sweat production. When you are cold, reduced blood flow to the skin conserves body heat and reduces sweat production.

- When you are cold, arrector pili contract to raise the angle of body hairs and cause goose bumps. The raised body hair reduces air microcirculation immediately above the skin's surface. The reduced air circulation helps impede the loss of body heat via the skin.

The skin forms a protective barrier for the internal organs. Keratin protects the body from heat, abrasion, chemicals, and microbes. Keratinocytes resist invasion by microbes, and the Langerhans cells alert the immune system to the invaders. Furthermore, the skin can usually repair itself following minor injuries, such as cuts and tears. See **Figures 4.6** and **4.7** for descriptions of how the epidermis and dermis self-repair—processes called *epidermal wound healing* and *deep wound healing*, respectively.

PROCESS DIAGRAM

Epidermal wound healing • Figure 4.6

THE PLANNER

An injury such as an abrasion removes the outer layers of the epidermis, possibly down to the dermis. The stratum basale cells repair the wound by moving over, filling the void, and making new epidermal layers.

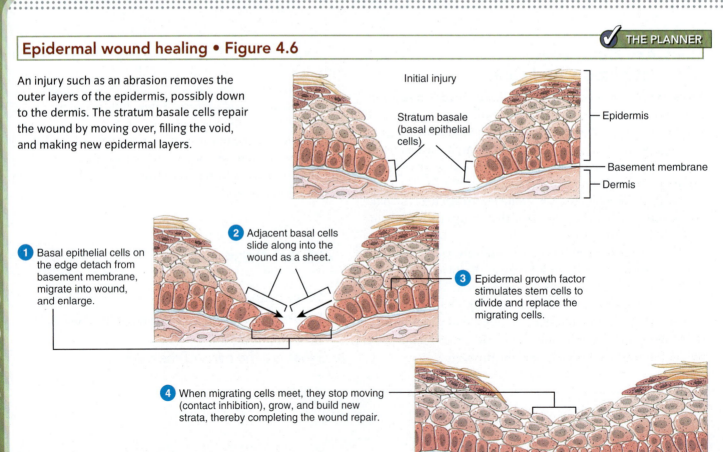

Initial injury

Stratum basale (basal epithelial cells)

Epidermis

Basement membrane

Dermis

1 Basal epithelial cells on the edge detach from basement membrane, migrate into wound, and enlarge.

2 Adjacent basal cells slide along into the wound as a sheet.

3 Epidermal growth factor stimulates stem cells to divide and replace the migrating cells.

4 When migrating cells meet, they stop moving (contact inhibition), grow, and build new strata, thereby completing the wound repair.

Deep wound healing • Figure 4.7

When a cut or tear extends into the dermis, blood vessels get torn. Once blood clots form, various cells move in to repair the damaged area by removing damaged tissue and microbes, building connective tissue matrix, and making new epidermal tissue.

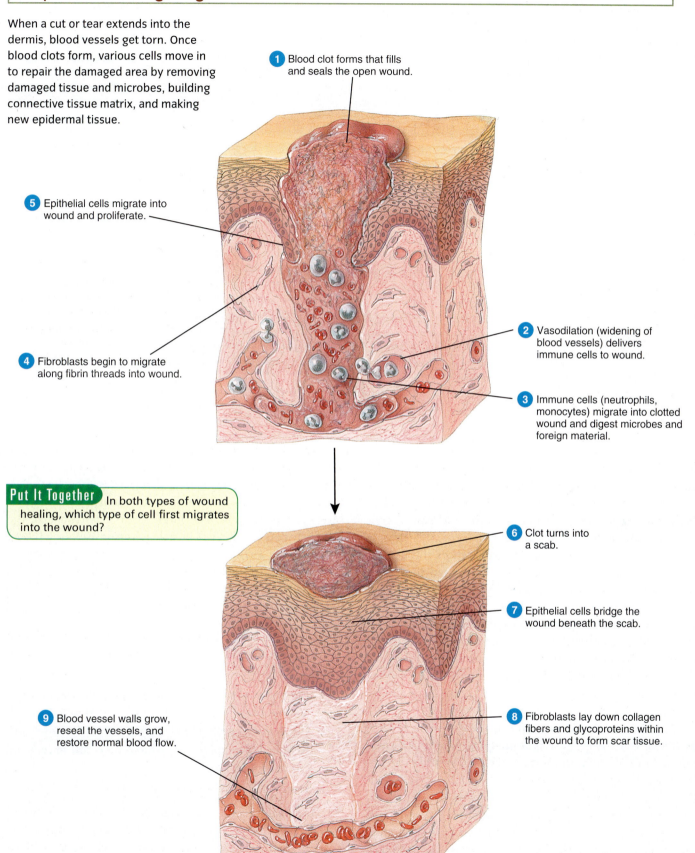

1 Blood clot forms that fills and seals the open wound.

5 Epithelial cells migrate into wound and proliferate.

4 Fibroblasts begin to migrate along fibrin threads into wound.

2 Vasodilation (widening of blood vessels) delivers immune cells to wound.

3 Immune cells (neutrophils, monocytes) migrate into clotted wound and digest microbes and foreign material.

Put It Together In both types of wound healing, which type of cell first migrates into the wound?

6 Clot turns into a scab.

7 Epithelial cells bridge the wound beneath the scab.

9 Blood vessel walls grow, reseal the vessels, and restore normal blood flow.

8 Fibroblasts lay down collagen fibers and glycoproteins within the wound to form scar tissue.

The Skin Plays a Number of Roles in the Body **101**

The skin senses the outside • Figure 4.8

Sensory receptors within the skin provide information about the outside environment through touch, temperature, pressure, and pain.

Nociceptors are free nerve endings that sense pain rapidly. Other free nerve endings sense tickle, itch, and some touch.

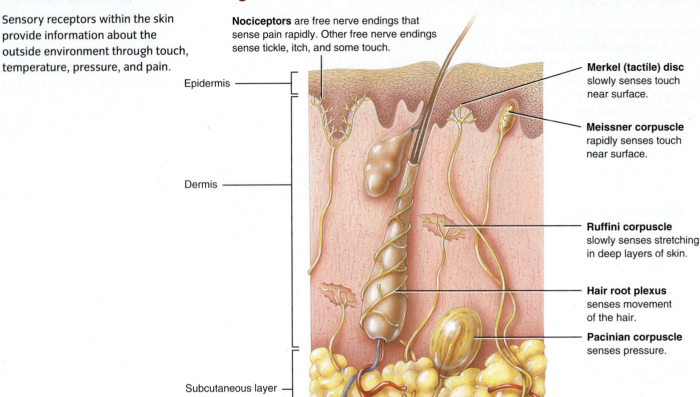

Epidermis

Dermis

Subcutaneous layer

Merkel (tactile) disc slowly senses touch near surface.

Meissner corpuscle rapidly senses touch near surface.

Ruffini corpuscle slowly senses stretching in deep layers of skin.

Hair root plexus senses movement of the hair.

Pacinian corpuscle senses pressure.

The skin is one of our connections to the outside world. Its specialized nerve receptors sense pressure, pain, changes in temperature, and things we touch. These receptors include mechanoreceptors, **Meissner corpuscles**, **Pacinian corpuscles**, thermoreceptors, and pain receptors called **nociceptors** (**Figure 4.8**).

The skin absorbs and excretes substances through its surface. Transdermal patches have been used to deliver various drugs such as nicotine for quitting smoking and estrogen derivatives for birth control. Chemicals such as those found in nail polish removers and cleaning products can also absorb readily through the skin and into the blood. Some of these products can be quite toxic if they enter the blood. In contrast, through sweat, the skin excretes water, small amounts of carbon dioxide, salts, ammonia, and urea.

The skin plays an important role in calcium homeostasis within your body. Your skin produces vitamin D when exposed to UV radiation. Vitamin D is converted to an active form by reactions in the liver and kidney. Just 5 to 30 minutes of sunlight exposure twice a week can provide you with an adequate amount of vitamin D. As you will see later, vitamin D gets processed further by other organs and ultimately acts in the intestines to assist absorption of calcium and phosphorus.

All of the skin's many functions may be disrupted if the skin is severely damaged. One source of damage is a burn, which is damage to proteins in tissues caused by heat, electricity, radiation, or corrosive chemicals, and can vary in their severity (**Figure 4.9**). The more tissue gets damaged, the less likely that the skin will be able to repair itself (see *What a Health Provider Sees*). Some burns are accompanied by **edema**. Severely and extensively burned tissue can be life threatening due to the following:

> **edema** (e-DE-ma) An abnormal accumulation of interstitial fluid.

- Because skin receives a large percentage of blood flow, there are several circulatory consequences that lead to fluid loss and shock:
  - Large loss of water, plasma, and proteins can result in edema; and hypovolemia (low blood volume), which can result in a sudden drop of blood pressure
  - Reduced blood pressure, which can result in reduced circulation of blood
  - Decreased urine production
- Microbial infection can occur because of the absence of the skin's protective covering and inability of immune cells to reach the invaders due to changes in blood pressure and blood flow.

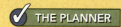
Burns vary in type and severity, depending on which skin layers are affected, the degree of tissue damage, and the extent of the burned surface area. First- and second-degree burns are often referred to as partial-thickness burns. Third- and fourth-degree burns are called full-thickness burns; fourth-degree burns (not shown in this figure) are burns that damage additional soft tissue underlying the skin, such as muscle and tendons.

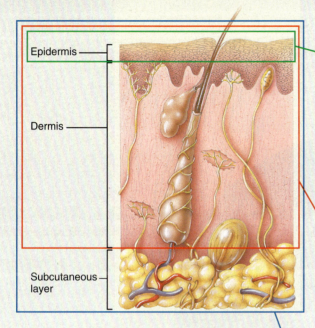

Epidermis

Dermis

Subcutaneous layer

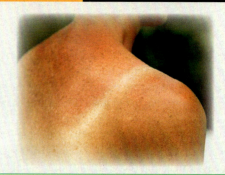

First-degree burn
- Mild pain
- Redness (no blisters)
- Skin functions normally
- Treatment: flush with cold water to relieve pain
- Heals within 3-6 days
- Example: sunburn

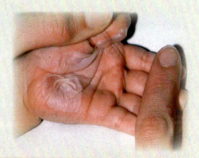

Second-degree burn
- Pain
- Redness
- Blisters (epidermis separates from underlying layers and fluid fills void)
- Edema
- Hair follicles and glands are not injured
- Some skin function is lost
- If there is no infection and no grafting is required, heals within 3–4 weeks

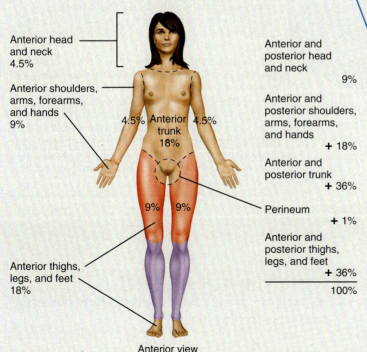

Rule of nines is used to estimate the surface area affected by burns in an adult

Anterior head and neck 4.5%

Anterior shoulders, arms, forearms, and hands 9%

4.5% Anterior trunk 18% 4.5%

9% 9%

Anterior thighs, legs, and feet 18%

Anterior view

Anterior and posterior head and neck   9%

Anterior and posterior shoulders, arms, forearms, and hands   + 18%

Anterior and posterior trunk   + 36%

Perineum   + 1%

Anterior and posterior thighs, legs, and feet   + 36%
_____
100%

Third-degree burn
- Severe pain (burned region is numb due to nerve damage)
- Marked edema
- Marble-white to black color
- Most skin functions are lost
- Tissue damage
- Susceptible to infection
- Slow healing
- May require skin graft to promote healing and minimize scarring

Ask Yourself

In which type of burn have more layers of the skin been damaged?
a. First-degree
b. Second-degree
c. Third-degree
d. All types of burns affect the same number of skin layers.

WHAT A HEALTH PROVIDER SEES

Skin Grafts and Artificial Skin

Extensive and severe burns can have grave consequences for a patient's life. Ultimately, the damaged skin must be removed, some underlying tissue may have to be reconstructed, and new skin might have to be sutured over the damaged area (skin graft). The skin graft **a.** uses donor skin, which may come from either the patient (autologous graft) or a cadaver (allograft). The major advantage of an autologous graft is that the graft will not be rejected by the patient's immune system. Recently, burn surgeons have developed artificial skin that serves as both a bandage and a scaffold for growing new skin. Artificial skin **b.** is a lower-layer matrix consisting of collagen and glycosaminoglycan (a carbohydrate). The matrix mimics the dermis, and the patient's own dermal cells migrate and grow into the burned area. The upper layer is made of silicone that mimics the epidermis. After 2 to 4 weeks, the surgeon removes the silicone layer and replaces it with a layer of epidermal cells from another area of the patient's body. In time, these cells form a normal epidermal layer without hair follicles.

a. Skin graft

b. Artificial skin

Think Critically

1. Following a severe burn, the skin that forms over the wounds will have many of the features of the skin that had been there prior to the burn, with the exception of hair follicles. Why will hair follicles be missing in the newly formed skin of the burn area?

2. Skin grafts are often taken from a small section of the skin of the thigh. The skin is run through a special instrument that makes tiny cuts in the graft tissue. The skin graft is then stretched, forming small gaps all through the tissue, and laid onto the burn wound. What would be the purpose of the small cuts in the skin graft? What will happen to the small gaps in the graft? Why would it be important to remove all the damaged tissue from the wound prior to placing the graft?

CONCEPT CHECK

1. **How** does the skin help to regulate body temperature?

2. **What** events happen in the healing of an epidermal wound?

3. **Which** receptors detect touch?

4. **What** type of burn is characterized by blisters, and why do they form?

Aging and Skin Cancer

LEARNING OBJECTIVES

1. **Describe** the effects of aging on the integumentary system.

2. **Identify** the three main types of skin cancer.

ome changes in the integumentary system occur with age, and some are caused by prolonged exposure to the sun. Let's take a closer look at these changes.

Aging Changes the Appearance and Quality of Skin and Its Associated Structures

Most age-related changes in skin do not occur until people reach their late 40s. **Figure 4.10** summarizes the changes that occur at various stages.

Skin Cancers Can Develop from Repeated Exposure to UV Radiation

Excessive exposure to ultraviolet (UV) radiation from the sun or tanning beds can cause damage to the skin and may lead to development of skin cancer. Long-wave ultraviolet light (UVA) and short-wave ultraviolet light (UVB) both contribute to the development of skin cancer.

There are three main types of skin cancer:

- **Basal cell carcinoma** (78% of all skin cancers)—Tumors arise from cells in the stratum basale of the epidermis and rarely **metastasize**.

> **metastasize**
> (me-TAS-ta-sīz) To spread to surrounding tissues (local) or to other body sites (distant).

- **Squamous cell carcinoma** (20% of all skin cancers)—Tumors arise from stratum spinosum cells in the epidermis and vary in their tendency to metastasize. These tumors tend to arise from preexisting lesions of sun-damaged skin.

Summary of age-related changes in the integumentary system • Figure 4.10

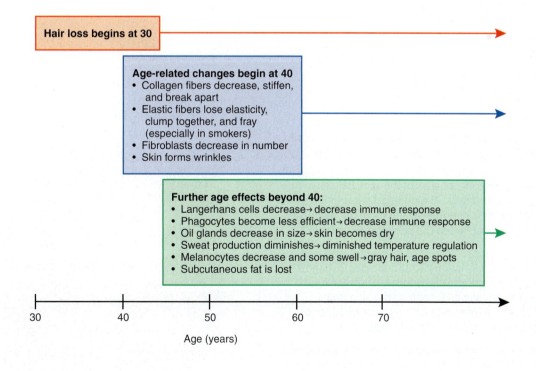

Hair loss begins at 30

Age-related changes begin at 40
- Collagen fibers decrease, stiffen, and break apart
- Elastic fibers lose elasticity, clump together, and fray (especially in smokers)
- Fibroblasts decrease in number
- Skin forms wrinkles

Further age effects beyond 40:
- Langerhans cells decrease→ decrease immune response
- Phagocytes become less efficient→ decrease immune response
- Oil glands decrease in size→ skin becomes dry
- Sweat production diminishes→ diminished temperature regulation
- Melanocytes decrease and some swell→gray hair, age spots
- Subcutaneous fat is lost

30   40   50   60   70

Age (years)

Comparison of a mole and a malignant melanoma • Figure 4.11

Malignant melanomas are the most aggressive form of skin cancer.

a. Normal nevus (mole)

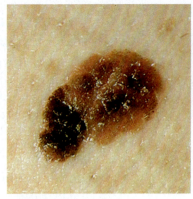

b. Malignant melanoma

ABCDE warning signs of malignant melanoma (MM)

A asymmetry - MMs lack symmetry
B border - MMs have irregular borders
C color - MMs have uneven coloration and may contain many colors
D diameter - MMs are larger than ordinary moles (>5 mm)
E evolving - MMs change in size, shape, or color

Risk factors include skin type (light =high risk), sun exposure (number of severe sunburns), family history of cancer, age, and immune status (immunosuppressed = high risk).

- **Malignant melanoma** (2% of all skin cancers)—Tumors arise from melanocytes, metastasize quickly, and can kill a patient within months of diagnosis (**Figure 4.11**).

The best way to prevent skin cancer is to reduce exposure to UV radiation from the sun or tanning beds. Tightly woven clothes keep the sun's rays out, and hats prevent the scalp from being exposed. Sunscreen lotions reduce the damaging effects of UV radiation. They should be consistently applied while outside and reapplied after swimming. Sunscreens are rated by their sun protection factor (SPF). SPF numbers refer to the amount of time it would take to burn if an individual were wearing sunscreen compared to not wearing sunscreen. Everyone should use sunscreen with a minimum SPF of 15. The fairer your skin, the higher the SPF should be: Fair-skinned individuals should probably apply sunscreen with an SPF of 55.

CONCEPT CHECK

1. **How** do skin wrinkles develop with age?
2. **What** is the most dangerous form of skin cancer?

 THE PLANNER ✓

Summary

1 The Integumentary System Is Composed of Skin, Glands, Hair, Nails, and Nerve Endings 92

- The **integumentary system** consists of the skin, hair, oil glands, sweat glands, and nails. The principal parts of the skin, as shown, are the superficial **epidermis** and deeper **dermis**, which overlies and is attached to the **subcutaneous layer**.

- The epidermis is composed of four or five layers containing many cells with various functions. Keratinocytes, the most common cells of the epidermis, produce the intracellular protein keratin. Other cells give color to the skin (melanocytes), participate in immunity (Langerhans cells), and provide **cutaneous** sensations (Merkel cells).

- The dermis consists of dense irregular connective tissue with collagen and elastic fibers. The papillary region contains thin collagen and fine elastic fibers. The reticular region contains thick collagen fibers, adipose tissue, hair follicles, nerves, oil glands, and sweat glands.

The skin • Figure 4.1a

EPIDERMIS

DERMIS

Subcutaneous layer

2 Accessory Structures Provide Protection and Help Regulate Body Temperature 96

- Hair, glands, and nails develop from the embryonic epidermis. Hairs consist of threads of fused, keratinized cells. A hair has a shaft above the surface, a root below, and a hair follicle in the subcutaneous layer. Smooth muscle bundles called **arrector pili** and oil glands are associated with hairs.

- Oil glands secrete sebum (oil) that softens the skin, prevents water loss, and inhibits bacterial growth. Eccrine sweat glands secrete sweat and participate in temperature regulation, while apocrine sweat glands open into hair follicles and secrete sweat during emotional states (for example, fear, embarrassment, sexual excitement). **Ceruminous glands**, which are found in the external ear canal, are modified sweat glands that secrete earwax (cerumen).

- As shown, **nails** are plates of dead, hard, keratinized cells that cover the ends of the fingers and toes. A nail has a nail body, free edge, root, lunula, cuticle, and nail matrix. The nail matrix is the portion of the nail root that contains the living, dividing nail cells.

Parts of the nail • Figure 4.5

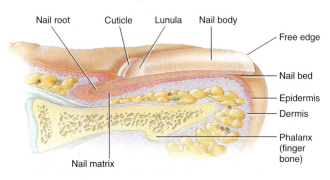

3 The Skin Plays a Number of Roles in the Body 100

- The skin covers a large surface area (~2 m², or 22 ft²) and acts as a large reservoir of blood (8%–10% total circulation) that makes it ideal for radiating heat. The skin removes excess body heat through radiation and by evaporating sweat. It can also conserve heat by restricting blood flow to the body surface.

- Throughout the skin, specialized nerve endings provide **cutaneous** sensations of touch, pressure, pain, heat, tickling,

stretching, and so on. The skin also makes vitamin D upon exposure to sunlight. The vitamin D gets processed elsewhere and ultimately stimulates calcium and phosphorus absorption from the intestine, so the skin plays an important role in calcium homeostasis.

- Within limits, the skin can repair cuts and abrasions. In epidermal wound healing, basal cells migrate into the wound area and begin to enlarge and grow, while adjacent stem cells divide to fill in the gaps. In deep wound healing, as shown, blood clots form to seal the wound, immune cells migrate into the area to fight bacteria and ingest foreign material, fibroblasts emerge to lay down collagen fibers that will form scar tissue, and epithelial cells move in to repair the epidermis. Eventually, the clot becomes a covering scab that sloughs off, the epidermis gets reformed, and tougher scar tissue fills the gap.

Deep wound healing • Figure 4.7

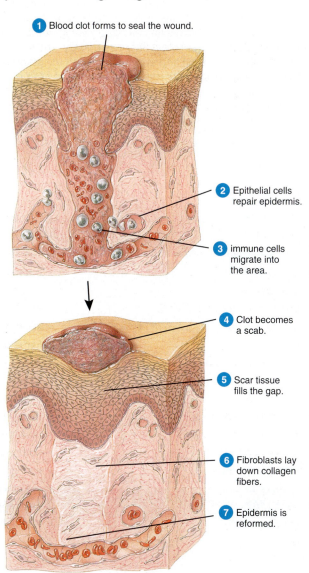

1 Blood clot forms to seal the wound.

2 Epithelial cells repair epidermis.

3 immune cells migrate into the area.

4 Clot becomes a scab.

5 Scar tissue fills the gap.

6 Fibroblasts lay down collagen fibers.

7 Epidermis is reformed.

- Burns can be extensive injuries. First-degree burns involve only the epidermis and are characterized by redness and mild pain. Second-degree burns involve both the epidermis and dermis and are characterized by pain and blistering. Third-degree burns involve extensive damage to epidermis, dermis, and subcutaneous tissues. Fourth-degree burns damage additional soft tissue underlying the skin, such as muscle and tendons. Severely burned patients are susceptible to circulatory shock and infection and may require skin grafting to repair the lost tissue. The extent of burns on the body is assessed using the rule of nines.

- Prolonged exposure to UV radiation from the sun or tanning beds over a lifetime has cumulative effects on the skin. UV rays can depress the immune system, cause sunburn, and damage the protein fibers in the dermis, and contribute to the development of skin cancer.

- Skin cancer manifests itself in three forms: **basal cell carcinoma** (most common), **squamous cell carcinoma** (less common), and **malignant melanoma** (the rarest but also the deadliest). Malignant melanomas, as shown, arise from melanocytes and **metastasize** quickly.

Malignant melanoma • Figure 4.11b

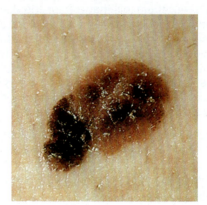

4 Aging and Skin Cancer 105

- Age-related changes in the skin begin around age 40. The protein fibers in the dermis become stiff and brittle and begin to break down. Most skin cells decrease in number and do not lay down new protein fibers as quickly. The secretions of oil (sebaceous) and sweat (sudoriferous) glands are reduced. Melanocytes decrease in number and stop producing melanin. These changes lead to wrinkled, dry skin, dotted with age spots. In addition, the skin is easily damaged and is slow to heal. Hair loss begins around age 30.

Key Terms

- albinism 95
- albino 95
- androgenic alopecia 97
- androgen 97
- apoptosis 94
- areolar connective tissue 93
- arrector pili 96
- axillary 97
- basal cell carcinoma 105
- bulb 96
- carotene 96
- ceruminous gland 98
- cutaneous membrane 93
- dermal papilla 95
- dermatology 92
- dermis 93
- edema 102
- epidermis 93
- hair 96

- hair follicle 96
- hair matrix 96
- hair root plexus 96
- hemoglobin 95
- hirsutism 97
- integumentary system 92
- keratin 94
- keratinization 95
- lunula 99
- male-pattern baldness 97
- malignant melanoma 106
- Meissner corpuscle 102
- melanin 95
- metastasize 105
- nail 99
- nevus 95
- nociceptors 102
- Pacinian corpuscle 102

- papilla of the hair 96
- pili 96
- root 96
- sebaceous gland 98
- shaft 96
- skin 93
- squamous cell carcinoma 105
- stem cell 94
- stratum basale 94
- stratum corneum 94
- stratum granulosum 94
- stratum lucidum 94
- stratum spinosum 94
- subcutaneous (subQ) layer 93
- sudoriferous gland 98
- thick skin 94
- thin skin 94
- vasodilation 97
- vitiligo 95

Critical and Creative Thinking Questions

1. While grading papers, Mr. Jackson gets a minor paper cut. Although the cut is irritating, it does not bleed. How will Mr. Jackson's cut heal? Will it leave a scar? Explain your answers.

2. In a beauty salon, Mrs. Reid is having her eyebrows plucked during a facial. She winces when each hair is pulled and thinks to herself, "I am killing my hairs." Is Mrs. Reid correct? Explain your answer.

3. In a dry-cleaning store, Joe spills some cleaning fluid on his arm. It immediately turns red but does not blister. What type of burn is Joe's injury, how severe is it, and how should he treat it?

4. Mrs. Stephenson, an elderly female, notices that some moles on her arm have changed color and become irregularly shaped. She shrugs them off as age spots and ignores them. Is Mrs. Stephenson doing the right thing? Why or why not?

5. Jill is nearly 13 years old. She begins to notice hair growing on her underarms and in her pubic area. She is worried that hair is spreading from her head to the rest of her body and tells her mother. What should her mother say to calm Jill?

What is happening in this picture?

Think Critically This woman is applying lotion to her exposed skin in preparation for a day at the beach.
1. How will exposure to sunlight damage her skin? What else can she do to prevent or minimize the damage?
2. What is one advantage of her being out in the sun for even a short period of time?

Self-Test

(Check your answers in Appendix C.)

1. A patient is admitted to the emergency room with second-degree burns on the anterior surfaces of both legs, including the thighs, lower legs, and feet. As the attending nurse, you must fill out the burn assessment. What percentage of the patient's body is burned?

   a. 9%

   b. 18%

   c. 27%

   d. 36%

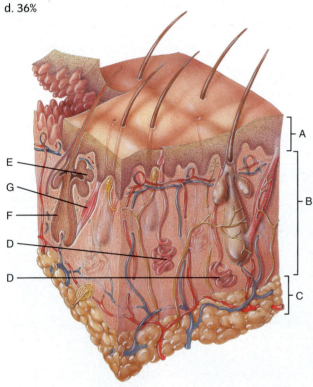

Use this figure for questions 2–5.

2. A third-degree burn involves which of the following layers?

   a. A only

   b. A and B

   c. B and C

   d. A, B, and C

3. Which of the following structures plays an important role in body temperature regulation?

   a. D

   b. E

   c. F

   d. G

4. The stratum corneum is located in which layer?

   a. B

   b. A and B

   c. A

   d. C

5. The structure labeled C on this diagram is the _____.

   a. epidermis

   b. dermis

   c. subcutaneous layer

   d. stratum granulosum

6. From which of the following layers of skin do cells slough off?

   a. stratum lucidum

   b. stratum basale

   c. stratum granulosum

   d. stratum spinosum

7. Age spots are caused by _____.

   a. keratinocytes

   b. Merkel cells

   c. Langerhans cells

   d. melanocytes

8. What is the major protein in the epidermis?

   a. collagen

   b. keratin

   c. myosin

   d. elastin

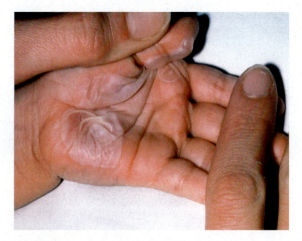

Use this figure for questions 9–10.

9. The burn shown in this photo involves the _____.

　a. epidermis, dermis, and hypodermis

　b. epidermis only

　c. epidermis and dermis

　d. dermis and hypodermis

10. What is the classification of the burn shown here?

　a. first-degree

　b. second-degree

　c. third-degree

　d. fourth-degree

11. Which of the following cells are most important in forming scar tissue?

　a. macrophage

　b. Langerhans cell

　c. keratinocyte

　d. fibroblast

Use this figure for question 12.

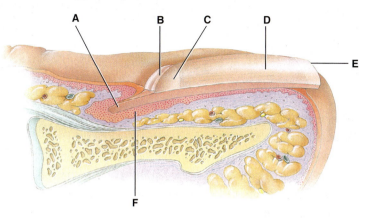

12. Fingernail cells arise from which portion of the nail?

　a. F　b. D　c. B　d. C

13. Which portion of the hair is involved in the "goose bumps" response to cold?

　a. shaft

　b. papilla

　c. arrector pili

　d. bulb

14. Which of the following nerve endings are found in the reticular region of the dermis?

　a. Meissner corpuscles

　b. Pacinian corpuscles

　c. Merkel cells

　d. nociceptors

Use this figure for question 15.

15. The structure shown is made of _____.

　a. keratinocytes

　b. Langerhans cells

　c. lamellar granules

　d. melanocytes

THE PLANNER

Review your Chapter Planner on the chapter opener and check off your completed work.

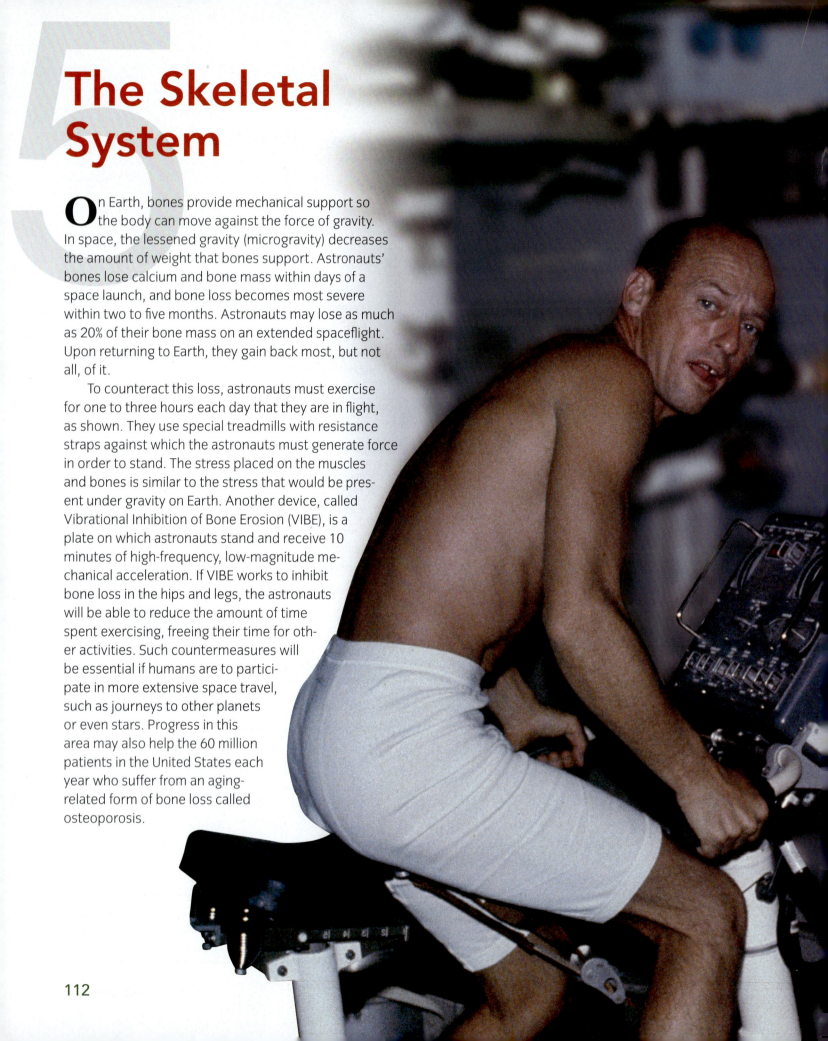

The Skeletal System

5

On Earth, bones provide mechanical support so the body can move against the force of gravity. In space, the lessened gravity (microgravity) decreases the amount of weight that bones support. Astronauts' bones lose calcium and bone mass within days of a space launch, and bone loss becomes most severe within two to five months. Astronauts may lose as much as 20% of their bone mass on an extended spaceflight. Upon returning to Earth, they gain back most, but not all, of it.

To counteract this loss, astronauts must exercise for one to three hours each day that they are in flight, as shown. They use special treadmills with resistance straps against which the astronauts must generate force in order to stand. The stress placed on the muscles and bones is similar to the stress that would be present under gravity on Earth. Another device, called Vibrational Inhibition of Bone Erosion (VIBE), is a plate on which astronauts stand and receive 10 minutes of high-frequency, low-magnitude mechanical acceleration. If VIBE works to inhibit bone loss in the hips and legs, the astronauts will be able to reduce the amount of time spent exercising, freeing their time for other activities. Such countermeasures will be essential if humans are to participate in more extensive space travel, such as journeys to other planets or even stars. Progress in this area may also help the 60 million patients in the United States each year who suffer from an aging-related form of bone loss called osteoporosis.

CHAPTER OUTLINE

CHAPTER PLANNER ✓

- ❏ Study the picture and read the opening story.
- ❏ Scan the Learning Objectives in each section:
  p. 114 ❏   p. 122 ❏   p. 130 ❏   p. 132 ❏   p. 136 ❏   p. 144 ❏
- ❏ Read the text and study all visuals. Answer any questions.

Analyze key features

- ❏ Process Diagram, p. 118 ❏   p. 121 ❏
- ❏ InSight, pp. 124–125 ❏   pp. 128–129 ❏   pp. 130–131 ❏
  pp. 134–135 ❏   p. 137 ❏
- ❏ What a Health Provider Sees, p. 143 ❏
- ❏ Stop: Answer the Concept Checks before you go on:
  p. 121 ❏   p. 127 ❏   p. 132 ❏   p. 133 ❏   p. 143 ❏   p. 145 ❏

End of chapter

- ❏ Review the Summary and Key Terms.
- ❏ Answer the Critical and Creative Thinking Questions.
- ❏ Answer What is happening in this picture?
- ❏ Complete the Self-Test and check your answers.

The Structure of Bone Controls Function and Growth

LEARNING OBJECTIVES

1. **Outline** the six functions of bone and the skeletal system.
2. **Describe** the parts of a long bone and the microscopic structure of bone.
3. **Explain** the formation of bone during different phases of a person's lifetime.
4. **Discuss** how bones heal after fractures.

The skeletal system consists of all the bones attached at joints and the cartilage between the joints. The functions of the skeletal system include support, protection, movement, mineral homeostasis, blood cell protection, and triglyceride storage. Bones are classified according to shape, with a bone's structure determining its function. Throughout life, bone is constantly made and destroyed in an ongoing process of bone remodeling.

The Skeleton Is More Than Just a Supportive Framework for the Body

The adult skeletal system is made of 206 different bones. During development, some bones (such as those that form the skull) fuse to create a solid unit, while others connect to the adjacent bones at **articulations** that offer varying degrees of movement. Bones and associated tissues perform several basic functions:

> **articulation** (ar-tik′-ū-LĀ-shun) A location at which two or more bones make contact. Also known as a **joint**.

- *Support*—The skeleton provides a scaffold or framework to support soft tissues and points of attachment for skeletal muscles.
- *Protection*—The skeleton protects many internal organs from injury. For example, the skull protects the brain, and the rib cage protects the heart and lungs.
- *Movement*—Muscles provide the force, while bones serve as the levers. Together, bones and muscles produce movement of the body in its environment.
- *Mineral homeostasis*—Bone tissue stores calcium and phosphorus. When needed, endocrine glands secrete hormones, which act on bone to either release calcium and phosphorus into the blood or store excess minerals in the bone extracellular matrix.
- *Blood cell production*—**Red bone marrow**, a connective tissue within bone, produces red blood cells, white blood cells, and platelets in a process called **hemopoiesis**.

> **hemopoiesis** (hē-mō-poy-Ē-sis) Blood cell production.

- *Triglyceride storage*—**Yellow bone marrow** within bone is composed mostly of adipose cells, which store triglycerides (fats). In a newborn, all bone marrow is red, but it eventually changes into yellow marrow with increasing age.

A Bone's Structure Determines Its Function

The skeletal system contains four types of bones, based on shape:

- **Long bones** are longer than they are wide and have knobby ends where the articulations form. Their slightly curved structure gives them strength. Long bones include those of the arms, legs, fingers, and toes.
- **Short bones** are equal in length and width, making them nearly cube-shaped. Examples include most bones of the ankles and the wrists.
- **Flat bones** are thin and provide both protection and surfaces for muscle attachments. The bones of the skull, sternum, and ribs are all flat bones.
- **Irregular bones** have complex shapes, such as those of the face and vertebral column.

Let's take a look at the structure of bones, from gross (macroscopic) to microscopic.

Long bones are typically used as examples when studying gross bone structure (**Figure 5.1**). Long bones are hollow in the middle and more solid on the ends. Typically, the ends of an adult's bones contain the red marrow, while the hollow cavity contains the yellow marrow. The outside of the bone has a connective tissue covering known as the periosteum. The ends of a bone that form movable joints are covered with a layer of articular cartilage. As we will discuss later, long bones lengthen by adding new bone to an area known as the epiphyseal plate that lies between the epiphysis and diaphysis at each end of the bone.

Parts of a long bone • Figure 5.1

A typical long bone, such as the humerus from the arm, exemplifies the characteristics of bone structure.

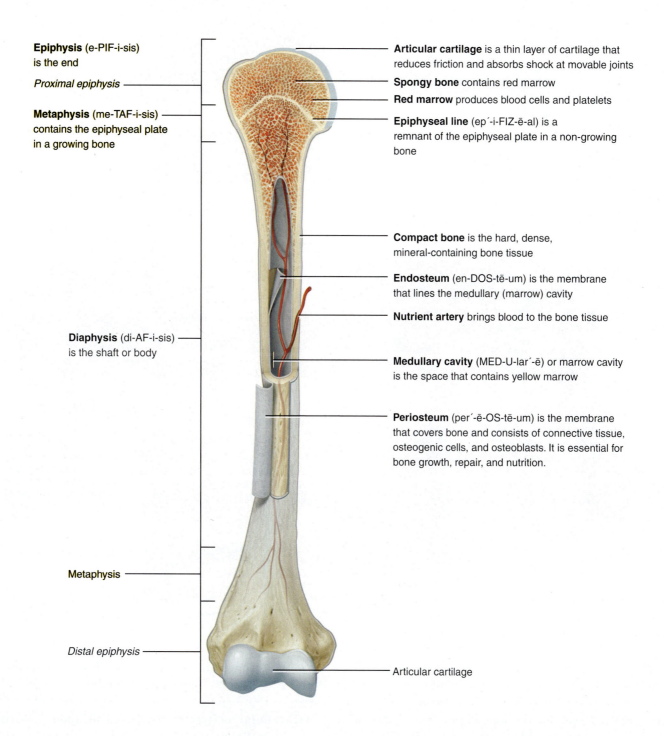

Epiphysis (e-PIF-i-sis) is the end

Proximal epiphysis

Metaphysis (me-TAF-i-sis) contains the epiphyseal plate in a growing bone

Diaphysis (di-AF-i-sis) is the shaft or body

Metaphysis

Distal epiphysis

Articular cartilage is a thin layer of cartilage that reduces friction and absorbs shock at movable joints

Spongy bone contains red marrow

Red marrow produces blood cells and platelets

Epiphyseal line (ep´-i-FIZ-ē-al) is a remnant of the epiphyseal plate in a non-growing bone

Compact bone is the hard, dense, mineral-containing bone tissue

Endosteum (en-DOS-tē-um) is the membrane that lines the medullary (marrow) cavity

Nutrient artery brings blood to the bone tissue

Medullary cavity (MED-U-lar´-ē) or marrow cavity is the space that contains yellow marrow

Periosteum (per´-ē-OS-tē-um) is the membrane that covers bone and consists of connective tissue, osteogenic cells, and osteoblasts. It is essential for bone growth, repair, and nutrition.

Articular cartilage

Bone tissue consists of numerous types of cells arranged in organized structural units. Each cell has a specific role in building or destroying bone tissue. Osteoblasts build bone, osteoclasts destroy bone, osteocytes maintain healthy bone, and osteogenic cells make new osteblasts.

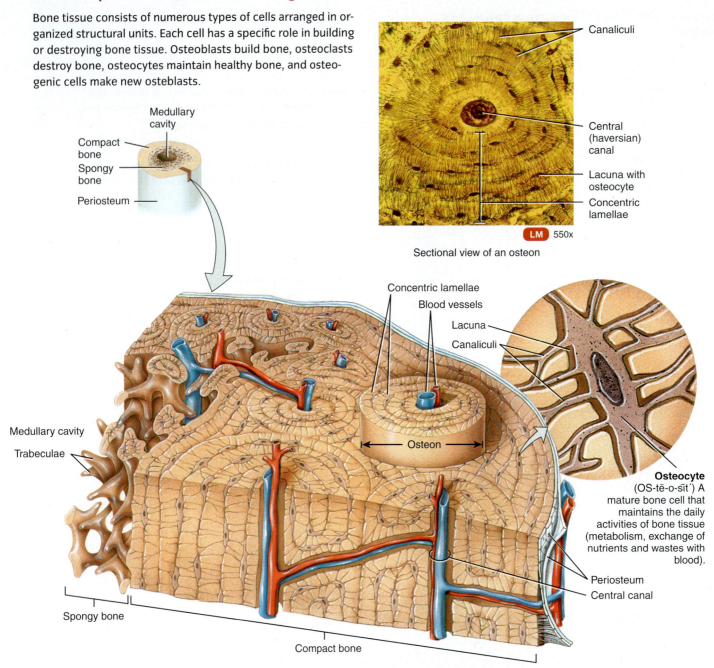

Sectional view of an osteon

Canaliculi

Central (haversian) canal

Lacuna with osteocyte

Concentric lamellae

LM 550x

Concentric lamellae
Blood vessels
Lacuna
Canaliculi

Medullary cavity
Trabeculae

Osteon

Spongy bone

Compact bone

Osteocyte (OS-tē-o-sīt´) A mature bone cell that maintains the daily activities of bone tissue (metabolism, exchange of nutrients and wastes with blood).

Periosteum
Central canal

Medullary cavity
Compact bone
Spongy bone
Periosteum

Bones also contain compact and spongy bone and a periosteum. Flat bones such as the scapula, or shoulder blade, form "plates" with none of the distinguishing features that are seen on the long bones. This is because long bones are designed for movement while flat bones serve for protection of delicate organs. Now let's take a look at the microscopic structure of bone (**Figure 5.2**).

Like other connective tissues, **bone**, or **osseous tissue** (OS-ē-us), contains abundant extracellular matrix that surrounds widely separated cells. The extracellular matrix is composed of about 25% water, 25% collagen fibers, and 50% crystallized mineral salts. Cells called **osteoblasts** deposit these mineral salts in the framework formed by the collagen fibers of the extracellular matrix during a process called **calcification** (kal-si-fi-KĀ-shun). The mineral salts crystallize and the tissue hardens.

A bone's hardness is determined by the degree of calcification, but its flexibility is determined by the collagen fibers. These fibers, along with other organic molecules, provide *tensile strength* (resistance to being stretched). Besides osteoblasts, three other cell types help maintain bone tissue: **osteoclasts**, osteocytes, and **osteogenic cells** (see Figure 5.2).

Bone is not solid but contains numerous spaces for blood vessels and storage areas for marrow. Depending on the size and distribution of the spaces, bone can be classified as **compact bone** or **spongy bone** (see Figure 5.2). Features of compact bone include the following:

- Compact bone is strong and dense, provides protection and support, and resists the stresses produced by weight and movement.

- Compact bone is found beneath the periosteum of all bones and makes up the bulk of long bones.

- Compact bone is made of cylindrical units called **osteons** (OS-tē-ons).

- Each osteon consists of **concentric lamellae** (la-MEL-ē), concentric layers that surround a **central canal**, or **haversian canal**, containing blood and lymph vessels.

- Between the lamellae are spaces called **lacunae** (la-KOO-nē), which contain osteocytes, and smaller channels called **canaliculi** (kan'-a-LIK-ū-lī) that radiate out from the lacunae. These canaliculi allow nutrients and wastes to be passed more easily from one osteocyte to another within the osteon.

In contrast to compact bone, spongy bone is lightweight. Spongy bone, also known as cancellous bone, contains irregular lattices of thin bone columns called **trabeculae** (tra-BEK-ū-lē). Trabeculae form a supportive framework that is firm but not exceedingly strong. This tissue must be covered by compact bone or cartilage because it could be damaged easily if exposed. The spaces between the trabeculae of some bones are filled with red bone marrow; in such cases, the functions of the trabeculae are to support and protect the red bone marrow. Spongy bone is found mostly in short, flat, and irregular bones. In long bones, spongy tissue forms the majority of the epiphyses and is also found around the inner rim of the diaphysis.

Now that we have examined the microscopic structure of bones, let's see how they form.

Bone Is Formed During Ossification and Maintained by Remodeling

The process of bone formation, called **ossification** (os'-i-fi-KĀ-shun), occurs in four situations:

- Initial formation of bones in the embryo and fetus

- Bone growth during infancy, childhood, and adolescence prior to adulthood

- Bone remodeling, which occurs as old bone tissue is replaced with new bone tissue throughout life

- Repair of broken bones (fractures) throughout life

Bones form initially in the embryo by two processes. In the first process, called **intramembranous ossification** (in'-tra-MEM-bra-nus), bone forms directly from **mesenchyme**. Intramembranous ossification occurs in the flat bones of the skull, mandible, and clavicle. In the second

> **mesenchyme**
> (MEZ-en-kīm) Embryonic connective tissue.

process, called **endochondral ossification** (en'-dō-KON-dral), bone forms within and replaces cartilage (**Figure 5.3** on the next page). Intramembranous ossification is the simpler of these two processes.

During infancy, childhood, and adolescence, long bones grow in both length and thickness:

- Growth in length:
  - Within the epiphyseal plate (cartilage) are chondrocytes that divide and form additional cartilage.
  - New chondrocytes form on the epiphyseal side, while the cartilage on the diaphyseal side is replaced by bone.
  - The thickness of the epiphyseal plate remains the same, but the bone lengthens.
  - Cartilage growth stops at adulthood, and bone replaces the remaining cartilage to form the epiphyseal line.

- Growth in thickness:
  - As the bone lengthens, it also thickens.
  - Cells in the perichondrium differentiate into osteoblasts, which secrete extracellular matrix that calcifies.
  - Osteoblasts differentiate into **osteocytes** as new lamellae are formed.
  - **Osteoclasts** break down the inner surface of the medullary cavity but at a slower rate than the bone forms on the outer surface. So the medullary cavity grows in diameter as the bone thickens.

Endochondral ossification • Figure 5.3

Most bones in the body are formed through endochondral ossification, replacing cartilage with bone.

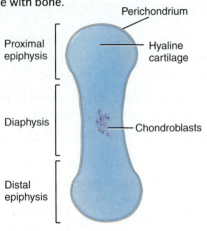

Proximal epiphysis

Perichondrium

Hyaline cartilage

Diaphysis

Chondroblasts

Distal epiphysis

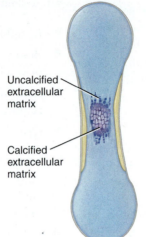

Uncalcified extracellular matrix

Calcified extracellular matrix

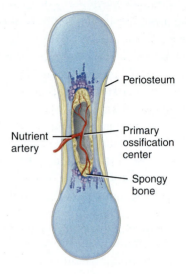

Periosteum

Nutrient artery

Primary ossification center

Spongy bone

1 Development of cartilage model:
- Mesenchymal cells cluster and differentiate into **chondroblasts**
- Chondroblasts secrete hyaline cartilage
- **Perichondrium** (a membrane) forms around the cartilage

2 Growth of cartilage model:
- Chondroblasts differentiate into **chondrocytes**
- Chondrocytes in the middle grow and surrounding matrix calcifies
- Inner chondrocytes die and form spaces (lacunae)

3 Development of primary ossification center:
- Nutrient artery penetrates the center of the cartilage model
- Osteogenic cells in perichondrium differentiate into osteoblasts
- Perichondrium forms bone and becomes periosteum
- In primary ossification center, cartilage breaks down, blood vessels expand, and osteoblasts deposit bone over cartilage remnants

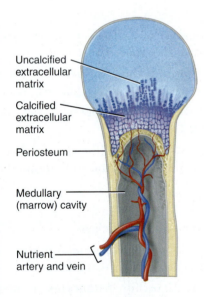

Uncalcified extracellular matrix

Calcified extracellular matrix

Periosteum

Medullary (marrow) cavity

Nutrient artery and vein

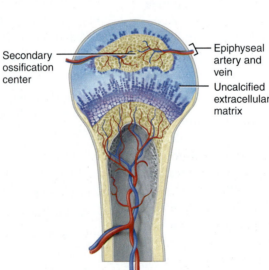

Secondary ossification center

Epiphyseal artery and vein

Uncalcified extracellular matrix

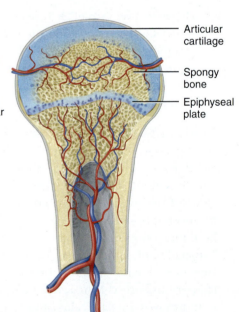

Articular cartilage

Spongy bone

Epiphyseal plate

4 Development of medullary cavity:
- Primary ossification center grows toward ends
- Osteoclasts break down some spongy bone trabeculae, leaving a marrow cavity
- Most of the diaphysis wall becomes compact bone

5 Development of secondary ossification centers:
- Blood vessels enter epiphyses and form secondary ossification centers
- Secondary ossification proceeds outward
- No cavities form

6 Formation of articular cartilage and epiphyseal plate:
- Hyaline cartilage at ends become articular cartilage
- Hyaline cartilage remaining between the ossification centers (diaphysis and epiphyses) becomes epiphyseal growth plate

Throughout life, bone is constantly made and destroyed in an ongoing process called **bone remodeling**. At various stages of life, osteoblasts and osteoclasts regulate bone formation and bone destruction. During some stages, such as childhood, formation exceeds destruction, allowing for growth of the skeleton. For a short period of time after ossification processes cease, formation and destruction are fairly equivalent. As a person ages, destruction often exceeds formation, leading to bone loss. Rates of remodeling vary depending on the location within the body. The remodeling process allows for mineral homeostasis, as well as repair of damaged bone tissue. Remodeling is affected by many factors:

- Hormones:
  - *Parathyroid hormone (PTH)*—Stimulates osteoclasts
  - *Calcitonin*—Stimulates osteoblasts
  - *Human growth hormone (hGH)*—Stimulates cartilage and bone growth
  - *Insulin-like growth factor (IGF)*—Stimulates cartilage and bone growth
  - *Sex hormones*—Sex-related differences in skeletal growth
- Minerals—Availability of calcium, magnesium, and phosphorus
- Vitamins A, C, and D
- Activity level—Active versus sedentary lifestyle
- Diet

Bone remodeling plays an active role in calcium homeostasis because well over 90% of the calcium in the body is located in the bone extracellular matrix. A certain level of ionized calcium (Ca^{2+}) in the blood (1.3 mM) is necessary for proper functioning of nerve and muscle cells. **Parathyroid glands** detect decreased Ca^{2+} in the blood and secrete PTH, which acts on bone (stimulating osteoclasts to resorb matrix), on kidneys (to prevent calcium loss into the urine), and indirectly on intestines (to help absorb calcium from food following the formation of **calcitriol**), bringing about an increase in blood levels of Ca^{2+} (**Figure 5.4**).

> **Parathyroid glands** (par'-a-THĪ-royd) Four small endocrine glands embedded in the posterior surfaces of the lateral lobes of the thyroid gland.
>
> **calcitriol** The active form of vitamin D.

The parafollicular cells of the thyroid gland produce the hormone calcitonin, which is also involved in calcium homeostasis. The parafollicular cells monitor blood calcium levels, releasing calcitonin if the blood calcium increases above normal. This hormone stimulates osteoblasts to increase formation of bone extracellular matrix and the kidneys to eliminate excess calcium, thus lowering the calcium to more normal levels.

Calcium homeostasis • Figure 5.4

When blood Ca^{2+} level falls, PTH stimulates osteoclasts to resorb bone and the kidneys to retain calcium and make calcitriol, which acts on the intestines to absorb Ca^{2+}. These actions restore Ca^{2+} to normal levels.

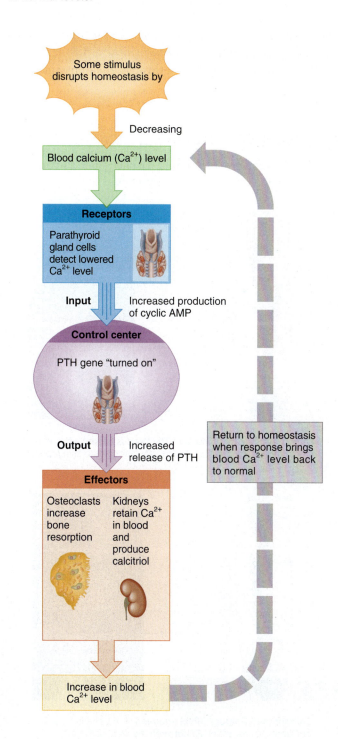

Some stimulus disrupts homeostasis by

Decreasing

Blood calcium (Ca^{2+}) level

Receptors

Parathyroid gland cells detect lowered Ca^{2+} level

Input — Increased production of cyclic AMP

Control center

PTH gene "turned on"

Output — Increased release of PTH

Effectors

Osteoclasts increase bone resorption

Kidneys retain Ca^{2+} in blood and produce calcitriol

Increase in blood Ca^{2+} level

Return to homeostasis when response brings blood Ca^{2+} level back to normal

Bones Repair After Fracture by a Four-stage Process

Bones can reform after being fractured (broken). The classification of fractures depends on whether the bone is splintered (partial), entirely broken (complete), or shattered (*comminuted*, KOM-i-noo-ted). If the broken bone protrudes through the skin, it is an *open fracture*. If it remains within the skin, it is a *closed fracture* (**Figure 5.5**).

Broken bones are repaired through the four-stage process outlined in **Figure 5.6**.

To treat fractures, the broken ends of the bones must be brought together and aligned in a process called **reduction**. The ends must then be immobilized for healing. Reduction can be accomplished by manipulating the broken ends of the bone without rupturing the skin (*closed reduction*) or by surgical insertion of screws, pins, and/or plates (*open reduction*). Bones take many months to heal completely because the fracture interferes with the blood supply to the bone, and the bone's strength comes from its extracellular matrix that has also been disrupted by the fracture.

Bone fractures • Figure 5.5 _____

Bone fractures come in many forms, including open fracture; greenstick fracture; comminuted fracture; and impacted fracture.

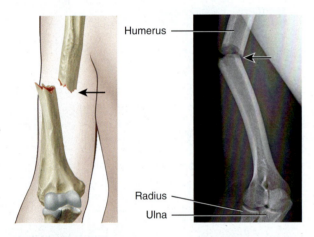

Open fracture (also complete) The bone is broken into two or more pieces (complete) and protrudes through the skin (open)

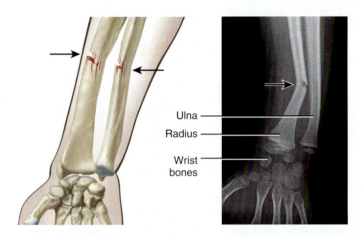

Greenstick fracture The bone is partially broken and one piece bends; only occurs in children

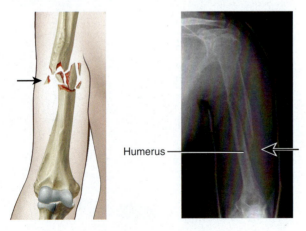

Comminuted fracture (KOM-i-noo-ted) The bone is crushed into small pieces, which lie between the ends of the main broken bone fragments

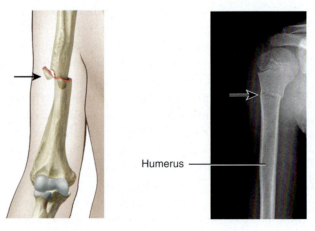

Impacted fracture One broken end is forced into the interior of the other

Repairing broken bones • Figure 5.6

The healing process takes place in four stages and can take up to several months.

1 Formation of fracture hematoma (several weeks)
- Broken blood vessel leaks blood, which clots
- Nearby bone cells die
- Phagocytes and osteoclasts digest dead cells, debris, and bone tissue in and around hematoma

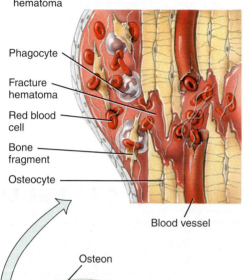

Phagocyte
Fracture hematoma
Red blood cell
Bone fragment
Osteocyte
Blood vessel

Osteon
Periosteum
Compact bone
Spongy bone
Fracture hematoma

2 Fibrocartilaginous callus formation (3 weeks)
- Fibroblasts from periosteum invade site and produce collagen fibers
- Chondroblasts from periosteum produce fibrocartilage
- These events produce fibrocartilaginous callus that bridges the broken ends of bone

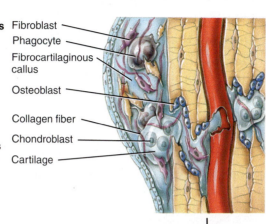

Fibroblast
Phagocyte
Fibrocartilaginous callus
Osteoblast
Collagen fiber
Chondroblast
Cartilage

3 Bony callus formation (3 – 4 months)
- In areas close to healthy bone, osteogenic cells become osteoblasts
- Osteoblasts produce spongy bone trabeculae, which joins living bone and dead bone fragments
- Fibrocartilage becomes spongy bone (bony callus)

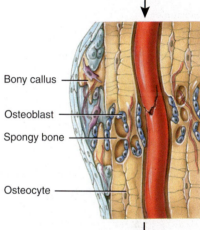

Bony callus
Osteoblast
Spongy bone
Osteocyte

4 Bone remodeling
- Osteoclasts resorb dead fragments of original bone
- Compact bone replaces spongy bone around the fracture periphery
- Healed fracture line may be visible in X-ray; if not, a thickened area on the surface remains where fracture occurred.

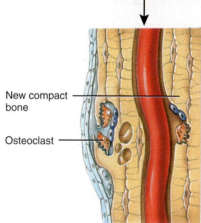

New compact bone
Osteoclast

Ask Yourself

Which of the following sequences correctly describes the steps in the bone healing process after a fracture?
a. Fibrocartilaginous callus; bony callus; remodeling; hematoma
b. Hematoma; bony callus; fibrocartilaginous callus; remodeling
c. Remodeling; fibrocartilaginous callus; bony callus; hematoma
d. Hematoma; fibrocartilaginous callus; bony callus; remodeling

---

CONCEPT CHECK STOP

1. **What** are the functions of bone tissue and the skeletal system?

2. **How** does compact bone differ from spongy bone?

3. **What** are the stages of intramembranous ossification?

4. **How** does an open, complete fracture differ from a closed, comminuted one?

The Structure of Bone Controls Function and Growth **121**

The Axial Skeleton Is Composed of 80 Bones

LEARNING OBJECTIVES

1. **Name** the cranial and facial bones and describe their locations.
2. **Describe** the unique features of the skull.
3. **Identify** the regions and structural features of the vertebral column.

The 206 bones in the adult human skeleton are divided into two categories: the **axial skeleton** (skull, vertebral column, and thorax) and the **appendicular skeleton** (shoulders, upper limbs, pelvis, and lower limbs)(**Figure 5.7**). Together, the bones of the skull and vertebral column make up 55 of the 80 bones in the axial skeleton. Let's look first at the bones of the skull.

Axial and appendicular skeletons • Figure 5.7

The bones of the skull, vertebral column, and thorax form the axial skeleton. The bones of the shoulders, upper limbs, pelvis, and lower limbs form the appendicular skeleton.

Division of the Skeleton	Structure	Number of Bones	Division of the Skeleton	Structure	Number of Bones
Axial Skeleton	**Skull**		**Appendicular Skeleton**	**Pectoral (shoulder) girdles**	
	Cranium	8		Clavicle	2
	Face	14		Scapula	2
	Hyoid	1		**Upper limbs**	
	Auditory ossicles	6		Humerus	2
	Vertebral column	26		Ulna	2
	Thorax			Radius	2
	Sternum	1		Carpals	16
	Ribs	24		Metacarpals	10
				Phalanges	28
				Pelvic (hip) girdle	
				Hip or pelvic bone	2
				Lower limbs	
				Femur	2
				Patella	2
				Fibula	2
				Tibia	2
				Tarsals	14
				Metatarsals	10
				Phalanges	28
		Subtotal = 80			Subtotal = 126
					Total in an adult skeleton = 206

The Head Is Formed by the Skull and Hyoid Bones

The skull, which consists of 22 bones, rests on top of the vertebral column. It has two groups of bones: 8 **cranial bones** and 14 **facial bones**. The cranial bones protect the brain and form attachment points for the **meninges** (me-NIN-jēz) on the interior and the muscles that move the head on the exterior. The facial bones house the openings to the airways and the digestive system, protect the sensory organs (eyes, ears, nose, and taste buds), and provide attachments for facial muscles.

> **meninges** (me-NIN-jēz) Three membranes that cover the brain and spinal cord.

Cranial bones The cranium consists of the following bones (**Figure 5.8**, on the next page):

- The **frontal bone** forms the forehead, roofs of the eye sockets, and front part of the cranial floor. The mucous membrane–lined spaces (frontal sinuses) deep within it resonate sound.
- The **parietal** (pa-RĪ-e-tal) **bones** (2) form the sides and roof of the cranium
- The **temporal bones** (2) form the lower side of the cranium and part of the cranial floor. The temporal bones have several features:
  - They form joints with the jawbone (mandible) called the **temporomandibular joints (TMJ)**.
  - The **external auditory meatus** is the canal that leads to the middle ear.
  - The **mastoid process** is a point of attachment for some of the muscles involved in head movement.

  > **foramen** (fō-RĀ-men) A hole in a bone for passage of vessels or nerves (plural is *foramina*).

  - The carotid artery passes through a **foramen** called the **carotid foramen**.
  - The **styloid process** serves as a point of attachment for the tongue and neck muscles.
- The **occipital bone** (ok-SIP-i-tal) forms the back part of the skull and most of the cranial floor. The medulla, spinal cord, and vertebral and spinal arteries all pass through its **foramen magnum**. The first cervical vertebra attaches to the occipital bone at two processes called **occipital condyles**.
- The **sphenoid bone** (SFĒ-noyd) is in the middle of the cranial floor and is where all the other cranial bones attach, like the keystone joining two arches to form a doorway. It contains **sphenoidal sinuses**, which drain into the nasal cavity. The pituitary gland sits in a depression of the sphenoid bone called the **sella turcica**. The optic nerve passes through its **optic foramen**, and the mandibular nerve passes through its **foramen ovale**.
- The **ethmoid bone** (ETH-moid) forms the anterior part of the cranial floor, the medial part of the eye sockets, and superior portions of the nasal cavity. It has 3 to 18 **ethmoidal sinuses** (air spaces) and mucus-lined **conchae** that warm and moisten inhaled air and trap foreign particles. The **crista galli**, a ridge on the superior portion of the ethmoid bone, serves as an attachment point for the meninges. This is surrounded by the **cribriform plate** through which the nerves associated with the receptors for smell pass from the nose into the brain.

Facial bones Your face changes and grows from the time you are born until around age 16. Your teeth form and erupt, the cranial bones grow, and the paranasal sinuses expand. Your facial bones include the following:

- **Nasal bones** (2) form the bridge of the nose.
- **Maxillae** (mak-SIL-ē; singular is *maxilla*) (2) form the upper jawbone and join with all the other facial bones except the mandible (lower jawbone):
  - Each maxilla has a **maxillary sinus** that empties into the nasal cavity.
  - The **alveolar process** (al-VĒ-ō-lar) forms the arch that contains the sockets (alveoli) for the teeth.
  - The maxillae form the anterior three-fourths of the roof of the mouth (hard palate).
- **Palatine bones** (PAL-a-tīn) (2) form the posterior portion of the hard palate, part of the lower eye sockets, and part of the floor and the sides of the nasal cavity.
- The **mandible** is the largest, strongest facial bone and the only one that moves:
  - Each **condylar process** (KON-di-lar) forms a temporomandibular joint with each temporal bone.
  - Like the maxillae, the mandible has an alveolar process for the lower teeth.

The skull consists of cranial (8) and facial (14) bones that surround and protect the brain, provide attachments for muscles and meninges, form the framework for the nasal and oral cavities, and have passages for major nerves and blood vessels.

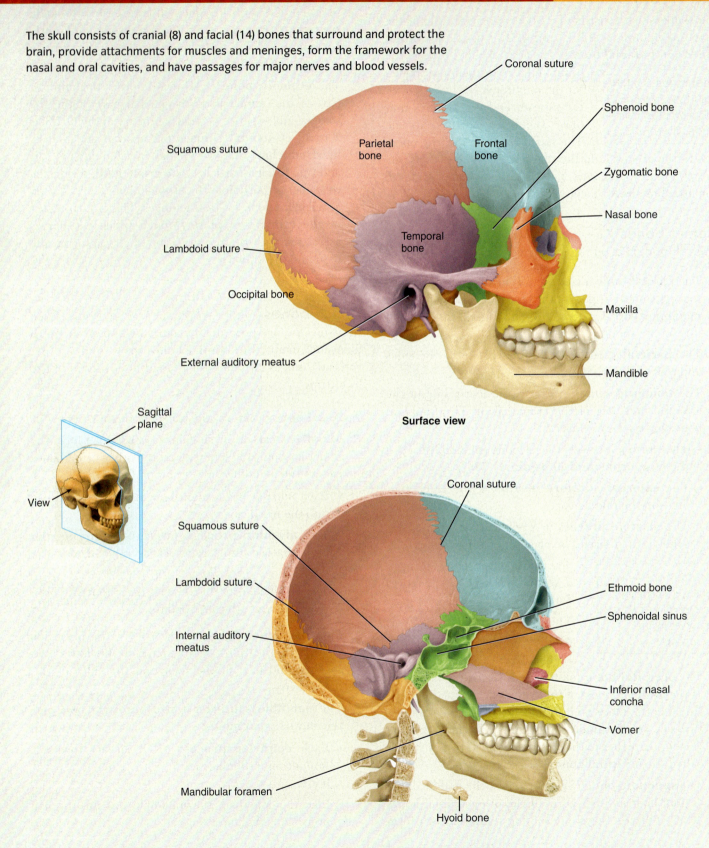

Coronal suture

Sphenoid bone

Zygomatic bone

Nasal bone

Parietal bone

Frontal bone

Squamous suture

Temporal bone

Lambdoid suture

Occipital bone

Maxilla

Mandible

External auditory meatus

Surface view

Sagittal plane

View

Coronal suture

Squamous suture

Lambdoid suture

Ethmoid bone

Sphenoidal sinus

Internal auditory meatus

Inferior nasal concha

Vomer

Mandibular foramen

Hyoid bone

Medial view of sagittal plane

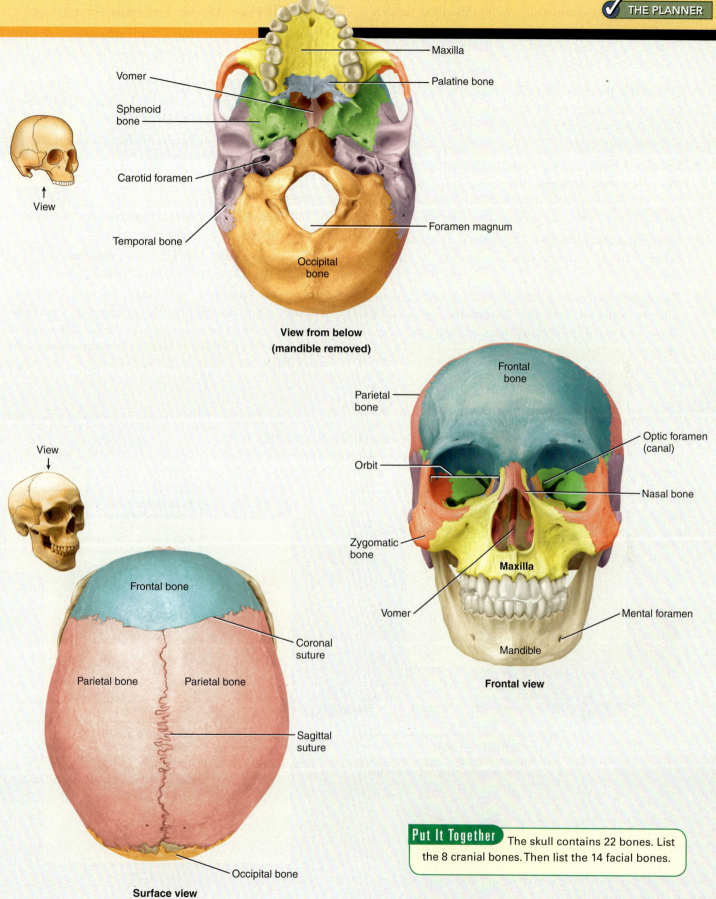

View

Maxilla

Vomer

Palatine bone

Sphenoid
bone

Carotid foramen

Foramen magnum

Temporal bone

Occipital
bone

**View from below
(mandible removed)**

View

Frontal
bone

Parietal
bone

Optic foramen
(canal)

Orbit

Nasal bone

Zygomatic
bone

Maxilla

Vomer

Mental foramen

Mandible

Frontal view

Frontal bone

Coronal
suture

Parietal bone

Parietal bone

Sagittal
suture

Occipital bone

Surface view

Put It Together The skull contains 22 bones. List
the 8 cranial bones. Then list the 14 facial bones.

The Axial Skeleton Is Composed of 80 Bones   125

- The **mental foramina** are holes that allow passage of the mental nerve. Dentists use the mental foramina as landmarks to inject anesthetics into the mental nerve.

- **Zygomatic bones** (2), or cheekbones, form the cheek prominences and part of the wall of the eye sockets. They form joints with the frontal, maxilla, sphenoid, and temporal bones.

- **Lacrimal bones** (LAK-ri-mal) (2) are the smallest, thinnest bones on the medial eye socket. They house the tear ducts, which tunnel through to the nasal cavity. This is why your nose runs when you cry.

- **Inferior nasal conchae** (2) project into the nasal cavity to filter air before it passes toward the trachea and lungs.

- The **vomer** (VŌ-mer) joins with the maxillae and the palatine bones to form the floor of the nasal cavity. Along with cartilage and the ethmoid bone, the single vomer forms the nasal septum, which divides the nasal cavity into right and left sides.

The Skull Has Many Unique Features

The skull has unique features, such as sutures, sinuses, and fontanels (soft spots at birth and early infancy):

- A **suture** (SOO-chur) is a special type of immovable joint that joins most of the skull bones. There are 4 major sutures in the skull (Figure 5.8):
  - The **coronal** (kō-RŌ-nal) **suture** unites the frontal bone and two parietal bones.
  - The **sagittal** (SAJ-i-tal) **suture** attaches the two parietal bones.
  - The **lambdoid** (LAM-doyd) **suture** joins the parietal bones to the occipital bone.
  - The **squamous** (SKWĀ-mus) **sutures** seal the parietal bones to the temporal bones.

- **Paranasal sinuses** are found in the sphenoid, frontal, ethmoid, and maxillary bones. They produce mucus, lighten the weight of the skull, and serve as echo chambers, which produce the unique sounds of your voice.

Fontanels • Figure 5.9

The fontanels are the soft spots on a baby's head. They allow the bones of the skull to compress during passage through the birth canal and provide room for the brain to grow after birth.

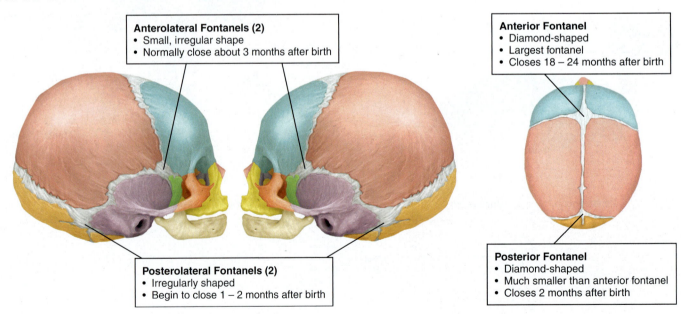

Anterolateral Fontanels (2)
- Small, irregular shape
- Normally close about 3 months after birth

Anterior Fontanel
- Diamond-shaped
- Largest fontanel
- Closes 18 – 24 months after birth

Posterolateral Fontanels (2)
- Irregularly shaped
- Begin to close 1 – 2 months after birth

Posterior Fontanel
- Diamond-shaped
- Much smaller than anterior fontanel
- Closes 2 months after birth

- **Fontanels** (fonta-NELZ) are mesenchyme-filled spaces between the cranial bones of infants at birth (**Figure 5.9**). These soft spots compress as the baby passes through the birth canal. For a short time after birth, the fontanels also provide room for the brain to grow. Within the first two years of life, they are replaced by bone via intramembranous ossification. As the fontanels close, the bones of the skull fuse to form the sutures.

- The **hyoid bone** (HĪ-oyd) is located in the neck, between the mandible and larynx (see Figure 5.8). It is suspended from the styloid process of each temporal bone by ligaments and muscle. It supports the tongue, stabilizes the airways, and provides attachment points for tongue, neck, and pharyngeal muscles.

The Vertebral Column Contains 26 Vertebrae

The vertebral column (also called the *spine, spinal column,* or *backbone*) protects the spinal cord, supports the head and neck, permits movement, and provides attachment points for the back muscles, ribs, and pelvis. The vertebral column consists of 26 bones called **vertebrae** (VER-te-brē; singular is vertebra). Vertebrae have the following general structures (see **Figure 5.10** on the next page):

- The **body** is the thick, disc-shaped anterior portion that bears weight.
- The **vertebral arch** extends posteriorly from the body. It consists of two short, thick processes called **pedicles** (PED-i-kuls) that project backward and join with two flat parts called **laminae** (LAM-i-nē). The **vertebral foramen** is an opening through which the spinal cord passes.
- Seven processes arise from the vertebral arch:
  - *Spinous process* (1) projects from the laminae; it serves as attachment point for muscles.
  - *Transverse processes* (2) are lateral extensions that serve as attachment points for muscles.
  - *Superior articular processes* (2) attach to vertebra above.
  - *Inferior articular processes* (2) attach to vertebra below.

The exact shape and structure of the vertebrae vary with the region where they are located:

- **Cervical vertebrae** (7) are in the neck region. Each cervical vertebra has three openings (foramina): a larger, central opening (vertebral foramen) for the spinal cord, and two transverse foramina, passages for blood vessels and nerves.
- **Thoracic vertebrae** (12) are posterior to the chest cavity and serve as attachments for the ribs.
- **Lumbar vertebrae** (5) form the lower back.
- The **sacrum** (SĀ-krum) consists of 5 fused vertebrae and forms the posterior wall of the pelvis. Blood vessels and nerves pass through the openings.
- The **coccyx** (KOK-siks), sometimes referred to as the tailbone, consists of 4 fused vertebrae.

Note that the adult vertebral column has four curved regions: cervical, thoracic, lumbar, and sacral. The curves develop from a single, concave curve in the fetus. When the infant begins holding its head erect (at approximately three months of age), the cervical curve develops. The lumbar curve develops later, when the child starts sitting up, standing, and walking. Sometimes, abnormal curvatures develop in the spine due to uneven growth or weakening of the bones and/or musculature associated with the spine (see Figure 5.10):

- *Scoliosis*—A lateral curvature that causes the spine to "lean" to one side more than the other. This condition is seen more commonly in females than in males.
- *Kyphosis*—An exaggeration of the thoracic curve that forms a "humpback" appearance.
- *Lordosis*—An exaggeration of the lumbar curve that causes a "sway back."

CONCEPT CHECK

1. **Which** cranial bones form the floor of the cranium?
2. **How** do sutures form from the anterior fontanel?
3. **Compare** and contrast the five types of vertebrae.

The vertebral column (spine) functions as a strong, flexible rod that can rotate and move anteriorly (forward), posteriorly (backward), and laterally (sideways). Each type of vertebra has a slightly different shape. Intervertebral discs form cartilage cushions between the vertebrae.

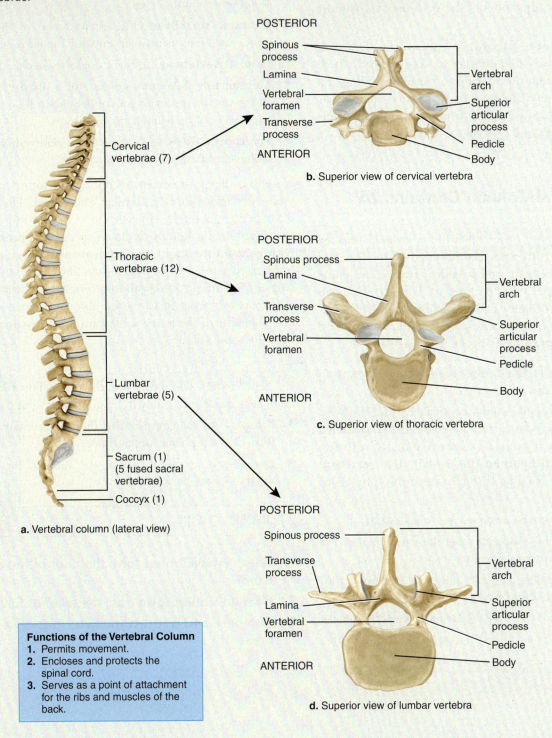

POSTERIOR

Spinous process

Lamina

Vertebral foramen

Transverse process

ANTERIOR

Vertebral arch

Superior articular process

Pedicle

Body

b. Superior view of cervical vertebra

POSTERIOR

Spinous process

Lamina

Transverse process

Vertebral foramen

ANTERIOR

Vertebral arch

Superior articular process

Pedicle

Body

c. Superior view of thoracic vertebra

POSTERIOR

Spinous process

Transverse process

Lamina

Vertebral foramen

ANTERIOR

Vertebral arch

Superior articular process

Pedicle

Body

d. Superior view of lumbar vertebra

Cervical vertebrae (7)

Thoracic vertebrae (12)

Lumbar vertebrae (5)

Sacrum (1) (5 fused sacral vertebrae)

Coccyx (1)

a. Vertebral column (lateral view)

Functions of the Vertebral Column
1. Permits movement.
2. Encloses and protects the spinal cord.
3. Serves as a point of attachment for the ribs and muscles of the back.

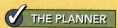

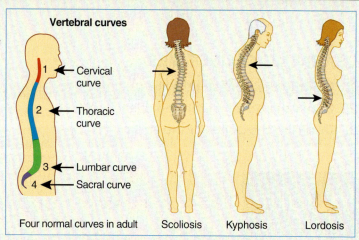

Vertebral curves

1 → Cervical curve

2 → Thoracic curve

3 → Lumbar curve

4 → Sacral curve

Four normal curves in adult   Scoliosis   Kyphosis   Lordosis

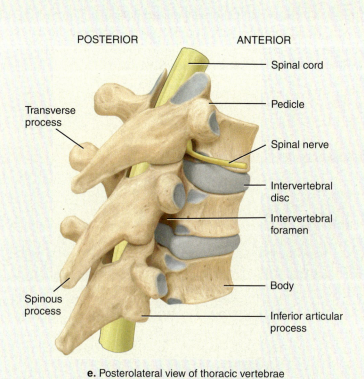

POSTERIOR   ANTERIOR

Spinal cord

Transverse process

Pedicle

Spinal nerve

Intervertebral disc

Intervertebral foramen

Body

Spinous process

Inferior articular process

e. Posterolateral view of thoracic vertebrae

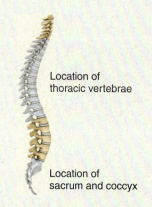

Location of thoracic vertebrae

Location of sacrum and coccyx

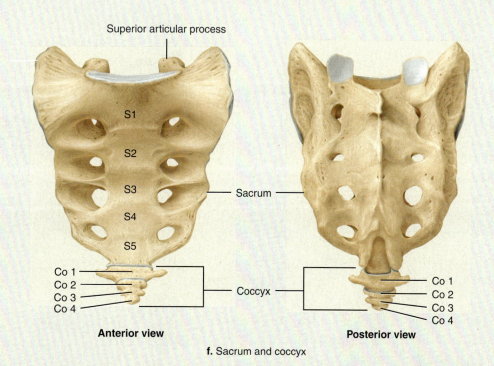

Superior articular process

S1

S2

S3

S4

S5

Sacrum

Co 1
Co 2
Co 3
Co 4

Coccyx

Co 1
Co 2
Co 3
Co 4

Anterior view   **Posterior view**

f. Sacrum and coccyx

Bones of the Upper Body Form the Thorax and Arms

LEARNING OBJECTIVES

1. **Identify** the bones of the thorax.
2. **Name** the bones of the shoulder.
3. **Describe** the bones of the arm.

I n this section you will learn about the bones of the thorax, which are part of the axial skeleton, and those of the arms, which along with the hips, legs, and feet form the appendicular skeleton (see Figure 5.7). The thorax consists of the **sternum** and 12 pairs of **ribs** (**Figure 5.11a**).

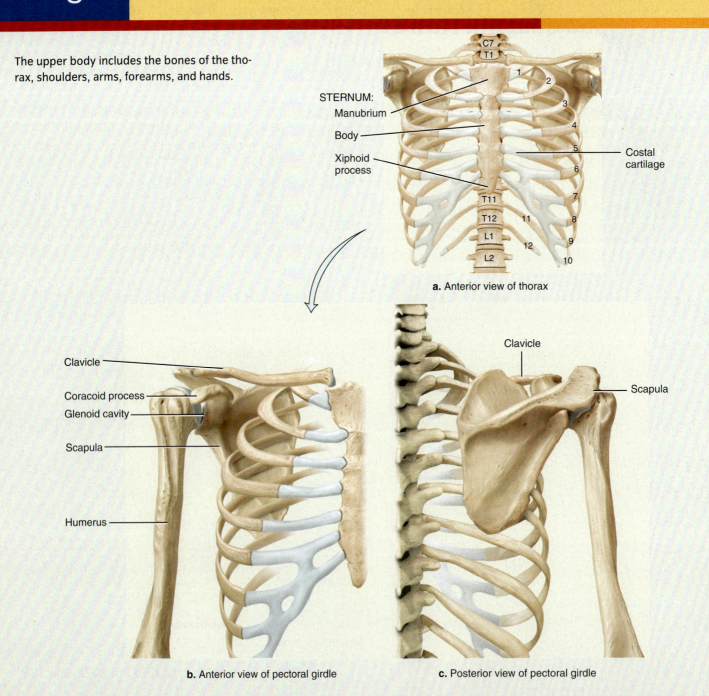

The upper body includes the bones of the thorax, shoulders, arms, forearms, and hands.

C7
T1
1
2
3
4
5
6
7
8
9
10
11
12
T11
T12
L1
L2

STERNUM:
Manubrium
Body
Xiphoid process
Costal cartilage

a. Anterior view of thorax

Clavicle
Coracoid process
Glenoid cavity
Scapula
Humerus

Clavicle
Scapula

b. Anterior view of pectoral girdle

c. Posterior view of pectoral girdle

The Ribs and Sternum Form the Framework for the Thorax

The sternum is made of three parts that fuse by age 25:

- The **manubrium** (ma-NOO-brē-um) articulates with the collarbone, or clavicle, and the first rib.
- The body articulates with part of the second rib and ribs 3 through 10.
- The **xiphoid process** (ZĪ-foyd) is made of cartilage that ossifies by age 40. No ribs attach to this pointed

structure, which rescuers use to locate the proper hand position for cardiopulmonary resuscitation (CPR).

The sternum attaches directly to the first through seventh pairs of ribs by a form of hyaline cartilage called *costal cartilage*. The remaining pairs of ribs either attach indirectly to the sternum (pairs 8–12) or do not attach at all (pairs 11–12). Ribs are named based on how they attach to the sternum; rib pairs 1 through 7 are called **true ribs**, rib pairs 8 through 12 are called **false ribs**, and rib pairs 11 and 12 are referred to as **floating ribs**. The bones of the thorax

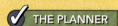

THE PLANNER

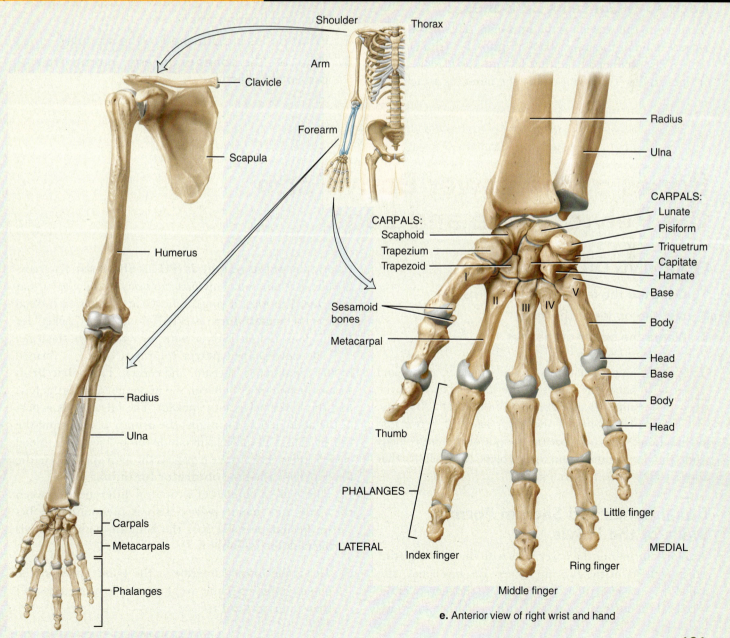

d. Anterior view of upper limb

e. Anterior view of right wrist and hand

protect the heart and lungs and provide attachment points for muscles, especially those involved in breathing.

The Scapula and Clavicle Form the Pectoral Girdle

The next set of bones in the upper body is the shoulder girdles or **pectoral girdles** (PEK-tō-ral), which attach the upper limbs to the axial skeleton (see **Figure 5.11b,c**). The **clavicle** (KLAV-i-kul), or collarbone, attaches to the manubrium of the sternum and the **scapula**, or shoulder blade. The **coracoid process** (KOR-a-koyd) of the scapula serves as a point of attachment for muscles and its **glenoid cavity** forms the shoulder joint with the head of the humerus (upper arm bone).

Each Arm Contains 22 Bones

Attached to the scapula is the **humerus** (HŪ-mer-us), the longest bone in the upper body (**Figure 5.11d**). The humerus has a rounded head that fits into the glenoid cavity of the scapula. The distal end of the humerus attaches to the two bones of the forearm, the **ulna** and **radius**. The ulna is medial to the radius. The proximal end of the radius is rounded and articulates with the humerus, to allow approximately 180° of forearm rotation. Distal to the ulna and radius are the bones of the wrist, the **carpals** (8 bones), which are arranged in two rows of 4 bones each. The distal row attaches to the bones of the palm of the hand (**metacarpals**), whose distal heads form the knuckles. Finally, the metacarpals attach to the bones of the fingers, the **phalanges** (fa-LAN-jēz). There are 2 phalanges in the thumb and 3 in each finger (Figure 5.11).

CONCEPT CHECK

1. **What** is the difference between true ribs, false ribs, and floating ribs?
2. **What** bones are attached to the scapula?
3. **What** are the bones of the arm, in order from proximal to distal?

Bones of the Lower Body Form the Pelvic Girdle and Legs

LEARNING OBJECTIVES

1. **Describe** the features of the pelvis.
2. **List** the bones of the legs.
3. **Name** the bones of the feet.

A long with those of the shoulder, arms, and hand, the bones of the lower body form the appendicular skeleton—a total of 126 bones. The bones of the lower body have two major functions: They support the weight of the entire body while you are standing, and they provide the mobility that allows you to walk, run, and jump.

Coxal Bones and Sacrum Form Walls of the Pelvis

We start our tour of the lower body with the **pelvic girdle**, which consists of the two hip bones. The hip bones, also called the *coxal bones* (or *os coxa*), attach to the sacrum of the vertebral column posteriorly and with each other anteriorly to form the **pubic symphysis** (PŪ-bik SIM-fi-sis). Each coxal bone is composed of an *ilium*, an *ischium*, and a *pubis* that have fused to form a single unit. Centrally, on each coxal bone is the **acetabulum** (as-e-TAB-ū-lum), the "socket" for the hip joint that articulates with the femur from the leg.

The bowl-shaped **pelvis** (plural is *pelves*) is formed by the coxal bones, sacrum, and coccyx. The **pelvic brim** forms the boundary between the upper pelvis (false pelvis) and the lower pelvis (true pelvis). The false pelvis is part of the abdomen and contains the urinary bladder and the uterus. The true pelvis surrounds the pelvic cavity. Blood vessels and nerves to the legs pass through openings in the lower pelvis called the **obturator foramina**.

There are a number of structural differences between the male and female pelves. This is predominantly because the female must pass the baby through the pelvis during childbirth (**Table 5.1**):

- The angle formed inferior to the pubic bones at the pubic symphysis (pubic arch) is wider in women (> 90°) than in men (< 90°).

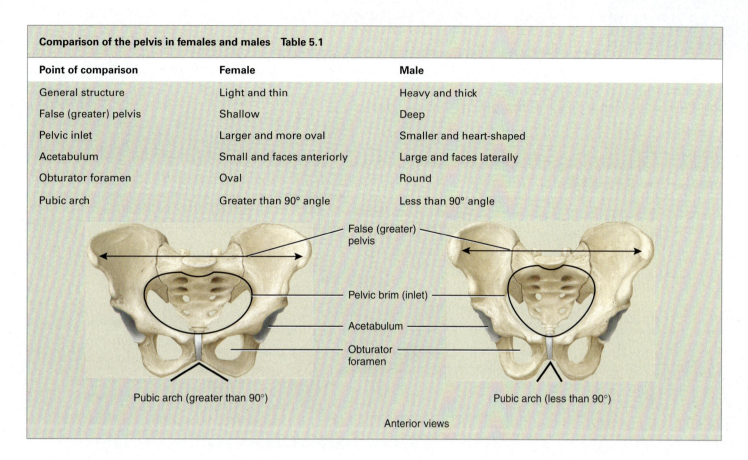

Comparison of the pelvis in females and males Table 5.1		
Point of comparison	Female	Male
General structure	Light and thin	Heavy and thick
False (greater) pelvis	Shallow	Deep
Pelvic inlet	Larger and more oval	Smaller and heart-shaped
Acetabulum	Small and faces anteriorly	Large and faces laterally
Obturator foramen	Oval	Round
Pubic arch	Greater than 90° angle	Less than 90° angle

False (greater) pelvis

Pelvic brim (inlet)

Acetabulum

Obturator foramen

Pubic arch (greater than 90°)

Pubic arch (less than 90°)

Anterior views

- The pelvic inlet—the superior opening into the bony pelvis—is heart-shaped in males and oval in females.
- The sacrum is wider and more posteriorly placed in the female than in the male.

All bones of the lower body are shown in **Figure 5.12** on the next page.

The Design of the Lower Limb Is Similar to That of the Upper Limb

Like the upper limb, the lower limb is composed of a single bone proximally with increasing numbers of bones as one moves distally. Attached to each coxal bone is a **femur** (thigh bone), the longest single bone in the body (see Figure 5.12a). Like the humerus, the femur has a rounded, proximal head where it fits into the acetabulum of the pelvis. The femur bends medially and attaches distally to the **patella** (knee cap) and the tibia.

The **tibia** is the large medial bone of the shin and bears most of the weight. It attaches proximally with the femur and fibula and distally with the fibula and the talus of the ankle. The patellar ligament, which holds the patella in place, attaches to the tibia at the **tibial tuberosity**.

The lateral bone of the shin is called the **fibula**. The proximal end of the fibula articulates with the tibia and

its rounded distal end and forms a joint with the talus of the ankle. The ankle bones, or **tarsals**, consist of 7 bones. Two of these bones, the **talus** (TĀ-lus) and the **calcaneus** (kal-KĀ-nē-us), are on the posterior part of the foot. The talus is part of the ankle joint, while the calcaneus forms the heel of the foot. The rest of the tarsals are anterior to these and attach to the metatarsals.

Foot Structure Helps Disperse Body Weight and Absorb Shock

The **metatarsals** (5) are like the metacarpals of the hand. Attached to the metatarsals are the phalanges (toes), which have a structure similar to the phalanges in the hand (Figure 5.12d). The calcaneus forms the posterior portion of the foot. The foot has two raised bends, or **arches**, in it (Figure 5.12e). The *longitudinal arch* spreads from anterior to posterior, while the *transverse arch* goes from side to side. The arches are flexible and springy. They absorb shocks, distribute body weight over the foot, and provide leverage while walking.

CONCEPT CHECK

1. **What** is the pubic symphysis?
2. **How** does the thigh attach to the leg?
3. **What** are the functions of the arches of the foot?

The bones of the lower body include the pelvis, femur, patella, tibia, fibula, tarsals, metatarsals, and phalanges.

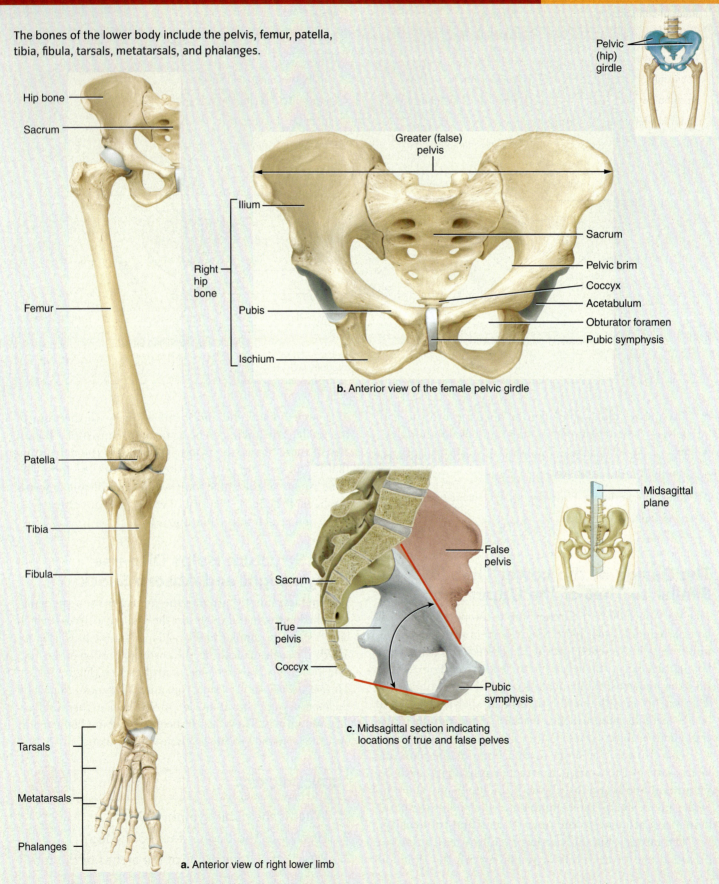

Pelvic (hip) girdle

Hip bone

Sacrum

Femur

Patella

Tibia

Fibula

Tarsals

Metatarsals

Phalanges

a. Anterior view of right lower limb

Greater (false) pelvis

Right hip bone

Ilium

Pubis

Ischium

Sacrum

Pelvic brim

Coccyx

Acetabulum

Obturator foramen

Pubic symphysis

b. Anterior view of the female pelvic girdle

Midsagittal plane

Sacrum

True pelvis

Coccyx

False pelvis

Pubic symphysis

c. Midsagittal section indicating locations of true and false pelves

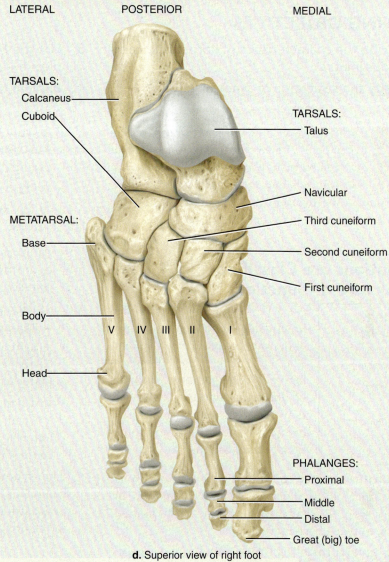

LATERAL   POSTERIOR   MEDIAL

TARSALS:
Calcaneus
Cuboid

TARSALS:
Talus

Navicular

Third cuneiform

METATARSAL:
Base

Second cuneiform

First cuneiform

Body

V   IV   III   II   I

Head

PHALANGES:
Proximal
Middle
Distal
Great (big) toe

d. Superior view of right foot

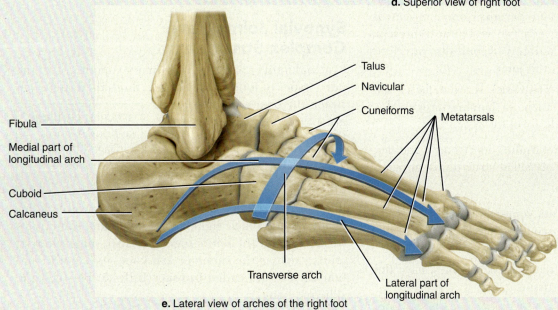

Talus
Navicular
Cuneiforms   Metatarsals

Fibula

Medial part of
longitudinal arch

Cuboid

Calcaneus

Transverse arch

Lateral part of
longitudinal arch

e. Lateral view of arches of the right foot

Bones of the Lower Body Form the Pelvic Girdle and Legs **135**

Articulations Form Where Bones Join Together

LEARNING OBJECTIVES

1. **Describe** the characteristics of fibrous and cartilaginous joints.
2. **Identify** the structures of synovial joints.
3. **List** the types of synovial joints and give examples of each.
4. **Explain** the various types of joint movements.

An *articulation* or joint is a point of contact between bones, between cartilage and bone, or between teeth and bones. There are two classification schemes for joints; one is based on how the bones come together and the other on how the bones work together to make the body move.

Articulations are Classified by Structure or Function

> **ligaments** (LIG-a-ments) Dense regular connective tissue that attaches bone to bone.
>
> **tendons** (TEN-dons) White fibrous cords of dense regular connective tissue that attach muscles to bones.

An articulation can be classified functionally based on the range of movement possible by the joint. The shape of the bones, the placement of the **ligaments**, and the flexibility of **tendons** are important for determining movement in a joint. There are several functional classifications:

- A **synarthrosis** (sin'-ar-THRŌ-sis) is a very strong, tightly fitted joint that permits virutally no movement. The sutures of the skull are an example of this limited-movement type of articulation.

- An **amphiarthrosis** (am'-fē-ar-THRŌ-sis) is looser than a synarthrosis and permits some movement. The joint formed between two vertebrae represents an amphiarthrosis. The intervertebral disc interferes with free movement of this joint.

- A **diarthrosis** (dī'-ar-THRŌ-sis) is even looser and more freely movable than an amphiarthrosis. Most articulations of the body are of this type.

The structure of a joint determines its strength and flexibility. Joints can also be classified structurally in one of three categories:

- **Fibrous** (FĪ-brus) **joints** contain varying densities of irregular dense collagenous connective tissue (**Figure 5.13**):
  - Sutures, like those that join skull bones, have thin layers of connective tissue.

- A **syndesmosis** has more space between the bones and more connective tissue than a suture. One example is the joint that forms between the tibia and fibula distally (distal tibiofibular joint) at the ankle.

Another example is a *gomphosis*, the attachment of a tooth into its socket.

- **Interosseous membranes** are large sheets of dense connective tissue that span greater distances and connect long bones, such as the ulna and radius.

- As their name implies, **cartilaginous joints** (kar-ti-LAJ-i-nus) are connected by cartilage (see Figure 5.13):
  - A **synchondrosis** (sin'-kon-DRŌ-sis) has hyaline cartilage and is immovable, like the epiphyseal plates within long bones.
  - In a **symphysis** (SIM-fi-sis), the bones are joined by a broad, flat disc of fibrocartilage—such as on the anterior pelvis where the pubic bones articulate—so they are only slightly movable.

- The **synovial joints** (si-NŌ-vē-al) have numerous components and are the most common type of articulation. These joints are described in more detail in the next section (Figure 5.13).

Synovial Joints Have Complex Structures

Synovial joints are structurally more complex than the other types of joints. They have a fluid-filled cavity between the bones, and the bones are covered by **articular cartilage**. A **fibrous capsule** lined with a **synovial membrane** surrounds the cavity, which is filled with **synovial fluid**. The

> **articular cartilage** (ar-TIK-ū-lar) Hyaline cartilage attached to articular bone surfaces.

fluid and cartilage lubricate this freely movable joint. Synovial joints have ligaments outside and sometimes within to hold the joints together. Also, some synovial joints, such as the shoulder and knee, have additional lubricating sacs called **bursae** (BER-sē) between the skin and bone.

Several types of joints connect bones in the skeleton.
Joints vary in their structure (fibrous, cartilaginous,
synovial), and degree of movement.

Synovial joints
- Fluid-filled cavity between bones
- Encapsulated
- Bone ends covered with cartilage
- Held together by ligament
- Six types
- Flexible and freely-movable joints

Cartilaginous joints
- Held together by hyaline cartilage or fibrocartilage
- Permit little or no movement
- Types include synchondroses and symphyses

Fibrous joints
- Held together by irregular connective tissue
- Types include sutures, syndesmoses, and interosseous membranes
- Permit little or no movement

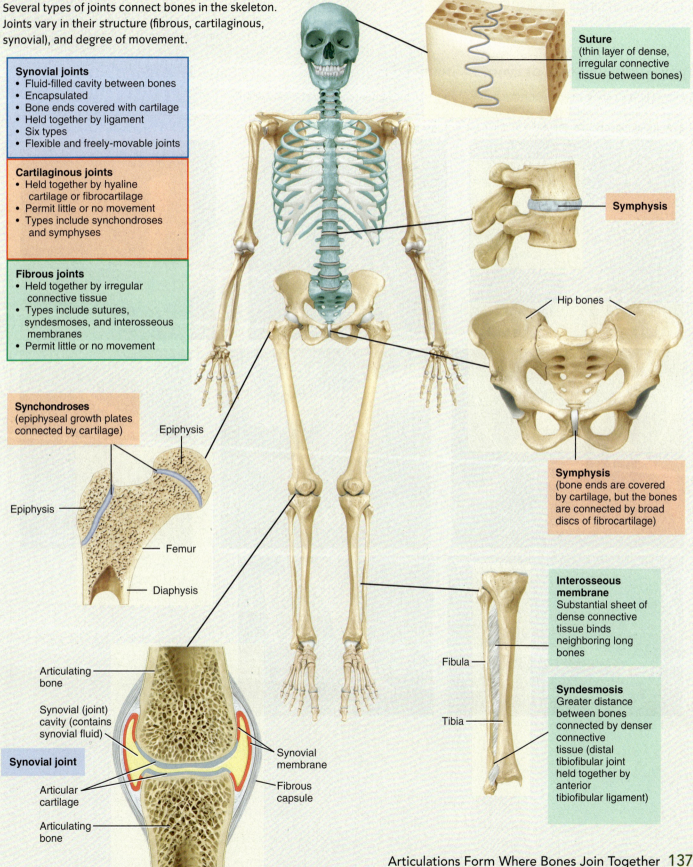

Suture
(thin layer of dense, irregular connective tissue between bones)

Symphysis

Hip bones

Symphysis
(bone ends are covered by cartilage, but the bones are connected by broad discs of fibrocartilage)

Synchondroses
(epiphyseal growth plates connected by cartilage)

Epiphysis

Epiphysis

Femur

Diaphysis

Interosseous membrane
Substantial sheet of dense connective tissue binds neighboring long bones

Fibula

Tibia

Syndesmosis
Greater distance between bones connected by denser connective tissue (distal tibiofibular joint held together by anterior tibiofibular ligament)

Articulating bone

Synovial (joint) cavity (contains synovial fluid)

Synovial joint

Articular cartilage

Articulating bone

Synovial membrane

Fibrous capsule

Freely Movable Joints Are Capable of Various Motions

Many types of movements occur at freely movable synovial joints (**Figure 5.14**):

- **Gliding**—Flat bone surfaces move back-and-forth and side-to-side relative to one another, like the movements of your carpal bones against one another that allow you to move your wrist. Gliding movements are limited to loose-fitting structures of the articular capsule and associated ligaments and bones (not shown in Figure 5.14).

Types of movements at synovial joints • Figure 5.14

Synovial joints can have angular movements, rotational movements, or special movements.

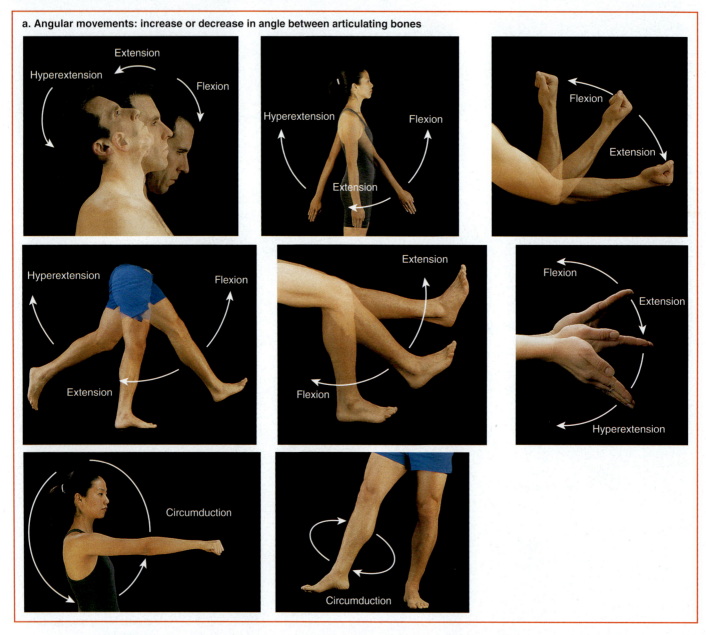

a. Angular movements: increase or decrease in angle between articulating bones

- **Angular movements**—These movements are increases or decreases in the angle of articulating bones. Angular movements include flexion, extension, hyperextension, abduction, adduction, and circumduction (Figure 5.14a).
- **Rotation**—In this type of movement, a bone revolves around its own longitudinal axis. *Medial rotation* (internal rotation) turns the body part towards the midline, while *lateral rotation* (external rotation) turns it away from the midline (Figure 5.14b).
- **Special movements**—These movements occur only at certain joints. Such movements include elevation, depression, protraction, retraction, inversion, eversion, dorsiflexion, plantar flexion, supination, and pronation (Figure 5.14c).

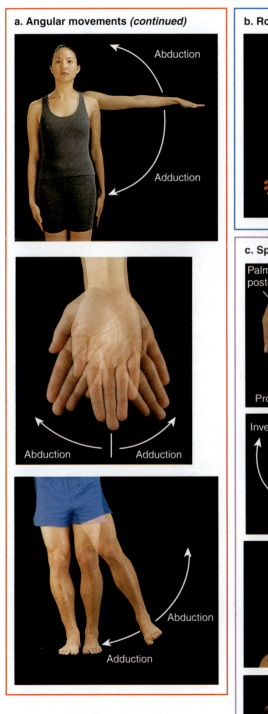

a. Angular movements (continued)

Abduction

Adduction

Abduction | Adduction

Abduction

Adduction

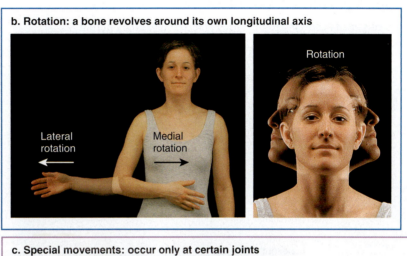

b. Rotation: a bone revolves around its own longitudinal axis

Lateral rotation

Medial rotation

Rotation

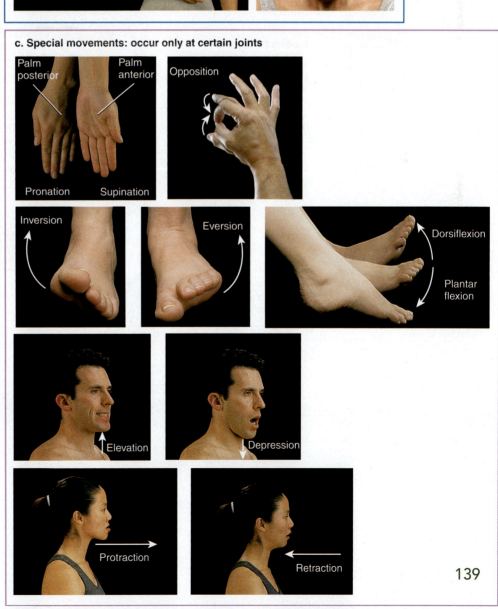

c. Special movements: occur only at certain joints

Palm posterior

Palm anterior

Pronation   Supination

Opposition

Inversion

Eversion

Dorsiflexion

Plantar flexion

Elevation

Depression

Protraction

Retraction

139

Synovial Joints Are Classified by Type of Movement

Six types of synovial joints make the various types of movement possible:

- **Plane**—Back-and-forth, side-to-side. Many plane joints are biaxial (movement in two axes). Some are triaxial (movement in three axes).
- **Hinge**—Angular opening and closing movements (uniaxial)

Types of synovial joints • Figure 5.15

Synovial joints are classified according to the types of movements they permit.

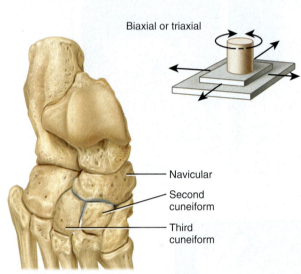

Biaxial or triaxial

Navicular
Second cuneiform
Third cuneiform

a. **Plane joint between navicular and second and third cuneiforms of tarsus in foot**

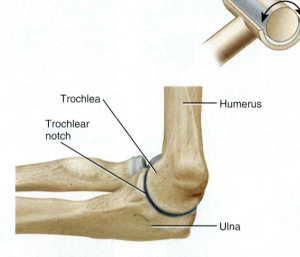

Uniaxial

Trochlea
Humerus
Trochlear notch
Ulna

b. **Hinge joint between trochlea of humerus and trochlear notch of ulna at the elbow**

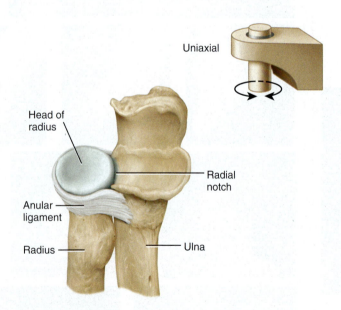

Uniaxial

Head of radius
Radial notch
Anular ligament
Radius
Ulna

c. **Pivot joint between head of radius and radial notch of ulna**

- **Pivot**—Rotation around longitudinal axis (uniaxial)
- **Condyloid** (KON-di-loyd)—Rotation around two axes (biaxial)
- **Saddle**—Rotation around two axes (biaxial)

- **Ball-and-socket**—Rotation around three axes (triaxial)

Figure 5.15 shows examples and types of permissible movement for each type of synovial joint.

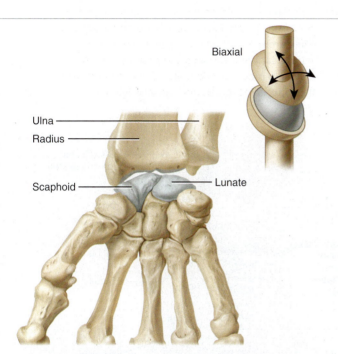

d. Condyloid joint between radius and scaphoid and lunate bones of carpus (wrist)

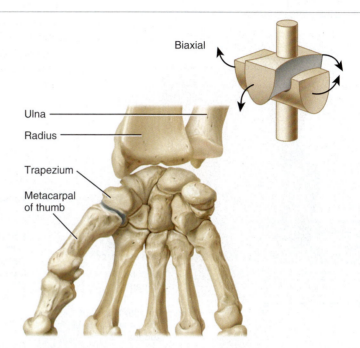

e. Saddle joint between trapezium of carpus (wrist) and metacarpal of thumb

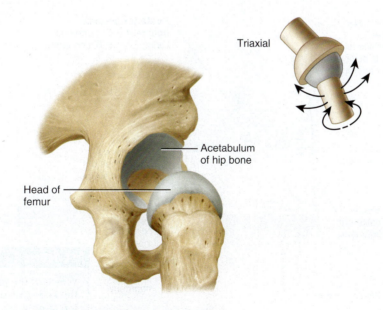

f. Ball-and-socket joint between head of femur and acetabulum of hip bone

The knee, the most complex synovial joint in the body (**Figure 5.16**), is held together with many ligaments that stabilize it. The sudden changes of direction of movement or impacts to the legs that cause unnatural movements of the joint often lead to knee injuries, especially in young athletes (see *What a Health Provider Sees*). The most common knee injury is tearing of the anterior cruciate ligament (ACL), which occurs in about 70% of all serious knee injuries. The ACL is often damaged if the knee joint is twisted or hyperextended.

Anatomy of the knee • Figure 5.16

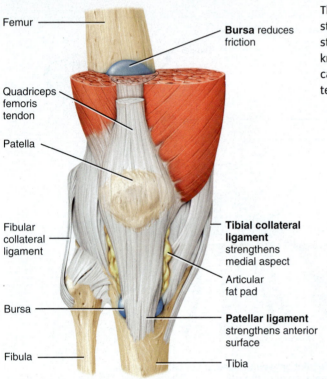

a. Anterior superficial view

The knee is a synovial joint that is reinforced by several ligaments that stabilize it in various directions. This helps prevent hyperextension and stops the tibia from sliding around on the femur. The menisci of the knee help even out the irregular bone surfaces and protect the articular cartilage, while its bursae help reduce friction. Most knee injuries involve tearing of the ligaments, especially the anterior cruciate ligament.

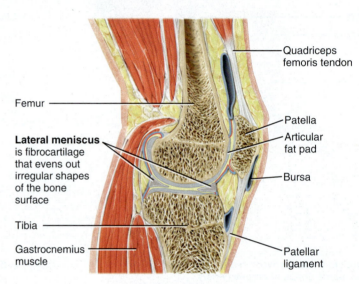

b. Sagittal section

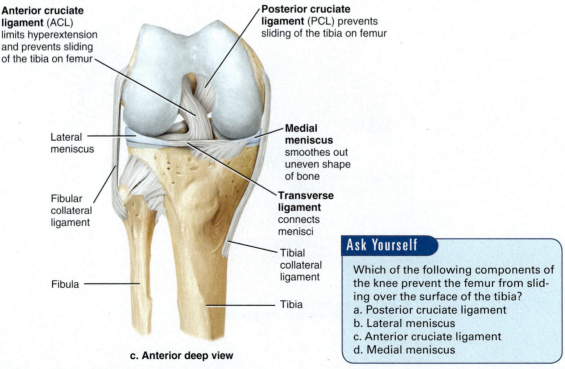

c. Anterior deep view

Ask Yourself

Which of the following components of the knee prevent the femur from sliding over the surface of the tibia?
a. Posterior cruciate ligament
b. Lateral meniscus
c. Anterior cruciate ligament
d. Medial meniscus

WHAT A HEALTH PROVIDER SEES

Knee Injuries

The knee is the most complex and vulnerable joint in your body. Whether you are a professional athlete or a "weekend warrior," knee injuries are common in activities that place stress on the knee joint. Knee injuries tend to be most common in sports where there is a lot of lateral motion, twisting, and pulling, such as basketball, tennis, and football. Knees can be hyper-extended or twisted, which can lead to tears of the ligaments or cartilage. The result can be severe pain, swelling, and loss of mobility. Some knee injuries can ruin promising athletic careers.

Pittsburgh Steelers quarterback Dennis Dixon injured his knee in a 2010 game against the Tennessee Titans (**Figure a**). At one time, the only way to repair a knee injury was to fully expose the knee joint. This type of surgery was not always successful with respect to complete repair of the injury and full recovery of function, and recovery times were generally as long as four to six months

Today, many knee injuries can be treated using arthroscopic knee surgery, in which a flexible microscope and micro-precision tools are placed into the knee joint through tiny incisions. The surgeons can remove, repair, or replace injured tissue, such as torn cartilage (**Figure b**). Recovery, which is usually complete within a month, must still be followed by two to four weeks of physical therapy to ensure that the patient regains full mobility.

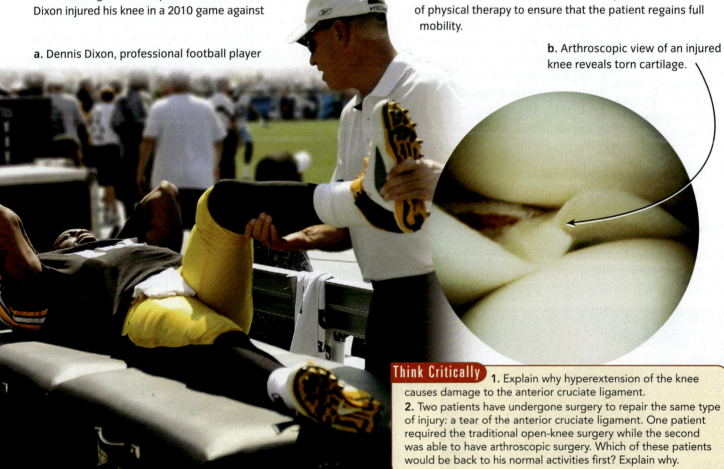

a. Dennis Dixon, professional football player

b. Arthroscopic view of an injured knee reveals torn cartilage.

Think Critically
1. Explain why hyperextension of the knee causes damage to the anterior cruciate ligament.
2. Two patients have undergone surgery to repair the same type of injury: a tear of the anterior cruciate ligament. One patient required the traditional open-knee surgery while the second was able to have arthroscopic surgery. Which of these patients would be back to his normal activities first? Explain why.

CONCEPT CHECK STOP

1. **Which** type of fibrous joint has the thickest layer of connective tissue?

2. **What** is the function of articular cartilage in a synovial joint?

3. **Which** type of synovial joint permits movement along two axes?

4. **Where** in the body does flexion occur? What type of movement opposes it?

Skeletal Structure Changes with Aging

LEARNING OBJECTIVES

1. **Describe** the effects of aging on bones.
2. **Explain** what happens as joints age.

Bone constantly remodels itself. At any given time, the amount of bone mass in your body is related to the rates at which your osteoblasts make bone and your osteoclasts resorb bone.

Bone Mass Decreases As You Get Older

From birth through adolescence, you make more bone than you lose. Young adults make as much bone as they lose. From middle age onward, you begin to lose more bone than you replace; this is especially true for women due to the loss of sex hormones such as estrogen after **menopause**. These changes in the balance of bone creation and resorption result in two major effects of aging on bone:

> **menopause** (MEN-ō-pawz) The termination of the menstrual or reproductive cycles.

- **Bones become more brittle**—Rates of protein synthesis and production of collagen fibers decrease, probably due to diminished growth hormone production. (Recall that protein fibers give bone its flexibility.)

- **Bones lose mass**—Increased osteoclast activity with age causes bones to lose minerals, mostly calcium. The mineral extracellular matrix gives bone its strength. This usually begins around age 30, increases around age 45, and continues throughout the rest of life. By age 70, as much as 30% of bone mass is lost.

Age-related demineralization of bone affects 60 million people in the United States each year, mostly middle-aged and elderly women. In the condition known as **osteoporosis** (os′-tē-ō-pō-RŌ-sis), so much mineral has been removed from the bone matrix that the bones become porous (**Figure 5.17**). This loss of minerals makes the bones weak and susceptible to fractures. Weight-bearing bones such as the spine, hip, and femur are especially prone to injury.

A diagnosis of osteoporosis is usually made based on a combination of factors: family history of the condition, small skeletal stature, and a bone mineral density (BMD) test indicating a decreased mineral composition of the bone. Treatment includes diets high in calcium and vitamin D, exercise, and medications.

Osteoporosis • Figure 5.17

Comparing spongy bone tissue from a normal young adult (**a**) with that of a person with osteoporotic bone (**b**) shows how the bone tissue has lost minerals and become porous. Osteoporosis affects both spongy bone tissue and compact bone tissue.

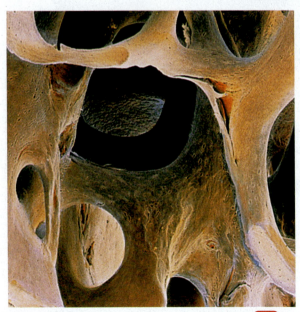

SEM 30x

a. Normal bone

SEM 30x

b. Osteoporotic bone

Knee replacement • Figure 5.18

In a surgical procedure called arthroplasty, an injured knee is replaced surgically with an artificial one.

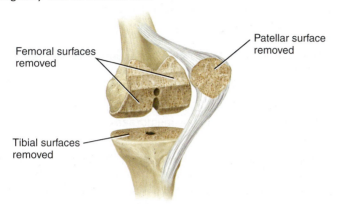

Femoral surfaces removed

Patellar surface removed

Tibial surfaces removed

a. Preparation for total knee replacement

Plastic spacer

Femoral component   Tibial component   Patellar component

b. Components of artificial knee joint prior to implantation

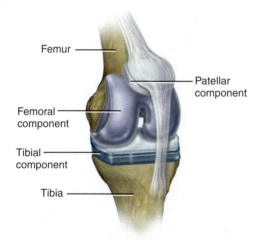

Femur

Patellar component

Femoral component

Tibial component

Tibia

c. Implanted components of a total knee replacement

Arthritis Develops In Joints Due to Wear and Tear

Joints also show wear and tear with age. The following changes in joints occur as you get older:

- Synovial joints decrease production of synovial fluid.
- Cartilage becomes thinner and worn.
- Ligaments shorten and become less flexible.

The result is pain, swelling, and loss of mobility. Age-related changes can begin as early as age 20, and by age 80 most peo-

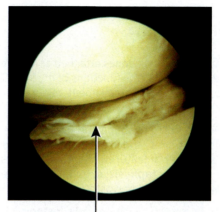

Arthroscopic view of a knee whose medial meniscus show signs of degeneration

ple show signs of degeneration in the knees (**Figure 5.18**), elbows, hips, shoulders, and even the vertebral column.

A degenerative joint disease associated with aging is called **osteoarthritis** (os′-tē-ō-ar-THRĪ-tis). In this condition, the joint cartilage deteriorates because with use, the cartilage thins. As the cartilage disappears from the joint, the bone surfaces have more friction, which leads to irritation and inflammation of joints, muscle weakness, and further wear and abrasion. Osteoarthritis is the most common type of **arthritis**. Osteoarthritis may be treated by injecting hyaluronic acid into the affected joint to improve lubrication. Severe osteoarthritis may require surgery to replace the degenerated joint with an artificial one, such as a knee replacement (see Figure 5.18) or hip replacement. Knee and hip replacements are usually accompanied by months of physical therapy to ensure that the patient regains normal mobility.

> **arthritis** (ar-THRĪ-tis) Inflammation of a joint.

A second type of arthritis, **rheumatoid arthritis**, also leads to deterioration of the joint surfaces. This condition is caused by an abnormal reaction of the immune system known as **autoimmunity**. The immune system appears to attack the tissues of the joint, causing deterioration and inflammation of the articulation. Unlike osteoarthritis, rheumatoid arthritis tends to be a disease condition that starts at a much younger age (often in the teens).

CONCEPT CHECK 🛑 STOP

1. **What** are the two major effects of aging on bone?

2. **What** happens to joints as they age?

Summary

1 The Structure of Bone Controls Function and Growth 114

- The skeletal system consists of all bones attached at joints and the cartilage between the joints. The functions of the skeletal system include support, protection, movement, mineral homeostasis, blood cell production, and triglyceride storage.

- Bones are classified according to shape and include long bones, short bones, flat bones, and irregular bones. A typical long bone has a number of features, including **diaphysis** (shaft), **epiphyses** (ends), **metaphysis**, **articular cartilage**, **medullary (marrow) cavity**, and **periosteum** (covering).

- As shown, bone tissue consists of various cells (**osteogenic cells**, **osteocytes**, **osteoblasts**, and **osteoclasts**) surrounded by a protein and mineral extracellular matrix. Compact bone tissue has arrays of **osteons** with little space between them, while spongy bone tissue consists of **trabeculae** with red marrow–filled spaces.

Microscopic structure of bone • Figure 5.2

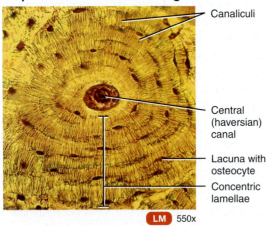

Canaliculi

Central (haversian) canal

Lacuna with osteocyte

Concentric lamellae

LM 550x

Sectional view of an osteon

- Bone formation (ossification) occurs via two processes: intramembranous ossification (in which mesenchyme is replaced by bone tissue) and endochondral ossification (in which cartilage is replaced by bone tissue).

- Bone grows near the ends (epiphyseal plates) and increases in length. Bone is constantly destroyed by osteoclasts and remade by osteoblasts (in a process called bone remodeling). Fractures (broken bones) are healed by bone remodeling and ossification.

- Bone growth depends on the presence of minerals, vitamins, and hormones. Bones store and release calcium, magnesium, and phosphate under the control of **parathyroid hormone** and **calcitonin** and thereby regulate the calcium concentration in the blood.

2 The Axial Skeleton Is Composed of 80 Bones 122

- The skull consists of 8 **cranial** bones and 14 **facial** bones, which protect the brain, provide attachment points for muscles, and have openings for blood vessels and nerves.

- Bones of the skull have air-filled cavities (**paranasal sinuses**) that are lined with mucous membranes and open into the nasal cavity.

- As shown, the bones of the vertebral column include 7 cervical vertebrae, 12 thoracic vertebrae, 5 lumbar vertebrae, the sacrum (5 fused vertebrae), and the coccyx (4 fused vertebrae).

The vertebral column • Figure 5.10a

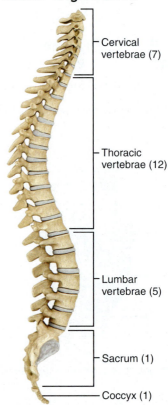

Cervical vertebrae (7)

Thoracic vertebrae (12)

Lumbar vertebrae (5)

Sacrum (1)

Coccyx (1)

- The vertebral column is curved and provides strength, support, and balance. It protects the spinal cord and enables movement.

- Each vertebra consists of a body, a vertebral arch, and seven processes. The vertebrae vary in size and shape, depending on where in the spinal column they are located.

3 Bones of the Upper Body Form the Thorax and Arms 130

- The thoracic bones consist of the **sternum** and 12 pairs of **ribs**, as shown. The thoracic cage protects the heart and lungs and provides attachment points for muscles.

The thorax
• Figure 5.11a

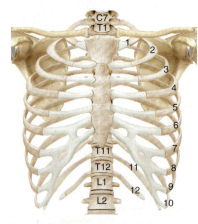

- The shoulder consists of the **clavicle** and **scapula**, which form the pectoral girdle and attach the upper limb to the trunk.

- The 30 bones of the upper limb consist of the humerus, ulna, radius, carpals (8), metacarpals (5), and phalanges (14).

4 Bones of the Lower Body Form the Pelvic Girdle and Legs 132

- The pelvis consists of coxal (hip) bones attached to the sacrum and to each other. Each coxal bone is composed of a fused ilium, ischium, and pubis.

- As shown, the 30 bones of the lower limb include the femur, tibia, fibula, tarsals (8), metatarsals (5), and phalanges (14).

- The medial and longitudinal arches of the foot distribute weight and provide support and leverage.

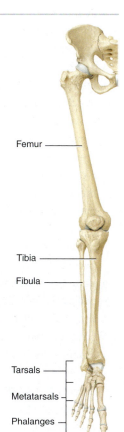

The lower body • Figure 5.12a

5 Articulations Form Where Bones Join Together 136

- An articulation (joint) is where two bones, cartilage and bone, or teeth and bone make contact. The structure of a joint determines its strength and flexibility.

- Structurally, joints are classified as **fibrous**, **cartilaginous**, and **synovial**. Functionally, they can be immovable (**synarthroses**), partially movable (**amphiarthroses**), or freely movable (**diarthroses**).

- Fibrous joints have no cavity and the bones are held together by irregular connective tissue. The types of fibrous joints depend on the thickness of the connective tissue: **sutures** (immovable), **syndesmoses** (slightly movable), or **interosseous membranes** (slightly movable).

- Cartilaginous joints have no cavity, and the bones are held together by cartilage. As shown, they include immovable **synchondroses** with hyaline cartilage and slightly movable **symphyses** with fibrocartilage.

Joints • Figure 5.13

Synchondrosis
(epiphyseal growth plates
connected by cartilage)

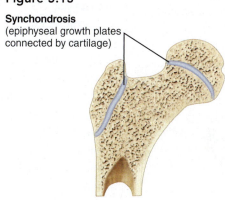

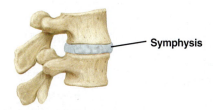

Symphysis

- Synovial joints contain a fluid-filled cavity and are freely movable. Their structures are complex, with several parts, including **articular cartilage**, **articular capsule**, **synovial fluid**, **accessory discs**, and **ligaments**.

- Six types of synovial joints allow various types of movements.

6 Skeletal Structure Changes with Aging 144

- As you age, your bones lose calcium, which can result in demineralization and decreased production of extracellular matrix proteins (for example, collagen). If this is severe, it can lead to osteoporosis, which causes bones to become brittle and susceptible to fractures.

- As you get older, synovial joints produce less fluid, articular cartilage thins, and ligaments become less flexible.

- Severe degeneration of hip and knee joints can be treated using hip and knee replacement; shown are implanted components of a total knee replacement.

Knee replacement • Figure 5.18c

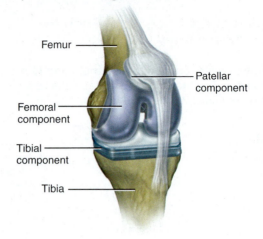

Key Terms

- acetabulum 132
- alveolar process 123
- amphiarthrosis 136
- angular movement 139
- appendicular skeleton 122
- arch 133
- arthritis 145
- articular cartilage 136
- articulation 114
- autoimmunity 145
- axial skeleton 122
- ball-and-socket 141
- body 127
- bone 116
- bone remodeling 119
- bursae 136
- calcaneus 133
- calcification 117
- calcitriol 119
- canaliculus 117
- carotid foramen 123
- carpal 132
- cartilaginous joint 136
- central canal 117
- cervical vertebrae 127
- clavicle 132
- coccyx 127
- compact bone 117
- concentric lamellae 117
- conchae 123
- condylar process 123
- condyloid 141

- coracoid process 132
- coronal suture 126
- cranial bone 123
- cribriform plate 123
- crista galli 123
- diarthrosis 136
- endochondral ossification 117
- ethmoid bone 123
- ethmoidal sinus 123
- external auditory meatus 123
- facial bone 123
- false rib 131
- femur 133
- fibrous capsule 136
- fibrous joint 136
- fibula 133
- flat bone 114
- floating rib 131
- fontanel 127
- foramen 123
- foramen magnum 123
- foramen ovale 123
- frontal bone 123
- glenoid cavity 132
- gliding 138
- haversian canal 117
- hemopoiesis 114
- hinge 140
- humerus 132
- hyoid bone 127
- inferior nasal conchae 126
- interosseous membrane 136

- intramembranous ossification 117
- irregular bone 114
- joint 114
- lacrimal bone 126
- lacuna 117
- lambdoid suture 126
- laminae 127
- ligament 136
- long bone 114
- lumbar vertebrae 127
- mandible 123
- manubrium 131
- mastoid process 123
- maxillae 123
- maxillary sinus 123
- meninges 123
- menopause 144
- mental foramina 126
- mesenchyme 117
- metacarpal 132
- metatarsal 133
- nasal bone 123
- obturator foramina 132
- occipital bone 123
- occipital condyle 123
- optic foramina 123
- osseous tissue 116
- ossification 117
- osteoarthritis 145
- osteoblast 116
- osteoclast 117

- osteocyte 117
- osteogenic cell 117
- osteon 117
- osteoporosis 144
- palatine bone 123
- paranasal sinus 126
- parathyroid gland 119
- parietal bone 123
- patella 133
- pectoral girdle 132
- pedicle 127
- pelvic brim 132
- pelvic girdle 132
- pelvis 132
- phalanges 132
- pivot 141
- plane 140
- pubic symphysis 132
- radius 132
- red bone marrow 114
- reduction 120
- rheumatoid arthritis 145
- rib 130
- rotation 139
- sacrum 127
- saddle 141
- sagittal suture 126
- scapula 132
- sella turcica 123
- short bone 114
- special movement 139
- sphenoid bone 123

Critical and Creative Thinking Questions

1. Mrs. Thompson is 78-year-old woman. Since her 45th birthday, she has been getting measurably shorter every 5 years. Her back has become hunched, she can no longer stand, and she has constant pain in her back and pelvis. What is happening to Mrs. Thompson?

2. Jimmy jumped off the garage roof, onto the hard pavement of his driveway. When his mother arrived, she saw the end of a bone sticking out of his arm. The end of the bone looked jagged but not shattered. (An X-ray later confirmed this finding.) What bone did Jimmy break, and what type of fracture did he have?

3. Mr. Sanders is suffering from a disease in which the levels of all of his blood cells are decreased. What part of his bones is likely affected? Explain.

4. Mrs. Hill has been having trouble walking. She complains of pain and swelling in her right knee. Images of her knee show that her medial meniscus is worn down to less than 5% of its normal thickness. Her doctor recommends that she have a total knee replacement. Provide a simple explanation of what will happen during the surgery.

5. Kelly has just been diagnosed with osteoporosis at age 25. An MRI revealed a tumor on her parathyroid gland, and her blood levels of PTH are two to three times normal. How can she have osteoporosis at such a young age?

What is happening in this picture?

Many expectant couples are excited to see their unborn baby's image in a sonogram. At about 16 weeks, the skeleton looks like this.

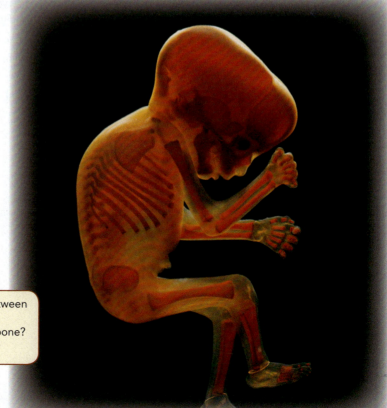

Think Critically
1. What is in the gaps between the bones?
2. How do these gaps eventually become bone?
3. How did the fetus's skull form?

Self-Test

(Check your answers in Appendix C.)

1. Which of the following is a flat bone?

    a. tibia

    b. carpal

    c. scapula

    d. thoracic vertebra

Use this diagram to answer questions 2–4.

2. In which part of the bone shown, would blood cells be produced?

    a. A   b. B   c. C   d. D

3. What is the part shown in D?

    a. diaphysis

    b. articulating cartilage

    c. epiphysis

    d. metaphysis

4. Which part is the diaphysis?

    a. A   b. B   c. C   d. D

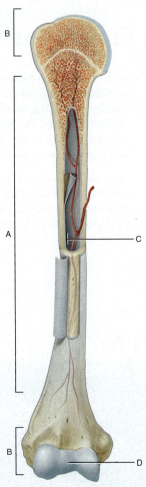

5. An immovable joint that has thin, irregular connective tissue is a _____.

    a. synchondrosis

    b. symphysis

    c. syndesmosis

    d. suture

Use this figure for questions 6–8.

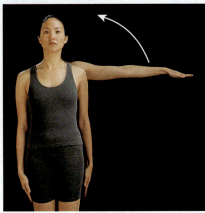

6. The movement indicated by the arrow in this figure is _____.

    a. flexion

    b. extension

    c. opposition

    d. abduction

7. The joint that is moving in this figure involves the _____.

    a. scapula and clavicle

    b. humerus and ulna

    c. humerus and scapula

    d. humerus and tibia

8. What is the opposing movement to the one shown in the figure?

    a. abduction

    b. circumduction

    c. depression

    d. adduction

9. Which type of cell is stimulated by parathyroid hormone?

    a. osteoclast

    b. osteogenic cell

    c. osteoblast

    d. osteocyte

10. The _____ joint is freely movable.

    a. synovial

    b. symphysis

    c. syndesmosis

    d. suture

Use this diagram to answer questions 11–12.

11. The region numbered 3 in the diagram corresponds to the _____ spine.

    a. sacral

    b. cervical

    c. lumbar

    d. thoracic

12. Which of the numbered regions follows the original fetal curve?

    a. 1

    b. 2

    c. 3

    d. 4

13. In the cranium, the spinal cord passes through the _____.

    a. olfactory foramen in the ethmoid bone

    b. foramen ovale in the sphenoid bone

    c. jugular foramen in the temporal bone

    d. foramen magnum in the occipital bone

14. Bursae are typically associated with a _____.

    a. cartilaginous joint

    b. fibrous joint

    c. symphysis

    d. synovial joint

15. What is the most common knee injury?

    a. torn bursa

    b. torn ACL

    c. worn medial meniscus

    d. torn patellar ligament

Use this figure to answer questions 16–17.

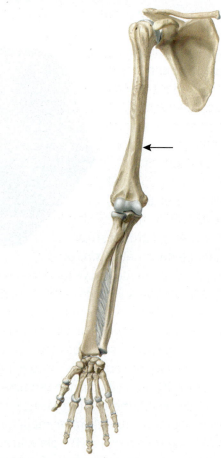

16. The bone indicated by the arrow is which type of bone?

    a. Flat bone

    b. Short bone

    c. Long bone

    d. Irregular bone

17. What is the bone indicated by the arrow?

    a. pelvis

    b. humerus

    c. clavicle

    d. phalanges

THE PLANNER ✔

Review your Chapter Planner on the chapter opener and check off your completed work.

The Muscular System

Body-building and body-sculpting both build up muscles in selected areas of the body, resulting in a generally pleasing shape or appearance. Muscle makes up almost 40–50% of the total mass of your body. Many of these primary engines of motion are intimately linked to the skeletal system, but there are many muscles that you don't even think about. For example, smooth muscle tissue forms the walls of your blood vessels, gastrointestinal tract, and urinary bladder, helps control blood pressure, circulate blood, move food through the digestive tract, and receive and store urine for periods of time. Your heart, made of cardiac muscle tissue, works constantly throughout your life, without a rest.

Muscle is one of the few tissue types that you can actually modify, but how? The answer is through rigorous and regular exercise. You can change the size and type of muscle fibers (cells), thereby increasing strength and endurance. In fact, your muscle mass readily adjusts to your lifestyle and exercise regimen. People with active lifestyles tend to have more muscle mass than sedentary individuals. Patients on long-term bed rest tend to lose muscle mass quickly. Maintaining your muscle health is important to maintaining your overall health.

Let's take a closer look at these remarkable tissues.

NATIONAL GEOGRAPHIC

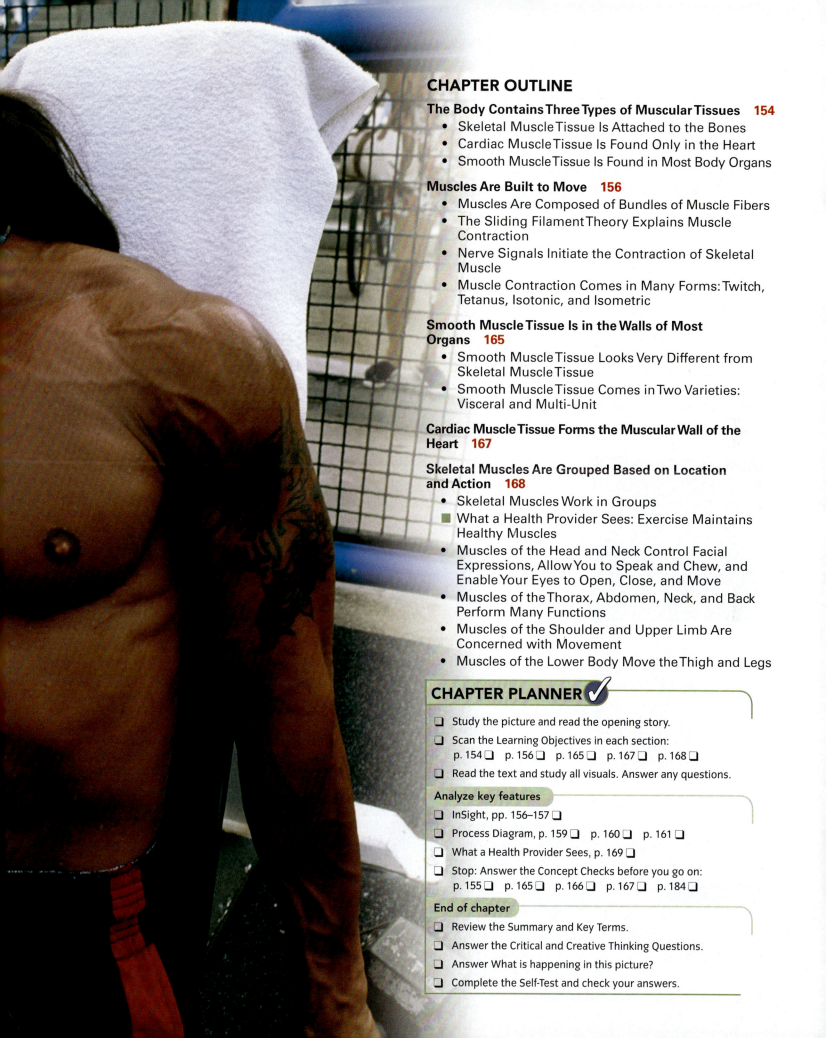

CHAPTER OUTLINE

CHAPTER PLANNER ✔

- ❏ Study the picture and read the opening story.
- ❏ Scan the Learning Objectives in each section:
  p. 154 ❏   p. 156 ❏   p. 165 ❏   p. 167 ❏   p. 168 ❏
- ❏ Read the text and study all visuals. Answer any questions.

Analyze key features

- ❏ InSight, pp. 156–157 ❏
- ❏ Process Diagram, p. 159 ❏   p. 160 ❏   p. 161 ❏
- ❏ What a Health Provider Sees, p. 169 ❏
- ❏ Stop: Answer the Concept Checks before you go on:
  p. 155 ❏   p. 165 ❏   p. 166 ❏   p. 167 ❏   p. 184 ❏

End of chapter

- ❏ Review the Summary and Key Terms.
- ❏ Answer the Critical and Creative Thinking Questions.
- ❏ Answer What is happening in this picture?
- ❏ Complete the Self-Test and check your answers.

The Body Contains Three Types of Muscular Tissues

LEARNING OBJECTIVES

1. **Describe** the types of muscular tissue.
2. **List** the various locations of the types of muscular tissue.
3. **Outline** the functions of muscular tissue.

M uscle accounts for 40–50 percent of your body mass. Muscle allows the body to move around in its environment or the organs to move within the body. Compared to other tissues in the body, muscular tissue has a limited capacity to divide and regenerate. There are three types of muscular tissue (**Figure 6.1**): skeletal, cardiac, and smooth.

Skeletal Muscle Tissue Is Attached to the Bones

There are about 700 **skeletal muscles** in your body. Skeletal muscles produce body movements, stabilize the skeleton, and produce much of the heat that helps maintain body temperature. Movements of skeletal muscle can be *voluntary*—you can knowingly contract and relax them—or *involuntary*. Skeletal muscle tissue is banded, or **striated**; the striations can be seen only under a microscope. The ability of skeletal muscle to regenerate is somewhat limited and involves satellite cells, a type of stem cell that becomes activated to form new skeletal muscle cells.

Types of muscular tissue • Figure 6.1

Muscular tissues consist of three types.

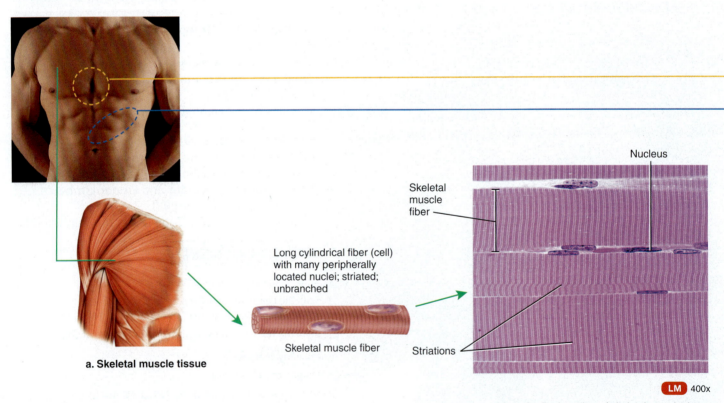

a. Skeletal muscle tissue

Long cylindrical fiber (cell) with many peripherally located nuclei; striated; unbranched

Skeletal muscle fiber

Nucleus

Skeletal muscle fiber

Striations

LM 400x

Longitudinal section of skeletal muscle tissue

Ask Yourself

Which type of muscle tissue is composed of striated cells that are interconnected by intercalated discs?
a. Smooth muscle
b. Skeletal muscle
c. Cardiac muscle

Cardiac Muscle Tissue Is Found Only in the Heart

Cardiac muscle tissue makes up the walls of the heart and generates the force necessary to pump your blood. Cardiac muscle contractions are *involuntary*: You don't think about contracting and relaxing this muscle. Unlike most other muscle tissue, cardiac muscle tissue has the ability to contract without the assistance of the nervous system. Like skeletal muscle, cardiac muscle is striated. The regeneration ability of cardiac muscle is minimal.

Smooth Muscle Tissue Is Found in Most Body Organs

Smooth muscle tissue forms the walls of hollow organs such as blood vessels, airways, the stomach, the intestines, and the uterus. The smooth muscle of these organs helps to store and move substances within the body and regu-

lates organ volume. Smooth muscle cells are considerably smaller than other muscle cells and are not striated. Like cardiac muscle, the contraction and relaxation of smooth muscle is *involuntary*. Of the three types of musclar tissue, smooth muscle tissue regenerates most easily, most likely because this type of muscle has a less complex structure than that of the striated cardiac or skeletal muscle tissues.

CONCEPT CHECK

1. **Which** type of muscular tissue is striated and voluntary?

2. **What** type of muscular tissue will an obstetrician be cutting through to deliver a baby via cesarean section?

3. **Which** types of muscular tissue play a role in generating body heat?

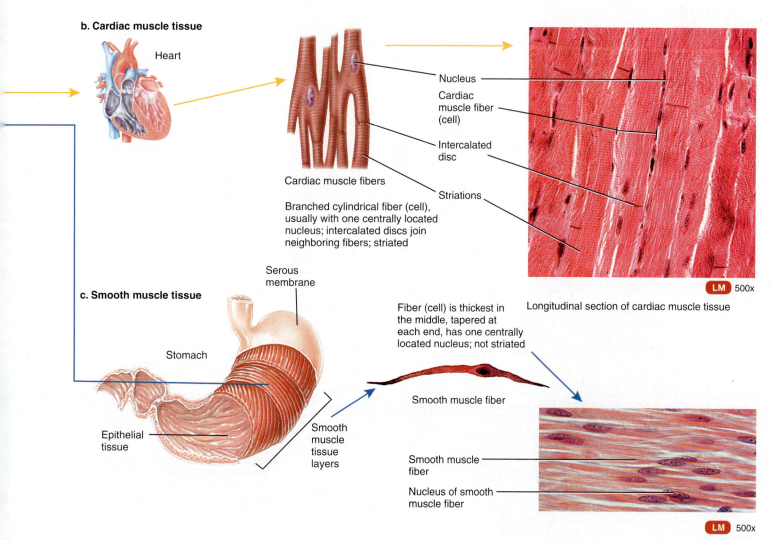

b. Cardiac muscle tissue

Heart

Cardiac muscle fibers

Branched cylindrical fiber (cell), usually with one centrally located nucleus; intercalated discs join neighboring fibers; striated

Nucleus

Cardiac muscle fiber (cell)

Intercalated disc

Striations

LM 500x

Longitudinal section of cardiac muscle tissue

c. Smooth muscle tissue

Serous membrane

Stomach

Epithelial tissue

Smooth muscle tissue layers

Fiber (cell) is thickest in the middle, tapered at each end, has one centrally located nucleus; not striated

Smooth muscle fiber

Smooth muscle fiber

Nucleus of smooth muscle fiber

LM 500x

Longitudinal section of smooth muscle tissue

Muscles Are Built to Move

LEARNING OBJECTIVES

1. **Describe** the structure of skeletal muscle tissue.
2. **Explain** how skeletal muscle tissue shortens.
3. **Outline** the sequence of events of skeletal muscle contraction (excitation–contraction coupling).
4. **Associate** the type of muscle fiber (cell) with the source of energy in given situations.

The structure of a skeletal muscle and its connective tissue coverings: A macroscopic view • **Figure 6.2**

The cylindrical muscle fibers (cells) contain **myofibrils**, consisting of overlapping thin filaments and thick filaments that are arranged into **sarcomeres**. This arrangement of overlapping filaments allows the muscle to shorten, thereby generating force and movements.

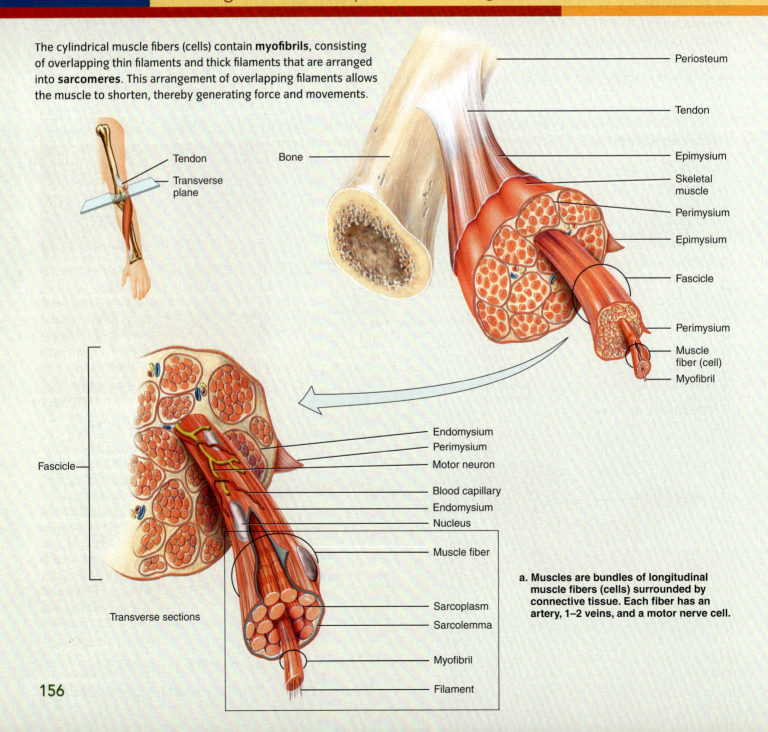

Tendon
Transverse plane

Bone

Periosteum
Tendon
Epimysium
Skeletal muscle
Perimysium
Epimysium
Fascicle
Perimysium
Muscle fiber (cell)
Myofibril

Fascicle

Endomysium
Perimysium
Motor neuron
Blood capillary
Endomysium
Nucleus
Muscle fiber
Sarcoplasm
Sarcolemma
Myofibril
Filament

Transverse sections

a. Muscles are bundles of longitudinal muscle fibers (cells) surrounded by connective tissue. Each fiber has an artery, 1–2 veins, and a motor nerve cell.

S keletal muscles allow us to move. They can shorten (contract), thereby generating the force for movement. Muscles can also relax and return to their original length. The specific structure and arrangement of skeletal muscle allows its muscle cells to work the way they do. Let's take a closer look.

Muscles Are Composed of Bundles of Muscle Fibers

Muscles consist of bundles of muscle fibers or muscle cells wrapped in connective tissue (**Figure 6.2**). (Note that we use the terms *muscle fibers* and *muscle cells* interchangeably; they mean the same thing.) Each wrapped bundle of fibers is called a **fascicle**. The individual muscle cells are

fascicle (FAS-i-kul) A small bundle or cluster of skeletal muscle fibers.

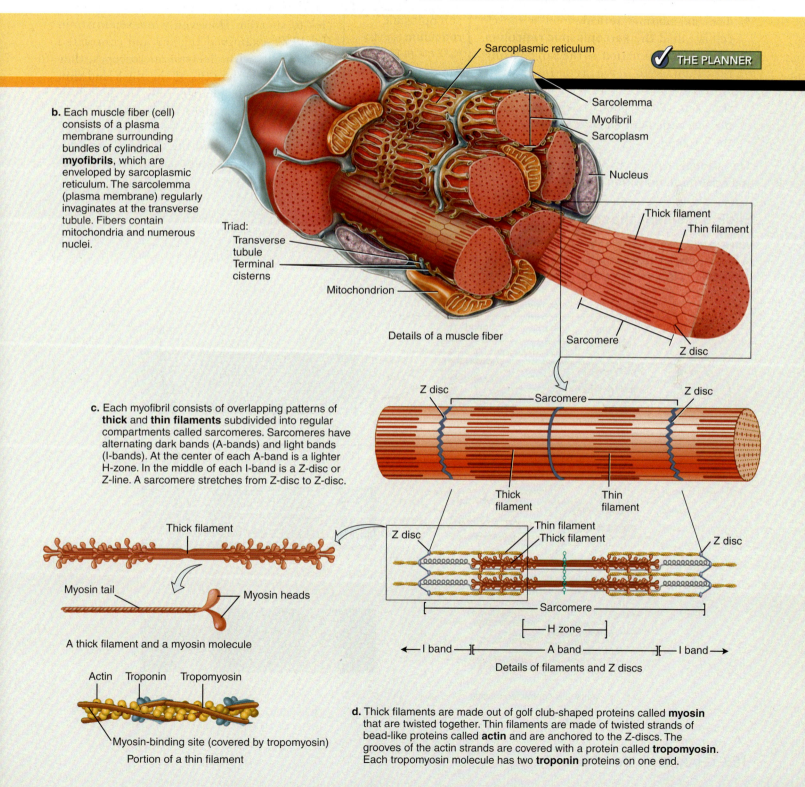

THE PLANNER

b. Each muscle fiber (cell) consists of a plasma membrane surrounding bundles of cylindrical **myofibrils**, which are enveloped by sarcoplasmic reticulum. The sarcolemma (plasma membrane) regularly invaginates at the transverse tubule. Fibers contain mitochondria and numerous nuclei.

Sarcoplasmic reticulum

Sarcolemma
Myofibril
Sarcoplasm

Nucleus

Thick filament
Thin filament

Triad:
Transverse tubule
Terminal cisterns

Mitochondrion

Details of a muscle fiber

Sarcomere
Z disc

c. Each myofibril consists of overlapping patterns of **thick** and **thin filaments** subdivided into regular compartments called sarcomeres. Sarcomeres have alternating dark bands (A-bands) and light bands (I-bands). At the center of each A-band is a lighter H-zone. In the middle of each I-band is a Z-disc or Z-line. A sarcomere stretches from Z-disc to Z-disc.

Z disc
Sarcomere
Z disc

Thick filament
Thin filament

Thick filament

Myosin tail
Myosin heads

A thick filament and a myosin molecule

Z disc
Thin filament
Thick filament
Z disc

Sarcomere
H zone
I band
A band
I band

Details of filaments and Z discs

Actin   Troponin   Tropomyosin

Myosin-binding site (covered by tropomyosin)

Portion of a thin filament

d. Thick filaments are made out of golf club-shaped proteins called **myosin** that are twisted together. Thin filaments are made of twisted strands of bead-like proteins called **actin** and are anchored to the Z-discs. The grooves of the actin strands are covered with a protein called **tropomyosin**. Each tropomyosin molecule has two **troponin** proteins on one end.

cylindrical. Like other cells, muscle fibers have mitochondria to provide energy and a plasma membrane to separate the inside of the cell from the outside environment. However, skeletal muscle cells differ from other cells in several ways:

- A muscle cell has more than one nucleus.

- A muscle cell's plasma membrane invaginates regularly into the deep parts of the muscle fiber to form a **transverse tubule**, or **T-tubule**.

- The endoplasmic reticulum of a muscle cell is called the **sarcoplasmic reticulum**. The primary function of the sarcoplasmic reticulum is to serve as a reservoir to store calcium ions between muscle contractions. The regularly structured sarcoplasmic reticulum ends as small sacs near the T-tubules.

> **sarcoplasmic reticulum** (sar'-kō-PLAZ-mik re-TIK-ū-lum) A network of saccules and tubes surrounding myofibrils of a muscle fiber.

The arrangement of **myofilaments** (threadlike proteins) and other proteins within the muscle fibers allows them to shorten. The proteins and the surrounding connective tissue are stretchable and allow muscle cells to lengthen. Let's take a closer look at this arrangement of the myofilaments, which allows muscle cells to both contract and relax.

The Sliding Filament Theory Explains Muscle Contraction

Early on, scientists thought that muscle cells fold when they contract, thereby causing muscles to shorten. However, when scientists in the 1950s examined relaxed and contracted muscle cells with electron microscopes, they found that the lengths of the thick and thin filaments do not change; only the length of the sarcomere changes. Physiologists A. L. Hodgkin and H. E. Huxley explained these findings with a hypothesis called the *sliding filament theory*, in which **thin filaments** slide past the **thick filaments (Figure 6.3)**.

Sliding filament theory of muscle contraction • Figure 6.3

When muscles contract, thin filaments slide along the thick filaments toward the M-line, thereby shortening the sarcomere. When the thick filaments butt up against the Z discs, the sarcomere cannot shorten anymore.

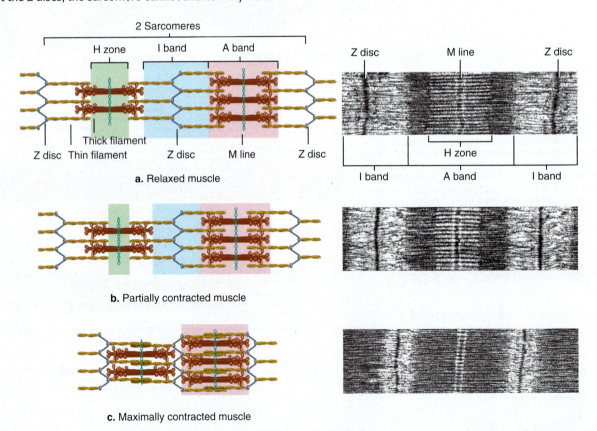

a. Relaxed muscle

b. Partially contracted muscle

c. Maximally contracted muscle

The contraction (cross-bridge) cycle • Figure 6.4

✓ THE PLANNER

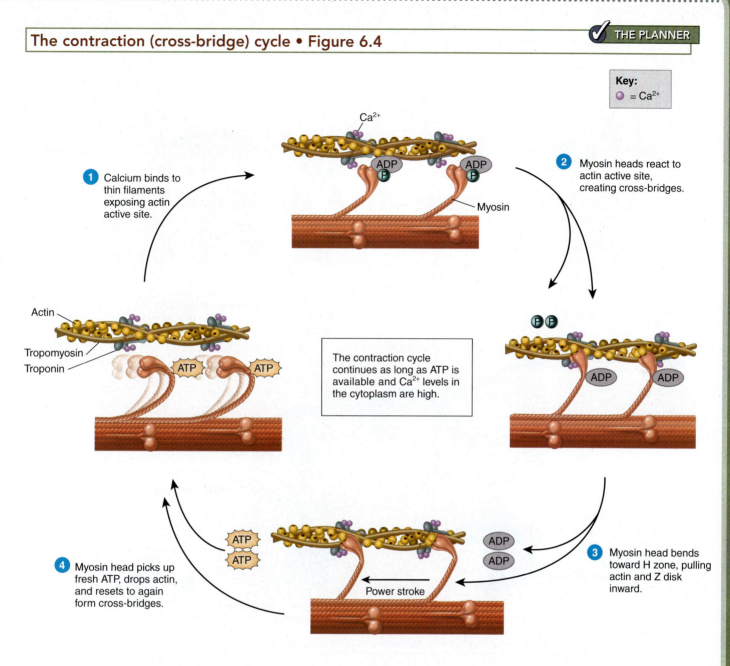

Key:
● = Ca^{2+}

1 Calcium binds to thin filaments exposing actin active site.

Ca^{2+}

ADP P

ADP P

Myosin

2 Myosin heads react to actin active site, creating cross-bridges.

P P

ADP

ADP

Actin

Tropomyosin

Troponin

ATP

ATP

The contraction cycle continues as long as ATP is available and Ca^{2+} levels in the cytoplasm are high.

ATP

ATP

4 Myosin head picks up fresh ATP, drops actin, and resets to again form cross-bridges.

Power stroke

ADP

ADP

3 Myosin head bends toward H zone, pulling actin and Z disk inward.

The myosin heads grab onto the binding sites of the actin filaments to form cross-bridges and then pull the actin filaments along in a repeating process called the cross-bridge cycle or contraction cycle (**Figure 6.4**). This contraction cycle requires calcium ions (Ca^{2+}) and ATP. The contraction cycle repeats as long as ATP and Ca^{2+} are available in the sarcoplasm. At any one instant, some of the myosin heads are attached to actin, forming cross-bridges, and generating force, and other myosin heads are detached from actin and getting ready to bind again. How is muscle contraction initiated? How is it stopped? This process, called *excitation–contraction coupling*, begins with a nerve impulse. Let's take a closer look.

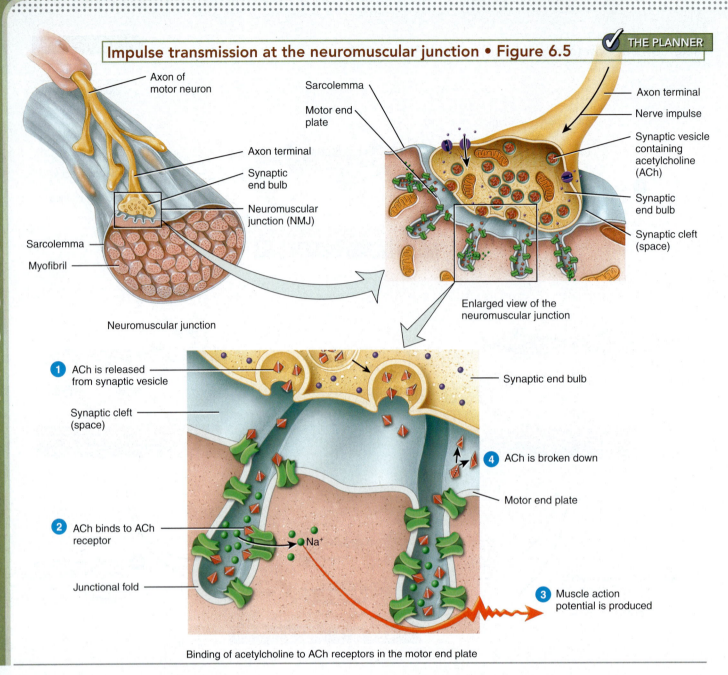

Impulse transmission at the neuromuscular junction • Figure 6.5

✔ THE PLANNER

- Axon of motor neuron
- Axon terminal
- Synaptic end bulb
- Neuromuscular junction (NMJ)
- Sarcolemma
- Myofibril

Neuromuscular junction

- Sarcolemma
- Motor end plate
- Axon terminal
- Nerve impulse
- Synaptic vesicle containing acetylcholine (ACh)
- Synaptic end bulb
- Synaptic cleft (space)

Enlarged view of the neuromuscular junction

1 ACh is released from synaptic vesicle

- Synaptic end bulb

Synaptic cleft (space)

4 ACh is broken down

Motor end plate

2 ACh binds to ACh receptor

Na⁺

Junctional fold

3 Muscle action potential is produced

Binding of acetylcholine to ACh receptors in the motor end plate

Nerve Signals Initiate the Contraction of Skeletal Muscle

As mentioned earlier, skeletal muscle contraction can be voluntary. If you want to pick up a pencil, your brain initiates that voluntary muscle contraction using a nerve impulse. So how is the nerve impulse transmitted from the nerve cell to the muscle cell? At a specialized meeting place between the two cells called a **neuromuscular junction**, the neuron sends a chemical message (acetylcholine [ACh]) that starts an electrical impulse in the muscle cell (**Figure 6.5**).

neuromuscular junction (noo-rō-MUS-kū-lar) A synapse (functional junction) between the axon terminals of a motor neuron and the sarcolemma of a muscle fiber.

This impulse (muscle action potential) reverses the normal electrical state of the membrane (**depolarization**) and initiates contraction along the length of the muscle fiber.

The muscle action potential passes along the cell membrane and down into the T-tubules (**Figure 6.6**). The action potential causes calcium ions to be released through calcium channels from their storage site inside the sarcoplasmic reticulum. The rise in calcium ion level in the sarcoplasm (cytoplasm) initiates the contraction cycle. The muscle continues to contract until enough calcium ions get pumped back inside the sarcoplasmic reticulum. Once the calcium levels in the cytoplasm return to normal, the muscle relaxes. This process of returning to the resting state is referred to as **repolarization**.

Excitation–contraction coupling in skeletal muscle • Figure 6.6

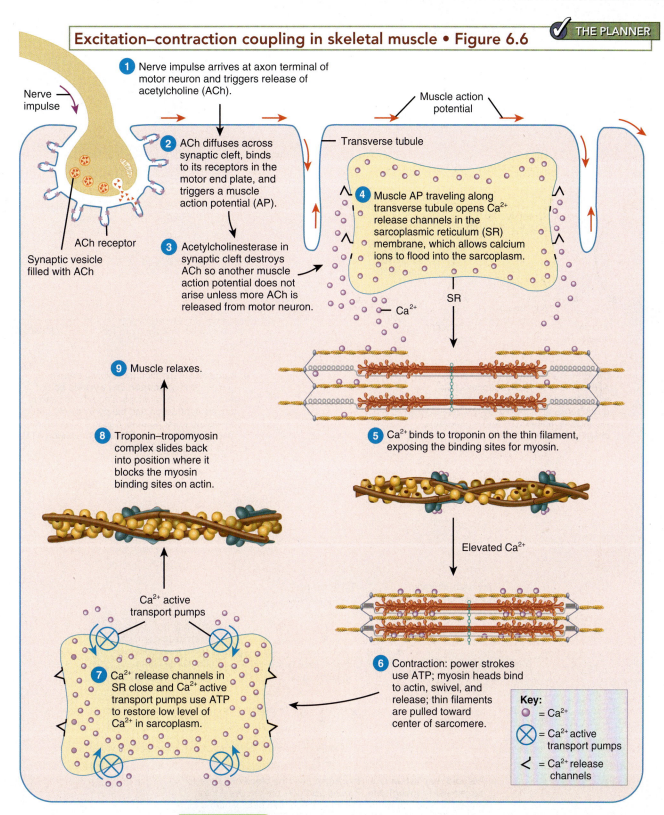

1 Nerve impulse arrives at axon terminal of motor neuron and triggers release of acetylcholine (ACh).

Nerve impulse

2 ACh diffuses across synaptic cleft, binds to its receptors in the motor end plate, and triggers a muscle action potential (AP).

ACh receptor

Synaptic vesicle filled with ACh

3 Acetylcholinesterase in synaptic cleft destroys ACh so another muscle action potential does not arise unless more ACh is released from motor neuron.

Muscle action potential

Transverse tubule

4 Muscle AP traveling along transverse tubule opens Ca^{2+} release channels in the sarcoplasmic reticulum (SR) membrane, which allows calcium ions to flood into the sarcoplasm.

Ca^{2+}

SR

9 Muscle relaxes.

8 Troponin–tropomyosin complex slides back into position where it blocks the myosin binding sites on actin.

5 Ca^{2+} binds to troponin on the thin filament, exposing the binding sites for myosin.

Elevated Ca^{2+}

Ca^{2+} active transport pumps

7 Ca^{2+} release channels in SR close and Ca^{2+} active transport pumps use ATP to restore low level of Ca^{2+} in sarcoplasm.

6 Contraction: power strokes use ATP; myosin heads bind to actin, swivel, and release; thin filaments are pulled toward center of sarcomere.

Key:
⬤ = Ca^{2+}
⊗ = Ca^{2+} active transport pumps
< = Ca^{2+} release channels

Put It Together

1. What makes the muscle contract in step 6?
2. What kind of contraction is shown in step 6? (a) a partial contraction (b) a maximum contraction
3. Why does the muscle relax in step 9?
4. Which numbered steps in this figure are part of the contraction cycle?

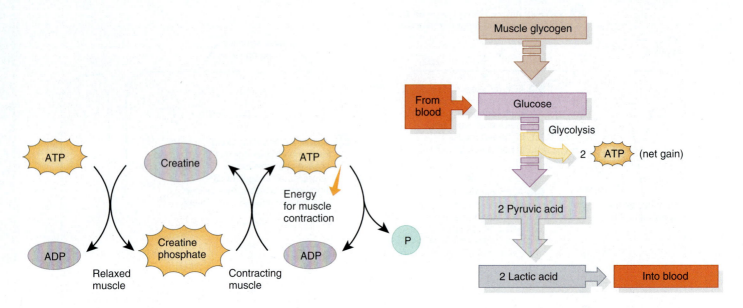

a. **Creatine-phosphate** transfers phosphate from ADP to make ATP. This process gets exhausted within 15 seconds, so it is good for quick bursts of muscle activity. Once at rest, the muscle makes ATP and regenerates creatine phosphate.

15 seconds

b. Anaerobic respiration (glycolysis) breaks glucose from **glycogen** into **pyruvic acid** through a series of steps. The ATP is used in muscle contraction and the pyruvic acid gets converted to **lactic acid**. Lactic acid is what makes muscles sore. Glycolysis provides ATP for a further 30–60 seconds.

30-60 seconds

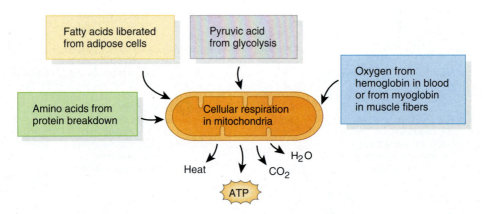

c. **Aerobic respiration** requires oxygen from the blood or a muscle protein called **myoglobin**. Pyruvic acid gets broken down into carbon dioxide and water. The energy released gets captured to make ATP and some gets lost as heat. Aerobic respiration can also use amino acids and fatty acids to make ATP. This process provides more ATP than the others and can last a long time (e.g., minutes to hours of muscle activity).

1-2⁺ hours

Muscle contraction requires lots of energy. The energy comes from ATP, but stores of ATP can be exhausted within seconds. The body must be able to make ATP quickly to keep up with the energy demand of the working muscle. ATP can come from three sources: **creatine phosphate**, anaerobic respiration or **glycolysis**, and aerobic respiration (**Figure 6.7**). **Anaerobic respiration** is respiration that does not require oxygen; oxygen is required for **aerobic respiration**.

> **glycolysis** (glī-KOL-i-sis) A series of anaerobic chemical reactions in the cytosol of a cell in which a molecule of glucose is split into two molecules of pyruvic acid, with the net production of two ATPs.

When muscles can no longer contract, they become fatigued. Although muscle fatigue can come from a decrease in the release of calcium (calcium depletion), it most often results from insufficient supplies of ATP (due to depleted amounts of creatine phosphate and **glycogen**, inadequate delivery of oxygen to the muscle, or build-up of **lactic acid**). Sometimes, as ATP levels decrease and lactic acid levels increase, muscle cramping may also occur because muscles cells cannot relax. This is because the active transport process that moves calcium out of the cell also requires ATP.

After strenuous exercise, the oxygen consumption by working muscles continues to be elevated. This elevated oxygen consumption is referred to as oxygen debt. Following vigorous physical activity, the heart rate and breathing rate generally remain elevated for a period of time to move more oxygen into the tissue, which helps "repay" this oxygen debt. The excess oxygen delivered to the tissues converts lactic acid back to glucose and glycogen, remakes creatine phosphate, and replaces the oxygen lost from myoglobin during aerobic respiration. (**Myoglobin** is a muscle protein that stores oxygen within the muscle cell.) Oxygen is also used for the following processes:

- Increased body heat speeds up other reactions in the body that use ATP, so oxygen is used to replenish these stores.
- Oxygen feeds hard-working heart and breathing muscles, which use more ATP.
- Oxygen replenishes ATP used in tissue repair processes.

Muscle Contraction Comes in Many Forms: Twitch, Tetanus, Isotonic, and Isometric

There are two categories of muscle contraction: twitch and tetanus (contracture). A **twitch** is a single muscle contraction in response to a single nerve impulse. It consists of three periods: the **latent period**, **contraction period**, and **relaxation period**. Each period is associated with the events of excitation–contraction coupling (**Figure 6.8**).

Phases of a muscle twitch • Figure 6.8

A **myogram** records the properties of a twitch in response to a single stimulus (arrow), starting with the latent period, moving to the contraction period, and then to the relaxation period. Notice that the twitch lasts only about 50 ms, or 0.05 s.

1. **Latent period.** There is a time delay during which the muscle action potential travels down the muscle and calcium ions get released from the sarcoplasmic reticulum.
2. **Contraction period.** Contraction cycles generate force.
3. **Relaxation period.** Ionized calcium levels in the cell return to normal and contraction cycles decrease.

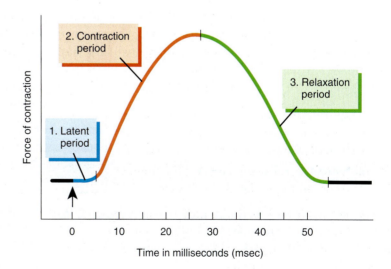

Mechanisms for controlling muscle twitch and tetanus • Figure 6.9

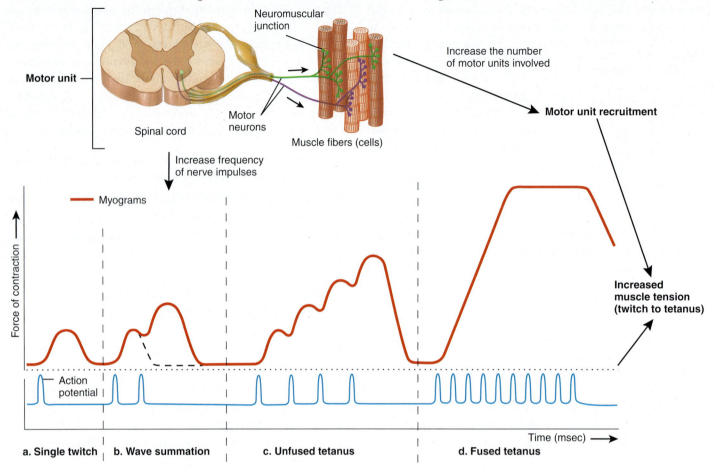

Neuromuscular junction

Motor unit

Spinal cord

Motor neurons

Muscle fibers (cells)

Increase the number of motor units involved

Motor unit recruitment

Increase frequency of nerve impulses

— Myograms

Force of contraction

Increased muscle tension (twitch to tetanus)

Action potential

Time (msec) →

a. Single twitch
b. Wave summation
c. Unfused tetanus
d. Fused tetanus

Effects of summation on contraction:
a. **Single twitch**
b. **Wave summation** — when subsequent stimuli arrive before the first wave is finished, the tension is higher
c. **Unfused tetanus** — continued wave summations add together, but each stimulus has a partial relaxation (20-30 impulses/s)
d. **Fused tetanus** — contraction force is steady and sustained (80-100 impulses/s)

As each subsequent stimulus arrives before the previous wave returns to normal, the levels of cytoplasmic Ca^{2+} accumulate and produce greater contractions up to a maximum when all motor units are recruited

A twitch lasts only a short time and does not produce much tension. In contrast, **tetanus** is a longer, sustained contraction that develops more force than a twitch. There are two ways to increase the amount of tension and transition from a twitch to tetanus (**Figure 6.9**):

- **Temporal summation** (or **wave summation**) increases the frequency of nerve impulses (that is, the number of impulses per second).

- **Motor unit** recruitment increases the number of motor units stimulated at one time.

Most muscles use both methods to control and maintain sustained contractions. Sustained contractions are also maintained by asynchronous (unsynchronized) stimulation of various motor units.

Muscle contraction can be isotonic or isometric. In **isotonic contractions**, muscle length changes. For example,

when you move your forearm toward your arm, the biceps muscle shortens, moving the forearm. In contrast, **isometric contraction** results in a change of the tension that is generated but does not cause motion. For example, when you stand, your back muscles must contract continually to maintain your upright posture.

Even when a whole muscle is not contracting, a small number of its motor units are involuntarily activated to produce a sustained contraction of their muscle fibers. This process, which results in **muscle tone**, occurs in both skeletal and smooth muscle. For example, the smooth muscles in your arteries and veins always have some degree of tone to maintain your blood pressure.

motor unit A motor neuron together with the muscle fibers it stimulates.

muscle tone A state of continued partial contraction.

Based on the type of metabolism and the rate at which tension develops, skeletal muscle fibers can be classified into three types:

- **Slow oxidative (SO) fibers** have small diameters, contain many large mitochondria, and appear red because they contain large amounts of myoglobin. They make ATP mainly by aerobic respiration, develop tension relatively slowly, resist fatigue, and are capable of prolonged, sustained contractions. Such muscle fibers are found in large numbers in the muscles associated with posture. These are also the most important muscle fiber type for long-distance runners, who require endurance.

- **Fast oxidative-glycolytic (FOG) fibers** are intermediate in diameter. They have large amounts of myoglobin and therefore appear red. They make ATP through both aerobic respiration and glycolysis, so they are moderately resistant to fatigue. Most of the muscle fibers associated with large motor skills are of this type. They contract and relax faster than SO fibers but less quickly than the FG fiber type, providing an intermediate level of contraction sustainability. Middle-distance runners rely heavily on this type of muscle fiber.

- **Fast glycolytic (FG) fibers** are large and white, with considerable amounts of glycogen, but no myoglobin. Because they make ATP mainly through glycolysis, they fatigue rapidly. However, they contract and relax quickly, providing a short surge of power. Many of these fibers are found in areas where fine motor skills occur. Sprinters need lots of this muscle fiber type, as these muscles allow rapid reactions and short bursts of speed.

Most skeletal muscles have all three types of fibers. About 50 percent of the fiber composition is SO fibers. Depending on the muscle, the rest of the composition varies. For example, arm and shoulder muscles have high proportions of FG fibers, while leg muscles have large numbers of both SO and FOG fibers. Regardless of the overall mixture of muscle fibers within any given muscle, a single motor unit consists of identical fiber types. Therefore, there are SO motor units, FOG motor units, and FG motor units. These motor units are recruited in specific orders, depending on whether speed, force, or duration of contraction is required in the muscle action.

CONCEPT CHECK

1. **Which** skeletal muscle organelle stores and releases calcium?

2. **What** does ATP do in the contraction, or cross-bridge, cycle?

3. **Through** what steps is the action potential stimulus translated into sliding filament contractions?

4. **Which** type of muscle fibers would dominate the movements of your arm muscles in performing short but powerful contractions, as in throwing a ball or swinging a tennis racket?

Smooth Muscle Tissue Is in the Walls of Most Organs

LEARNING OBJECTIVES

1. **Explain** the structure of smooth muscle.
2. **Differentiate** the classes of smooth muscle and describe where they are found.

Smooth muscles are found in the walls of all hollow organs, such as blood vessels, the stomach, and the intestines. (They are also found in the skin, attached to hair follicles.) Because of their unique structure, they are capable of being stretched much more than skeletal or cardiac muscle and can maintain steady levels of contractions for long periods of time.

Smooth Muscle Tissue Looks Very Different from Skeletal Muscle Tissue

Smooth muscle cells are much smaller than cells of other muscle types, and they are spindle-shaped. Smooth muscle differs from skeletal muscle in a number of ways:

- Each smooth muscle cell has a single nucleus and no striations because the thin and thick filaments are not arranged regularly.

- Because smooth muscle cells are small and thin, there are no transverse tubules.

- Each cell contains **intermediate filaments** and **dense bodies**, which are equivalent to the Z-discs in

Types and structure of smooth muscle cells • Figure 6.10

Whether arranged as single- or multi-unit smooth muscle tissue, the structure of individual smooth muscle cells is the same.

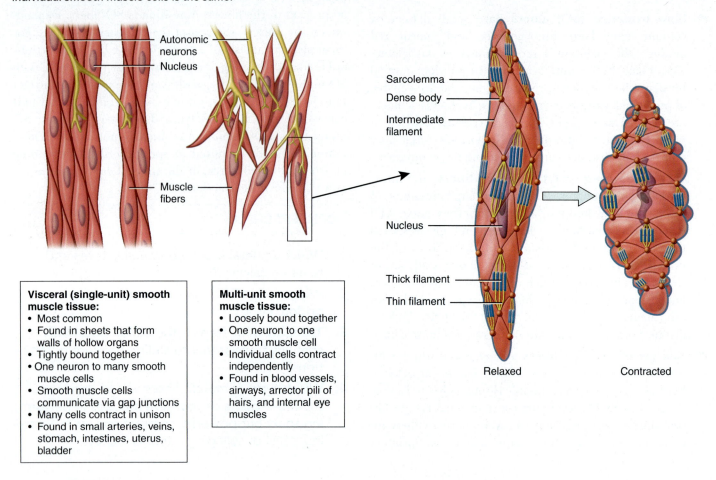

Autonomic neurons
Nucleus
Muscle fibers

Sarcolemma
Dense body
Intermediate filament
Nucleus
Thick filament
Thin filament

Relaxed   Contracted

Visceral (single-unit) smooth muscle tissue:
- Most common
- Found in sheets that form walls of hollow organs
- Tightly bound together
- One neuron to many smooth muscle cells
- Smooth muscle cells communicate via gap junctions
- Many cells contract in unison
- Found in small arteries, veins, stomach, intestines, uterus, bladder

Multi-unit smooth muscle tissue:
- Loosely bound together
- One neuron to one smooth muscle cell
- Individual cells contract independently
- Found in blood vessels, airways, arrector pili of hairs, and internal eye muscles

skeletal muscle. Some of the dense bodies are within the cytoplasm, and others are attached to the cell membrane. This arrangement gives the intermediate filament network a net-like appearance.

- Smooth muscle cells do not have a well-developed sarcoplasmic reticulum.

Smooth Muscle Tissue Comes in Two Varieties: Visceral and Multi-Unit

There are two types of smooth muscle tissue: **visceral (single-unit) muscle tissue** and **multi-unit smooth muscle tissue** (**Figure 6.10**). The smooth muscle cells of both types have the structural features listed above, but the appearance of the contraction differs between the two. Visceral smooth muscle cells are interconnected by gap junctions and function as a group to produce a wave-like contraction known as *peristalsis*. Multi-unit smooth muscle does not contain gap junctions, so the cells function independently, allowing more pinpoint control of the contraction.

Smooth muscles contract or relax involuntarily in response to stimulation by neurons of the autonomic (involuntary) nervous system. Smooth muscles also respond to hormones and local events, such as changes in pH, carbon dioxide levels, temperature, and ion concentrations.

CONCEPT CHECK STOP

1. **Why** do smooth muscle cells lack striations?

2. **Which** type of smooth muscle tissue would you find in the walls of your stomach?

Cardiac Muscle Tissue Forms the Muscular Wall of the Heart

LEARNING OBJECTIVES

1. **Describe** the structure of cardiac muscle tissue.

2. **Compare** cardiac muscle, skeletal muscle, and smooth muscle.

The heart is composed mostly of cardiac muscle tissue. This tissue has some of the characteristics seen in skeletal muscle and some of those seen in smooth muscle tissue, creating a unique type of muscular tissue.

Cardiac muscle has a structure that is intermediate between skeletal muscle and smooth muscle (**Figure 6.11**). Cardiac cells are larger than smooth muscle cells but smaller than skeletal muscle cells. Like skeletal muscle, they have cylindrical myofibrils, sarcomeres, T-tubules, and a fairly well-developed sarcoplasmic reticulum. Like single-unit smooth muscle cells, cardiac cells communicate with one another through gap junctions in the

intercalated disc
(in-TER-ka-lăt-ed)
An irregular transverse thickening of sarcolemma where two cardiac cells come together. The intercalated disc is composed of desmosomes, which hold cardiac muscle fibers together, and gap junctions, which aid in conduction of muscle action potentials from one fiber to the next.

intercalated discs between them. Cardiac cells are unique in that they are often branched.

Cardiac cells respond to signals from specialized heart muscle cells called *pacemaker cells*, which in turn are influenced by the autonomic nervous system. Pacemaker cells are **autorhythmic**; they can contract on their own, without nervous system stimulation. Via pacemaker cells, the autonomic nervous system can speed up or slow down the contraction of cardiac cells as needed. Unlike other types of muscle cells, cardiac cells can produce only twitch contractions, and they do so at a rate of 70 to 80 contractions per minute. The actions of cardiac muscle are discussed more fully in Chapter 11.

CONCEPT CHECK STOP

1. **What** is the function of intercalated discs?

2. **What** is the source of action potentials in cardiac muscle? in skeletal muscle?

Structure of cardiac muscle cells • Figure 6.11

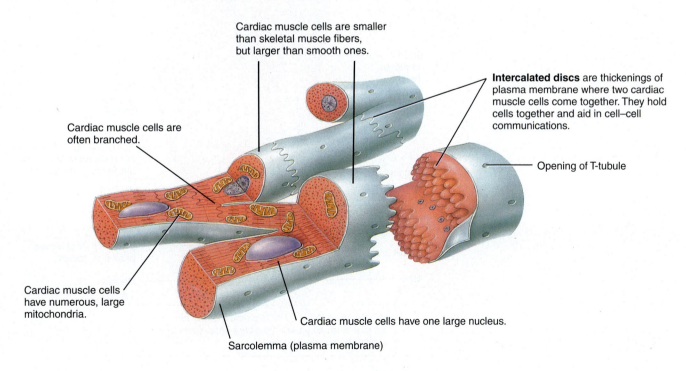

Cardiac muscle cells are smaller than skeletal muscle fibers, but larger than smooth ones.

Cardiac muscle cells are often branched.

Intercalated discs are thickenings of plasma membrane where two cardiac muscle cells come together. They hold cells together and aid in cell–cell communications.

Opening of T-tubule

Cardiac muscle cells have numerous, large mitochondria.

Cardiac muscle cells have one large nucleus.

Sarcolemma (plasma membrane)

Skeletal Muscles Are Grouped Based on Location and Action

LEARNING OBJECTIVES

1. **Explain** how skeletal muscles make bones move.
2. **Identify** major muscle groups of the upper body.
3. **Describe** major muscle groups of the lower body.

The body consists of around 700 skeletal muscles that are grouped based on their locations and actions. Muscles of the upper body control functions of the head and neck, shoulder, and arms, thorax, and abdomen. Muscles of the lower body move the legs and pelvis.

Skeletal Muscles Work in Groups

Skeletal muscles attach to bones via bands of connective tissue called *tendons*. Two bones are usually involved in the movement of a joint: one is stationary, and the other one moves. The location where a muscle attaches to a stationary bone via a tendon is called the **origin**. The other end of the muscle is attached by means of a tendon to the movable bone at a point called the **insertion**. The fleshy portion of the muscle, between the tendons of the origin and insertion, is called the **belly** (**Figure 6.12**).

A good analogy is a spring on a door. The part of the spring attached to the door represents the insertion, the part attached to the frame is the origin, and the coils of the spring are the belly.

Relationships of muscles to bones • Figure 6.12

The ends of muscles have different names, depending on where they are attached. They work in groups or pairs that have opposing actions. For movements of the elbow, the biceps and triceps are a muscle pair with opposing actions.

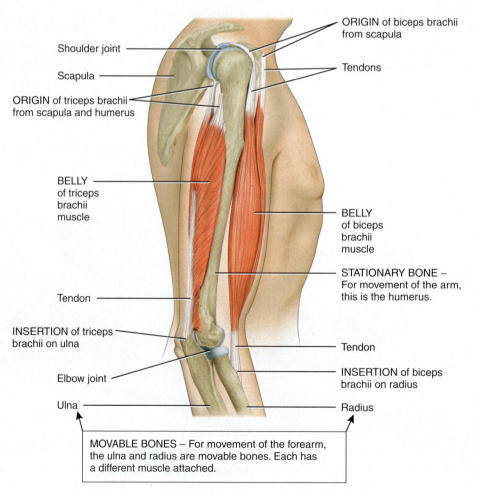

- Shoulder joint
- Scapula
- ORIGIN of triceps brachii from scapula and humerus
- BELLY of triceps brachii muscle
- Tendon
- INSERTION of triceps brachii on ulna
- Elbow joint
- Ulna
- ORIGIN of biceps brachii from scapula
- Tendons
- BELLY of biceps brachii muscle
- STATIONARY BONE – For movement of the arm, this is the humerus.
- Tendon
- INSERTION of biceps brachii on radius
- Radius

MOVABLE BONES – For movement of the forearm, the ulna and radius are movable bones. Each has a different muscle attached.

WHAT A HEALTH PROVIDER SEES

Exercise Maintains Healthy Muscles

Stretching prior to exercise is as important as exercise in maintaining your muscle health. Proper stretching (**Figure a**) lengthens the connective tissue associated with the muscles and improves flexibility. Stretching should not only be done prior to exercise but also on a regular, daily basis. Frequent stretching can reduce muscle tension, improve agility, and increase range of joint motion. Stretching should be done slowly, with the application of gentle force to prevent injury to the muscle or associated connective tissues.

Exercise does not change the total number of muscle fibers but can affect the distribution of muscle fiber types within working muscles. Although your ratio of FG fibers to SO fibers is determined by genetics, you can change the proportion of FOG fibers with exercise. For example, aerobic/endurance exercises such as running (**Figure b**) or swimming can cause some FG fibers to change into FOG fibers. The new FOG fibers are larger and have increased numbers of mitochondria, increased blood supply, and strength. The increased size (hypertrophy) is caused by increased synthesis of additional **myosin**, **actin**, **troponin**, and **tropomyosin**.

In contrast, lack of exercise causes a loss of muscle mass. This is especially true in bedridden patients and astronauts in microgravity (see the opening to Chapter 5). The muscle fibers decrease in size (atrophy) as the muscle proteins degrade.

a.

b.

WILEY PLUS Video

Think Critically

1. Soft tissue injuries (sprains or muscle pulls) are common in athletes. The soft tissues generally involved in these injuries are the ligaments and tendons. Muscle strains (aches) can also occur if muscles are stretched too much. One important part of the treatment for such injuries is rest. Why would rest be an integral part of the therapy for healing soft tissue and muscle injuries?

2. Sam just started an exercise regime. At first, he was stiff, and his muscles fatigued easily. After several weeks of exercising, however, Sam notices that he is more flexible and can tolerate longer and longer sessions. What is happening here?

---

prime mover
The muscle directly responsible for producing a desired motion. Also called the *agonist*.

Muscles work in groups, usually as pairs of muscles with opposing actions. The muscle that starts the desired action is the **prime mover**, or **agonist**. As the agonist contracts, the muscle with an opposing action, called the **antagonist**, relaxes. For example, when you flex (bend) your arm at the elbow, the biceps is the agonist and the triceps is the antagonist. When you extend your arm, the two muscles switch roles.

Most movements involve additional muscles called **synergists** (SIN-er-gists). A synergist helps the agonist function more efficiently by reducing unnecessary move-ments. Some muscles in a group act as **fixators**, stabilizing one of the bones so the agonist can move the other bone more efficiently. During arm movement, for example, muscles associated with the scapula serve as fixators to stabilize the scapula while the humerus moves. Depending on the movement and the conditions, many muscles can switch roles, acting as agonists, antagonists, synergists, or fixators at different times. Exercise is important to maintaining healthy muscles (see *What a Health Provider Sees*).

As mentioned earlier, there are about 700 skeletal muscles, each with a different name. For an overview of selected superficial muscles of the human body, (**Figure 6.13** on the following pages.)

Skeletal Muscles Are Grouped Based on Location and Action **169**

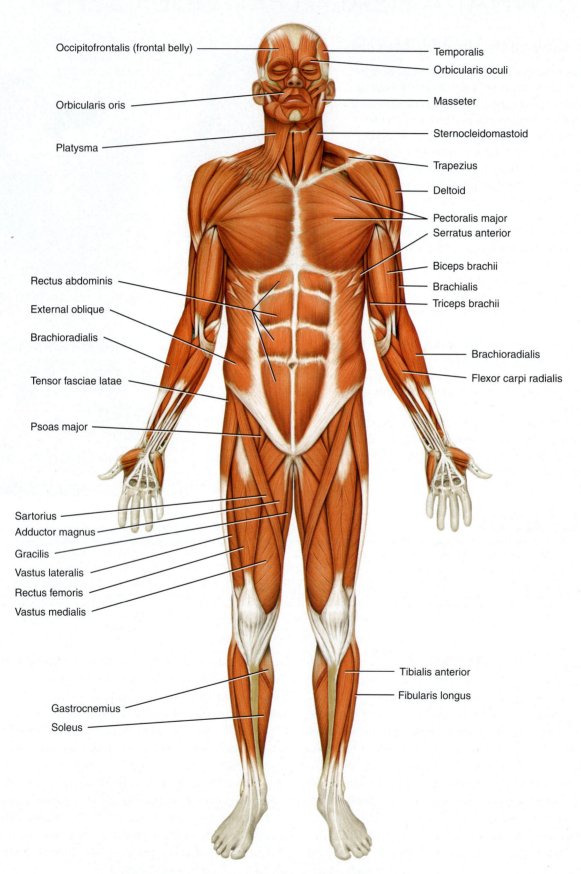

Occipitofrontalis (frontal belly)

Orbicularis oris

Platysma

Rectus abdominis

External oblique

Brachioradialis

Tensor fasciae latae

Psoas major

Sartorius

Adductor magnus

Gracilis

Vastus lateralis

Rectus femoris

Vastus medialis

Gastrocnemius

Soleus

Temporalis

Orbicularis oculi

Masseter

Sternocleidomastoid

Trapezius

Deltoid

Pectoralis major

Serratus anterior

Biceps brachii

Brachialis

Triceps brachii

Brachioradialis

Flexor carpi radialis

Tibialis anterior

Fibularis longus

a. Anterior view

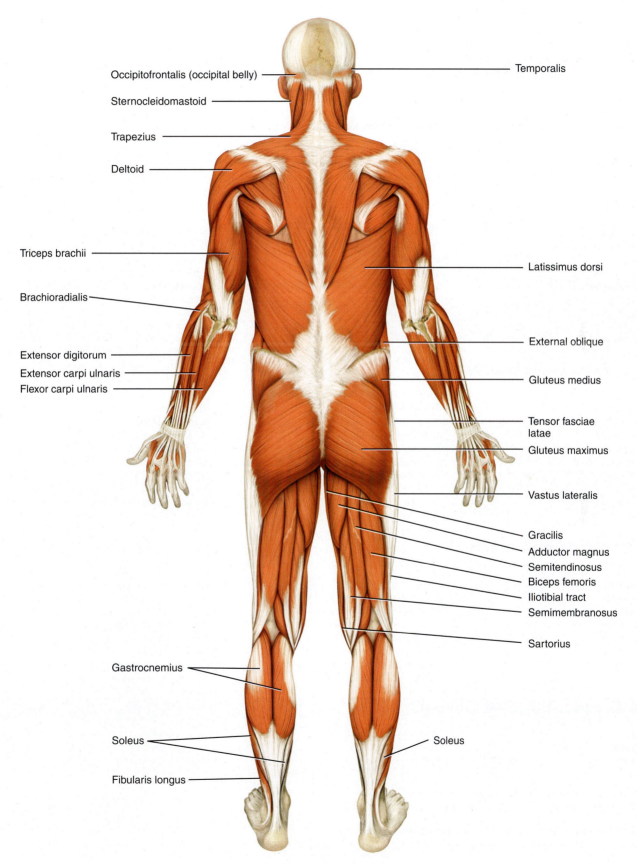

Occipitofrontalis (occipital belly)

Sternocleidomastoid

Trapezius

Deltoid

Triceps brachii

Brachioradialis

Extensor digitorum

Extensor carpi ulnaris

Flexor carpi ulnaris

Temporalis

Latissimus dorsi

External oblique

Gluteus medius

Tensor fasciae latae

Gluteus maximus

Vastus lateralis

Gracilis

Adductor magnus

Semitendinosus

Biceps femoris

Iliotibial tract

Semimembranosus

Sartorius

Gastrocnemius

Soleus

Soleus

Fibularis longus

b. Posterior view

Skeletal Muscles Are Grouped Based on Location and Action **171**

The names of the skeletal muscles often relate to muscle characteristics such as fiber direction, size, shape, action, number of origins, and location (see **Table 6.1**).

Characteristics used to name skeletal muscles Table 6.1		
Name	**Meaning**	**Example**
Direction: **Orientation of muscle fibers relative to the body's midline**		
Rectus	Parallel to midline	Rectus abdominis
Transverse	Perpendicular to midline	Transverse abdominis
Oblique	Diagonal to midline	External oblique
Size: **Relative size of the muscle**		
Maximus	Largest	Gluteus maximus
Minimus	Smallest	Gluteus minimus
Longus	Longest	Adductor longus
Latissimus	Widest	Latissimus dorsi
Longissimus	Longest	Longissimus muscles
Magnus	Large	Adductor magnus
Major	Larger	Pectoralis major
Minor	Smaller	Pectoralis minor
Vastus	Great	Vastus lateralis
Shape: **Relative shape of the muscle**		
Deltoid	Triangular	Deltoid
Trapezius	Trapezoid	Trapezius
Serratus	Saw-toothed	Serratus anterior
Rhomboid	Diamond-shaped	Rhomboid major
Orbicularis	Circular	Orbicularis oculi
Pectinate	Comblike	Pectineus
Piriformis	Pear-shaped	Piriformis
Platys	Flat	Platysma
Quadratus	Square	Quadratus lumborum
Gracilis	Slender	Gracilis
Action: **Principal action of the muscle**		
Flexor	Decreases joint angle	Flexor carpi radialis
Extensor	Increases joint angle	Extensor carpi ulnaris
Abductor	Moves bone away from midline	Adductor longus
Levator	Produces superior movement	Levator scapulae
Depressor	Produces inferior movement	Depressor labii inferioris
Supinator	Turns palm anteriorly	Supinator
Pronator	Turns palm posteriorly	Pronator teres
Sphincter	Decreases size of opening	External anal sphincter
Tensor	Makes a body part rigid	Tensor fasciae latae
Number of Origins: **Number of tendons of origin**		
Biceps	Two origins	Biceps brachii
Triceps	Three origins	Triceps brachii
Quadriceps	Four origins	Quadriceps femoris
Location: **Structure near which a muscle is found** *Example:* Temporalis, a muscle near the temporal bone.		
Origin and Insertion: **Sites where muscle originates and inserts** *Example:* Brachioradialls, originating on the humerus and inserting on the radius (*brachii* = arm).		

Muscles of the Head and Neck Control Facial Expressions, Allow You to Speak and Chew, and Enable Your Eyes to Open, Close, and Move

Muscles of the head and neck produce facial expressions that give humans the ability to express a wide variety of emotions, such as happiness, displeasure, fear, and surprise. Muscles that move the mandible (lower jaw) are involved in biting and chewing. Because of this, they are also known as muscles of *mastication* (chewing). These muscles also help in speech. The muscles that control facial expression, speech, and chewing are illustrated in **Figure 6.14**.

Muscles of the head and neck • Figure 6.14

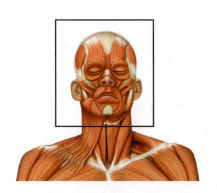

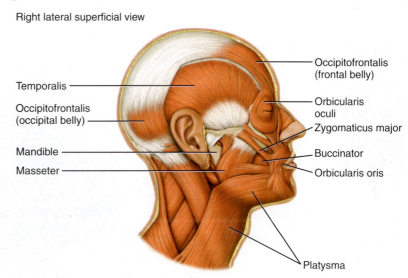

Right lateral superficial view

- Temporalis
- Occipitofrontalis (occipital belly)
- Mandible
- Masseter
- Occipitofrontalis (frontal belly)
- Orbicularis oculi
- Zygomaticus major
- Buccinator
- Orbicularis oris
- Platysma

Muscle	Origin	Insertion	Action
Muscles of the head and neck that produce facial expressions			
Occipitofrontalis (ok-sip-i-tō-frun-TĀ-lis)—divided into frontal belly and occipital belly			
Frontal belly	Epicranial aponeurosis (flat tendon attaching frontalis and occipitalis)	Skin superior to orbit	Draws scalp forward, as in frowning; raises eyebrows; and wrinkles skin of forehead horizontally, as in a look of surprise
Occipital belly	Occipital and temporal	Epicranial aponeurosis	Draws scalp backward
Orbicularis oris (or-bi'-kū-LAR-is OR-is)	Muscle fibers surrounding the opening of the mouth	Skin at the corner of the mouth	Closes and protrudes lips, as in kissing; compresses lips against teeth; and shapes lips during speech
Zygomaticus major (zī-gō-MA-ti-kus)	Zygomatic bone	Skin at the angle of the mouth and orbicularis oris	Draws angle of mouth upward and outward, as in smiling or laughing
Buccinator (BUK-si-nā'-tor)	Maxilla and mandible	Orbicularis oris	Presses cheeks against teeth and lips, as in whistling, blowing, and sucking; draws corner of mouth laterally; assists in chewing by keeping food between teeth
Platysma (pla-TIZ-ma)	Fascia over deltoid and pectoralis major	Mandible, muscles around angle of mouth, skin of lower face	Draws outer part of lower lip downward and backward, as in pouting; depresses mandible
Orbicularis oculi (OK-ū-lī)	Medial wall of orbit	Circular path around orbit	Closes eye
Muscles of the head and neck that control speech and chewing			
Masseter (MA-se-ter)	Maxilla and zygomatic arch	Mandible	Elevates mandible, as in closing mouth
Temporalis (tem'-por-Ā-lis)	Temporal bone	Mandible	Elevates and retracts mandible

The muscles that move the eyeballs are *extrinsic* muscles because they originate outside the eyeball and are inserted on the outer surface (*ex-* = out of). They move the eyeballs in various directions.

These muscles are among the fastest contracting and most precisely controlled skeletal muscles in the body. These muscles are illustrated in **Figure 6.15**, which also provides their respective origins, insertions, and actions.

Muscles of the head that move the eyeballs and upper eyelids • Figure 6.15

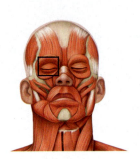
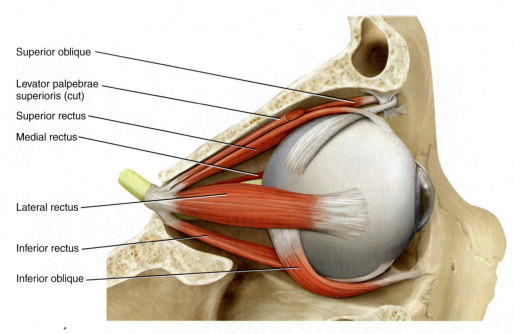

Superior oblique

Levator palpebrae superioris (cut)

Superior rectus

Medial rectus

Lateral rectus

Inferior rectus

Inferior oblique

Lateral view of right eyeball

Muscle	Origin	Insertion	Action
Muscles of the head that move the eyeballs and upper eyelids			
Superior rectus (REK-tus)	Tendinous ring attached to bony orbit around optic foramen	Superior and central part of eyeball	Moves eyeball upward (elevation) and medially (adduction) and rotates it medially
Lateral rectus	Same as above	Lateral side of eyeball	Moves eyeball laterally (abduction)
Medial rectus	Same as above	Medial side of eyeball	Moves eyeball medially (adduction)
Superior oblique (ō-BLĒK)	Same as above	Eyeball between superior and lateral recti; the muscle moves through a ring of fibrocartilaginous tissue called the trochlea	Moves eyeball downward (depression) and laterally (abduction) and rotates it medially
Inferior oblique	Maxilla	Eyeball between inferior and lateral recti	Moves eyeball upward (elevation) and laterally (abduction) and rotates it laterally
Levator palpebrae superioris (le-VĀ-tor PAL-pe-brē soo-per'-ē-OR-is)	Roof of orbit	Skin of upper eyelid	Elevates upper eyelid (opens eye)

Muscles of the Thorax, Abdomen, Neck, and Back Perform Many Functions

Muscles of the thorax (chest) have three functions: They are involved in breathing, movement of the shoulder, and movement of the humerus. The chest muscles involved in breathing alter the size of the thoracic cavity. During inhalation (breathing in), the cavity increases in size; during exhalation (breathing out), the cavity decreases in size. Muscles that help move the shoulder also act to hold the scapula in place so that it is a stable point of origin for the muscles that move the humerus. These muscles are illustrated in **Figure 6.16**, which also provides their respective origins, insertions, and actions.

Muscles of the thorax involved in breathing • Figure 6.16

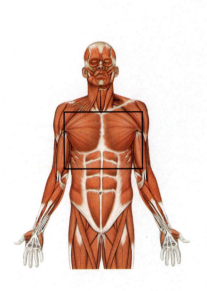

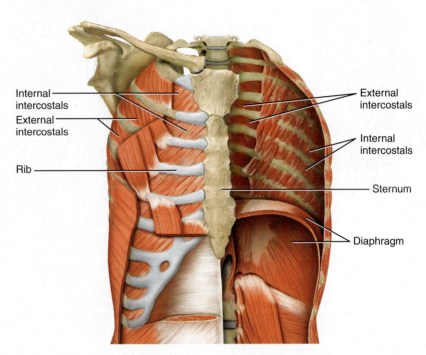

Internal intercostals

External intercostals

Rib

External intercostals

Internal intercostals

Sternum

Diaphragm

Anterior superficial view   Anterior deep view

Muscle	Origin	Insertion	Action
Muscles of the thorax involved in breathing			
Diaphragm (DĪ-a-fram)	Xiphoid process of sternum, costal cartilages of inferior six ribs, lumbar vertebrae, and their intervertebral discs	Central tendon (strong aponeurosis near the center of the diaphragm)	Contraction causes it to flatten and increases the vertical (top-to-bottom) dimension of the thoracic cavity, resulting in inhalation; relaxation causes it to move superiorly and decreases the vertical dimension of the thoracic cavity, resulting in exhalation
External intercostals	Inferior border of rib above	Superior border of rib below	Contraction elevates the ribs and increases the anteroposterior and lateral dimensions of the thoracic cavity, resulting in inhalation; relaxation depresses the ribs and decreases the anteroposterior and lateral dimensions of the thoracic cavity, resulting in exhalation
Internal intercostals	Superior border of rib below	Inferior border of rib above	Contraction draws adjacent ribs together to further decrease the anteroposterior (front-to-back) and lateral (side-to-side) dimensions of the thoracic cavity during forced exhalation

Muscles of the thorax that move the shoulder originate on the axial skeleton and insert on the clavicle or scapula. **Figure 6.17** illustrates and describes these muscles.

Muscles of the thorax that move the shoulder • Figure 6.17

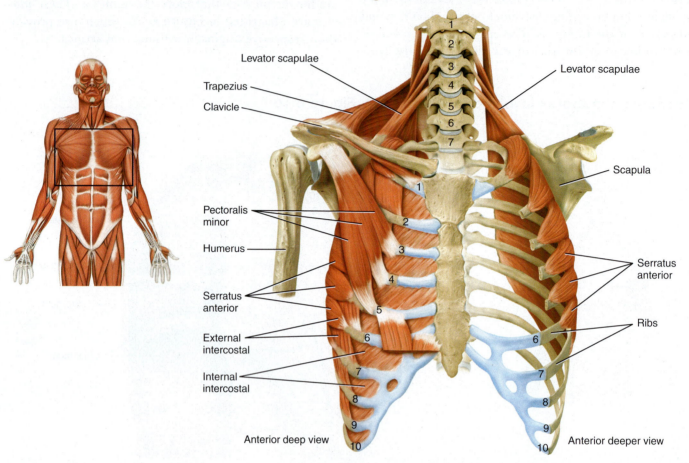

Anterior deep view

Anterior deeper view

Muscle	Origin	Insertion	Action
Muscles of the thorax that move the shoulder			
Pectoralis minor (pek'-tor-A-lis; *pect-* = breast, chest, thorax; *minor* = lesser)	Second through fifth, third through fifth, or second through fourth ribs	Scapula	Abducts scapula and rotates it downward (movement of glenoid cavity upward); elevates third through fifth ribs during forced inhalation when scapula is fixed
Serratus anterior (ser-A-tus; *serratus* = saw-toothed; *anterior* = front)	Upper eight or nine ribs	Scapula	Abducts scapula and rotates it upward (movement of glenoid cavity downward); elevates ribs when scapula is fixed; known as "boxer's muscle" because it is important in horizontal arm movements such as punching and pushing
Trapezius (tra-PE-ze-us; *trapeze-* = trapezoid-shaped)	Occipital bone and spines of seventh cervical and all thoracic vertebrae	Clavicle and scapula	Superior fibers elevate scapula; middle fibers adduct scapula; inferior fibers depress and upward rotate scapula; superior and inferior fibers together rotate scapula upward; stabilizes scapula
Levator scapulae (le-VA-tor SKA-pu-le; *levator* = to raise; *scapulae* = of the scapula)	Upper four cervical vertebrae	Scapula	Elevates and adducts scapula and rotates it downward

Four pair of muscles of the abdomen protect the abdominal organs. They include the rectus abdominis, external oblique, internal oblique, and transverse abdominis. **Figure 6.18** illustrates and describes these muscles.

Muscles of the abdomen that protect the abdominal organs • Figure 6.18

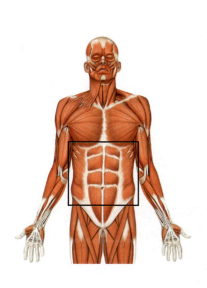

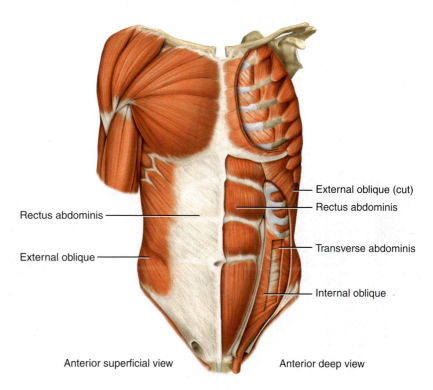

Rectus abdominis

External oblique

External oblique (cut)

Rectus abdominis

Transverse abdominis

Internal oblique

Anterior superficial view   Anterior deep view

Muscle	Origin	Insertion	Action
Muscles of the abdomen that protect the abdominal organs			
Rectus abdominis (REK-tus ab-DOM-in-is)	Pubis and pubic symphysis	Cartilage of fifth to seventh ribs and xiphoid process of sternum	Flexes vertebral column and compresses abdomen to aid in defecation, urination, forced exhalation, and childbirth
External oblique	Lower eight ribs	Crest of ilium and linea alba (a tough connective tissue band that runs from xiphoid process to pubic symphysis)	Contraction of both external obliques compresses abdomen and flexes vertebral column; contraction of one side alone bends vertebral column laterally and rotates it
Internal oblique	Ilium, inguinal ligament, and thoracolumbar fascia	Cartilage of last three or four ribs and linea alba	Contraction of both internal obliques compresses abdomen and flexes vertebral column; contraction of one side alone bends vertebral column laterally and rotates it
Transverse abdominis	Ilium, inguinal ligament, lumbar fascia, and cartilages of last six ribs	Xiphoid process, linea alba, and pubis	Compresses abdomen

Muscles of the neck and back are responsible for movement of the spine. These include the erector spinae, sternocleidomastoid, quadratus lumborum, and psoas major. The erector spinae is the largest muscular mass of the back and consists of three groups of overlapping muscles: the iliocostalis group, longissimus group, and spinalis group. These groups are illustrated in **Figure 6.19**. See Figure 6.13b for an illustration of the sternocleidomastoid, and see Figure 6.23 for illustrations of the quadratus lumborum and psoas major.

Muscles of the neck and back that move the spine • Figure 6.19

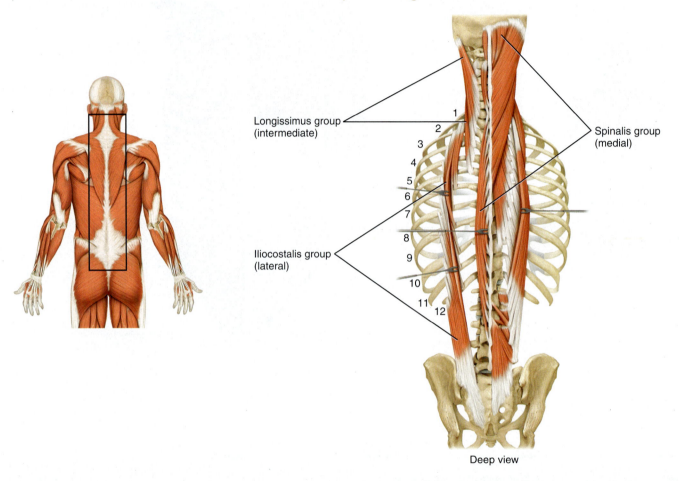

Longissimus group (intermediate)

Spinalis group (medial)

Iliocostalis group (lateral)

1
2
3
4
5
6
7
8
9
10
11
12

Deep view

Muscle	Origin	Insertion	Action
Muscles of the neck and back that move the spine			
Erector spinae (e-REK-tor SPĪ-nē; iliocostalis group, longissimus group, and spinalis group)	All ribs plus cervical, thoracic, and lumbar vertebrae	Occipital bone, temporal bone, ribs, and vertebrae	Extends head; extends and laterally flexes vertebral column
Sternocleidomastoid (ster'-nō-klī-dō-MAS-toid) (see Figure 6.13b)	Sternum and clavicle	Temporal bone	Contractions of both muscles flex cervical part of vertebral column and flex head; contraction of one muscle rotates head toward side opposite contracting muscle
Quadratus lumborum (kwod-RĀ-tus lum-BOR-um) (see Figure 6.23)	Ilium	Twelfth rib and upper four lumbar vertebrae	Contractions of both muscles extend lumbar part of vertebral column; contraction of one muscle flexes lumbar part of vertebral column
Psoas major (SŌ-as) (see Figure 6.23)	Lumbar vertebrae	Femur	Flexes vertebral column; flexes and rotates thigh laterally at hip joint

Muscles of the Shoulder and Upper Limb Are Concerned with Movement

You learned in Chapter 5 that the pectoral girdle consists of the clavicle and scapula. The region of the muscles that attach to these bones and to the bones of the arm and forearm are also referred to as the *pectoral girdle* or *shoulder girdle*. The muscles of the shoulder include the deltoid, subscapularis, supraspinatus, infraspinatus, teres major, and teres minor. **Figure 6.20** illustrates and describes these muscles.

Muscles of the shoulder and upper limb • Figure 6.20

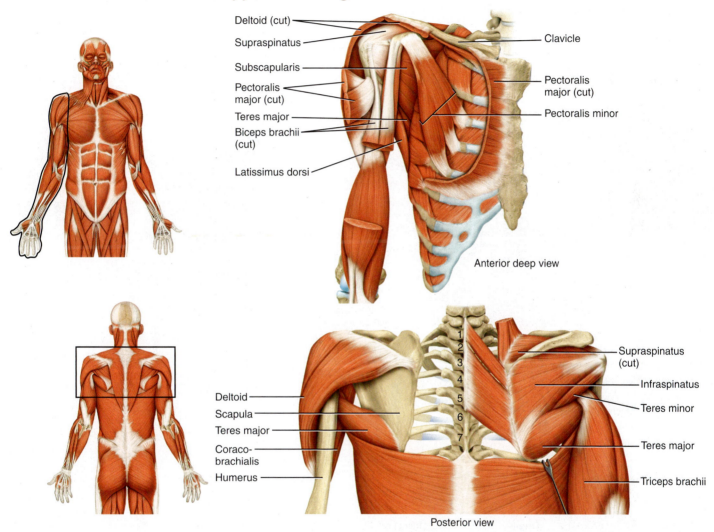

Anterior deep view

Posterior view

Muscle	Origin	Insertion	Action
Muscles of the shoulder			
Deltoid (DEL-toyd)	Clavicle and scapula	Humerus	Abducts, flexes, extends, and rotates arm at shoulder joint
Subscapularis (sub-scap'-ū-LĀ-ris)	Scapula	Humerus	Rotates arm medially at shoulder joint
Supraspinatus (soo'-pra-spi-NĀ-tus)	Scapula	Humerus	Assists deltoid muscle in abducting arm at shoulder joint
Infraspinatus (in'-fra-spi-NĀ-tus)	Scapula	Humerus	Rotates arm laterally at shoulder joint
Teres major (TE-rēz)	Scapula	Humerus	Extends arm at shoulder joint; assists in adduction and rotation of arm medially at shoulder joint
Teres minor	Scapula	Humerus	Rotates arm laterally and extends arm at shoulder joint

The muscles of the arm have three functions: flexing the forearm, extending the forearm, and rotating the forearm. The biceps brachii, brachialis, and brachioradialis flex the forearm. The triceps brachii extends the forearm. The muscles responsible for the rotation of the forearm are the supinator and the pronator teres. **Figure 6.21** illustrates and describes these muscles.

The two functions of the muscles of the forearm are moving the wrist and moving the hand and fingers. The flexor carpi radialis, flexor carpi ulnaris, extensor carpi radialis longus, and extensor carpi ulnaris are all involved in the wrist motion required to toss a Frisbee. The muscles that let you grasp an object with your hand or move your fingers to play the piano include the palmaris longus, flexor digitorum superficialis, flexor digitorum profundus, and extensor digitorum. **Figure 6.22** illustrates all of these muscles and includes descriptions of their origins, insertions, and actions.

Muscles of the arm that move the radius and ulna • Figure 6.21 _____

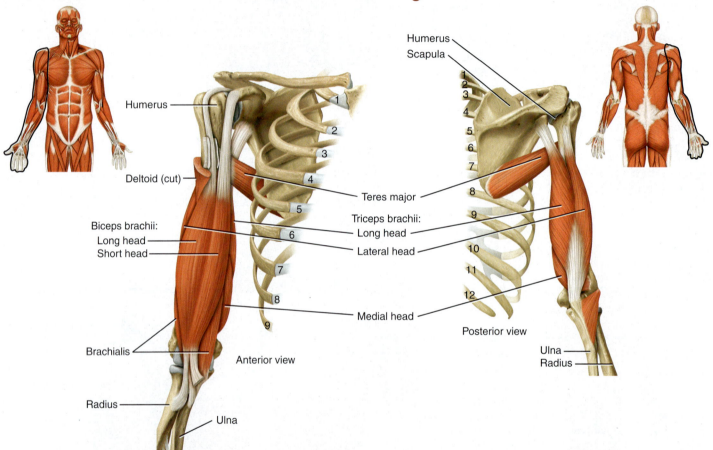

Muscle	Origin	Insertion	Action
Muscles of the arm			
Biceps brachii (BĪ-ceps BRĀ-kē-ī)	Scapula	Radius	Flexes and supinates forearm at elbow joint; flexes arm at shoulder joint
Brachialis (brā′-kē-Ā-lis)	Humerus	Ulna	Flexes forearm at elbow joint
Brachioradialis (brā′-kē-ō-rā′-dē-Ā-lis)	Humerus	Radius	Flexes forearm at elbow joint
Triceps brachii (TRĪ-ceps BRĀ-kē-ī)	Scapula and humerus	Ulna	Extends forearm at elbow joint; extends arm at shoulder joint
Supinator (SOO-pi-nā-tor)	Humerus and ulna	Radius	Supinates forearm
Pronator teres (PRŌ-nā-tor TE-rēz)	Humerus and ulna	Radius	Pronates forearm

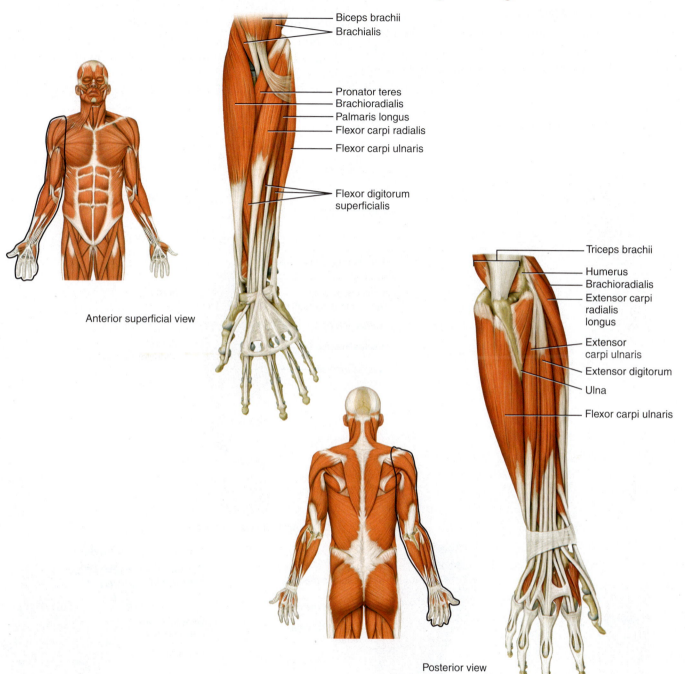

Biceps brachii
Brachialis

Pronator teres
Brachioradialis
Palmaris longus
Flexor carpi radialis
Flexor carpi ulnaris

Flexor digitorum superficialis

Anterior superficial view

Triceps brachii
Humerus
Brachioradialis
Extensor carpi radialis longus
Extensor carpi ulnaris
Extensor digitorum
Ulna
Flexor carpi ulnaris

Posterior view

Muscle	Origin	Insertion	Action
Muscles of the forearm			
Flexor carpi radialis (FLEK-sor KAR-pē rā′-dē-Ā-lis)	Humerus	Second and third metacarpals	Flexes and abducts hand at wrist joint
Flexor carpi ulnaris (ul-NAR-is)	Humerus and ulna	Pisiform, hamate, and fifth metacarpal	Flexes and adducts hand at wrist joint
Extensor carpi radialis longus (eks-TEN-sor)	Humerus	Second metacarpal	Extends and abducts hand at wrist joint
Extensor carpi ulnaris	Humerus and ulna	Fifth metacarpal	Extends and adducts hand at wrist joint

Muscles of the Lower Body Move the Thigh and Legs

The muscles of the lower body include those of the gluteal region, thigh, and leg. These muscles are larger and more powerful than those of the upper limbs. They pro- vide stability, locomotion, and maintenance of posture. **Figure 6.23** lists their origins, insertions, and actions. Gluteal muscles move the thigh and include the gluteus maximus, gluteus medius, and piriformis. Thigh muscles can be divided into five groups, based on their functions:

Muscles of the gluteal region that move the thigh • Figure 6.23

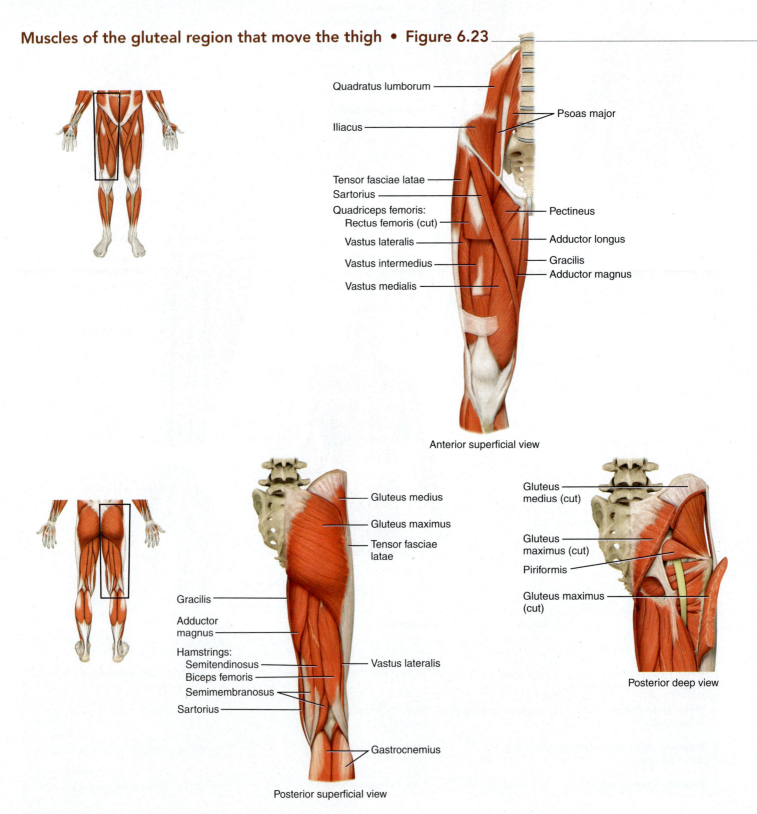

Quadratus lumborum

Psoas major

Iliacus

Tensor fasciae latae

Sartorius

Quadriceps femoris:
Rectus femoris (cut)

Pectineus

Vastus lateralis

Adductor longus

Vastus intermedius

Gracilis

Adductor magnus

Vastus medialis

Anterior superficial view

Gluteus medius

Gluteus maximus

Tensor fasciae latae

Gracilis

Adductor magnus

Hamstrings:
Semitendinosus

Biceps femoris

Semimembranosus

Sartorius

Vastus lateralis

Gastrocnemius

Posterior superficial view

Gluteus medius (cut)

Gluteus maximus (cut)

Piriformis

Gluteus maximus (cut)

Posterior deep view

1. The adductor group moves the thigh medially and includes the adductor magnus, adductor longus, pectineus, and gracilis.

2. The quadriceps group extends the leg and includes the vastus lateralis, vastus medialis, vastus intermedius, and rectus femoris.

3. The hamstring group flexes the leg and includes the biceps femoris, semitendinosus, and semimembranosus.

4. The sartorius helps you cross your legs by weakly flexing the leg and the knee joint and flexing, abducting, and laterally rotating the thigh at the hip joint.

5. The tensor fasciae latae moves the thigh laterally.

Muscle	Origin	Insertion	Action
Muscles of the gluteal region			
Gluteus maximus (GLOO-tē-us MAK-si-mus)	Ilium, sacrum, coccyx, and aponeurosis of sacrospinalis	Iliotibial tract of fascia lata and femur	Extends and rotates thigh laterally at the hip joint; helps lock knee in extension
Gluteus medius (MĒ-dē-us)	Ilium	Femur	Abducts and rotates thigh medially at the hip joint
Piriformis (pir-i-FOR-mis)	Sacrum	Femur	Rotates thigh laterally and abducts it at the hip joint
Muscles of the thigh			
Adductor group			
Adductor magnus (MAG-nus)	Pubis and ischium	Femur	Adducts, rotates, flexes, and extends thigh at hip joint
Adductor longus (LONG-us)	Pubis and pubic symphysis	Femur	Adducts, rotates, and flexes thigh at hip joint
Pectineus (pek-TIN-ē-us)	Pubis	Femur	Flexes and adducts thigh at hip joint
Gracilis (gras-I-lis)	Pubis	Tibia	Adducts and medially rotates thigh at hip joint, flexes leg at knee joint
Quadriceps group			
Vastus lateralis (VAS-tus lat′-er-Ā-lis)	Femur	Patella by means of quadriceps tendon and then tibial tuberosity by means of patellar ligament	Extend leg at knee joint
Vastus medialis (mē-dē-Ā-lis)	Femur	Patella by means of quadriceps tendon and then tibial tuberosity by means of patellar ligament	Extend leg at knee joint
Vastus intermedius (in′-ter-MĒ-dē-us)	Femur	Patella by means of quadriceps tendon and then tibial tuberosity by means of patellar ligament	Extend leg at knee joint
Rectus femoris (REK-tus FEM-or-is)	Ilium	Patella by means of quadriceps tendon and then tibial tuberosity by means of patellar ligament	Extend leg at knee joint; flex thigh at hip joint
Hamstrings			
Biceps femoris (BĪ-ceps FEM-or-is)	Ischium and femur	Fibula and tibia	Flexes leg at knee joint; extends thigh at hip joint
Semitendinosus (sem′-ē-TEN-di-nō′-sus)	Ischium	Tibia	Flexes leg at knee joint; extends thigh at hip joint
Semimembranosus (sem′-ē-MEM-bra-nō′-sus)	Ischium	Tibia	Flexes leg at knee joint; extends thigh at hip joint
Sartorius (sar-TOR-ē-yus)	Ilium	Tibia	Weakly flexes leg at knee joint; flexes, abducts, and laterally rotates thigh at hip joint, thus crossing leg
Tensor fasciae latae (TEN-sor FA-shē-ē LĀ-tē)	Ilium	Tibia by means of iliotibial tract	Flexes and abducts thigh at hip joint; helps lock knee in extension

The leg muscles can be divided into two groups: those that move the feet and those that allow you to wiggle your toes. The tibialis anterior, fibularis longus, gastrocnemius, soleus, and tibialis posterior move the ankle. The extensor digitorum longus and flexor digitorum longus allow movement of the toes. See **Figure 6.24** for an illustration of these muscles and details about their origins, insertions, and actions.

CONCEPT CHECK STOP

1. **During** movement, what is the role of the agonist?
2. **Which** muscle groups are responsible for breathing?
3. **Where** would you find the gastrocnemius muscle?

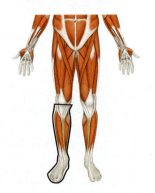

Muscles of the leg that move the feet and toes • Figure 6.24

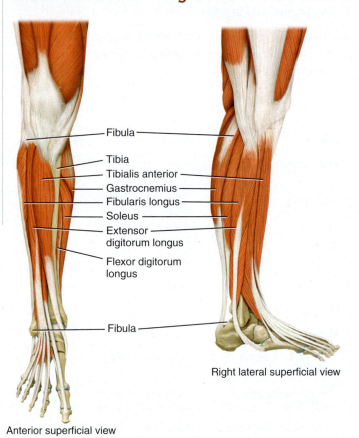

Right lateral superficial view

Anterior superficial view

Muscle	Origin	Insertion	Action
Muscles of the leg that move the feet			
Tibialis anterior (tib'-ē-Ā-lis)	Tibia	First metatarsal and first cuneiform	Dorsiflexes and inverts (supinates) foot
Fibularis longus (fib-ū-LAR-is LON-gus)	Fibula and tibia	First metatarsal and first cuneiform	Plantar flexes and everts (pronates) foot
Gastrocnemius (gas'-trok-NĒ-mē-us)	Femur	Calcaneus by means of calcaneal (Achilles) tendon	Plantar flexes foot; flexes leg at knee joint
Soleus (SŌ-lē-us)	Fibula and tibia	Calcaneus by means of calcaneal (Achilles) tendon	Plantar flexes foot
Tibialis posterior (tib'-ē-Ā-lis)	Tibia and fibula	Second, third, and fourth metatarsals; navicular; all three cuneiforms; and cuboid	Plantar flexes and inverts foot
Muscles of the leg that move the toes			
Extensor digitorum longus (eks-TEN-sor di'-ji-TOR-um LON-gus)	Tibia and fibula	Middle and distal phalanges of each toe (except great toe)	Dorsiflexes and everts foot; extends toes
Flexor digitorum longus (FLEK-sor)	Tibia	Distal phalanges of each toe (except great toe)	Plantar flexes foot; flexes toes

Summary

1 The Body Contains Three Different Types of Muscular Tissue 154

• There are three types of muscular tissue: skeletal, cardiac, and smooth, as shown.

Types of muscular tissues • Figure 6.1

Skeletal muscle fiber

Long cylindrical fiber with many peripherally located nuclei; striated; unbranched

Cardiac muscle fibers

Branched cylindrical fiber, usually with one centrally located nucleus; intercalated discs join neighboring fibers; striated

Smooth muscle fiber

Fiber is thickest in the middle, tapered at each end; has one centrally located nucleus; not striated

• Skeletal muscle is attached to bones; it is striated and can be voluntarily controlled.

• Smooth muscle is located in the walls of hollow organs; it is non-striated and involuntary.

• Cardiac muscle is found in the heart; it is striated and involuntary.

• Muscular tissue contracts and relaxes, thereby producing body movements, stabilizing body positions, regulating organ volume, moving substances within the body, and producing heat.

2 Muscles Are Built to Move 156

• As shown, skeletal muscles consist of bundles of fibers (cells), wrapped in connective tissue. The fibers are large, cylindrical, and multi-nucleated. The cell membrane has tunnel-like T-tubules extending into the center of the cell; there are numerous mitochondria; and the endoplasmic reticulum (**sarcoplasmic reticulum**) is well developed and specialized for storing calcium. The cells are filled with arrays of myofibrils that consist of thick and thin filaments that are arranged into sarcomeres. Thick filaments are made of the protein myosin, and thin filaments are made of actin, tropomyosin, and troponin.

The structure of a skeletal muscle and its connective tissue coverings • Figure 6.2

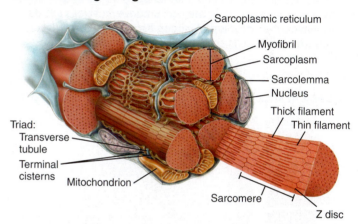

Sarcoplasmic reticulum
Myofibril
Sarcoplasm
Sarcolemma
Nucleus
Thick filament
Thin filament
Triad:
Transverse tubule
Terminal cisterns
Mitochondrion
Sarcomere
Z disc

• Muscles shorten when the myosin in thick filaments pulls on the actin of thin filaments, thereby causing them to slide past one another, toward the center of the sarcomere. This process, called a contraction cycle, or cross-bridge cycle, is triggered by a rise in the concentration of calcium ions in the cytoplasm.

• In a process called excitation–contraction coupling, a nerve impulse sets up an electrochemical impulse that gets conducted along the muscle fiber. This impulse, or **action potential,** is conducted down the T-tubules and causes the release of calcium ions from the sarcoplasmic reticulum. The calcium ions trigger the contraction cycle. When the cell repolarizes, calcium gets pumped back into the sarcoplasmic reticulum, the contraction cycle stops, and the muscle relaxes.

• The ATP that provides the energy for muscle contraction comes from several sources, depending on the amount of time the muscle remains contracted: ATP and creatine phosphate reserves (0–15 s), anaerobic respiration or glycolysis (15–30 s), and aerobic respiration (less than 30 s). Working muscles need good supplies of oxygen and blood for sustained activity. When the muscle cannot contract forcefully, it is said to be fatigued.

- A single muscle contraction is called a twitch, while a sustained contraction is called tetanus. The tension generated by a muscle can be increased by increasing the frequency of nerve impulses to the **motor unit** (via temporal summation or wave summation) or by increasing the number of motor units that are contracting at one time (recruitment).

- Smooth muscles can be stretched considerably and still maintain the ability to contract, as shown. The contraction and relaxation is slower, and this helps maintain muscle tone. Smooth muscles contract in response to nerve impulses, hormones, stretching, and local factors.

3 Smooth Muscle Tissue Is in the Walls of Most Organs 165

- Smooth muscle tissue is either visceral (single-unit) or multi-unit, as shown. Smooth muscle cells are much smaller, thinner, and more spindle shaped than skeletal muscle. In addition to thin and thick filaments, smooth muscle cells have intermediate filaments attached to dense bodies, forming a net-like network in the cell. Smooth muscles cells do not have a well-developed sarcoplasmic reticulum. Visceral smooth muscle cells communicate with each other through gap junctions.

4 Cardiac Muscle Tissue Forms the Muscular Wall of the Heart 167

- As shown, the structure of cardiac muscle has some characteristics of smooth muscle and some of skeletal muscle. It is intermediate in size between the other muscle types and has cylindrical, branched fibers that are interconnected by **intercalated discs** that contain gap junctions. Cardiac muscle cells have large mitochondria, one nucleus, and bundles of myofibrils arranged into sarcomeres. They have well-developed sarcoplasmic reticulum and T-tubule systems, like skeletal muscle.

Structure of cardiac muscle cells • Figure 6.11

Types and structure of smooth muscle cells • Figure 6.10

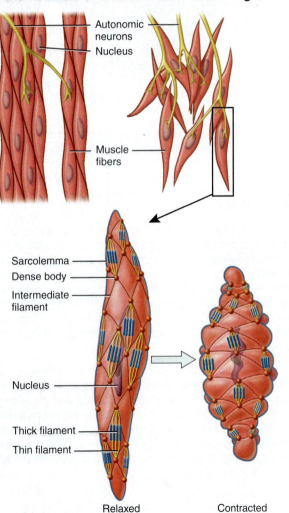

Autonomic neurons
Nucleus
Muscle fibers

Sarcolemma
Dense body
Intermediate filament
Nucleus
Thick filament
Thin filament

Relaxed   Contracted

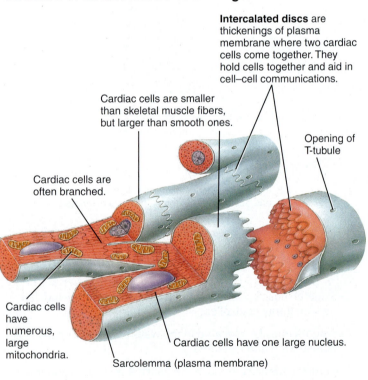

Intercalated discs are thickenings of plasma membrane where two cardiac cells come together. They hold cells together and aid in cell–cell communications.

Cardiac cells are smaller than skeletal muscle fibers, but larger than smooth ones.

Opening of T-tubule

Cardiac cells are often branched.

Cardiac cells have numerous, large mitochondria.

Cardiac cells have one large nucleus.

Sarcolemma (plasma membrane)

- Cardiac cells respond to signals from specialized heart muscle cells called *pacemaker cells*. Via pacemaker cells, the autonomic nervous system can speed up or slow down the contraction of cardiac cells as needed.

5 Skeletal Muscles Are Grouped Based on Location and Action 168

• Skeletal muscles are attached to bones with tendons. The attachment of a muscle to a stationary bone is its origin, the attachment to a movable bone is its insertion, and the part of the muscle in between is the belly, as shown.

• Muscles usually work in pairs that have opposing actions. The muscle acting as the **agonist (prime mover)** contracts to produce the desired movement, and the **antagonist** relaxes. The synergist works with the prime mover to cause the action and help reduce unnecessary movement, and the fixator stabilizes the origin of the prime mover so it can work more efficiently.

• The principal skeletal muscles are grouped according to region. Their names indicate specific characteristics, which are based on descriptive categories such as direction of fibers, location, size, number of origins, shape, origin and insertion points, and action.

ORIGIN of biceps brachii from scapula

Shoulder joint

Tendons

Scapula

ORIGIN of triceps brachii from scapula and humerus

BELLY of triceps brachii muscle

BELLY of biceps brachii muscle

STATIONARY BONE – For movement of the arm, this is the humerus.

Tendon

INSERTION of triceps brachii on ulna

Tendon

INSERTION of biceps brachii on radius

Elbow joint

Ulna

Radius

Movable bones

Key Terms

- actin 169
- aerobic respiration 163
- agonist 169
- anaerobic respiration 163
- antagonist 169
- autorhythmic 167
- belly 168
- cardiac muscle tissue 155
- contraction period 163
- creatine phosphate 163
- dense body 165
- depolarization 160
- fascicle 157
- fast glycolytic (FG) fiber 165
- fast oxidative-glycolytic (FOG) fiber 165
- fixator 169
- glycogen 163
- glycolysis 163
- insertion 168

- intercalated disc 167
- intermediate filament 165
- isometric contraction 164
- isotonic contraction 164
- lactic acid 163
- latent period 163
- motor unit 164
- multi-unit smooth muscle tissue 166
- muscle tone 164
- myofibril 156
- myofilament 158
- myoglobin 163
- myosin 169
- neuromuscular junction 160
- origin 168
- prime mover 169
- relaxation period 163
- repolarization 160
- sarcomere 156

- sarcoplasmic reticulum 158
- skeletal muscles 154
- slow oxidative (SO) fiber 165
- smooth muscle 155
- striated 154
- synergist 169
- temporal summation 164
- tetanus 164
- thick filament 158
- thin filament 158
- transverse tubule (T-tubule) 158
- tropomyosin 169
- troponin 169
- twitch 163
- visceral (single-unit) muscle tissue 166
- wave summation 164

Critical and Creative Thinking Questions

1. Rigor mortis is a stiffening of all the muscles in the body that occurs shortly after death. It happens when all the ATP in the muscles gets used up. Explain how this stiffening occurs with respect to aspects of the contraction cycle.

2. Drugs called calcium channel blockers help relax the walls of arteries and veins, thereby lowering high blood pressure. How does blocking a calcium channel cause the muscles of the arteries and veins to relax?

3. The night before racing, most marathon runners eat meals that are high in carbohydrates, like pasta. Explain how this "carbing up" helps their muscles during the race.

4. Jim lifts a hand weight by moving his elbow and subsequently lowers it. List the muscles involved in each motion and explain which ones are the prime movers, antagonists, synergists, and fixators.

5. Celia is training to run a 10k race. She stretches and exercises daily for months. Explain why she must stretch. What changes will occur to the structure of her leg muscles as a result of her training?

What is happening in this picture?

Many athletes complain about soreness after an extensive workout. Sometimes massage can help.

Think Critically

1. What should athletes do before exercise to prevent or minimize injury?
2. What biochemical or physiological aspects of muscle contraction cause muscle soreness following exercise?

Self-Test

(Check your answers in Appendix C.)

1. The first energy source used in a working muscle is _____.
   a. aerobic respiration
   b. glycolysis
   c. ATP
   d. creatine phosphate

   Use this figure to answer questions 2–4.

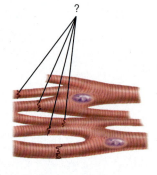

2. This figure shows _____ muscle tissue.
   a. skeletal   c. smooth
   b. cardiac   d. synergist

3. What are the structures indicated by the ? in the figure?
   a. Z-lines   c. nuclei
   b. striations   d. intercalated discs

4. The cells in the figure are capable of _____.
   a. twitch only   c. fused tetanus
   b. unfused tetanus   d. wave summation

5. Which of the following is not true of smooth muscle tissue?
   a. It can maintain steady contractions for a long time.
   b. Contractions are involuntary.
   c. It has multiple nuclei.
   d. It is found in the walls of all hollow organs.

6. The biceps brachii muscle helps to flex the elbow. The end of this muscle that attaches to the humerus is called the _____.
   a. insertion
   b. origin
   c. belly
   d. tendon

7. What is the largest skeletal muscle fiber?
   a. slow oxidative fiber
   b. fast glycolytic fiber
   c. slow glycolytic fiber
   d. fast oxidative-glycolytic fiber

Use this figure to answer questions 8–10.

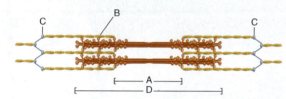

8. In which labeled structure would you find actin?
   a. A   c. C
   b. B   d. D

9. Which structure contains the Z-disc?
   a. A   c. C
   b. B   d. D

10. Which label represents the A-band?
    a. A   c. C
    b. B   d. D

Use this figure to answer questions 11–12.

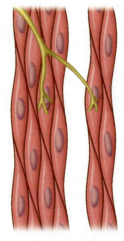

11. The cells in the figure regulate contraction by placing a phosphate group on _____.
    a. myosin
    b. actin
    c. troponin
    d. tropomyosin

12. Where would you find the arrangement of muscle fibers shown in this figure?
    a. biceps muscle
    b. heart
    c. bronchial airway
    d. urinary bladder

13. The _____ stabilizes the insertion point of a skeletal muscle.
    a. agonist muscle
    b. antagonist muscle
    c. synergist
    d. fixator

14. The _____ assists in breathing.
    a. soleus
    b. latissimus dorsi
    c. external intercostal
    d. extensor digitorum

15. The _____ has white fibers.
    a. uterus
    b. heart
    c. biceps
    d. bladder

THE PLANNER ✔

Review your Chapter Planner on the chapter opener and check off your completed work.

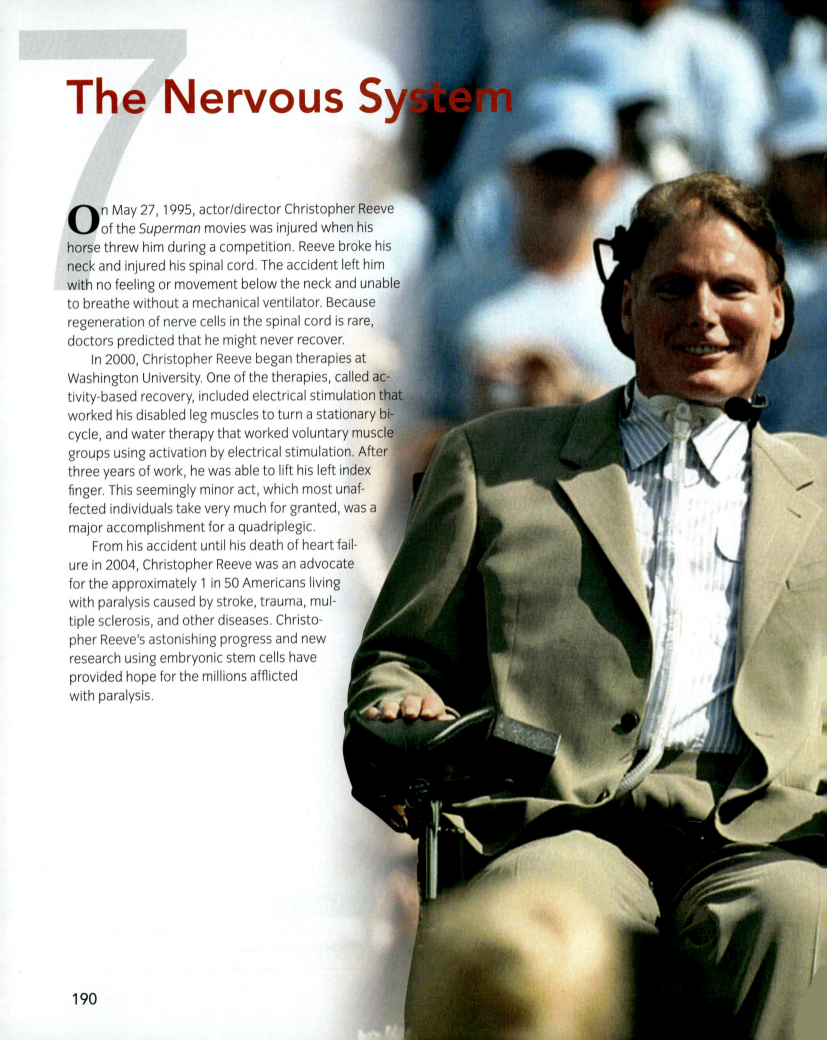

The Nervous System

On May 27, 1995, actor/director Christopher Reeve of the *Superman* movies was injured when his horse threw him during a competition. Reeve broke his neck and injured his spinal cord. The accident left him with no feeling or movement below the neck and unable to breathe without a mechanical ventilator. Because regeneration of nerve cells in the spinal cord is rare, doctors predicted that he might never recover.

In 2000, Christopher Reeve began therapies at Washington University. One of the therapies, called activity-based recovery, included electrical stimulation that worked his disabled leg muscles to turn a stationary bicycle, and water therapy that worked voluntary muscle groups using activation by electrical stimulation. After three years of work, he was able to lift his left index finger. This seemingly minor act, which most unaffected individuals take very much for granted, was a major accomplishment for a quadriplegic.

From his accident until his death of heart failure in 2004, Christopher Reeve was an advocate for the approximately 1 in 50 Americans living with paralysis caused by stroke, trauma, multiple sclerosis, and other diseases. Christopher Reeve's astonishing progress and new research using embryonic stem cells have provided hope for the millions afflicted with paralysis.

CHAPTER OUTLINE

CHAPTER PLANNER ✔

- ❏ Study the picture and read the opening story.
- ❏ Scan the Learning Objectives in each section:
  p. 192 ❏   p. 200 ❏   p. 208 ❏   p. 214 ❏
- ❏ Read the text and study all visuals. Answer any questions.

Analyze key features

- ❏ Process Diagram, p. 197 ❏   p. 198 ❏   p. 199 ❏   p. 219 ❏
- ❏ InSight, pp. 202–203 ❏   p. 206 ❏
- ❏ What a Health Provider Sees, p. 204 ❏
- ❏ Stop: Answer the Concept Checks before you go on:
  p. 199 ❏   p. 208 ❏   p. 213 ❏   p. 219 ❏

End of chapter

- ❏ Review the Summary and Key Terms.
- ❏ Answer the Critical and Creative Thinking Questions.
- ❏ Answer What is happening in this picture?
- ❏ Complete the Self-Test and check your answers.

Nerve Cells "Talk" to Each Other

LEARNING OBJECTIVES

1. **List** the structures and basic functions of the nervous system.

2. **Identify** the cells of the nervous system and their functions.

3. **Outline** the events in an action potential.

4. **Describe** the process of synaptic transmission.

Consider a computer. It has a central processing unit (microchip) with memory, ways to receive information (keyboard, mouse, disk drives, ports, modem), and ways to send information (disk drives, ports, modem). It receives information, processes it, and sends it out. Your nervous system does basically the same thing.

The Nervous System Has Sensory, Motor, and Integrative Functions

Instead of being made of silicon, wires, and circuit boards, your nervous system consists of billions of nerve cells called **neurons** and supporting cells called **neuroglia** (noo-RŌG-lē-a) or glial cells. Cells of the nervous system either lie within the **central nervous system (CNS)**, which includes the brain and spinal cord (see **Figure 7.1**), or outside the CNS in the **peripheral nervous system (PNS)**, which includes the cranial and spinal nerves as well as cells in the

peripheral nervous system (PNS) The portion of the nervous system that consists of nervous tissue structures that lie outside the brain and spinal cord.

central nervous system (CNS) The portion of the nervous system that is composed of the brain and spinal cord.

Structural overview of the nervous system • Figure 7.1

The nervous system is divided into two parts. The peripheral nervous system (PNS) senses changes in the environment, sends information to the central nervous system (CNS), and receives information from the CNS. The central nervous system processes information and makes decisions.

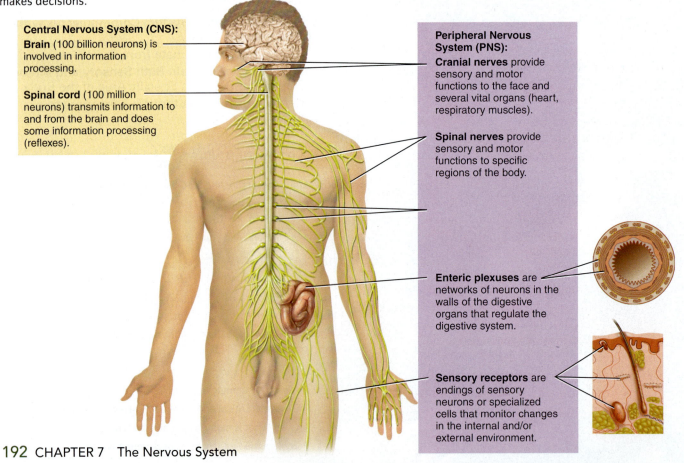

Central Nervous System (CNS):
Brain (100 billion neurons) is involved in information processing.

Spinal cord (100 million neurons) transmits information to and from the brain and does some information processing (reflexes).

Peripheral Nervous System (PNS):

Cranial nerves provide sensory and motor functions to the face and several vital organs (heart, respiratory muscles).

Spinal nerves provide sensory and motor functions to specific regions of the body.

Enteric plexuses are networks of neurons in the walls of the digestive organs that regulate the digestive system.

Sensory receptors are endings of sensory neurons or specialized cells that monitor changes in the internal and/or external environment.

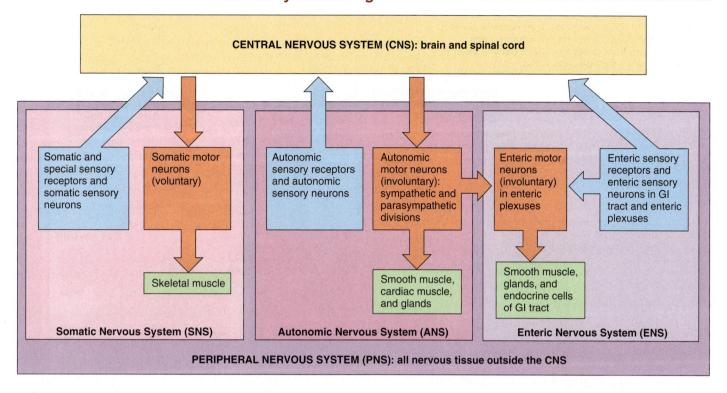

walls of the digestive organs and the sensory organs. Not to be confused with neurons, **nerves** are bundles of nerve fibers in the peripheral nervous system. The PNS is subdivided further according to functions of groups of sensory and motor neurons (**Figure 7.2**):

- The **afferent (sensory) nervous system** consists of a variety of nerve receptors and their associated nerve fibers:

  - *Somatosensory receptors* are associated with the muscles, joints, and skin.
  - *Special sense receptors* are found in the ear, eye, nose, and tongue.
  - *Autonomic sensory receptors* are found in the internal organs.

- The **efferent (motor) nervous system** is composed of motor nerve fibers that regulate the activities of muscle and glandular tissues throughout the body. This system can be subdivided into three sections:

  - The **somatic nervous system (SNS)** deals with initiating voluntary (under conscious control) skeletal muscle actions that move the body around in space.
  - The **autonomic nervous system (ANS)** regulates involuntary functions (such as heart rate, breathing rate, and body temperature) involving cardiac muscle, smooth muscle, and glandular tissue. This

system consists of two divisions, sympathetic and parasympathetic, which have opposite effects.

- The **enteric nervous system (ENS)** is an intricate network of nerve fibers within the digestive organs that regulates the involuntary functions of the digestive system and interacts with the ANS.

Like a computer, your nervous system accomplishes the following functions:

1. *Sensory function (information input).* Neurons in the peripheral nervous system (PNS) sense changes in internal and external environments, such as changes in blood pressure, injuries, touch, and pain, and send this information to the central nervous system (CNS).

2. *Integrative function (information processing).* Neurons in the CNS analyze sensory information and make decisions unconsciously or consciously. Conscious decisions require perception (or mental awareness) and higher-level processing.

3. *Motor function (information output).* After processing information and making decisions, CNS neurons send commands to muscular or glandular effectors that carry out responses such as muscle contraction/relaxation or increased/decreased secretion of substances such as oil or sweat.

Let's take a closer look at the cells that make up the nervous system.

a. Regardless of class (see part b.), all neurons consist of a cell body, dendrites, and axon. Information (arrows) flows from dendrites → cell body → axon → axon terminal → synaptic end bulbs.

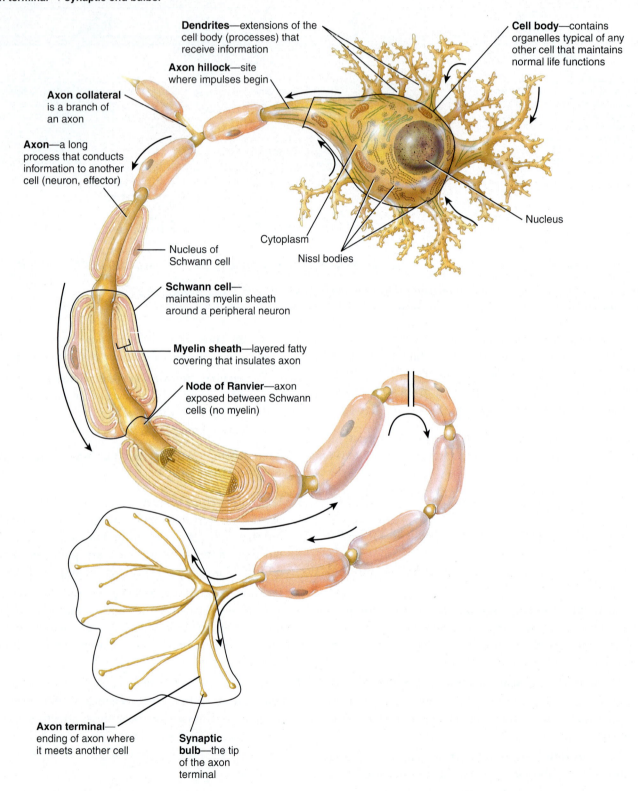

Dendrites—extensions of the cell body (processes) that receive information

Axon hillock—site where impulses begin

Axon collateral is a branch of an axon

Axon—a long process that conducts information to another cell (neuron, effector)

Nucleus of Schwann cell

Schwann cell—maintains myelin sheath around a peripheral neuron

Myelin sheath—layered fatty covering that insulates axon

Node of Ranvier—axon exposed between Schwann cells (no myelin)

Cell body—contains organelles typical of any other cell that maintains normal life functions

Nucleus

Cytoplasm

Nissl bodies

Axon terminal—ending of axon where it meets another cell

Synaptic bulb—the tip of the axon terminal

b. Neurons are classified according to number of processes extending from the cell body:

Multipolar neurons have several dendrites and one axon. They are found in the brain and spinal cord.

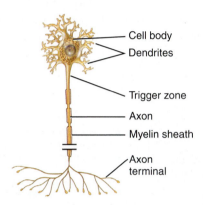

- Cell body
- Dendrites
- Trigger zone
- Axon
- Myelin sheath
- Axon terminal

Bipolar neurons have one dendrite and one axon. They are found in the retina of the eye, the inner ear, and olfactory areas of the brain.

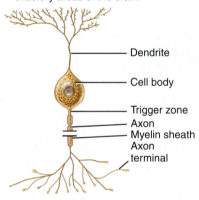

- Dendrite
- Cell body
- Trigger zone
- Axon
- Myelin sheath
- Axon terminal

Unipolar neurons have dendrites and axons fused into one process. Most sensory neurons are unipolar.

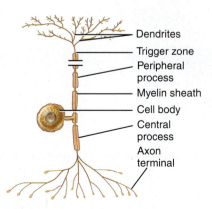

- Dendrites
- Trigger zone
- Peripheral process
- Myelin sheath
- Cell body
- Central process
- Axon terminal

Neurons Are Electrically Excitable Cells Designed for Transmitting Information

Like other cells, neurons have organelles such as endoplasmic reticulum (ER), mitochondria, and nuclei in their **cell bodies** that maintain normal cell functions, including protein synthesis and metabolism. Like muscle cells, neurons are electrically excitable and specialized for transmitting electrical impulses (**Figure 7.3a**).

Unlike most other cells, neurons have cellular extensions that are commonly referred to as **nerve fibers**. There are two different types of nerve fibers. A neuron always has one **axon**. This fiber conducts electrical impulses away from the cell body and is involved in signalling motor activity in other cells. The other fiber is called a **dendrite**. This fiber conducts impulses toward the cell body and is associated with sensory nerve receptors. A neuron can have multiple dendrites.

Neurons are classified in different ways. Structurally they are labeled as multipolar, bipolar, or unipolar, depending on the number of processes extending from the cell body (**Figure 7.3b**). Functionally, neurons fall into one of three classes:

- **Sensory neurons**, or **afferent neurons** (AF-er-ent NOO-ronz), either have specialized receptors themselves or are connected to separate **sensory receptors**. The somatosensory and autonomic sensory neurons are **unipolar neurons**, while the special sense receptors are attached to **bipolar neurons**. Sensory neurons transmit impulses toward the CNS through spinal or cranial nerves.

- **Motor neurons**, or **efferent neurons** (EF-er-ent NOO-ronz), are mostly **multipolar neurons** that carry information from the CNS to effectors through cranial or spinal nerves.

- **Interneurons** (in'-ter-NOO-ronz) extend only for short distances and contact nearby neurons in the brain, spinal cord, or a ganglion. A **ganglion** (plural is *ganglia*) is a collection of nerve cell bodies found in the peripheral nervous system. Similar collections of nerve cell bodies in the CNS are referred to as **nuclei**. Interneurons are mostly multipolar and comprise the vast majority of neurons in the body.

Neurons can conduct action potentials over a range of distances from just a few millimeters for an interneuron to over a meter for some of the sensory and motor neurons. To enable the impulses to be conducted rapidly over these distances, most neuronal fibers are covered by layers of **myelin**, which serves to insulate the axon and speed the conduction of impulses.

> **myelin** (MĪ-e-lin)
> A multilayered axon covering that consists of lipid and protein.

Now let's take a closer look at the supporting cells of the nervous system, neuroglia or glial cells.

Neuroglia Protect Neurons and Help Them Do Their Jobs

As mentioned previously, neuroglia, or glial (GLĒ-al) cells, provide various support functions for neurons. Neuroglia are found in both the CNS and PNS:

CNS neuroglia (**Figure 7.4**) include the following:

- **Microglia** provide protection against microbial organisms.
- **Astrocytes** form the **blood–brain barrier** to protect neurons from harmful chemicals.

> **blood–brain barrier** A barrier consisting of specialized brain capillaries and astrocytes that prevents the passage of materials from the blood to the cerebrospinal fluid and brain.

- **Oligodendrocytes** produce and maintain myelin.
- **Ependymal cells** produce cerebrospinal fluid (CSF).

PNS neuroglia include the following:

- **Schwann cells** produce myelin around a single axon and help in neuron regeneration (see Figure 7.3).
- **Satellite** (SAT-i-līt) **cells** support cells within the ganglia that regulate the exchange of material between the neurons and the surrounding interstitial fluid.

Now that we have looked at the cells that comprise the nervous system, let's see how they communicate with each other.

Neuroglia • Figure 7.4

Neuroglia provide many supporting functions that assist neurons in both the CNS and PNS.

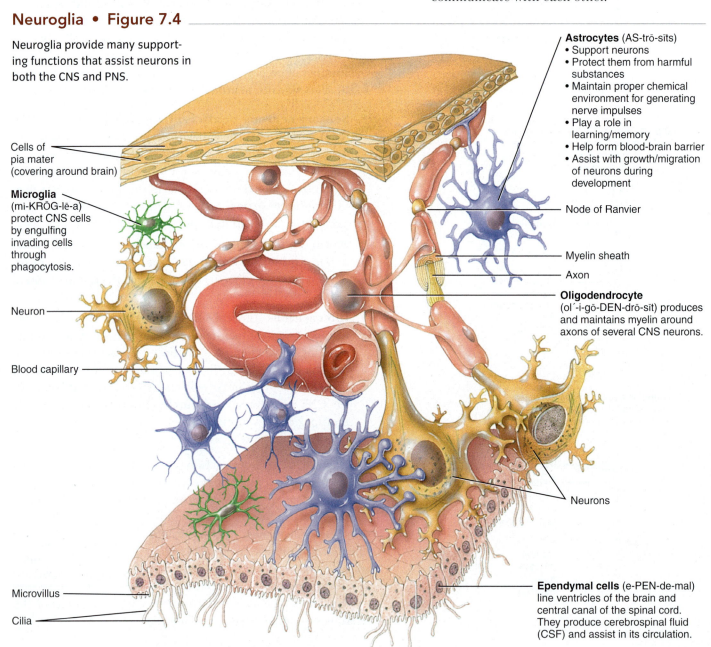

Cells of pia mater (covering around brain)

Microglia (mi-KRŌG-lē-a) protect CNS cells by engulfing invading cells through phagocytosis.

Neuron

Blood capillary

Astrocytes (AS-trō-sīts)
- Support neurons
- Protect them from harmful substances
- Maintain proper chemical environment for generating nerve impulses
- Play a role in learning/memory
- Help form blood-brain barrier
- Assist with growth/migration of neurons during development

Node of Ranvier

Myelin sheath

Axon

Oligodendrocyte (ol´-i-gō-DEN-drō-sīt) produces and maintains myelin around axons of several CNS neurons.

Neurons

Microvillus

Cilia

Ependymal cells (e-PEN-de-mal) line ventricles of the brain and central canal of the spinal cord. They produce cerebrospinal fluid (CSF) and assist in its circulation.

Ventricle (space) of brain

Events in the action potential • Figure 7.5

✓ THE PLANNER

Sodium (Na⁺) and potassium (K⁺) channels open and close in sequence during the phases of the action potential.

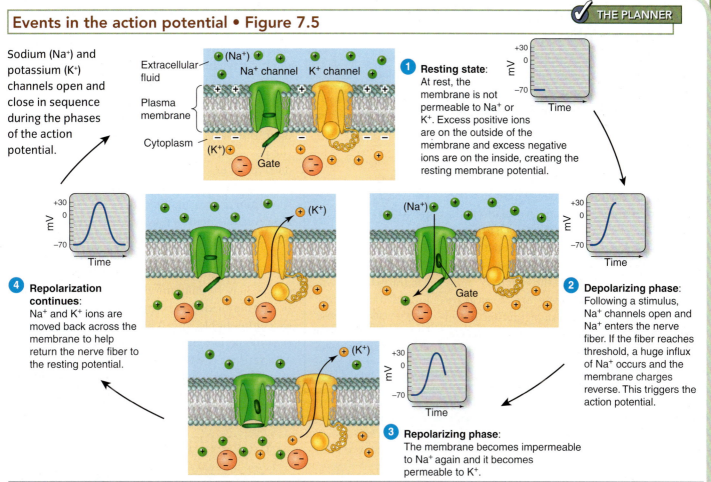

1 Resting state: At rest, the membrane is not permeable to Na⁺ or K⁺. Excess positive ions are on the outside of the membrane and excess negative ions are on the inside, creating the resting membrane potential.

2 Depolarizing phase: Following a stimulus, Na⁺ channels open and Na⁺ enters the nerve fiber. If the fiber reaches threshold, a huge influx of Na⁺ occurs and the membrane charges reverse. This triggers the action potential.

3 Repolarizing phase: The membrane becomes impermeable to Na⁺ again and it becomes permeable to K⁺.

4 Repolarization continues: Na⁺ and K⁺ ions are moved back across the membrane to help return the nerve fiber to the resting potential.

Action Potentials Help Propagate Nerve Impulses Along the Nerve Fiber

Like muscle cells, neurons are electrically excitable. They have a voltage difference across their membranes, called a **resting membrane potential,** which is similar to the voltage difference across the terminals of a battery. The resting membrane potential is about −70 mV (significantly smaller than that of a battery) and is caused by several factors related to the distribution of electrically charged sodium (Na⁺) and potassium (K⁺) ions across the membrane.

The membrane potential of a neuron can change slightly. When **stimuli** (detectable changes in the internal or external environment) open or close small ion channels at dendrites, the membrane potential can become more positive (depolarization) or more negative (hyperpolarization). These small changes, called **graded potentials**, can add together to depolarize the membrane to some critical level called the **threshold**. Once a threshold is reached, the membrane potential changes dramatically; various ion channels open and close in sequence to form an **action potential** (**Figure 7.5**).

action potential
An electrical signal that propagates along the membrane of a neuron or muscle fiber (cell).

The action potential is transmitted rapidly (meters per second) along the nerve fiber as the inflow of sodium ions in one portion of the axon stimulates the same sequence of events in adjacent portions. The action potential is divided into a depolarizing phase and a repolarizing phase. The repolarizing phase of the action potential "resets" the membrane potential so that the neuron can transmit another action potential. It also limits the transmission of the action potential to one direction.

The size of every action potential, called the magnitude of depolarization, is always the same. The intensity of a stimulus is not encoded by the size of the action potential but rather by the frequency of action potentials. The greater the stimulus, the more action potentials occur in a given period of time.

The same sequence of ionic events occurs in unmyelinated and myelinated nerve fibers. However, because action potentials propagate sequentially down portions of the membranes of unmyelinated axons and move from node to node in myelinated axons, myelinated axons transmit action potentials faster than unmyelinated axons.

Now let's see how an action potential moves from one cell to another.

Synapses Help Neurons Communicate with Other Cells

As you have just learned, a stimulus evokes an action potential in a neuron. The neuron conducts the action potential down its axon to the **axon terminal** (see Figure 7.3a), and to another cell (neuron, effector) at a junction called the **synapse**. The neuron that sends the message is called the *presynaptic neuron,* and the one that receives the message is called the *postsynaptic neuron*. Some synapses are considered to be *electrical synapses*. These junctions are generally found between two neurons that lie very close to one another. The plasma membranes of the two cells are connected by gap junctions, and the ionic signal passes through these gaps from one cell to the other.

Most synapses, however, are *chemical synapses*. The electrical message (action potential) in the presynaptic neuron gets converted to a chemical message with the release of a **neurotransmitter**. This chemical messenger drifts across the synapse, binds to receptors on the postsynaptic cell, and transmits information to the postsynaptic cell (**Figure 7.6**).

There are several different neurotransmitters in the human body, including **acetylcholine (ACh)**, dopamine, serotonin, and norepinephrine. They are stored in structures in synaptic bulbs called synaptic vesicles. Some neurotransmitters cause depolarizations in the postsynaptic cell, while others cause hyperpolarizations in the postsynaptic cell. A single neuron produces only one type of neurotransmitter.

Generally, synaptic transmission goes only one way, from a presynaptic cell to a postsynaptic cell. However, one neuron may receive input from several others. For example, a typical CNS neuron receives input from 1,000 to 10,000 synapses. Some synapses may be excitatory and cause depolarizations, while others may be inhibitory and cause hyperpolarizations. Many toxins and drugs affect synaptic transmission either by blocking it altogether or by modulating the amount of neurotransmitter released.

Neurons can play different roles in neural pathways. They can send, receive, or process information. Sensory neurons help relay information from

> **synapse** (SYN-aps) The functional junction between two neurons or between a neuron and an effector, such as a muscle or gland. A synapse may be electrical or chemical.
>
> **neurotransmitter** One of a variety of molecules within axon terminals that are released into the synaptic cleft in response to a nerve impulse and that change the membrane potential of the postsynaptic cell.

PROCESS DIAGRAM

Synaptic transmission • Figure 7.6

✓ THE PLANNER

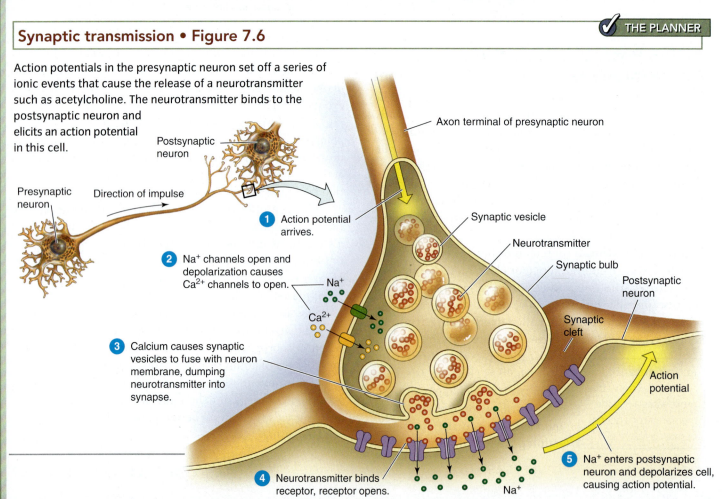

Action potentials in the presynaptic neuron set off a series of ionic events that cause the release of a neurotransmitter such as acetylcholine. The neurotransmitter binds to the postsynaptic neuron and elicits an action potential in this cell.

Postsynaptic neuron

Presynaptic neuron

Direction of impulse

1 Action potential arrives.

2 Na⁺ channels open and depolarization causes Ca²⁺ channels to open.

Na⁺

Ca²⁺

3 Calcium causes synaptic vesicles to fuse with neuron membrane, dumping neurotransmitter into synapse.

4 Neurotransmitter binds receptor, receptor opens.

Axon terminal of presynaptic neuron

Synaptic vesicle

Neurotransmitter

Synaptic bulb

Postsynaptic neuron

Synaptic cleft

Action potential

Na⁺

5 Na⁺ enters postsynaptic neuron and depolarizes cell, causing action potential.

Overview of neuron communication within the nervous system • Figure 7.7

 THE PLANNER

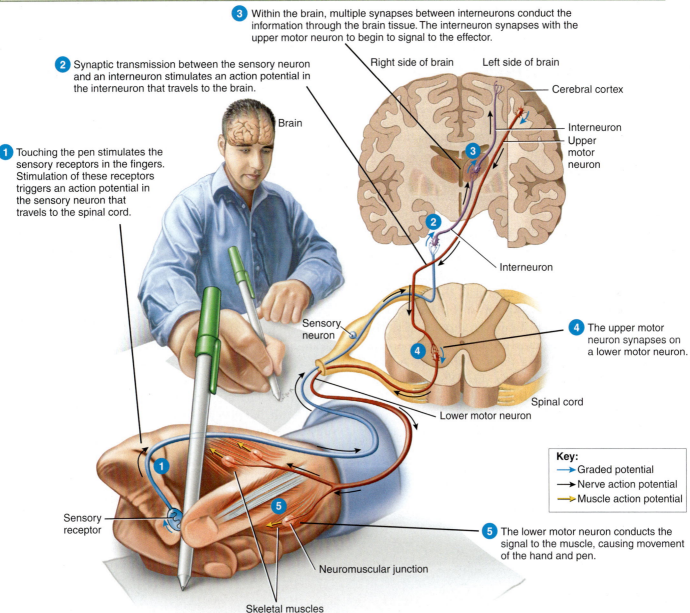

3 Within the brain, multiple synapses between interneurons conduct the information through the brain tissue. The interneuron synapses with the upper motor neuron to begin to signal to the effector.

2 Synaptic transmission between the sensory neuron and an interneuron stimulates an action potential in the interneuron that travels to the brain.

Right side of brain   Left side of brain

Cerebral cortex

Brain

1 Touching the pen stimulates the sensory receptors in the fingers. Stimulation of these receptors triggers an action potential in the sensory neuron that travels to the spinal cord.

Interneuron
Upper motor neuron

3

2

Interneuron

Sensory neuron

4 The upper motor neuron synapses on a lower motor neuron.

4

Spinal cord

Lower motor neuron

Key:
→ Graded potential
→ Nerve action potential
→ Muscle action potential

Sensory receptor

1

5

5 The lower motor neuron conducts the signal to the muscle, causing movement of the hand and pen.

Neuromuscular junction

Skeletal muscles

sensory receptors that gather data about the internal or external environment of the body. Interneurons link the sensory neurons with motor neurons and assist with information processing. Motor neurons are involved in sending signals out to effector organs (such as muscles or glands) to produce a response. In motor neuron connections to skeletal muscles, two neurons form the motor pathway. The upper motor neuron begins at the brain and relays the message to the lower motor neuron in the spinal cord. The lower motor neuron then sends the signal to the muscle cell. All neurons involved in such a pathway use action potentials and synaptic transmissions from one neuron to another to move information along (**Figure 7.7**).

CONCEPT CHECK STOP

1. **What** is the relationship between the enteric nervous system and the autonomic nervous system?

2. **What** is the function of the dendrites of a neuron?

3. **What** ionic changes occur in each of the phases of the action potential?

4. **How** does an action potential in a presynaptic cell elicit an action potential in a postsynaptic cell?

The Central Nervous System Coordinates All Nervous Activity

LEARNING OBJECTIVES

1. **Name** the coverings of the brain and spinal cord.
2. **Explain** the circulation of cerebrospinal fluid.
3. **List** the parts of the brain and describe the function of each.
4. **Describe** the major parts of the spinal cord and their functions.
5. **Compare** the ascending and descending spinal tracts.

As you have already learned, the central nervous system (CNS) consists of the brain and spinal cord. The brain is composed of 100 billion neurons and 10 to 50 trillion neuroglia. It has a mass of about 1.3 kg (about 3 lb). The "wiring" of the brain and spinal cord allows the CNS to coordinate and process nervous activity. Because different specialized areas of the brain receive and process different types of sensory information, knowledge of the structures of the brain and spinal cord is essential to an understanding of their functions. Let's take a closer look.

The Central Nervous System Requires Protection

As shown in **Figure 7.8**, the brain lies within the cranial cavity of the skull and is covered by three connective tissue membranes called **meninges**. The brain and spinal cord are bathed in cerebrospinal fluid (CSF), which is derived from the blood in spaces found inside the brain tissue called **ventricles** (**Figure 7.9**). The CSF circulates through the ventricles and around the brain and spinal cord, cushioning the CNS. The composition of CSF is very different from that of the blood. This separation of CSF from the blood (called the blood–brain barrier) protects the brain cells from harmful substances and toxins.

> **meninges** (me-NIN-jēz) Three membranes covering the brain and spinal cord, called the dura mater, arachnoid mater, and pia mater.
>
> **ventricles** (VEN-tri-kuls) Cavities in the brain filled with cerebrospinal fluid.

The coverings of the brain • Figure 7.8

The skull and cranial meninges protect the brain.

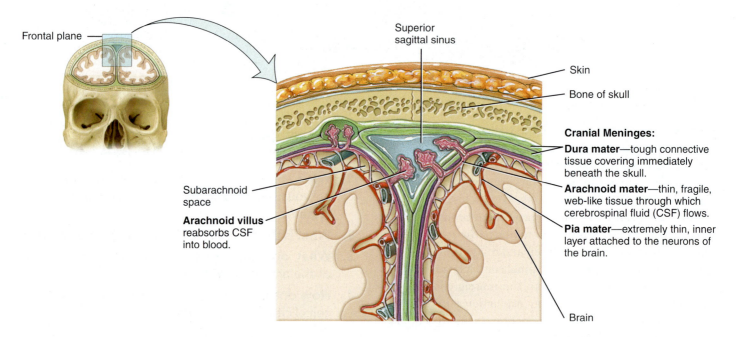

Frontal plane

Superior sagittal sinus

Skin

Bone of skull

Cranial Meninges:
Dura mater—tough connective tissue covering immediately beneath the skull.
Arachnoid mater—thin, fragile, web-like tissue through which cerebrospinal fluid (CSF) flows.
Pia mater—extremely thin, inner layer attached to the neurons of the brain.

Subarachnoid space

Arachnoid villus reabsorbs CSF into blood.

Brain

Circulation of the cerebrospinal fluid • Figure 7.9

CSF is produced in the ventricles and circulates through the ventricles and around the brain and spinal cord.

a. View of the ventricles (through a transparent brain)

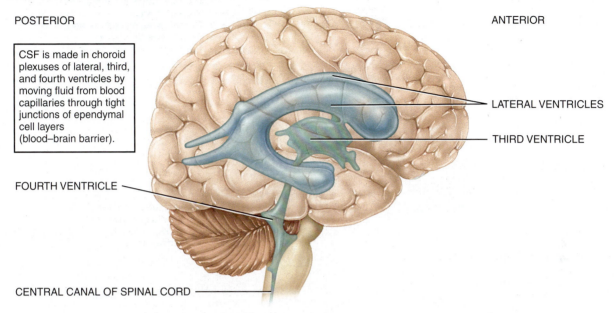

POSTERIOR

ANTERIOR

CSF is made in choroid plexuses of lateral, third, and fourth ventricles by moving fluid from blood capillaries through tight junctions of ependymal cell layers (blood–brain barrier).

LATERAL VENTRICLES

THIRD VENTRICLE

FOURTH VENTRICLE

CENTRAL CANAL OF SPINAL CORD

Right lateral view of brain

b. Flow of CSF through the brain and spinal cord

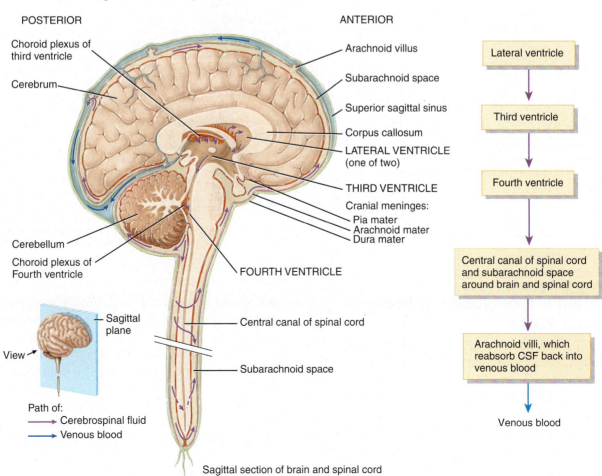

POSTERIOR

ANTERIOR

Choroid plexus of third ventricle

Cerebrum

Cerebellum

Choroid plexus of Fourth ventricle

Arachnoid villus

Subarachnoid space

Superior sagittal sinus

Corpus callosum

LATERAL VENTRICLE (one of two)

THIRD VENTRICLE

Cranial meninges:
Pia mater
Arachnoid mater
Dura mater

FOURTH VENTRICLE

Sagittal plane

View

Central canal of spinal cord

Subarachnoid space

Path of:
→ Cerebrospinal fluid
→ Venous blood

Sagittal section of brain and spinal cord

Lateral ventricle

↓

Third ventricle

↓

Fourth ventricle

↓

Central canal of spinal cord and subarachnoid space around brain and spinal cord

↓

Arachnoid villi, which reabsorb CSF back into venous blood

↓

Venous blood

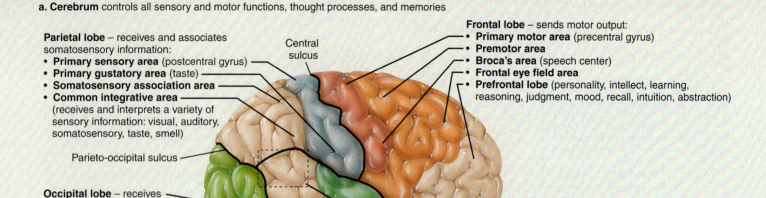

a. Cerebrum controls all sensory and motor functions, thought processes, and memories

Parietal lobe – receives and associates somatosensory information:
- **Primary sensory area** (postcentral gyrus)
- **Primary gustatory area** (taste)
- **Somatosensory association area**
- **Common integrative area**
  (receives and interprets a variety of sensory information: visual, auditory, somatosensory, taste, smell)

Central sulcus

Frontal lobe – sends motor output:
- **Primary motor area** (precentral gyrus)
- **Premotor area**
- **Broca's area** (speech center)
- **Frontal eye field area**
- **Prefrontal lobe** (personality, intellect, learning, reasoning, judgment, mood, recall, intuition, abstraction)

Parieto-occipital sulcus

Occipital lobe – receives visual information

Lateral cerebral sulcus

Temporal lobe – receives auditory information:
- Auditory association area (green)
- Wernicke's area interprets speech (partially located in parietal lobe)

b. Diencephalon, cerebellum, and brainstem

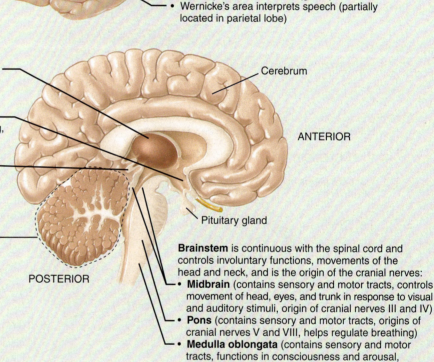

Diencephalon:
- **Thalamus** (relays sensory information to sensory cortex, motor information from cerebellum and basal ganglia to frontal cortex, role in conciousness)
- **Hypothalamus** (controls ANS and pituitary gland, emotional states, eating, drinking, sexual activity, circadian rhythms, sleep/wake cycles)
- **Pineal gland** (secretes melatonin)

Cerebrum

ANTERIOR

Pituitary gland

Cerebellum is attached to the brainstem by two bundles of axons called **cerebellar peduncles**, one from each hemisphere. The cerebellum smoothes and coordinates voluntary movements of skeletal muscles, regulates posture and balance, and may have a role in cognition and language processing.

POSTERIOR

Brainstem is continuous with the spinal cord and controls involuntary functions, movements of the head and neck, and is the origin of the cranial nerves:
- **Midbrain** (contains sensory and motor tracts, controls movement of head, eyes, and trunk in response to visual and auditory stimuli, origin of cranial nerves III and IV)
- **Pons** (contains sensory and motor tracts, origins of cranial nerves V and VIII, helps regulate breathing)
- **Medulla oblongata** (contains sensory and motor tracts, functions in consciousness and arousal, regulates vital functions (heart rate, breathing, blood vessel diameter), origins of cranial nerves VIII - XII)

The Brain Is Composed of Many Different Parts

The brain itself consists of four parts: brainstem, diencephalon, cerebellum, and cerebrum (**Figure 7.10**). The **cerebrum** (**Figure 7.10a**) is located superior to the other parts of the brain and is divided into two hemispheres, which communicate through a tract of axons called the **corpus callosum** (**Figure 7.10b**). The **brainstem** (**Figure 7.10a**) is continuous with the spinal cord and controls involuntary movements of the head and neck, and is the origin of the cranial nerves. In addition, the brainstem regulates many important physiological activities that keep you alive, including breathing and blood pressure. The **diencephalon** (**Figure 7.10a**) contains a relay station that coordinates sensory pathways to the brain from the periphery and motor pathways among various regions within the brain and an area which controls many vital functions, includ-

c. Basal ganglia and corpus callosum

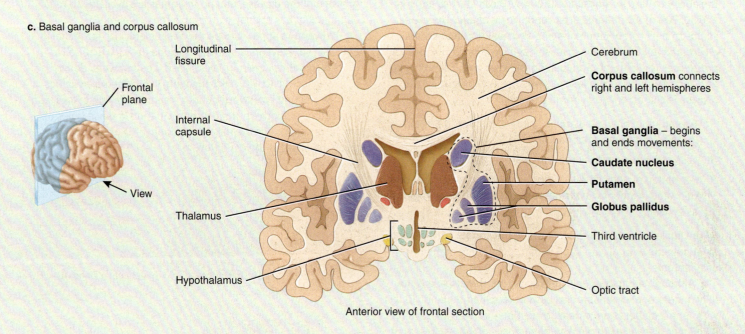

Longitudinal fissure

Frontal plane

View

Internal capsule

Thalamus

Hypothalamus

Cerebrum

Corpus callosum connects right and left hemispheres

Basal ganglia – begins and ends movements:

Caudate nucleus

Putamen

Globus pallidus

Third ventricle

Optic tract

Anterior view of frontal section

The cerebral surface consists of **gray matter** (cell bodies and dendrites), while underneath is **white matter** (axons). The surface is highly folded; each fold is called a **gyrus**. The fissures between the gyri (*pl.*) are called **sulci** (*singular:* sulcus) and several large sulci divide the cerebrum into four lobes.

ing eating, drinking, and body temperature, as well as the autonomic and endocrine systems. The **cerebellum** (**Figure 7.10a**) is located posterior to the brainstem, controls the stability of muscle movements, and initiates subconscious skeletal muscle movements associated with posture, balance, and rhythmic, repetitive activities (like walking or running). Each of the two hemispheres is in turn divided into four **lobes** (frontal, parietal, temporal, occipital).

The cerebrum also contains the basal ganglia (**Figure 7.10c**) and neurons associated with the limbic system. The cerebrum controls all sensory and voluntary motor functions, thought processes, and memories through sensory areas, motor areas, and association areas within the lobes. A number of diseases are caused by impairment of brain functions such as memory and reasoning; among these is Alzheimer disease, which causes extensive damage to the brain (see *What a Health Provider Sees* on the next page).

WHAT A HEALTH PROVIDER SEES

Alzheimer Disease

Alzheimer disease (ALTZ-hī-mer) or **(AD)** is a disabling senile dementia that results in the loss of reasoning, memory, and the ability to care for oneself. Although AD can start as early as age 30, 90% of AD patients are over 65 years of age. The cause seems to be a combination of aging and genetic, lifestyle, and environmental factors. The disease begins with memory loss and progresses to disorientation, personality changes, loss of cognition and recognition, and loss of language. Pathologically, the disease is characterized by three changes:

- Formation of amyloid plaques, made of β-amyloid protein and bits of degenerated neurons

- Loss of neurons associated with learning and memory

- Formation of neurofibrillary tangles (NFT), abnormal bundles of protein filaments inside neurons of affected brain regions, as shown.

By the final stage of AD, the brain tissue has shrunk considerably. Most patients with AD are dead within seven years of the onset of symptoms.

Although the pathogenic mechanism for the disease is not fully understood, recent studies have determined that individuals with Type II diabetes are at least three times as likely to develop Alzheimer disease as those who do not have diabetes. Research studies have also isolated three genes involved in the process; current research is focused on determining how the amyloid

plaques and NFTs form and identifying risk factors for AD. Medical therapies have not been entirely effective in slowing the progression of the disease, and no therapies are available to repair damage that has already occurred in these patients.

Prevention appears to be the best "medicine" for Alzheimer's disease. Preventive activities include eating a diet rich in fruits, vegetables, grains, olive oil, fish, and red wine; doing regular physical activity; and engaging in intellectual activities such as working crossword puzzles, reading, playing games, and interacting with others. Regular participation in these types of activities can reduce the risk of developing AD or will at least delay the onset of the condition.

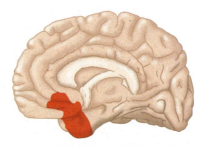

Very early stage AD

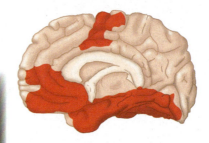

Mild to moderate stage AD

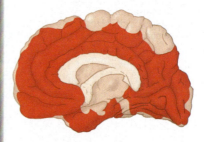

Final stage (severe) AD

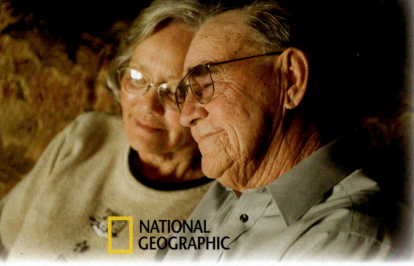

NATIONAL GEOGRAPHIC

Think Critically **1.** AD is associated with loss of memory and cognition. Which part of the brain is most affected? **2.** How would advanced AD affect motor skills such as writing? Explain your answer. In your explanation, refer to the neural pathways involved.

In the cerebrum, the sensory and motor areas have groups of neurons devoted to specific parts of the body (mouth, hands, feet, visual fields, auditory fields, and so on). Essentially, these areas have "maps" of the body

wired into them. **Figure 7.11** shows the maps called **homunculi** (singular is *homunculus*) of the primary somatosensory area and the primary motor area of the body. Although the maps appear distorted, they show that ar-

Maps of the body within the brain • Figure 7.11

Your brain has distorted maps within its sensory and motor areas called homunculi that show the amount of cells dedicated to any given area of the body.

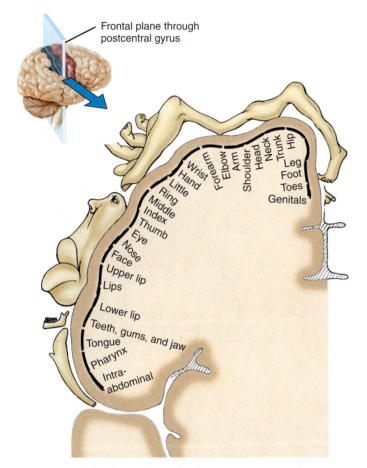

a. Frontal section of primary somatosensory area in right cerebral hemisphere

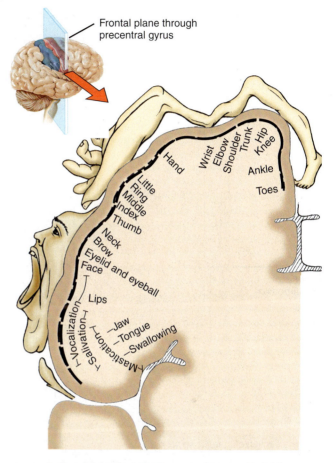

b. Frontal section of primary motor area in right cerebral hemisphere

eas such as the hands and mouth have more cells devoted to them than do the legs or arms. This makes sense because we have fine motor skills and need detailed control of muscular movements in our fingers and hands and to make sounds with our mouths. Motor units in these areas are quite small, so numerous neurons would be required to control the muscle cells. The large motor skills associated with moving our legs or arms do not require such fine control. The motor units in these areas of the body are quite large, requiring fewer neurons to regulate the muscle movements.

The hemispheres of the brain are symmetrical but vary slightly. Both have sensory and motor functions. Although information is generally processed by both hemispheres, the focus tends to vary a bit, depending on which hemisphere is dominant. The left hemisphere is inclined to address information from a more logical, linear reasoning perspective; the right hemisphere tends to introduce more of an artistic, imaginative perspective. For many

individuals, the primary speech centers associated with vocabulary, grammar, and speech production are found in the left hemisphere, but the right hemisphere helps to add intonation and contextual qualities to our speech. Most individuals use one hemisphere slightly more than the other to process information. This is referred to as *brain lateralization* or, more commonly, *brain dominance*.

Despite the fact that we have the ability to control muscle functions on both sides of the body, we usually elect to choose a particular side to lead the activity. About three-quarters of the population are right-handed, with most of the remaining individuals being left-handed. A very small percentage of the population is ambidextrous (equal dominance)—a trait that is often "learned" rather than inborn. For many of us, attempting to use the non-dominant hand to perform a task easily done by the dominant hand can require great concentration and often frustration.

Now let's turn to the spinal cord to see how the brain communicates with the outside world.

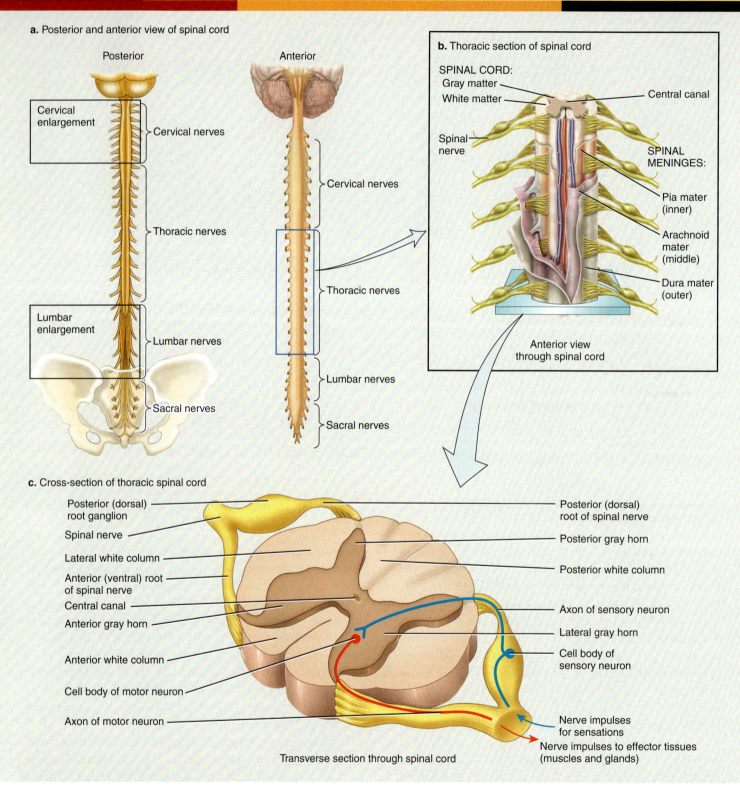

a. Posterior and anterior view of spinal cord

Posterior

Anterior

Cervical enlargement

Cervical nerves

Thoracic nerves

Lumbar enlargement

Lumbar nerves

Sacral nerves

Cervical nerves

Thoracic nerves

Lumbar nerves

Sacral nerves

b. Thoracic section of spinal cord

SPINAL CORD:
Gray matter
White matter

Central canal

Spinal nerve

SPINAL MENINGES:

Pia mater (inner)

Arachnoid mater (middle)

Dura mater (outer)

Anterior view through spinal cord

c. Cross-section of thoracic spinal cord

Posterior (dorsal) root ganglion

Spinal nerve

Lateral white column

Anterior (ventral) root of spinal nerve

Central canal

Anterior gray horn

Anterior white column

Cell body of motor neuron

Axon of motor neuron

Posterior (dorsal) root of spinal nerve

Posterior gray horn

Posterior white column

Axon of sensory neuron

Lateral gray horn

Cell body of sensory neuron

Nerve impulses for sensations

Nerve impulses to effector tissues (muscles and glands)

Transverse section through spinal cord

The Spinal Cord Helps the Brain Communicate with the Environment

The **spinal cord** (**Figure 7.12**) lies in the vertebral canal and is 42 to 45 cm (16 to 18 in.) long. It is continuous with the **medulla oblongata** of the brainstem. The spinal cord is divided into segments that correspond to the segments of the bony vertebral column (**Figure 7.12a**). However, because the spinal cord is shorter than the vertebral column, the lumbar, sacral, and coccygeal nerves do not exit the cord at the same levels where they exit the column. Furthermore, the spinal cord is enlarged in the cervical and lumbar areas where nerves from the upper and lower limbs connect.

Like the brain, the spinal cord is covered by meninges (**Figure 7.12b**) and bathed in CSF. The nerve fibers of the spinal nerves enter and exit the cord at each segment. The cell bodies of the sensory neurons lie outside the cord, in ganglia. The cell bodies of the motor neurons lie in the butterfly-shaped gray area (**Figure 7.12c**) at the center of the spinal cord. Surrounding the **gray matter** is the **white matter**. The white matter contains bundles of myelinated nerve fibers ascending and descending through the spinal cord in separate tracts. The shape and amount of white and gray matter varies with the spinal cord segment. The lower segments have less white matter than upper segments because there are fewer ascending and descending pathways at the lower levels of the spinal cord.

The spinal cord is a relay and processing center for neural input from the periphery. Nerve fibers from receptors of sensory neurons throughout the body enter the spinal cord at various levels depending upon the receptor's location. The pathway from there to the brain can involve several synapses with interneurons prior to the message reaching its destination, where analysis of the sensation will occur (**Figure 7.13**). There are a number of ascending spinal tracts that conduct information through the spinal cord. Some of the tracts will cross to opposite sides and some remain on the same side of the body as they make their way to the brain.

Ascending and descending pathways through the brain and spinal cord • Figure 7.13

a. Sensory information is carried from outside the body through the spinal cord to the brain.

Ascending (sensory) pathway
1. Incoming sensory neuron (first neuron) has axon that extends up to medulla.
2. Axon of second neuron crosses midline and extends to thalamus.
3. Third neuron extends to primary somatosensory cortex.

b. Motor pathways descend from the brain through the spinal cord to the outside.

Descending (motor) pathway
1. Upper motor neuron extends from primary motor cortex into spinal cord.
2. Upper motor neuron synapses on lower motor neuron in spinal cord. If the neuron decussates (crosses over) as shown, the synapse will be on the opposite side of the spinal cord. Note that some motor neurons do not cross over.

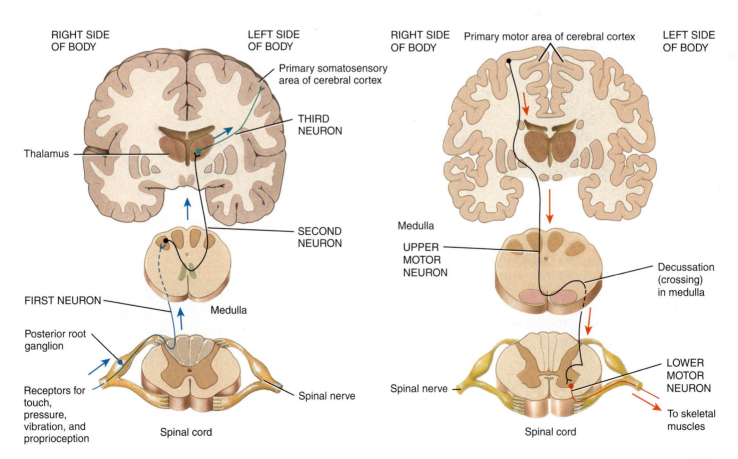

Similarly, descending tracts pass through the spinal cord in different ways. The axons of some motor neurons cross the midline at the medulla, allowing the right hemisphere of your cerebrum to control motor functions on the left side of your body, and vice versa. The fibers of some tracts do not cross over so some functions are controlled by the same side of the brain. The descending tracts have a two-neuron pathway that leads from the brain to the muscle. The upper motor neuron conducts the message from the brain to the lower motor neuron, which then relays that message to the muscle.

CONCEPT CHECK　　STOP

1. **Cerebrospinal** fluid flows between which layers of the meninges?

2. **Where** is cerebrospinal fluid produced?

3. **What** cerebral lobe is primarily responsible for processing visual information?

4. **Where** are the cell bodies of motor neurons located?

5. **Which** spinal tracts contain the upper motor neurons?

The Autonomic Nervous System Controls the Activities of Smooth Muscle, Cardiac Muscle, and Glands

LEARNING OBJECTIVES

1. **Outline** the parts of the autonomic nervous system.

2. **Compare** the main structures and functions of the somatic and autonomic nervous systems.

3. **Describe** the functions of the sympathetic and parasympathetic divisions of the autonomic nervous system.

While crossing the street to go to your car after lunch, a bus careens around the corner, barely missing you. Your heart starts racing, you are breathing rapidly, you begin to sweat, and your lunch feels like a lump in your stomach. What is going on here? These and many other involuntary functions are controlled by your **autonomic nervous system (ANS)**.

> **autonomic nervous system (ANS)**
> An automatic or self-regulating system of autonomic motor neurons, both sympathetic and parasympathetic, that conduct nerve impulses from the central nervous system to smooth muscle, cardiac muscle, and glands.

The ANS Uses a Two-Neuron Pathway to Communicate with the Effectors

The ANS is divided into two divisions, sympathetic and parasympathetic. Like the somatic nervous system (SNS), the ANS has sensory and motor neurons. The sensory neurons are located in the walls of organs and blood vessels. The ANS has motor neurons that regulate cardiac muscles, smooth muscles, and glands. However, unlike the somatic nervous system, the ANS has two neurons that link the CNS to the peripheral effector (**Figure 7.14**). The cell bodies for the second motor neurons in ANS motor pathways are located in ganglia. These ganglia are outside the spinal cord, either in the walls of the organs or organized into chains. The chains can run alongside the vertebral column or be positioned near the large abdominal arteries (see **Table 7.1** on page 210).

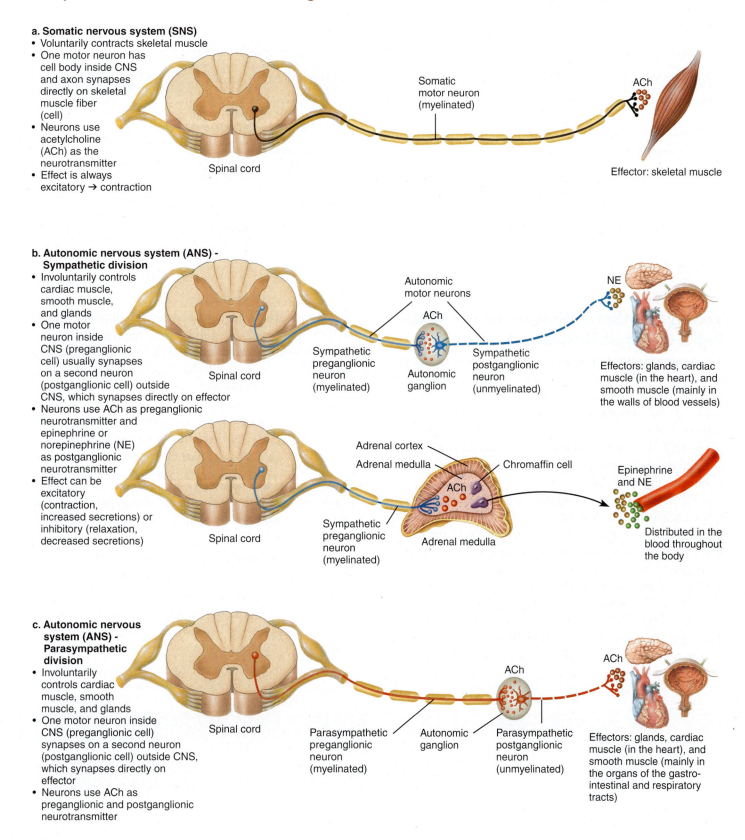

a. Somatic nervous system (SNS)
- Voluntarily contracts skeletal muscle
- One motor neuron has cell body inside CNS and axon synapses directly on skeletal muscle fiber (cell)
- Neurons use acetylcholine (ACh) as the neurotransmitter
- Effect is always excitatory → contraction

Spinal cord

Somatic motor neuron (myelinated)

ACh

Effector: skeletal muscle

b. Autonomic nervous system (ANS) - Sympathetic division
- Involuntarily controls cardiac muscle, smooth muscle, and glands
- One motor neuron inside CNS (preganglionic cell) usually synapses on a second neuron (postganglionic cell) outside CNS, which synapses directly on effector
- Neurons use ACh as preganglionic neurotransmitter and epinephrine or norepinephrine (NE) as postganglionic neurotransmitter
- Effect can be excitatory (contraction, increased secretions) or inhibitory (relaxation, decreased secretions)

Spinal cord

Autonomic motor neurons

ACh

Sympathetic preganglionic neuron (myelinated)

Autonomic ganglion

Sympathetic postganglionic neuron (unmyelinated)

NE

Effectors: glands, cardiac muscle (in the heart), and smooth muscle (mainly in the walls of blood vessels)

Adrenal cortex

Adrenal medulla

Chromaffin cell

ACh

Sympathetic preganglionic neuron (myelinated)

Adrenal medulla

Epinephrine and NE

Distributed in the blood throughout the body

c. Autonomic nervous system (ANS) - Parasympathetic division
- Involuntarily controls cardiac muscle, smooth muscle, and glands
- One motor neuron inside CNS (preganglionic cell) synapses on a second neuron (postganglionic cell) outside CNS, which synapses directly on effector
- Neurons use ACh as preganglionic and postganglionic neurotransmitter

Spinal cord

Parasympathetic preganglionic neuron (myelinated)

Autonomic ganglion

ACh

Parasympathetic postganglionic neuron (unmyelinated)

ACh

Effectors: glands, cardiac muscle (in the heart), and smooth muscle (mainly in the organs of the gastro-intestinal and respiratory tracts)

Comparison of somatic and autonomic nervous systems Table 7.1		
Property	**Somatic**	**Autonomic**
Effectors	Skeletal muscles	Cardiac muscle, smooth muscle, and glands
Type of control	Mainly voluntary	Mainly involuntary
Neural pathway	One motor neuron extends from CNS and synapses directly with a skeletal muscle fiber	One motor neuron extends from the CNS and usually synapses with another motor neuron in a ganglion; the second motor neuron synapses with an autonomic effector
Neurotransmitter	Acetylcholine	Acetylcholine or norepinephrine
Action of neurotransmitter on effector	Always excitatory (causing contraction of skeletal muscle)	May be excitatory (causing contraction of smooth muscle, increased heart rate, increased force of heart contraction, or increased secretion from glands) or inhibitory (causing relaxation of smooth muscle, decreased heart rate, or decreased secretions from glands)

The Sympathetic Division of the ANS: Fight-or-Flight Responses

The **sympathetic division** of the ANS is also called the *thoracolumbar division* because the sympathetic nerves emerge from the thoracic and lumbar segments of the spinal cord. The ganglia lie on either side of the vertebral column and near large abdominal arteries. A single sympathetic preganglionic axon may synapse with 20 or more postganglionic neurons and can therefore elicit a number of effects (**Figure 7.15**).

The neurons of the sympathetic division use two neurotransmitters. Like cells of the somatic nervous system, preganglionic sympathetic neurons release acetylcholine (ACh) at their synapses in the sympathetic ganglia. ACh gets destroyed in the synaptic cleft by the enzyme acethylcholinesterase (AChE), so synaptic transmission is short-lived. The sympathetic postganglionic cells release norepinephrine (NE) at their synapses with target organs. NE is inactivated more slowly than ACh, so its effects last longer. The adrenal medulla releases both NE and epinephrine into the blood so they circulate throughout the body and have an even longer-lasting effect.

The sympathetic division responds to physical or emotional stressors such as exercise, emergency, excitement, or embarrassment. Your body's reaction to sympathetic stimulation is often called the "fight-or-flight" response because it prepares your body for the physical exertion required to fight or run from danger (Figure 7.15). Remember the bus incident described at the beginning of this section? The reactions that your body experienced were due to activation of the sympathetic nervous system:

- Increased heart rate, force of contraction, and blood pressure, as well as vasodilation, increase blood flow to organs and muscles. This helps deliver oxygen and nutrients to working muscles.
- Dilating the airways and increasing the rate of breathing increase air flow and bring more oxygen into the body.
- Breaking down glycogen and fatty acids and releasing glucose into the blood provide energy for working muscles.
- Inhibiting digestion keeps blood from being diverted from muscles.

Structure and functions of the sympathetic division of the ANS • Figure 7.15

Cell bodies of the sympathetic preganglionic neurons are located in the gray matter of spinal segments T1–T12 and L1–L2. Outflow from these neurons has multiple effects on various target organs.

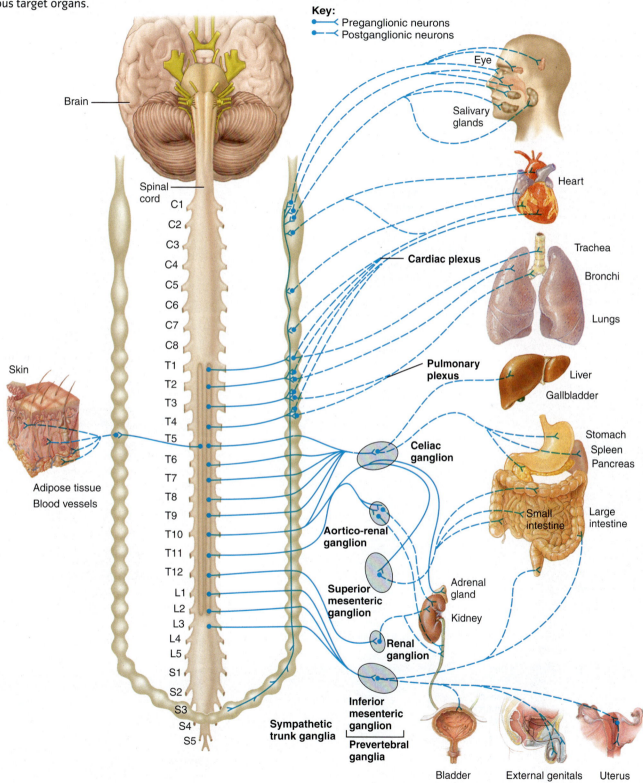

Key:
- Preganglionic neurons
- Postganglionic neurons

Brain

Spinal cord

C1
C2
C3
C4
C5
C6
C7
C8
T1
T2
T3
T4
T5
T6
T7
T8
T9
T10
T11
T12
L1
L2
L3
L4
L5
S1
S2
S3
S4
S5

Skin

Adipose tissue
Blood vessels

Eye

Salivary glands

Heart

Trachea
Bronchi
Lungs

Liver
Gallbladder

Stomach
Spleen
Pancreas

Small intestine
Large intestine

Adrenal gland
Kidney

Bladder
External genitals
Uterus

Cardiac plexus

Pulmonary plexus

Celiac ganglion

Aortico-renal ganglion

Superior mesenteric ganglion

Renal ganglion

Inferior mesenteric ganglion

Sympathetic trunk ganglia

Prevertebral ganglia

Cell bodies of parasympathetic preganglionic neurons are located in brainstem nuclei and in the gray matter of spinal segments S2-S4. Outflow from these neurons has multiple effects on various target organs (See Table 7.2).

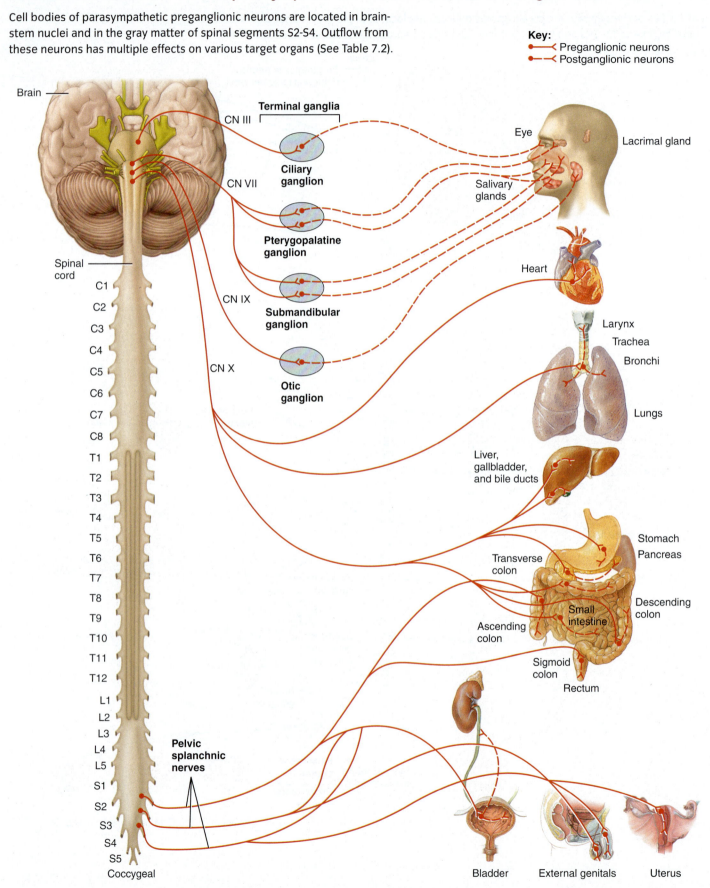

Key:
- Preganglionic neurons
- Postganglionic neurons

The Parasympathetic Division of the ANS: A Rest-and-Digest Response

The **parasympathetic division** of the ANS is also called the *craniosacral division* because the parasympathetic nerves emerge from some of the cranial nerves and the sacral segments of the spinal cord (**Figure 7.16**). Unlike their location in the sympathetic division, the cell bodies of parasympathetic neurons lie in **terminal ganglia** (also called *intramural ganglia*), which are close to the target organ or within the walls of the target organ. So, compared to sympathetic pathways, the parasympathetic preganglionic cells are longer, but the parasympathetic postganglionic cells are shorter. Also, parasympathetic preganglionic neurons synapse with only four or five postganglionic neurons. Parasympathetic effects therefore tend to be limited to specific target organs. Both pre- and postganglionic parasympathetic neurons release ACh at their synapses.

Because parasympathetic activities dominate periods of relative inactivity, the reaction of your body has been referred to as the "rest-and-digest" response:

- Stimulation of the digestive tract leads to increased gastric and pancreatic secretions, which increase digestion.
- Vasoconstriction decreases blood flow to working muscle and shunts it toward the digestive tract.
- Relaxing urinary bladder muscles and anal muscles leads to increased urination and defecation.

A common mnemonic (a device that helps recall) for remembering parasympathetic activities is SLUDD, which stands for salivation, lacrimation, urination, digestion, and defecation. **Table 7.2** compares the effects of stimulation of the sympathetic and parasympathetic divisions of the ANS.

CONCEPT CHECK

1. **Where** do you find the cell bodies of postsynaptic neurons involved in the ANS?
2. **Contrast** the neurons of the ANS with those of the somatic nervous system.
3. **What** are the body's responses to sympathetic stimulation? To parasympathetic stimulation?

Effects of stimulation of the sympathetic and parasympathetic divisions of the autonomic nervous system Table 7.2

Target	Effect of Sympathetic Stimulation	Effect of Parasympathetic Stimulation
Sweat gland	Increases secretion of antidiuretic hormone (ADH)	No known effect
Skin	Increases sweating; contraction of hair follicles to produce goosebumps	No known effect
Adrenal medullae	Breaks down triglycerides and releases fatty acids	No known effect
Blood vessels	Dilation; increases blood flow	No known effect
Eye	Dilation of pupil; relaxation of ciliary muscle	Constriction of pupil
Lacrimal glands (tears)	Slight secretion	Secretion of tears
Salivary glands	Decrease secretion	Increase secretion
Heart	Increases rate and force of contraction; increases blood flow to heart muscle	Decreases heart rate and force of contraction, decreases blood flow to heart muscle
Lungs	Widening of the airways	Narrowing of the airways
Liver	Breaks down glycogen into glucose, makes new glucose, releases glucose into blood, and decreases bile secretion	Increases glycogen synthesis; increases release of bile into small intestine
Digestive system	Decreases movement through GI tract, inhibits gastric secretions, inhibits insulin secretion, and increases glucagon secretion	Increases pancreatic secretions of digestive enzymes and insulin, increases movement through GI tract, and increases gastric secretions
Adrenal glands	Increase secretions of epinephrine and norepinephrine	No known effect
Uterus	Inhibits contraction in non-pregnant women and stimulates contraction in pregnant women	Minimal effect
External genitals	Causes ejaculation (males)	Erection of penis (males) and clitoris (females)
Kidney	Decreases urine production	No known effect
Urinary bladder	Relaxation of bladder muscle wall, and contraction of internal bladder sphincter muscle	Contraction of muscular wall and relaxation of internal sphincter

The Peripheral Nervous System Communicates with the Outside World

LEARNING OBJECTIVES

1. **Describe** the parts of a nerve.

2. **Identify** the 12 pairs of cranial nerves and the type of information carried by each.

3. **Outline** the distribution of spinal nerves.

4. **Explain** how a reflex arc works.

The CNS communicates with the outside world through the peripheral nervous system (PNS), which consists of cranial nerves and peripheral, or spinal, nerves. Nerves are collections of neuronal axons wrapped into bundles by various connective tissue coverings. These bundles are in turn wrapped together again, much like wires in a conduit (**Figure 7.17**). Most nerves carry both sensory and

Parts of a nerve • Figure 7.17

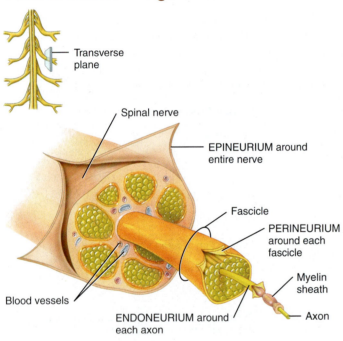

- Transverse plane
- Spinal nerve
- EPINEURIUM around entire nerve
- Fascicle
- PERINEURIUM around each fascicle
- Myelin sheath
- Axon
- ENDONEURIUM around each axon
- Blood vessels

Transverse section showing the parts and coverings of a nerve

Cranial nerves • Figure 7.18

The cranial nerves provide sensory and motor functions for the head, special senses (sight, hearing, smell, taste), and parasympathetic outputs to the body.

NERVE	FUNCTION
I	Sensory for smell
II	Sensory for vision
III	Motor for eye movement and focusing
IV	Motor for eye movement
V	Sensory from face and motor for chewing
VI	Motor for eye movement
VII	Sensory for taste and motor for facial movement
VIII	Sensory for hearing and equilibrium
IX	Sensory for taste and blood pressure and motor for swallowing and speech
X	Sensory and motor associated with thoracic and abdominal organs
XI	Motor to muscles of head, neck, and shoulders
XII	Motor to tongue

Mnemonic for remembering the names of the cranial nerves
Oh, **O**h, **O**h, **T**o **T**ouch **A**nd **F**eel **V**ery **G**reen **V**egetables **AH**!

motor neurons (mixed); however, some cranial nerves carry sensory neurons only (sensory nerves) or motor neurons only (motor nerves). Let's start by looking at cranial nerves.

Cranial Nerves Originate from Brain Tissue

There are 12 pairs of cranial nerves. Each pair of nerves is numbered with a Roman numeral that designates the order in which it emerges from the brainstem. Cranial nerves I, II, and VIII are sensory nerves, and cranial nerves III, IV, VI, XI, and XII are motor nerves; see **Figure 7.18** for the specific functions of each pair. The cell bodies of sensory neurons in the cranial nerves lie in ganglia outside the brain, while those of motor neurons are located in nuclei within the brain. The major functions of the cranial nerves are sensory and motor functions of the face and head, special senses, and ANS output (Figure 7.18).

Now let's examine spinal nerves.

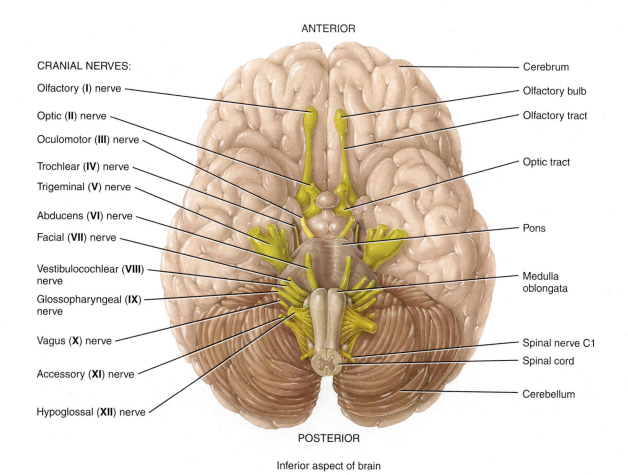

Inferior aspect of brain

Spinal Nerves Contain Both Sensory and Motor Fibers That Supply a Specific Area of the Body

The **spinal nerves** are all mixed nerves. The cell bodies of sensory neurons lie in the posterior root ganglia (dorsal root ganglia), while those of motor neurons lie in the anterior horns of the spinal cord (see Figure 7.12c). After emerging from the spinal cord, many spinal nerves divide into several branches that form networks of neurons called *plexuses* (singular is **plexus**; **Figure 7.19**). After the nerves pass through the plexus, the emerging nerves carry neurons from several adjacent spinal segments.

Spinal nerves • Figure 7.19

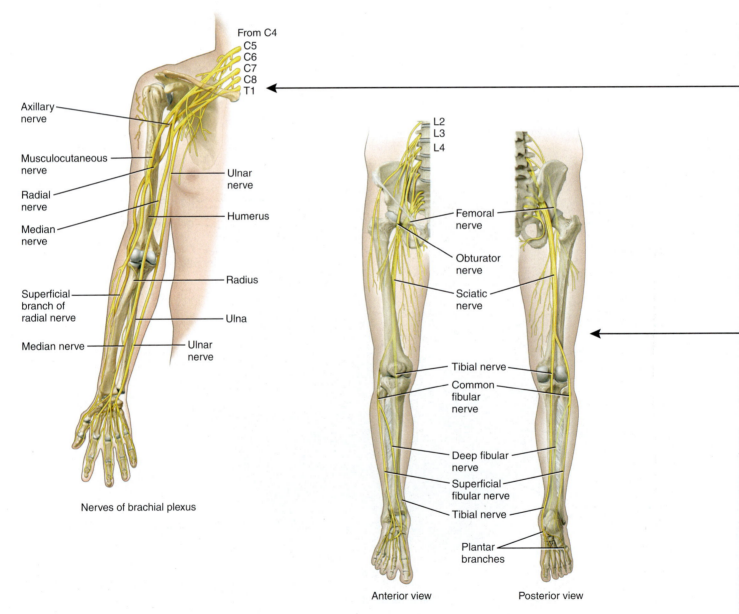

From C4
C5
C6
C7
C8
T1

Axillary nerve

Musculocutaneous nerve

Radial nerve

Median nerve

Superficial branch of radial nerve

Median nerve

Ulnar nerve

Humerus

Radius

Ulna

Ulnar nerve

Nerves of brachial plexus

L2
L3
L4

Femoral nerve

Obturator nerve

Sciatic nerve

Tibial nerve

Common fibular nerve

Deep fibular nerve

Superficial fibular nerve

Tibial nerve

Plantar branches

Anterior view

Posterior view

Nerves of lumbar plexus

In contrast, the spinal nerves T2 to T11 (**intercostal nerves**) do not form plexuses; instead, they extend directly to the structures they supply.

The sensory neurons within each spinal nerve receive input from a specific area of skin called a **dermatome**. The dermatomes overlap somewhat due to the mixing of spinal nerves in plexuses, but they are arranged in an orderly fashion. Physicians can assess nerve damage by touching specific dermatomes and determining whether the patient can feel the stimulation. Due to the overlap of dermatomes, anesthesiologists may have to block several nerves to prevent pain perception in a given area.

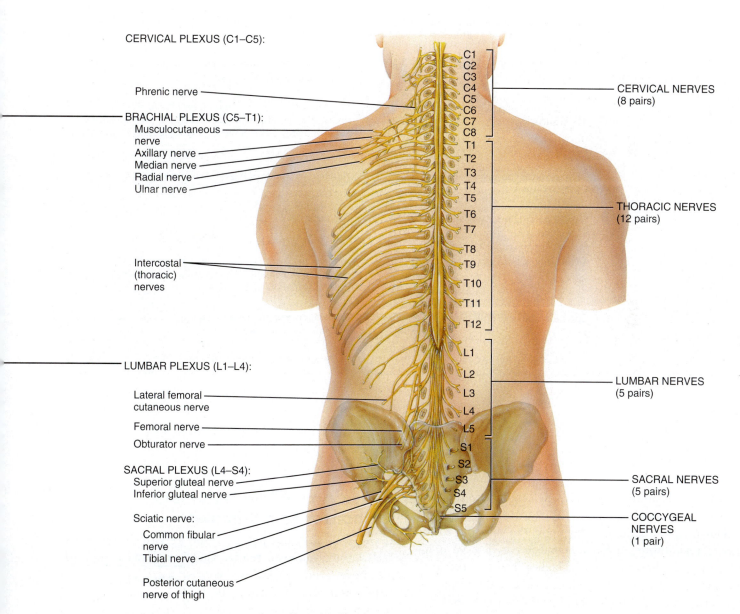

CERVICAL PLEXUS (C1–C5):

Phrenic nerve

BRACHIAL PLEXUS (C5–T1):
Musculocutaneous nerve
Axillary nerve
Median nerve
Radial nerve
Ulnar nerve

Intercostal (thoracic) nerves

LUMBAR PLEXUS (L1–L4):

Lateral femoral cutaneous nerve

Femoral nerve

Obturator nerve

SACRAL PLEXUS (L4–S4):
Superior gluteal nerve
Inferior gluteal nerve

Sciatic nerve:

Common fibular nerve

Tibial nerve

Posterior cutaneous nerve of thigh

C1
C2
C3
C4
C5
C6
C7
C8
T1
T2
T3
T4
T5
T6
T7
T8
T9
T10
T11
T12
L1
L2
L3
L4
L5
S1
S2
S3
S4
S5

CERVICAL NERVES (8 pairs)

THORACIC NERVES (12 pairs)

LUMBAR NERVES (5 pairs)

SACRAL NERVES (5 pairs)

COCCYGEAL NERVES (1 pair)

Posterior view of entire spinal cord and portions of spinal nerves

Selected nerves of the spinal nerve plexuses Table 7.3

Plexus	Origin	Nerve	Distribution
Cervical	C1–C5	Phrenic (FREN-ik; origin between C3 and C5)	Diaphragm
Brachial	C5–C8 and T1	Musculocutaneous (mus'-kū-lō-kū-TĀN-ē-us; origin between C5 and C7)	Muscles of arm
		Axillary (AK-si-lar-ē; origin between C5 and C6)	Deltoid and teres minor muscles; skin over deltoid and superior posterior aspect of arm
		Median (origin between C5 and T1)	Flexors of forearm; skin of palm of hand and fingers
		Radial (origin between C5 and T1)	Triceps and extensor muscles of forearm; skin of posterior arm, forearm, hand, and fingers
		Ulnar (origin between C8 and T1)	Flexor muscles of forearm, and most muscles of hand; skin of hand and some fingers
Lumbar	L1–L4	Femoral (including lateral and anterior cutaneous branches) (origin between L2 and L4)	Largest nerve arising from lumbar plexus; flexor muscles of hip and extensor muscles of knee; skin over anterior and medial aspect of thigh and medial side of leg and foot
		Obturator (OB-too-rā-tor; origin between L2 and L4)	Adductor muscles of hip joint; skin over medial aspect of thigh
Sacral	L4–L5 and S1–4	Superior and inferior gluteal (origin between L4 and S2)	Gluteus muscles
		Sciatic (origin between L4 and S3)	Actually two nerves—tibial and common fibular—bound by a common sheath of connective tissue; splits into two divisions, usually at the knee
		Common fibular (origin between L4 and S2)	Divides into superficial and deep branches; serves lateral aspects of leg and foot
		Tibial (origin between L4 and S3)	Posterior muscles of leg and foot

Table 7.3 summarizes some selected nerves of the spinal nerve plexuses.

During an episode of chickenpox, the varicella zoster virus can sometimes infect a spinal nerve. The virus invades the peripheral nerve cell bodies in the dorsal root ganglion. After a period of lying dormant, the virus can reactivate, causing shingles. Shingles is characterized by a painful rash in the skin of the associated dermatome.

The PNS and spinal cord process certain information without input from the brain. This is done via reflexes. Let's take a closer look to see how this is accomplished.

Reflex Arcs Allow Automatic Responses to a Stimulus

If you pick up a hot pot, you may release it automatically before you even realize that your hand is injured. Similarly, you withdraw your hand immediately when it is pricked by a sharp object before you consciously feel the pain. These responses, called **reflexes**, are rapid and involuntary, and they are caused by specific

reflex A fast response to a change (stimulus) in the internal or external environment that attempts to restore homeostasis.

Reflex arc • Figure 7.20

When you tap the patellar tendon with a rubber hammer, your knee jerks involuntarily in a reflex arc called the patellar reflex.

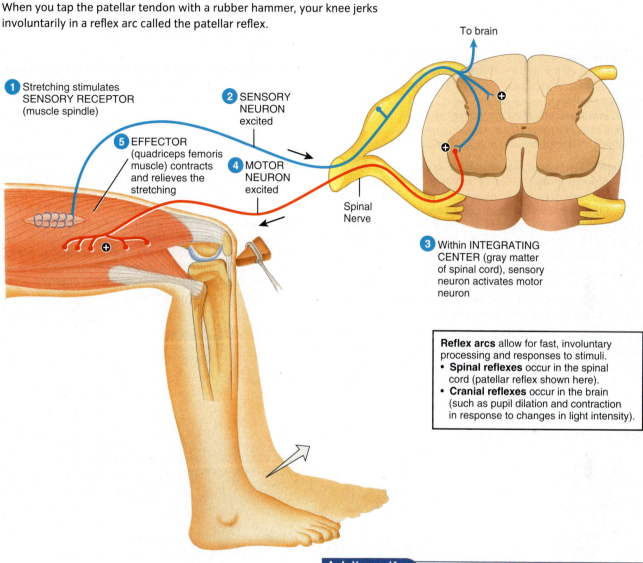

1 Stretching stimulates SENSORY RECEPTOR (muscle spindle)

2 SENSORY NEURON excited

5 EFFECTOR (quadriceps femoris muscle) contracts and relieves the stretching

4 MOTOR NEURON excited

To brain

Spinal Nerve

3 Within INTEGRATING CENTER (gray matter of spinal cord), sensory neuron activates motor neuron

Reflex arcs allow for fast, involuntary processing and responses to stimuli.
- **Spinal reflexes** occur in the spinal cord (patellar reflex shown here).
- **Cranial reflexes** occur in the brain (such as pupil dilation and contraction in response to changes in light intensity).

Ask Yourself

Suppose that the individual in this figure has suffered an injury to the spinal cord segment that is shown, and the integrating center is damaged. Explain why the leg would not move when the tendon is struck with the hammer.

neuronal circuits called reflex arcs (**Figure 7.20**). A **reflex arc** consists of a sensor receptor, a sensory neuron, an integrating center (the brain or spinal cord), a motor neuron, and an effector. Some reflexes also involve inhibitory interneurons. As the introductory examples imply, some reflex arcs are important mechanisms that protect us from harm. Others, such as the patellar reflex, help us maintain our balance when we stand up (Figure 7.20).

CONCEPT CHECK

1. **What** is a bundle of axons in the peripheral nervous system called?
2. **Which** cranial nerve is a sensory nerve for sight?
3. **Which** spinal nerves do not have plexuses?
4. **Where** is the integrating center for a cranial nerve reflex?

Summary

1 Nerve Cells "Talk" to Each Other 192

- The nervous system is made of **neurons** and **neuroglia**. Neurons have cell bodies, dendrites, axons, and axon terminals and can transmit information in the form of electrochemical signals. A layer of **myelin**, which helps to speed up nerve impulse transmissions, covers some **nerve fibers**. Neuroglial cells (astrocytes, microglia, Schwann cells, and ependymal cells) support and maintain neurons.

- Like other excitable cells, neurons have a resting membrane potential (approximately –70 mV), which is caused by an imbalance of ions across the cell membrane. Various stimuli cause sodium or potassium channels to open, which can change the membrane potential slightly. If enough sodium channels open, the cell will reach the threshold potential. Then the membrane will depolarize rapidly. This triggers openings of potassium channels, which eventually repolarize the membrane. An **action potential** is transmitted rapidly down the length of the neuron, causing other nerve segments to go through the same process.

- At the **synapse**, the action potential triggers the release of a chemical called a **neurotransmitter** into the gap or cleft between neurons, as shown. The neurotransmitter diffuses across the cleft, binds to receptors on the postsynaptic cell, and triggers an action potential.

Synaptic transmission • Figure 7.6

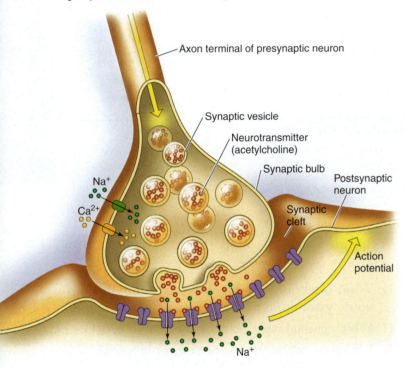

2 The Central Nervous System Coordinates All Nervous Activity 200

- The **central nervous system (CNS)** consists of the brain and spinal cord. The brain lies in the cranial cavity, is encased in the protective **meninges**, and is bathed in cerebrospinal fluid. The **spinal cord** is connected to the brain and lies in the vertebral column, where it is also encased in meninges and bathed in cerebrospinal fluid. The CNS tissues are isolated from harmful chemicals by the **blood–brain barrier**.

- As shown, the **brain** is composed of four parts: the brainstem, diencephalon, cerebellum, and cerebrum. The brain receives sensory information, processes it, and sends out motor commands using various pathways. It also takes care of involuntary functions (breathing, adjustments to heart rate, and so on) and voluntary muscle actions, processes information from special senses, controls emotions, and allows thought processes to occur. Each different part of the brain performs designated tasks.

The brain • Figure 7.10

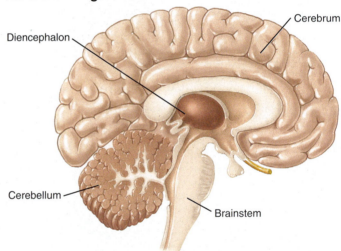

- The spinal cord is divided into segments. Sensory and motor neurons enter and leave the various segments through spinal nerves. Nerves are conduits that consist of wrapped bundles of nerve fibers (axons, dendrites, or a combination of both). **Spinal nerves** make up pathways that relay sensory information to the brain through ascending spinal tracts and receive motor information from the brain via descending spinal tracts. **Interneurons** within the **gray matter** of the CNS help process information and relay it to the appropriate portion of the brain.

3 The Autonomic Nervous System Controls the Activities of Smooth Muscle, Cardiac Muscle, and Glands 208

- The **autonomic nervous system (ANS)**, as shown, regulates involuntary functions that are vital to life. It consists of the sympathetic division and the parasympathetic division.

Autonomic nervous system—sympathetic division • Figure 7.14

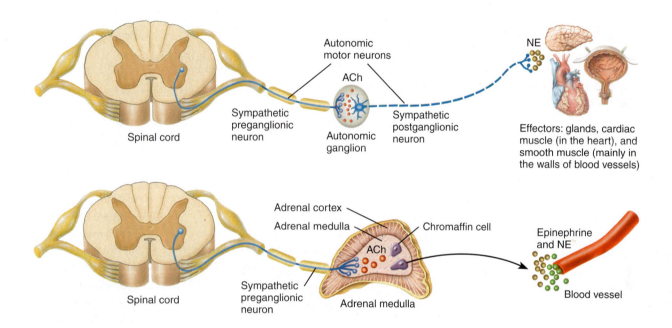

- The motor nerves of the autonomic nervous system usually consist of a preganglionic neuron that releases acetylcholine and a postganglionic neuron that releases the neurotransmitters epinephrine or norepinephrine (sympathetic) or **acetylcholine** (parasympathetic) to the **effector cells** (cardiac muscle, smooth muscle, or glands).

- Because of the use of different neurotransmitters, the sympathetic and parasympathetic divisions have opposite effects. The **sympathetic division** produces "fight-or-flight" responses in target organs, while the **parasympathetic division** produces "rest-and-digest" responses.

4 The Peripheral Nervous System Communicates with the Outside World 214

- **The peripheral nervous system** relays information to the **central nervous system** through cranial and spinal nerves. Cranial and spinal nerves can contain only sensory neurons, only motor neurons, or both.

- **Cranial nerves** consist of 12 pairs of nerves that govern sensory and motor information in the head, transmit information to and from special sensory organs (eyes, nose, ears, mouth, and so on), and provide output of some parasympathetic functions. **Spinal nerves** like the one shown contain sensory and **motor neurons** that transmit information from specific spinal cord segments to different areas of the body.

- **Reflexes** produce involuntary responses to various stimuli. The response is the result of passing impulses through reflex arcs, which consist of sensory receptors, sensory neurons, integrating centers, motor neurons, and effectors.

Parts of a nerve • Figure 7.17

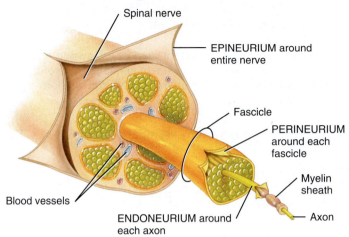

Spinal nerve

EPINEURIUM around entire nerve

Fascicle

PERINEURIUM around each fascicle

Myelin sheath

Axon

ENDONEURIUM around each axon

Blood vessels

Key Terms

- acetylcholine (ACh) 198
- action potential 197
- afferent neuron 195
- afferent (sensory) nervous system 193
- Alzheimer disease (AD) 204
- arachnoid villus 200
- astrocyte 196
- autonomic nervous system (ANS) 208
- axon collateral 194
- axon hillock 194
- axon 195
- axon terminal 198
- basal ganglia 203
- bipolar neuron 195
- blood–brain barrier 196
- brain 192
- brainstem 202
- cell body 195
- central nervous system (CNS) 192
- cerebellar peduncles 202

- cerebellum 203
- cerebrum 202
- corpus callosum 202
- cranial nerves 192
- dendrite 195
- dermatome 217
- diencephalon 202
- efferent (motor) nervous system 193
- efferent neuron 195
- enteric nervous system (ENS) 193
- enteric plexuses 192
- ependymal cell 196
- frontal lobe 202
- ganglion 195
- graded potential 197
- gray matter 207
- gyrus 203
- homunculus 204
- intercostal nerve 217
- interneuron 195
- lobe 203

- medulla oblongata 206
- meninges 200
- microglia 196
- midbrain 202
- motor neuron 195
- multipolar neuron 195
- myelin 195
- nerve 193
- nerve fiber 195
- neuroglia 192
- neuron 192
- neurotransmitter 198
- node of Ranvier 194
- nucleus 195
- occipital lobe 202
- oligodendrocyte 196
- parasympathetic division 213
- parietal lobe 202
- peripheral nervous system (PNS) 192
- pineal gland 202
- plexus 216
- pons 202

- reflex 218
- reflex arc 219
- resting membrane potential 197
- satellite cell 196
- Schwann cell 196
- sensory neuron 195
- sensory receptor 195
- somatic nervous system (SNS) 193
- spinal cord 206
- spinal nerve 216
- stimulus 197
- sulcus 203
- sympathetic division 210
- synapse 198
- temporal lobe 202
- terminal ganglion 213
- thalamus 202
- threshold 197
- unipolar neuron 195
- ventricle 200
- white matter 207

Critical and Creative Thinking Questions

1. Multiple sclerosis is a disease that causes destruction of the myelin sheaths of CNS neurons. Explain the effects of this disease on the conduction of action potentials.

2. Head trauma suffered in a car accident left Mrs. Anderson blind, even though she did not injure her eyes. Where on her head did she suffer trauma, and what part of her brain was most likely affected?

3. Jack fell off a ladder and injured his back. He could not move his legs and had no feeling below his waist, but he did respond to pin pricks from his waist upward. Which spinal cord segment did he injure in his fall?

4. Julia was startled when her boyfriend sneaked up behind her and yelled loudly. Her heart was racing, and she was breathing rapidly. Explain her responses and trace the neural pathways involved.

5. A dentist wishes to numb a patient's upper teeth for a root canal procedure. Near which cranial nerve should he inject the anesthetic?

What is happening in this picture?

Consuming alcohol slows reflexes, impairs judgment, and causes drowsiness. Great quantities of γ-aminobutyric acid (GABA), an inhibitor that floods synapses, are released from certain neurons in the brain when a person drinks.

Think Critically 1. Explain how GABA influences synaptic transmission to produce the effects associated with alcohol.
2. If coffee or caffeine stimulates the same neural pathways as those affected by alcohol, then what would be the effects of excessive caffeine?

Self-Test

(Check your answers in Appendix C.)

1. Which part of the brain is responsible for processing visual information?

   a. parietal lobe

   b. occipital lobe

   c. temporal lobe

   d. frontal lobe

Use this figure to answer questions 2–4.

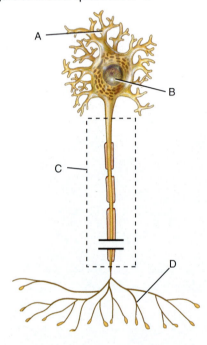

2. Which labeled part contains receptors for a neurotransmitter?

   a. A

   b. B

   c. C

   d. D

3. Which part of the neuron has myelin?

   a. A

   b. B

   c. C

   d. D

4. Which labeled part contains high concentrations of neurotransmitter?

   a. A

   b. B

   c. C

   d. D

5. The cell bodies of motor neurons are in the _____ of the spinal cord.

   a. posterior gray horn

   b. lateral corticospinal tract

   c. spinothalamic tract

   d. anterior gray horn

6. Sodium channels inactivate _____ of the action potential.

   a. during the stimulus phase

   b. during the depolarizing phase

   c. during the repolarizing phase

   d. after the hyperpolarizing phase

Use this figure to answer questions 7–9.

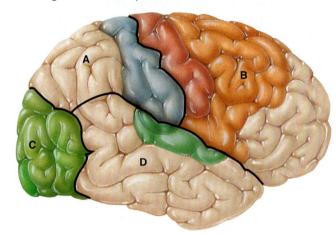

7. Which labeled part contains the somatosensory cortex?

   a. A

   b. B

   c. C

   d. D

8. Which labeled part receives auditory information?

   a. A

   b. B

   c. C

   d. D

9. The figure shows the _____ of the brain.

   a. cerebellum

   b. brainstem

   c. cerebrum

   d. diencephalon

10. The _____ nerve is *not* part of a plexus.

    a. intercostal

    b. cervical

    c. radial

    d. sciatic

11. The flow of CSF starts in the _____.

    a. subarachnoid space

    b. lateral ventricle

    c. third ventricle

    d. fourth ventricle

Use this figure for questions 12–13.

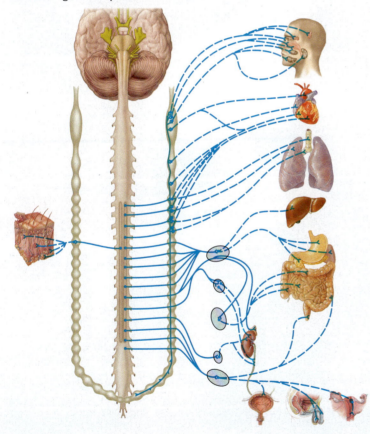

12. The postganglionic cells of the division shown in the figure use the neurotransmitter _____.

    a. dopamine

    b. norepinephrine

    c. acetylcholine

    d. serotonin

13. Which part of the nervous system is shown in the figure?

    a. central nervous system

    b. somatic nervous system

    c. peripheral nervous system

    d. autonomic nervous system

14. Which part of the brain is continuous with the spinal cord?

    a. cerebellum

    b. brainstem

    c. cerebrum

    d. diencephalon

15. In which of the following is the limbic system involved?

    a. anger

    b. balance

    c. hearing

    d. taste

THE PLANNER ✔

Review your Chapter Planner on the chapter opener and check off your completed work.

Somatic Senses and Special Senses

You are seated in your favorite restaurant, at a table not far from the kitchen. You can feel the fine linen of the tablecloth and napkins. You can hear the bustling sounds from the kitchen, such as the clanking of pots and cooking utensils, the orders shouted by the chefs, and the sounds of the dishes hitting the metal shelves and counter tops. You can also smell the simmering pasta sauce, the grilling steaks, the sautéing seafood, and the aromas of herbs.

Your waiter places your food in front of you. You can see the bright colors and artistic presentation of each dish. You take a bite and savor the flavor of the ingredients the chef has combined in the entrée. Perhaps you even enjoy your meal with a glass of wine; its fragrant bouquet and taste complement your food perfectly. We can appreciate all these facets of the dining experience because we have somatic senses of touch and special senses of sight, hearing, taste, and smell.

Let's take a look at each of these wonderful senses.

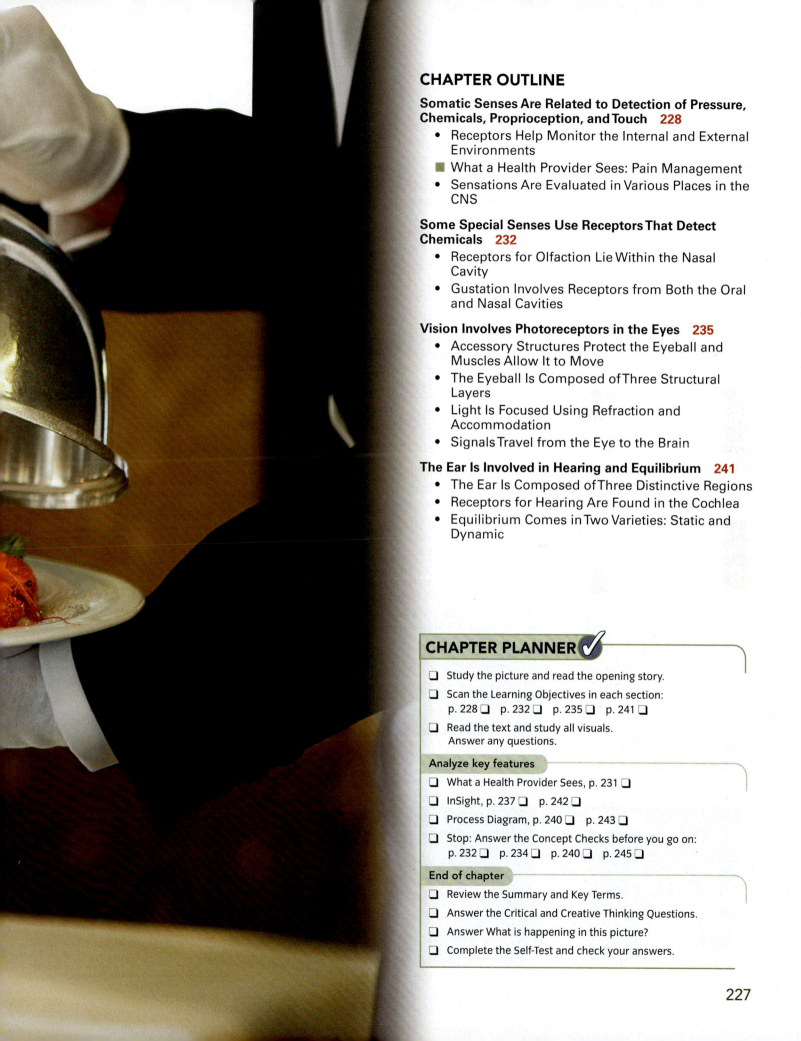

CHAPTER OUTLINE

CHAPTER PLANNER ✔

- ❑ Study the picture and read the opening story.
- ❑ Scan the Learning Objectives in each section:
  p. 228 ❑ p. 232 ❑ p. 235 ❑ p. 241 ❑
- ❑ Read the text and study all visuals. Answer any questions.

Analyze key features

- ❑ What a Health Provider Sees, p. 231 ❑
- ❑ InSight, p. 237 ❑   p. 242 ❑
- ❑ Process Diagram, p. 240 ❑   p. 243 ❑
- ❑ Stop: Answer the Concept Checks before you go on:
  p. 232 ❑ p. 234 ❑ p. 240 ❑ p. 245 ❑

End of chapter

- ❑ Review the Summary and Key Terms.
- ❑ Answer the Critical and Creative Thinking Questions.
- ❑ Answer What is happening in this picture?
- ❑ Complete the Self-Test and check your answers.

Somatic Senses Are Related to Detection of Pressure, Chemicals, Proprioception, and Touch

LEARNING OBJECTIVES

1. **Describe** the location and function of receptors for the somatic senses (tactile, thermal, pain).

2. **Identify** the receptors for proprioception and describe their functions.

3. **Describe** the neural pathways involved in somatic senses.

You know about the world around you through **sensations** and **perceptions**. Sensations are detected by specialized nerve receptors

sensation Conscious or subconscious awareness of changes in internal or external environment; can occur in all parts of the CNS.

perception Conscious awareness and interpretation of sensations; occurs only in the cerebral cortex.

that detect a change in the body's internal or external environment. Each receptor is specific to a particular type of sensation. When a receptor is stimulated, it initiates a signal in the associated dendrite, which conducts impulses related to the sensation to the central nervous system (CNS). Somatic sensations include tactile, pain, temperature, and proprioception. Receptors for the somatic senses are spread diffusely around the body in structures such as the skin, mucous membranes, muscles, tendons, and joints (**Figure 8.1a**).

Visceral senses have receptors within the walls of the internal organs that detect pain and

Overview of sensations • Figure 8.1

The neural pathway for sensations consists of the following: Stimulus → sensory receptor → neural pathway → brain (integrate nerve impulses) → sensation. For example, when the skin of the cheek is touched, sensory receptors are stimulated, and a nerve impulse is transmitted through the trigeminal nerve (cranial nerve V) to the cerebrum for integration leading to perception of the sensation.

a. Somatic senses

Tactile sensations: touch, pressure, vibration

Thermal sensations: hot, cold

Nociception (pain)

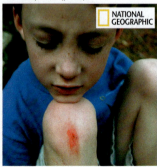

Proprioception: joint and muscle position, movements of head and limbs

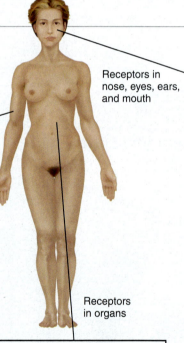

Receptors in nose, eyes, ears, and mouth

Receptors in skin, muscles, and joints

Receptors in organs

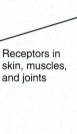

b. Visceral senses (conditions within the organs)

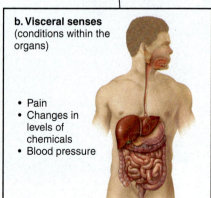

- Pain
- Changes in levels of chemicals
- Blood pressure

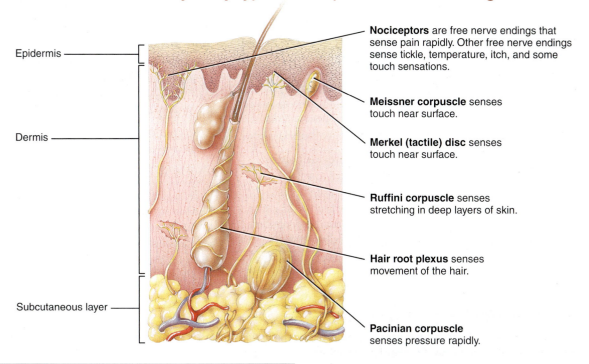

Epidermis

Dermis

Subcutaneous layer

Nociceptors are free nerve endings that sense pain rapidly. Other free nerve endings sense tickle, temperature, itch, and some touch sensations.

Meissner corpuscle senses touch near surface.

Merkel (tactile) disc senses touch near surface.

Ruffini corpuscle senses stretching in deep layers of skin.

Hair root plexus senses movement of the hair.

Pacinian corpuscle senses pressure rapidly.

c. Special senses

Smell

Sight

Hearing and balance

Taste

changes in blood pressure and chemical levels (**Figure 8.1b**). Like somatic senses, visceral senses have receptors that are spread throughout the organs of the body. In contrast, the special senses, such as smell, sight, hearing, balance, and taste, have receptors that are localized in the nose, eyes, ears, and mouth (**Figure 8.1c**). Let's look first at the somatic senses.

Receptors Help Monitor the Internal and External Environments

The dermis of the skin contains various tactile receptors, which include **encapsulated nerve endings** and free nerve endings (bare dendrites) located at various levels (**Figure 8.2**). With prolonged stimulation, **adaptation** occurs, causing a loss of sensation. Some receptors can adapt rapidly (desensitize quickly) and some adapt more slowly (take longer to adjust). Either type of adaptation results in desensitization to a stimulus. Four types of receptors sense touch: **Meissner corpuscles**, **hair root plexuses**, **Merkel (tactile) discs**, and **Ruffini corpuscles**. Pressure is sensed by rapidly adapting **Pacinian corpuscles** (pa-SIN-ē-an), which are widely distributed throughout the body. Pacinian corpuscles also sense high-frequency

> **encapsulated nerve ending** A receptor enclosed in a connective tissue capsule.
>
> **adaptation** A decrease in the strength of a sensation during a prolonged stimulus.

vibrations, while Meissner corpuscles sense low-frequency vibrations.

Free nerve endings sense the following:

- *Itch.* Itch is caused by local inflammatory responses to certain chemicals.
- *Tickle.* Tickle is a complex sensation.
- *Cold.* Rapidly adapting cold receptors in the epidermis respond to temperatures between 10°C and 40°C.
- *Warmth.* Rapidly adapting warm receptors in the dermis respond to temperatures between 32°C and 48°C.
- *Pain.* **Nociceptors** are found in tissues throughout the body except the brain. Unlike

> **nociceptor** (nō′-sē-SEP-tor) A free (naked) nerve ending that detects painful stimuli.
>
> **proprioceptor** (prō′-prē-ō-SEP-tor) A receptor that provides information about body position and movements.

other sensory receptors, nociceptors do not adapt. Fast (acute) pain receptors respond within 0.1 s and result in a sharp or pricking pain. Slow (chronic) pain receptors respond after 1 s or more and result in a burning, aching, or throbbing pain. Often, slow pain or pain from internal organs (visceral pain) is felt over a larger area than the one stimulated; this phenomenon is called **referred pain** (**Figure 8.3**). (See *What a Health Provider Sees* for information on the management of the different types of pain.)

Proprioceptors are found in skeletal muscles, tendons, and joints throughout the body. Proprioceptors adapt slowly and only slightly, so the brain constantly receives information about head and limb position as well as movements.

Distribution of referred pain • Figure 8.3

Areas of referred pain for particular visceral organs are indicated by the various colors. The referred pain usually arises because the innervations of these areas of the skin share the same spinal cord segment as those of the corresponding visceral organ.

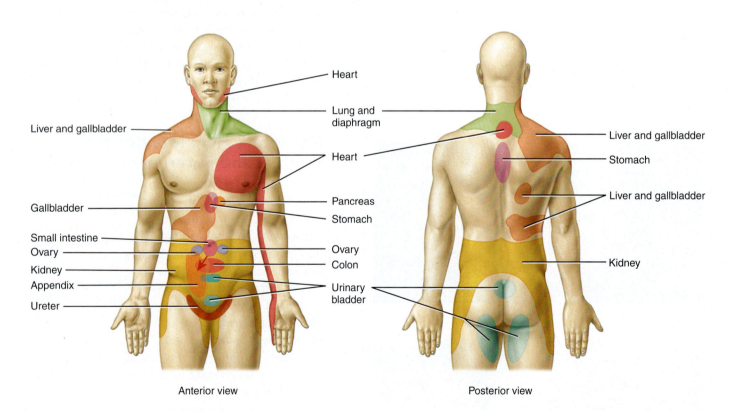

Anterior view

Posterior view

WHAT A HEALTH PROVIDER SEES

Pain Management

Health providers vary their treatment of pain, depending on the type of pain being experienced. Most pain can be managed with medications (analgesics), which act at different places in the somatosensory pathway:

- When an injury causes pain, damaged cells or immune cells release certain products (such as bradykinin, prostaglandins, and histamines) that activate nearby nociceptors. *Anti-inflammatory analgesics,* such as aspirin, ibuprofen, acetaminophen, or naproxen, interfere with the enzymes that make these products.

- A class of neurotransmitters called opioids (such as endorphin, dynorphin, and enkephalin) is responsible for synaptic transmission in CNS pain pathways. At the synapse, *opioid analgesics* (such as morphine, meperidine, oxycodone, and codeine) bind to the opioid receptors but do not activate them. Health providers use opioid analgesics to treat high levels of pain. Opioid analgesics can become addictive and are easily overdosed; therefore, health providers must closely monitor their use.

- Pain relievers called *adjuvant analgesics (co-analgesics)* are used to treat other conditions and also help relieve pain. For example, *anti-epileptic drugs* (such as phenytoin) reduce the ability of CNS neurons to conduct action potentials. *Tricyclic antidepressants* (such as amitriptyline) block synaptic transmission involving the neurotransmitter serotonin. *Anesthetics* (such as lidocaine and benzocaine) block sodium and potassium channels, thereby preventing propagation of action potentials; most are applied topically or injected locally to relieve pain.

Other therapies can also be used to manage chronic pain without the use of medications or in conjunction with smaller doses of pain medications:

- Physical therapy involves a series of exercises, massage, thermal stimulation, and electrotherapy to release muscle tension and build muscle strength that can help to relieve pain caused by pressure on peripheral nerves.

- Transcutaneous electric nerve stimulator (TENS) units produce electrical pulses via the skin to stimulate the release of endorphins and enkephalins within the CNS. These neurotransmitters help to block pain signals from reaching the brain.

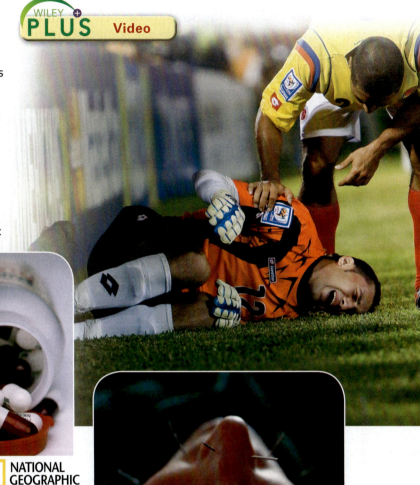

WILEY PLUS Video

NATIONAL GEOGRAPHIC

- Acupuncture and acupressure techniques involve the insertion of needles or application of pressure to strategic parts of the body, which helps to relieve pain in other parts of the body. The areas stimulated by the techniques share the spinal segments with the area that will ultimately be "treated" by the procedure.

- Biofeedback can sometimes be used to help patients manage chronic pain by helping individuals learn to control the impact of the pain on their daily lives.

Think Critically 1. Migraine headaches are thought to involve areas of the brain where the primary neurotransmitter is serotonin. Which pain medication might work best for this condition?
2. Jenny is 10 years old and has a cold. She fell from a step and twisted her ankle, and her ankle is beginning to swell. Pediatricians do not recommend giving children aspirin as it has been implicated in a deadly condition called Reye's syndrome. What medication might be best to give Jenny for the pain?

Sensations Are Evaluated in Various Places in the CNS

Receptors for the somatic senses communicate with the somatosensory cortex in the parietal lobe of the cerebrum through pathways in the spinal cord and brain (see Figure 7.13a). Touch, pressure, vibration, and proprioception are communicated via one pathway, and pain, cold, warmth, and itch are transmitted via a separate pathway. In addition, proprioceptors also pass impulses to the cerebellum and contribute to its role in coordinating movements.

CONCEPT CHECK STOP

1. **Which** specific receptors sense touching of the skin?
2. **Where** would you find proprioceptors, and what do they do?
3. **Where** is the somatosensory cortex located in the brain?

Some Special Senses Use Receptors That Detect Chemicals

LEARNING OBJECTIVES

1. **Identify** the structures involved in smell (olfaction).
2. **Describe** the processes involved in perceiving odors.
3. **List** the structures involved in taste (gustation).
4. **Outline** the processes involved in perceiving tastes.

T he special senses include smell, taste, sight, hearing, and balance. Some special senses detect physical phenomena such as light, sound, or gravity. Because two of the special senses require the interaction of chemicals with sensory receptors—**olfaction** (smell) and **gustation**—they are sometimes referred to as the *chemical senses*. Also, because the nose and mouth are connected, the senses of smell and taste are associated. For example, when you have a stuffy nose and cannot smell, you often cannot taste foods. Let's start our examination of the chemical senses with olfaction.

> **olfaction** (ōl-FAK-shun) The sense of smell.
>
> **gustation** (gus-TĀ-shun) The sense of taste.

Receptors for Olfaction Lie Within the Nasal Cavity

The nose contains 10 to 100 million receptors for the sense of smell. These **olfactory receptors** are located within the olfactory epithelium and are surrounded by other cells, including **supporting cells** and **basal cells** (**Figure 8.4**). Olfactory receptors bind and recognize almost 10,000 different molecules, which are collectively called **odorants**. The odorants elicit action potentials within the receptors, and the information is sent via the olfactory nerves (cranial nerve I) to the olfactory areas of the brain (temporal and frontal lobes of the cerebrum, the limbic system, and the hypothalamus) (see Figure 7.10). Because some of the nerve impulses are passed to the limbic system, certain odors and tastes evoke strong emotional responses and memories.

Olfactory receptors adapt rapidly to odors. Within the first second or so of exposure to an odor, approximately 50% of the receptors adapt. The remaining receptors adapt more slowly. When you first put on perfume or aftershave in the morning, there is a strong odor to it, but that odor seems to "disappear" very quickly. The chemical is still there, but the nasal receptors lose their ability to react effectively to it.

Structures of the olfactory epithelium • Figure 8.4

The olfactory epithelium is located within the nose. The epithelium contains olfactory receptors, several supporting cells, and neurons of the olfactory nerves.

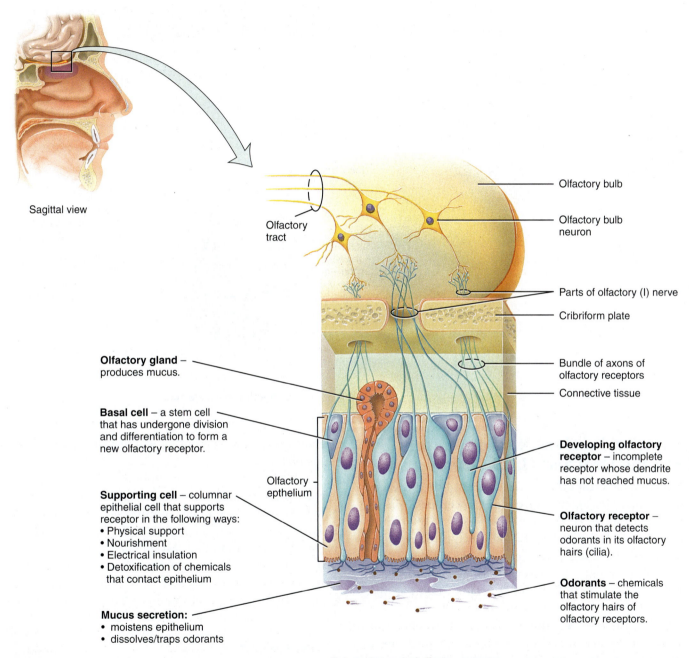

Sagittal view

Olfactory tract

Olfactory bulb

Olfactory bulb neuron

Parts of olfactory (I) nerve

Cribriform plate

Olfactory gland – produces mucus.

Bundle of axons of olfactory receptors

Connective tissue

Basal cell – a stem cell that has undergone division and differentiation to form a new olfactory receptor.

Developing olfactory receptor – incomplete receptor whose dendrite has not reached mucus.

Olfactory epthelium

Supporting cell – columnar epithelial cell that supports receptor in the following ways:
• Physical support
• Nourishment
• Electrical insulation
• Detoxification of chemicals that contact epithelium

Olfactory receptor – neuron that detects odorants in its olfactory hairs (cilia).

Odorants – chemicals that stimulate the olfactory hairs of olfactory receptors.

Mucus secretion:
• moistens epithelium
• dissolves/traps odorants

Enlarged aspect of olfactory receptors

Olfaction, like all the special senses, has a low threshold. As little as a few molecules of certain substances can be detected in the air. A good example is the chemical methyl mercaptan, which smells like rotten cabbage and can be detected in very low concentrations. Because natural gas is odorless, but a leak is potentially lethal, this chemical is added in very small amounts to the natural gas used for cooking and heating.

Structures of the tongue involved in the sense of taste • Figure 8.5 _____

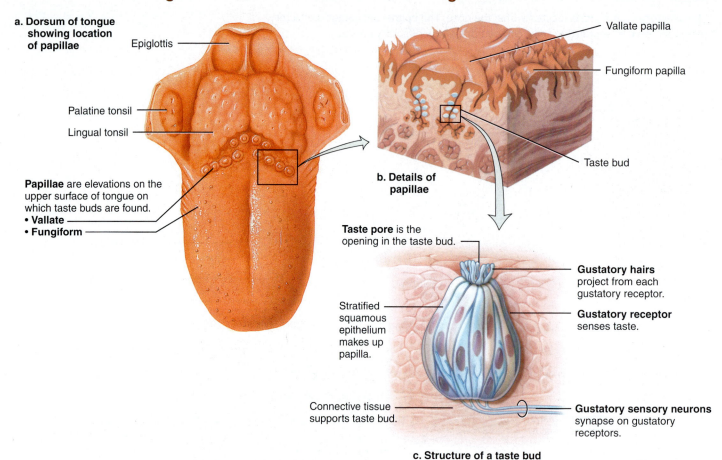

a. Dorsum of tongue showing location of papillae

Epiglottis

Palatine tonsil

Lingual tonsil

Papillae are elevations on the upper surface of tongue on which taste buds are found.
• **Vallate**
• **Fungiform**

Vallate papilla

Fungiform papilla

Taste bud

b. Details of papillae

Taste pore is the opening in the taste bud.

Stratified squamous epithelium makes up papilla.

Connective tissue supports taste bud.

Gustatory hairs project from each gustatory receptor.

Gustatory receptor senses taste.

Gustatory sensory neurons synapse on gustatory receptors.

c. Structure of a taste bud

Gustation Involves Receptors from Both the Oral and Nasal Cavities

Like odors, molecules that are collectively called **tastants** bind to taste receptors (**gustatory receptors**) on the tongue (**Figure 8.5**). While we can detect 10,000 different odors, we can detect only about five primary tastes: sour, sweet, bitter, salty, and umami (ū-MAM-ē). Umami is described as "meaty" or "savory." In the gustatory center, combinations of these five primary tastes and the associated smells allow detection of hundreds of flavors.

Unlike the nose, the tongue has no "sticky" mucus to trap tastant molecules. Instead, the tastants dissolve in saliva, enter **taste pores**, and contact the hairs of gustatory receptors on the tongue. The gustatory receptors transmit their information via several cranial nerves (VII, IX, X; see Figure 7.18) to the **primary gustatory area** in the parietal lobe of the cerebrum, the limbic system, and the hypothalamus (see Figure 7.10). Because gustatory information is sent to the limbic system, some tastes may evoke strong memories. Some memories may be strong and unpleasant and cause you to avoid certain tastes (taste aversion); others may evoke more positive responses, such as reminding you of your mother's delicious pie.

Individual gustatory receptors may respond to one of the five primary tastes and may adapt completely to a specific taste within one to five minutes of continuous stimulation. There is a tendency for the first bite of food to be more "flavorful" than that last bite on the plate. Different tastes arise from different patterns of activation of various taste buds as well as activation of the various olfactory receptors. In the absence of smell (such as when your nose is stuffy with a cold), flavors tend to "taste" differently.

CONCEPT CHECK   STOP

1. **Which** cells in the olfactory pathway detect odors?

2. **How** are odorants transformed into neural signals?

3. **What** structures are involved in detecting a tastant?

4. **Through** what cranial nerves do nerve impulses from taste buds pass?

Vision Involves Photoreceptors in the Eyes

LEARNING OBJECTIVES

1. **Identify** the accessory structures of the eye and their functions.
2. **Describe** the interior structure of the eyeball.
3. **Explain** how receptors detect light.
4. **Describe** several common vision problems.
5. **Outline** the processes involved in vision, including the neural pathways involved.

So far we have talked about senses in which the stimulus either touches or chemically binds to the receptors. The remaining special senses—sight, hearing, and balance—deal with intangible physical phenomena—light, sound, and gravity, respectively. The structures that change these stimuli into nerve impulses are quite complex. Let's look first at our window to the world, the eye.

Accessory Structures Protect the Eyeball and Muscles Allow It to Move

Because sight is our dominant sense, the human eye is a complex sense organ. Beginning at the surface of the face, the eye has several accessory structures that protect it (**Figure 8.6**). The eyebrows and eyelashes protect the eyeballs from foreign objects. The eyelids cover, protect, and help prevent the eyes from drying out via the blink reflex. The lacrimal apparatus secretes and drains the fluid that moistens the eye. **Lacrimal fluid**, also referred to as *tears*, is a watery solution that contains salts, mucus, and a bacteria-killing enzyme called **lysozyme**. Typically, the tears drain into the nasal cavity, and when we cry, parasympathetic activity stimulates the **lacrimal glands** to produce excessive tears, which spill over the

> **lysozyme** (LĪ-sō-zīm)
> A bactericidal enzyme found in tears, saliva, perspiration, nasal secretions, and tissue fluids.

Accessory structures of the eye • Figure 8.6

The eye has several external structures to protect it, including eyebrows, eyelashes, and eyelids. Lacrimal structures also protect the eye by keeping it moist.

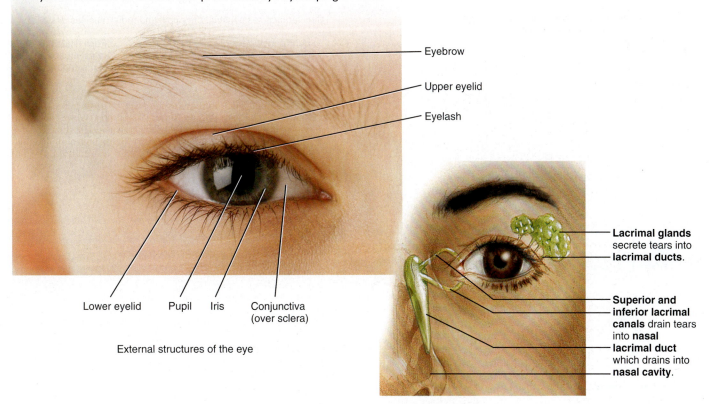

Eyebrow

Upper eyelid

Eyelash

Lower eyelid   Pupil   Iris   Conjunctiva (over sclera)

External structures of the eye

Lacrimal glands secrete tears into **lacrimal ducts**.

Superior and inferior lacrimal canals drain tears into **nasal lacrimal duct** which drains into **nasal cavity**.

Lacrimal structures of the eye

eyelids and drain into the nasal cavity, thereby producing a runny nose.

The eye has external and internal muscles (**Figure 8.7**). Contraction of different combinations of the six external muscles enables the eye to move rapidly in many directions (also see Figure 6.15). Internal muscles of the **iris** control the size of the **pupil**, regulating the amount of light that enters the eye. Other internal muscles change the shape of the lens to focus light refracted by the lens onto the retina.

The Eyeball Is Composed of Three Structural Layers

The eyeball is a layered structure that consists of the outer coat or **fibrous tunic**, a middle **vascular tunic** (vascular layer) that supplies blood to the eye's tissues and secretes fluids, and an inner **neural tunic (retina)**, which **transduces** incoming light to nerve impulses (**Figure 8.8**). Inside the eye are

> **transduce** To change from one form to another.

two chambers, the **anterior cavity** in front of the **lens** and the **posterior cavity** behind it. The anterior cavity is filled with a liquid called **aqueous humor**, which helps maintain the shape of the eyeball and nourishes the lens and **cornea**. Aqueous humor has a composition similar to that of cerebrospinal fluid. The posterior cavity is filled with a jelly-like substance called the **vitreous body**, which prevents the eyeball from collapsing and holds the retina in place.

The retina has layers of connecting **ganglionic cells**, interneurons, supporting cells, and **photoreceptors**. The photoreceptors consist of two types, **rods** and **cones**, which allow us to see color and to see in different light levels.

> **aqueous humor** (AK-wē-us HŪ-mer) The watery fluid that fills the anterior cavity of the eye between the cornea and the lens.
>
> **vitreous body** (VIT-rē-us) A soft, jellylike substance that fills the posterior cavity of the eyeball between the lens and the retina.
>
> **photoreceptor** A receptor that detects light shining on the retina of the eye.

Muscles of the eye • Figure 8.7

The eye has exterior and interior muscles to allow it to move and to control the entry of light.

Exterior Eye Muscles

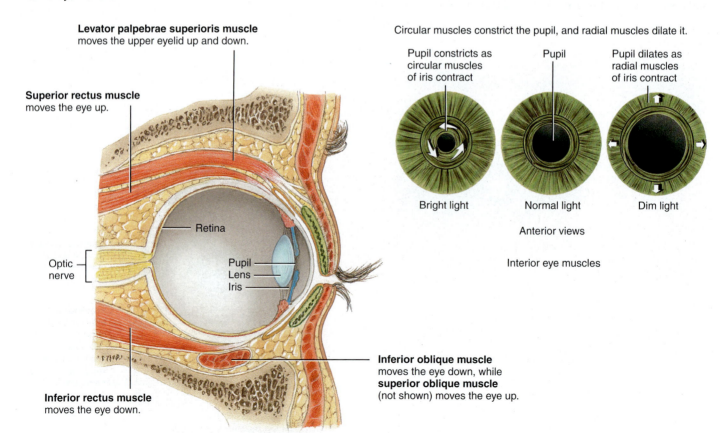

Levator palpebrae superioris muscle
moves the upper eyelid up and down.

Superior rectus muscle
moves the eye up.

Retina

Optic nerve

Pupil
Lens
Iris

Inferior rectus muscle
moves the eye down.

Inferior oblique muscle
moves the eye down, while **superior oblique muscle** (not shown) moves the eye up.

Circular muscles constrict the pupil, and radial muscles dilate it.

Pupil constricts as circular muscles of iris contract

Pupil

Pupil dilates as radial muscles of iris contract

Bright light

Normal light

Dim light

Anterior views

Interior eye muscles

InSight   The layers of the eyeball • Figure 8.8

The layers of the eyeball provide protection, nourishment, and the "screen" that transduces light images into nerve impulses.

Layers of the eye

Cornea admits and bends light.

Light

Visual axis

Anterior cavity contains aqueous humor that maintains shape and supplies oxygen and nutrients to lens and cornea.

Pupil is the opening.

Iris regulates the size of the pupil and the amount of light entering the eye.

Lens bends light.

Ciliary body secretes aqueous humor and alters shape of the lens for near or far vision.

Retina converts light into nerve impulses (output to brain).

Choroid provides blood supply and absorbs scattered light.

Medial rectus muscle moves eye medially.

Sclera provides shape and protects inner parts.

Inside is the **vitreous chamber**, which contains the vitreous body that maintains shape and keeps the retina attached.

MEDIAL

Blood vessels

Lateral rectus muscle moves eye laterally.

Optic nerve carries axons of ganglion cells to the brain.

Optic disc is site where axons exit and has no rods or cones (blind spot).

Fovea centralis is the center of the retina and has the highest concentrations of cone cells, which makes it the area of highest visual acuity.

| Fibrous tunic consists of cornea and sclera. |
| Vascular tunic ciliary body, lens, and choroid. |
| Neural tunic consists of the retina. |

Microscopic Structure of the Retina

Retinal blood vessel

Optic nerve axons

Ganglion cells are optic nerve fibers.

Ganglionic cell layer

Interneurons connect photoreceptors with ganglionic cells.

Bipolar cell layer

Photoreceptors detect light:
1. **Rods** detect shades of gray in dim light.
2. **Cones** detect bright light and colors.

Photoreceptor layer

Pigmented layer – melanin-containing epithelial cells between retina and choroids absorb stray light.

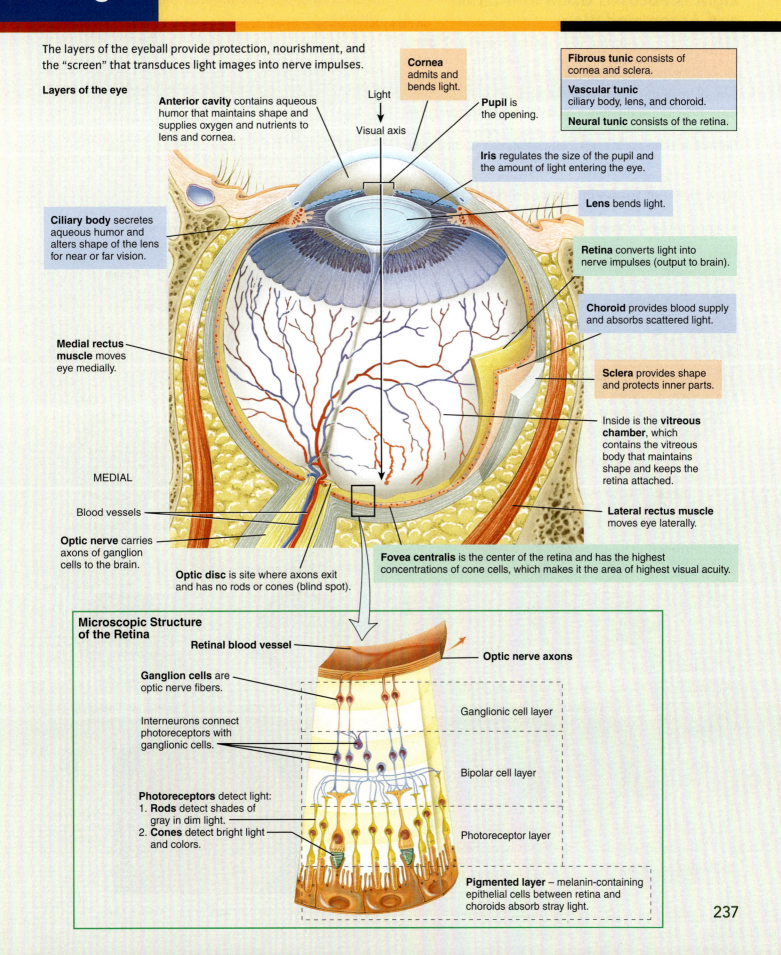

237

Light Is Focused Using Refraction and Accommodation

The eye is very similar to a digital camera (**Figure 8.9**). A digital camera has a lens that can be adjusted to focus the image onto a photosensitive electronic chip. The chip has an array of picture elements that transduce the light into electrical impulses that are combined to form an electronic photograph. In the eye, the image is focused on the retina, which is packed with an array of photoreceptors that convert light to action potentials.

Two processes are involved in focusing an image on the retina. **Refraction** or bending of light occurs as the light passes through areas with different densities. (The various tissues of the eye have different densities.) Much of the refraction occurs as the light enters the eyeball through the cornea. The amount of refraction is determined by the contour of the cornea. Most of the remaining refraction occurs at the lens. Unlike the immobile cornea, the lens can be reshaped

How the eye forms images and detects light • Figure 8.9

a. Image formation in the eye

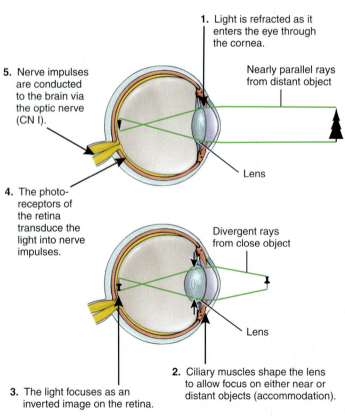

Light ray before refraction

Air

Water

Light ray after refraction

Refraction of light rays

1. Light is refracted as it enters the eye through the cornea.

Nearly parallel rays from distant object

Lens

5. Nerve impulses are conducted to the brain via the optic nerve (CN I).

4. The photoreceptors of the retina transduce the light into nerve impulses.

Divergent rays from close object

Lens

3. The light focuses as an inverted image on the retina.

2. Ciliary muscles shape the lens to allow focus on either near or distant objects (accommodation).

b. Normal and abnormal refractions in the eyeball

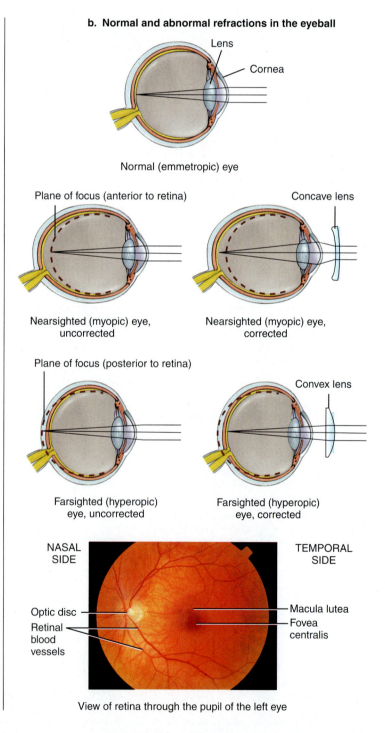

Lens

Cornea

Normal (emmetropic) eye

Plane of focus (anterior to retina)

Concave lens

Nearsighted (myopic) eye, uncorrected

Nearsighted (myopic) eye, corrected

Plane of focus (posterior to retina)

Convex lens

Farsighted (hyperopic) eye, uncorrected

Farsighted (hyperopic) eye, corrected

NASAL SIDE

TEMPORAL SIDE

Optic disc

Retinal blood vessels

Macula lutea

Fovea centralis

View of retina through the pupil of the left eye

by the **ciliary body** to adjust the refraction of the light passing through the lens. This process of adjusting the shape of the lens (anywhere from very flat to round) for focusing is called **accommodation**. Accommodation can keep an image focused on the retina, regardless of whether you are looking at a textbook on your desk or an instructor at the front of a lecture hall. Abnormally shaped eyeballs can lead to vision problems, which can be corrected with eyeglasses, contact lenses, or surgery.

> **photopigment** A substance that can absorb light and undergo structural changes that can lead to the development of a receptor potential.

How do your eyes provide visual information to the brain? In the retinal photoreceptors, light interacts with a **photopigment** called **rhodopsin**. Light causes rhodopsin to break up into two pieces, **retinal** and **opsin**. The chemical changes in rhodopsin lead to electrochemical changes in the photoreceptor membrane, which in turn lead to action potentials. The rhodopsin molecule must be reassembled by an enzyme before it can respond to light again (see Figure 8.9c).

c. Stimulation of a nerve impulse by light

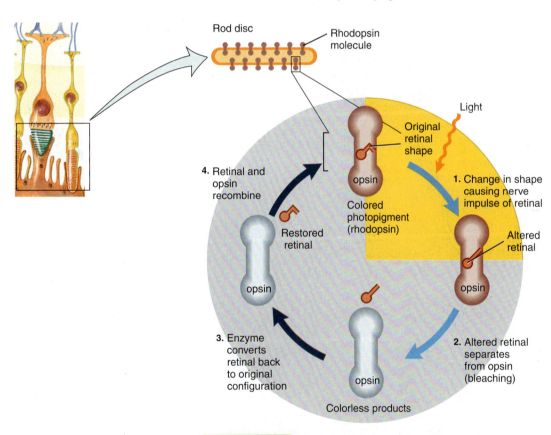

Vision Involves Photoreceptors in the Eyes **239**

Tracing the visual pathway from the eye to the brain • Figure 8.10

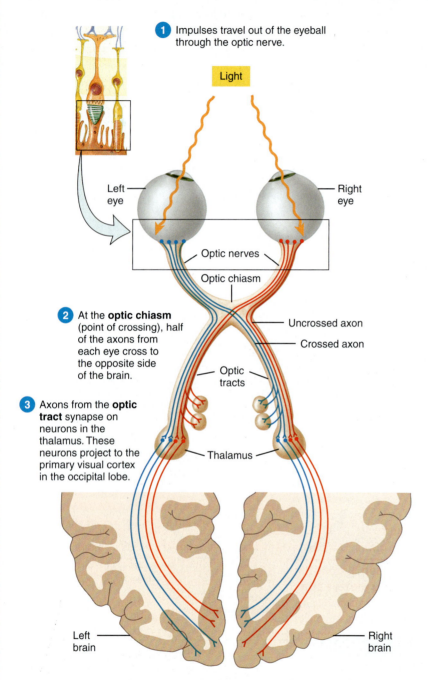

1. Impulses travel out of the eyeball through the optic nerve.

Light

Left eye

Right eye

Optic nerves

Optic chiasm

Uncrossed axon

Crossed axon

Optic tracts

Thalamus

2. At the **optic chiasm** (point of crossing), half of the axons from each eye cross to the opposite side of the brain.

3. Axons from the **optic tract** synapse on neurons in the thalamus. These neurons project to the primary visual cortex in the occipital lobe.

Left brain

Right brain

4. **Primary visual areas** in the occipital lobes receive input from both eyes. Because of the crossing in the optic chiasm, the following occurs:
   • Left side of brain interprets visual information from right side of an object.
   • Right side of brain interprets visual information from left side of an object.

Signals Travel from the Eye to the Brain

The action potentials follow a pathway from the receptor through several neurons into the optic nerves. Along the way to the brain, the optic nerves cross at a point called the **optic chiasm**. In the optic chiasm, some nerve fibers cross to the other side of the brain and subsequently to the brain's visual cortex in the occipital lobe of the cerebrum (**Figure 8.10**). There, the impulses are integrated to form a representative image of the visual field of each eye. Some information is also relayed to the cerebellum to coordinate eye movements and balance cues.

CONCEPT CHECK   STOP

1. **How** do tears flow over the eye?

2. **What** part of the eyeball supplies blood and nutrients?

3. **Which** photoreceptors respond to colors?

4. **What** is nearsightedness, and how can it be corrected?

5. **Where** in the brain is visual information processed?

The Ear Is Involved in Hearing and Equilibrium

LEARNING OBJECTIVES

1. **Identify** structures of the ear, including those of the outer ear, middle ear, and inner ear.

2. **Explain** how the ear converts sound waves into nerve impulses and the path those impulses take to the brain.

3. **Describe** the structures of the ear involved in static and dynamic equilibrium.

4. **Outline** the neural pathways involved in equilibrium.

Consider a ballet dancer. She must hear the music so that she knows her cues as well as the proper timing and rhythm for the execution of her steps. Her body must be aware at all times of the positions of her limbs with respect to gravity and make minute changes in muscles so she can stay balanced on her toes. The receptors for both the sense of hearing and the sense of balance, or equilibrium, are located in the ears. Let's take a closer look at the structures of the ear.

The Ear Is Composed of Three Distinct Regions

The ear consists of three parts: the outer ear, middle ear, and inner ear (**Figure 8.11**). The outer ear mainly receives and directs sound waves into the ear canal. The middle ear conveys sound vibrations to the inner ear, which houses the receptors for hearing and balance. The middle ear and inner ear are located within the temporal bone of the skull. The spaces of the outer ear and middle ear are filled with air, which allows the **eardrum (tympanic membrane)** and auditory **ossicles** to vibrate rapidly and freely.

ossicles (OS-si-kuls)
Small bones of the middle ear.

The ear consists of three parts: the outer ear, the middle ear, and the inner ear • Figure 8.11

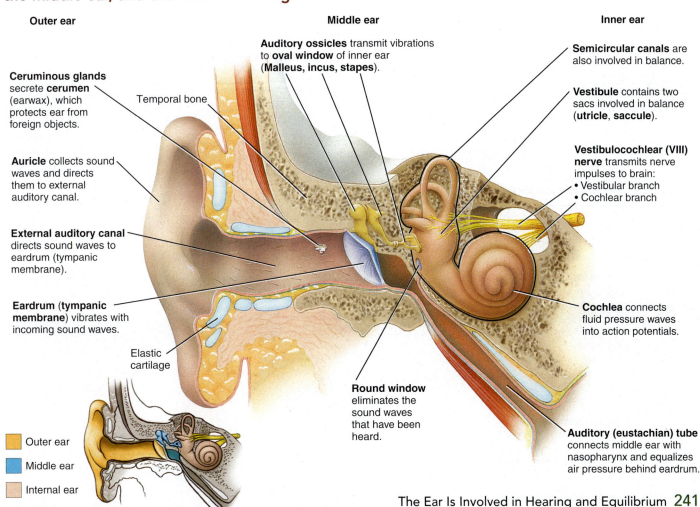

Outer ear

Ceruminous glands secrete **cerumen** (earwax), which protects ear from foreign objects.

Temporal bone

Auricle collects sound waves and directs them to external auditory canal.

External auditory canal directs sound waves to eardrum (tympanic membrane).

Eardrum (tympanic membrane) vibrates with incoming sound waves.

Elastic cartilage

Middle ear

Auditory ossicles transmit vibrations to **oval window** of inner ear (**Malleus, incus, stapes**).

Round window eliminates the sound waves that have been heard.

Inner ear

Semicircular canals are also involved in balance.

Vestibule contains two sacs involved in balance (**utricle, saccule**).

Vestibulocochlear (VIII) nerve transmits nerve impulses to brain:
• Vestibular branch
• Cochlear branch

Cochlea connects fluid pressure waves into action potentials.

Auditory (eustachian) tube connects middle ear with nasopharynx and equalizes air pressure behind eardrum.

- Outer ear
- Middle ear
- Internal ear

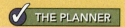
The inner ear is divided into the outer bony labyrinth and inner membranous labyrinth. The bony labyrinth contains the cochlea, the sense organ for hearing, and the vestibule and semicircular canals, the sense organs for equilibrium and balance.

The spiral organ (organ of Corti) is the organ of hearing and consists of supporting cells and hair cells, the receptors for auditory sensations.

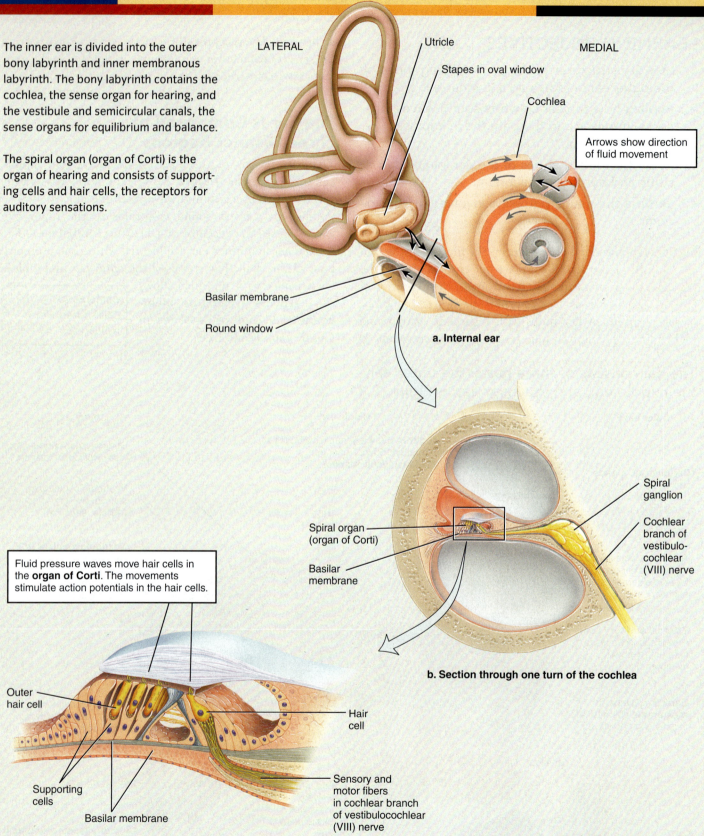

LATERAL

Utricle

Stapes in oval window

MEDIAL

Cochlea

Arrows show direction of fluid movement

Basilar membrane

Round window

a. Internal ear

Spiral organ (organ of Corti)

Basilar membrane

Spiral ganglion

Cochlear branch of vestibulo-cochlear (VIII) nerve

b. Section through one turn of the cochlea

Fluid pressure waves move hair cells in the **organ of Corti**. The movements stimulate action potentials in the hair cells.

Outer hair cell

Hair cell

Supporting cells

Basilar membrane

Sensory and motor fibers in cochlear branch of vestibulocochlear (VIII) nerve

c. Spiral organ (organ of Corti)

In contrast, the inner ear is filled with fluid, which translates vibrations into pressure waves that move rapidly through the channels in the cochlea and stimulate auditory receptors. The inner ear also contains the **vestibular apparatus** (**semicircular canals**, utricles, and saccules), which senses positions and movements of the head. Let's take a closer look at how these structures of the ear work together to perceive the ballerina's music (**Figure 8.12**).

Receptors for Hearing Are Found in the Cochlea

The outer ear directs incoming sound waves to the ear canal and eardrum, which vibrates with the sound (**Figure 8.13**). Vibrations of the eardrum cause the auditory ossicles to vibrate and move the **oval window** of the inner ear. The vibrations set up pressure waves in the fluid of the **cochlea**, which moves hair cells in the **organs of Corti** and elicits action potentials in sensory neurons. **Hair cells** are the receptors for hearing. Nerve impulses travel via the **vestibulo-cochlear nerve (cranial nerve VIII)** to the brain's auditory cortex in the temporal lobe.

> **cochlea** (KŌK-lē-a) A winding, cone-shaped tube that forms a portion of the inner ear.

Along the middle of the cochlea, there runs a membrane called the basilar membrane. Sound waves of various frequencies cause certain regions of the basilar membrane within the cochlea to vibrate more than others. Each segment of the **basilar membrane** is "tuned" for a particular pitch. Higher-frequency sounds vibrate the membrane near the oval window, while lower-frequency sounds vibrate the membrane near the tip of the cochlea. Louder sounds cause larger vibrations of the membrane and higher-frequency nerve impulses reaching the brain.

The mechanism of hearing • Figure 8.13

THE PLANNER

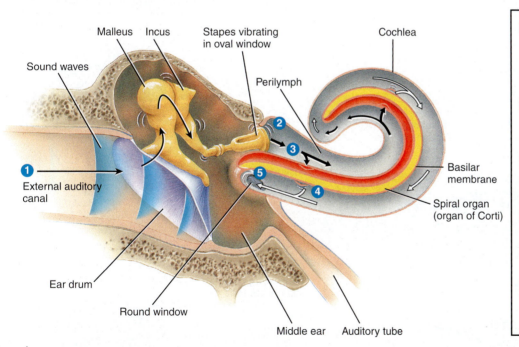

1. Sound waves hit the eardrum, causing the auditory ossicles to move. The ossicles pull the oval window in and out, thereby setting up fluid pressure waves in the cochlea.

2. The pressure waves move through the upper chamber of the cochlea and stimulate the hair cells in the spiral organs.

3. The stimulation of the hair cells initiates action potentials in the cochlear branch of the vestibulocochlear nerve (CN VIII).

4. The pressure waves shift into the lower chamber of the cochlea and travel to the round window where they are eliminated from the inner ear.

5. The vestibulocochlear nerve conducts the impulses to the medulla, midbrain, thalamus, and temporal lobes.

Ask Yourself

Suppose you had an ear infection and had fluid in your middle ear that was pushing on the eardrum and making it somewhat stiff. Why would you have trouble hearing?

The Ear Is Involved in Hearing and Equilibrium **243**

Equilibrium Comes in Two Varieties: Static and Dynamic

Sometimes ballet dancers stumble, but on these rare occasions, they usually regain their balance (equilibrium) quickly. There are two types of equilibrium: static and dynamic. **Static equilibrium** maintains body position, specifically the position of the head, relative to the direction of gravity, as when a ballerina is holding a pose. **Dynamic equilibrium** maintains head and body positions when the body is in motion (accelerating or decelerating), as might occur when a ballerina is doing a pirouette. Different parts of the vestibular apparatus are involved in the different types of equilibrium.

The saccule and utricle of the vestibule are involved in static equilibrium. They contain structures called *maculae*, which contain hair cells that are covered by a crystal and gelatinous layer (**Figure 8.14a**). The force of gravity pulls on the crystal layer and bends the hair cells, which elicit action potentials relaying position information.

The *ampulla* at the base of each semicircular canal is involved in dynamic equilibrium. Within each ampulla is a structure called a **crista** (KRIS-ta), which has embedded sensory hair cells (**Figure 8.14b**). When you move your head, the change of position causes the fluid within the semicircular canal to move and bend the hair cells, which elicit action potentials conveying movement information. The semicircular canals are oriented at right angles to one another in three dimensions, so that you can sense motion in all three dimensions of space. For example, if you spin around and then stop, you may feel

Hair cells within structures of the vestibular apparatus sense equilibrium • Figure 8.14 _____

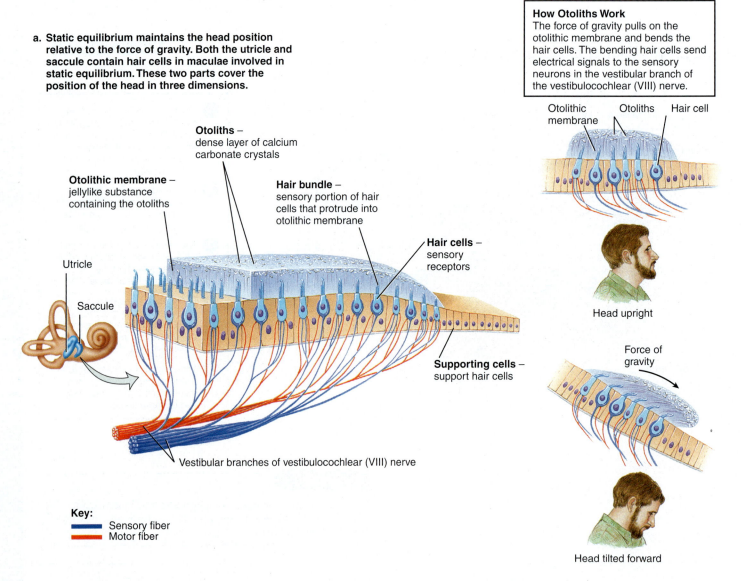

a. Static equilibrium maintains the head position relative to the force of gravity. Both the utricle and saccule contain hair cells in maculae involved in static equilibrium. These two parts cover the position of the head in three dimensions.

How Otoliths Work
The force of gravity pulls on the otolithic membrane and bends the hair cells. The bending hair cells send electrical signals to the sensory neurons in the vestibular branch of the vestibulocochlear (VIII) nerve.

Otolithic membrane   Otoliths   Hair cell

Head upright

Force of gravity

Head tilted forward

Otoliths –
dense layer of calcium carbonate crystals

Otolithic membrane –
jellylike substance containing the otoliths

Hair bundle –
sensory portion of hair cells that protrude into otolithic membrane

Hair cells –
sensory receptors

Utricle

Saccule

Supporting cells –
support hair cells

Vestibular branches of vestibulocochlear (VIII) nerve

Key:
— Sensory fiber
— Motor fiber

dizzy. Although you have stopped, the fluid in the semi-circular canal continues moving for a while. It stimulates hair cells and gives you the impression that you are still in motion. When the fluid stops moving, you feel stationary again.

The movements of hair cells within the vestibular apparatus generate action potentials and send nerve impulses through the vestibular branch of the vestibulocochlear nerve to various parts of the brain including the medulla, pons, cerebellum, and vestibular area of the cortex in the parietal lobe. The **vestibular nuclei** in the pons also receive and integrate sensory information from the eyes and proprioceptors in the head and neck muscles. This information goes to the cortex and cerebellum, which coordinate muscle movements.

CONCEPT CHECK

1. **Which** structure in the ear first vibrates with incoming sound waves?
2. **How** does the ear change sound waves into nerve impulses?
3. **What** does the otolithic membrane do?
4. **What** type of equilibrium would be used to determine the head position when an individual is doing a somersault?

b. **Dynamic equilibrium** maintains the head position relative to rotational acceleration or deceleration. The ampullae within the semicircular canals contain sites of hair cells in cristae involved in dynamic equilibrium. The semicircular canals cover acceleration in three dimensions.

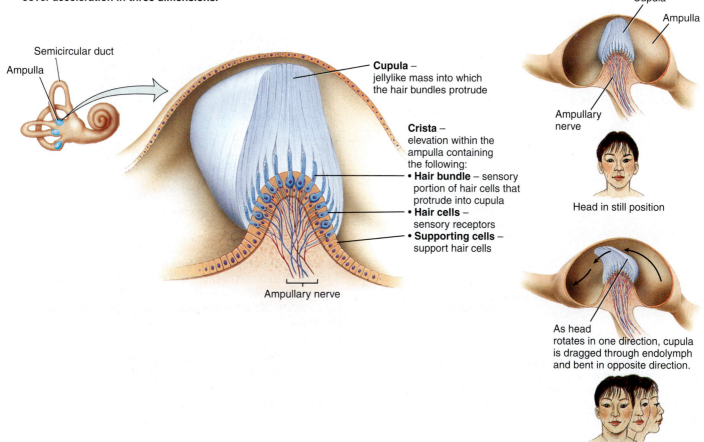

How Ampullae Work
Turning the head accelerates fluid within the semicircular canals. The fluid drags the cupula, which bends the hair cells. The bending hair cells send electrical signals to the sensory neurons in the vestibular branch of the vestibulocochlear (VIII) nerve.

Cupula

Ampulla

Ampullary nerve

Head in still position

As head rotates in one direction, cupula is dragged through endolymph and bent in opposite direction.

Head rotating

Semicircular duct

Ampulla

Cupula – jellylike mass into which the hair bundles protrude

Crista – elevation within the ampulla containing the following:
• **Hair bundle** – sensory portion of hair cells that protrude into cupula
• **Hair cells** – sensory receptors
• **Supporting cells** – support hair cells

Ampullary nerve

Summary

1 Somatic Senses Are Related to Detection of Pressure, Chemicals, Proprioception, and Touch 228

- **Sensation** is the conscious or subconscious awareness of changes in the external or internal environment. A stimulus causes a receptor to evoke nerve impulses that travel along specific neural pathways to the brain. Each receptor is specific for a particular type of stimulus. The brain perceives the stimulus as a sensation.

- As shown, the skin has various tactile receptors for touch, pressure, stretching, warmth, cold, and pain. Upon stimulation, these receptors evoke action potentials in sensory neurons. The action potentials propagate through ascending pathways in the spinal cord to the somatosensory cortex.

Skin receptors • Figure 8.2

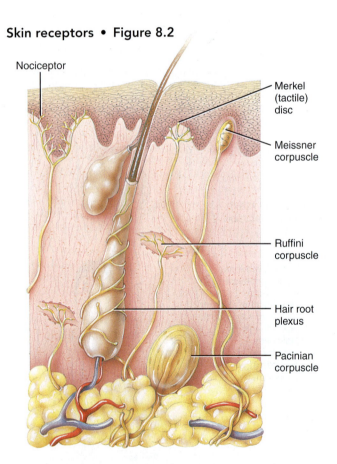

Nociceptor

Merkel (tactile) disc

Meissner corpuscle

Ruffini corpuscle

Hair root plexus

Pacinian corpuscle

- **Proprioceptive** sensations allow us to know where our head and limbs are located or moving. Proprioceptors in muscle, tendons, and joints sense the positions of the head and limbs. The vestibular apparatus in the inner ear senses the orientation of the head and neck. These receptors feed information to the somatosensory cortex and to the cerebellum.

2 Some Special Senses Use Receptors That Detect Chemicals 232

- The chemical senses include smell (**olfaction**) and taste (**gustation**). Specific chemicals bind to receptors in the nose and on the tongue, where they elicit nerve impulses that travel to the brain.

Structures of the olfactory epithelium • Figure 8.4

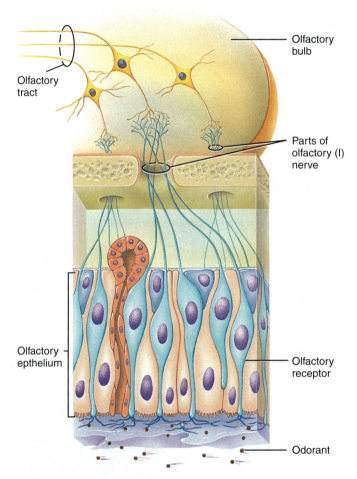

Olfactory bulb

Olfactory tract

Parts of olfactory (I) nerve

Olfactory epthelium

Olfactory receptor

Odorant

- As shown, **olfactory receptors** are located in the olfactory epithelium and can detect 10,000 different odorants. Nerve impulses from olfactory receptors travel through the olfactory tract to the temporal lobe, limbic system, and hypothalamus.

- **Gustatory receptors** are located in taste buds on the tongue and detect five classes of tastes: sweet, sour, bitter, salty, and umami (meaty, or savory). Nerve impulses evoked from taste buds travel through cranial nerves VII, IX, and X to the parietal lobe, limbic system, and hypothalamus.

- The gustatory center, which evaluates taste, combines information provided by both the nasal and oral receptors to determine flavor. If you have a cold and cannot smell, you usually cannot taste foods very well either.

- Light is refracted (bent) as it passes through the tissues of the eyeball. The **lens** focuses an inverted image on the retina. Ciliary muscles change the shape of the lens, depending on the distance from the objects being viewed, a process known as accommodation. Changes in the shape of the eyeball influence the focal plane of the lens. Abnormalities can be corrected with contact lenses, eyeglasses, or surgery.

3 Vision Involves Photoreceptors in the Eyes 235

- The eye has accessory structures to protect it (eyebrows, eyelashes, and eyelids), lacrimal apparatus to moisten it, and six external muscles to move it.

The layers of the eyeball • Figure 8.8

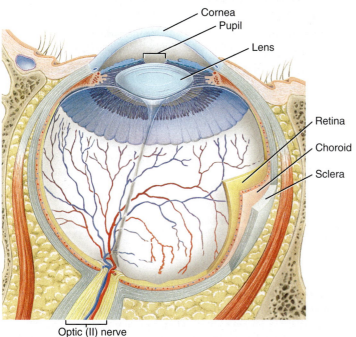

Cornea
Pupil
Lens
Retina
Choroid
Sclera
Optic (II) nerve

- The eyeball, as shown, is a three-layered structure (outer protective layer, middle vascular layer, and an inner neural layer). **Iris** muscles control the opening that lets light through (pupil). The **ciliary body** controls the shape of the lens for focusing the light on photoreceptors in the retina.

- **Photoreceptors** (rods and cones) contain photopigments that elicit nerve impulses in the retina and optic nerve. The axons from the lateral portions of each eye cross in the optic chiasm. The visual pathway ends in the occipital lobe. The right occipital lobe interprets information about the left halves of the visual fields of both eyes, while the left occipital lobe interprets input from the right halves of the visual fields.

4 The Ear Is Involved in Hearing and Equilibrium 241

- The ear is composed of the outer, middle, and inner ears. The outer ear, including the **auricle** and **external auditory canal** shown in the figure below, collects and directs sound waves to the middle ear, where they are converted to mechanical vibrations. Vibrations of the middle ear cause pressure waves in the fluid-filled inner ear. The pressure waves move hair cells in the **organs of Corti** that send impulses to the brain through the vestibulocochlear nerve (CN VIII) to the temporal lobe of the cerebral cortex.

- There are two types of equilibrium: static and dynamic. **Static equilibrium** senses the position of the head, and **dynamic equilibrium** senses movements of the head. Both are detected by the vestibular apparatus, which is located in the inner ear. In static equilibrium, gravity pulls on sensory hairs in the saccule and utricle and evokes nerve impulses. In dynamic equilibrium, the fluid inside the vestibular apparatus moves hair cells within the semicircular canals and evokes nerve impulses.

- Nerve impulses from the **vestibular apparatus** travel through the **vestibulocochlear nerve (CN VIII)** to the **pons**, somatosensory cortex, and cerebellum. The **vestibular nuclei** in the pons integrate this information along with visual information and **proprioceptive** information to sense equilibrium.

The inner ear • Figure 8.11

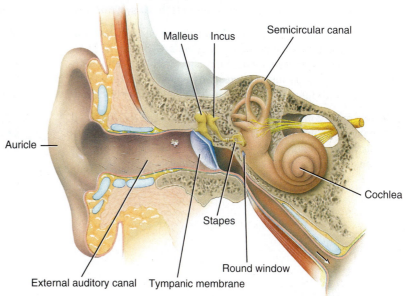

Malleus   Incus   Semicircular canal
Auricle
Cochlea
Stapes
Round window
External auditory canal   Tympanic membrane

Key Terms

- accommodation 239
- adaptation 229
- anterior cavity 236
- aqueous humor 236
- auditory (Eustacian) tube 241
- auditory ossicles 241
- auricle 241
- basal cell 232
- basilar membrane 243
- ceruminous gland 241
- choroid 237
- ciliary body 239
- cochlea 243
- cone 236
- cornea 236
- crista 244
- cupula 245
- dynamic equilibrium 244
- eardrum (tympanic membrane) 241
- encapsulated nerve ending 229
- external auditory canal 241
- fibrous tunic 236

- fovea centralis 237
- ganglionic cell 236
- gustation 232
- gustatory hairs 234
- gustatory receptor 234
- gustatory sensory neurons 234
- hair bundle 245
- hair cell 243
- hair root plexus 229
- incus 241
- iris 236
- lacrimal fluid 235
- lacrimal gland 235
- lens 236
- lysozyme 235
- malleus 241
- Meissner corpuscle 229
- Merkel (tactile) disc 229
- nasal cavity 235
- nasal lacrimal duct 235
- neural tunic 236
- nociceptor 230

- odorant 232
- olfaction 232
- olfactory receptor 232
- opsin 239
- optic chiasm 240
- optic disc 237
- optic nerve 237
- optic tract 240
- organ of Corti 243
- ossicle 241
- otolithic membrane 244
- otolith 244
- oval window 243
- Pacinian corpuscle 229
- papillae 234
- perception 228
- photopigment 239
- photoreceptor 236
- posterior cavity 236
- primary gustatory area 234
- primary visual area 240
- proprioceptor 230
- pupil 236

- referred pain 230
- refraction 238
- retinal 239
- rhodopsin 239
- rod 236
- round window 241
- Ruffini corpuscle 229
- sclera 237
- semicircular canal 243
- sensation 228
- stapes 241
- static equilibrium 244
- supporting cell 232
- tastant 234
- taste pore 234
- transduce 236
- vascular tunic 236
- vestibular apparatus 243
- vestibular nucleus 245
- vestibule 241
- vestibulocochlear nerve (CN VIII) 243
- vitreous body 236
- vitreous chamber 237

Critical and Creative Thinking Questions

1. Cindy likes to spin around. When she spins rapidly to the right three times and stops, she feels dizzy. However, if she spins rapidly to the right three times and then rapidly to the left three times and stops, she does not feel as dizzy. How can you explain this phenomenon with respect to the vestibular system?

2. After years of working around machinery in a noisy factory, John cannot hear high-frequency sounds. However, he can hear low-frequency sounds. Tests show that his vestibulocochlear nerves are functioning at all frequencies. Explain what is wrong with John's hearing and where the damage is located.

3. One patient in the ER has cut his hand and can feel localized pain. A second patient is having kidney stones and feels pain distributed around his back and waist. How can you explain the differences in the patients' experiences of pain?

4. Jim is a student in middle school who sits in the rear of the classroom. He complains to his mother that he can no longer read things written on the board or overhead projector. Any writing looks blurry to him. What type of visual abnormality does Jim have, and how could it be corrected?

5. Rita has allergies that give her a stuffy nose and painful sinus pressure. When she has these allergic episodes, she cannot smell anything, and her sense of taste seems to be altered as well. Explain what is happening to her to diminish her senses of smell and taste.

6. Robert was denied admission to the U.S. Naval Academy because he could not see the color red. What part of his eyes is affected by this abnormality?

What is happening in this picture?

Wines have distinct smells and subtle flavors. The olfactory sense can identify 10,000 different odors, but the gustatory sense can classify only 5 tastes.

NATIONAL GEOGRAPHIC

Think Critically 1. How can a wine taster distinguish between different varieties of wine, such as a Merlot and a Cabernet?
2. Explain why a wine taster would have to drink water in between wines or have a cracker to cleanse the palate. What do these actions do for the olfactory or gustatory process? How would they help the wine taster's ability to distinguish wines?

Self-Test

(Check your answers in Appendix C.)

1. Astronauts spend vast amounts of time in weightlessness, where the effect of gravity is minimal. Which sense would most likely be affected by this environment?

   a. hearing

   b. smell

   c. touch

   d. equilibrium

Use this figure for questions 2–4.

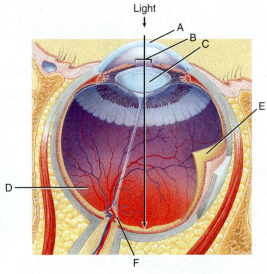

2. The structure labeled E _____.

   a. supplies blood to the eye

   b. focuses light

   c. converts light to nerve impulses

   d. regulates the amount of light entering the eye

3. Which labeled structure is the cornea?

   a. A

   b. B

   c. C

   d. D

4. Which labeled structure is the pupil?

   a. A

   b. B

   c. C

   d. D

5. _____ are receptors that detect pain.

   a. Proprioceptors

   b. Pacinian corpuscles

   c. Nociceptors

   d. Meissner corpuscles

6. What part of the ear changes sound waves to vibrations?

   a. auditory ossicles

   b. tympanic membrane

   c. cochlea

   d. oval window

Use this figure for questions 7–9.

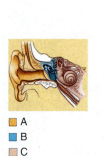
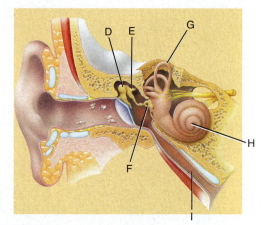

7. The area indicated by the letter C _____.

   a. collects and transmits sound

   b. converts sound waves to vibrations

   c. converts pressure waves to nerve impulses

   d. relieves pressure

8. Which labeled part is the stapes?

   a. D

   b. E

   c. F

   d. G

9. Which labeled part senses accelerations?

   a. D

   b. E

   c. F

   d. G

10. The _____ is the tactile receptor that detects pressure.
    a. Meissner corpuscle
    b. Ruffini corpuscle
    c. Merkel disc
    d. Pacinian corpuscle

11. Axons of the visual pathways cross at the _____.
    a. optic nerve
    b. optic chiasm
    c. thalamus
    d. optic tract

12. Which cells are the stem cells that eventually become olfactory receptors?
    a. basal cells
    b. supporting cells
    c. olfactory gland cells
    d. cribiform plate cells

Use this figure for questions 13–15.

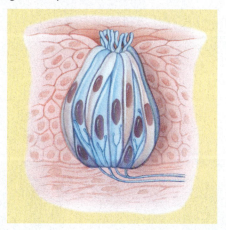

13. Where would you find the structure shown in the figure?
    a. inner ear
    b. tongue
    c. nose
    d. retina

14. This structure would detect _____.
    a. high-frequency sounds
    b. red light
    c. umami
    d. pressure

15. To what part of the brain do nerve impulses from the structure travel?
    a. hypothalamus
    b. limbic system
    c. parietal lobe
    d. all of the above

Use this figure for questions 16.

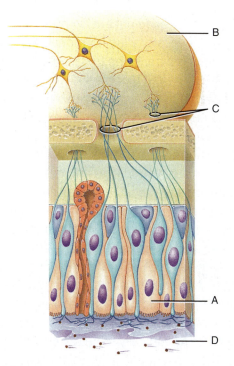

16. Which of the structures shown is the primary receptor for smell?
    a. A
    b. B
    c. C
    d. D

THE PLANNER ✓

Review your Chapter Planner on the chapter opener and check off your completed work.

The Endocrine System

A racing biker feels that he needs an edge so that he can get a better time on each leg of the Tour de France. A baseball player wants to hit more home runs. A championship track star wants to break a world record. In many instances, athletes such as these have turned to performance-enhancing drugs to achieve their goals.

Many performance-enhancing drugs are actually natural chemicals called hormones. For example, erythropoietin is a hormone that signals the bone marrow to make more red blood cells, which increases the ability of the blood to carry oxygen to working muscles. Because it increases oxygen-carrying capacity, it enhances an athlete's endurance. Human growth hormone and steroid hormones, such as testosterone, increase the growth of skeletal muscles. Increased muscle mass enables athletes to run faster and become stronger.

Unfortunately, these performance-enhancing drugs can have harmful side effects, not to mention the unfair competitive advantage they may give athletes. Many athletic agencies such as the International Olympic Committee, National Baseball League, and National Football League forbid players from using these substances. Furthermore, athletes must submit blood and urine samples periodically to be tested for banned substances. The consequences for using these substances can be severe. In some cases, athletes even face criminal charges for using and distributing these substances.

In addition to the hormones that enhance performance, the body produces many hormones that are necessary for normal functions. Let's look at the various hormones that the body produces, how the endocrine system uses them, and their physiological functions.

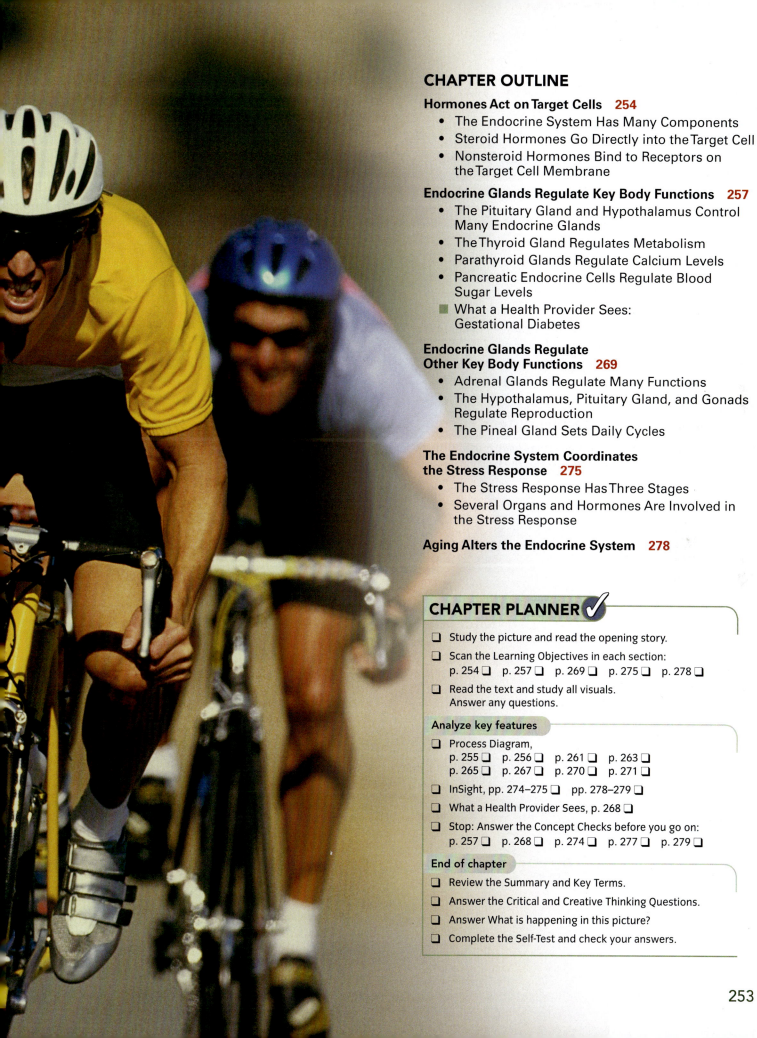

CHAPTER OUTLINE

CHAPTER PLANNER ✔

- ❏ Study the picture and read the opening story.
- ❏ Scan the Learning Objectives in each section:
  p. 254 ❏   p. 257 ❏   p. 269 ❏   p. 275 ❏   p. 278 ❏
- ❏ Read the text and study all visuals. Answer any questions.

Analyze key features

- ❏ Process Diagram,
  p. 255 ❏   p. 256 ❏   p. 261 ❏   p. 263 ❏
  p. 265 ❏   p. 267 ❏   p. 270 ❏   p. 271 ❏
- ❏ InSight, pp. 274–275 ❏   pp. 278–279 ❏
- ❏ What a Health Provider Sees, p. 268 ❏
- ❏ Stop: Answer the Concept Checks before you go on:
  p. 257 ❏   p. 268 ❏   p. 274 ❏   p. 277 ❏   p. 279 ❏

End of chapter

- ❏ Review the Summary and Key Terms.
- ❏ Answer the Critical and Creative Thinking Questions.
- ❏ Answer What is happening in this picture?
- ❏ Complete the Self-Test and check your answers.

Hormones Act on Target Cells

LEARNING OBJECTIVES

1. **Define** endocrine glands, hormones, hormone receptors, and target cells.

2. **Describe** how steroid hormones act.

3. **Describe** how nonsteroid hormones act.

To maintain a steady state, the body must regulate many parameters, such as heart rate, breathing, blood glucose, fluid balance, and so on. The nervous system regulates many of these on a moment-by-moment basis and is ready to rapidly respond to changes within seconds. Besides the nervous system, the **endocrine system** regulates many parameters chemically. Let's look at the components of the endocrine system.

The Endocrine System Has Many Components

The endocrine system consists of several components (see **Figure 9.1a**):

- **Endocrine glands** are glands that secrete a chemical signal directly into the bloodstream rather than through a duct or tube. For example, the salivary gland is an exocrine gland; it secretes saliva through a duct. In contrast, the pituitary, thyroid, parathyroid, adrenal, and pineal glands are endocrine glands because they secrete chemicals into the bloodstream.

> **endocrine system** A system of glands and hormone-secreting cells that regulate body functions through chemical messages (hormones).

Components of the endocrine system • Figure 9.1

This figure shows the major endocrine glands (red), organs that contain endocrine cells (blue), and some nearby organs for reference (a), and indicates the functional components of the endocrine system (b).

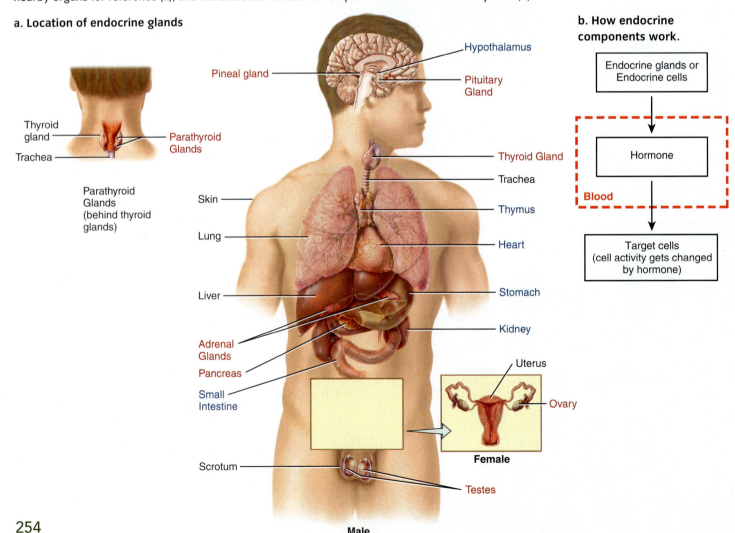

a. Location of endocrine glands

Thyroid gland
Trachea
Parathyroid Glands

Parathyroid Glands (behind thyroid glands)

Pineal gland
Hypothalamus
Pituitary Gland
Thyroid Gland
Trachea
Thymus
Heart
Stomach
Kidney
Uterus
Ovary
Skin
Lung
Liver
Adrenal Glands
Pancreas
Small Intestine
Scrotum
Testes

Female

Male

b. How endocrine components work.

Endocrine glands or Endocrine cells

Hormone

Blood

Target cells (cell activity gets changed by hormone)

How steroid hormones work • Figure 9.2

Steroid hormones go directly into the target cell and stimulate the making of new proteins, which alter the cell's activity. The entire process of steroid hormone action takes a significant amount of time (minutes to hours).

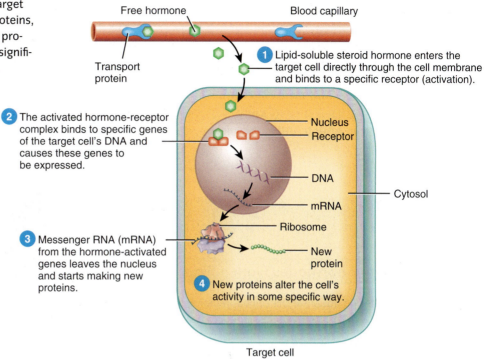

1 Lipid-soluble steroid hormone enters the target cell directly through the cell membrane and binds to a specific receptor (activation).

2 The activated hormone-receptor complex binds to specific genes of the target cell's DNA and causes these genes to be expressed.

3 Messenger RNA (mRNA) from the hormone-activated genes leaves the nucleus and starts making new proteins.

4 New proteins alter the cell's activity in some specific way.

Free hormone

Blood capillary

Transport protein

Nucleus

Receptor

DNA

Cytosol

mRNA

Ribosome

New protein

Target cell

- Some organs—such as the hypothalamus, pancreas, thymus, ovaries, testes, heart, liver, and kidneys—contain endocrine cells.

- Some tissues, such as adipose tissue, secrete hormones.

The endocrine system also controls long-term changes, such as growth and the onset of puberty.

The components of the endocrine system function in this general manner. In direct response to a change in some physiological parameter or by stimulation from the nervous system, an endocrine cell secretes a chemical signal called a **hormone** into the bloodstream. Although the hormone circulates throughout the body, it acts only on specific cells, called **target cells**. The target cells contain the appropriate receptor for a secreted hormone, called a **hormone receptor**; imagine the hormone receptor as a lock that can be opened only by a specific key (the hormone). The hormone–receptor complex elicits some action within the target cells that restores the changed parameter to normal levels. Once normal levels are restored, the secretion of the endocrine cells is usually stopped via negative feedback (see Chapter 1). In this way, the endocrine system helps maintain homeostasis.

Hormones come in two major types, steroid and nonsteroid. **Steroid hormones** dissolve in fats or lipids, while **nonsteroid hormones** dissolve in water. Steroid hormones are made from cholesterol, while nonsteroid hormones are usually made from small amino acids, peptides, or large

proteins. Steroid hormones can go directly through the target cell membrane, which is made of lipids, while nonsteroid hormones usually must bind to a hormone receptor on the surface of the cell membrane to enter the cell. These two modes of entry lead to entirely different mechanisms of action. Let's look at steroid hormones first.

Steroid Hormones Go Directly into the Target Cell

Steroid hormones pass directly through the cell membrane and stimulate specific genes to make new proteins, which alter the activity of the target cell (**Figure 9.2**). Because new proteins must be made, steroid hormones take longer to act than their nonsteroid counterparts. However, the effects of steroid hormones are generally longer lasting.

Different endocrine cells produce various steroid hormones, which act on a variety of targets. For example, the ovaries, testes, and adrenal glands produce androgens and estrogens, which act on many cells to produce male and female characteristics. The adrenal cortex secretes mineralocorticoids, which act on the kidneys to increase sodium and water reabsorption and potassium excretion. The kidneys produce calcitriol, the active form of vitamin D, which acts on the intestinal lining to promote absorption of calcium and phosphate.

Now, let's look at how nonsteroid hormones work.

Nonsteroid Hormones Bind to Receptors on the Target Cell Membrane

Nonsteroid hormones are water-soluble and cannot pass through the cell membrane. Instead, they bind to receptors on the surface of the cell. The hormone–receptor complex elicits the formation of a **second messenger** inside the cell, which alters the target cell's activity (**Figure 9.3**). Second messengers include **cyclic AMP (cAMP)** and calcium. cAMP is made from adenosine triphosphate (ATP) but does not participate in energy transformations. Instead, it is a chemical second message inside target cells. In some cells, calcium can act as a second messenger. The binding of a nonsteroid hormone to its receptor opens calcium channels in the membrane that allow calcium to flow into the cell, alter the activities of enzymes, and elicit responses.

> **cyclic AMP (cAMP)** A form of adenosine monophosphate in which the phosphate is in a ring structure.

PROCESS DIAGRAM

How nonsteroid hormones work • Figure 9.3

THE PLANNER

Nonsteroid hormones do not enter the cell but rather act through second messengers, such as cAMP, inside the cell. They achieve their physiological effects by activating existing proteins rather than making new ones. Their effects are relatively short–lived.

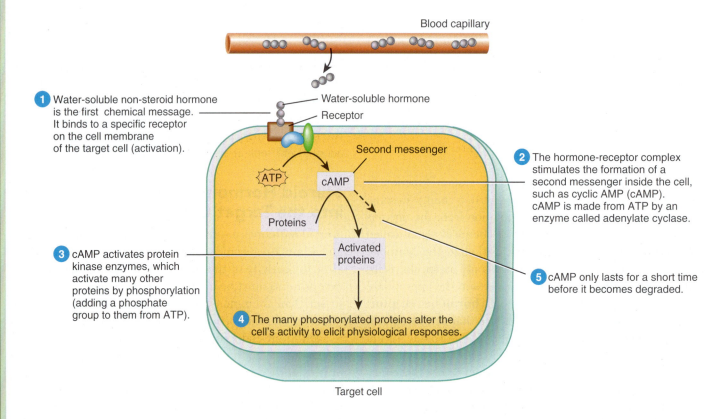

1 Water-soluble non-steroid hormone is the first chemical message. It binds to a specific receptor on the cell membrane of the target cell (activation).

Water-soluble hormone

Receptor

Second messenger

ATP

cAMP

Proteins

Activated proteins

2 The hormone-receptor complex stimulates the formation of a second messenger inside the cell, such as cyclic AMP (cAMP). cAMP is made from ATP by an enzyme called adenylate cyclase.

3 cAMP activates protein kinase enzymes, which activate many other proteins by phosphorylation (adding a phosphate group to them from ATP).

5 cAMP only lasts for a short time before it becomes degraded.

4 The many phosphorylated proteins alter the cell's activity to elicit physiological responses.

Blood capillary

Target cell

Because second messengers elicit a multistep process, their effects are amplified. This means that a small amount of nonsteroid hormone can produce an exaggerated and greatly diversified set of responses in the target cells. Note that this amplification occurs in each step depicted in Figure 9.3: One hormone molecule can elicit changes in many proteins. Also, because nonsteroid hormones do not involve making new proteins, they generally act faster, but their effects are more short-lived than those of steroid hormones.

A variety of endocrine cells make nonsteroid hormones that act on many target cells. For example, the adrenal medulla makes epinephrine and norepinephrine, which are also called adrenaline and noradrenaline, respectively. These hormones increase heart rate, contract blood vessels to increase blood pressure, and stimulate the liver to break down glycogen into glucose. The hypothalamus makes antidiuretic hormone (ADH), which is released from the pituitary gland and acts on the kidneys to reabsorb water. The pituitary gland also produces follicle-stimulating hormone (FSH), which stimulates sex cell production in the ovaries and testes. The pancreas secretes insulin, which acts on most cells in the body to take up glucose and store lipids.

CONCEPT CHECK

1. **How** does an endocrine gland communicate with a target cell?
2. **How** does a steroid hormone exert its action?
3. **What** is the role of the second messenger?

Endocrine Glands Regulate Key Body Functions

LEARNING OBJECTIVES

1. **Identify** the hormones secreted by the pituitary gland and their physiological effects.
2. **Outline** the processes of secretion and the functions of thyroid hormones.
3. **Explain** the roles of parathyroid glands in regulating calcium metabolism.
4. **Describe** the role of the pancreatic islets in glucose homeostasis.

The endocrine glands, organs containing endocrine cells, and endocrine tissues secrete approximately 30 different hormones. They regulate and control many body functions, including chemical composition and volume of blood, metabolism, contractions of smooth and heart muscles, secretions of endocrine glands, the immune system, growth and development, reproduction, and daily rhythms (circadian rhythms).

Diseases or endocrine disorders may involve diminished secretions of endocrine glands (hyposecretion) or increased secretions of endocrine glands (hypersecretion). Some hormone secretions are associated with more than one endocrine gland, as you will see with the hypothalamic–pituitary–thyroid axis. Diminished hormonal secretion by an endocrine gland may have one of two causes:

- *Primary hyposecretion.* This is a defect in a gland that directly secretes a hormone.

- *Secondary hyposecretion.* This is a defect in a gland that provides a stimulating hormone or releasing hormone to the gland that directly secretes a hormone.

Let's look at the functions of four major elements of the endocrine system: the hypothalamus/pituitary gland, thyroid gland, parathyroid gland, and pancreas.

The Pituitary Gland and Hypothalamus Control Many Endocrine Glands

The **pituitary gland**, which is about the size of a grape, is located just beneath the brain (**Figure 9.4**). It consists of two parts, the larger **anterior pituitary lobe** and the smaller **posterior pituitary lobe**; the lobes of the pituitary are often referred to merely as the anterior pituitary and the posterior pituitary. A third region of the pituitary gland, called the pars intermedia (intermediate lobe), atrophies during human fetal development and ceases to exist as a separate lobe in adults; however, some of its cells migrate into adjacent parts of the anterior pituitary, where they persist.

The structure and blood supply of the pituitary gland • Figure 9.4

The hypothalamus controls the secretions of the pituitary gland into the blood. Some hypothalamic neurosecretory cells make up the posterior pituitary, while others influence the anterior pituitary by secreting releasing and inhibiting hormones into a common blood supply.

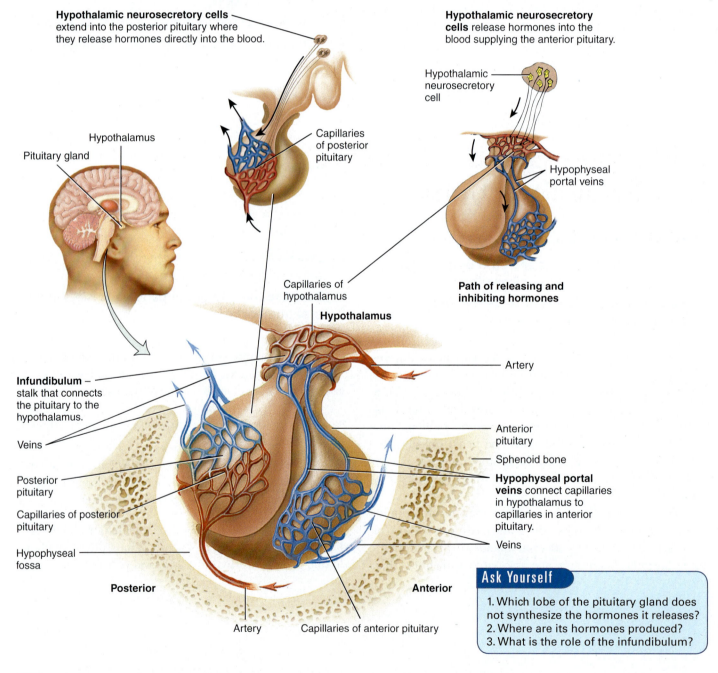

Hypothalamic neurosecretory cells extend into the posterior pituitary where they release hormones directly into the blood.

Hypothalamus

Pituitary gland

Capillaries of posterior pituitary

Hypothalamic neurosecretory cells release hormones into the blood supplying the anterior pituitary.

Hypothalamic neurosecretory cell

Hypophyseal portal veins

Path of releasing and inhibiting hormones

Capillaries of hypothalamus

Hypothalamus

Artery

Infundibulum – stalk that connects the pituitary to the hypothalamus.

Veins

Posterior pituitary

Capillaries of posterior pituitary

Hypophyseal fossa

Posterior

Artery

Capillaries of anterior pituitary

Anterior

Anterior pituitary

Sphenoid bone

Hypophyseal portal veins connect capillaries in hypothalamus to capillaries in anterior pituitary.

Veins

Ask Yourself

1. Which lobe of the pituitary gland does not synthesize the hormones it releases?
2. Where are its hormones produced?
3. What is the role of the infundibulum?

The pituitary gland secretes nine nonsteroid hormones, many of which control other endocrine glands. However, the pituitary gland itself is controlled by the brain, specifically the hypothalamus. The *posterior pituitary* is essentially an extension of the brain—axons and terminals of specialized nerve cells from the hypothalamus called **neurosecretory cells** make up the posterior pituitary lobe. The cells dump their contents directly into the blood vessels supplying the posterior pituitary lobe. The *anterior pituitary* shares a blood supply with the hypothalamus—some neurosecretory cells within the hypothalamus synapse on blood vessels that are connected to blood vessels of the anterior pituitary lobe. Secretions from these neurosecretory cells influence the cells

> **neurosecretory cells** A specialized form of nerve cells that secrete a neurotransmitter into the bloodstream rather than into a synaptic cleft.

of the anterior pituitary lobe. The hypothalamic secretions can either stimulate the pituitary gland (releasing hormones) or inhibit the pituitary gland (inhibiting hormones).

Hormones of the anterior pituitary regulate growth, metabolism, sexual maturation and reproduction, milk production, glucocorticoid production, and melanocyte activity (**Table 9.1**). Four of the hormones secreted by the anterior pituitary (TSH, FSH, LH, and ACTH; see Table 9.1) alter body functions indirectly by influencing the hormone secretion of other endocrine glands (the thyroid gland, ovaries, testes, and adrenal gland). The secretions of the anterior pituitary are controlled by hypothalamic releasing hormones (GHRH, TRH, GnRH, CRH, and PRH; see Table 9.1) and hypothalamic inhibiting hormones (GHIH and PIH; see Table 9.1). Remnant cells from the *pars intermedia* secrete melanocyte-stimulating hormone (MSH), but its role in humans is not known.

Hypothalamic hormones and corresponding anterior pituitary hormones   Table 9.1

Hypothalamic hormone	Anterior pituitary hormone	Target cell	Action
Growth-hormone releasing hormone (GHRH) promotes growth hormone secretion. Growth-hormone inhibiting hormone (GHIH) inhibits growth hormone secretion*	Human growth hormone (hGH)	Various tissues (e.g., liver, muscle, bones, cartilage)	Makes insulinlike growth factor (IGF) in target cells to control growth (stimulate cell division in muscle, bone, and cartilage)
Thyrotropin-releasing hormone (TRH)†	Thyroid-stimulating hormone (TSH)	Thyroid follicle cells	Secretes thyroid hormones (T_3, T_4) to control metabolism
Gonadotropin-releasing hormone (GnRH)	Follicle-stimulating hormone (FSH)	Ovaries, testes	Promotes egg and sperm development
	Luteinizing hormone (LH)	Ovaries, testes	Controls ovulation and production of estrogen, progesterone, and testosterone
Prolactin-releasing hormone (PRH) stimulates prolactin release. Prolactin-inhibiting hormone (PIH) inhibits prolactin release.	Prolactin (PRL)	Breast cells	Produces milk
Corticotropin-releasing hormone (CRH)	Adrenocorticotropic hormone (ACTH)	Adrenal cortex cells	Produces glucocorticoid to break down proteins and fats, form glucose, and reduce inflammation
	Melanocyte stimulating hormone (MSH); excessive CRH can stimulate MSH release	Melanocytes	Have an unknown function in humans, but it may influence brain activity

Notes

* Hypothalamic-inhibitory hormones are denoted in blue and have opposite effects of the releasing hormones.

† The terms (or endings) *tropic* (TRŌ-pik) *hormones, trophic* (TRŌ-fik) *hormones,* and *tropins* refer to hormones that act on other endocrine glands.

Pituitary dwarfism, gigantism, and acromegaly are caused by abnormal levels of secretion of human growth hormone • Figure 9.5

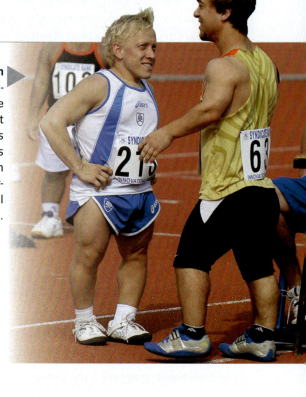

a. Pituitary Dwarfism
When the levels of secretion of human growth hormone (hGH) by the anterior pituitary are insufficient prior to puberty, bone growth is impaired and the individual does not grow to normal height. A person with pituitary dwarfism has normal body proportions but overall shorter-than-normal height.

b. Gigantism
A tumor of the anterior pituitary existing prior to puberty causes secretion of too much human growth hormone, resulting in gigantism. The acceleration of bone growth in this condition results in a person with normal proportions but taller-than-normal height.

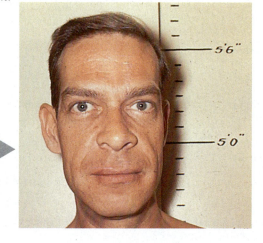

c. Acromegaly
A tumor of the anterior pituitary after puberty causes excess secretion of human growth hormone, resulting in acromegaly, a condition in which long bones can no longer grow. Instead, the bones of the hands, feet, face, and jaw thicken and grow larger.

When the anterior pituitary secretes either too little or too much human growth hormone (hGH), several disorders are manifested. These include **pituitary dwarfism** (**Figure 9.5a**), **gigantism** (**Figure 9.5b**) and **acromegaly** (**Figure 9.5c**). Abnormal secretions that result in dwarfism or gigantism usually occur prior to puberty, while acromegaly results from abnormal secretions after puberty. When a tumor is the cause of excess secretion, the condition may be treated with either surgery or chemotherapy to shrink the tumor.

The anterior pituitary may be involved in endocrine disorders of other endocrine glands. We will discuss these diseases when we discuss those glands.

Hypothalamic neurosecretory cells make and secrete **oxytocin** through the posterior pituitary. Oxytocin influences birth and milk release (lactation). During birth, oxytocin stimulates muscle contractions in the uterus. After birth, oxytocin stimulates milk release or let-down from breast tissue when the infant suckles.

Other hypothalamic neurosecretory cells make and secrete **antidiuretic hormone (ADH)** through the posterior pituitary (**Figure 9.6**). These cells sense increases in the salt concentration in the blood (osmolarity) and release ADH. ADH acts on the kidneys, sweat glands, arterioles, and brain to conserve body water and increase blood pressure.

The most common pituitary disorder is **diabetes insipidus**, which is caused by a lack of ADH secretion. Diabetes insipidus is most often caused by brain tumors, head trauma, or damage to the pituitary gland, pituitary stalk, or hypothalamus during surgery. Diabetes insipidus leads to excessive thirst, frequent urination, and large volumes of urine. While a normal person can tolerate dehydration for several days, individuals with diabetes insipidus cannot tolerate dehydration for even a day.

> **antidiuretic hormone (ADH)**
> A hormone secreted through the posterior pituitary lobe that conserves body water and increases blood pressure.

Effect of ADH secretions on body water and blood pressure levels • Figure 9.6

 THE PLANNER

The hypothalamus senses changes in osmolarity and directs neurosecretory cells to secrete ADH through the posterior pituitary.

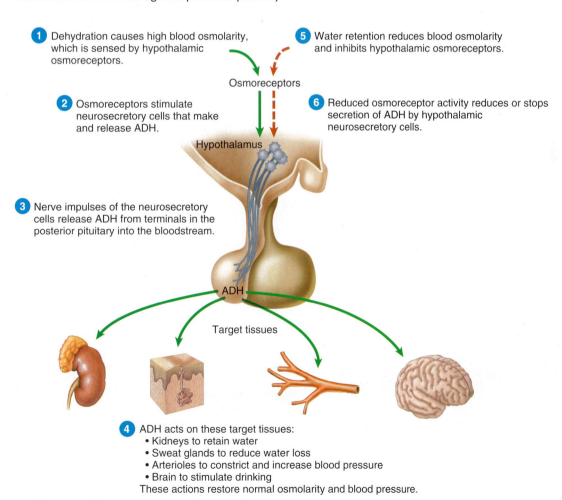

1. Dehydration causes high blood osmolarity, which is sensed by hypothalamic osmoreceptors.

2. Osmoreceptors stimulate neurosecretory cells that make and release ADH.

3. Nerve impulses of the neurosecretory cells release ADH from terminals in the posterior pituitary into the bloodstream.

5. Water retention reduces blood osmolarity and inhibits hypothalamic osmoreceptors.

6. Reduced osmoreceptor activity reduces or stops secretion of ADH by hypothalamic neurosecretory cells.

Osmoreceptors

Hypothalamus

ADH

Target tissues

4. ADH acts on these target tissues:
   • Kidneys to retain water
   • Sweat glands to reduce water loss
   • Arterioles to constrict and increase blood pressure
   • Brain to stimulate drinking
   These actions restore normal osmolarity and blood pressure.

Endocrine Glands Regulate Key Body Functions **261**

The Thyroid Gland Regulates Metabolism

The thyroid gland wraps around the trachea just below the voice box (**Figure 9.7**). The spongy thyroid tissue is composed of many small sacs called **follicles**. In the walls of the follicles, **follicular cells** make thyroid hormones, **thyroxine** and **triiodothyronine** (trī-ī-ō-dō-THĪ-rō-nēn), which are stored in the cavities of the follicles. Located between the follicles are the **parafollicular cells**, which make the hormone **calcitonin**.

Thyroid hormones are made from the amino acid tyrosine and have three or four iodine atoms added to them (thyroxine [T_4] has four iodines, and triiodothyronine [T_3] has three). Even though they are not lipid-soluble, the thyroid hormones are small enough to pass through the lipids of the target cell membrane without a surface receptor. Inside the cell, T_4 gets converted to the active form T_3. T_3 binds to the

calcitonin A hormone secreted by the parafollicular cells of the thyroid gland that inhibits osteoclast activity and lowers blood calcium levels.

Location and structure of the thyroid gland • Figure 9.7

Follicular cells of the thyroid gland make thyroid hormones and store them in cavities. Parafollicular cells lie between the follicles and make the hormone calcitonin.

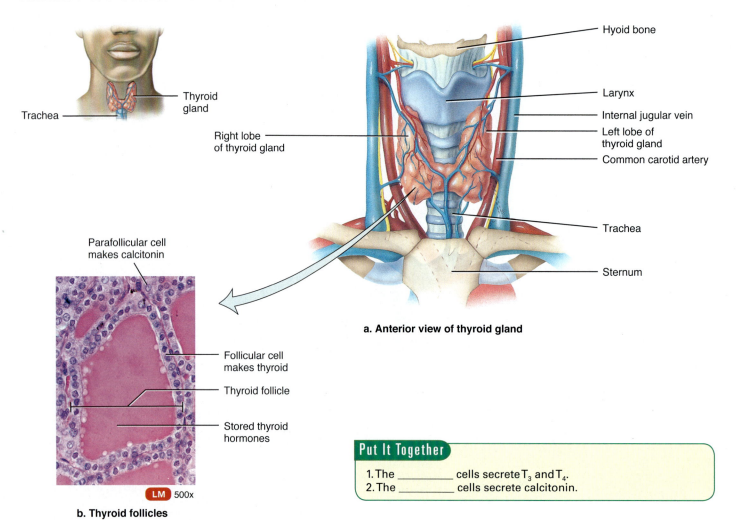

Trachea

Thyroid gland

Right lobe of thyroid gland

Hyoid bone

Larynx

Internal jugular vein

Left lobe of thyroid gland

Common carotid artery

Trachea

Sternum

a. Anterior view of thyroid gland

Parafollicular cell makes calcitonin

Follicular cell makes thyroid

Thyroid follicle

Stored thyroid hormones

LM 500x

b. Thyroid follicles

Put It Together

1. The _____ cells secrete T_3 and T_4.
2. The _____ cells secrete calcitonin.

Regulation of thyroid hormone secretion by the hypothalamus and pituitary gland • Figure 9.8

The hormones of the hypothalamus, pituitary gland, and thyroid make up a negative feedback loop that controls thyroid hormone levels in the blood.

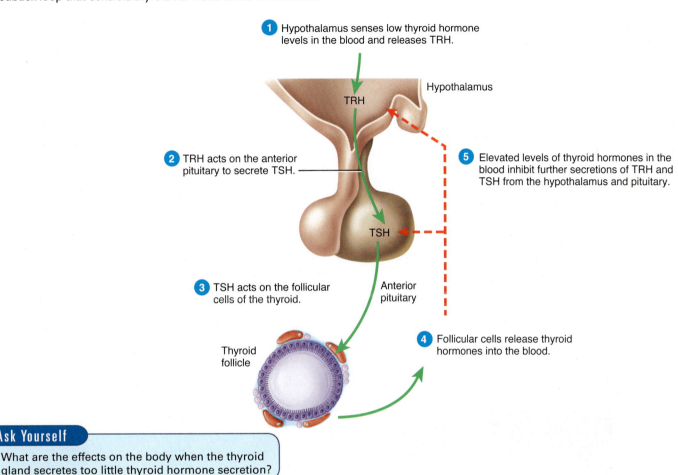

1. Hypothalamus senses low thyroid hormone levels in the blood and releases TRH.

2. TRH acts on the anterior pituitary to secrete TSH.

3. TSH acts on the follicular cells of the thyroid.

4. Follicular cells release thyroid hormones into the blood.

5. Elevated levels of thyroid hormones in the blood inhibit further secretions of TRH and TSH from the hypothalamus and pituitary.

Hypothalamus

TRH

TSH

Anterior pituitary

Thyroid follicle

Ask Yourself

What are the effects on the body when the thyroid gland secretes too little thyroid hormone secretion?

---

thyroid hormone receptor, which is located on nuclear DNA. Once T_3 is bound to the receptor, new proteins get made. T_3 has the following effects on target cells, which are most cells in the body:

- It stimulates the breakdown of glucose and fatty acids for ATP production (increasing the basal metabolism rate).

- During ATP production, cells use more oxygen and release more heat.

- Protein synthesis increases.

thyrotropin-releasing hormone (TRH) A hormone secreted by the hypothalamus that stimulates the anterior pituitary gland to secrete thyroid-stimulating hormone (TSH)

Thyroid hormones increase the excretion of cholesterol, thereby reducing its level in the blood. Together with insulin and hGH, thyroid hormones stimulate body growth.

The hypothalamus and anterior pituitary gland regulate the secretions of thyroid hormones by using a negative-feedback loop involving **thyrotropin-releasing hormone (TRH)**, thyroid-stimulating hormone (TSH), and thyroxine (**Figure 9.8**).

Parafollicular cells of the thyroid secrete calcitonin, which inhibits a type of bone cell called **osteoclasts**. Osteoclasts "eat" bone mineral and release the calcium into the blood. By inhibiting the activity of osteoclasts, calcitonin can decrease the levels of calcium in the blood. However, the normal physiological role of calcitonin remains a mystery because calcitonin can be present in excess or completely absent without causing any abnormal physiology.

When the thyroid gland does not secrete enough thyroid hormone, a condition called **hypothyroidism** results. If it occurs at birth, it is called **congenital hypothyroidism**; if it occurs later in life, it is referred to as **myxedema**. Congenital (or primary) hypothyroidism causes abnormal development and mental retardation, but it can be treated with oral doses of thyroxine; by law, newborns must be screened for proper thyroid function. Myxedema (secondary hypothyroidism) is characterized by puffiness in the face, slow heart rate, sensitivity to hot and cold, and muscle weakness. Primary hypothyroidism is caused by defects in the follicular cells of the thyroid, while secondary hypothyroidism can be caused by damage or defects in the hypothalamus (diminished TRH) or the anterior pituitary lobe (diminished TSH).

When the thyroid gland secretes too much thyroxine, **hyperthyroidism** occurs. In the most common form of this condition, called **Graves disease**, the immune system produces antibodies that bind to the TSH receptor, mimic TSH, and overstimulate the thyroid. Patients with Graves disease may have an enlarged thyroid called a goiter (**Figure 9.9a**) and/or peculiar puffiness in the eyes called **exophthalmos** (**Figure 9.9b**). Graves disease can be treated with drugs that block the synthesis of thyroid hormone, by destroying thyroid tissue with radiation, or by surgically removing thyroid tissue.

Parathyroid Glands Regulate Calcium Levels

The **parathyroid glands** are small glands embedded in the posterior side of the thyroid gland (**Figure 9.10**). The parathyroid glands secrete a protein hormone called **parathyroid hormone (PTH)**, which acts on bone cells (osteoclasts) to release calcium and on kidney cells to reduce calcium excretion and to release **calcitriol**.

Graves disease • Figure 9.9

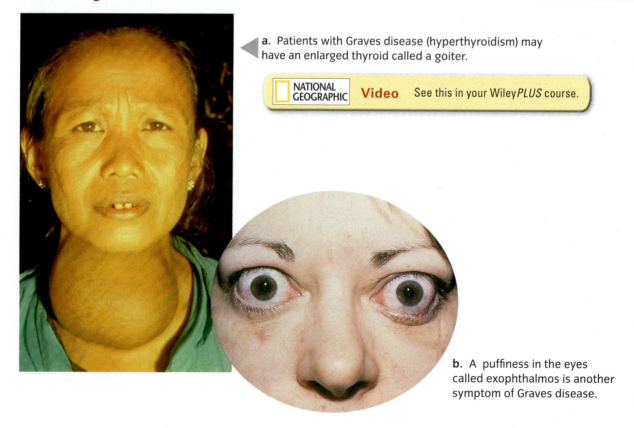

a. Patients with Graves disease (hyperthyroidism) may have an enlarged thyroid called a goiter.

NATIONAL GEOGRAPHIC **Video** See this in your Wiley*PLUS* course.

b. A puffiness in the eyes called exophthalmos is another symptom of Graves disease.

The parathyroid glands and the parafollicular cells of the thyroid gland are thought to work together to control blood calcium levels (**Figure 9.11**). The calcium level in the blood must be maintained within narrow limits for proper functioning of nerves and muscles, particularly the heart muscle, vascular smooth muscle, and intestinal smooth muscles.

When the parathyroid glands are damaged, which happens most often during surgery on the thyroid, the circulating levels of PTH fall. Hypoparathyroidism leads to a fall in blood calcium levels, which especially affects nerve and muscle cells. It can often be treated with calcium and vitamin D supplements.

Location of the parathyroid glands • Figure 9.10

The parathyroid glands are small glands embedded in the posterior side of the thyroid gland.

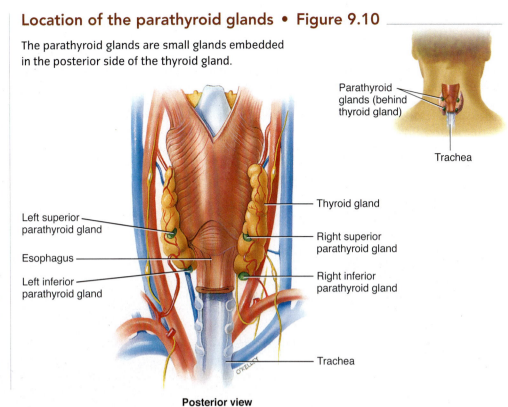

Parathyroid glands (behind thyroid gland)

Trachea

Thyroid gland

Left superior parathyroid gland

Right superior parathyroid gland

Esophagus

Right inferior parathyroid gland

Left inferior parathyroid gland

Trachea

O'KELLY

Posterior view

The activity of calcitonin and parathyroid hormone • Figure 9.11

☑ THE PLANNER

Calcitonin and parathyroid hormone act in concert to regulate blood calcium levels.

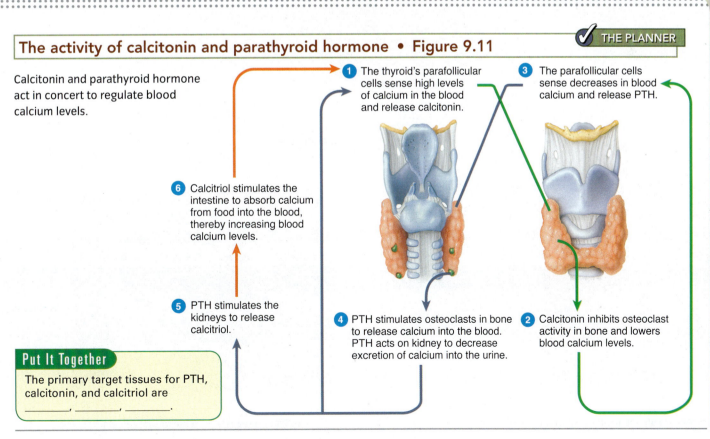

1 The thyroid's parafollicular cells sense high levels of calcium in the blood and release calcitonin.

3 The parafollicular cells sense decreases in blood calcium and release PTH.

6 Calcitriol stimulates the intestine to absorb calcium from food into the blood, thereby increasing blood calcium levels.

5 PTH stimulates the kidneys to release calcitriol.

4 PTH stimulates osteoclasts in bone to release calcium into the blood. PTH acts on kidney to decrease excretion of calcium into the urine.

2 Calcitonin inhibits osteoclast activity in bone and lowers blood calcium levels.

Put It Together

The primary target tissues for PTH, calcitonin, and calcitriol are _____, _____, _____.

PROCESS DIAGRAM

Pancreatic Endocrine Cells Regulate Blood Sugar Levels

The pancreas, which is both an exocrine gland and an endocrine gland, is located off the first section of the small intestine, called the duodenum (**Figure 9.12**). Its exocrine function concerns digestion, which you will learn more about in Chapter 14. In this chapter, we are concerned with its endocrine functions, which occur in the cells of the **pancreatic islets**, or **islets of Langerhans**.

The structure of the endocrine pancreas • Figure 9.12

The pancreas lies in the abdomen next to the small intestine.

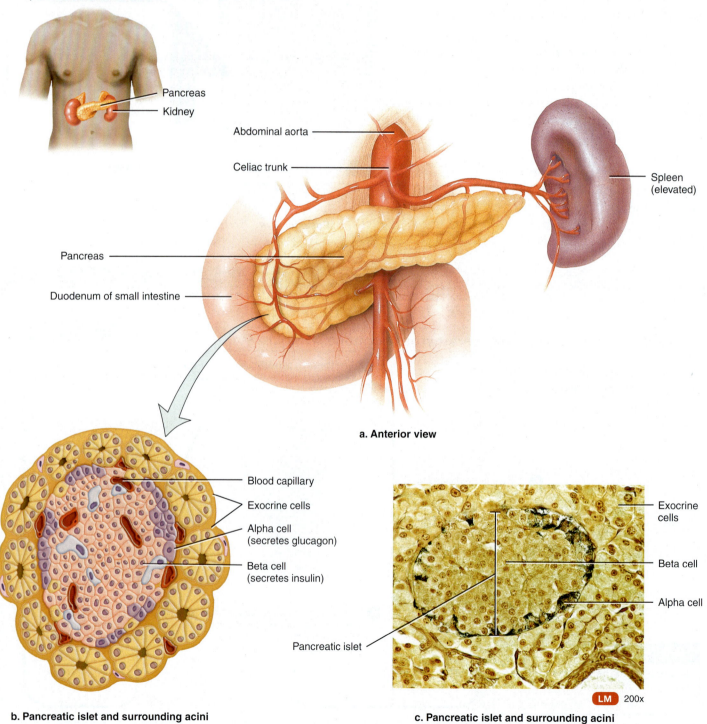

Pancreas
Kidney

Abdominal aorta
Celiac trunk
Spleen (elevated)

Pancreas
Duodenum of small intestine

a. Anterior view

Blood capillary
Exocrine cells
Alpha cell (secretes glucagon)
Beta cell (secretes insulin)

b. Pancreatic islet and surrounding acini

Exocrine cells
Beta cell
Alpha cell
Pancreatic islet

LM 200x

c. Pancreatic islet and surrounding acini

Within the islets are two types of cells: alpha cells, which secrete **glucagon**, and beta cells, which secrete **insulin**. Both glucagon and insulin are protein hormones that act mainly on the liver and skeletal muscle; they have opposite effects on the levels of glucose in the blood (**Figure 9.13**). Maintenance of normal blood glucose levels is important for the proper functioning of the nervous system.

The secretions of glucagon and insulin are coordinated to prevent blood glucose from rising too much after a meal and falling too much between meals. Abnormal levels of insulin, glucagon, and/or glucose in the blood can indicate **hypoglycemia**, **hyperglycemia**, or the presence of diseases such as diabetes mellitus, the most common disorder.

> **hypoglycemia**
> A condition in which blood glucose levels are below normal.
>
> **hyperglycemia**
> A condition in which blood glucose levels are above normal

There are two types of **diabetes mellitus,** which affects about 7.8% of the U.S. population:

- *Type I*. Type I diabetes involves a lack of insulin because the patient's immune system has destroyed the beta cells of the pancreas. Type I diabetes usually develops early in life and has been called juvenile diabetes.

- *Type II*. Type II diabetes involves decreased sensitivity of target cells to insulin. Type II diabetes is often associated with obesity, usually develops later in life, and has been called adult-onset diabetes. Patients with type II diabetes may have nearly normal levels of insulin in their blood. This type of diabetes is a growing epidemic in America, mainly due to the rise in obesity.

Both types of diabetes are characterized by elevated blood glucose levels and impaired glucose tolerance (see *What a Health Provider Sees*), presence of glucose in the urine, excessive thirst, excessive hunger and eating, and frequent urination. In both type I and type II diabetes, the body acts as if it is starving. Elevated glucagon and cortisol levels cause the breakdown of liver

Regulator of blood glucose by glucagon and insulin • Figure 9.13

THE PLANNER

Glucagon and insulin have opposite effects on skeletal muscle to regulate blood glucose levels.

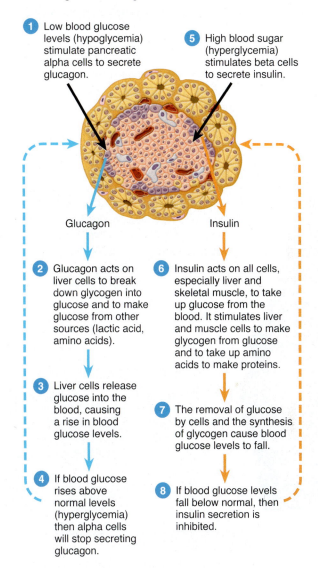

1. Low blood glucose levels (hypoglycemia) stimulate pancreatic alpha cells to secrete glucagon.

5. High blood sugar (hyperglycemia) stimulates beta cells to secrete insulin.

Glucagon

Insulin

2. Glucagon acts on liver cells to break down glycogen into glucose and to make glucose from other sources (lactic acid, amino acids).

6. Insulin acts on all cells, especially liver and skeletal muscle, to take up glucose from the blood. It stimulates liver and muscle cells to make glycogen from glucose and to take up amino acids to make proteins.

3. Liver cells release glucose into the blood, causing a rise in blood glucose levels.

7. The removal of glucose by cells and the synthesis of glycogen cause blood glucose levels to fall.

4. If blood glucose rises above normal levels (hyperglycemia) then alpha cells will stop secreting glucagon.

8. If blood glucose levels fall below normal, then insulin secretion is inhibited.

glycogen to release glucose, triglycerides to release fatty acids, and muscle proteins to release amino acids. Because most cells of the body do not take up excess glucose from the blood, the blood becomes hypertonic (Chapter 3). This condition, along with increased excretion by the kidneys, leads to dehydration, circulatory problems, and tissue damage. Diabetes mellitus often leads to death due to cardiovascular damage and kidney failure. Individuals with type I diabetes can be treated with blood glucose monitoring, changes in diet, exercise, and administration of insulin. Individuals with type II diabetes can be treated with blood glucose monitoring, changes in diet, exercise, and medications that stimulate insulin secretion.

Endocrine Glands Regulate Key Body Functions **267**

WHAT A HEALTH PROVIDER SEES

Gestational Diabetes

Jenna is pregnant and has her blood glucose checked during her regular obstetrical visits. During one of her tests, her fasting blood glucose is greater than normal. She fears that she might have **gestational diabetes**, a type of diabetes that affects about 4% of pregnant women, representing about 135,000 new cases in the United States each year. Although the exact cause is unknown, it's thought that hormones released by the placenta during pregnancy block the action of the mother's insulin (that is, cause insulin resistance). The mother's pancreas secretes more insulin than usual but not enough to remove glucose adequately. The increased blood glucose and other nutrients stimulate the baby's pancreas to hypersecrete insulin. As a result, the baby gains additional fat during development, which presents complications for delivery, low blood glucose at birth, and risk for obesity and type II diabetes in the mother.

Jenna's physician orders a one-hour oral glucose tolerance test (OGTT), usually done between weeks 24 to 28. To prepare for this test, Jenna fasts for 8 to 12 hours prior to the test. A technician takes a blood sample and then gives her a high-glucose drink (~50 g). She has blood drawn 1 hour later. Her blood glucose level is higher than normal (see the table below), so the physician orders a 3-hour OGTT. Jenna prepares for this test in the same way but drinks a higher-glucose drink (100 g) and has blood drawn at 1-hour intervals for 3 hours. When her test results show higher-than-normal responses, her physician diagnoses gestational diabetes.

For treatment, Jenna goes on a restricted carbohydrate diet, begins a new exercise program, and uses a blood glucose meter to monitor her blood glucose daily. Also, because Jenna may require two or three times the normal insulin levels, she requires daily insulin injections. This treatment will continue until she gives birth. For many women, gestational diabetes goes away after birth, but some women become diabetic permanently.

Test	Normal blood glucose (mg/dL)		Other blood glucose (mg/dL)	Diagnosis or treatment
Fasting	< 100		100–126	Administer GTT
1-hour OGTT (50 g glucose → blood sample in 1 hour)	< 140		>140	Gestational diabetes; may take 3-hour test and/or retake the test in 4 weeks
3-hour OGTT (100 g glucose → blood sample at 1, 2, and 3 hours)	Fast 1 2 3	< 95 < 180 < 155 < 140	Values greater than normal	Gestational diabetes

Think Critically

1. Why does hypersecretion of insulin by the fetus cause the baby to gain excess weight?
2. Why might Jenna require two or three times the normal insulin levels?

CONCEPT CHECK STOP

1. **How** is human growth hormone secreted, and what are its effects?

2. **How** do the hypothalamus and pituitary gland interact to control the secretion of thyroid hormone?

3. **What** is the effect of parathyroid hormone on blood calcium levels?

4. **How** does the pancreas help control blood glucose levels?

Endocrine Glands Regulate Other Key Body Functions

LEARNING OBJECTIVES

1. **List** the major hormones of the adrenal gland and explain their effects on the body.

2. **Describe** the influence of the hypothalamus and pituitary gland on the hormones of the gonads.

3. **Identify** the hormone secreted by the pineal gland and its function.

4. **List** the hormones secreted by endocrine cells not located in endocrine organs and their subsequent effects.

In addition to the hypothalamus/pituitary gland, thyroid gland, parathyroid gland, and pancreas, other endocrine glands control several body functions, including chemical composition and volume of blood, metabolism, contractions of smooth muscles and heart muscles, reproduction, and daily rhythms (circadian rhythms). These glands include the adrenal glands, the gonads (ovaries and testes), and the pineal gland, along with endocrine cells contained within other organs. Let's start with the adrenal glands.

Adrenal Glands Regulate Many Functions

An **adrenal gland** sits on top of each of the kidneys (**Figure 9.14**). Each gland consists of an outer **cortex** that surrounds an inner **medulla**. The cortex secretes three types of steroid hormones—mineralocorticoids, glucocorticoids, and androgens—each from different layers. The medulla secretes epinephrine and norepinephrine. Let's take a closer look at each of these hormones, their effects, and their secretion.

The structure of the adrenal glands • Figure 9.14

The adrenal glands rest on top of the kidneys. Each gland has two parts (cortex, medulla) surrounded by a capsule. Cells within each part secrete different hormones.

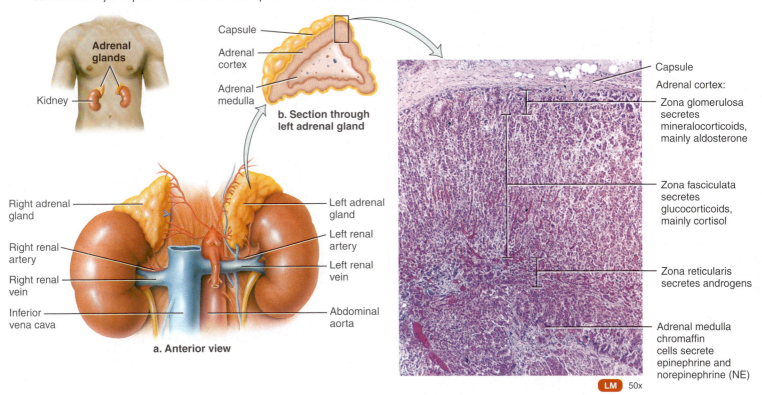

Adrenal glands

Kidney

Capsule

Adrenal cortex

Adrenal medulla

b. Section through left adrenal gland

Right adrenal gland

Right renal artery

Right renal vein

Inferior vena cava

Left adrenal gland

Left renal artery

Left renal vein

Abdominal aorta

a. Anterior view

Capsule

Adrenal cortex:

Zona glomerulosa secretes mineralocorticoids, mainly aldosterone

Zona fasciculata secretes glucocorticoids, mainly cortisol

Zona reticularis secretes androgens

Adrenal medulla chromaffin cells secrete epinephrine and norepinephrine (NE)

LM 50x

c. Subdivisions of the adrenal gland

Mineralocorticoids regulate mineral composition of the blood Cells in the outer zone of the adrenal cortex secrete steroid hormones called **mineralocorticoids**. One of these, **aldosterone**, acts on the kidneys to reabsorb sodium into the blood and

mineralocorticoids Steroid hormones secreted by the adrenal cortex that regulate the mineral composition of the blood.

excrete potassium in response to dehydration, blood loss, or sodium deficiency. Aldosterone acts in concert with the hormones renin and angiotensin to change the mineral composition of the blood and influence both blood volume and blood pressure (**Figure 9.15**).

Renin–angiotensin–aldosterone system • Figure 9.15

THE PLANNER

In a complex series of events involving several organs, the renin–angiotensin–aldosterone system restores blood volume and blood pressure.

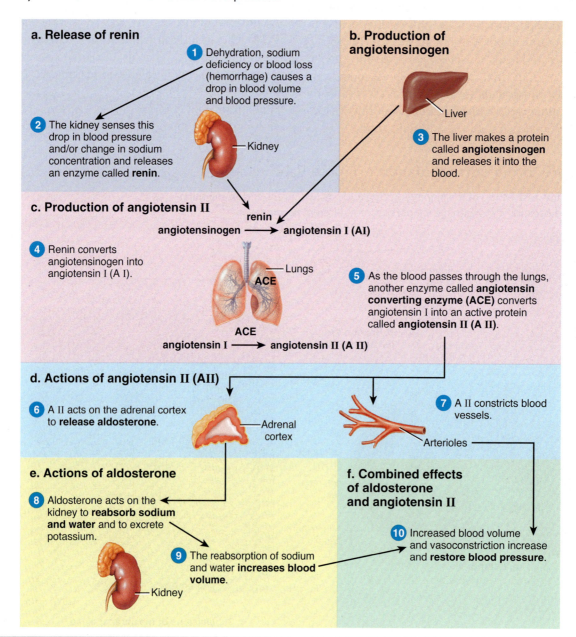

a. Release of renin

1. Dehydration, sodium deficiency or blood loss (hemorrhage) causes a drop in blood volume and blood pressure.

2. The kidney senses this drop in blood pressure and/or change in sodium concentration and releases an enzyme called **renin**.

— Kidney

b. Production of angiotensinogen

— Liver

3. The liver makes a protein called **angiotensinogen** and releases it into the blood.

c. Production of angiotensin II

renin
angiotensinogen ⟶ angiotensin I (AI)

4. Renin converts angiotensinogen into angiotensin I (A I).

— Lungs
ACE

5. As the blood passes through the lungs, another enzyme called **angiotensin converting enzyme (ACE)** converts angiotensin I into an active protein called **angiotensin II (A II)**.

ACE
angiotensin I ⟶ angiotensin II (A II)

d. Actions of angiotensin II (AII)

6. A II acts on the adrenal cortex to **release aldosterone**.

— Adrenal cortex

7. A II constricts blood vessels.

— Arterioles

e. Actions of aldosterone

8. Aldosterone acts on the kidney to **reabsorb sodium and water** and to excrete potassium.

9. The reabsorption of sodium and water **increases blood volume**.

— Kidney

f. Combined effects of aldosterone and angiotensin II

10. Increased blood volume and vasoconstriction increase and **restore blood pressure**.

Control of glucocorticoid secretion by the hypothalamus and pituitary gland • Figure 9.16

The hypothalamus and pituitary gland control secretion of glucocorticoids from the adrenal gland through negative feedback.

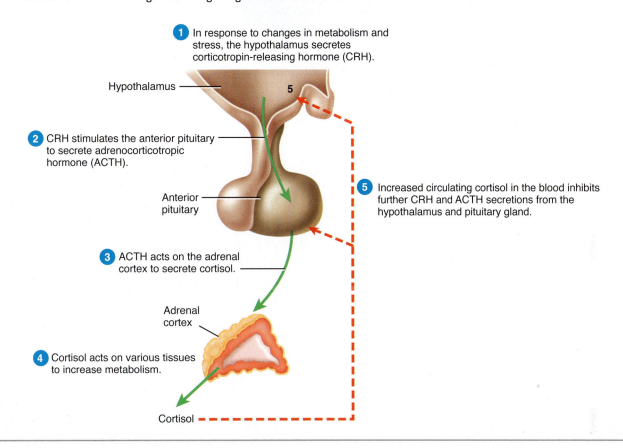

1 In response to changes in metabolism and stress, the hypothalamus secretes corticotropin-releasing hormone (CRH).

Hypothalamus

2 CRH stimulates the anterior pituitary to secrete adrenocorticotropic hormone (ACTH).

Anterior pituitary

5 Increased circulating cortisol in the blood inhibits further CRH and ACTH secretions from the hypothalamus and pituitary gland.

3 ACTH acts on the adrenal cortex to secrete cortisol.

Adrenal cortex

4 Cortisol acts on various tissues to increase metabolism.

Cortisol

If blood pressure is restored to normal, then renin and aldosterone secretions stop. With severe blood loss, the renin–angiotensin–aldosterone system cannot fully compensate for the drop in blood volume and pressure. Secretions of renin and aldosterone will stop only when the blood loss is fixed (for example, when a wound is repaired) and when blood volume is restored by other means (such as transfusion).

Glucocorticoids regulate energy balance

The cells in the middle layer of the adrenal cortex secrete **glucocorticoids**, mainly **cortisol**. Cortisol acts on many tissues to mobilize energy resources:

- *Skeletal muscle*. Cortisol causes the breakdown of muscle proteins into amino acids, which get released into the blood.
- *The liver*. Cortisol stimulates liver cells to convert amino acids into glucose, which is released into the blood.

- *Fat (adipose) tissue*. Cortisol stimulates fat cells to break down triglycerides into fatty acids, which are released into the blood. Other cells can use the fatty acids to make ATP.

> **glucocorticoids**
> Steroid hormones secreted by the adrenal cortex that regulate energy balance.

Glucocorticoids also suppress white blood cells from participating in the inflammatory response. In high doses, cortisol suppresses immune responses. Therefore, cortisol (hydrocortisone) is often prescribed as an anti-inflammatory treatment for rashes and allergic responses. In addition, cortisol is used to treat chronic inflammations, such as arthritis, and to decrease the risk of tissue rejection by the immune system in organ transplant recipients.

The hypothalamus and pituitary gland control secretion of glucocorticoids from the adrenal gland through a negative feedback loop involving corticotropin-releasing hormone (CRH), adrenocorticotropic hormone (ACTH), and cortisol (**Figure 9.16**).

One symptom of Cushing's syndrome, which results from excess cortisol, is a round, flushed face.

Secretion of too much cortisol by the adrenal gland causes a condition called **Cushing's syndrome**. Cushing's syndrome can be caused by adrenal tumors or by pituitary tumors that result in oversecretion of ACTH. Cushing's syndrome can also result from using high doses of cortisone for treating arthritis. Overstimulation by cortisol causes muscles to break down and fat to be redistributed. One obvious symptom is a round, flushed-looking face (**Figure 9.17**), along with a humped back and hanging abdomen. Complications include increased blood pressure, elevated blood glucose, weakness, brittle bones, and decreased resistance to stress. Depending on the cause, Cushing's syndrome can be treated using drugs that interfere with cortisol synthesis or by surgical removal of the adrenal glands or pituitary gland.

When the adrenal cortex does not secrete enough cortisol or aldosterone, the result is a rare disease called **Addison's disease**. Addison's disease can be caused by antibodies attacking the adrenal tissue or damage to the pituitary gland due to surgery or trauma that leads to insufficient ACTH secretion. Due to the loss of cortisol and aldosterone, Addison's disease patients often have symptoms including low blood sugar, weight loss, nausea/vomiting, muscle weakness, low blood pressure, dehydration, and heart problems (for example, arrhythmias, low cardiac output). Addison's disease is usually treated with artificial glucocorticoids and mineralocorticoids.

Adrenal androgens influence sexual characteristics and behaviors

In both males and females, the inner zone of the adrenal cortex secretes adrenal **androgens**. Before puberty, adrenal androgens contribute to growth of arm hair and pubic hair in males and pre-pubertal growth spurts in females. Adrenal androgens have little influence after puberty in males because the testes secrete far more androgens (testosterone) than the adrenal gland. Adrenal androgens have greater effects in females after puberty; they influence sex drive and become converted to estrogens (female hormones). While a female is fertile (between puberty and menopause), the ovaries are the organs primarily responsible for the secretion of estrogens. (You will learn more about the ovaries shortly.) After menopause, the conversion of adrenal androgens into estrogens becomes the only source of natural estrogen.

Epinephrine and norepinephrine regulate the body's response to stress and exercise

The cells of the adrenal medulla are like post-ganglionic nerve cells of the autonomic nervous system. Upon nervous stimulation, they secrete the hormones **epinephrine** (adrenaline) and **norepinephrine** (noradrenaline). These hormones act on many tissues to accomplish the following functions:

- Increase heart rate and force of contraction, which increases blood pressure and blood flow
- Constrict blood vessels and raise blood pressure
- Increase blood flow to the heart, liver, and skeletal muscles
- Dilate airways of the lung, which increases ventilation
- Break down liver glycogen into glucose, which is released into the blood
- Break down adipose tissue fats into fatty acids, which are released into the blood

These responses help the body during activities such as exercise by mobilizing energy reserves, bringing more oxygen into the body through the lungs, and increasing oxygen delivery to working muscles.

The Hypothalamus, Pituitary Gland, and Gonads Regulate Reproduction

The gonads, or sex-cell-producing organs, are the ovaries, which are located in the pelvis of the female, and the testes, which lie in the scrotal sac of the male. The ovaries produce the hormones **estrogen** and **progesterone**. These female sex hormones stimulate the development of female sex characteristics at puberty (for example, female body shape, pubic hair growth, breast development), regulate the menstrual cycle, and maintain pregnancy. The ovaries also produce other hormones, inhibin and relaxin, that have roles in pregnancy.

The testes produce **testosterone**, which is responsible for development of the male sex characteristics at puberty (for example, muscle growth, thickening of the vocal cords, increased facial and pubic hair) and sperm development. The testes also produce inhibin, which inhibits secretion of FSH from the pituitary gland.

Let's take a brief look at how the hypothalamus, anterior pituitary, and gonads work together to control the secretions of sex hormones, determine the onset of puberty, and maintain normal sexual functions (**Figure 9.18**). We will return to this topic in greater detail in Chapter 16.

- Neurosecretory cells in the hypothalamus of both males and females secrete pulses of **gonadotropin-releasing hormone (GnRH)** into the pituitary blood supply at regular intervals.

- GnRH stimulates the anterior pituitary lobe to secrete similar pulses of follicle-stimulating hormone (FSH) and luteinizing hormone (LH) into the bloodstream.

- In the male, FSH and LH stimulate sperm development and testosterone production by the testes. Testosterone in the blood inhibits secretions of GnRH from the hypothalamus and LH and FSH secretions from the pituitary gland.

- In the female, LH and FSH cause the egg cells to mature and to produce estrogen:

  - As the estrogen level increases and an egg develops, there comes a point when a massive release of LH and FSH from the pituitary gland occurs, leading to ovulation.

- After ovulation, the remnants of the follicle (corpus luteum) continue to produce estrogen and progesterone, which suppress GnRH secretion from the hypothalamus and LH and FSH secretions from the pituitary gland.

- Inhibin produced by the ovaries helps to suppress FSH secretion from the pituitary gland and inhibit the development of other egg cells.

- During pregnancy, the ovaries produce relaxin to increase the flexibility of uterine and vaginal smooth muscle in preparation for childbirth.

> **gonadotropin-releasing hormone (GnRH)**
> A hormone released by the hypothalamus that stimulates the anterior pituitary gland to secrete hormones—specifically luteinizing hormone (LH) and follicle stimulating hormone (FSH)—that act on the ovaries and testes.

Regulation of sex hormone secretions by the hypothalamus, pituitary gland, and gonads • Figure 9.18

The coordination of GnRH, FSH, and LH secretion, along with the secretion of sex hormones by the gonads, signal the onset of puberty and gamete production in both males and females. In addition, this process signals ovulation and maintenance of pregnancy in females.

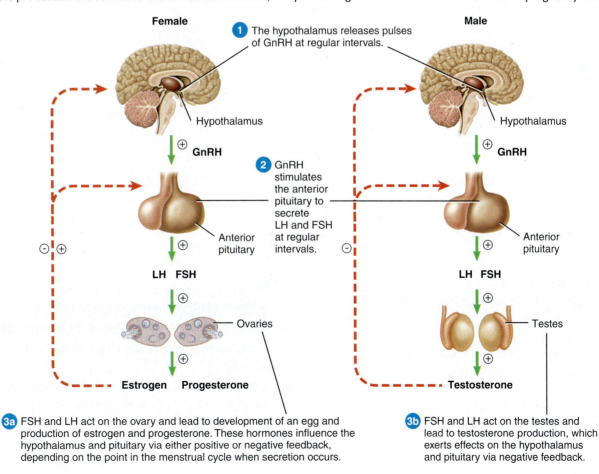

Female **Male**

1 The hypothalamus releases pulses of GnRH at regular intervals.

Hypothalamus — GnRH ⊕

2 GnRH stimulates the anterior pituitary to secrete LH and FSH at regular intervals.

Anterior pituitary

LH FSH ⊕

⊖ ⊕

Ovaries

Testes

Estrogen Progesterone

Testosterone

3a FSH and LH act on the ovary and lead to development of an egg and production of estrogen and progesterone. These hormones influence the hypothalamus and pituitary via either positive or negative feedback, depending on the point in the menstrual cycle when secretion occurs.

3b FSH and LH act on the testes and lead to testosterone production, which exerts effects on the hypothalamus and pituitary via negative feedback.

There are many cells in various non-endocrine organs that secrete hormones.

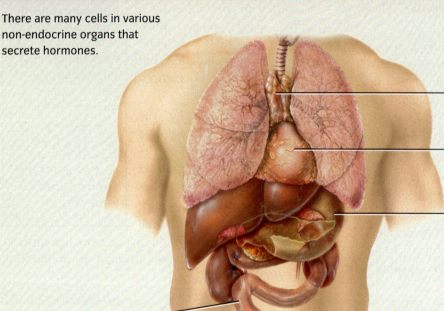

Thymus – Releases **thymosin**, which helps T-cell lymphocytes to mature and slow the aging process.

Heart – When blood volume increases, endocrine cells in the heart secrete **atrial natruretic peptide**, which stimulates sodium excretion by the kidney, reducing blood volume and pressure.

Stomach – Stretching and food in the stomach causes it to release **gastrin**, which stimulates gastric glands to secrete gastric juice and stomach muscles to increase motility.

Small intestine secretes several hormones:
• The duodenum secretes **glucose-dependent insulinotropic hormone (GIP)**, which stimulates the beta cells of the pancreas to secrete insulin. GIP coordinates the timing of intestinal glucose absorption with the insulin-evoked glucose uptake, so that the rise in blood glucose during the digestion of a meal is minimized.
• The mucosa secretes **secretin**, which stimulates the secretion of bicarbonate-rich pancreatic juices.
• The mucosa also secretes **cholecystokinin (CKK)**, which stimulates secretion of pancreatic juices and ejection of bile from the gallbladder, and inhibits gastric emptying.

The Pineal Gland Sets Daily Cycles

The **pineal** gland (PĪN-ē-al) is attached to the roof of the third ventricle in the midline of the brain. The pineal gland consists of masses of neuroglia and secretory cells called **pinealocytes** (pin-ē-AL-ō-sītz). This gland secretes the hormone **melatonin**, which is derived from the amino acid serotonin. Melatonin is thought to be involved in setting the body's daily clock. The pineal gland receives input from the eyes, indicating that light and dark may influence its activities. In support of this hypothesis, melatonin secretion is greatest during sleep and in darkness (almost 10 times greater than at other times). For patients who have trouble falling asleep, small doses of melatonin may be taken orally to promote sleep.

Overproduction of melatonin may be the cause of a type of depression called **seasonal affective disorder (SAD)**. SAD afflicts some people during the winter months, when day length is short. To provide relief from this disorder, full-spectrum bright-light therapy (that is, repeated doses of several hours of exposure to artificial light as bright as sunlight) is used. Another disorder that may be similar to SAD is jet lag, fatigue suffered by travelers who quickly cross several time zones. Three to six hours of exposure to bright light appears to help speed recovery from jet lag.

There are cells in non-endocrine organs that secrete hormones. See **Figure 9.19** to learn about some important endocrine functions in these non-endocrine glands.

CONCEPT CHECK

1. **What** is the role of aldosterone in the renin–angiotensin–aldosterone system?
2. **How** is ovulation controlled by hormones?
3. **What** is thought to be the function of the pineal gland?
4. **What** are the functions of hormones secreted by endocrine cells that are not located in endocrine glands?

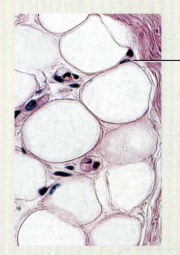

Adipose tissue releases **leptin**, which suppresses appetite.

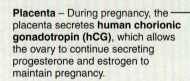

Placenta – During pregnancy, the placenta secretes **human chorionic gonadotropin (hCG)**, which allows the ovary to continue secreting progesterone and estrogen to maintain pregnancy.

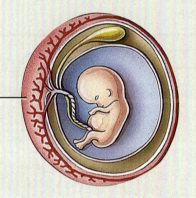

All cells except red blood cells secrete two hormones that are derived from fatty acids: **leukotrienes** (loo-ko-TRI-ens) and **prostaglandins** (pros'-ta-GLAN-dins). Unlike most other hormones, these act locally (that is, close to their sites of release), so they are not found in significant quantities in the blood. Leukotrienes stimulate the movement of white blood cells and mediate inflammation. The prostaglandins alter smooth muscle contraction, glandular secretions, blood flow, reproductive processes, platelet function, respiration, nerve impulse transmission, fat metabolism, and immune responses. Prostaglandins also have roles in inflammation, promoting fever, and intensifying pain.

The Endocrine System Coordinates the Stress Response

LEARNING OBJECTIVES

1. **Describe** the stages of the stress response.
2. **List** the glands of the endocrine system involved in the stress response.

Imagine that you are driving to class and you narrowly avoid a collision with another vehicle. The sound of the other car's horn startles you. You immediately realize what happened. Your heart beats faster and harder. You are breathing more heavily. You may be sweating. When you get to class, your professor springs a pop quiz for which you are unprepared, and the same symptoms appear again. These events are stressors, and we encounter such events every day. Some other examples of stress include fear, anxiety, other strong emotions, exposure to toxins or allergens, changes in temperature, exercise, injury trauma, and disease.

The body has a series of physiological responses to stress, which is called the **stress response**, or **general adaptation syndrome**. The stress response, which has three stages, strives to maintain homeostasis.

The Stress Response Has Three Stages

Whatever the cause of stress, the body responds in stages. The *fight-or-flight stage* prepares the body for immediate physical activity. The *resistance stage* involves prolonged resistance to the stressor. The *exhaustion stage* features an inability to resist prolonged exposure to the stressor.

Let's look at the responses in each stage.

The stress response • Figure 9.20

By secreting norepinephrine and various hypothalamic releasing factors, the sympathetic nervous system and hypothalamus coordinate the secretions of other glands and activation of various processes to make an immediate and sustained response to stress. (Green arrows represent nervous activity, red arrows represent primary hormone secretions, and black arrows indicate subsequent hormone secretions.)

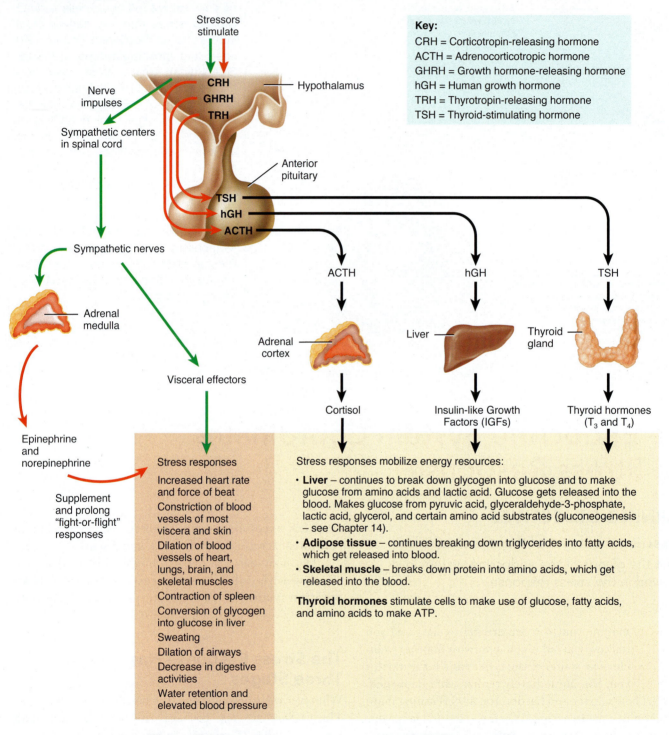

Key:
CRH = Corticotropin-releasing hormone
ACTH = Adrenocorticotropic hormone
GHRH = Growth hormone-releasing hormone
hGH = Human growth hormone
TRH = Thyrotropin-releasing hormone
TSH = Thyroid-stimulating hormone

Stressors stimulate

Nerve impulses

CRH
GHRH
TRH

Hypothalamus

Sympathetic centers in spinal cord

Anterior pituitary

TSH
hGH
ACTH

Sympathetic nerves

ACTH   hGH   TSH

Adrenal medulla

Adrenal cortex   Liver   Thyroid gland

Visceral effectors

Epinephrine and norepinephrine

Supplement and prolong "fight-or-flight" responses

Cortisol   Insulin-like Growth Factors (IGFs)   Thyroid hormones (T_3 and T_4)

Stress responses

Increased heart rate and force of beat

Constriction of blood vessels of most viscera and skin

Dilation of blood vessels of heart, lungs, brain, and skeletal muscles

Contraction of spleen

Conversion of glycogen into glucose in liver

Sweating

Dilation of airways

Decrease in digestive activities

Water retention and elevated blood pressure

Stress responses mobilize energy resources:

• **Liver** – continues to break down glycogen into glucose and to make glucose from amino acids and lactic acid. Glucose gets released into the blood. Makes glucose from pyruvic acid, glyceraldehyde-3-phosphate, lactic acid, glycerol, and certain amino acid substrates (gluconeogenesis – see Chapter 14).

• **Adipose tissue** – continues breaking down triglycerides into fatty acids, which get released into blood.

• **Skeletal muscle** – breaks down protein into amino acids, which get released into the blood.

Thyroid hormones stimulate cells to make use of glucose, fatty acids, and amino acids to make ATP.

a. "Fight-or-flight" responses   **b. Resistance reaction**

Fight-or-flight stage In the **fight-or-flight stage**, the sympathetic nervous system (Chapter 7) kicks in and stimulates heart, lungs, blood vessels, and the adrenal medulla (see **Figure 9.20**):

- Sympathetic stimulation of the heart and blood vessels increases heart rate, blood pressure, and blood flow, especially to skeletal muscles.
- Sympathetic stimulation of the lungs dilates airways and increases the rate of breathing.
- Sympathetic stimulation of the adrenal medulla releases epinephrine and norepinephrine.

These coordinated responses increase the availability of oxygen and fuel to working muscles. However, this initial stage cannot be sustained for long.

Resistance stage The **resistance stage** takes over when the fight-or-flight stage is almost complete. In this stage, the body's energy reserves are tapped to provide energy to working muscles or other tissues, such as sites of injury. The resistance stage involves the hypothalamus, pituitary gland, adrenal cortex, and thyroid gland. Depending on the type and severity of the stress, the resistance stage may be successful in combating the stressor and returning the body to normal. However, when an individual cannot defeat a powerful stressor, such as severe trauma, the body becomes exhausted.

Exhaustion stage The **exhaustion stage** occurs when the body's energy resources become depleted, muscles waste away, and the immune system is suppressed.

Several Organs and Hormones Are Involved in the Stress Response

The sympathetic nervous system and the hypothalamus direct the stress response by stimulating other organs and glands, such as the adrenal glands, the pituitary gland, the thyroid gland, the liver, and skeletal muscle (see Figure 9.20). Let's discuss them at each stage of the stress response.

In the fight-or-flight stage, the sympathetic nervous system is activated and releases norepinephrine, which:

- Stimulates the heart, which increases heart rate, blood pressure, and blood flow.
- Constricts blood vessels to visceral organs, such as the digestive system and skin. This reduces digestion and

diverts blood flow to working skeletal muscles, the heart, and the lungs.

- Stimulates sweat glands in the skin, which leads to increased sweating.
- Dilates the airways of the lungs and increases the rate of breathing.
- Stimulates release of epinephrine and norepinephrine from the adrenal medulla, enhancing the effects of the sympathetic nervous system and causing the liver and adipose tissue to begin mobilizing energy reserves. In the liver, glycogen breaks down into glucose, which is released into the blood. In adipose tissue, triglycerides break down into fatty acids, which are released into the blood.

In the resistance stage, the hypothalamus stimulates the pituitary gland, the thyroid gland, the adrenal glands, the liver, and muscle, and adipose tissues (see Figure 9.18). The hypothalamus secretes CRH, GHRH, and TRH, which act on the anterior pituitary lobe (see Table 9.1). The anterior pituitary secretes TSH, hGH, and ACTH. TSH acts on the thyroid gland to release thyroid hormones, which increase metabolism. hGH acts on the liver and other tissues through IGFs to sustain the breakdown of glycogen and triglycerides. ACTH stimulates the adrenal cortex to release cortisol, the major stress hormone. Cortisol acts on the liver, adipose tissue, and skeletal muscle to maintain the mobilization of energy reserves and reduce inflammation. So, the resistance stage mobilizes energy sources (glucose, fatty acids, amino acids) that tissues can use to make ATP. Because the resistance stage relies on chemical messages, it takes longer to initiate, but it can be sustained for a long period of time to combat a stressor.

Exhaustion occurs due to prolonged exposure to cortisol, which, among other effects, depletes the body's energy resources, wastes away muscles, and suppresses the immune system. Prolonged stress or frequent stresses may weaken the body and make it susceptible to other diseases. However, the exact role of stress in diseases is not known.

CONCEPT CHECK

1. **What** are the stages of the stress response?
2. **How** are the components of the endocrine system and their target organs involved in each stage of the stress response?

Aging Alters the Endocrine System

LEARNING OBJECTIVE

1. **Describe** the effects of aging on the endocrine system.

Aging alters the secretions of various endocrine glands. Because several endocrine functions involve more than one endocrine gland through negative-feedback loops, aging may cause different hormonal changes; some glands may increase secretions, while others may decrease secretions.

Most often, as we age, the secretions of major endocrine glands diminish, and some endocrine glands may shrink.

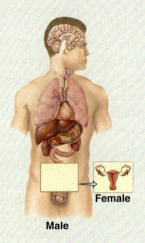

Male

Female

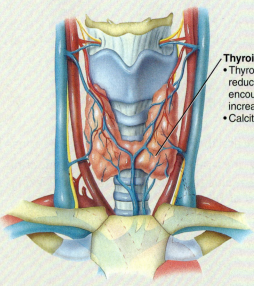

Hypothalamus
• GHRH secretion from the hypothalamus may also diminish with age.
• TRH secretion may be elevated due to failure of negative feedback by thyroid hormones.
• Slightly higher levels of CRH due to failure of negative feedback by cortisol.
• GnRH secretion decreases.

Pituitary gland
• hGH secretion decreases, leading to muscle atrophy and weakness.
• TSH secretion may be elevated due to negative feedback by thyroid hormones.
• Slightly higher levels of ACTH due to failure of negative feedback by cortisol.
• LH, FSH secretions decrease probably due to decreasing GnRH from hypothalamus.

Pineal gland – Secretion of melatonin is reduced with age, which may disrupt normal sleep-wake cycles.

Thyroid gland
• Thyroid hormone secretions diminish, resulting in reduced rate of metabolism. A reduction in metabolism encourages the formation of more body fat and increases sensitivity to cold temperatures.
• Calcitonin secretion diminishes.

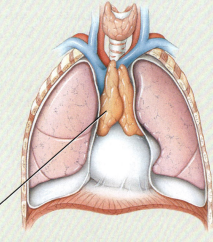

The **thymus gland** shrinks significantly, resulting in decreased immune cell production and decreased resistance to disease.

In general, the size or secretions of many endocrine glands diminish as the body ages (**Figure 9.21**). Hormonal changes include diminished output by endocrine glands, diminished sensitivity of target tissues to hormonal stimulation, and increased secretions of some glands as a consequence of failed negative-feedback loops within the endocrine system. For example, the cessation of secretions of estrogen and progesterone from the ovaries causes one of the most dramatic changes in the female endocrine system—menopause (see Chapter 16). Also, insufficient dietary calcium leads to elevated PTH secretion and diminished secretions of calcitonin and calcitriol; these events lead to osteoporosis, weakening of bone and increased occurrences of bone fractures (see Chapter 5).

CONCEPT CHECK STOP

1. **How** does aging affect the ovaries in women?

 THE PLANNER

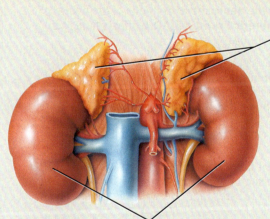

Adrenal glands – The adrenal cortex tissues become more fibrous with age, so the secretions of cortisol and aldosterone diminish, which impairs the body's ability to respond to stress and to regulate the mineral composition of the blood. For example, lack of aldosterone causes the kidney to excrete sodium.

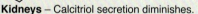

Kidneys – Calcitriol secretion diminishes.

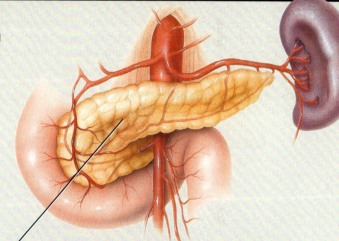

Pancreas – Insulin secretion and insulin sensitivity of the target tissues diminishes with age, which affects changes in blood glucose; increases occur faster and the return to normal glucose levels occurs more slowly.

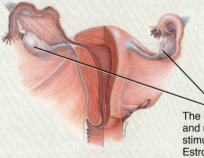

The **ovaries** shrink with age and no longer respond to stimulation by FSH and LH. Estrogen and progesterone secretions diminish greatly, which causes ovulation and the menstrual cycle to stop (menopause).

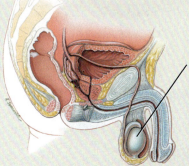

The **testes** continue to produce sperm and secrete testosterone as men age, but the secretions of testosterone diminish. Large decreases in testosterone production do not occur until well into old age. The decrease in testosterone reduces sex drive and can lead to increased body fat.

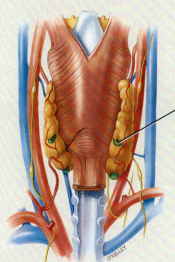

Parathyroid glands – Increase PTH secretion due to insufficient calcium in the diet. Elevated levels of PTH cause increased bone resorption by osteoclasts, which leads to weakening of the bones and susceptibility to fractures (osteoporosis).

Summary

1 Hormones Act on Target Cells 254

- The endocrine system consists of both endocrine glands and endocrine cells of other organs. The endocrine system regulates many physiological parameters chemically.

- Endocrine cells secrete hormones that bind to receptors in target cells and cause some actions in the target cells to evoke physiological responses.

- Hormones can be classified as steroid hormones or nonsteroid hormones.

- Steroid hormones can go directly through the target cell membrane, and stimulate specific genes to make new proteins, which alter the activity of the target cell.

- As shown, nonsteroid hormones, usually must bind to a hormone receptor on the surface of the cell membrane, and elicit the formation of a "second messenger" inside the cell (for example, **cyclic AMP**, calcium), which alters the target cell's activity.

Nonsteroid hormones • Figure 9.3

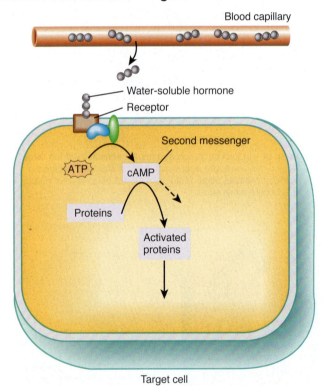

- Nonsteroid hormones generally act faster, but their effects are shorter than those of steroid hormones.

2 Endocrine Glands Regulate Key Body Functions 257

- The endocrine glands and organs containing endocrine cells secrete approximately 30 different hormones that regulate and control many body functions, including chemical composition and volume of blood, metabolism, contractions of smooth and cardiac muscles, secretions of endocrine glands, growth, development, and reproduction.

- Endocrine disorders involve either hyposecretion or hypersecretion of hormones. Causes of disorders can be either primary (within the affected gland) or secondary (within a gland that regulates the affected gland).

- The hypothalamus and pituitary gland regulate many endocrine glands and body functions. The hypothalamus secretes hormones (ADH, oxytocin) directly into the bloodstream through the **neurosecretory cells** of the posterior pituitary. **Antidiuretic hormone (ADH)** affects fluid reabsorption in the kidneys.

- The hypothalamus secretes releasing hormones (GnRH, GHRH, **thyrotropin-releasing hormone (TRH)**, CRH, PRH) and inhibiting hormones (GHIH, PIH) into a common blood supply, where they affect the anterior pituitary.

- The anterior pituitary secretes many hormones that affect human growth.

- Other anterior pituitary hormones (TSH, ACTH, FSH, LH, PRL) control the secretions of the thyroid gland, the adrenal glands, the ovaries/testes, and breast development and milk production.

- As shown here, the thyroid gland secretes thyroid hormones (thyroxine, triiodothyronine) that increase basal metabolism, and **calcitonin**, which decreases blood calcium levels by inhibiting the activity of a type of bone cell called **osteoclasts** that release calcium into the blood.

Thyroid hormone levels • Figure 9.8

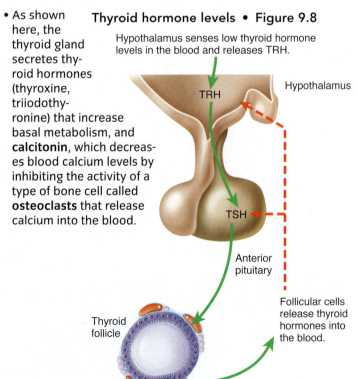

- The parathyroid glands secrete parathyroid hormone, which mobilizes calcium from bone, increases calcium reabsorption in the kidneys, and stimulates calcitriol secretion by the kidneys. **Calcitriol** stimulates calcium absorption by the small intestine. Parathyroid hormone and calcitonin act in opposite ways to control blood calcium levels.

- The pancreas, as seen here, secretes glucagon and insulin. Both glucagon and insulin act mainly on the liver and skeletal muscle; they have opposite effects on blood glucose. In response to low blood sugar levels (**hypoglycemia**), glucagon mobilizes energy sources to raise low blood glucose by stimulating the breakdown of liver glycogen into glucose, adipose tissue triglycerides into fatty acids, and muscle proteins into amino acids. In response to high blood sugar levels (**hyperglycemia**), insulin lowers high blood glucose by stimulating all cells to take up glucose from the blood, and it has the opposite effect of glucagon on liver, adipose tissue, and skeletal muscle.

The pancreas • Figure 9.12a

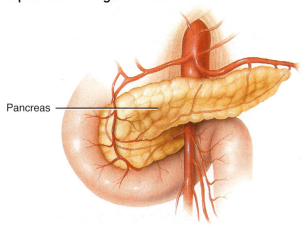

Pancreas

by hypothalamic CRH secretion and pituitary ACTH secretion.

- The adrenal medulla secretes epinephrine and norepinephrine, which increase heart rate, blood pressure, and blood flow and mobilize energy sources.

- The ovaries secrete estrogen and progesterone. Secretions are controlled by the hypothalamus via **gonadotropin-releasing hormone (GnRH)** and the pituitary gland via follicle-stimulating hormone (FSH) and luteinizing hormone (LH). These hormones control the menstrual cycle, initiate ovulation, and maintain pregnancy.

- The testes secrete testosterone under control of the hypothalamus (GnRH) and pituitary gland (LH, FSH). Testosterone supports sperm development and maintains male sexual characteristics.

- Other body tissues (for example, heart, kidney, intestine) secrete different hormones that have a variety of functions (for example, control blood volume, increase red blood cell production, stimulate insulin secretion).

Control of glucocorticoid secretion • Figure 9.16

Corticotropin releasing hormone (CRH)

Corticotropin (ACTH)

Cortisol

3 Endocrine Glands Regulate Other Key Body Functions 269

- The adrenal cortex secretes a **mineralocorticoid** called aldosterone, which affects sodium reabsorption and potassium excretion in the kidneys. Aldosterone secretion, which is controlled by the renin–angiotensin–aldosterone pathway, influences blood volume and pressure.

- As shown, the adrenal cortex also secretes a **glucocorticoid** called cortisol, which mobilizes energy sources from skeletal muscle, the liver, and adipose tissue. Cortisol also reduces inflammation. Its secretion is controlled

4 The Endocrine System Coordinates the Stress Response 275

- As shown here, the stress response has three stages: fight-or-flight, resistance, and exhaustion.

The stress response • Figure 9.20

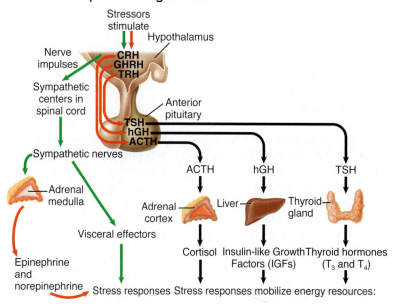

Stressors stimulate

Hypothalamus

Nerve impulses

CRH
GHRH
TRH

Sympathetic centers in spinal cord

Anterior pituitary

TSH
hGH
ACTH

Sympathetic nerves

ACTH   hGH   TSH

Adrenal medulla

Adrenal cortex   Liver   Thyroid gland

Visceral effectors

Cortisol   Insulin-like Growth Factors (IGFs)   Thyroid hormones (T_3 and T_4)

Epinephrine and norepinephrine

Stress responses   Stress responses mobilize energy resources:

Supplement and prolong "fight-or-flight" responses

Summary **281**

- The sympathetic nervous system starts the fight-or-flight stage. The adrenal medulla secretes epinephrine and nor-epinephrine, which increase blood flow and mobilize energy sources.

- Activity of the hypothalamus and pituitary gland continue during the resistance stage by stimulating secretion of cortisol, growth hormone, and thyroid hormones. These hormones mobilize energy sources (glucose, fatty acids, amino acids) that tissues can use to make ATP. Resistance can be sustained for a long period of time to combat a stressor.

- The exhaustion stage occurs when resistance can no longer combat the stressor and body reserves are depleted.

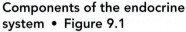

5 Aging Alters the Endocrine System 278

- The size or secretions of many endocrine glands, as shown here, diminish with aging.

- Secretions of growth hormone, thyroid hormones, cortisol, aldosterone, estrogen, and progesterone all decrease with aging.

Components of the endocrine system • Figure 9.1

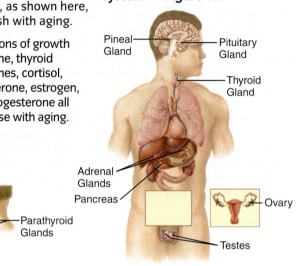

- TSH, PTH, LH, and FSH secretions increase with aging.

- As a body ages, the pancreas secretes insulin more slowly, and tissues take longer to respond to insulin.

Key Terms

- acromegaly 260
- Addison's disease 272
- adrenal gland 269
- aldosterone 270
- androgen 272
- anterior pituitary lobe 258
- antidiuretic hormone (ADH) 261
- calcitonin 262
- calcitriol 264
- congenital hypothyroidism 264
- cortex 269
- cortisol 271
- Cushing's syndrome 272
- cyclic AMP (cAMP) 256
- diabetes insipidus 261
- diabetes mellitus 267
- endocrine gland 254
- endocrine system 254
- epinephrine 272
- estrogen 272
- exhaustion stage 277
- exophthalmos 264
- fight-or-flight stage 277

- follicle 262
- follicular cell 262
- general adaptation syndrome 275
- gestational diabetes 268
- gigantism 260
- glucagon 267
- glucocorticoid 271
- gonadotropin-releasing hormone (GnRH) 273
- Graves disease 264
- hormone 255
- hormone receptor 255
- hyperglycemia 267
- hyperthyroidism 264
- hypoglycemia 267
- hypothyroidism 264
- insulin 267
- medulla 269
- melatonin 274
- mineralocorticoid 270
- myxedema 264
- neurosecretory cell 259
- nonsteroid hormone 255
- norepinephrine 272

- osteoclast 264
- oxytocin 260
- pancreatic islets or islets of Langerhans 266
- parafollicular cell 262
- parathyroid gland 264
- parathyroid hormone (PTH) 264
- pineal 274
- pinealocyte 274
- pituitary dwarfism 260
- pituitary gland 258
- posterior pituitary lobe 258
- progesterone 272
- resistance stage 277
- seasonal affective disorder (SAD) 274
- second messenger 256
- steroid hormone 255
- stress response 275
- target cell 255
- testosterone 272
- thyrotropin-releasing hormone (TRH) 263
- thyroxine 262
- triiodothyronine 262

Critical and Creative Thinking Questions

1. Two teenage boys, Bob and Bill, are diagnosed with diabetes. Bob is slightly underweight for his age, while Bill is overweight, possibly obese. Bob must take daily insulin injections to control his diabetes, while Bill must watch his diet and take oral medications. Identify the reasons for the different treatments and explain what is going on in each of their bodies.

2. Joe's wife of 20 years has noticed changes in his appearance. When they were married at age 18, Joe was a tall, handsome teenager. As Joe has aged, his hands, feet, and jaws have thickened, and his head has become elongated. What might be happening to Joe? What tests might you conduct to support your hypothesis?

3. A scientist has two cultures of the same type of cell. She administers Hormone A to one culture and Hormone B to the other. She notices that Hormone A has an immediate effect on the culture, while the culture with Hormone B takes about 90 minutes to show any activity. When the scientist examines protein synthesis in both cultures, she notices that new proteins are made only in the culture exposed to Hormone B. Explain the differences between the two hormones and indicate what types they might be.

4. Laura recently suffered a head injury in a car accident. As she is recovering, she notices that she is always thirsty, urinates frequently, and passes large volumes of urine. If she does not drink often, she feels weak, faint, and nauseated. What might have happened to Laura, and what is going on in her body that would explain these signs and symptoms?

5. Michael is a 45-year-old man who was recently diagnosed with high blood pressure. His physician prescribed a drug called an angiotensin-converting enzyme (ACE) inhibitor. How will the drug affect Michael's ability to withstand dehydration? Explain.

What is happening in this picture?

All these girls are 15 years old, but one of them looks very different from the others. At only 23.5 inches tall, Jyoti Amge is allegedly the world's smallest person.

Think Critically

1. What endocrine problem might make Jyoti so much smaller than the other girls?
2. How would it affect her size?

Self-Test

(Check your answers in Appendix C.)

1. Which of the following does a body need for successful endocrine communication?
   a. endocrine cell
   b. hormone
   c. target cell
   d. all of the above

2. The letter C indicates the _____.
   a. adrenal glands
   b. thyroid gland
   c. pineal gland
   d. pituitary gland

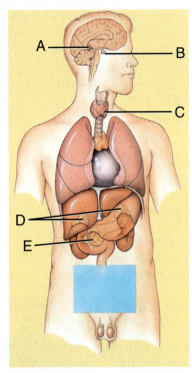

3. In the above figure, the endocrine gland that controls reproductive functions is labeled _____.
   a. A   d. D
   b. B   e. E
   c. C

4. Of the glands labeled in the figure above, the one that is most involved in the resistance phase of the stress response is _____.
   a. A   d. D
   b. B   e. E
   c. C

5. The hormone _____ would be elevated in an individual with primary thyroid hormone deficiency.
   a. ADH   c. TSH
   b. GnRH   d. ACTH

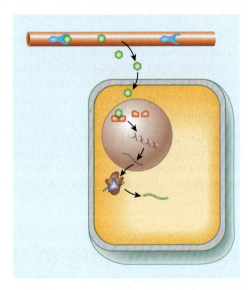

6. The figure above demonstrates the action of the hormone _____.
   a. cortisol   c. epinephrine
   b. insulin   d. parathyroid hormone

7. The hormone _____ would require a second messenger for its action.
   a. cortisol   c. aldosterone
   b. epinephrine   d. estrogen

8. The hormone _____ is essential to reproduction.
   a. CRH   c. ACTH
   b. hGH   d. GnRH

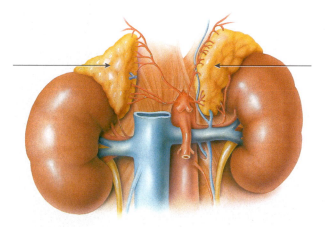

9. The endocrine gland indicated by arrows in the figure above regulates _____.
   a. the growth of long bones
   b. blood calcium levels
   c. blood volume
   d. basal metabolism

10. The gland pictured in the figure above also plays a role in _____.
    a. reproduction
    b. stress response
    c. type I diabetes
    d. gigantism

11. The mineralocorticoid secreted from the gland in the figure above originates in the _____.
    a. hypothalamus
    b. anterior pituitary lobe
    c. thyroid gland
    d. kidneys

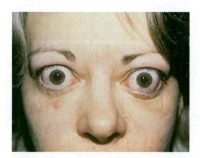

12. The person shown in the photo above likely _____.
    a. lacks insulin
    b. secretes too much aldosterone
    c. secretes too much thyroid hormone
    d. lacks parathyroid hormone

13. The hormone _____ has the opposite effects of glucagon.
    a. cortisol
    b. insulin
    c. epinephrine
    d. ADH

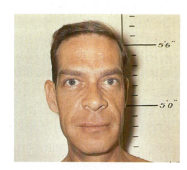

14. What might be the cause of the condition shown in the photo above?
    a. secretion of too much hGH during adulthood
    b. lack of hGH secretion during childhood
    c. secretion of too much hGH during childhood
    d. secretion of too much cortisol during adulthood

15. Elderly people often have osteoporosis. What might contribute to this condition?
    a. not enough thyroid hormone
    b. secretion of too much calcitonin
    c. high levels of parathyroid hormone
    d. high levels of cortisol

THE PLANNER ✔

Review your Chapter Planner on the chapter opener and check off your completed work.

10 The Cardiovascular System: Blood

Blood has been sort of a mystery for much of human history. Because our ancestors observed that people could die from the loss of too much blood, it was considered a "life force." Despite this observation, physicians from ancient times through the 18th century practiced the technique of bloodletting, either by directly cutting into a vein or by using leeches. The idea was to balance the "humors," or vital fluids, that traveled in the blood. As the science of medicine advanced, physicians realized that blood played an essential role in carrying oxygen to body tissues. Giving blood to trauma or surgical patients via transfusion replaced bloodletting. With further medical advances, scientists discovered that individuals had different blood types and that donor blood had to match the recipient's type for the transfusion to be successful; transfusions became safer and saved countless lives. Today, millions of individuals safely donate blood to others.

Despite the safety of transfusions, many places experience shortages of available blood for health care. Many people who would qualify do not participate in blood donation. A single blood donation can benefit up to four recipients, since the blood can be separated to provide the recipients with only the blood components they need most. Because of blood shortages, along with the limited time that blood can be stored, scientists have been working to create artificial blood.

Let's take a closer look at this vital substance.

286

CHAPTER OUTLINE

CHAPTER PLANNER ✔

- ❑ Study the picture and read the opening story.
- ❑ Scan the Learning Objectives in each section:
  p. 288 ❑   p. 291 ❑   p. 294 ❑   p. 296 ❑   p. 299 ❑
- ❑ Read the text and study all visuals.
  Answer any questions.

Analyze key features

- ❑ InSight, p. 289 ❑   pp. 296–297 ❑
- ❑ Process Diagram, p. 293 ❑   p. 295 ❑
- ❑ What a Health Provider Sees, p. 298 ❑
- ❑ Stop: Answer the Concept Checks before you go on:
  p. 290 ❑   p. 292 ❑   p. 294 ❑   p. 299 ❑   p. 300 ❑

End of chapter

- ❑ Review the Summary and Key Terms.
- ❑ Answer the Critical and Creative Thinking Questions.
- ❑ Answer What is happening in this picture?
- ❑ Complete the Self-Test and check your answers.

Blood Functionally Connects the Body Organ Systems

LEARNING OBJECTIVES

1. **Describe** the functions of blood.
2. **Outline** the components of plasma.
3. **Identify** the blood cells that make up formed elements.

While blood looks like a **homogeneous** liquid, it is actually a liquid connective tissue that consists of cells surrounded by a liquid extracellular matrix. Blood is thicker than water and has a temperature of about 37°C and a pH of 7.35 to 7.45. An average adult male has about 5 to 6 L (1.5 gal) of circulating blood volume, while a female has about 4 to 5 L (1.2 gal); the difference in blood volume is mainly due to differences in average body size. Blood makes up 8% of your body weight, while other fluids and tissues make up 92%.

homogeneous (hō-mo-JĒ-nē-us) Uniform in structure or composition throughout.

Blood has three major functions:

- *Transportation*—Blood delivers oxygen from the lungs to the cells of the body, moves carbon dioxide from the cells to the lungs, and carries nutrients, waste products, and hormones to various destinations.

- *Regulation*—Blood helps to maintain a steady pH of body fluids. It also distributes heat, thereby adjusting body temperature. Blood's osmotic pressure influences the water content of cells and tissues.

- *Protection*—Blood forms seals or clots in response to injury, thereby preventing blood loss and maintaining cardiovascular function. White blood cells protect against disease by ingesting invading bacteria and producing antibodies. Blood also contains proteins that protect against disease.

plasma A liquid extracellular matrix in blood that contains dissolved substances.

formed elements Cells and cell fragments in blood.

Blood has two components: a liquid component called **plasma** and cellular components called **formed elements**. These two components can be separated by placing a container of whole blood into a centrifuge, which spins the blood at high speed. The formed elements sink to the bottom because they are denser than the plasma. Let's first take a look at plasma.

Plasma Is the Liquid Portion of Blood

When you separate whole blood into its components (**Figure 10.1**), you find that over half of it is plasma. Most of plasma is water, with some proteins and other solutes mixed in. The proteins, including albumins, globulins, and fibrinogen, are made mainly in the liver. These proteins help maintain osmotic pressure (albumins), defend against foreign substances (globulins), and help form blood clots (fibrinogen). Other solutes include various salts or electrolytes, nutrients, wastes, hormones, and gases.

Formed Elements Consist of the Many Types of Blood Cells

The remaining 45% of blood by weight consists of cells that are collectively called formed elements. The formed elements consist primarily of red blood cells and, to a much smaller extent, platelets and white blood cells. The different types of cells have different functions. Let's look at each one.

Red blood cells (RBCs), or **erythrocytes** (e-RITH-rō-sīts), are about 7–8 μm in diameter and have a unique **biconcave**, disc-like shape (see Figure 10.1). This shape provides a large surface area for the exchange of gases, specifically oxygen and carbon dioxide. Red blood cells have no nucleus and few organelles. Their flexible plasma membrane allows them to easily maneuver through the various vessels of the circulation. The percentage of total blood volume occupied by red blood cells is termed the **hematocrit** (he-MAT-ō-krit).

biconcave (bī-KON-kāv) Inwardly curved on both sides or surfaces.

Blood has both liquid and cellular parts. Blood cells come in various types and densities. Here, the density of blood cells is expressed as the number of cells per microliter (µL). One drop of blood equals about 50 µL.

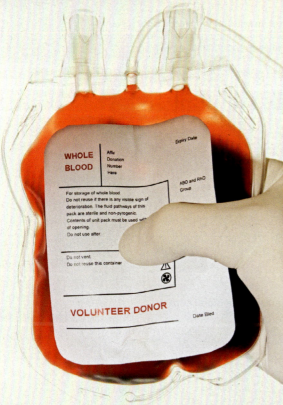

Plasma components (by weight) and their functions

Water (91.5%)	Water acts as a solvent for carrying other substances and absorbs heat
Proteins (7%)	Albumin, globulins, and fibrinogen maintain osmotic balance and pH buffering, help with blood clotting, and help transport antibodies and lipids
Other solutes (1.5%)	Other solutes include: • Electrolytes (Na^+, K^+, and Cl^-), which help with osmotic balance, pH buffering, and regulation of membrane permeability • Nutrients (glucose and fatty acids) • Respiratory gases (O_2, CO_2, N_2) • Regulatory substances, such as hormones • Waste products (urea, uric acid)

Formed elements and their functions
(number per microliter of blood)

The buffy coat is composed of white blood cells and platelets

White blood cells (5,000–10,000/µL)	White blood cells fight invading microbes and eliminate cell debris; they make up less than 0.2% of formed elements

 Lymphocytes (20–25% of total white blood cell (WBC) count; 6–16µm in diameter) have a round nucleus and attack viruses and bacteria

 Eosiniphils have a nucleus with 2 lobes and are involved in allergic reactions, phagocytosis of antigen-antibody complexes, and defense against worms

 Neutrophils have a nucleus with 2–5 lobes and destroy bacteria by phagocytosis and chemicals

 Basophils have a nucleus with 2 lobes and secrete heparin, histamine, and serotonin that intensify allergic reactions

 Monocytes have a horseshoe-shaped nucleus and destroy microbes and cell debris by phagocytosis

Platelets (150,000–400,000/µL)		Platelets are cell fragments that are important in blood clotting, although they make up less than 1% of formed elements
Red blood cells (4–6 million/µL)		These disc-shaped cells contain a protein called hemoglobin and transport most of the O_2 and CO_2 in the blood; they make up most of the formed elements

Centrifuge

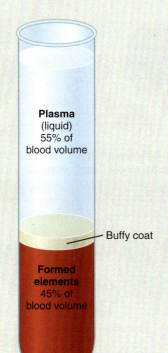

Plasma (liquid) 55% of blood volume

Buffy coat

Formed elements 45% of blood volume

The structure of red blood cells • Figure 10.2 _____

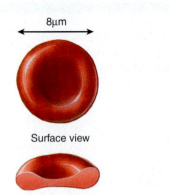

8μm

Surface view

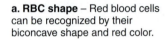

Sectioned view

a. RBC shape – Red blood cells can be recognized by their biconcave shape and red color.

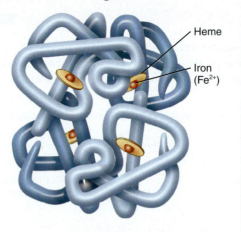

Heme

Iron (Fe²⁺)

b. Hemoglobin molecule – Hemoglobin consists of four peptide chains, each with an iron-containing center, or heme.

The RBC is filled with a protein called **hemoglobin** (hē′-mō-GLŌ-bin), which gives these cells their red color. Hemoglobin is an iron-based protein that binds most of the oxygen and a small portion of the carbon dioxide and allows the RBCs to transport these gases through the blood (**Figure 10.2**).

The next major formed element is the **platelet** (PLĀT-let)(see Figure 10.1). Platelets are actually tiny fragments of larger cells called **megakaryocytes** (meg-a-KAR-ē-ō-sīts), found only in the red bone marrow. Platelets consist of a little cytoplasm enclosed within a piece of cell membrane. As you will see later, platelets are important for blood clotting.

The last category of formed elements is **white blood cells (WBCs)**, or **leukocytes** (LOO-kō-sīts). WBCs comprise less than 0.2% of the formed elements and are made up of five types of cells: **neutrophils** (NOO-trō-fils), **lymphocytes** (LIM-fō-sīts), **monocytes** (MON-ō-sīts′), **eosinophils** (ē′-ō-SIN-ō-fils), and **basophils** (BĀ-sō-fils)(see Figure 10.1). WBCs are very different in appearance from one to another and from RBCs and platelets. One major distinction is that, unlike RBCs and platelets, WBCs have nuclei. Each type of WBC has characteristic features when stained and viewed under a microscope. As a result, they can be classified as either *granular* (having cytoplasmic granules) or *agranular* (no cytoplasmic granules) (**Figure 10.3**). All white blood cells are involved in protecting the body by engulfing invading bacterial cells or viruses, releasing chemicals that initiate immune responses, or producing antibodies. You will learn more about these functions in Chapter 12.

CONCEPT CHECK STOP

1. **What** are the three major functions of blood?
2. **How** is blood plasma different from water?
3. **Which** type of formed element consists of pieces of another cell?

Appearance of blood cells in a blood smear • Figure 10.3 _____

a. Blood smear stained with Wright's stain – Note that there are few white blood cells compared to platelets and red blood cells.

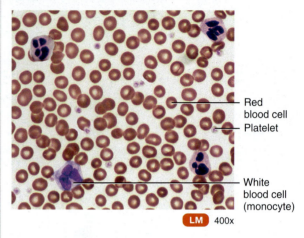

Red blood cell

Platelet

White blood cell (monocyte)

LM 400x

b. Granular white blood cells

Eosinophil

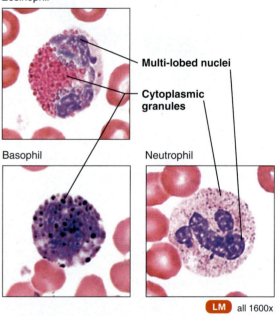

Multi-lobed nuclei

Cytoplasmic granules

Basophil

Neutrophil

LM all 1600x

c. Agranular white blood cells – Single nucleus, no cytoplasmic granules

Monocyte

Lymphocyte

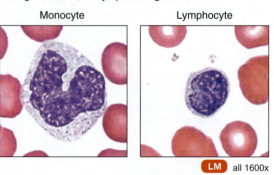

LM all 1600x

Blood Cells Are Created in the Red Bone Marrow

LEARNING OBJECTIVES

1. **Outline** the events in the life span of a red blood cell.

2. **Describe** the life span of a white blood cell.

3. **Identify** the steps in the formation and destruction of red blood cells.

4. **Explain** how red blood cell formation is regulated.

As you learned in Chapter 5, blood cells are formed in red bone marrow. They travel throughout the body via large arteries and veins and narrow arterioles, capillaries, and venules. As they move through these vessels, they bend, stretch, twist, get compressed, and collide with each other and with the vessel walls, experiencing quite a bit of wear and tear along the way. As a result, RBCs last only about 120 days. In contrast, most WBCs last only a few days because they eventually engulf enough foreign bacteria and cell debris to interfere with their own life functions. In times of severe infection, WBCs may last only a few hours. In contrast, some types of lymphocytes that provide immunity can last for years.

All the Formed Elements Develop from Red Bone Marrow Stem Cells

The process of making formed elements of blood is called **hemopoiesis**, or *hematopoiesis* (hem′-a-tō-poy-Ē-sis). Before birth, hemopoiesis first occurs in the yolk sac of the embryo and then in the liver, thymus, and lymph nodes. Eventually the process is moved to red bone marrow, starting about 3 months before birth through the end of life. In response to hormones, **pluripotent stem cells** (0.05–0.1% of red bone marrow cells) differentiate into two other types, **lymphoid stem cells** and **myeloid stem cells**. These stem cells further differentiate into the formed elements (**Figure 10.4**).

> **hemopoiesis** (hē-mō-poy-Ē-sis) Blood cell production.

Origin and development of blood cells • Figure 10.4

The various blood cells form from stem cells in the red bone marrow and, in some cases, lymphoid tissue through cell division and differentiation. Some of these blood cells eventually migrate into the tissues and receive new names (for example, monocytes become macrophages).

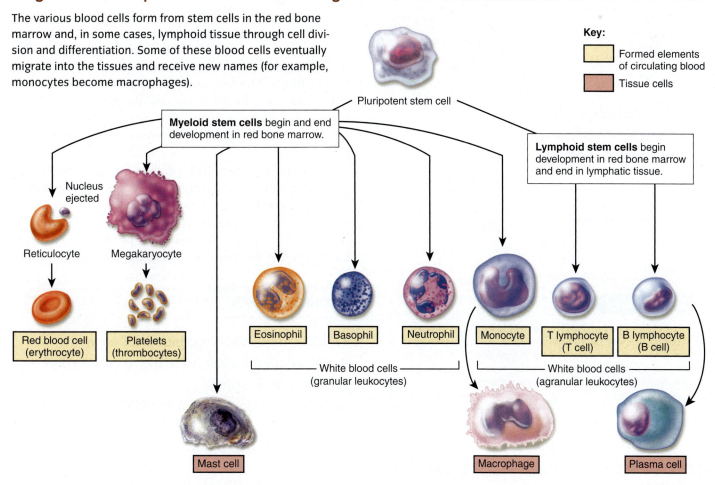

Key:
- ▢ Formed elements of circulating blood
- ▢ Tissue cells

Pluripotent stem cell

Myeloid stem cells begin and end development in red bone marrow.

Lymphoid stem cells begin development in red bone marrow and end in lymphatic tissue.

Nucleus ejected

Reticulocyte   Megakaryocyte

Red blood cell (erythrocyte) | Platelets (thrombocytes) | Eosinophil | Basophil | Neutrophil | Monocyte | T lymphocyte (T cell) | B lymphocyte (B cell)

White blood cells (granular leukocytes)

White blood cells (agranular leukocytes)

Mast cell   Macrophage   Plasma cell

Regulation of red blood cell formation • Figure 10.5

Red blood cell formation is controlled through a negative feedback system involving the kidney and red bone marrow that is mediated by the hormone erythropoietin.

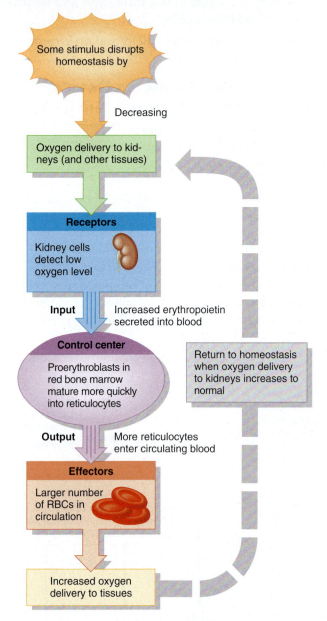

Some stimulus disrupts homeostasis by

Decreasing

Oxygen delivery to kidneys (and other tissues)

Receptors

Kidney cells detect low oxygen level

Input Increased erythropoietin secreted into blood

Control center

Proerythroblasts in red bone marrow mature more quickly into reticulocytes

Return to homeostasis when oxygen delivery to kidneys increases to normal

Output More reticulocytes enter circulating blood

Effectors

Larger number of RBCs in circulation

Increased oxygen delivery to tissues

During red blood cell formation, also called **erythropoiesis**, the pre–red blood cell, or **reticulocyte**, ejects its nucleus to form the biconcave erythrocyte. Therefore, the mature RBC can no longer reproduce and will die within 120 days. Platelets, which are fragments of the bone marrow's megakaryocytes, are also unable to reproduce.

Besides routine maintenance, the major stimulus for producing RBCs is reduced oxygen delivery to tissues (*tissue hypoxia*). The kidneys sense hypoxia and release a hormone called **erythropoietin**, which stimulates the red bone marrow to produce additional RBCs (**Figure 10.5**).

> **erythropoietin**
> (e-rith′-rō-POY-e-tin)
> A hormone released by the kidneys that stimulates red blood cell production.

Like RBCs, granular leukocytes develop in red bone marrow from myeloid stem cells, while lymphocytes develop in the red bone marrow from lymphoid stem cells. Some lymphocytes later develop in lymphoid tissues through mitosis. An increase in WBCs called **leukocytosis** (loo′-kō-sī-TŌ-sis) is a normal response to infections, strenuous exercise, anesthesia, and surgery.

When Blood Cells Are Destroyed, Many of Their Components Are Recycled

What happens to red blood cells after 120 days? Old, worn-out RBCs are dismantled in the spleen, liver, and red bone marrow (see Figure 10.4). Hemoglobin breaks apart. The iron is recycled to red bone marrow, where it ends up back in new RBCs. The non-iron portions of heme get broken down, processed in the liver, and ultimately excreted by the kidneys and intestines. The globin protein degrades into amino acids, which are released into the bloodstream and reused.

Usually, the rate of RBC production equals the rate of RBC destruction. To maintain normal numbers of RBCs, new mature cells must enter the bloodstream at an astonishing rate of at least 2 million per second. However, if tissue hypoxia occurs, then RBC production is stimulated at a rate faster than its destruction (**Figure 10.6**).

CONCEPT CHECK **STOP**

1. **How** long does a red blood cell survive?
2. **What** causes most white blood cells to die?
3. **What** is bilirubin?
4. **Which** hormone stimulates RBC production?

Formation and destruction of red blood cells • Figure 10.6

Red blood cells are destroyed by many organs. Their hemoglobin is broken down, and some parts are recycled. The remainder of the hemoglobin is excreted in a process that involves the liver, intestines, and kidney.

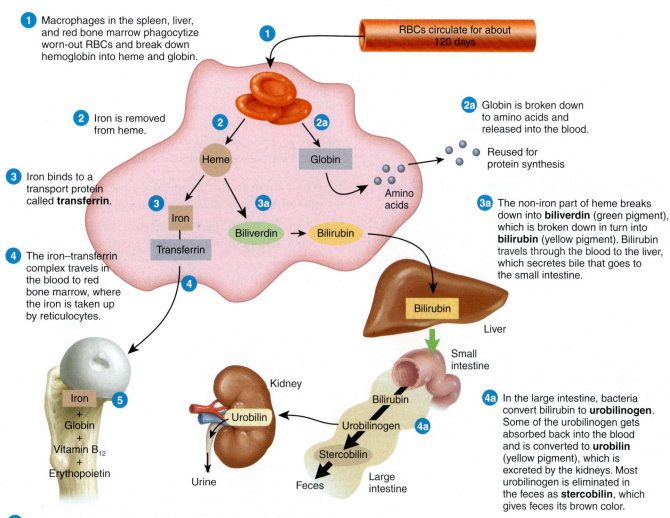

1 Macrophages in the spleen, liver, and red bone marrow phagocytize worn-out RBCs and break down hemoglobin into heme and globin.

2 Iron is removed from heme.

3 Iron binds to a transport protein called **transferrin**.

4 The iron–transferrin complex travels in the blood to red bone marrow, where the iron is taken up by reticulocytes.

2a Globin is broken down to amino acids and released into the blood.

Reused for protein synthesis

3a The non-iron part of heme breaks down into **biliverdin** (green pigment), which is broken down in turn into **bilirubin** (yellow pigment). Bilirubin travels through the blood to the liver, which secretes bile that goes to the small intestine.

4a In the large intestine, bacteria convert bilirubin to **urobilinogen**. Some of the urobilinogen gets absorbed back into the blood and is converted to **urobilin** (yellow pigment), which is excreted by the kidneys. Most urobilinogen is eliminated in the feces as **stercobilin**, which gives feces its brown color.

RBCs circulate for about 120 days

Heme
Globin
Amino acids
Iron
Transferrin
Biliverdin → Bilirubin
Bilirubin
Liver
Small intestine
Kidney
Urobilin
Bilirubin
Urobilinogen
Stercobilin
Urine
Feces
Large intestine
Iron + Globin + Vitamin B$_{12}$ + Erythropoietin

5 Under stimulation by the hormone erythropoietin, erythropoiesis occurs in red bone marrow. Iron combines with globin protein to make hemoglobin in reticulocytes, which differentiate into RBCs that are released into the blood.

Ask Yourself

Sometimes too much bilirubin and its by-products can build up in the body, causing a yellowing of the tissues—a condition known as *jaundice*. Which of the following would *not* be a likely cause of jaundice?

a. Reduced production of red blood cells.
b. Liver disease that is reducing liver function.
c. Excess destruction of RBCs.
d. Kidney disease that is reducing kidney function.
e. Blockage of the tubing that connects the liver to the intestine.

Blood Clotting Controls Bleeding

LEARNING OBJECTIVES

1. **Describe** the three mechanisms that stop bleeding.
2. **Identify** the steps involved in forming a blood clot.
3. **Explain** how your body minimizes blood clots on a daily basis.

While shaving in the morning, you nick your face, leg, or underarm with a razor. You begin to bleed. You instinctively apply pressure to the wound to stop the bleeding, and your body does something that eventually seals the wound. But how exactly does your body prevent you from bleeding to death?

Hemostasis Helps Minimize Blood Loss

> **hemostasis** (hē-MŌ-stā-sis) The stoppage of bleeding.

The series of physiological responses to stop bleeding when blood vessels are injured is called **hemostasis**. The major hemostatic mechanisms that stop large amounts of blood loss, or **hemorrhage** (HEM-o-rij), include the following (**Figure 10.7**):

- **Vascular spasm**, or *vasospasm*, restricts blood flow in damaged vessels.
- Platelets assemble to form a **platelet plug** to seal the damaged blood vessel. (This works well for small holes.)
- **Blood clotting**, or **coagulation**, is a more complex mechanism that ultimately closes up the wound. This process requires chemicals called **clotting factors** and **fibrinogen**.

Depending on the extent of damage, these mechanisms take seconds to minutes to seal the wound.

Fibrinolysis Creates Space for the Final Repair of a Damaged Vessel

Once a blood clot has sealed a wound, platelets within the clot pull on the **fibrin** threads and contract the clot. The contracting clot removes excess fluid and brings the damaged surfaces close together so that permanent wound healing can occur. Later, clots get dissolved in a process called **fibrinolysis** that makes room for a more permanent patch of connective tissue to be placed on the vessel and a new endothelium to be formed. Fibrinolysis involves the following steps:

> **fibrinolysis** (fī-bri-NOL-i-sis) Dissolution of a blood clot by the action of a protein-digesting enzyme.

1. When a clot forms, an inactive plasma enzyme called **plasminogen** incorporates into the clot.
2. Various factors from blood and tissues, such as thrombin, factor XII, and tissue plasminogen activator (tPA), convert inactive plasminogen into active **plasmin**.
3. Plasmin digests fibrin threads and dissolves the clot after the damage is repaired.

Occasionally a small clot called a **thrombus** (plural is *thrombi*) will form in an unbroken vessel and dissolve spontaneously. However, if the thrombus remains intact, it can interfere with blood flow to the organ "downstream" from the clot. Sometimes a clot can break off and travel through the cardiovascular system, eventually blocking small blood vessels. This moving clot is called an **embolus**. Emboli can lodge in various places and cause problems, such as pulmonary embolism (lungs), heart attack (heart), stroke (brain), or kidney failure (kidney). To treat such embolic events, physicians try to dissolve the clots by injecting tissue plasminogen activator (tPA) into the patient. If this is done soon after the event begins, blood flow can be re-established through the vessel, and the damage to the blocked organ can be minimized.

> **embolus** A blood clot, a bubble of air or fat from broken bones, a mass of bacteria, or other debris or foreign material transported by the blood. The plural is *emboli*.

CONCEPT CHECK STOP

1. **What** are the steps in the process of platelet plug formation?
2. **What** does prothrombinase do?
3. **How** can tissue plasminogen activator (tPA) be used to treat a heart attack?

Hemostatic mechanisms involved in preventing blood loss • Figure 10.7

Three processes that prevent blood loss—**a.** vascular spasm, **b.** platelet plug formation, and **c.** blood clotting—occur on the order of seconds to minutes after damage to a blood vessel. Each varies in its effectiveness, depending on the extent of the damage.

Damage to blood vessel

Immediate   **Within seconds**   **Seconds to minutes**

a. Vascular spasm
- Smooth muscles in the vessel wall contract, thereby reducing blood flow and loss.
- Vascular spasm lasts minutes to hours.

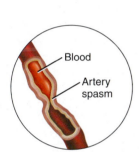

Blood

Artery spasm

b. Platelet plug formation can temporarily stop blood loss completely if the hole in the vessel is small enough.

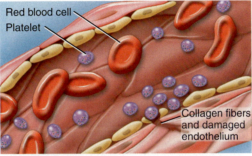

Red blood cell
Platelet
Collagen fibers and damaged endothelium

1 Platelets stick to collagen fibers at damage site.

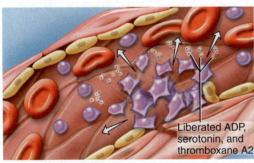

Liberated ADP, serotonin, and thromboxane A2

2 Platelets secrete chemicals (ADP, serotonin, thromboxane A2) to activate nearby platelets.

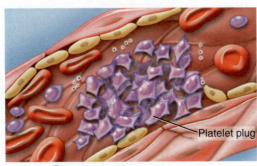

Platelet plug

3 Activated platelets stick together to form a plug that blocks blood flow in the vessel.

c. Clotting – Blood clot forms and seals the wound.

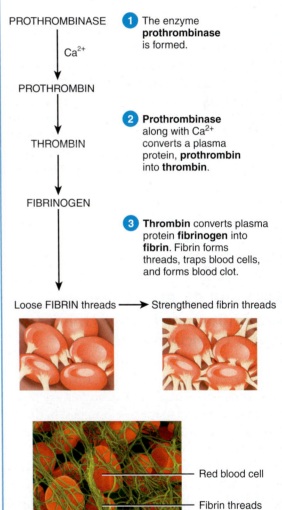

PROTHROMBINASE

Ca^{2+}

PROTHROMBIN

THROMBIN

FIBRINOGEN

1 The enzyme **prothrombinase** is formed.

2 **Prothrombinase** along with Ca^{2+} converts a plasma protein, **prothrombin** into **thrombin**.

3 **Thrombin** converts plasma protein **fibrinogen** into **fibrin**. Fibrin forms threads, traps blood cells, and forms blood clot.

Loose FIBRIN threads ⟶ Strengthened fibrin threads

Red blood cell

Fibrin threads

SEM 1600x

Put It Together

When several small blood vessels in the skin get cut with a sharp knife, what is the first event that leads to hemostasis and eventual repair of the damage?
a. Platelet plug forms.
b. Common pathway occurs.
c. Vascular spasm occurs.
d. Fibrin clot is created.

Matching the ABO Group Allows Safe Transfusions

LEARNING OBJECTIVES

1. **Outline** the components of the ABO blood group.
2. **Describe** the Rh blood group.
3. **Explain** the consequences of ABO blood typing for transfusions.

The surfaces of RBCs and of many other cells contain genetically determined molecules called **antigens**, which are composed of glycolipids and glycoproteins. These antigens occur in various combinations. Based on the presence or absence of these antigens, blood can be categorized into different **blood types** (blood cells that have the same antigens). There are at least 24 **blood groups** (classification system based on hereditary characteristics of the blood) and more than 100 antigens on RBCs. The most common blood groups are the ABO group and the Rh group (**Figure 10.8**).

> **antigen** (AN-ti-jen)
> A substance that has the ability to provoke an immune response.

ABO and Rh Blood Groups Are Important in Determining Blood Compatibility

The basis for the blood groups is the interactions between the antigens on the surface of an RBC and the antibodies circulating in the plasma. **Antibodies** are proteins in the blood that can bind to specific antigens. In the ABO group, there are two antigens (A and B) and two corresponding antibodies (anti-A and anti-B) (see Figure 10.8b). People with Type A blood usually have anti-B antibodies, and people with Type B blood usually have anti-A antibodies. When RBCs with a specific antigen come in contact with a corresponding antibody (for example, A antigen with anti-A antibody), the RBCs either clump together (**agglutinate**) or burst (**hemolysis**). This reaction would happen if blood from a type B donor was given to a recipient with type A blood with anti-B antibodies.

The antigen–antibody reaction forms the basis of the ABO blood typing test (Figure 10.8a). One drop of the patient's blood is mixed with anti-A serum and another is mixed with anti-B serum. Both spots are examined for clumping, and the appropriate blood type is assigned based on the results.

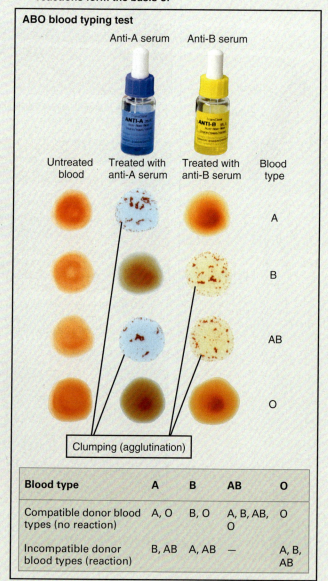

InSight Blood groups and blood types
• **Figure 10.8**

Antigen–antibody reactions are essential in determining the compatibility of various blood types and in the tests for blood types.

Antigen–antibody reactions:

- **Antigens** are proteins on the RBC membrane
- **Antibodies** are proteins in the blood that can bind to specific antigens
- If an antigen combines with the corresponding antibody (e.g., RBCs with A antigen combine with anti-A antibodies), then the RBCs will clump together (agglutinate) and burst, thereby releasing hemoglobin (**hemolysis**).

a. Antigen–antibody reactions form the basis of

ABO blood typing test

Anti-A serum   Anti-B serum

Untreated blood	Treated with anti-A serum	Treated with anti-B serum	Blood type
			A
			B
			AB
			O

Clumping (agglutination)

Blood type	A	B	AB	O
Compatible donor blood types (no reaction)	A, O	B, O	A, B, AB, O	O
Incompatible donor blood types (reaction)	B, AB	A, AB	—	A, B, AB

b. ABO blood groups

BLOOD TYPE	TYPE A	TYPE B	TYPE AB	TYPE O
	A antigen	B antigen	Both A and B antigens	Neither A nor B antigen

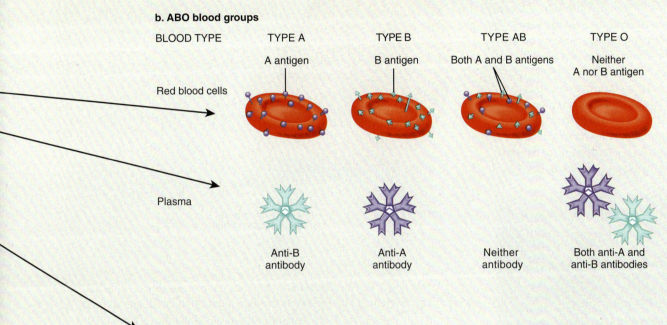

Red blood cells

Plasma

Anti-B antibody	Anti-A antibody	Neither antibody	Both anti-A and anti-B antibodies

c. Antigen–antibody reactions cause problems between mother and child with

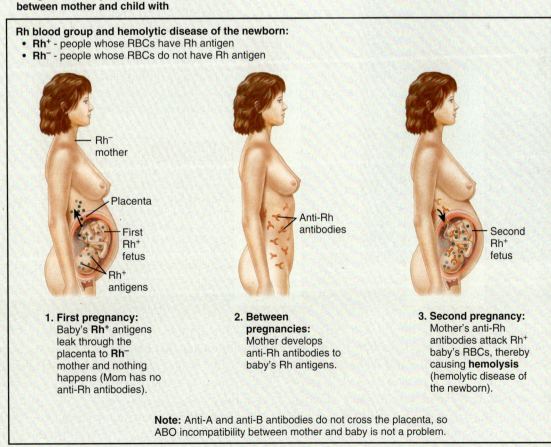

Rh blood group and hemolytic disease of the newborn:
- **Rh⁺** - people whose RBCs have Rh antigen
- **Rh⁻** - people whose RBCs do not have Rh antigen

Rh⁻ mother

Placenta

First Rh⁺ fetus

Rh⁺ antigens

Anti-Rh antibodies

Second Rh⁺ fetus

1. First pregnancy: Baby's **Rh⁺** antigens leak through the placenta to **Rh⁻** mother and nothing happens (Mom has no anti-Rh antibodies).

2. Between pregnancies: Mother develops anti-Rh antibodies to baby's Rh antigens.

3. Second pregnancy: Mother's anti-Rh antibodies attack Rh⁺ baby's RBCs, thereby causing **hemolysis** (hemolytic disease of the newborn).

Note: Anti-A and anti-B antibodies do not cross the placenta, so ABO incompatibility between mother and baby is not a problem.

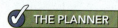

WHAT A HEALTH PROVIDER SEES

Artificial Blood

With the risks of bloodborne diseases, such as human immunodeficiency virus (HIV) and hepatitis B, many hospitals, clinics, and blood banks are experiencing a shortage of available blood for transfusions. Another challenge is that blood has a relatively short useful shelf life (less than 120 days), after which time it must be thrown away. To counter these problems, some pharmaceutical companies have developed artificial blood (**a.**). Artificial blood must be osmotically compatible with real blood and must have a high carrying capacity for oxygen and carbon dioxide. Two types have been devel-

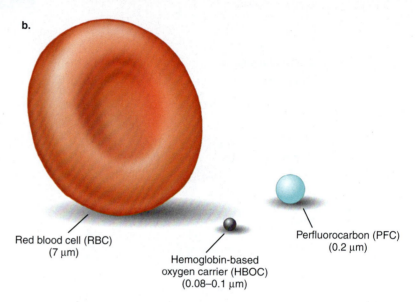

b.

Red blood cell (RBC)
(7 μm)

Hemoglobin-based
oxygen carrier (HBOC)
(0.08–0.1 μm)

Perfluorocarbon (PFC)
(0.2 μm)

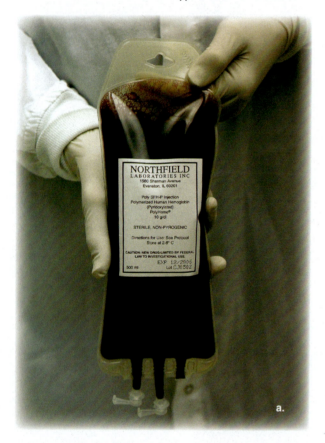

a.

oped for testing. One type uses hemoglobin-based oxygen carriers (HBOCs), forms of hemoglobin that are chemically modified to prevent degradation by macrophages (see Figure 10.6). The other type uses perfluorocarbons (PFCs), fluorine-based organic compounds. Both types of artificial blood can carry oxygen and carbon dioxide efficiently. Their components are also smaller than RBCs, so they can reach into small vessels more easily (**b.**). Both types are being evaluated in clinical trials for their effectiveness compared to whole blood transfusions as well as possible side effects. Some reported side effects of HBOCs include abdominal cramping and elevated blood pressure. If approved for medical use, artificial blood would be especially helpful in trauma centers, on battlefields, and during surgical procedures.

Think Critically **1.** How would artificial blood improve the storage of blood for transfusions?
2. In certain trauma and battle situations, there is little time to type the patient's blood before giving a transfusion. Sometimes the blood is administered, and the health provider hopes that he or she has made a proper selection so there will not be a transfusion reaction. Would it be more likely that an individual would have a transfusion reaction to real blood or to artificial blood? Explain.

Similar antigen–antibody reactions occur with the Rh blood group, or Rh factor, which was named for the rhesus monkey, where it was first found. People are said to be either Rh-negative (not have Rh factor) or Rh-positive (have Rh factor). This blood antigen must also be carefully matched during a transfusion. When selecting donor blood for a transfusion, one must be careful to ensure that no "new" antigens are being given to the recipient. For example, if you are an Rh-negative individual, you cannot receive Rh-positive blood because you would be getting a new antigen (Rh factor). Foreign antigens can stimulate a transfusion reaction that can sometimes be fatal.

Researchers are attempting to address the challenges of blood group compatibility in their efforts to develop artificial blood (see *What a Health Provider Sees*).

Hemolytic Disease of a Newborn Is Due to an Rh Incompatibility

The Rh blood antigen can cause problems during pregnancy if the mother is Rh-negative (see Figure 10.8c). (Rh-positive mothers do not need to be concerned about this condition.) If an Rh-negative mother gives birth to an Rh-positive baby, some of the baby's blood can move into the mother's blood during the birth process. The mother's body responds to this foreign antigen by creating anti-Rh antibodies. Although the recently delivered baby will be unaffected, subsequent pregnancies involving an Rh⁺ infant can be affected. During a subsequent pregnancy, these antibodies can cross over the placenta and get into the baby's circulation, causing incompatibility and damage to the baby's RBCs—a condition known as hemolytic disease of the newborn. This condition can be prevented by injecting the mother with anti-Rh antibodies after delivery of the first baby and during the pregnancy and delivery of subsequent children. These antibodies destroy any Rh antigens that manage to get into the mother's circulation and prevent the mother from developing anti-Rh antibodies. (See Chapter 12 for more about immunity.) Without anti-Rh antibodies in the mother's system, no antibody is passed to the future babies, and hemolytic disease of the newborn will not occur.

CONCEPT CHECK

1. **What** antigen(s) does a person with type B-positive blood have on her RBCs?

2. **How** does hemolytic disease of the newborn occur?

3. **Which** blood types are compatible with type O?

Analysis of Blood Components Can Tell Much About an Individual's Health

LEARNING OBJECTIVES

1. **Identify** the locations in the body where you can obtain blood samples.

2. **Describe** what is measured in blood tests, including hematocrit, reticulocyte count, differential WBC count, and complete blood count.

3. **Explain** the condition(s) each blood test might indicate.

Because the composition of blood at any given time can be an indicator of good health or disease, blood tests are common in clinical settings. Blood samples can be taken from multiple places, and common blood tests can be run for diagnostic purposes.

Blood for Testing Can Be Obtained in Several Ways

The first step in any blood test is to obtain a sample of blood. A test may require only a drop or two of blood (as in blood typing), or it may require volumes as large as one or more 5–10 mL tubes (as in measuring various chemicals in blood). Blood samples can usually be obtained by one of the following methods:

- *Fingerstick*—A sterile needle or lancet is used to prick a finger, an earlobe, or a heel to obtain a drop or two of capillary blood.

- *Venipuncture*—A sterile needle and syringe can puncture a surface vein, usually the median cubital vein in front of the elbow. A tourniquet is used to stop the blood flow from the veins and makes them easier to see. This method is used to obtain several milliliters of blood.

- *Arterial stick*—A sterile needle and syringe are used to sample blood from an artery, usually the radial artery at the wrist or the femoral artery at the thigh. Like venipuncture, this method is used to obtain several milliliters of blood.

Depending on the type of tests that will be done, the blood will either be maintained in a fluid form or will be allowed to clot. Tests that are evaluating the formed elements or the clotting capability of the blood require the blood to be in a fluid form. Tests that are determining the chemical composition of the blood work with the liquid component and may require either fluid blood or clotted blood.

Blood Chemical Composition Testing Uses Either Plasma or Serum

The liquid used for blood chemical composition tests is referred to either as *plasma* or *serum*, depending on how the sample is collected. Plasma is extracted from blood that is still in a fluid form. In the blood vessels, the blood remains in fluid form because it is constantly moving. To keep blood in a fluid form after it is removed from the body, the collection tube must contain *anticoagulant* chemicals.

If blood removed from the body is not stored with anticoagulants, the blood will quickly solidify (clot). The liquid in a blood sample that has clotted is called **serum**. Serum is quite similar in composition to plasma but is missing one of the proteins (fibrinogen) found in plasma. Extracting the liquid from the blood, whether fluid or clotted, requires that the tube be spun in a centrifuge. As we mentioned before, cells fall to the bottom, and the plasma or serum ends up at the top of the tube (see Figure 10.1).

Blood Tests Are Often Used to Diagnose Disease

Once a blood sample has been collected, it can be used for a number of different tests that study the formed elements of the blood. These tests include the hematocrit, reticulocyte count, differential white blood cell count, and/or complete blood count (**Figure 10.9**). These tests can provide a wide range of information regarding blood conditions, such as the following:

> **polycythemia**
> (pol′-ē-sī-THĒ-mē-a) A disorder characterized by an above-normal hematocrit (above 55%).
>
> **anemia** (a-NĒ-mē-a) A condition of the blood in which the number of functional red blood cells or their hemoglobin content is below normal.

1. **Polycythemia**. Too many RBCs (or *polycythemia vera*, too many formed elements of all types).

2. **Anemia** due to too few RBCs:
   - *Pernicious anemia*. Insufficient hemopoiesis
   - *Hemorrhagic anemia*. Excessive loss of blood through bleeding resulting from large wounds, stomach ulcers, or especially heavy menstruation.
   - *Hemolytic anemia*. Premature rupture of RBC plasma membranes; may be caused by inherited defects or by outside agents such as parasites, toxins, or antibodies from transfused blood.
   - *Aplastic anemia*. Destruction of red bone marrow due to toxins, gamma radiation, and certain medications

3. Anemia due to deficiency of functional hemoglobin in each RBC:
   - *Iron-deficiency anemia*. Inadequate intake or absorption of iron or excessive loss of iron (This is the most common type of anemia.) Women are at greater risk for this type of anemia because of monthly menstrual blood loss.
   - *Sickle cell anemia*. Genetic disorder in which a defective hemoglobin is formed in the RBC. (This abnormal hemoglobin causes the RBCs to form a crescent shape, which rupture easily. Even though the loss of RBCs stimulates erythropoiesis, it cannot keep pace with hemolysis.)

   > **Video**

   - *Thalassemia*. A group of hereditary hemolytic anemias in which there is an abnormality in one or more of the four polypeptide chains of the hemoglobin molecule. Thalassemia occurs primarily in populations from countries bordering the Mediterranean Sea.

4. **Leukemia**. Red bone marrow cancers in which abnormal white blood cells multiply uncontrollably.

   > **leukemia** (loo-KĒ-mē-a) A malignant disease of the blood-forming tissues.

5. **Infections**. Can be bacterial, fungal, viral, or parasitic.

Plasma samples can be used to determine the patient's ability to clot the blood effectively. A variety of tests using serum can also be done to study the chemical composition of the blood. These tests can determine such things as the level of various metabolites that may be present in the blood, including glucose, cholesterol, urea, and enzymes. Such tests can determine whether internal organs are functioning correctly or whether there are abnormal levels of electrolytes and waste materials that could impact body organs. Because the blood is such an important transport medium, it can serve as a "window" to your health.

CONCEPT CHECK 🛑 STOP

1. **Which** part of the body would you most likely use to get a few drops of blood?

2. **What** does a hematocrit measure?

3. **Which** blood test would be the most helpful in determining whether a patient has an infection?

Common medical blood tests • Figure 10.9

Values and interpretations of four common tests are shown.
(Note that not all components of a complete blood count (CBC) have been included.)

HEMATOCRIT TESTS

Measures: Percentage of red blood cells
Uses: Diagnosis of anemia, polycythemia, or abnormal states of hydration
Normal values: 38–42% (avg 42) in females; 40–54% (avg 47) in males

Abnormal indicators

High	Low
polycythemia	anemia

RETICULOCYTE TESTS

Measures: Percentage of reticulocytes
Uses: Assessment of rate of erythropoiesis
Normal values: 0.5–1.5%

Abnormal indicators

High	Low
Malfunctioning red bone marrow (with anemia), nutritional deficiency, leukemia, or pernicious anemia	Bleeding, hemolysis, or iron deficiency

DIFFERENTIAL WBC COUNT

Measures: Percentage of each type of WBC in sample of 100
Uses: Assesses immune status

Type WBC	%	Abnormal indicators High	Low
Neutrophils	60–70	Bacterial infection, burn, stress	Radiation, drugs, vitamin B_{12} deficiency
Eosinophils	2–4	Parasitic infection, allergic reaction, autoimmune disease, or adrenal insufficiency	Drugs, stress, Cushing's syndrome
Basophils	0.5–1	Allergy, leukemia, cancer	Pregnancy, stress, hyper-thyroidism
Lymphocytes	20–25	Viral infection, leukemia, immune disease	Immuno-suppression, prolonged severe illness, high steroid levels
Monocytes	3–8	Viral/fungal infection, tuberculosis, leukemia	Rare

COMPLETE BLOOD COUNT

Levels of various formed elements: Information about formed elements
Uses: Diagnosis of various conditions

Cell type/ units	Value	Abnormal indicators High	Low
RBC count (#/μl)	Males: 5.4 million Female: 4.8 million	Polycythemia, hypoxia, congenital heart disease*	Hemorrhage, anemia[†]
Hemoglobin (g/dl)	Males: 14–18 Females: 12–16	Polycythemia, hypoxia, congenital heart disease*	Hemorrhage, anemia[†]
WBC count (#/μl)	5000–10,000	Acute or chronic infections, trauma, leukemia, stress[‡]	Anemia, viral infections[‡]
Platelets (#/μl)	150,000–400,000	Cancer, trauma, cirrhosis	Anemia, allergic conditions, hemorrhage
Hematocrit (%)	(See panel at top left)		
Differential WBC	(See panel at bottom left)		

*indicated by increased RBC, hemoglobin, and hematocrit
[†]indicated by decreased RBC, hemoglobin, and hematocrit
[‡]see notes under differential WBC

Summary

1 Blood Functionally Connects the Body Organ Systems 288

- **Blood** is a liquid connective tissue that consists of a liquid component called **blood plasma** and a cellular component called the **formed elements**. Blood plasma consists mostly of water, proteins, and other substances. As shown, the formed elements include red blood cells, white blood cells, and platelets.

Appearance of blood cells in a blood smear • Figure 10.3

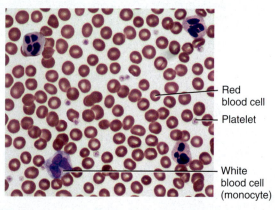

Red blood cell

Platelet

White blood cell (monocyte)

- Blood transports oxygen and carbon dioxide as well as nutrients, heat, and waste products. It regulates the pH of body fluids and helps manage body temperature. Blood also protects the body from infection, disease, and excessive blood loss during injury.

- Red blood cells transport oxygen and carbon dioxide; platelets are involved in blood clotting; and white blood cells (five different types) are involved in fighting infections, producing chemicals involved in the immune reaction, and removing microbes and cell debris.

2 Blood Cells Are Created in the Red Bone Marrow 291

- Within red bone marrow, pluripotent stem cells differentiate into **myeloid** and **lymphoid stem cells**, which in turn differentiate into all of the formed elements. Myeloid stem cells form the RBCs, platelets, and granulocytes. Lymphoid stem cells form the agranulocytes.

- Once produced, red blood cells live about 120 days. Old, worn-out red cells are destroyed in the spleen, liver, and red bone marrow. The hemoglobin is broken down into various parts. The amino acids and iron are recycled, and the non-iron heme pigments are broken down and excreted.

- As shown here, **erythropoiesis** is regulated by a negative feedback system involving the kidney, and red bone marrow via the hormone erythropoietin. The trigger for this system is a decrease in oxygen delivery to the kidney and other tissues.

Regulation of red blood cell formation • Figure 10.5

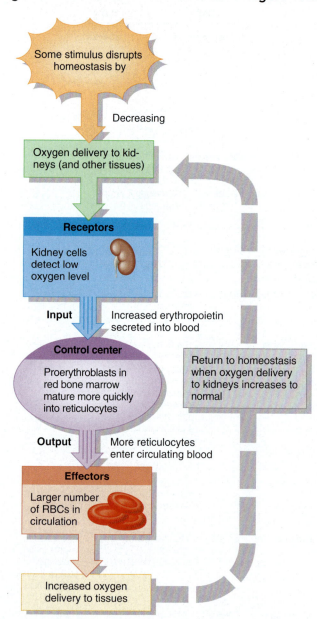

Some stimulus disrupts homeostasis by

Decreasing

Oxygen delivery to kidneys (and other tissues)

Receptors

Kidney cells detect low oxygen level

Input Increased erythropoietin secreted into blood

Control center

Proerythroblasts in red bone marrow mature more quickly into reticulocytes

Return to homeostasis when oxygen delivery to kidneys increases to normal

Output More reticulocytes enter circulating blood

Effectors

Larger number of RBCs in circulation

Increased oxygen delivery to tissues

3 Blood Clotting Controls Bleeding 294

- When blood vessels are damaged, three hemostatic mechanisms take place: **vascular spasm**, **platelet plug** formation (shown here), and blood clotting.

Platelet plug formation • Figure 10.7b

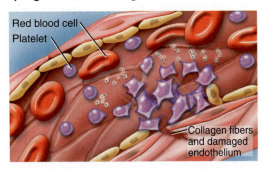

Red blood cell
Platelet
Collagen fibers and damaged endothelium

- Blood clotting is a complex process in which insoluble fibrin threads form to seal a tear or hole. Blood clotting involves a number of chemical reactions and ends with the formation of fibrin strands. Clotting involves various clotting factors, ionized calcium, plasma proteins, and enzymes.

- Clots get dissolved during a process called **fibrinolysis**. A connective tissue covering is then created to permanently seal the wound.

- A clot, called a **thrombus**, can form in an intact vessel. A clot that breaks can travel though the bloodstream and clog smaller blood vessels in various organs, thereby causing heart attacks, strokes, pulmonary embolism, and kidney failure.

4 Matching the ABO Group Allows Safe Transfusions 296

- Red blood cells have various molecules called **antigens** on their surfaces. Blood plasma contains antibodies to various antigens circulating within it. During a transfusion, if red blood cells containing a specific antigen are introduced into a person who has antibodies to that specific antigen, the antibodies attack and destroy the foreign red blood cells in what is called a transfusion reaction.

- The most common **blood group** is the ABO group, where red blood cells can have either A-antigens (type A), B-antigens (type B), both antigens (type AB), or neither antigen (type O); blood type testing (as shown) is based on these characteristics.

ABO blood typing test • Figure 10.8a

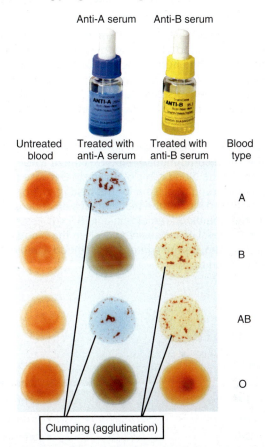

Anti-A serum   Anti-B serum

| Untreated blood | Treated with anti-A serum | Treated with anti-B serum | Blood type |

Clumping (agglutination)

Blood type	A	B	AB	O
Compatible donor blood types (no reaction)	A, O	B, O	A, B, AB, O	O
Incompatible donor blood types (reaction)	B, AB	A, AB	—	A, B, AB

- Within the Rh group, blood is either Rh-positive (has Rh antigen) or Rh-negative (doesn't have Rh antigen). This group is especially crucial to pregnant women who are Rh-negative because Rh incompatibility between the mother's blood and baby's blood can lead to hemolytic disease of the newborn. In this condition, the mother's antibodies, produced during the delivery of the first child, can attack the blood of a later baby who has an incompatible blood type.

5 Analysis of Blood Components Can Tell Much About an Individual's Health 299

- Blood samples can be taken from multiple places by finger-stick, venipuncture, or arterial stick.

- Four common blood tests to study the formed elements can be conducted on blood samples for diagnostic purposes: hematocrit (described here), reticulocyte count, differential white blood cell count, and complete blood count (CBC). These tests are used to screen for a variety of conditions, including **anemia**, **polycythemia**, **leukemia**, and various infections.

- Tests can also be done to determine the chemical composition of the liquid component of blood. Some of these tests check whether the blood can clot properly. Because these clotting studies require plasma for the testing, the blood must be collected into a tube that contains an anticoagulant to ensure that the blood remains in a fluid form. Tests that determine the level of various chemicals in the blood are done using **serum**. Serum comes from a blood sample that has been allowed to clot.

A common medical blood test • Figure 10.9

HEMATOCRIT TESTS

Measures:	Percentage of red blood cells
Uses:	Diagnosis of anemia, polycythemia, or abnormal states of hydration
Normal values:	38–42% (avg 42) in females; 40–54% (avg 47) in males

Abnormal indicators

High	Low
polycythemia	anemia

Key Terms

- agglutinate 296
- anemia 300
- antibody 296
- antigen 296
- basophil 290
- biconcave 288
- blood clotting 294
- blood group 296
- blood type 296
- clotting factor 294
- coagulation 294
- embolus 294
- eosinophil 290
- erythrocyte 288
- erythropoiesis 292
- erythropoietin 292

- fibrin 294
- fibrinogen 294
- fibrinolysis 294
- formed elements 288
- hemoglobin 290
- hemolysis 296
- hemopoiesis 291
- hemorrhage 294
- hemostasis 294
- homogeneous 288
- leukemia 300
- leukocyte 290
- leukocytosis 292
- lymphocyte 290
- lymphoid stem cell 291
- megakaryocyte 290

- monocyte 290
- myeloid stem cell 291
- neutrophil 290
- plasma 288
- plasmin 294
- plasminogen 294
- platelet 290
- platelet plug 294
- pluripotent stem cell 291
- polycythemia 300
- red blood cell (RBC) 288
- reticulocyte 292
- serum 300
- thrombus 294
- vascular spasm 294
- white blood cell (WBC) 290

Critical and Creative Thinking Questions

1. Jane feels tired all the time. She is often short of breath when she tries to exercise, but her lung function tests come back normal. Her doctor wants to assess whether she is suffering from anemia. What type of test should he order? If Jane has anemia, what might the expected test results be?

2. During an occupational accident in a nuclear lab, Jim was exposed to high doses of gamma radiation. He feels tired and weak, and he is especially prone to any infection that comes along. His CBC panel shows that numbers of all formed elements are well below normal. Which of his cells were damaged by the radiation exposure?

3. John is a 60-year-old male who presented at the emergency room with a heart attack. The physicians quickly gave him an intravenous drip containing tissue plasminogen activator (tPA). Within minutes, his coronary artery was clear, and blood flow had been re-established to the heart. What caused the blockage, and how did tPA administration solve the problem?

4. Julia is expecting her first child, and you are her health provider. While Julia is Rh-negative, there is a possibility that her baby is Rh⁺ positive. She asks you to explain hemolytic disease of the newborn and whether her baby is in danger. What is your response?

5. The Pittsburgh Steelers are preparing for a game against the Denver Broncos at Denver in seven days. Because the atmospheric pressure is lower in Denver than in Pittsburgh, there are lower amounts of gases in the air. This can affect the performance of athletes, so the Steelers travel to Denver immediately to begin practice. At the end of the week, the team doctor takes blood samples from random players and finds that, in all of them, the hematocrits are increased and the erythropoietin levels are higher than normal. Explain the mechanism that would account for these results.

What is happening in this picture?

It is essential that a donor and recipient be closely matched in the millions of blood transfusions that occur each year. The blood being administered here is type A.

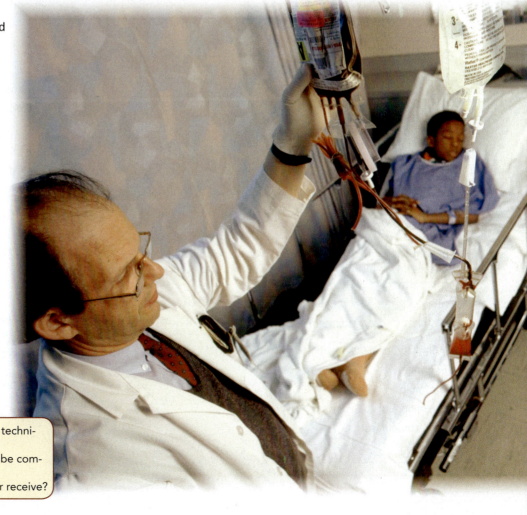

Think Critically 1. Describe how the technician determined the blood type.
2. Which recipient blood types would be compatible with this blood?
3. What blood type(s) could this donor receive?

Self-Test

(Check your answers in Appendix C.)

1. Which organ produces formed elements?

    a. spleen

    b. red bone marrow

    c. liver

    d. kidney

Use this figure to answer questions 2–4.

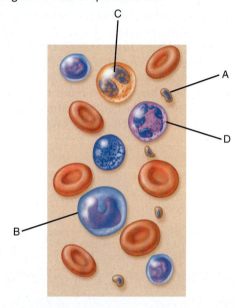

2. Which of the labeled cells is an agranular white blood cell?

    a. A   b. B   c. C   d. D

3. Which of the labeled cells is an eosinophil?

    a. A   b. B   c. C   d. D

4. Which of the formed elements labeled is involved in hemostasis?

    a. A   b. B   c. C   d. D

5. A patient has a lung disease in which oxygen delivery to the tissues is low. Which formed element would be elevated in response to this condition?

    a. monocyte   c. neutrophil

    b. platelet   d. red blood cell

6. Clotting factors activate which of the following?

    a. fibrinogen

    b. thrombin

    c. plasminogen

    d. prothrombinase

7. Which of the following tests would you use to diagnose an infection?

    a. reticulocyte count

    b. differential WBC

    c. hematocrit

    d. blood typing

8. Bilirubin is a by-product from the breakdown of which substance?

    a. fibrin

    b. globin

    c. hemoglobin

    d. thrombin

Use this figure to answer questions 9–10.

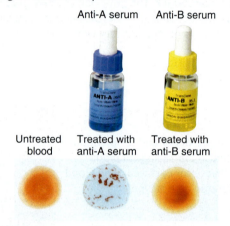

9. What is the blood type of the sample shown in this figure?

    a. B   b. A   c. AB   d. O

10. Which blood type would be the **best** recipient of the donor whose test results are shown?

    a. A   b. B   c. AB   d. O

11. A multi-lobed nucleus is characteristic of which formed element?

    a. platelet

    b. agranular white blood cell

    c. granular white blood cell

    d. red blood cell

12. What blood protein transports iron freely through the blood?

    a. hemoglobin

    b. transferrin

    c. prothrombin

    d. fibrin

Use this figure for questions 13–14.

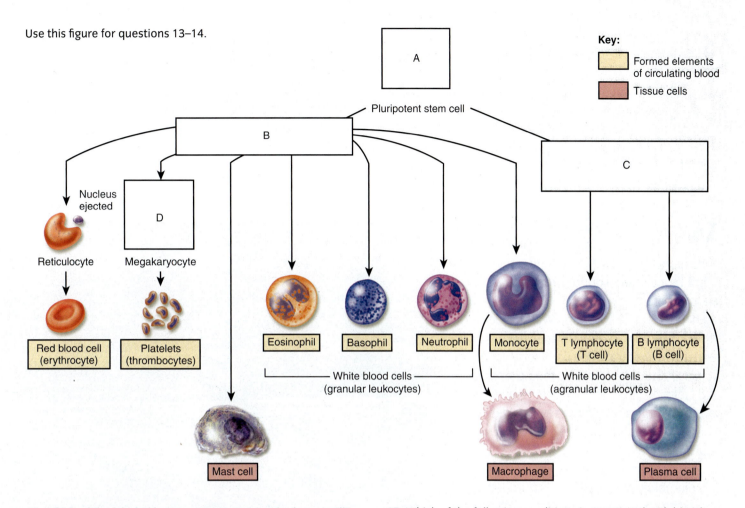

13. Which of the labeled boxes represents a myeloid stem cell?

 a. A

 b. B

 c. C

 d. D

14. Which of the labeled boxes represents a lymphoid stem cell?

 a. A

 b. B

 c. C

 d. D

15. Which of the following conditions is associated with blood type incompatibility?

 a. pernicious anemia

 b. polycythemia

 c. hemolytic disease of the newborn

 d. leukemia

16. A patient suffers from polycythemia vera, a condition in which levels of all the formed elements are elevated. What blood test is done to confirm this diagnosis?

 a. hematocrit

 b. reticulocyte count

 c. CBC

 d. differential WBC count

THE PLANNER

Review your Chapter Planner on the chapter opener and check off your completed work.

11 The Cardiovascular System: Heart, Blood Vessels, and Circulation

Water is delivered to your house through pipes that originate from pumping stations many miles away. Within your house, the water pipes branch out to provide water to many different rooms (such as the bathroom, kitchen, and laundry room). Water is delivered to you and your family members at your faucet. The water can be used for a variety of purposes (such as drinking, food preparation, and cleaning you and your belongings). Then waste water flows through sewer pipes to the waste-water treatment plant (typically far from your home) for processing.

Your cardiovascular system is much like this plumbing system. Blood is pumped by the heart (the "pumping station") through arteries (the "pipes") to your tissues (your "house"). Within the tissues, the blood is dispersed, using arterioles, to all of the cells ("your family"). Materials are delivered to the individual cells through an exchange at the capillaries (your "sink"). Blood is then moved back to the heart through the venules and veins (the "sewer pipes"), eventually being directed to various places (such as the lungs, liver, and kidneys—the "waste-water treatment plants") for disposal of the wastes that are being carried in the blood. Ensuring proper delivery and removal of blood from the tissues and ensuring that blood travels to all areas of the body are essential for survival.

Let's look at how this amazing pumping system works.

Aerial view of downtown New Orleans flooded after Hurricane Katrina Aug. 31, 2005

CHAPTER OUTLINE

CHAPTER PLANNER ✔

- ❑ Study the picture and read the opening story.
- ❑ Scan the Learning Objectives in each section:
  p. 310 ❑   p. 320 ❑   p. 332 ❑
- ❑ Read the text and study all visuals.
  Answer any questions.

Analyze key features

- ❑ InSight, p. 313 ❑   pp. 322–323 ❑   p. 329 ❑
- ❑ Process Diagram, p. 314 ❑   p. 315 ❑   p. 317 ❑
- ❑ What a Health Provider Sees, p. 335 ❑
- ❑ Stop: Answer the Concept Checks before you go on:
  p. 319 ❑   p. 331 ❑   p. 337 ❑

End of chapter

- ❑ Review the Summary and Key Terms.
- ❑ Answer the Critical and Creative Thinking Questions.
- ❑ Answer What is happening in this picture?
- ❑ Complete the Self-Test and check your answers.

The Heart Pumps Blood Through Blood Vessels to All Tissues

LEARNING OBJECTIVES

1. **Identify** the structural features of the heart.
2. **Outline** the flow of blood through the heart.
3. **Explain** how the electrical conduction system of the heart works.
4. **Identify** parts of an electrocardiogram and possible abnormalities.
5. **Describe** each phase of the cardiac cycle.
6. **Define** cardiac output and identify the factors that regulate it.

Like city pumping stations, which circulate water to your house and off to the waste-water treatment plant, your body's cardiovascular system circulates blood, gases, nutrients, and wastes throughout your body (**Figure 11.1**). This system consists of a single pump (your heart), and between 60,000 to 100,000 km (36,000 to 60,000 miles) of blood vessels. Your **heart** actually consists of two pumps in one. The right side collects oxygen-poor blood from the body and pumps it to the lungs. At the same time, the left side collects oxygen-rich blood from the lungs and propels it through the rest of the body. This essential muscular organ contracts about 100,000 times every day, which adds up to 35 million beats in a year. In total, the heart pumps over 8640 L (2282 gal) of blood each day and 3.2 million L (833,000 gallons) in a year.

Overview of the cardiovascular system • Figure 11.1

The cardiovascular system consists of two divisions: the pulmonary circulation (between the heart and the lungs) and the systemic circulation (between the heart and the body).

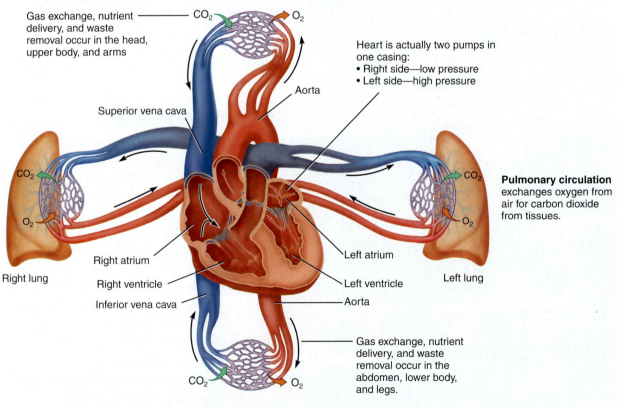

Gas exchange, nutrient delivery, and waste removal occur in the head, upper body, and arms

CO_2 · O_2

Superior vena cava

Aorta

Heart is actually two pumps in one casing:
• Right side—low pressure
• Left side—high pressure

CO_2 · O_2

Pulmonary circulation exchanges oxygen from air for carbon dioxide from tissues.

Right lung

Right atrium

Right ventricle

Inferior vena cava

Left atrium

Left ventricle

Aorta

Left lung

CO_2 · O_2

Gas exchange, nutrient delivery, and waste removal occur in the abdomen, lower body, and legs.

Systemic circulation exchanges carbon dioxide from tissues of the body for oxygen from the lungs. Systemic circulation also delivers nutrients to and removes waste from tissues.

Let's take a closer look at the operation of this amazing pump.

The Heart Is Located in the Thoracic Cavity

For all its might, the heart is relatively small, roughly the same size (but not the same shape) as your closed fist. It is about 12 cm (5 in.) long, 9 cm (3.5 in.) wide at its widest point, and 6 cm (2.5 in.) thick, with an average mass of 250 g (8 oz.) in adult females and 300 g (10 oz.) in adult males. The heart rests on the diaphragm, near the midline of the thoracic cavity. It lies in the **mediastinum** (mē-dē-a-STĪ-num), an anatomical region that extends from the sternum to the vertebral column, the first rib to the diaphragm, and between the lungs. About two-thirds of the mass of the heart lies to the left of the body's midline (**Figure 11.2**). You can visualize the heart as a cone lying on its side. Its pointed **apex** rests on the diaphragm and is directed toward the left hip. The **base** of the heart is its posterior surface, formed by the upper chambers of the heart.

The location of the heart • Figure 11.2

The heart is located in the thoracic cavity, with two-thirds of its mass to the left of the midline

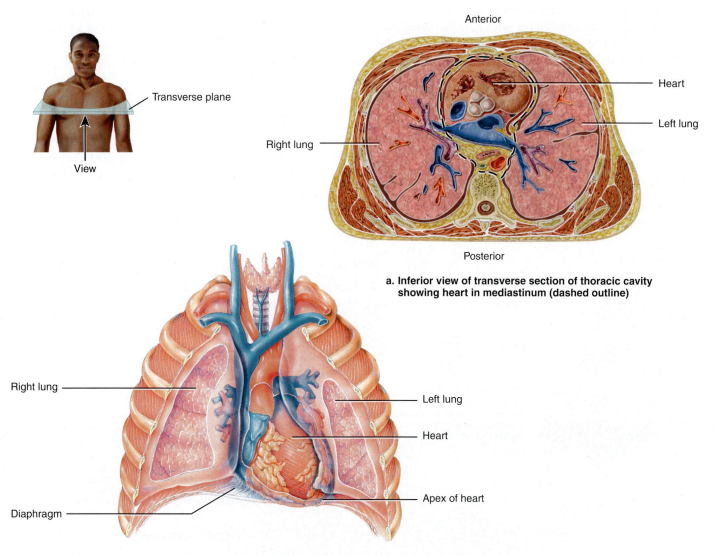

Transverse plane

View

Anterior

Heart

Left lung

Right lung

Posterior

a. Inferior view of transverse section of thoracic cavity showing heart in mediastinum (dashed outline)

Right lung

Left lung

Heart

Apex of heart

Diaphragm

b. Anterior view of heart in thoracic cavity

Valves Keep the Blood Flowing the Right Direction Through the Heart

The heart is enclosed in a protective sac called the **pericardium** (**Figure 11.3**). Each side has a receiving chamber called an **atrium** that collects the blood from the circulation (**right atrium** from systemic circulation, **left atrium** from the pulmonary circulation) and a pump called a **ventricle** that delivers blood to the circulation (**right ventricle** to the pulmonary circulation, **left ventricle** to the systemic circulation). A large muscular interventricular septum divides the two sides, which operate simultaneously (**Figure 11.4**).

Each side of the heart has **valves** that help to keep the blood flowing in the proper direction. One set of valves (atrioventricular) separates the atrium from the ventricle on each side of the heart, and another (semilunar) separates each ventricle from its major artery. The atrioventricular (AV) valves are the *triscuspid valve* (on the right side) and the *bicuspid*, or mitral, *valve* (on the left side). The semilunar valves are the *pulmonic valve* (on the right) and the *aortic valve* (on the left). The AV valves are designed to withstand the very high pressures produced by the pumping ventricles. The semilunar valves deal with much smaller volumes of blood (see Figure 11.4).

Cardiac muscle tissue of varying thicknesses forms the **myocardium**. The intercalat-

myocardium The middle layer of the heart wall, composed of cardiac muscle.

ed discs of cardiac muscle allow the myocardium to function as a unit by transmitting signals rapidly from cell to cell. The myocardium is sandwiched between the inner *endocardium* and the outer *epicardium*, both of which are composed of thin layers of connective and epithelial tissues (see Figure 11.3; for a review of muscle tissue and cardiac muscle contraction, see Chapter 6).

The walls of both atria are thin because they are mainly collection chambers and need to pump blood only a short distance into the ventricles. The walls of the left ventricle are thicker than those of the right because the left ventricle must do more work to drive blood flow through the entire body; the right ventricle transports blood a shorter distance, to the lungs.

Several large blood vessels called **veins** and **arteries** carry blood toward and away from the heart, respectively (see Figure 11.4b). The blood supply to the heart itself comes off the **aorta** via the **coronary arteries** and drains through the **coronary sinus** into the right atrium.

veins Blood vessels that convey blood from tissues back to the heart.

arteries Blood vessels that carry blood away from the heart.

Here is a simple mnemonic for remembering the differences between arteries and veins. The "A" in *ar*tery stands for "away" from the heart, while the "in" in v*ein* stands for "*in*to" the heart.

Let's take a closer look at how blood flows through the cardiovascular system.

Microscopic structure of the heart • Figure 11.3

The heart is composed mostly of cardiac muscle tissue, with connective tissue and epithelial linings inside and out.

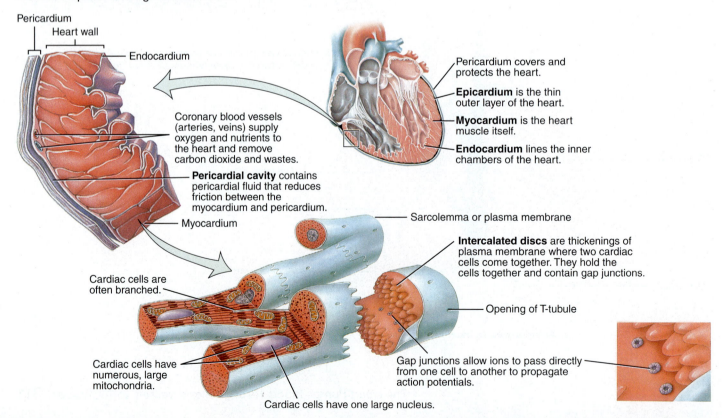

Pericardium
Heart wall
Endocardium

Coronary blood vessels (arteries, veins) supply oxygen and nutrients to the heart and remove carbon dioxide and wastes.

Pericardial cavity contains pericardial fluid that reduces friction between the myocardium and pericardium.

Myocardium

Cardiac cells are often branched.

Cardiac cells have numerous, large mitochondria.

Pericardium covers and protects the heart.

Epicardium is the thin outer layer of the heart.

Myocardium is the heart muscle itself.

Endocardium lines the inner chambers of the heart.

Sarcolemma or plasma membrane

Intercalated discs are thickenings of plasma membrane where two cardiac cells come together. They hold the cells together and contain gap junctions.

Opening of T-tubule

Gap junctions allow ions to pass directly from one cell to another to propagate action potentials.

Cardiac cells have one large nucleus.

The heart is a four-chambered pump with valves to keep blood flowing through it in only one direction. Several large vessels deliver blood to the heart and remove blood from it.

WILEY PLUS   Video

a. Major parts of the heart

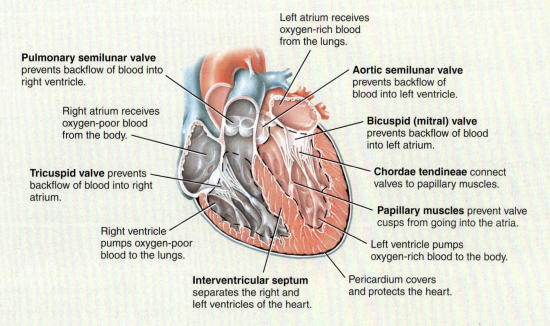

Left atrium receives oxygen-rich blood from the lungs.

Pulmonary semilunar valve prevents backflow of blood into right ventricle.

Aortic semilunar valve prevents backflow of blood into left ventricle.

Right atrium receives oxygen-poor blood from the body.

Bicuspid (mitral) valve prevents backflow of blood into left atrium.

Tricuspid valve prevents backflow of blood into right atrium.

Chordae tendineae connect valves to papillary muscles.

Papillary muscles prevent valve cusps from going into the atria.

Right ventricle pumps oxygen-poor blood to the lungs.

Left ventricle pumps oxygen-rich blood to the body.

Interventricular septum separates the right and left ventricles of the heart.

Pericardium covers and protects the heart.

b. Major blood vessels of the heart

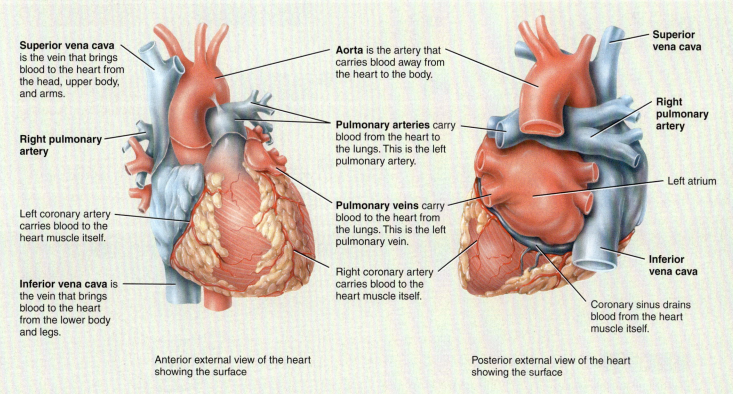

Superior vena cava is the vein that brings blood to the heart from the head, upper body, and arms.

Aorta is the artery that carries blood away from the heart to the body.

Superior vena cava

Right pulmonary artery

Pulmonary arteries carry blood from the heart to the lungs. This is the left pulmonary artery.

Right pulmonary artery

Left coronary artery carries blood to the heart muscle itself.

Pulmonary veins carry blood to the heart from the lungs. This is the left pulmonary vein.

Left atrium

Inferior vena cava is the vein that brings blood to the heart from the lower body and legs.

Right coronary artery carries blood to the heart muscle itself.

Inferior vena cava

Coronary sinus drains blood from the heart muscle itself.

Anterior external view of the heart showing the surface

Posterior external view of the heart showing the surface

Blood Flows Through the Heart Because of Pressure Gradients

The cardiovascular system is a closed-circuit system, meaning that the blood flows in a loop through the system, never leaving the vessels (**Figure 11.5**). Contractions of atria or ventricles increase the pressure on the fluid inside the chamber, forming a pressure gradient (a difference in pressures from one area to another) and causing blood to flow from the area of higher pressure to the area of lower pressure. When the atria contract, the pressure of the blood builds up on the atrial side of the AV valves, causing them to open and allowing the blood to move into the ventricles. When the ventricles contract, the pressure of the blood increases against the ventricular side of the AV valves, causing them to close. This pressure also forces the semilunar valves to open, permitting the blood to enter the arteries.

Once the left ventricle has pumped blood out, the blood will flow through the systemic circulation, driven by a pressure difference. This pressure difference is between the aorta (highest pressure at the beginning of the systemic circulation) and the right atrium (lowest pressure

PROCESS DIAGRAM

Blood flow through the heart • Figure 11.5

Follow the path of blood through the heart and circulation as it enters the right atrium of the heart (**1**) and flows through the rest of the cardiovascular system to ultimately return through the venae cavae (**10**). Oxygen-poor (deoxygenated) blood is shown in blue, while oxygen-rich (oxygenated) blood is shown in red.

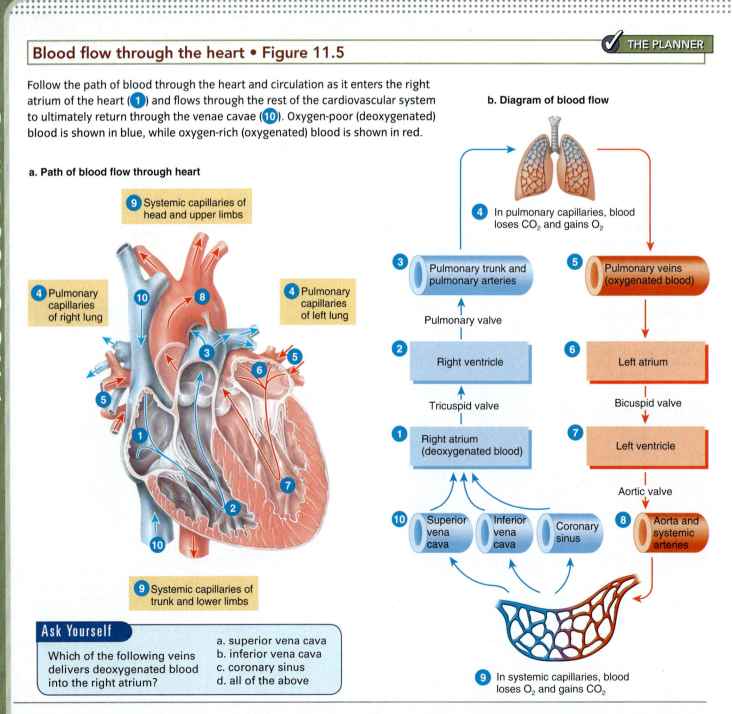

a. Path of blood flow through heart

9 Systemic capillaries of head and upper limbs

4 Pulmonary capillaries of right lung

4 Pulmonary capillaries of left lung

9 Systemic capillaries of trunk and lower limbs

b. Diagram of blood flow

4 In pulmonary capillaries, blood loses CO_2 and gains O_2

3 Pulmonary trunk and pulmonary arteries

5 Pulmonary veins (oxygenated blood)

Pulmonary valve

2 Right ventricle

6 Left atrium

Tricuspid valve

Bicuspid valve

1 Right atrium (deoxygenated blood)

7 Left ventricle

Aortic valve

10 Superior vena cava — Inferior vena cava — Coronary sinus

8 Aorta and systemic arteries

9 In systemic capillaries, blood loses O_2 and gains CO_2

Ask Yourself

Which of the following veins delivers deoxygenated blood into the right atrium?

a. superior vena cava
b. inferior vena cava
c. coronary sinus
d. all of the above

at the end of the systemic circulation). Similarly, the right ventricle will pump the blood into the pulmonary circulation, where the blood will flow from the pulmonary artery (higher pressure) to the left atrium (lower pressure).

The contractions of atria and ventricles, which generate pressure and force blood to flow through the system, are coordinated: Both atria contract together, and later, both ventricles contract together. At the "heart" of this coordination is an elaborate electrical system that stimulates muscle cells to contract. Let's take a look.

Electrical Signals Control Heart Rate

About 1% of cardiac muscle cells differ from the rest in that they can generate action potentials over and over

again in a rhythmic pattern. These cells, the heart's natural pacemaker, control the rate at which the heart beats and form an electrical conduction system called the **cardiac conduction system**. The electrical activity begins in the sinoatrial node (SA node), passes over the atria to the atrioventricular node (AV node), and spreads via the Purkinje fibers throughout the ventricles, using the intercalated discs to pass the electrical signals (**Figure 11.6**). This conduction system coordinates the contraction of the atria and the later contraction of the ventricles in a series of events referred to as the *cardiac cycle*.

The cardiac conduction system sets the heart rate but can be influenced by the sympathetic and parasympathetic divisions of the autonomic nervous system. Left alone, the heart would beat at a rate between 60 to 100 times per

PROCESS DIAGRAM

The heart's electrical activity • Figure 11.6

✔ THE PLANNER

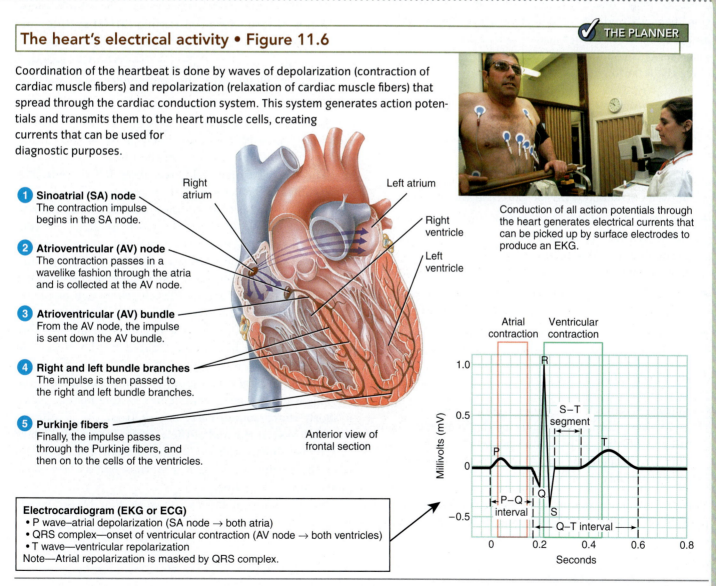

Coordination of the heartbeat is done by waves of depolarization (contraction of cardiac muscle fibers) and repolarization (relaxation of cardiac muscle fibers) that spread through the cardiac conduction system. This system generates action potentials and transmits them to the heart muscle cells, creating currents that can be used for diagnostic purposes.

Conduction of all action potentials through the heart generates electrical currents that can be picked up by surface electrodes to produce an EKG.

1 **Sinoatrial (SA) node**
The contraction impulse begins in the SA node.

2 **Atrioventricular (AV) node**
The contraction passes in a wavelike fashion through the atria and is collected at the AV node.

3 **Atrioventricular (AV) bundle**
From the AV node, the impulse is sent down the AV bundle.

4 **Right and left bundle branches**
The impulse is then passed to the right and left bundle branches.

5 **Purkinje fibers**
Finally, the impulse passes through the Purkinje fibers, and then on to the cells of the ventricles.

Anterior view of frontal section

Electrocardiogram (EKG or ECG)
• P wave–atrial depolarization (SA node → both atria)
• QRS complex—onset of ventricular contraction (AV node → both ventricles)
• T wave—ventricular repolarization
Note—Atrial repolarization is masked by QRS complex.

minute, with the typical rate at about 70 beats per minute. We will look at the regulation of the cardiovascular system in more detail later in the chapter.

The coordinated action potentials among various heart muscle cells create electrical currents that can be detected by electrodes attached to the surface of the chest. The record of the electrical changes accompanying the heartbeat is called an **electrocardiogram** (**EKG** or **ECG**). **Cardiologists** analyze the EKG pattern to diagnose problems with the heart (see Figure 11.6).

> **cardiologists** (kar-dē-OL-ō-jists) Physicians who study the heart and diseases associated with it.

When the normal rhythm established by the SA node (the normal sinus rhythm) is disrupted by a defect in the conduction system, a condition called **arrhythmia** occurs. There may be several causes of arrhythmias, including disease (such as coronary artery disease, hypertension, myocardial infarction, hyperthyroidism, or defective heart valves), stress, and drugs and other chemicals (such as caffeine, cocaine, nicotine, or alcohol). Some examples of arrhythmias include the following:

> **arrhythmia** (a-RITH-mē-a) An irregular heart rhythm.

- *Supraventricular tachycardia*—A rapid but regular heart rate (160–200 beats/min) that originates in the atria. Episodes begin and end suddenly and may last minutes to hours.

- *Heart block*—A blockage of the electrical pathways between atria and ventricles, often at the AV node.

- *Premature ventricular contraction*—An early heartbeat in the ventricular myocardium that briefly interrupts the normal rhythm. A similar arrhythmia, called atrial premature contraction, can occur in the atrial myocardium.

- *Atrial flutter*—Rapid but regular atrial contractions (240–360 beats/min) accompanied by AV block.

- *Ventricular tachycardia*—A rapid heart rhythm (at least 120 beats/min) that originates in one of the ventricles.

- *Atrial fibrillation*—Asynchronous contractions (300–600 beats/min) in which atrial contraction stops altogether. A similar, but more deadly, condition called ventricular fibrillation occurs in the ventricles. Ventricular fibrillation is treated by passing an electrical current through the heart to stop the contractions (defibrillation).

Now that we have seen how the heart is structured, the path that blood flows through it, and the way in which the electrical activity coordinates contractions, let's put it all together and examine the events of the cardiac cycle.

The Cardiac Cycle Alternates Between Systole and Diastole

With each heartbeat, there is a cycle of events called the **cardiac cycle** (see **Figure 11.7**). During each cardiac cycle, the cardiac chambers alternate between a phase of contraction, known as **systole**, and a phase of relaxation, called **diastole**.

> **systole** (SIS-tō-lē) The contraction phase of the cardiac cycle in which blood moves out of the heart chamber.
>
> **diastole** (dī-AS-tō-lē) The relaxation phase of the cardiac cycle in which the heart chamber is resting and filling with blood.

1. During diastole, or the relaxation period, the chamber of the heart fills with blood.

2. At systole, the chamber empties. The systole phase begins with the atria contracting (**atrial systole**) and forcing blood into the ventricles.

3. Then the ventricles contact (**ventricular systole**), pumping blood into the pulmonary and systemic circulations.

Each chamber returns to diastole following contraction and begins filling again. Each phase of the cardiac cycle corresponds to one wave of the EKG.

The entire cardiac cycle lasts about 0.8 seconds. The atria contract for 0.1 s and then rest for 0.7 s. In contrast, the ventricles contract for about 0.3 s and then rest for 0.5 s. For 0.4 s of the cycle, both the atria and ventricles are relaxed. When the heart rate increases, this relaxation time becomes shorter. Because most of the filling occurs during the first 0.1 s, as long as the chamber spends at least that long in diastole, there will be plenty of blood to pump out at systole.

The sound of the heartbeat comes primarily from the turbulence in blood flow created by the closure of the valves, not from the contraction of the heart muscle. The first sound, *lubb*, is a long, booming sound from the AV valves closing after ventricular systole begins. Next, a short, sharp sound, *dubb*, comes from the semilunar valves closing at the end of the ventricular systole. There is a pause during the relaxation period. Thus you can hear the cardiac cycle with a stethoscope as *lubb, dubb*, pause; *lubb, dubb*, pause.

So, how do we measure the work done by the heart? Read on to find out.

The phases of the cardiac cycle • Figure 11.7

1 Relaxation period (diastole)
   a. Ventricles relax isovolumetrically (no change in volume) and repolarize (relaxation of cardiac muscle fibers)(T wave of EKG).
   b. Blood from body and lungs fills atria.
   c. When ventricular pressure falls below atrial pressure, AV valves open and ventricles fill with blood.
   d. At end of this phase, ventricles are about 3/4 full and the atria begin to depolarize (P wave of EKG).

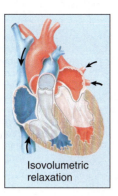

Isovolumetric relaxation

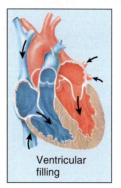

Ventricular filling

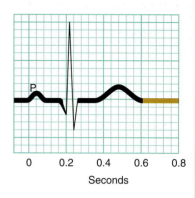

Seconds

2 Atrial contraction (atrial systole)
   a. As P wave spreads over atria, they contract and force the final 25% of blood into the ventricles (ventricles are full ~130 mL).
   b. AV valves are open and semilunar valves are closed.
   c. End of atrial systole = end of ventricular relaxation

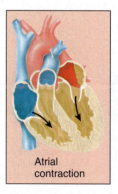

Atrial contraction

Seconds

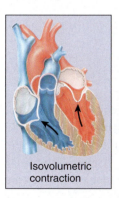

Isovolumetric contraction

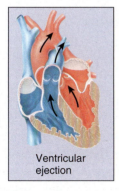

Ventricular ejection

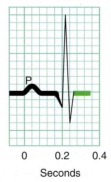

Seconds

3 Ventricular contraction (ventricular systole)
   a. Ventricular depolarization begins (QRS complex of EKG begins).
   b. Ventricular fibers contract, but do not shorten (isovolumetric contraction). Ventricular pressure increases.
   c. When ventricular pressure exceeds atrial pressure, the AV valves close. Ventricular pressure continues to increase.
   d. When ventricular pressure exceeds pressure in aorta and pulmonary arteries, the semilunar valves open.
   e. Ventricles pump about 70 mL of blood out (ventricular ejection) into aorta and pulmonary arteries.
   f. As blood leaves the ventricles, the ventricular pressure falls below aortic and pulmonary artery pressures and the semilunar valves close.
   g. Ventricular systole ends with the onset of the T wave.

Put It Together

Which of the following correctly pairs events of the cardiac cycle and the EKG?
a. QRS complex, ventricular systole, opening of the AV valves
b. P-wave, atrial diastole, flow of blood into the ventricles
c. T wave, closing of the AV valves, flow of blood into the ventricles
d. QRS complex, ventricular systole, opening of the semilunar valves
e. P wave, atrial systole, closing of the AV valves

Several factors influence cardiac output by changing stroke volume and/or heart rate.

a. Autonomic nervous system activity changes heart rate and stroke volume

During exercise or emotional stress, cardiac output rises to supply working tissues with increased amounts of oxygen and nutrients.

Baroreceptors sense changes in blood pressure and send signals to the brain.

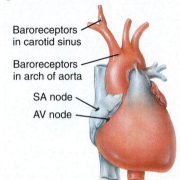

Baroreceptors in carotid sinus

Baroreceptors in arch of aorta

SA node

AV node

Cardiac accelerator nerves increase heart rate. Cardiac accelerator nerves in the sympathetic nervous system stimulate the SA and AV nodes of the heart to increase heart rate. In addition, these nerves increase the contractility of the atria and ventricles, which increases stroke volume.

Vagus nerve activity slows heart rate. Parasympathetic nerves, primarily the vagus nerves, decrease the stimulation of the nodes when the demand for cardiac output decreases.

Cardiac output (CO)		Stroke Volume (SV)		Heart Rate (HR)
Measured in L/min	=	Measured in mL/beat	×	Measured in beats/min

b. Three factors regulate stroke volume under normal conditions:

Degree of stretch in heart before contraction (preload)	Force of contraction	Pressure required to eject blood from ventricles (afterload):
• Increased blood flow into the heart increases the force of contraction (Frank-Starling Law of the heart) and increases SV	• Increased force of contraction increases SV • Increased Ca^{2+} increases contractility • Hormones (epinephrine and norepinephrine) increase contractility	• Increased aortic and/or pulmonary artery pressures reduce the time that semilunar valves are open, reducing SV

| • Decreased blood flow has the opposite effect | • Decreased force of contraction decreases SV
• Increased Na^+ and K^+ decrease contractility
• Decreased temperature decreases contractility | • Decreased pressures have the opposite effect |

Many Factors Affect Cardiac Output

The work done by the heart, called the **cardiac output (CO)**, is defined as the volume of blood ejected from the ventricle per minute. The same amount of blood is ejected from the right and left ventricles. Cardiac output is the product of **stroke volume** (**SV**; the amount of blood ejected during each beat) and the **heart rate** (**HR**; number of beats per minute). For example, in a resting adult, the stroke volume is typically 70 mL per beat and the heart rate is 72 beats/min. So, the cardiac output is as follows:

$$CO = SV \times HR$$
$$= 70 \text{ mL/beat} \times 72 \text{ beats/min}$$
$$= 5040 \text{ mL/min,}$$

or slightly over 5 L/min. You may recall that this is the typical blood volume for an individual.

Cardiac output is affected by factors that influence stroke volume and/or heart rate (**Figure 11.8**). Stroke volume can be influenced by the amount of blood present in the ventricle at the end of diastole (*preload*), the force of contraction (*contractility*), or the vascular pressures that the ventricles must overcome to push the blood out of the chamber (*afterload*). The force of contraction is influenced by how much the heart has been stretched before it contracts. The more blood that enters the heart during diastole, the stronger the subsequent contraction will be; this principle is called the Frank-Starling Law of the heart. While the heart rate is set by an intrinsic pacemaker called the sinoatrial node, which initiates regular waves of depolarization across the atria, it can also be influenced by other factors:

- *Autonomic nervous system.* In times of physical or emotional stress, the ANS can stimulate the heart to beat faster.
- *Hormones.* For example, epinephrine and norepinephrine released from the adrenal medullae during exercise and stress increase both heart rate and contractility.
- *Ions.* Elevated levels of sodium and potassium decrease heart rate and contractility. A moderate increase in calcium ions increases heart rate.
- *Body temperature.* Increased body temperature, as occurs during a fever or strenuous exercise, increases heart rate.
- *Age.* Newborns have higher heart rates (120 beats/min) than adults. The heart rate slows down gradually during childhood, until it reaches the adult rate of around 72 beats/min.
- *Gender.* Females tend to have higher resting heart rates than males.
- *Physical fitness.* Regular exercise tends to slow the resting heart rate. Well-trained athletes have slower resting heart rates (40–60 beats/min) but have normal resting cardiac outputs because their hearts are enlarged, giving them larger stroke volumes.

What Can Go Wrong with Blood Flow Through the Heart?

In many organs and tissues, including the heart, the blood vessels branch so that blood has many alternate routes to reach the cells within. These branches, called **anastomoses**, help maintain sufficient blood supply to an organ when one of the branches is blocked. However, when blood flow in the coronary arteries gets partially blocked (**myocardial ischemia**) or totally blocked (**myocardial infarction**), the heart muscle gets starved for oxygen and may weaken or die. Such obstructions may occur due to blood clots, coronary artery disease, or narrowing of the arteries. **Atherosclerosis**, for example, is an inflammatory disease condition in which fatty materials (mostly cholesterol and triglycerides) build up in the walls of the blood vessels, forming lesions called **atherosclerotic plaques**. These plaques narrow the diameter of the vessel and roughen the inner surface, which can attract platelets and lead to the formation of a thrombus (**Figure 11.9**).

> **anastomoses**
> (a-nas-tō-MŌ-sēs)
> Anatomical connections between tubular structures, especially blood vessels.

CONCEPT CHECK STOP

1. **What** is the function of the atria?
2. **Blood** returning to the heart from the body enters which of the chambers?
3. **What** does the atrioventricular node do?
4. **What** is the significance of the QRS complex?
5. **What** happens during ventricular systole?
6. **How** does preload influence cardiac output?

Normal and obstructed arteries • Figure 11.9

Photomicrographs of transverse sections of **a.** a normal artery and **b.** an artery partially obstructed by an atherosclerotic plaque.

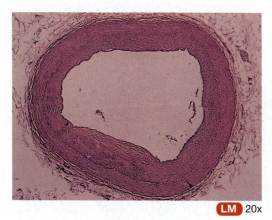

LM 20x

a. Normal artery

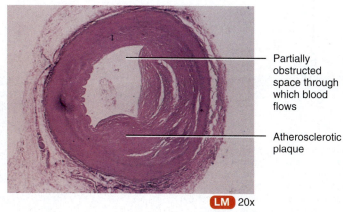

Partially obstructed space through which blood flows

Atherosclerotic plaque

LM 20x

b. Obstructed artery

Blood Vessels Are the Body's Plumbing

LEARNING OBJECTIVES

1. **Compare** the structure and function of the different types of blood vessels.

2. **Describe** how capillary exchange works.

3. **Explain** how venous blood returns to the heart.

4. **Outline** the pulmonary and hepatic portal circulations.

While the job of the heart is to pump the blood and generate pressure for blood flow, the blood vessels transport and distribute blood to and from the various tissues. From the heart to the tissues, various types of blood vessels form a series of branching tubes where single large vessels divide into numerous smaller ones (**Figure 11.10**). From the cells to the heart, small narrow blood vessels combine into larger ones. The types of vessels include arteries, arterioles, capillaries, venules, and veins.

Overview of systemic blood vessels • Figure 11.10

There are five types of blood vessels that carry blood away from the heart to the tissues and return it to the heart. The systemic blood vessels transport blood to tissues, but the distribution of blood between the heart and the various types of blood vessels is uneven.

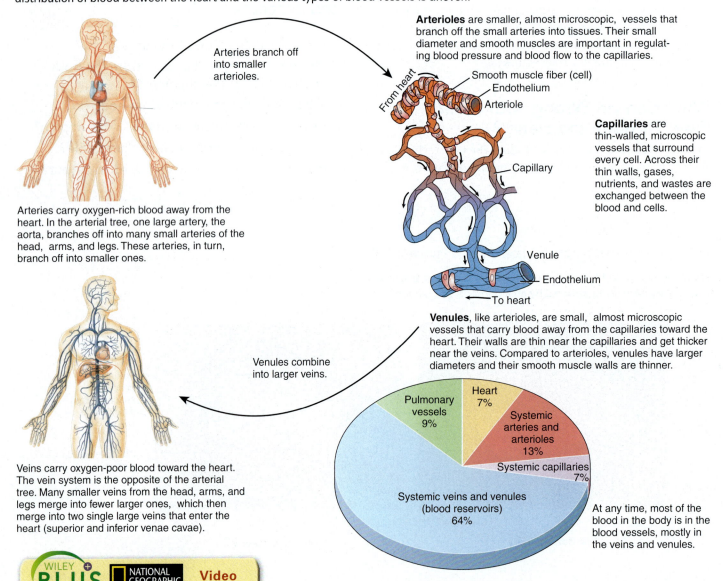

Arteries branch off into smaller arterioles.

Arteries carry oxygen-rich blood away from the heart. In the arterial tree, one large artery, the aorta, branches off into many small arteries of the head, arms, and legs. These arteries, in turn, branch off into smaller ones.

Arterioles are smaller, almost microscopic, vessels that branch off the small arteries into tissues. Their small diameter and smooth muscles are important in regulating blood pressure and blood flow to the capillaries.

From heart

Smooth muscle fiber (cell)
Endothelium
Arteriole

Capillaries are thin-walled, microscopic vessels that surround every cell. Across their thin walls, gases, nutrients, and wastes are exchanged between the blood and cells.

Capillary

Venule

Endothelium

To heart

Venules, like arterioles, are small, almost microscopic vessels that carry blood away from the capillaries toward the heart. Their walls are thin near the capillaries and get thicker near the veins. Compared to arterioles, venules have larger diameters and their smooth muscle walls are thinner.

Venules combine into larger veins.

Veins carry oxygen-poor blood toward the heart. The vein system is the opposite of the arterial tree. Many smaller veins from the head, arms, and legs merge into fewer larger ones, which then merge into two single large veins that enter the heart (superior and inferior venae cavae).

Heart
7%

Pulmonary vessels
9%

Systemic arteries and arterioles
13%

Systemic capillaries
7%

Systemic veins and venules (blood reservoirs)
64%

At any time, most of the blood in the body is in the blood vessels, mostly in the veins and venules.

In brief, blood vessels have three functions.

1. Blood vessels form a closed system of tubes that carries blood away from the heart (in arteries), transports it through the tissues of the body (in arterioles, capillaries, and venules) and then returns it to the heart (in veins).

2. Exchange of substances between the blood and body tissue cells occurs as blood flows through the capillaries.

3. Nutrients and oxygen diffuse from the blood through interstitial fluid into tissue cells. Waste products, including carbon dioxide, diffuses from tissue cells through interstitial fluid into the blood.

Let's start by looking at arteries.

Arteries and Arterioles Are Thick-Walled Vessels

The walls of arteries and **arterioles** are thick and have three layers of tissues, or tunics. These three layers surround the **lumen**, or hollow area through which the blood flows (**Figure 11.11**). Under control of the autonomic nervous system and/or local chemicals, smooth muscles in the arterial walls can contract and narrow the lumen (**vasoconstriction**) or relax and enlarge the lumen (**vasodilation**). Vasoconstriction and vasodilation are important in maintaining blood pressure and controlling blood flow, as we shall discuss later in the chapter.

The major systemic arteries are shown in **Figure 11.12** on the next page. The aorta is the largest artery of the body, with a diameter of 2–3 cm (about 1 in.). Its four principal divisions are as follows.

- The *ascending aorta* emerges from the left ventricle posterior to the pulmonary trunk. The ascending aorta gives off two coronary artery branches that supply the myocardium of the heart.

- The *arch of the aorta* is formed when the ascending aorta branches to the left. The arch of the aorta descends and ends at the level of the intervertebral disc between the fourth and fifth thoracic vertebrae.

- The *thoracic aorta* is the section of the aorta between the arch of the aorta and the diaphragm.

- The *abdominal aorta* is the section of the aorta between the diaphragm and the common iliac arteries, which carry blood to the lower limbs.

Structure of an artery • Figure 11.11

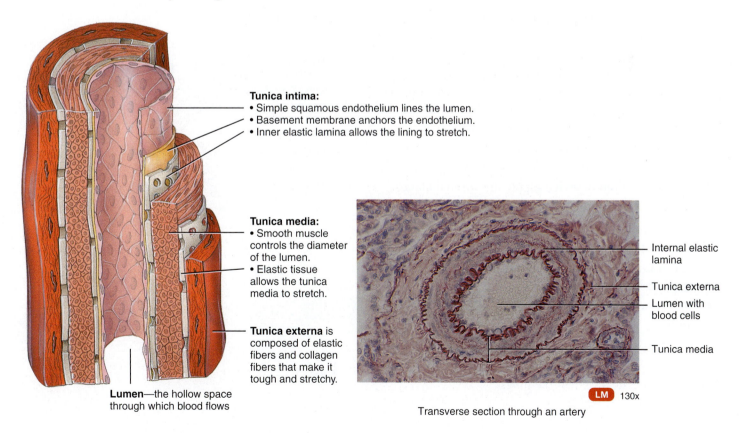

Tunica intima:
- Simple squamous endothelium lines the lumen.
- Basement membrane anchors the endothelium.
- Inner elastic lamina allows the lining to stretch.

Tunica media:
- Smooth muscle controls the diameter of the lumen.
- Elastic tissue allows the tunica media to stretch.

Tunica externa is composed of elastic fibers and collagen fibers that make it tough and stretchy.

Lumen—the hollow space through which blood flows

Internal elastic lamina

Tunica externa

Lumen with blood cells

Tunica media

LM 130x

Transverse section through an artery

The major arteries of your body branch off from the aorta and lead to the head, arms, and trunk. The remaining arteries branch off from there.

The ascending aorta branches into the right and left coronary arteries, which supply the heart.

The arch of the aorta branches into three divisions:

- The **brachiocephalic trunk** (brāke̅-ō-se-FAL-ik; *brachio* = arm; *-cephalic* = head) leads to the right common carotid artery, which supplies the right side of the head and neck, and to the **right subclavian artery**, which supplies the right upper limb.
- The **left common carotid artery** supplies the left side of the head and neck.
- The **left subclavian artery** supplies the left upper limb.

The thoracic aorta has four branches:

- The **bronchial arteries** supply the bronchi of the lungs.
- The **esophageal arteries** supply the esophagus.
- The **posterior intercostal arteries** supply the intercostal and chest muscles.
- The **superior phrenic arteries** supply the superior and posterior surfaces of the diaphragm.

The abdominal aorta branches into eight divisions:

- The **inferior phrenic arteries** supply the inferior surface of the diaphragm.
- The **celiac trunk** leads to the common hepatic artery, supplying the liver, stomach, duodenum, and pancreas. It also leads to the left gastric artery, which supplies the stomach and esophagus, and to the splenic artery, which supplies the spleen, pancreas, and stomach.
- The **superior mesenteric artery** supplies the small intestine, cecum, parts of the colon, and pancreas.
- The **suprarenal arteries** supply the adrenal glands.
- The **renal arteries** supply the kidneys.
- The **gonadal arteries** supply the testes (males) and ovaries (female).
- The **inferior mesenteric artery** supplies the rectum and parts of the colon.
- The **common iliac arteries** carry blood to the lower limbs.

Within the organs the arteries divide into arterioles and then into capillaries that service the systemic tissues (all tissues except the alveoli of the lungs).

InSight

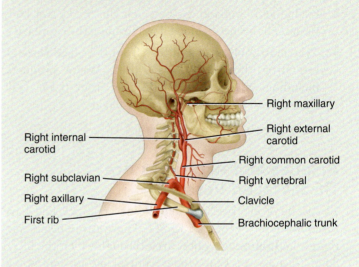

a. Right lateral view of branches of brachiocephalic trunk in neck and head

Right internal carotid
Right subclavian
Right axillary
First rib
Right maxillary
Right external carotid
Right common carotid
Right vertebral
Clavicle
Brachiocephalic trunk

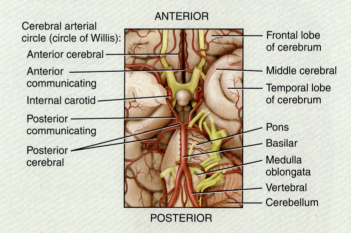

b. Inferior view of base of brain showing cerebral arterial circle

ANTERIOR

Cerebral arterial circle (circle of Willis):
Anterior cerebral
Anterior communicating
Internal carotid
Posterior communicating
Posterior cerebral

Frontal lobe of cerebrum
Middle cerebral
Temporal lobe of cerebrum
Pons
Basilar
Medulla oblongata
Vertebral
Cerebellum

POSTERIOR

Major arteries in the systemic circulation • Figure 11.12

c. Overall anterior view of the principal branches of the aorta

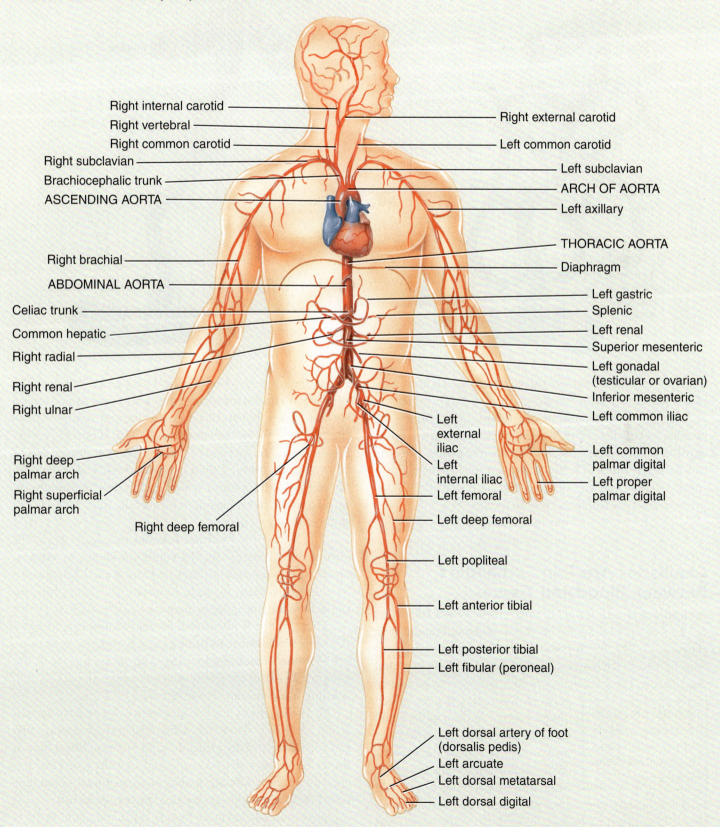

Right internal carotid
Right vertebral
Right common carotid
Right subclavian
Brachiocephalic trunk
ASCENDING AORTA

Right external carotid
Left common carotid
Left subclavian
ARCH OF AORTA
Left axillary

Right brachial
ABDOMINAL AORTA

THORACIC AORTA
Diaphragm

Celiac trunk
Common hepatic
Right radial
Right renal
Right ulnar

Left gastric
Splenic
Left renal
Superior mesenteric
Left gonadal
(testicular or ovarian)
Inferior mesenteric
Left common iliac

Right deep
palmar arch
Right superficial
palmar arch

Left
external
iliac
Left
internal iliac
Left femoral

Left common
palmar digital
Left proper
palmar digital

Right deep femoral

Left deep femoral

Left popliteal

Left anterior tibial

Left posterior tibial
Left fibular (peroneal)

Left dorsal artery of foot
(dorsalis pedis)
Left arcuate
Left dorsal metatarsal
Left dorsal digital

a. Sphincters relaxed: blood flowing through capillaries

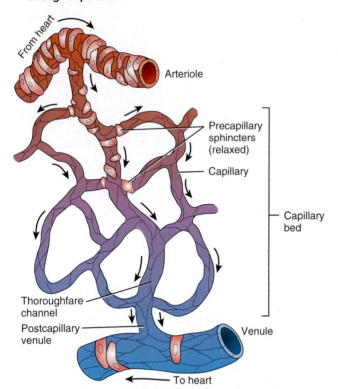

b. Sphincters contracted: blood flowing through thoroughfare channel

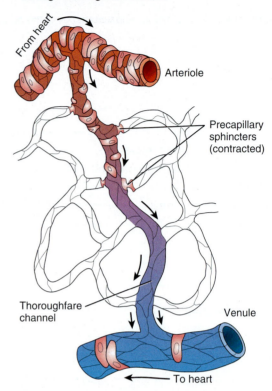

Arterioles are smaller, thick-walled vessels that vary in size from 15 µm to 300 µm in diameter. They have an endothelial layer covered by one or two layers of smooth muscle. At the distal end, the smooth muscle forms a **precapillary sphincter**, which controls blood flow into the capillaries (**Figure 11.13**). Like arteries, arteriolar vasoconstriction and vasodilation can help maintain blood pressure and control blood flow to organs.

Next, let's take a closer look at capillaries.

Capillaries Are Thin Exchangers Between Blood and Tissues

A **capillary** is a thin-walled, microscopic vessel consisting of only an endothelial layer and a basement membrane (**Figure 11.14**). Blood flows slowly through these narrow vessels, where exchange of gases, nutrients, wastes, and fluid occurs. The exchange of fluid is governed by a balance between *hydrostatic pressure*, which drives fluid out of the capillary, and **osmotic pressure**, which draws it back in. Nutrient-rich fluid moves out of the capillary via filtration, while waste-laden fluid is returned to the capillary, using absorption. In the arteriolar end, there is net filtration because blood hydrostatic pressure is

> **osmotic pressure**
> The pressure of a fluid due to its solute concentration, mostly the protein content.

the dominant force. However, as blood moves through the capillary, hydrostatic pressure diminishes and osmotic pressure increases due to the decrease in fluid volume within the capillary. So, in the venule end, there is net absorption because osmotic pressure is the dominant force. About 85–90% of the filtered fluid gets absorbed. The remaining filtered fluid drains into lymph capillaries, eventually returning to the blood plasma. The overall effect is that blood volume does not change.

The interstitial fluid also remains relatively consistent during this exchange. In the condition known as **edema**, however, the interstitial fluid level increases. Edema can be caused by many factors, but they generally fall into the following categories: increased hydrostatic pressure (causing too much filtration), decreased osmotic pressure (causing too little absorption), or a lymphatic system issue (leading to inability to return the excess filtrate to the system). Because excess fluid is left in the tissues, visible swelling is a common sign of edema.

At the capillary level, local conditions regulate vasodilation and vasoconstriction. Chemicals released from cells can vasodilate nearby arterioles and increase blood flow to the cells. Vasoconstrictors released by cells can have the opposite effects. The ability of tissues to adjust blood flow based on local conditions is called **autoregulation**.

Now let's take a closer look at veins and venules.

Structure and function of capillaries • Figure 11.14

a. Structure of capillaries

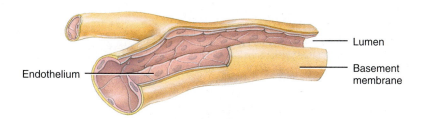

The thin capillary walls allow gases and other substances to pass through by diffusion.

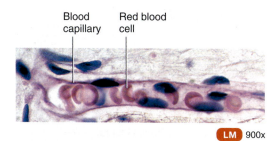

Capillaries are narrow (5–10μm in diameter). Blood flows through them slowly and blood cells move through them single file.

b. Capillary exchange

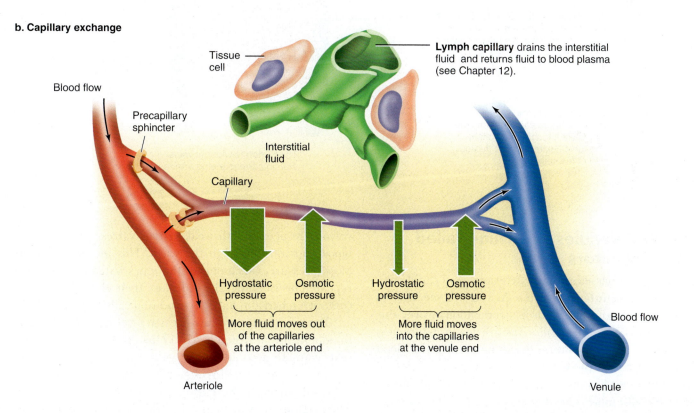

Lymph capillary drains the interstitial fluid and returns fluid to blood plasma (see Chapter 12).

In the capillaries, exchange of substances occurs as follows:
• Red blood cells exchange oxygen for carbon dioxide.
• Nutrients diffuse out of the blood into the interstitial fluid and into cells.
• Wastes diffuse from the cells into the interstitial fluid and into the capillaries
• Fluid moves out of the capillary at the arteriole end (net filtration) and into the capillary at the venule end (net absorption). The direction of net fluid movement is a balance between hydrostatic pressure and osmotic pressure.
• Interstitial fluid drains into lymph capillaries, which ultimately return the fluid to the blood plasma.

Although veins have a structure similar to arteries, they have larger diameters, thinner walls, and less smooth muscle and elastic tissue.

a. Structure of a vein

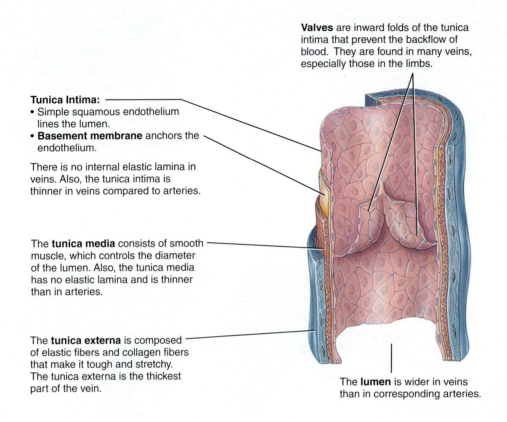

Valves are inward folds of the tunica intima that prevent the backflow of blood. They are found in many veins, especially those in the limbs.

Tunica Intima:
• Simple squamous endothelium lines the lumen.
• **Basement membrane** anchors the endothelium.

There is no internal elastic lamina in veins. Also, the tunica intima is thinner in veins compared to arteries.

The **tunica media** consists of smooth muscle, which controls the diameter of the lumen. Also, the tunica media has no elastic lamina and is thinner than in arteries.

The **tunica externa** is composed of elastic fibers and collagen fibers that make it tough and stretchy. The tunica externa is the thickest part of the vein.

The **lumen** is wider in veins than in corresponding arteries.

Veins and Venules Are Thin-Walled Blood Collectors

Once blood has passed through arterioles and capillaries, it reaches the **venules** and veins. As you will see later in this chapter, blood pressure drops tremendously over this distance. So the venules and veins are low-pressure vessels. This difference in pressure can be seen in the blood leaving a cut vessel. Blood flows from a cut vein slowly and evenly, but it gushes out of a cut artery in rapid spurts. When a blood sample is needed, it is usually collected from a vein because pressure is low in veins and because veins are close to the skin surface.

Structurally, veins and venules are similar to their arterial counterparts, with three layers, or tunics, but they have thinner walls and larger lumens (**Figure 11.15a**). These features make veins very expandable. Many veins have valves in them to prevent blood from flowing backward. All veins of the lower limbs have valves, which are more numerous than in veins of the upper limbs. (Arteries do not have

valves.) In people with weak venous valves, gravity forces blood backward through the valve. This increases venous blood pressure, which pushes the vein's wall outward. After repeated overloading, the walls lose their elasticity and become stretched and flabby, a condition called *varicose veins*.

Because veins are low-pressure vessels, blood tends to pool in them, which is why veins are the largest blood reservoir. To get adequate volumes of blood flowing back to the heart, veins depend on the continuous flow of blood through the cardiovascular system caused by contractions of the heart. In addition, contractions of the skeletal muscles and respiratory muscles help out by directing the blood in the veins back to the heart.

Let's discuss these two pumps in more detail.

The **skeletal muscle pump** acts in the following way to move blood back to the heart.

1. While you are standing at rest, as in **Figure 11.15b**, both the venous valve closer to the heart and the one farthest from the heart in this part of the leg are open,

b. Mechanisms of venous return

- Contractions of the heart pump blood through the entire cardiovascular system, including the veins.
- Contractions of skeletal muscles squeeze veins and move blood along, thereby acting as a **skeletal muscle pump** (shown here).
- Much like the skeletal muscle pump, pressure changes in the thoracic and abdominal cavities during breathing squeeze the abdominal veins and move blood through them to create a **respiratory pump**.

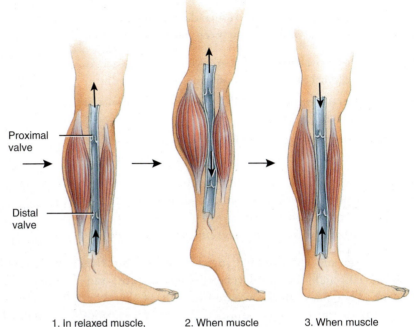

Proximal valve

Distal valve

1. In relaxed muscle, both proximal and distal valves are open and blood flows upward.

2. When muscle contracts, the vein gets compressed.
   - The proximal valve opens and blood gets squeezed upward.
   - The distal valve closes and pressure builds behind it.

3. When muscle relaxes again, the vein gets compressed.
   - Pressure behind the proximal valve drops and the valve closes.
   - Increased pressure behind the distal valve opens it and blood flows back into it.

and blood flows upward toward the heart. However, the pressure is barely enough to overcome the force of gravity pushing the blood back down.

2. Contraction of leg muscles, such as when you take a step, compresses the vein, which pushes the blood through the valve closer to the heart. At the same time, the valve farther from the heart in the uncompressed segment of the vein closes, as some blood is pushed against it. People who are immobilized through injury or disease lack these contractions of leg muscles. As a result, their venous return is slower and they may develop circulation problems.

3. Finally, just after muscle relaxation, pressure falls in the previously compressed section of vein, which causes the valve closer to the heart to close. The valve farther from the heart now opens because blood pressure in the foot is higher than in the leg, and the vein fills with blood from the foot.

The **respiratory pump** is also based on alternating compression and decompression of veins. During inhalation (breathing in), the diaphragm moves downward, which decreases the pressure in the thoracic cavity and increases the pressure in the abdominal cavity. As a result, abdominal veins are compressed, and a greater volume of blood moves from the compressed abdominal veins into the thoracic veins and then into the right atrium. When the pressure reverses during exhalation (breathing out), the valves in the veins prevent backflow of blood from the thoracic veins into the abdominal veins.

Veins may be superficial or deep. (Arteries are all deep below the skin.) You can see the superficial veins as they are located just beneath the skin; these are important as sites for withdrawing blood or giving injections. Deep veins generally travel alongside arteries and usually bear the same name.

The veins, which form from merging venules, drain the head, limbs, and trunk. The major systemic veins are shown in **Figure 11.16**. Although only one systemic artery, the aorta, takes oxygenated blood away from the heart (left ventricle), three systemic veins return deoxygenated blood to the heart (right atrium). These three veins are as follows.

- The **coronary sinus** is the main vein of the heart; it receives almost all venous blood from the myocardium.
- The **superior vena cava (SVC)** drains the head, neck, chest, and upper limbs.
- The **inferior vena cava (IVC)** is the largest vein in the body and drains the abdomen, pelvis, and lower limbs.

We'll start by identifying the veins that drain into the SVC from the head and neck. Keep in mind that these veins and the others we will identify are named in a distal-to-proximal direction, the same direction as the blood flows.

- The right and left **internal jugular veins** drain the brain (through the dural venous sinuses), face, and neck. They pass inferiorly on either side of the neck lateral to the internal carotid and common carotid arteries. They then unite with the subclavian veins to form the right and left **brachiocephalic veins**.
- The right and left **external jugular veins** empty into the subclavian veins and drain the scalp and superficial and deep regions of the face.
- The right and left **vertebral veins** empty into the brachiocephalic veins in the neck. They drain deep structures in the neck such as the cervical vertebrae, cervical spinal cord, and some neck muscles.

Blood from the upper limbs is returned to the SVC by both superficial and deep veins. The principal superficial veins that drain the upper limbs originate in the hand and convey blood from the smaller superficial veins into the axillary veins.

- The **cephalic veins** begin on the venous networks of the hands that drain the fingers. The cephalic veins drain blood from the lateral aspect of the upper limbs.
- The **basilic veins** begin on the medial aspects of the hands and drain blood from the medial aspects of the upper limbs. Anterior to the elbow, the basilic veins are connected to the cephalic veins by the **median cubital veins** (*cubitus* = elbow), which drain the forearm. If a vein must be punctured for an injection, transfusion, or removal of a blood sample, the median cubital vein is preferred. The basilic veins ascend until they merge with the brachial veins in the axillary area to form the axillary veins.
- The **median antebrachial veins** begin in the veins on the palms and ascend in the forearms to join the basilic or median cubital veins, sometimes both. They drain the palms and forearms.

Now to the deep veins that drain the upper limbs.

- The paired **radial veins** drain the lateral aspects of the forearms and pass alongside each radial artery. Just below the elbow, the radial veins unite with the ulnar veins to form the brachial veins.
- The paired **ulnar veins** drain the medial aspect of the forearms, pass alongside each ulnar artery, and join with the radial veins to form the brachial veins.
- The paired **brachial veins** accompany the brachial arteries. They drain the forearms, elbow joints, and arms. They join with the basilic veins to form the axillary veins.
- The **axillary veins** ascend to become the subclavian veins. They drain the arms, axillae, and upper part of the chest wall.
- The **subclavian veins** drain the arms, neck, and thoracic wall. They are continuations of the axillary veins that unite with the internal jugular veins to form the brachiocephalic veins. The brachiocephalic veins unite to form the SVC.

Now we turn to the veins that drain the lower limbs into the inferior vena cava. Again, we'll start with the superficial veins, which originate in the foot.

- The **great saphenous veins**, the longest veins in the body, begin at the medial side of the **dorsal venous arches** (VĒ-nus) of the foot, networks of veins on the top of the foot that collect blood from the toes. The great saphenous veins empty into the femoral veins and drain the leg and thigh, the groin, external genitals, and abdominal wall. Along their length, they have 10 to 20 valves, with more in the leg than the thigh. The great saphenous veins are often used for prolonged administration of intravenous fluids and as a source of vascular grafts, especially for coronary bypass surgery.
- The **small saphenous veins** begin at the lateral side of the dorsal venous arches of the foot. They drain the foot and leg and empty into the popliteal veins behind the knee. Along their length, they have 9 to 12 valves.

Now to the deep veins that drain the lower limbs.

- The paired **posterior tibial veins** drain the foot and posterior leg muscles. The paired **anterior tibial veins** drain the ankle joint, knee joint, tibiofibular joint, and anterior portion of the leg and unite with the posterior tibial veins to form the popliteal vein.
- The **popliteal veins**, formed by the union of the anterior and posterior tibial veins, drain the skin, muscles, and bones of the knee joint.

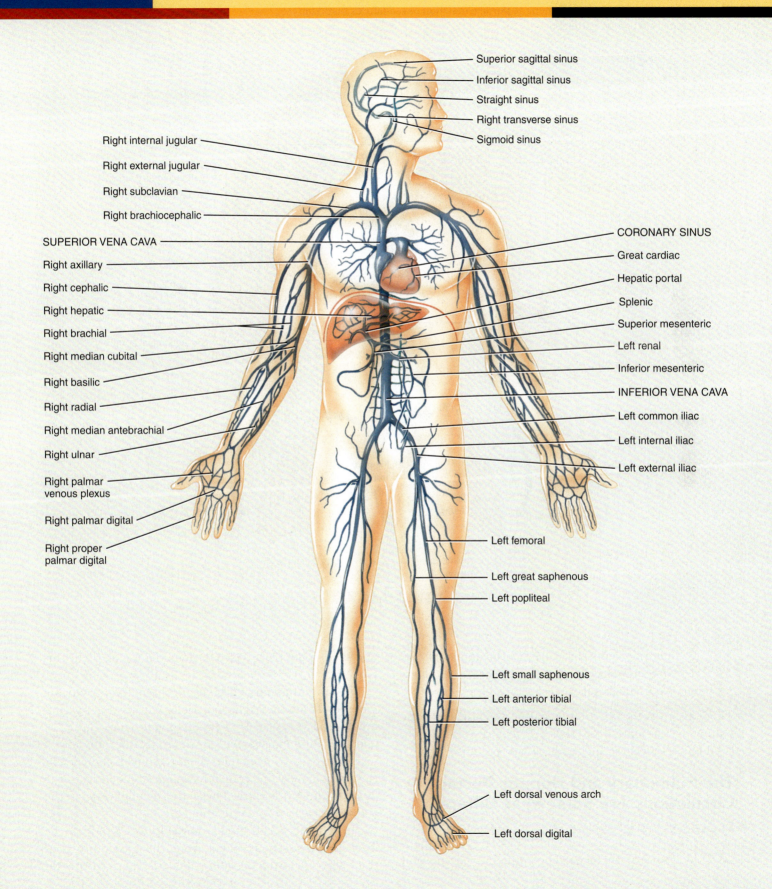

- Superior sagittal sinus
- Inferior sagittal sinus
- Straight sinus
- Right transverse sinus
- Sigmoid sinus

Right internal jugular
Right external jugular
Right subclavian
Right brachiocephalic
SUPERIOR VENA CAVA
Right axillary
Right cephalic
Right hepatic
Right brachial
Right median cubital
Right basilic
Right radial
Right median antebrachial
Right ulnar
Right palmar venous plexus
Right palmar digital
Right proper palmar digital

CORONARY SINUS
Great cardiac
Hepatic portal
Splenic
Superior mesenteric
Left renal
Inferior mesenteric
INFERIOR VENA CAVA
Left common iliac
Left internal iliac
Left external iliac

Left femoral
Left great saphenous
Left popliteal

Left small saphenous
Left anterior tibial
Left posterior tibial

Left dorsal venous arch
Left dorsal digital

Blood Vessels Are the Body's Plumbing **329**

The pulmonary circulation and the hepatic portal circulation have features that differ from systemic circulation patterns.

a. Pulmonary circulation

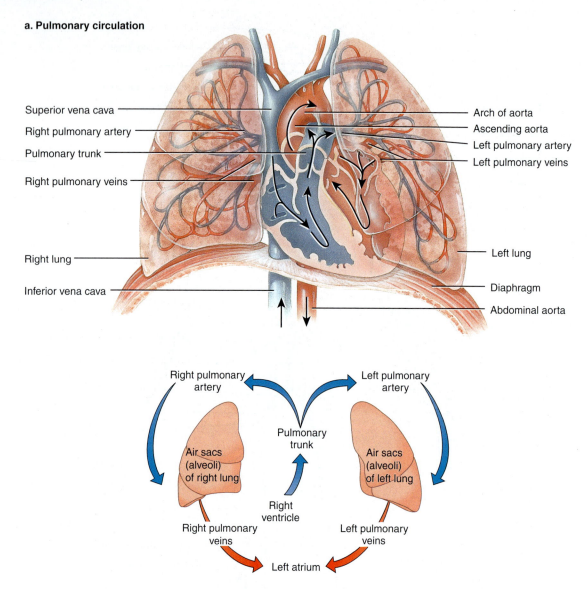

- The **femoral veins** are the continuations of the popliteal veins. They drain the muscles of the thigh, femur, external genitals, and superficial lymph nodes.

There are two other circulation pathways in the body: the pulmonary circulation and the hepatic portal circulation. Let's take a closer look at these two systems.

The Pulmonary and Hepatic Portal Circulations Are Somewhat Different

In the systemic circulation, arteries carry oxygenated blood away from the heart, and veins carry deoxygenated blood toward the heart. The pulmonary circulation and the hepatic portal circulation do not exactly fit this pattern.

In the pulmonary circulation, the pulmonary trunk emerges from the right ventricle, carrying deoxygenated blood (**Figure 11.17a**). It divides into the right and left pulmonary arteries, which are the only arteries in the body to carry deoxygenated blood. Conversely, the pulmonary veins return oxygenated blood from the lungs to the left atrium. These are the only veins in the body that carry oxygenated blood.

In hepatic portal circulation (**Figure 11.17b**), the liver receives venous blood from the gastrointestinal tract through a **portal vein** called the **hepatic portal**

b. Hepatic portal circulation

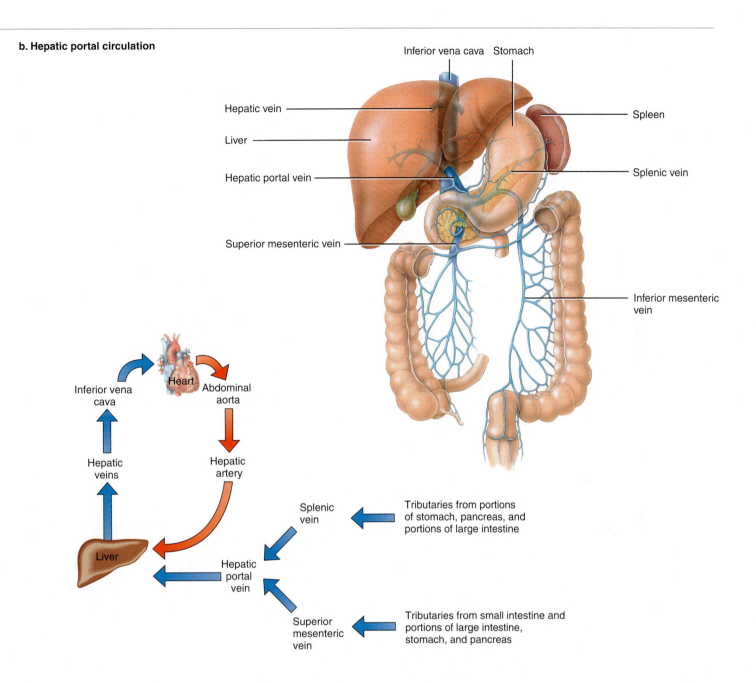

portal vein A vein that carries blood from one capillary network to another.

vein. This is an exception to the general principle that venous blood returns directly to the heart. However, because this venous blood is rich in substances absorbed from the gastrointestinal tract, this setup allows the liver to process the absorbed substances before they pass into the general circulation. This allows nutrient levels to be relatively stable throughout the body. The liver also receives oxygenated blood from the hepatic artery. Within the liver, oxygenated blood mixes with portal venous blood, and all venous blood exits via the hepatic veins, where the blood enters the general circulation.

CONCEPT CHECK STOP

1. **What** blood vessel structure is unique to veins?
2. **What** forces drive the exchange of fluid across the capillary wall?
3. **How** does the skeletal muscle pump work?
4. **How** is the pulmonary circulation different from the systemic circulation?

Maintaining Blood Pressure Is Critical for Survival

LEARNING OBJECTIVES

1. **Identify** the changes in blood pressure as blood flows through the cardiovascular system.
2. **Describe** the factors that influence blood pressure.
3. **Explain** how blood pressure is regulated.

 n many ways, the cardiovascular system is much like a water balloon:

- Water exerts outward pressure on the balloon's thin elastic walls just as the blood exerts pressure on the walls of the blood vessels.
- If you increase the volume of water in the balloon, you increase the pressure. Likewise, increasing blood volume increases blood pressure.
- If you squeeze the balloon, you increase the pressure inside. When the heart contracts, the fluid pressure inside increases.
- If you squeeze one end of the balloon (increase pressure), the water flows to the other side (low pressure). The heart increases the pressure at the beginning of the circulation, and the blood flows through the systemic and pulmonary circulations.
- If you constrict the balloon in the middle (increase resistance) and then squeeze one end, you have to squeeze harder (more pressure) to get the water to move than without the constriction. Similarly, narrowing the arteries increases the resistance to flow and yields higher pressure.

In addition, the amount of fluid that you put into the water balloon affects pressure, just as varying cardiac output alters the pressure within the vessels. Dilation of arterioles and precapillary sphincters allows fluid to move more quickly toward the venous circulation, also changing the pressure of the blood remaining in the arterial circulation. So blood flow in the cardiovascular system is subject to the same physical principles as the flow of any other fluid, or the flow of water in a balloon. Let's look at the dynamics of blood flow.

Pressure Drives the Flow of Blood

Like any other fluid, blood flows from areas of higher pressure to areas of lower pressure. Whether it's water in plumbing or blood in vessels, there is always some resistance to flow (such as friction of the fluid against the walls or width of the lumen). The general relationship describing blood flow is as follows:

$$\text{Blood flow} = \frac{\text{Change in blood pressure}}{\text{Vascular resistance}}$$

So the heart does not "push" the blood through the cardiovascular system but increases the pressure of the blood. The blood then flows according to the principle described above (**Figure 11.18**). The heart can also be compared to the chain that pulls a roller coaster car (stroke volume of blood) to the top of the first hill (aorta); once it reaches the top, the car proceeds through the rest of the ride on its own until it returns to the starting point (right atrium).

In the arteries, blood pressure fluctuates periodically with the systole and diastole phases of the cardiac cycle. The normal systolic pressure (SP) is about 120 mm Hg, and normal diastolic pressure (DP) is about 80 mm Hg. (Blood pressure is usually written as SP/DP and expressed as "systolic over diastolic," such as 120/80, or "120 over 80.") Often, this fluctuating pressure is expressed as **mean arterial pressure (MAP)**, which is calculated as follows: $\text{MAP} = \frac{1}{3}(\text{SP} - \text{DP}) + \text{DP}$. This calculation provides an "average" pressure at any given time and takes into consideration the fact that the heart remains at rest for a longer period of time than it is contracted.

Because the blood encounters narrower and narrower vessels as it proceeds through the circulation, the resistance increases, thereby causing blood pressure to drop along the way. In addition, the elasticity of the arteries causes the blood pressure to change to a more steady, non-fluctuating pressure, no longer showing systolic and diastolic pressure variations.

For the most part, the blood flow through the cardiovascular system (cardiac output) remains consistent. Therefore, the blood pressure remains steady. However, many factors can influence blood pressure. Let's take a closer look at a few of these.

Many Factors Influence Blood Pressure

Blood pressure is focused on the pressures within the arteries. Many factors, including the following, increase blood pressure:

- *Increased force of contraction (contractility).* Norepinephrine from sympathetic nerves to the heart and epinephrine

Changes in pressure as blood flows through the cardiovascular system • Figure 11.18 _____

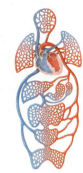

Systemic circulation

Like any other fluid, blood flows according to changes in pressure or pressure gradients. The blood vessels offer resistance to that flow, so pressure drops throughout the system. In addition, both pressure and resistance are affected by many factors.

How blood flows in the systemic circulation:
- The blood flow through the system is constant (steady state).
- Change in pressure (ΔP) drives the flow. (ΔP is directly proportional to flow.) So anything that increases blood pressure increases the flow.
- Vascular resistance (R) impedes the flow (R is inversely proportional to flow). For example, if R is increased, then flow is decreased.

What the heart does:
1. Takes in a small volume of blood during diastole
2. Applies a force (contraction) to the blood that raises its pressure during systole
3. Releases the blood into the systemic circulation (via the aorta)

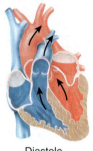

Diastole   Systole

Heart activity sets up the initial blood pressures at a steady flow rate (cardiac output, CO).

Elastic arteries have low R, so the systolic and diastolic pressure variations are maintained.

Arterioles, capillaries and venules have high R (narrow, rigid) so blood pressure drops and changes from fluctuating to steady.

Veins have low R (large diameters), so blood pressure drops only a little.

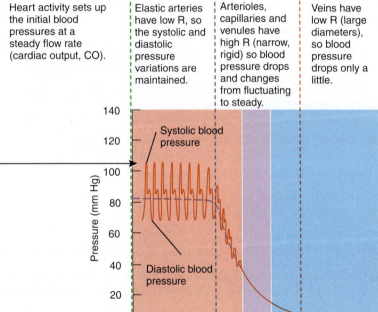

from the adrenal medullae directly increase the heart's contractility. Certain drugs, such as digitalis, also increase contractility.

- *Increased blood volume.* More blood exerts more pressure (like more water in the balloon). Also, increased blood volume increases the stroke volume. The increased stroke volume stretches the heart more, which generates a higher force of contraction and higher blood pressure.

- *Increased cardiac output.* In the cardiovascular system, venous return is equal to cardiac output. So, like an increase in blood volume, increasing the cardiac output returns more blood to the heart and increases blood pressure.

- *Increased vascular resistance.* Increased venous resistance prevents blood from moving into the capillaries, holding more blood in the artery and increasing the pressure. Several factors influence vascular resistance, including size of the lumen, viscosity, and total length of blood vessels. The size of the lumen may be affected by vasoconstriction or vasodilation resulting from autonomic nervous system activity. Hormones and drugs may also affect the size of the lumen. The thickness (or viscosity) of the blood may be affected by blood cell count, dehydration, or plasma protein levels. The total length of the blood vessels may vary, depending on the amount of fat.

The many factors that can contribute to increasing blood pressure are summarized in **Figure 11.19**. Addressing many of these factors can result in a decrease in blood pressure. Now that you understand the factors that can influence blood pressure, let's look at how it is regulated.

Blood Pressure Is Closely Regulated

To keep blood flowing at a constant rate, blood pressure must be kept at a steady level. This is especially important for the blood flowing against gravity to the brain. Your body has a neural mechanism and several hormonal mechanisms that regulate blood pressure. The neural mechanism involves the **cardiac** and **vasomotor centers** in the medulla oblongata. These centers receive and integrate input from several areas:

- **Baroreceptors**. Located in the carotid arteries, right atrium, and aortic arch; sense the stretch of these arteries due to the pressure of blood in the vessels and send a steady stream of nerve impulses to the cardiac and vasomotor centers. The frequency of nerve impulses is directly proportional to the pressure in these vessels.
- *Proprioceptors*. Located in limbs and joints; sense movements (such as those related to exercise).

> **baroreceptors**
> (bar'-ō-re-SEP-tors)
> Neurons capable of responding to changes in blood, air, or fluid pressure.

- *Chemoreceptors*. Located in the carotid arteries and aortic arch; sense chemicals in blood, such as the pH of blood and its levels of oxygen and carbon dioxide.
- *Higher brain centers*. Cortex, limbic system, hypothalamus; also integrate information from various sensory neurons and send commands to direct the cardiac center to alter heart rate and/or contractility and the vasomotor center to adjust vasoconstriction or vasodilation.

The cardiac and vasomotor centers integrate this information and send appropriate signals through the vagus nerve (parasympathetic nerve), the cardiac accelerator nerve to the heart (sympathetic nerve), or vasomotor nerves to the blood vessels (sympathetic nerves). Parasympathetic stimulation of the heart decreases heart rate, while sympathetic outflow causes increases in both heart rate and contractility. Sympathetic stimulation of the blood vessels causes vasoconstriction. So, parasympathetic stimulation tends to decrease cardiac output and blood pressure, and sympathetic stimulation has the opposite effect.

The neural regulatory system is known as the **baroreceptor reflex**. For example, when you stand up, gravity pulls blood away from the head and causes a drop in blood pressure. The baroreceptor reflex kicks in and restores blood pressure by increasing the heart rate, contractility, and blood pressure within a heart-

Summary of factors that increase blood pressure • Figure 11.19

Changes within green boxes increase cardiac output; changes within blue boxes increase systemic vascular resistance. Opposite changes lead to decreased blood pressure.

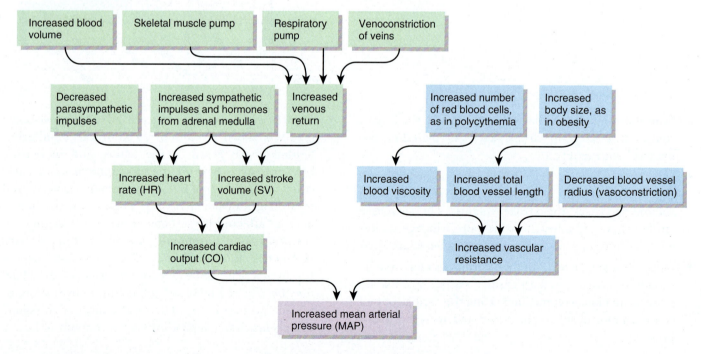

334 CHAPTER 11 The Cardiovascular System: Heart, Blood Vessels, and Circulation

WHAT A HEALTH PROVIDER SEES

Shock

Shock is a failure of the cardiovascular system to deliver enough oxygen and nutrients to meet cellular metabolic needs. The causes of shock are many and varied, but all are characterized by inadequate blood flow to body tissues. Here are some of the most common varieties:

- *Hypovolemic shock* is the result of excessive fluid loss, resulting in a decrease in the blood volume. Causes of hypovolemic shock include hemorrhage, dehydration, burns, excessive vomiting, persistent diarrhea, and excessive sweating.

- *Cardiogenic shock* is due to ineffective pumping of the ventricles of the heart, which reduces cardiac output.

- *Septic shock* is caused by an infection that has gotten into the bloodstream. As the body reacts to the infection, chemicals are released that cause vasodilation.

- *Anaphylactic shock* happens when there is an allergic reaction in which lots of histamine is released, causing widespread vasodilation.

- *Neurogenic shock* is the result of injury to the nervous system that interferes with sympathetic nervous system control of vessel size.

Despite multiple responses by the body to restore blood pressure, including the neurologic and hormonal responses discussed previously, with shock the body is unable to restore the blood pressure to a normal level. If shock persists, cells and organs become damaged, and cells may die unless proper treatment begins quickly.

The signs and symptoms of shock include the following, which may vary with the severity of the condition:

- Systolic blood pressure is lower than 90 mmHg.

- Resting heart rate is rapid due to sympathetic stimulation and increased blood levels of epinephrine and norepinephrine.

- Pulse is weak and rapid due to reduced cardiac output and fast heart rate.

- Skin is cool, pale, and clammy due to sympathetic constriction of skin blood vessels and sympathetic stimulation of sweating.

- Mental state is altered due to reduced oxygen supply to the brain.

- Urine formation is reduced due to increased levels of aldosterone and antidiuretic hormone (ADH).

- The person is thirsty due to loss of extracellular fluid.

- The pH of blood is low (acidosis) due to buildup of lactic acid.

- The person may have nausea because of impaired blood flow to the digestive organs from sympathetic vasoconstriction.

Immediate treatment includes keeping the patient warm with a blanket, elevating the feet above the heart and head to improve cardiac output, and administering intravenous fluid. It may be necessary to give medicines to increase blood pressure and cardiac output (such as epinephrine). Ultimately, blood volume must be restored by repairing the injury, administering fluids, or transfusing blood. If the condition is not treated quickly, death may ensue.

Think Critically 1. Explain how elevating the feet above the heart and head increases cardiac output in a patient suffering from hypovolemic shock.
2. Many people try to administer fluids by mouth to hypovolemic shock patients. This is not a recommended or advisable treatment. Why would this method of restoring fluid volume be ineffective?

---

beat or two. This important reflex prevents you from fainting every time you stand up. It also helps maintain blood pressure in other situations (such as exercise and hemor-

rhagic shock, one of several types of shock characterized by inadequate flow of blood to body tissues—see *What a Health Provider Sees*).

Mechanisms that maintain blood pressure and increase blood flow during exercise • Figure 11.20

Several factors can adjust blood pressure to maintain homoeostasis.

Working muscles require increased blood flow to meet oxygen demands.

Your muscles, brain, glands, heart, sympathetic nervous system, and systemic circulation work together to deliver the energy needed for exercise and to maintain blood pressure.

Muscles	Heart	Brain	Sympathetic nervous system	Glands	Systemic circulation
	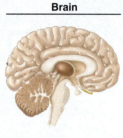		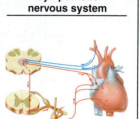		
Working muscles secrete metabolites that increase local vasodilation, which decreases blood pressure.	Baroreceptors in the heart sense the decrease in blood pressure caused by vasodilation and send a message to the brain.	Higher brain centers stimulate the ANS, including the CV center.	Increased sympathetic outflow to the cardiac accelerator nerve increases heart rate and contractility. Sympathetic outflow increases to the adrenal medullae and to vasomotor nerves supplying the blood vessels.	In response to increased sympathetic outflow to the adrenal medulla, the adrenal gland secretes epinephrine, which causes vasoconstriction.	Vasoconstriction caused by increased vasomotor nerve activity and increased epinephrine secretion results in increased blood volume in veins and venules, increased resistance in arteries and arterioles, and the diversion of blood flow from the digestive tract and skin to working muscles.

RESULTS: Cardiac output is increased in response to increased heart rate and increased blood volume.
Blood flow to working muscles increases dramatically.
Blood pressure remains the same or is elevated slightly.

Several hormones can be used to help maintain blood pressure, including epinephrine, norepinephrine, angiotensin II (AII), antidiuretic hormone (ADH), aldosterone, and atrial natriuretic peptide (ANP). Epinephrine, norepinephrine, and AII cause vasoconstriction and increase blood pressure. Aldosterone acts on the kidney to reabsorb sodium and water, thereby increasing the blood volume, cardiac output, and blood pressure. ADH impacts the kidneys to directly increase water reabsorption, leading to increased blood volume. In response to high blood pressure, the heart itself secretes ANP, which causes vasodilation, reduces blood volume, and lowers blood pressure.

To see how blood pressure is maintained under physiological stress, consider exercise (**Figure 11.20**). When you exercise, the working muscles use energy and generate metabolites that dilate the local arterioles (autoregulation). Local vasodilation increases tissue blood flow and decreases blood pressure, which causes the baroreceptor reflex to kick in to restore blood pressure. At the same time, the proprioceptors and higher brain centers signal the cardiac and vasomotor centers to increase sympathetic outflow to the heart, blood vessels, and adrenal medulla (which secretes epinephrine). All of these systems combine to increase heart rate, contractility, blood volume, cardiac output, and blood pressure. As a result, blood flow to working muscles is greatly increased to match the demand for oxygenated blood flow and to maintain normal or slightly elevated blood pressure. Once blood pressure is restored, the baroreceptor reflex shuts down, but the brain continues its increased sympathetic outflow to maintain the new steady state until you stop exercising.

hypertension (hī′-per-TEN-shun) Chronic high blood pressure.

Over 50 million Americans have **hypertension**, which can be caused by many factors, including heart disease, obesity, genetic factors, and kidney disease. Treatment includes lifestyle changes (such as exercise, weight loss, reduction of alcohol consumption and sodium intake) and medications (such as beta blockers, angiotensin-converting enzyme [ACE] inhibitors, and calcium channel blockers). The different drugs act using different mechanisms to reduce heart rate, contractility, and/or vasoconstriction so that blood pressure can be reduced. The consequences of hypertension include heart attack, heart failure, stroke, **glaucoma**, and kidney failure.

glaucoma (glaw-KŌ-ma) A progressive eye disorder in which there is increased intraocular pressure due to an excess of aqueous humor, resulting in irreversible loss of vision.

CONCEPT CHECK

1. **Where** does the greatest drop in blood pressure occur in the cardiovascular system?
2. **How** does angiotensin II alter blood pressure?
3. **How** does the baroreceptor reflex work?

THE PLANNER ✔

Summary

1 The Heart Pumps Blood Through Blood Vessels to All Tissues 310

- The **heart** is a four chambered pump that consists of two atria and two ventricles. Each chamber has an inner endocardium, a muscular **myocardium**, and an outer epicardium. The heart itself is covered in a tough protective sac called the **pericardium**. Several large vessels bring blood into the heart (inferior vena cava, superior vena cava, pulmonary vein) and take blood away (aorta, pulmonary artery).

- As shown, the right side of the heart collects oxygen-poor blood from the venous portion of the systemic circulation and pumps it to the pulmonary circulation. In contrast, the left side of the heart collects oxygen-rich blood from the pulmonary circulation and pumps it to the arterial portion of the systemic circulation. The four valves in the heart control the flow of blood between the atria and **ventricles** and between the ventricles and their associated arteries.

- The heart has an internal pacemaker, the sinoatrial node, which initiates regular waves of depolarization (contraction of cardiac muscle fibers) across the atria to the atrioventricular node. The atrioventricular node conducts the depolarization through a network of Purkinje fibers in the ventricles. This electrical conduction system coordinates contractions of both atria and subsequently both ventricles. The electrical activity of the heart

Overview of the cardiovascular system • Figure 11.1

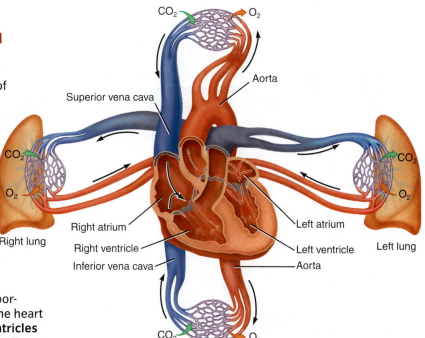

can be recorded by surface electrodes to produce an **electrocardiogram** (**EKG** or **ECG**). The EKG can be used diagnose abnormalities in the electrical conduction system (arrhythmias, blocks, and so on) and associated heart diseases.

- Each heartbeat recorded on an EKG is part of one **cardiac cycle**. The cardiac cycle consists of a relaxation phase (**diastole**) and a contraction phase (**systole**) for each of the heart chambers. During diastole, the chamber fills with blood. During systole, the atria contract first, followed by the ventricles. The electrical conduction system coordinates each contraction and relaxation phase. Specific pressure changes occur within each phase of the **cardiac cycle**.

- The work done by the heart, called the **cardiac output (CO)**, is the volume of blood pumped by the heart in one minute. Cardiac output is the product of the heart rate and stroke volume—factors affected by the brain's cardiovascular center working through the autonomic nervous system. Sympathetic outflow to the heart through the cardiac accelerator nerve increases the heart rate and force of contraction (contractility), while parasympathetic outflow through the vagus nerve decreases the heart rate. Sympathetic outflow to the blood vessels causes vasoconstriction, which increases venous return and stroke volume. These mechanisms strive to keep cardiac output in a steady state but also allow a shift from one steady state to another to adjust to changing physiological conditions.

2 Blood Vessels Are the Body's Plumbing 320

- Arteries and narrower **arterioles** are thick-walled vessels that have three layers: an inner endothelial layer, a middle smooth muscle layer, and an outer layer of elastic fibers. The smooth muscles of arteries and arterioles contract and relax to control the diameter of the opening (**lumen**). These ves-

How arterioles control blood flow into capillaries • Figure 11.13a

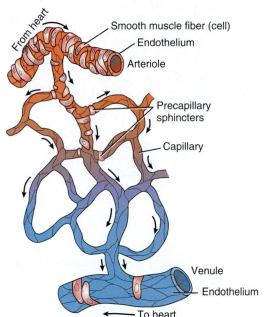

- sels, as shown, carry blood away from the heart and toward the tissues. **Precapillary sphincters** can restrict the blood flow into the tissue capillaries.

- **Capillaries** are thin-walled vessels where the exchange of gases, nutrients, and wastes occurs. Diffusion drives the movements of substances across the thin capillary walls. The movement of fluid is determined by the balance of hydrostatic pressure, which tends to move fluid out of the capillary (filtration), and osmotic pressure, which tends to move fluid into the capillary (absorption). At the arteriolar end, net fluid flow is out of the capillary, while at the venule end, net fluid flow is into the capillary. Excess filtrate will later be returned to the blood circulation via the lymphatic system.

- **Veins** and **venules** have structures similar to those of arteries and arterioles, but with larger lumens, thinner walls, and less smooth muscle. Many veins have valves in them that prevent the backflow of blood. These vessels are very elastic and hold almost two-thirds of the blood at any moment (blood reservoirs). Veins and venules carry blood from the tissues back to the heart. The pressure in the venous system is low, so skeletal muscles help pump venous blood back by squeezing the veins (skeletal muscle pump). Similarly, pressure changes in the abdominal cavity due to the breathing process help move venous blood along (respiratory pump).

- In the systemic circulation, oxygen-rich blood flows from the heart through arteries and arterioles to the tissue capillaries, and venules and veins carry deoxygenated blood back to the heart. However, there are two exceptions to this rule. In the pulmonary circulation, pulmonary arteries carry deoxygenated blood, and pulmonary veins carry oxygenated blood. In the hepatic portal circulation, blood comes into the liver from the veins of another vascular bed, the gastrointestinal tract, through the hepatic portal vein. This venous blood is rich in substances absorbed from the gut. The liver also receives oxygenated blood through the hepatic artery. All blood leaves the liver to join the vena cava through the hepatic vein.

3 Maintaining Blood Pressure Is Critical for Survival 332

- In the cardiovascular system, pressures drive blood through blood vessels that offer resistance to its flow, as shown. Because the cardiovascular system is closed, blood flow is constant, so the changes in pressure shown are directly related to changes in resistance. The heart increases the pressure in the ventricles on the blood, which then flows away.

- Many factors influence blood pressure by affecting the heart (altering contractility or heart rate), and/or the blood vessels (changing the resistance). These factors include alterations in blood volume and cardiac output, changes in blood vessel diameter, stimulation of sympathetic or parasympathetic nerve activity, release of various hormones, and administration of drugs.

- Blood pressure is maintained by neural and hormonal mechanisms. The neural mechanisms involve sensing changes in blood pressure (baroreceptors) and making adjustments. The **cardiac** and **vasomotor centers** alter heart rate, contractility, and blood vessel diameter through sympathetic or parasympathetic outflow. Various hormones affect blood pressure by causing increases in heart rate or contractility, vasoconstriction, vasodilation, or alterations in blood volume. These neurological and endocrine mechanisms interact to control blood pressure under changing physiological circumstances, such as exercise, trauma, or disease.

Changes in pressure as blood flows through the cardiovascular system • Figure 11.18

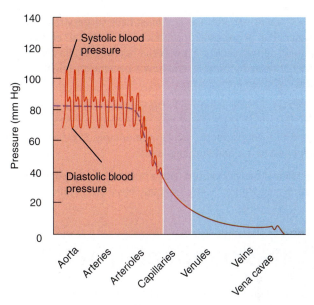

Key Terms

- anastomose 319
- anterior tibial veins 328
- aorta 312
- apex 311
- arrhythmia 316
- arteriole 321
- artery 312
- atherosclerosis 319
- atherosclerotic plaque 319
- atrial systole 316
- atrium 312
- autoregulation 324
- axillary veins 328
- baroreceptor 334
- baroreceptor reflex 334
- base 311
- basilic veins 328
- brachial veins 328
- brachiocephalic trunk 322
- brachiocephalic veins 328
- bronchial arteries 322
- capillary 324
- cardiac center 334
- cardiac conduction system 315
- cardiac cycle 316

- cardiac output (CO) 318
- cardiologist 316
- celiac trunk 322
- cephalic veins 328
- common iliac arteries 322
- coronary artery 312
- coronary sinus 312
- diastole 316
- dorsal venous arches 328
- edema 324
- electrocardiogram (EKG or ECG) 316
- esophageal arteries 322
- external jugular veins 328
- femoral vein 330
- glaucoma 337
- gonadal arteries 322
- great saphenous veins 328
- heart 310
- heart rate (HR) 318
- hepatic portal vein 330
- hypertension 337
- inferior mesenteric artery 322
- inferior phrenic arteries 322
- inverior vena cava (IVC) 328

- internal jugular veins 328
- left atrium 312
- left common carotid artery 322
- left subclavian artery 322
- left ventricle 312
- lumen 321
- mean arterial pressure (MAP) 332
- median antebrachial veins 328
- median cubital veins 328
- mediastinum 311
- myocardial infarction 319
- myocardial ischemia 319
- myocardium 312
- osmotic pressure 324
- pericardium 312
- popliteal vein 328
- portal vein 330
- posterior intercostal arteries 322
- posterior tibial veins 328
- precapillary sphincter 324
- radial veins 328
- renal arteries 322
- respiratory pump 327
- right atrium 312
- right subclavian artery 322

- right ventricle 312
- shock 335
- skeletal muscle pump 326
- small saphenous veins 328
- stroke volume (SV) 318
- subclavian veins 328
- superior mesenteric artery 322
- superior phrenic arteries 322
- superior vena cava (SVC) 328
- suprarenal arteries 322
- systole 316
- ulnar veins 328
- valve 312
- vasoconstriction 321
- vasodilation 321
- vasomotor center 334
- vein 312
- ventricle 312
- ventricular systole 316
- venule 326
- vertebral veins 328

Critical and Creative Thinking Questions

1. Juan is having arrhythmias and must have surgery to implant a pacemaker. A pacemaker produces an "artificial" electrical stimulus of muscle. Explain how the heart conducts electricity, what an arrhythmia is, and how a pacemaker might help a patient with an arrhythmia.

2. Anita had surgery in her neck. During the procedure, the sensory nerve from the baroreceptor was damaged. Now, every time Anita stands up, she faints momentarily. Explain what has happened to her and why she faints.

3. The Millers' baby was born with a hole in the interventricular septum (see Figure 11.4a). The baby has a blue skin color. Explain what's going on in this baby's heart to produce these symptoms and how this might be corrected.

4. Kim's doctor has prescribed a class of drugs called angiotensin-converting enzyme (ACE) inhibitors to treat her hypertension. ACE is used during the formation of angiotensin II (AII). How would ACE inhibitors help to lower blood pressure?

5. A cardiologist notices that the QRS complex of Cheryl's EKG is occasionally "missing" from the EKG pattern. What does the QRS complex represent, and what could be causing this unusual EKG pattern?

What is happening in this picture?

In adults of middle age and older, a heart attack or myocardial infarction typically occurs because a blood clot lodges in an artery of the heart already narrowed by atherosclerosis. During exercise, heart rate and blood pressure increase. Under this stress, an unstable plaque may rupture, stimulating the clotting process as the body tries to repair the damaged artery. Exercise stress tests are often performed to help diagnose coronary artery disease.

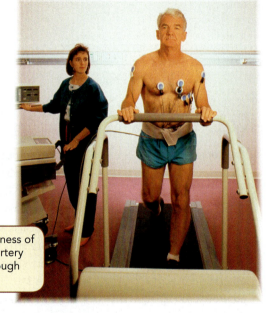

Think Critically Patients who experience chest pain and shortness of breath during exercise stress tests may be at risk for coronary artery disease. What do you think is happening to the blood flow through the heart that may cause such symptoms?

Self-Test

(Check your answers in Appendix C.)

1. A patient comes into the ER with a rapid pulse, mental disorientation, and cold, pale, sweaty skin. The patient is suffering from _____.

   a. heart attack   c. shock

   b. stroke   d. atherosclerosis

2. Which label indicates the chamber of the heart that pumps blood to the body?

   a. A   b. E   c. B   d. D

3. What is the function of the part of the heart that is labeled A?

   a. receives oxygenated blood from the lungs

   b. pumps deoxygenated blood to the lungs

   c. receives deoxygenated blood from the body

   d. pumps oxygenated blood to the body

Use this figure to answer questions 2–5.

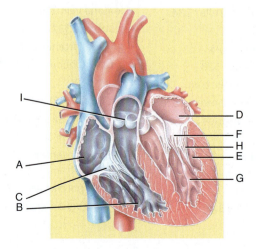

4. Which structure prevents backflow from the pulmonary circulation?

a. I   b. C   c. F   d. D

5. Which structure anchors the chordae tendineae to the papillary muscles??

a. H   b. C   c. E   d. G

6. In which branch of the systemic circulation does blood pressure drop the most?

a. arteries

b. veins

c. venules

d. arterioles

7. A subject takes a drug in clinical trials that lowers blood pressure, but his heart rate stays the same. Which of the following changes most likely explains the effects of the experimental drug?

a. The drug causes vasoconstriction in the systemic circulation.

b. The drug increases the stroke volume of the heart.

c. The drug causes vasodilation in the systemic circulation.

d. The drug increases sodium and water reabsorption in the kidney.

Use this figure to answer questions 8–10.

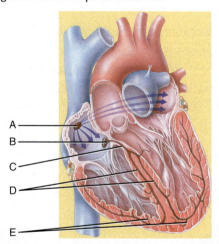

8. The signal for ventricular depolarization begins at which structure?

a. A   b. B   c. C   d. D

9. The function of the structures labeled E is which of the following?

a. originate the wave of depolarization in the ventricles

b. spread the wave of depolarization throughout the ventricles

c. originate the wave of depolarization in the atria

d. prevent the wave of depolarization from spreading backward in the ventricles

10. A heart block most commonly begins at the structure labeled _____.

a. B   b. A   c. D   d. E

11. What process occurs more at the venule end of a capillary than at the arteriole end?

a. oxygen diffusion into the capillary

b. fluid filtration

c. fluid absorption

d. oxygen diffusion out of the capillary

12. Which factor is most likely to cause a rise in blood pressure?

a. decreased stroke volume

b. vasoconstriction

c. vasodilation

d. decreased blood volume

13. Which of the following hormones would lower blood pressure?

a. AII   c. epinephrine

b. ADH   d. ANP

Use this figure to answer questions 14–15.

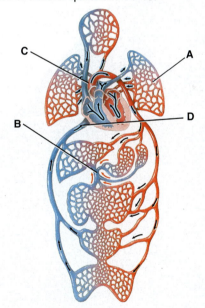

14. Which of the labeled structures is an artery?

a. A   b. B   c. C   d. D

15. Which of the labeled structures is a portal vein?

a. A   b. B   c. C   d. D

THE PLANNER ✓

Review your Chapter Planner on the chapter opener and check off your completed work.

The Lymphatic System and Immunity

Influenza, which comes in many varieties, can be devastating to the human population. In 1918, the Spanish flu pandemic killed somewhere between 30 and 50 million people worldwide. In 1997 and 2004/2005, bird flu originated in Hong Kong and spread quickly. Because the World Health Organization (WHO) and many nations provided enormous resources, these outbreaks were contained to parts of Southeast Asia. In early 2009, a strain of H1N1 influenza (swine flu) appeared in Mexico City and spread to many countries, mainly through commercial air travel. Some influenza viruses are "human" viruses that are harbored in other species, often causing no harm to that species. The most common forms of influenza (such as influenza A) are harbored in pigeons. Other forms of influenza are really "animal" viruses that can occasionally infect humans; thus many forms of influenza are referred to as a particular "animal" flu. Initially, these animal viruses can be passed only from the animal to the human. Of greater concern is the ability of a virus to be transmitted from one human to another, a transformation that may take several years to occur.

Vaccinations are often used to reduce the incidence of viral infections. Although researchers rushed to develop a vaccine for the H1N1 strain to be distributed throughout the United States, vaccination did not proceed as rapidly as the Centers for Disease Control and Prevention wished because of lack of vaccine and people's resistance to it. Fortunately, by first targeting those who were most susceptible, a full-blown epidemic was avoided.

Regular vaccinations for seasonal flu and for other childhood diseases (such as measles, mumps, rubella, and chicken pox) are essential for maintaining good health. But how do these vaccines protect your body from disease? They work with a powerful assortment of disease fighters collectively referred to as your *lymphatic system*.

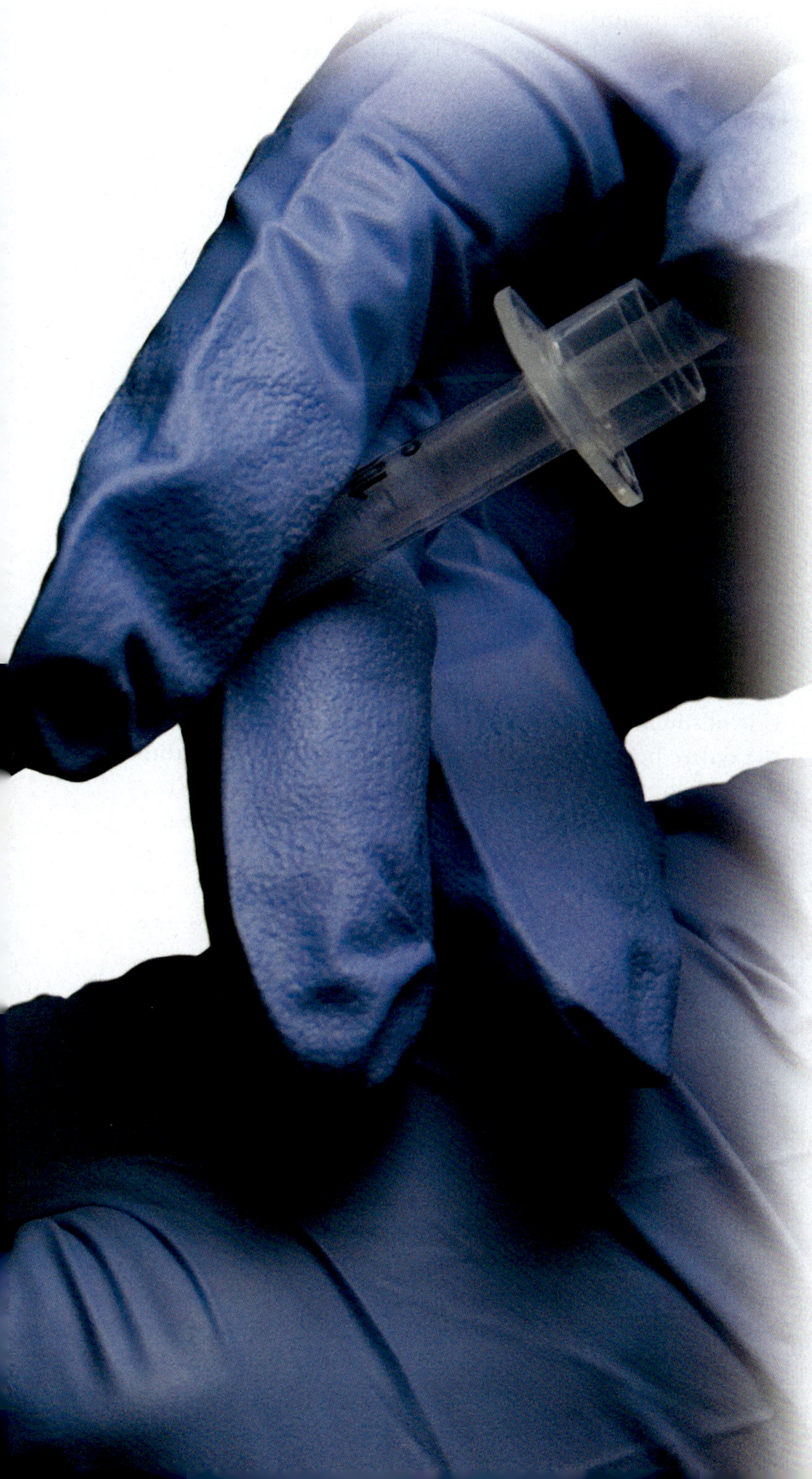

CHAPTER OUTLINE

CHAPTER PLANNER ✓

- ❏ Study the picture and read the opening story.
- ❏ Scan the L earning Objectives in each section:
  p. 344 ❏   p. 350 ❏   p. 362 ❏
- ❏ Read the text and study all visuals.
  Answer any questions.

Analyze key features

- ❏ InSight, p. 346 ❏   p. 363 ❏
- ❏ Process Diagram, p. 347 ❏   p. 356 ❏   p. 357 ❏   p. 358 ❏
- ❏ What a Health Provider Sees, p. 361 ❏
- ❏ Stop: Answer the Concept Checks before you go on:
  p. 349 ❏   p. 361 ❏   p. 364 ❏

End of chapter

- ❏ Review the Summary and Key Terms.
- ❏ Answer the Critical and Creative Thinking Questions.
- ❏ Answer What is happening in this picture?
- ❏ Complete the Self-Test and check your answers.

Components of the Lymphatic System Are Found Throughout the Body

LEARNING OBJECTIVES

1. **Identify** the components and functions of the lymphatic system.
2. **Describe** the flow of lymph.
3. **Compare** the structures and functions of the primary and secondary lymph organs.

Each day, about 20 liters of fluid filter out of the blood capillary walls to form *interstitial fluid*, the fluid that surrounds the cells of body tissues. Although chemically similar to blood plasma, interstitial fluid has less protein than blood plasma because most blood protein molecules are too large to filter through the capillary wall. About 17 L of this interstitial fluid are reabsorbed at the venous ends of the capillaries.

To maintain a constant blood volume, the remaining 3 L must also reenter the cardiovascular system. That is just one of the many jobs of your lymphatic system.

The Lymphatic System Has Three Functions: Drainage, Transport, and Immunity

The lymphatic system is basically a complex drainage system, much like a water sewage system in a city. Like sewage moving into and through the sewer pipes, interstitial fluid is carried through the system and eventually empties into the veins, where it again becomes part of the blood plasma.

The lymphatic system also helps transport lipids and lipid-soluble vitamins (A, D, E, and K) from the digestive tract to the blood. These materials do not dissolve well in the watery plasma, so sending them to the blood via the lymphatic system helps slow their entry into the blood supply and allows them to be dispersed more evenly throughout the body.

Finally, the lymphatic system contains various cells that participate in immune responses. These cells are lymphocytes, a type of white blood cell, as discussed in Chapter 10. Specifically, the lymph system uses T lymphocytes (T cells) and B lymphocytes (B cells) in its immune response to fight foreign cells (such as bacteria, fungi, and viruses), other foreign substances, and abnormal cells.

The Lymphatic System Consists of Lymph, Lymphatic Vessels, and Several Structures and Organs

Like the blood vessels of your cardiovascular system, the lymphatic system is a type of vascular system for moving fluid. It consists of lymph, lymphatic vessels, a number of structures containing lymphatic tissue, and red bone marrow. **Figure 12.1** is an overview of the system; we will discuss some of the structures shown in Figure 12.1 in more detail in the sections that follow.

Lymphatic vessels begin as **lymphatic capillaries** (see Figure 12.2). These tiny vessels begin in the tissues and carry the fluid that forms there. They are closed at one end and located in the spaces between cells. Lymphatic capillaries are slightly larger than blood capillaries and are uniquely structured to permit interstitial fluid to flow into—but not out—of them. The endothelial cells that make up the wall of a lymphatic capillary are not attached end to end; rather, the ends overlap, like roof shingles.

Just as blood capillaries unite to form venules and veins, lymphatic capillaries unite to form larger and larger **lymphatic vessels**. Lymphatic vessels resemble veins in structure but have thinner walls and more valves. Once interstitial fluid enters lymphatic vessels, it is called **lymph**.

From the lymphatic vessels, lymph eventually passes into one of two main channels. The **thoracic duct** is the main lymph-collecting duct. It receives lymph from the left side of the head, neck, and chest; the left upper limb; and the entire body below the ribs. The **right lymphatic duct** drains lymph from the upper right side of the body. Eventually the lymph again becomes part of the blood plasma.

Lymphatic organs and tissues, which are widely distributed throughout the body, are classified into two groups based on their functions: primary lymphatic organs and secondary lymphatic organs and tissues.

- The **primary lymphatic organs** include the **red bone marrow** and the **thymus**. These primary lymphatic organs are the sites where stem cells divide and develop into mature B cells and T cells.

- The **secondary lymphatic organs** and **tissues** include the **lymph nodes**, the **spleen**, and **lymphatic nodules**. Most immune responses occur in these sites. Lymphatic tissue is a specialized form of reticular connective tissue that contains large numbers of lymphocytes.

The lymphatic system consists of lymph, lymphatic vessels, lymphatic tissues, and red bone marrow.

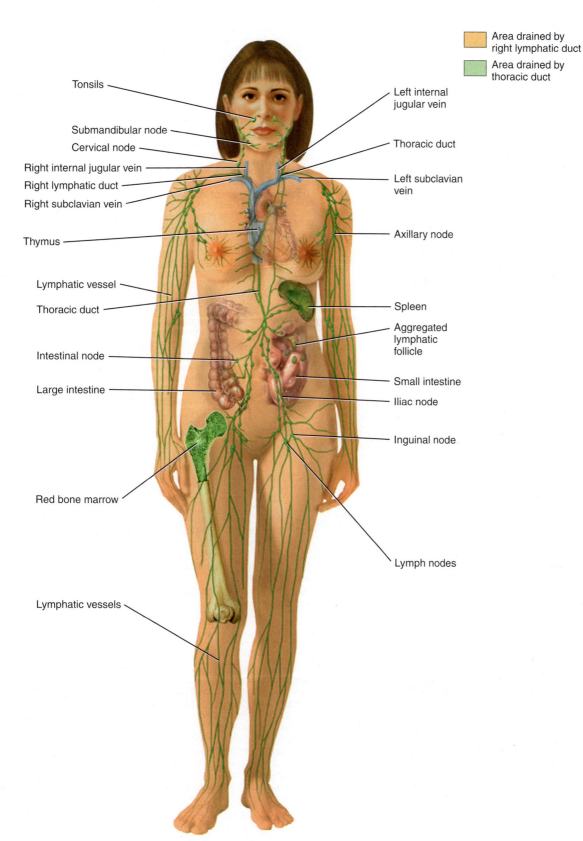

Area drained by right lymphatic duct

Area drained by thoracic duct

Tonsils

Submandibular node

Cervical node

Right internal jugular vein

Right lymphatic duct

Right subclavian vein

Thymus

Lymphatic vessel

Thoracic duct

Intestinal node

Large intestine

Red bone marrow

Lymphatic vessels

Left internal jugular vein

Thoracic duct

Left subclavian vein

Axillary node

Spleen

Aggregated lymphatic follicle

Small intestine

Iliac node

Inguinal node

Lymph nodes

Figure 12.2 provides an overview of the primary and secondary lymphatic organs. We will discuss the major organs and tissues in more detail soon, particularly the spleen and lymph nodes.

Primary and secondary organs and tissues of the lymphatic system • Figure 12.2

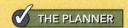 THE PLANNER

Primary organs and tissues

Red Bone Marrow
In red bone marrow, stem cells divide and develop into mature B and T lymphocytes (immune cells).

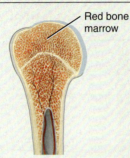

Red bone marrow

Thymus
Thymus is the organ where stem cells develop into mature T lymphocytes. It atrophies (shrinks) after puberty.

Secondary organs and tissues

Tonsils
Palatine, pharyngeal, and lingual tonsils are large aggregations of lymphatic nodules (not capsulated). They protect against inhaled or ingested foreign substances.

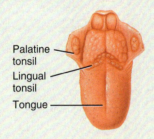

Palatine tonsil
Lingual tonsil
Tongue

Thoracic and Right Lymphatic Ducts
The thoracic duct and right lymphatic duct receive lymph from lymphatic vessels and empty into the junction between the jugular and subclavian veins.

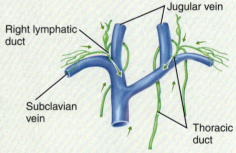

Jugular vein
Right lymphatic duct
Subclavian vein
Thoracic duct

Lymphatic Vessels
Although similar to veins, lymphatic vessels have thinner walls and more valves. They carry lymph away from tissues.

Lymphatic Capillaries
Lymphatic capillaries drain the interstitial spaces of excess fluid. This fluid, known as **lymph** once it enters the capillary, is passed on to the lymphatic vessels.

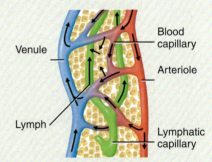

Venule
Blood capillary
Arteriole
Lymph
Lymphatic capillary

Spleen
The largest lymphatic organ is the spleen. It has macrophages and immune cells that destroy pathogens in the blood and remove worn-out blood cells. It also stores RBCs and platelets during fetal life, produces red blood cells.

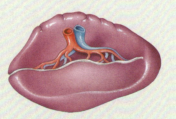

Lymph Nodes
Lymph nodes are bean-shaped, encapsulated tissues that filter lymph. Fibers trap foreign substances, which get broken down by macrophages and by immune responses involving lymphocytes. There are about 600 lymph nodes located mainly in the abdomen and thorax and near the head, axiliae, and groin.

Circulation of lymph • Figure 12.3

Lymph drains from interstitial fluid through lymphatic capillaries, which in turn feed into larger lymphatic vessels and ultimately flow back into the cardiovascular system. Lymph and interstitial fluid are chemically similar to blood plasma but with less protein. Each day, the lymphatic system returns about 3 L (about 15%) of fluid filtered from blood to maintain blood volume.

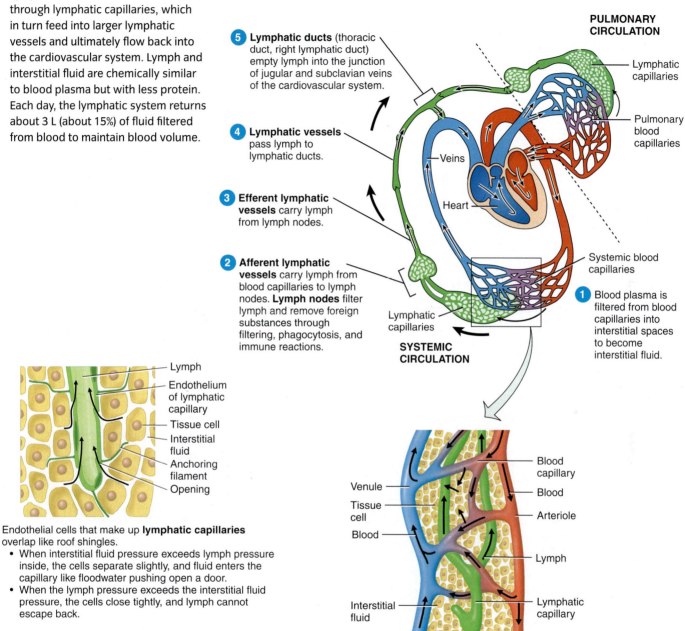

5 **Lymphatic ducts** (thoracic duct, right lymphatic duct) empty lymph into the junction of jugular and subclavian veins of the cardiovascular system.

4 **Lymphatic vessels** pass lymph to lymphatic ducts.

3 **Efferent lymphatic vessels** carry lymph from lymph nodes.

2 **Afferent lymphatic vessels** carry lymph from blood capillaries to lymph nodes. **Lymph nodes** filter lymph and remove foreign substances through filtering, phagocytosis, and immune reactions.

PULMONARY CIRCULATION

Lymphatic capillaries

Pulmonary blood capillaries

Veins

Heart

Systemic blood capillaries

1 Blood plasma is filtered from blood capillaries into interstitial spaces to become interstitial fluid.

Lymphatic capillaries

SYSTEMIC CIRCULATION

Lymph

Endothelium of lymphatic capillary

Tissue cell

Interstitial fluid

Anchoring filament

Opening

Endothelial cells that make up **lymphatic capillaries** overlap like roof shingles.

- When interstitial fluid pressure exceeds lymph pressure inside, the cells separate slightly, and fluid enters the capillary like floodwater pushing open a door.
- When the lymph pressure exceeds the interstitial fluid pressure, the cells close tightly, and lymph cannot escape back.

Venule

Tissue cell

Blood

Blood capillary

Blood

Arteriole

Lymph

Interstitial fluid

Lymphatic capillary

Let's take a closer look at the circulation of lymph in **Figure 12.3**, which shows the relationship of lymphatic vessels and lymph nodes to the cardiovascular system. Notice the close-up of the structure of lymphatic capillaries. This overlapping structure allows interstitial fluid to flow into the lymphatic capillaries—but not out. Figure 12.3 also shows that **afferent lymphatic vessels** carry lymph from capillaries to nodes (step **2**), and that **efferent lymphatic vessels** carry lymph away from a node (step **3**). Finally, the lymphatic vessels (step **4**) empty lymph into the junction of the jugular and subclavian veins of the cardiovascular system. (step **5**)

As in the venous system, the pressure responsible for lymph flow is generated by skeletal muscle pumps and respiratory pumps (see Chapter 11).

Immune Reactions Occur in the Lymph Nodes, Spleen, and Lymphatic Nodules

As you saw in Figure 12.2, the structures of the lymphatic system are divided into primary and secondary tissues. *Primary lymphatic tissues,* which include the thymus and red bone marrow, are responsible for making the cells that perform the immune functions of the lymphatic system.

These major immune cells are the B and T **lymphocytes** and phagocytic cells. The phagocytic cells destroy cellular debris and foreign cells that get into the system. Lymphocytes produce **antibodies** that help protect against foreign antigens (molecules that stimulate immune responses). You will learn more about these cells in the next section.

The *secondary lymphatic tissues* are locations throughout the body where various immune cells reside and where immune reactions occur. They include the lymph nodes, spleen, and lymphatic nodules. Let's take a look at the lymph nodes first (**Figure 12.4**).

Just as sewage goes through a filtration/treatment plant, the lymph is filtered and immunologically processed in the lymph nodes to remove foreign substances. Removal of debris and disease-causing organisms (also known as **pathogens**) from the lymph greatly reduces the incidence of disease. This ability to protect against disease is known as **immunity**. Because of these functions, the lymphatic system is sometimes referred to as the *immune system.*

> **lymphocytes** (LIM-fō-sītz) A type of white blood cell that helps carry out cell-mediated and antibody-mediated immune responses; found in blood and lymphatic tissues.

The structure and function of the lymph node • Figure 12.4 _____

Lymph nodes are multilayered structures that filter lymph.

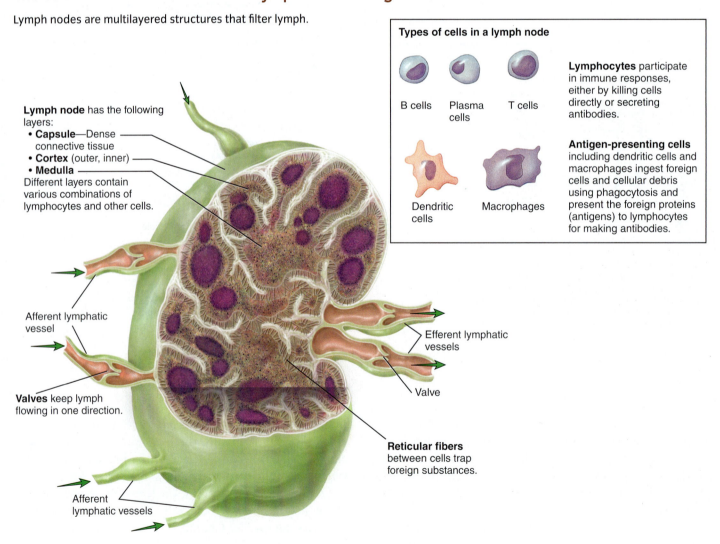

Lymph node has the following layers:
- **Capsule**—Dense connective tissue
- **Cortex** (outer, inner)
- **Medulla**
Different layers contain various combinations of lymphocytes and other cells.

Afferent lymphatic vessel

Valves keep lymph flowing in one direction.

Afferent lymphatic vessels

Efferent lymphatic vessels

Valve

Reticular fibers between cells trap foreign substances.

Types of cells in a lymph node

B cells   Plasma cells   T cells

Lymphocytes participate in immune responses, either by killing cells directly or secreting antibodies.

Dendritic cells   Macrophages

Antigen-presenting cells including dendritic cells and macrophages ingest foreign cells and cellular debris using phagocytosis and present the foreign proteins (antigens) to lymphocytes for making antibodies.

The structure and function of the spleen • Figure 12.5

The spleen filters blood and stores blood cells.

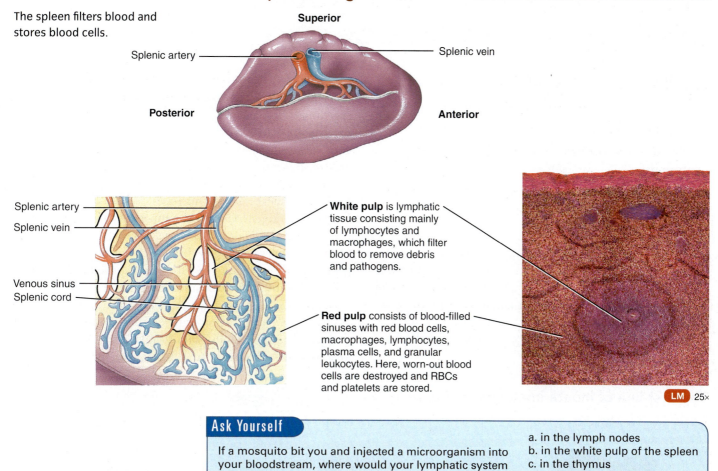

Superior

Splenic artery

Splenic vein

Posterior

Anterior

Splenic artery

Splenic vein

Venous sinus

Splenic cord

White pulp is lymphatic tissue consisting mainly of lymphocytes and macrophages, which filter blood to remove debris and pathogens.

Red pulp consists of blood-filled sinuses with red blood cells, macrophages, lymphocytes, plasma cells, and granular leukocytes. Here, worn-out blood cells are destroyed and RBCs and platelets are stored.

LM 25×

Ask Yourself

If a mosquito bit you and injected a microorganism into your bloodstream, where would your lymphatic system remove this organism to help reduce the incidence of disease?

a. in the lymph nodes
b. in the white pulp of the spleen
c. in the thymus
d. in the red pulp of the spleen
e. in the tonsils

The spleen is the largest single mass of lymphatic tissue in the body (**Figure 12.5**). It lies between the stomach and the diaphragm and is covered by a capsule of dense connective tissue. The spleen contains two types of tissue.

- White pulp is lymphatic tissue, consisting mostly of lymphocytes and macrophages.
- Red pulp consists of blood-filled venous sinuses and cords of splenic tissue consisting of red blood cells, macrophages, lymphocytes, plasma cells, and granular leukocytres.

Blood enters the spleen through the splenic artery and then enters the white pulp. B cells and T cells within the white pulp carry out immune responses, while macrophages destroy pathogens by phagocytosis. The red pulp then performs three functions related to blood cells:

1. The macrophages remove worn out or defective blood cells and platelets.

2. The red pulp stores platelets, perhaps up to one-third of the body's supply.

3. The red pulp produces blood cells during the growth of the fetus.

Finally, lymphatic nodules are egg-shaped masses of lymphatic tissue that are not surrounded by a capsule. Many are small and solitary, but some are grouped together. For example, the tonsils in your throat are a group of lymphatic nodules, strategically positioned to fight against inhaled or ingested foreign substances.

CONCEPT CHECK   STOP

1. **What** processes drive the flow of lymph through the lymphatic system?

2. **Which** glands are involved in the immune responses of the lymphatic system?

3. **What** is the major difference between primary and secondary lymphatic tissues?

Immune Reponses Help Protect the Body Against Disease

LEARNING OBJECTIVES

1. **Identify** the components of innate immunity.
2. **Explain** how cell-mediated immunity works.
3. **Outline** the process of antibody-mediated immunity.
4. **Describe** how vaccination protects the body from disease.

Microbes and pathogens are all around you; every day you come into contact with millions of them. While you're performing simple activities such as brushing your teeth, shaving, handling paper, or running your hand along a surface, pathogens can enter into your body. How does the lymphatic system protect you?

Your body has two types of immune responses that can defend against disease. **Innate immunity** is a series of nonspecific physical and chemical defenses. **Adaptive immunity** involves a specific reaction to pathogens called antigens. An **antigen** is any substance that the immune system recognizes as foreign (nonself). Entire microbes or parts of microbes may act as antigens. Chemical components of bacterial structures are antigenic, as are bacterial toxins and viral proteins. Other examples of antigens include chemical components of pollen, foods (such as egg white), drugs, incompatible blood cells, and transplanted tissues and organs.

Antigens induce plasma cells to secrete proteins known as *antibodies*, as we will see soon. In fact, the word antigen means *anti*body *gen*erator. In adaptive immunity, antigens are presented to lymphocytes that respond by

> **antigen** (AN-ti-jen)
> A substance that has the ability to provoke an immune response.

The first line of innate immunity • Figure 12.6

Skin

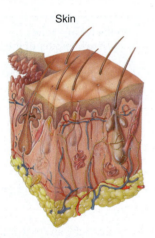

The epidermis is a physical barrier to microbes.

Skin's sebaceous (oil) glands form acidic film that prevents growth of microbes.

Skin's sweat glands produce sweat that flushes away microbes.

Nose and mouth

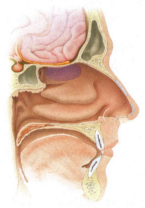

Hairs like those in the nose filter dust and microbes.

Mucus like that in the nose traps microbes, while ciliated cells move them away.

Saliva washes microbes from tooth surfaces and mucous membranes in the mouth.

Eyes

The lacrimal apparatus secretes tears that dilute and wash away irritants and microbes.

Physical barriers:
- The epidermis, hair, and mucus.
- Fluids that wash pathogens away.

Chemical barriers:
- Lysozyme, an antimicrobial agent that is present in sweat, nasal secretions, salvia, tears, and tissue fluids.
- The acidic film produced by the skin's sebaceous glands.

making killer cells and/or killer antibodies that destroy the antigens. Normally, a person's adaptive immune system cells recognize and do not attack his or her own tissues and chemicals. Let's look first at innate immunity.

Innate Immunity Includes Two Levels of Nonspecific Defense

The first line of defense that makes up innate immunity is a system of physical and chemical barriers that prevent pathogens from entering the body (**Figure 12.6**). The physical barriers include the skin and the various mucous membranes in the nose, upper respiratory tract, intestines, reproductive system (female), and urinary system. With its many layers of closely packed, keratinized cells, the epidermis (the outer epithelial layer of the skin) provides a formidable physical barrier to the entrance of microbes. In addition, continual shedding of the top epidermal cells helps remove microbes at the skin's surface. Bacteria rarely penetrate an intact and healthy epidermis.

The epithelial layer of mucous membranes secretes a fluid called mucus that lubricates and moistens the surface of a body cavity. Because mucus is sticky, it traps many microbes and foreign substances. The mucous membrane of the nose has mucous-coated hairs that trap and filter microbes, dust, and pollutants from inhaled air. The mucous membrane of the upper airways contain cilia, microscopic hairlike projections on the surface of the epithelial cells, which propel inhaled dust and microbes that have become trapped in mucus toward the throat.

Other fluids such as saliva, sweat, and tears can wash pathogens away from the skin and mucous membranes.

In addition to these physical barriers, chemicals within various fluids and secretions can slow the growth of microbes. Sebaceous (oil) glands of the skin secrete an oily substance called sebum that forms a protective film over the surface of the skin. Perspiration helps flush microbes from the surface of the skin and contains lysozyme, an enzyme capable of breaking down the cell walls of certain bacteria. Lysozyme is also found in tears, saliva, nasal secretions, and tissue fluids. Gastric juice, a mixture of hydrochloric acid, enzymes, and mucus in the stomach, destroys many bacteria and most bacterial toxins. Vaginal secretions are also slightly acidic, which discourages bacterial growth.

Digestive system	Vagina	Urethra
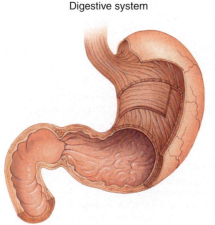		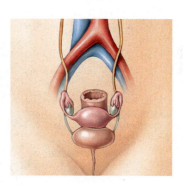
Acidic gastric juice destroys most bacteria and toxins in the stomach.	Vaginal secretions flush microbes out of the vagina.	Urine flow washes microbes from the urethra.
Vomiting and defecation expel microbes from the stomach and intestines.	Vaginal acidity discourages bacterial growth.	

Physical barriers:
• Vomiting and defecation.
• Vaginal secretions.
• Urine flow.

Chemical barriers:
• Acidic gastric juice.
• Vaginal acidity.

phagocytosis
(fag'-ō-sī-TŌ-sis) The process by which phagocytes ingest particulate matter; the ingestion and destruction of microbes, cell debris, and other foreign matter.

If pathogens penetrate the barriers that make up the first line of defense, several internal mechanisms serve as a second line of defense. These mechanisms include (1) antimicrobial chemicals, (2) **phagocytosis** (see Chapter 3), (3) inflammation, and (4) fever. **Figure 12.7** explains all four mechanisms in greater depth.

There are four types of antimicrobial substances (Figure 12.7a). **Complement proteins** from blood form holes in microbial membranes, which causes them to burst. **Interferons** are proteins produced by lymphocytes, macrophages, and fibroblasts that migrate to infected cells and interfere with viral replication. **Iron-binding proteins** inhibit the growth of certain bacteria by reducing the amount of available iron. **Antimicrobial proteins** are short peptides that kill microbes and attract other cells that participate in immune responses. When microbes penetrate the skin and mucous membranes or bypass the antimicrobial defenses in the blood, the next nonspecific defense consists of **phagocytes** and **T cells** and **B cells** (Figure 12.7b).

Inflammation (Figure 12.7c) is a defensive response of the body to tissue damage. Because inflammation is one of the body's innate defenses, the response of a tissue to a cut is similar to damage caused by burns, radiation,

The second line of innate immunity • Figure 12.7

Various mechanisms, including chemicals, killer cells, inflammation, and fever, make up a second line of defense when microbes get past surface barriers.

a. Antimicrobial substances

Complement
Complement proteins from blood become activated and form holes in microbial membranes, thereby causing them to burst (**cytolysis**). Complement also attracts phagocytes to a site (**chemotaxis**).

PHAGOCYTOSIS: Enhancement of phagocytosis by coating with C3b

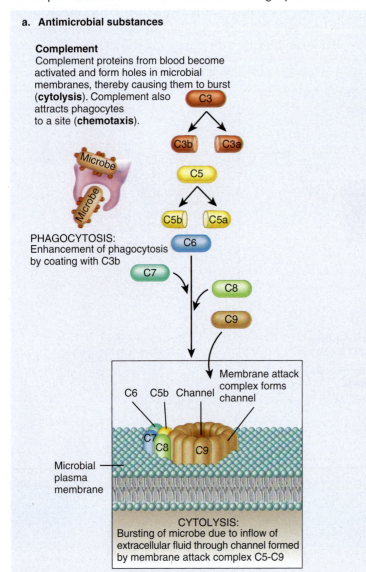

CYTOLYSIS: Bursting of microbe due to inflow of extracellular fluid through channel formed by membrane attack complex C5–C9

Interferons
Lymphocytes, macrophages, and fibroblasts infected with viruses produce proteins called interferons that interfere with viral replication. Interferons diffuse from one cell to another, rendering the receiving cells resistant to the viruses.

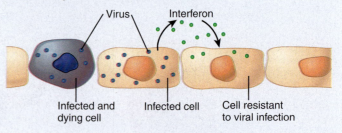

Antimicrobial proteins
Antimicrobial proteins (AMPs) are short peptides that kill microbes and attract mast cells and dendritic cells, which participate in immune responses. Microbes do not develop resistance to AMPs as they often do to antibiotics.

Iron-binding proteins
Iron-binding proteins reduce available iron and inhibit the growth of microbes. These proteins include:
• *Hemoglobin* in red blood cells
• *Transferrin* in blood and tissue fluids
• *Ferretin* in the liver, spleen, and red bone marrow
• *Lactoferrin* in milk, saliva, and mucus

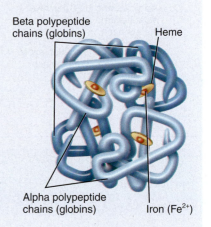

Hemoglobin molecule

or invasion by bacteria or viruses. The events of inflammation dispose of microbes, toxins, or foreign material at the site of the injury, prevent their spread to other tissues, and prepare the site for tissue repair. The four signs and symptoms of inflammation are redness, pain, heat, and swelling. Inflammation can also cause the loss of function in the injured area, depending on the site and extent of the injury.

From the events that occur during inflammation, it's easy to understand the signs and symptoms. Vasodilation of the arterioles causes increased blood flow to the area. You feel this as heat and see it as redness around the injured area. People often recognize redness as a sign of infection. The area swells because an increased amount of interstitial fluid has leaked out of the capillaries (*edema*). Pain results from injury to neurons, from toxic chemicals released by microbes, and from the increased pressure of edema.

Fever (Figure 12.7d) is an abnormally high body temperature that occurs because the hypothalamic thermostat is reset. It commonly occurs with infection and inflammation.

Inflammation and phagocytosis are always part of this second line of defense in response to any material that has crossed into the sterile body area. Inflammation causes swelling and redness, which people often recognize as symptoms of infection. Fever and the various antimicrobial chemicals are involved in only some situations.

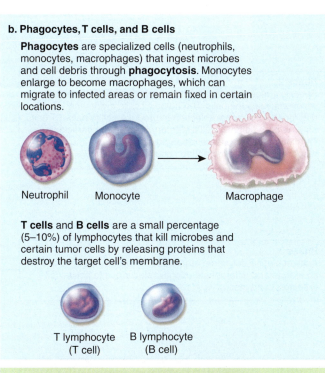

b. Phagocytes, T cells, and B cells

Phagocytes are specialized cells (neutrophils, monocytes, macrophages) that ingest microbes and cell debris through **phagocytosis**. Monocytes enlarge to become macrophages, which can migrate to infected areas or remain fixed in certain locations.

Neutrophil   Monocyte   Macrophage

T cells and **B cells** are a small percentage (5–10%) of lymphocytes that kill microbes and certain tumor cells by releasing proteins that destroy the target cell's membrane.

T lymphocyte   B lymphocyte
(T cell)   (B cell)

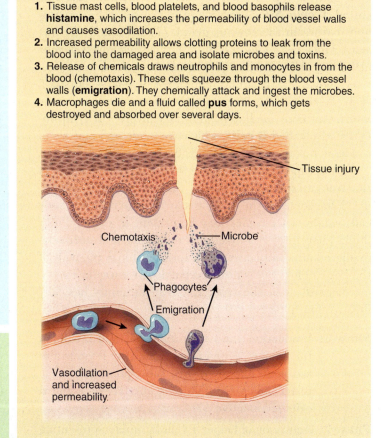

c. Inflammation
1. Tissue mast cells, blood platelets, and blood basophils release **histamine**, which increases the permeability of blood vessel walls and causes vasodilation.
2. Increased permeability allows clotting proteins to leak from the blood into the damaged area and isolate microbes and toxins.
3. Release of chemicals draws neutrophils and monocytes in from the blood (chemotaxis). These cells squeeze through the blood vessel walls (**emigration**). They chemically attack and ingest the microbes.
4. Macrophages die and a fluid called **pus** forms, which gets destroyed and absorbed over several days.

Tissue injury

Chemotaxis   Microbe

Phagocytes

Emigration

Vasodilation and increased permeability

d. Fever
Fever is an abnormally high body temperature. Many bacterial toxins elevate body temperature by triggering fever-causing substances (interleukin-1) from macrophages. The high temperature enhances the effects of interferons, inhibits growth of microbes, and speeds up repair reactions.

Adaptive Immunity Allows You to Respond to a Variety of Invaders

All innate defenses are *nonspecific*; they try to block everything that could possibly cause harm. However, your body also has the ability to adapt to specific types of infections, such as specific strains of viruses and bacteria, or specific antigens, including pollen, animal dander, and antibiotics. This ability is called **adaptive immunity**. Adaptive immunity can respond quickly—it can "remember" the particular antigens that have invaded your body—and can recognize the difference between foreign cells or substances and your own cells or substances.

There are two types of adaptive immunity: cell-mediated immunity and antibody-mediated immunity (**Figure 12.8**). **Cell-mediated immunity** is effective against intracellular pathogens (*intra-* = within), such as viruses, bacteria, and fungi located inside cells; it can also be used against some cancer cells and foreign tissue transplants. Cell-mediated immunity involves **T lymphocytes**, also called **T cells**, which are made in the red bone marrow and mature in the thymus gland. (The *T* stands for *thymus*.) Thus, cell-mediated immunity always involves cells attacking cells. **Antibody-mediated immunity** is effective against extracellular pathogens (*extra-* = outside), such as viruses, bacteria, and fungi in blood and body fluids. Antibody-mediated immunity involves **B lymphocytes**, also called **B cells**, which are made and mature in red bone marrow. (The *B* stands for *bone marrow*.)

When an antigen invades the body, many copies of that antigen spread throughout the body's tissues and fluids. Some copies of the antigen may be present inside body cells, while other copies may be present in extracellular fluid. Therefore, cell-mediated immunity and antibody-mediated immunity often work together to rid the body of the large number of copies of a particular antigen.

Regardless of the type, adaptive immunity basically involves four phases:

Phase 1: Production and maturation of T and B cells in the primary lymphatic organs (the thymus and red bone marrow), with subsequent migration to secondary lymphatic organs. These T and B cells are inactive.

Phase 2: Activation of helper T cells by antigen-presenting cells. This leads to **clonal selection**, a vast proliferation and differentiation of cells into active cells and memory cells.

Phase 3: Activation of B cells and cytotoxic T cells with the aid of active helper T cells. B cells and cytotoxic T cells also undergo clonal selection.

Overview of adaptive immune responses
• Figure 12.8

The phases of adaptive immune responses for cell-mediated immunity and antibody-mediated immunity are shown. Phase 1 occurs throughout life, while phases 2–4 occur only in response to presentations of various pathogens (such as infections, illnesses, and allergies).

Phase 1
T cells and B cells develop and mature in primary lymphatic organs. Antigen receptors are placed in their cell membranes. T cells and B cells migrate to secondary lymphatic organs, where they are inactive.

Phase 2
Specific antigens are presented to inactive helper T cells, which activates the cells. The helper T cells divide and differentiate via clonal selection to produce active helper T cells and memory helper T cells. Memory cells do not participate in immune reactions but are reserved so that later immune responses to the same antigen occur more quickly.

Phase 3
Active helper T cells help other cells (cytotoxic T cells, B cells) proliferate and differentiate (clonal selection) into active cells (effectors) and memory cells.

Phase 4
Active cytotoxic T cells kill invading antigens directly (cell-mediated immunity), while plasma cells (formed from B cells) secrete antibodies to do the job (antibody-mediated immunity).

Phase 4: Actions of active cytotoxic T cells and active B cells. **Cytotoxic T cells** are T cells that kill infected body cells. Active B cells differentiate into plasma cells, which secrete antibodies. Antibodies act on antigens in various ways, as you will learn shortly.

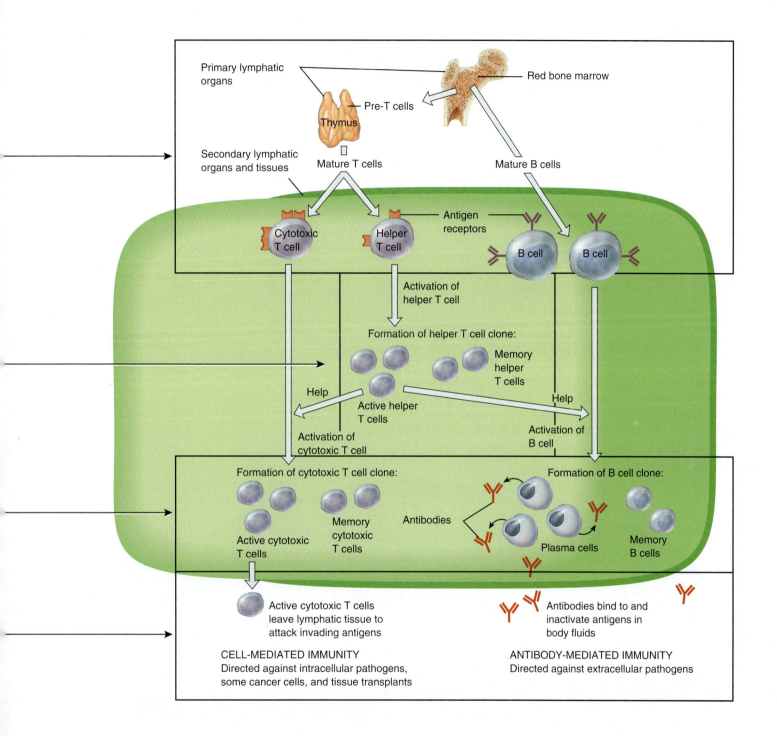

Primary lymphatic organs

Red bone marrow

Pre-T cells

Thymus

Secondary lymphatic organs and tissues

Mature T cells

Mature B cells

Cytotoxic T cell

Helper T cell

Antigen receptors

B cell

B cell

Activation of helper T cell

Formation of helper T cell clone:

Memory helper T cells

Help

Active helper T cells

Help

Activation of cytotoxic T cell

Activation of B cell

Formation of cytotoxic T cell clone:

Formation of B cell clone:

Antibodies

Active cytotoxic T cells

Memory cytotoxic T cells

Plasma cells

Memory B cells

Active cytotoxic T cells leave lymphatic tissue to attack invading antigens

Antibodies bind to and inactivate antigens in body fluids

CELL-MEDIATED IMMUNITY
Directed against intracellular pathogens, some cancer cells, and tissue transplants

ANTIBODY-MEDIATED IMMUNITY
Directed against extracellular pathogens

B cells can recognize and bind to antigens in lymph, interstitial fluid, or blood plasma, but T cells only recognize fragments of antigens that are processed and presented in a certain way. For adaptive immunity to work, T cells and B cells must be able to distinguish your body's own cells and antigens from foreign cells and antigens. To accomplish this, you have a unique set of proteins called the **major histocompatiblity complex (MHC)** on every cell in your body (except red blood cells). Unless you have an identical twin, your MHC proteins are unique.

When a foreign antigen enters the body, a cell called an **antigen-presenting cell (APC)** ingests it and processes the foreign substance (**Figure 12.9**). APCs include dendritic cells, macrophages, and B cells. They are strategically located in the skin, mucous membranes, and lymph nodes. The foreign antigens are bound to the MHC proteins in the APC and are moved to the APC's surface, where they are "presented" to the appropriate lymphocytes. Cells that have been infected by intracellular parasites can also present antigens in a similar manner.

In cell-mediated immunity, once the APCs present the antigen to helper T cells, the active helper T cells secrete interleukin and other proteins that stimulate cytotoxic T cells to undergo **clonal activation**. When an infected cell presents its antigen to the active cytotoxic T cell with the corresponding antigen receptor, the cytotoxic T cell destroys the infected cell (**Figure 12.10**).

Cytotoxic T cells kill infected target cells in one of two ways. In one method, protein-digesting enzymes called granzymes trigger apoptosis, the fragmentation of cellular contents. The cell is destroyed and releases the microbes, which are then killed by phagocytes. In the second method, the T cells release two proteins, perforin and granulysin, into infected cells; these proteins enter, destroy the microbes, and cause cytolysis (cell-bursting).

Processing and presentation of antigens • Figure 12.9

THE PLANNER

Antigens enter an antigen-presenting cell (APC), such as a dendritic cell, B cell, or macrophage. After processing, pieces of the antigen combine with major histocompatiblity complex (MHC) proteins. The antigen–MHC complex is inserted into the surface of the APC, where it can sensitize and activate a helper T cell. Infected body cells can also present antigens that have invaded them by going through a similar process using steps 2–5.

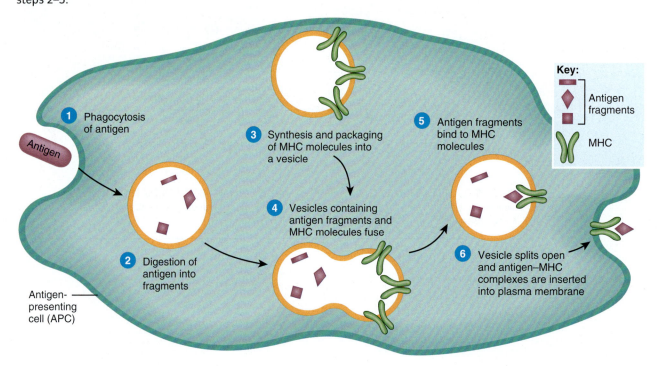

Cell-mediated immunity • Figure 12.10

THE PLANNER ✔

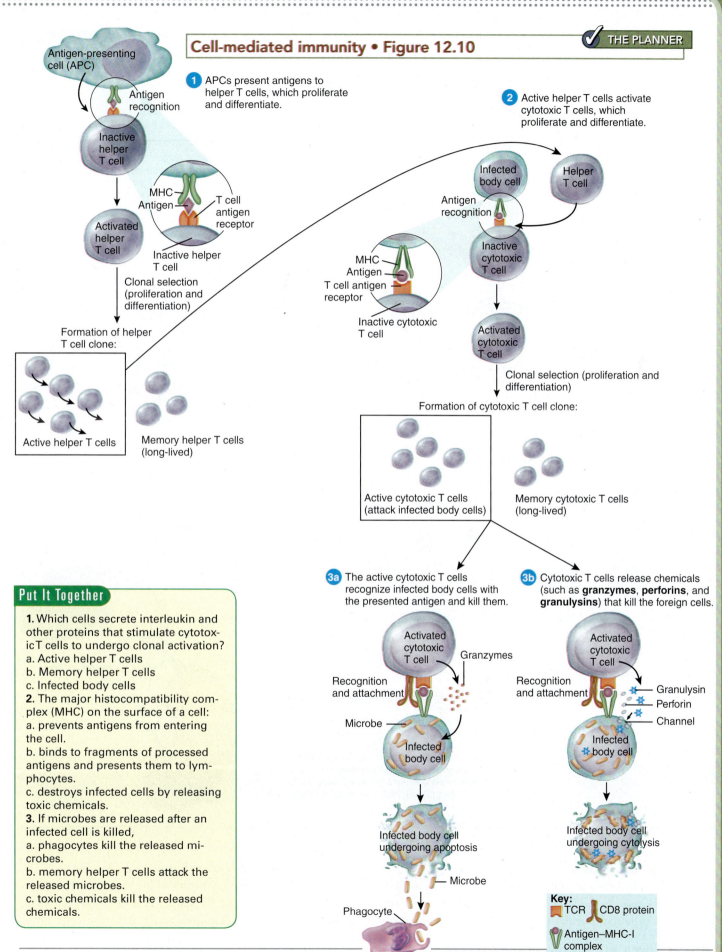

1 APCs present antigens to helper T cells, which proliferate and differentiate.

Antigen-presenting cell (APC)

Antigen recognition

Inactive helper T cell

MHC
Antigen
T cell antigen receptor

Inactive helper T cell

Activated helper T cell

Clonal selection (proliferation and differentiation)

Formation of helper T cell clone:

Active helper T cells

Memory helper T cells (long-lived)

2 Active helper T cells activate cytotoxic T cells, which proliferate and differentiate.

Infected body cell

Helper T cell

Antigen recognition

MHC
Antigen
T cell antigen receptor

Inactive cytotoxic T cell

Inactive cytotoxic T cell

Activated cytotoxic T cell

Clonal selection (proliferation and differentiation)

Formation of cytotoxic T cell clone:

Active cytotoxic T cells (attack infected body cells)

Memory cytotoxic T cells (long-lived)

3a The active cytotoxic T cells recognize infected body cells with the presented antigen and kill them.

Activated cytotoxic T cell

Granzymes

Recognition and attachment

Microbe

Infected body cell

Infected body cell undergoing apoptosis

Microbe

Phagocyte

3b Cytotoxic T cells release chemicals (such as **granzymes**, **perforins**, and **granulysins**) that kill the foreign cells.

Activated cytotoxic T cell

Recognition and attachment

Granulysin
Perforin
Channel

Infected body cell

Infected body cell undergoing cytolysis

Key:
TCR   CD8 protein
Antigen–MHC-I complex

Put It Together

1. Which cells secrete interleukin and other proteins that stimulate cytotoxic T cells to undergo clonal activation?
a. Active helper T cells
b. Memory helper T cells
c. Infected body cells
2. The major histocompatibility complex (MHC) on the surface of a cell:
a. prevents antigens from entering the cell.
b. binds to fragments of processed antigens and presents them to lymphocytes.
c. destroys infected cells by releasing toxic chemicals.
3. If microbes are released after an infected cell is killed,
a. phagocytes kill the released microbes.
b. memory helper T cells attack the released microbes.
c. toxic chemicals kill the released chemicals.

Antibody-mediated immunity • Figure 12.11

APC's active helper T cells activate B cells. B cells then proliferate and differentiate into memory B cells and plasma cells **1**. Plasma cells secrete antibodies, which bind to antigens and render them inactive in several ways **2**.

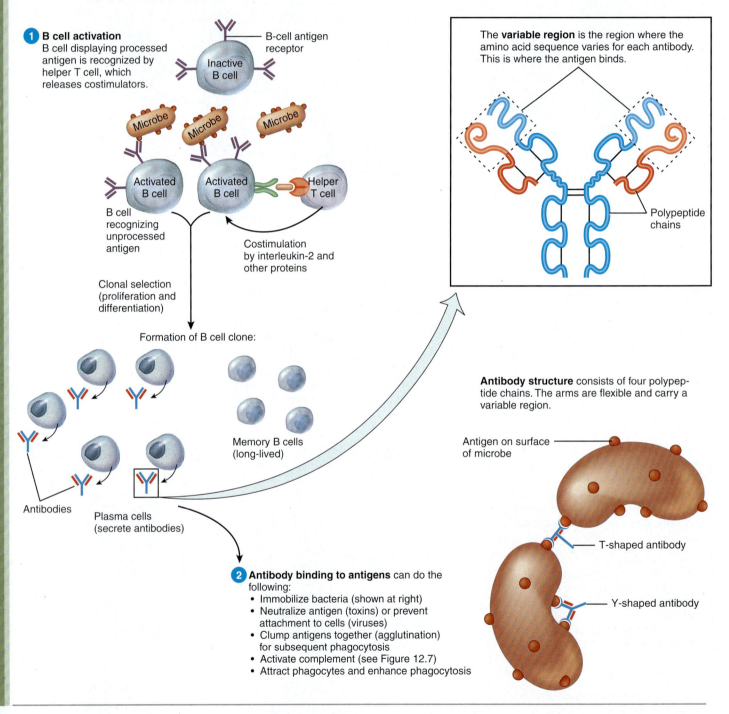

1 B cell activation
B cell displaying processed antigen is recognized by helper T cell, which releases costimulators.

B-cell antigen receptor

Inactive B cell

Microbe

Microbe

Microbe

Activated B cell

Activated B cell

Helper T cell

B cell recognizing unprocessed antigen

Costimulation by interleukin-2 and other proteins

Clonal selection (proliferation and differentiation)

Formation of B cell clone:

Memory B cells (long-lived)

Antibodies

Plasma cells (secrete antibodies)

2 Antibody binding to antigens can do the following:
• Immobilize bacteria (shown at right)
• Neutralize antigen (toxins) or prevent attachment to cells (viruses)
• Clump antigens together (agglutination) for subsequent phagocytosis
• Activate complement (see Figure 12.7)
• Attract phagocytes and enhance phagocytosis

The **variable region** is the region where the amino acid sequence varies for each antibody. This is where the antigen binds.

Polypeptide chains

Antibody structure consists of four polypeptide chains. The arms are flexible and carry a variable region.

Antigen on surface of microbe

T-shaped antibody

Y-shaped antibody

In antibody-mediated immunity, the B cells process the antigen and present it to helper T cells that then stimulate activated B cells to undergo clonal activation to produce memory B cells and plasma cells. The plasma cells subsequently produce antibodies (**Figure 12.11**). Because the antibody "arms" can move somewhat, an antibody can assume either a T shape or a Y shape. This flexibility allows an antibody to bind to two identical antigens at the same time.

Classes of immunoglobulins Table 12.1

Name and Structure	Characteristics	Functions
IgG	About 80% of all antibodies in the blood. Also found in lymph and the intestines. Only class of antibody to cross the placenta from mother to fetus, thereby conferring considerable immune protection to newborns.	Protects against bacteria and viruses by enhancing phagocytosis, neutralizing toxins, and triggering the complement system.
IgA	About 10% to 15% of all antibodies in the blood. Found mainly in sweat, tears, saliva, mucus, breast milk, and gastrointestinal secretions. Levels decrease during stress, lowering resistance to infection.	Provides localized protection against bacteria and viruses on mucous membranes.
IgM	About 5% to 10% of all antibodies in the blood. Also found in lymph. First antibody class to be secreted by plasma cells after an initial exposure to any antigen. In blood plasma, the anti-A and anti-B antibodies of the ABO blood group, which bind to A and B antigens during incompatible blood transfusions, are also IgM antibodies (see Figure 10.8).	Activates complement and causes agglutination and lysis of microbes.
IgD	About 0.2% of all antibodies in the blood. Also found in lymph and on the surfaces of B cells as antigen receptors.	Involved in activation of B cells.
IgE	Less than 0.1% of all antibodies in the blood. Also located on mast cells and basophils.	Involved in allergic and hypersensitivity reactions and provides protection against parasitic worms.

The structure of antibodies allows them to be specific for a particular antigen. Antibodies belong to a larger group of plasma proteins called **immunoglobulins** (Ig; im-ū-nō-GLOB-ū-lins). There are five classes of immunoglobulins, each with different functions (**Table 12.1**).

Immune Response Time Speeds Up After the Initial Exposure

During an initial infection, there are few lymphocytes (that is, helper T cells, cytotoxic T cells, and B cells) with the appropriate antigen receptor, so the *initial*, or *primary*, *immune response* may take several days or even weeks. During cell-mediated or antibody-mediated immunity, the clonal selection process described earlier produces numerous memory cells in each of those lymphocyte categories. Because these memory cells live for long periods of time (even decades), they can help your body respond to later infections with the same antigens. In a subsequent exposure to the same antigens, a *secondary immune response* occurs that is stronger and much more rapid (within hours or days) (**Figure 12.12**). The presence of long-lasting cells

and antibodies gives your lymphatic system **immunological memory**.

Immunological memory • Figure 12.12

The stronger and more rapid reaction to subsequent antigen exposures by memory cells can be seen in graphs of antibody concentrations in the blood.

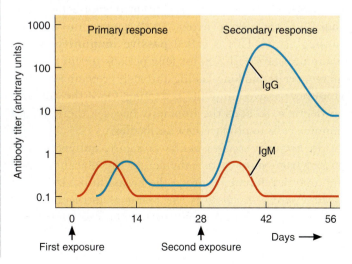

Allergies Are Caused by an Overreaction to an Antigen

An example of an overactive immunological memory is an **allergy**, or a *hypersensitivity*. Allergies develop when an individual is exposed to a type of antigen, referred to as an **allergen**. Examples of allergens include pet dander, foods, pollen, antibiotics, poison ivy, and bee venom. Upon initial exposure to the allergen, the lymphatic system produces IgE antibodies that bind to the surfaces of mast cells and basophils. Later, upon reexposure to the allergen, these cells secrete **histamine**, prostaglandins, and other chemicals, causing vasodilation, contraction of airway smooth muscle, and increased mucus secretion. These factors contribute to the allergy symptoms of runny nose, sneezing, congestion, and difficulty breathing. A severe, life-threatening allergic reaction in which the airways constrict so much that the person has extreme difficulty breathing is called **anaphylactic shock**. Anaphylactic shock can be treated by injecting epinephrine to dilate the airways and strengthen the heart.

Immunological memory forms the basis of vaccinations. Let's turn next to vaccinations and see how they work.

> **allergen** (AL-er-jen) An antigen that evokes a hypersensitivity reaction.
>
> **histamine** (HISS-ta-mēn) A substance found in many cells that is released when the cells are injured; results in vasodilation, increased permeability of blood vessels, and constriction of bronchioles.

There Are Many Ways to Develop Immunity

You can acquire immunity to a pathogen in a number of ways. In some of the mechanisms, you are exposed to the antigen, and your body develops antibodies in response to it. These processes result in **active immunity**. Other mechanisms involve receiving premade antibodies that your body can use to defend itself against the disease-causing agent. This is called **passive immunity**. Active immunity stays with you for long periods (or life), as the memory cells and long-lasting antibodies remain with you. However, passive immunity is fleeting. Once the antibodies degrade, so does the immunity because there are no immune cells to produce new antibodies.

In *naturally acquired passive immunity*, antibodies are passed from mother to baby across the placenta (during fetal development) or via the breast milk. *Artificially acquired passive immunity* is the result of intravenous injection of anti-serum or anti-toxin solutions containing antibodies. When you are exposed to a disease by coming in contact with someone else who has the disease, *naturally acquired active immunity* occurs: You react to that disease organism by making antibodies through the process described in the previous section. You can also be exposed to that pathogenic organism through the use of a vaccine. This type of exposure results in *artificially acquired active immunity*. See **Table 12.2** for a summary of the types of adaptive immunity.

While exposure to pathogens, getting sick, evoking a full immune response, and recovering after producing antibodies is the most effective process for developing immunity, some pathogens are fatal. A **vaccination** (receipt of a vaccine) can provide you with active immunity at greatly reduced risk. A **vaccine** consists of weakened or dead pathogens—either a whole organism or parts of an organism. When the vaccine is injected, your B and T cells are activated in a full immune response to this weakened opponent. Your body makes memory cells in response to the pathogen, so that when you encounter the pathogen at its full strength, the active secondary response should be enough to fend it off. Some vaccines may require booster doses to maintain adequate protection.

Most often, antibodies protect us from diseases, but sometimes they can attack our own cells and cause disease; such diseases are called *autoimmune diseases* (see *What a Health Provider Sees*).

Types of adaptive immunity Table 12.2

Type	How acquired
Naturally acquired active immunity	Following exposure to a microbe, antigen recognition by B cells and T cells and costimulation lead to the production of antibody-secreting plasma cells, cytotoxic T cells, and B and T memory cells.
Naturally acquired passive immunity	Transfer of IgG antibodies from mother to fetus across the placenta, or of IgA antibodies from mother to baby in milk during breastfeeding.
Artificially acquired active immunity	Antigens introduced during a vaccination stimulate cell-mediated and antibody-mediated immune responses, leading to production of memory cells. The antigens are pretreated to be immunogenic but not pathogenic; that is, they will trigger an immune response but not cause significant illness.
Artificially acquired passive immunity	Intravenous injection of immunoglobulins (antibodies).

WHAT A HEALTH PROVIDER SEES

When Your Own Immune System Attacks You

Your adaptive immune responses have ways of distinguishing your own cells and antigens from foreign ones. However, sometimes these recognition mechanisms go awry and your immune cells attack your own body cells. This lack of self-recognition by the immune cells leads to a variety of **autoimmune diseases**, also referred to as **autoimmunity**. The type of autoimmune disease depends on the tissue or system that is attacked. About three-quarters of the individuals with autoimmune disorders are women. Most are between the ages of 15 and 45. Autoimmune disorders that occur in men tend to be more severe than the same diseases in women. Here are a few of the autoimmune diseases:

> **autoimmunity**
> An immunological response against a person's own tissues.

- *Type I diabetes mellitus*—Insulin-producing cells of the pancreas are damaged, interfering with the individual's ability to maintain the proper level of glucose in the blood (see Chapter 9).

- *Multiple sclerosis*—Degrades the myelin sheaths of neurons. With the loss of the myelin sheath, neurons take longer to send nerve impulses, resulting in a variety of symptoms associated with slowed neurologic function. Many patients with MS eventually lose mobility and become wheelchair-bound.

- *Graves disease*—The individual produces antibodies that mimic the actions of thyroid hormone. This condition has dramatic effects on metabolic rate and cardiovascular function. Drugs such as propylthiouracil and methimazole are used to treat Graves disease because they block the production of thyroid hormone.

- *Systemic lupus erythematosus (SLE), or lupus*—This chronic autoimmune disease affects multiple body systems. Signs and symptoms include joint pain, slight fever, fatigue, oral ulcers, weight loss, enlarged lymph nodes and spleen, photosensitivity, rapid loss of large amounts of scalp hair, and sometimes an eruption across the bridge of the nose and cheeks , as shown, called a "butterfly rash." The name lupus comes from the term for wolf, as some of the skin lesions were thought to resemble the damage inflicted by the bite of a wolf. Kidney damage occurs as antigen–antibody complexes become trapped in kidney capillaries, obstructing blood filtering. Renal failure is the most common cause of death.

Treatments for autoimmune disorders depend on the individual disease but generally involve some type of immunosuppressive therapies (use of drugs to shut down the body's immune response), removal of the thymus gland, or plasmapheresis (filtering the blood to remove antibodies).

Think Critically
1. Explain the benefits and risks of using immunosuppressive therapies to help treat an autoimmune disease.
2. How might removing the thymus gland treat an autoimmune disease?

CONCEPT CHECK  STOP

1. **What** two parts of your body form the physical barriers to invading microbes in innate immunity?

2. **How** can a cytotoxic T cell kill an infected cell?

3. **How** does antigen presentation in B cell–mediated immunity differ from that in T cell–mediated immunity?

4. **How** does a vaccination provide immunity?

HIV Causes a Breakdown of the Immune Response

LEARNING OBJECTIVES

1. **Distinguish** between HIV and AIDS.
2. **Describe** the structure and mechanism of action of HIV.
3. **Outline** the symptoms and progression of AIDS.
4. **Identify** treatments for HIV infection.

Acquired immunodeficiency syndrome **(AIDS)** is a condition in which a person experiences an assortment of infections due to the progressive destruction of cells of the lymphatic system. The destruction is caused by a virus called **human immunodeficiency virus (HIV)**. Upon infection with HIV, the affected individual may experience no symptoms for many years but then deteriorate rapidly into a condition that is ultimately fatal. Since the first cases of AIDS were reported in 1981, more than 20 million people have died from the disease. Currently more than 40 million people worldwide are infected with HIV.

It Is Not Easy to Get Infected with HIV

HIV is present in the blood and body fluids of an infected person. The virus is transmitted from person to person through the exchange of bodily fluids (for example, blood, semen, vaginal fluid, breast milk). The most common routes of transmission include unprotected (without a condom) sexual intercourse, anal intercourse, oral sex, intravenous drug use that involves sharing of needles, and accidental needle sticks from HIV-contaminated needles. (The latter is most common among health care providers.) HIV is fragile and does not survive long outside the body. The virus cannot be spread through casual contact, such as hugging or sharing household items, and it can be easily eliminated from items with heat (57.2°C for 10 min), by disinfecting (using hydrogen peroxide, rubbing alcohol, bleach, or germicidal cleansers such as Betadine or Hibiclens), or by standard dishwashing or clothes washing. HIV cannot be transmitted through insect bites.

The Symptoms of HIV Infection Progress in Severity

HIV is a *retrovirus*—a virus that can convert its RNA into DNA (*retro-* = "backward"). This virus contains RNA and some proteins necessary to incorporate its nucleic acid into a host cell's DNA in a process called *reverse transcription*. The virus can remain dormant for an extended period of time, or it can force the host cell to make new particles of the virus. The virus attacks mainly helper T cells and can replicate more than 10 billion new viruses daily. Eventually, the host cell ruptures and dies, while new viruses infect other cells. Initially, helper T cells reproduce almost as fast as HIV destroys them, but after several years, the ability to replace T cells diminishes and the T cell concentration in the circulation gradually declines.

When the initial HIV infection occurs, the individual experiences flu-like symptoms (**Figure 12.13a**). During the initial infection, the virus spreads throughout the body. The initial infection is followed by a latency period, during which the T cell count is relatively stable but low (**Figure 12.13b**). Once the T cell count drops, however, the affected individual is unable to respond immunologically to even simple infections. At this point, the HIV infection has progressed to AIDS. The AIDS patient experiences *constitutional symptoms* including persistent fatigue, weight loss, enlarged lymph nodes, night sweats, rashes, diarrhea, and various lesions (of the mouth, gums, and skin). Death usually occurs due to one or more *opportunistic diseases* infections that the lymphatic system would have normally held in check (such as pneumonia).

If left untreated, people with HIV infections often begin showing symptoms of AIDS within 10 to 12 years. Once AIDS develops, untreated patients usually die within 8 to 12 months. Although there is no cure for HIV infection, some pharmaceutical therapies have been developed to prevent viral replication and help extend the lives of HIV-infected people (**12.13c**). **Reverse transcriptase inhibitors** block the action of HIV's reverse transcriptase enzyme from making DNA copies of the viral RNA. **Protease inhibitors** block the actions of HIV's protease, which cuts host proteins into pieces to assemble the coat of new HIV virus particles.

Highly active antiretroviral therapy (HAART) is a combination of two different reverse transcriptase inhibitors and one protease inhibitor that is effective but expensive (~$10,000 per year). In any patient, HAART drastically reduces levels of HIV, increases helper T cell counts, delays HIV progression to AIDS, and causes remission, or disappearance of opportunistic infections, restoring some semblance of normal health. However, HIV remains in the body and can be transmitted to others.

As HIV reproduces, the virus slowly destroys the body's T cells.

WILEY **PLUS** **Video**

a. Main symptoms of AIDS

Initially, HIV-infected patients exhibit flu-like symptoms, including the following:

- Fever
- Fatigue
- Headache
- Rash
- Sore throat
- Joint pain
- Swollen lymph nodes
- Night sweats (50% of cases)

When HIV progresses to AIDS, patients become susceptible to many opportunistic infections, tumors, and dementia.

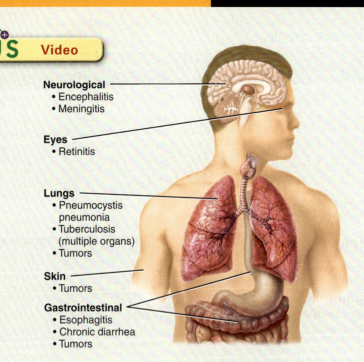

Neurological
- Encephalitis
- Meningitis

Eyes
- Retinitis

Lungs
- Pneumocystis pneumonia
- Tuberculosis (multiple organs)
- Tumors

Skin
- Tumors

Gastrointestinal
- Esophagitis
- Chronic diarrhea
- Tumors

b. HIV slowly destroys the body's T cells as it reproduces.

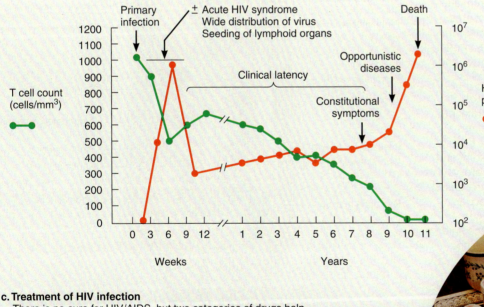

Primary infection

± Acute HIV syndrome
Wide distribution of virus
Seeding of lymphoid organs

Death

Clinical latency

Opportunistic diseases

Constitutional symptoms

T cell count (cells/mm³)

HIV RNA copies per ml plasma

Weeks   Years

c. Treatment of HIV infection

There is no cure for HIV/AIDS, but two categories of drugs help extend the lives of many HIV-infected people:

1. Reverse transcriptase inhibitors block the action of HIV's reverse transcriptase enzyme from making DNA copies of the viral RNA. These drugs include zidovudine (ZDV, previously AZT), didansoine (ddl), and stavudine (d4T). Trizivir is a combination of three reverse transcriptase inhibitors.
2. Protease inhibitors block the actions of HIV's protease, which cuts host proteins into pieces to assemble the coat of new HIV virus particles. Protease inhibitors include nelfinavir, saquinavir, ritinovir, and indinavir.

Highly active antiretroviral therapy (HAART) is an effective but expensive combination of two different reverse transcriptase inhibitors and one protease inhibitor.

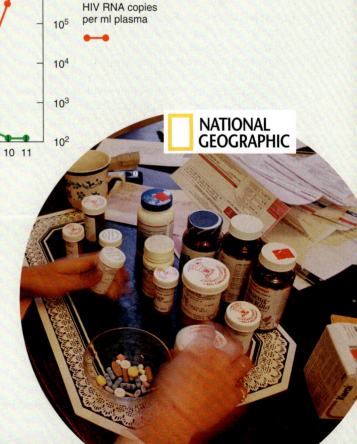

NATIONAL GEOGRAPHIC

With treatment, the onset of AIDS can sometimes be postponed for many years. Once AIDS develops in these individuals, continued treatment can help them survive with fewer complications for 4 to 10 years.

CONCEPT CHECK STOP

1. **What** is the difference between HIV and AIDS?
2. **How** does HIV replicate?
3. **What** are the major symptoms that a patient experiences during the initial infection with HIV?
4. **How** does a reverse transcriptase inhibitor treat HIV infection?

Summary

1 Components of the Lymphatic System Are Found Throughout the Body 340

• As shown, the **lymphatic system** consists of **primary organs** (red bone marrow and the thymus) where **lymphocytes** are produced and **secondary organs** (lymph nodes, spleen, and **lymphatic nodules**) where the lymphocytes and other cells reside and carry out immune reactions. The lymphatic system drains interstitial fluid, protects the body from invaders, and transports lipids and lipid-soluble vitamins.

• **Lymph** is interstitial fluid that drains into **lymph capillaries**. Lymph capillaries drain into **lymphatic vessels**. Lymph then drains into larger lymphatic vessels and lymphatic ducts, and it ultimately enters the venous system and general circulation. Skeletal and respiratory pumps drive lymph through the lymph vessels, while valves keep it moving in one direction. On the path back to the circulation, lymph passes through **lymph nodes**, where the fluid is filtered and immune reactions take place.

• The **spleen** is the largest lymphatic organ. It stores platelets and red blood cells, removes worn-out blood cells, and produces blood cells during fetal life.

The lymphatic system • Figure 12.1

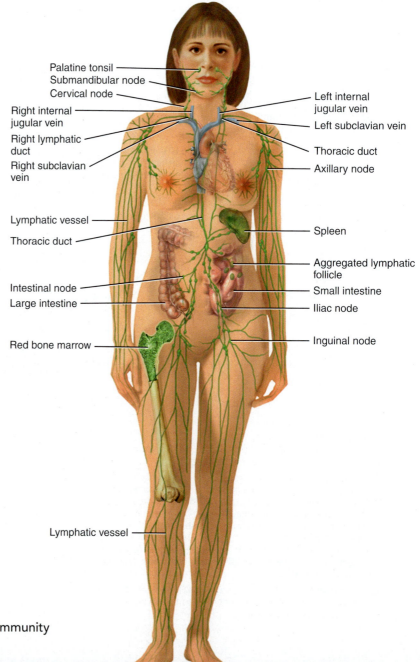

Palatine tonsil
Submandibular node
Cervical node
Right internal jugular vein
Right lymphatic duct
Right subclavian vein
Lymphatic vessel
Thoracic duct
Intestinal node
Large intestine
Red bone marrow
Lymphatic vessel

Left internal jugular vein
Left subclavian vein
Thoracic duct
Axillary node
Spleen
Aggregated lymphatic follicle
Small intestine
Iliac node
Inguinal node

---

2 Immune Reponses Help Protect the Body Against Disease 345

- Physical barriers such as the skin, mucus, and mucous membranes, as well as chemicals in various secretions, are the first line of nonspecific innate defenses. The second line of nonspecific innate defense includes internal defenses such as **antimicrobial substances**, macrophages, **natural killer** cells, inflammation (as shown at right), and **fever**. The lymphatic system also has the ability to adapt to specific pathogens through **adaptive immunity**.

- As shown below, adaptive immunity is of two types: **cell-mediated immunity** and **antibody-mediated immunity**. In cell-mediated immunity, T cells kill infected cells in tissues and release pathogens for subsequent phagocytosis. In antibody-mediated immunity, B cells differentiate to form plasma cells that secrete **antibodies**, which attack various pathogens. Antibody-mediated immunity occurs mostly in blood and fluids.

- In adaptive immunity, various cells present specific **antigens** to helper **T cells** and **B cells** that contain receptors for those antigens. These cells undergo **clonal activation**, where they proliferate and differentiate into active effector cells and memory cells. Helper T cells then activate **cytotoxic** T cells to undergo clonal selection. Cytotoxic T cells attack and kill infected cells (cell-mediated immunity). Plasma cells, which are derived from B cells, secrete antibodies (antibody-mediated immunity).

- Memory cells help the lymphatic system react more strongly and quickly when your body is subsequently reexposed to any given antigen. This **immunological memory** is the basis for vaccination.

- **Active immunity** produces antibodies following exposure to an antigen, while **passive immunity** supplies antibodies from another source without a full immune response. Passive immunity is temporary, but active immunity can last for decades.

Inflammation • Figure 12.7c

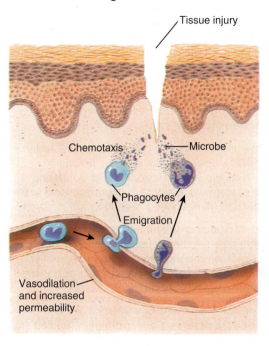

- Tissue injury
- Chemotaxis
- Microbe
- Phagocytes
- Emigration
- Vasodilation and increased permeability

Overview of adaptive immune responses • Figure 12.8

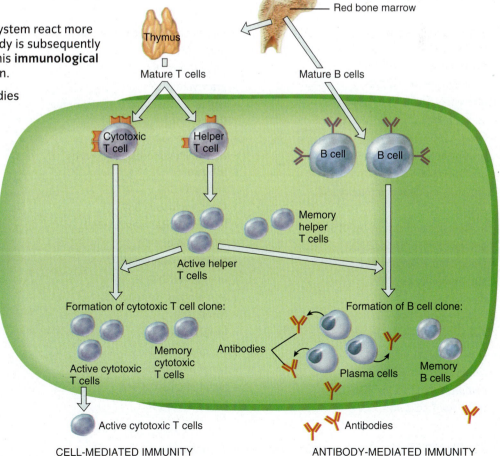

- Red bone marrow
- Thymus
- Mature T cells
- Mature B cells
- Cytotoxic T cell
- Helper T cell
- B cell
- B cell
- Memory helper T cells
- Active helper T cells
- Formation of cytotoxic T cell clone:
- Active cytotoxic T cells
- Memory cytotoxic T cells
- Formation of B cell clone:
- Antibodies
- Plasma cells
- Memory B cells
- Active cytotoxic T cells
- Antibodies

CELL-MEDIATED IMMUNITY

ANTIBODY-MEDIATED IMMUNITY

3 HIV Causes a Breakdown of the Immune Response 356

- **Human immunodeficiency virus (HIV)** is a retrovirus that attacks helper T cells of the lymphatic system. The virus incorporates its genetic information in the host's DNA and forces the host cell to make new viruses. Ultimately, the host cell dies and the virus spreads to other cells. In time, all the helper T cells are destroyed and the infection proceeds to **acquired immunodeficiency syndrome (AIDS)**.

- As shown, AIDS patients suffer from numerous constitutional symptoms, including chronic fatigue, weight loss, night sweats, skin rashes, and various lesions of the mouth and gums. They are susceptible to a host of opportunistic diseases that would be fended off easily by a healthy lymphatic system.

- HIV is present in blood and body fluids. It is passed from person to person through the exchange of body fluids (via unprotected sexual intercourse, anal intercourse, oral sex, and sharing of intravenous drug needles). It is not transmitted by insect bites or casual contact.

- While there is no cure for HIV/AIDS, patients have responded to combinations of **reverse transcriptase inhibitors** and **protease inhibitors**. While these drugs inhibit HIV replica-

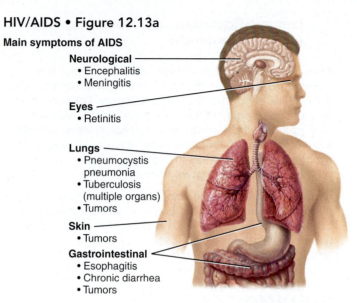

HIV/AIDS • Figure 12.13a

Main symptoms of AIDS

Neurological
- Encephalitis
- Meningitis

Eyes
- Retinitis

Lungs
- Pneumocystis pneumonia
- Tuberculosis (multiple organs)
- Tumors

Skin
- Tumors

Gastrointestinal
- Esophagitis
- Chronic diarrhea
- Tumors

tion, help restore T cell counts, and improve general health, an infected patient on this therapy can still transmit the virus to another person. Even with treatment, the disease is usually fatal.

Key Terms

- acquired immunodeficiency syndrome (AIDS) 362
- active immunity 360
- adaptive immunity 350
- afferent lymphatic vessels 347
- allergen 360
- allergy 360
- anaphylactic shock 360
- antibody 348
- antibody-mediated immunity 354
- antigen 350
- antigen-presenting cell (APC) 356
- antimicrobial proteins 352
- autoimmune disease 361
- autoimmunity 361
- B cell 352
- B lymphocyte 354
- cell-mediated immunity 354
- clonal activation 356
- clonal selection 354
- complement proteins 352
- cytotoxic T cells 354

- efferent lymphatic vessels 347
- fever 353
- granulysins 357
- granzymes 357
- highly active antiretroviral therapy (HAART) 362
- histamine 360
- human immunodeficiency virus (HIV) 362
- immunity 348
- immunoglobulin 359
- immunological memory 359
- inflammation 352
- innate immunity 350
- interferons 352
- iron-binding proteins 352
- lymph 344
- lymph node 344
- lymphatic capillary 344
- lymphatic nodule 344
- lymphatic vessel 344
- lymphocyte 348

- major histocompatibility complex (MHC) 356
- passive immunity 360
- pathogen 348
- phagocyte 352
- phagocytosis 352
- perforins 357
- primary lymphatic organs 344
- protease inhibitor 362
- red bone marrow 344
- reverse transcriptase inhibitor 362
- right lymphatic duct 344
- secondary lymphatic organs and tissues 344
- spleen 344
- T cell 352
- T lymphocyte 354
- thoracic duct 344
- thymus 344
- vaccination 360
- vaccine 360

Critical and Creative Thinking Questions

1. After a car accident, surgeons had to remove Saud's ruptured spleen. What is the role of the spleen, and what consequences will Saud have as a result of its loss?

2. Jill is allergic to bee venom. In case she gets stung, she carries injectable epinephrine with her to prevent anaphylactic shock. What is anaphylactic shock, and how does epinephrine help?

3. Rachel gets a seasonal flu shot each year to prevent influenza. For a couple days after her latest shot, she felt sick with flu-like symptoms. Explain how flu vaccines work to prevent this disease and why Rachel felt this way.

4. Robert was filing some papers and received a cut on his finger. Later, he noticed some pain, redness, swelling, and warmth around the area of the cut. Explain the physiological events that account for his symptoms.

5. Sarah's mom has been bedridden in the hospital for several months. The nurses come in and manually "exercise" her legs to prevent edema (swelling). Explain how manipulating her limbs would help inhibit edema.

What is happening in this picture?

This patient has developed a rare disease involving reddish brown skin tumors on his back and shoulders. He has also complained of fatigue, joint pain, weight loss, and night sweats. All of these symptoms could be related to a single disorder.

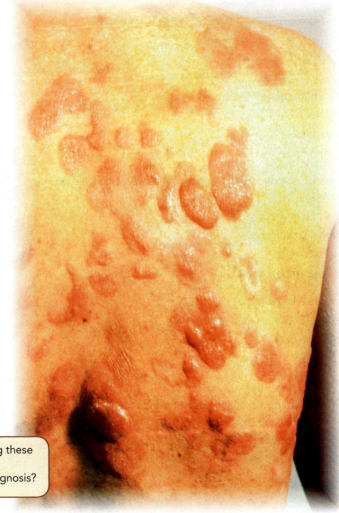

Think Critically 1. What is causing these symptoms?
2. How would you confirm your diagnosis?

Self-Test

(Check your answers in Appendix C.)

1. HIV specifically targets _____.
   a. cytotoxic T cells
   b. helper T cells
   c. plasma cells
   d. B cells

Use this figure to answer questions 2–4.

2. The organ shown is _____.
   a. a lymphatic vessel
   b. the spleen
   c. the thymus
   d. a lymph node

3. The vessels leading into the organ on the left are which of the following?
   a. afferent lymphatic vessels
   b. lymphatic capillaries
   c. lymphatic ducts
   d. efferent lymphatic vessels

4. What does this organ do?
   a. produce T cells
   b. produce B cells
   c. filter lymph
   d. pump lymph

5. Which of the following drains interstitial fluid from 75% of the body's area?
   a. inferior vena cava
   b. right lymphatic duct
   c. left lymphatic duct
   d. thoracic duct

6. An allergic response is an exaggerated example of which of the following?
   a. inflammation
   b. immunological memory
   c. innate defense
   d. passive immunity

7. Where do T cells form and mature?
   a. spleen
   b. lymph nodes
   c. thoracic duct
   d. thymus

8. Interferon is a defense against which of the following?
   a. bacteria
   b. virus
   c. fungi
   d. venom

Use this figure to answer questions 9–12.

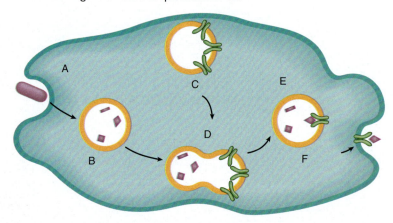

9. The cellular process shown in step A is which of the following?
   a. Pinocytosis
   b. Phagocytosis
   c. Receptor-mediated endocytosis
   d. Secretion

10. Which step isolates the foreign antigens?

    a. B   c. D

    b. C   d. F

11. In which step is the antigen–MHC complex presented?

    a. C   c. E

    b. D   d. F

12. What is the purpose of whole process shown?

    a. destroying foreign bacteria

    b. making lymphocytes

    c. distinguishing foreign cells from body cells

    d. creating immunological memory

Use this diagram to answer questions 13–14.

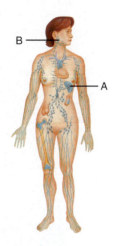

13. Which of the following is *not* one of the activities of the organ labeled A?

    a. pump lymph

    b. store platelets

    c. carry out immune responses

    d. destroy worn-out blood cells

14. The organ labeled B is which of the following?

    a. lymph node

    b. spleen

    c. tonsil

    d. aggregated lymphatic follicles

15. A reverse transcriptase inhibitor would be used to prevent the growth of _____.

    a. botulism toxin

    b. tetanus bacteria

    c. HIV

    d. influenza virus

16. Which cell allows for long-lasting immunization?

    a. cytotoxic T cell

    b. activated cytotoxic T cell

    c. activated helper T cell

    d. memory cytotoxic T cell

THE PLANNER ✔

Review your Chapter Planner on the chapter opener and check off your completed work.

The Respiratory System

Mountain climbing can be a dangerous sport. As you climb, ascending higher and higher through the Earth's atmosphere, the atmospheric pressure decreases. As the atmospheric pressure decreases, so does the concentration of oxygen in the air. Climbers can experience dizziness and may even faint if the oxygen concentration gets too low. To cope with the effects of high altitudes, climbers make extended stops in base camps located at various altitudes along the way to their destination. During these rest stops, their bodies adjust (acclimate) to the decreased oxygen levels. As they are acclimating, their breathing becomes rapid and their hearts beat faster. Climbers also use bottled oxygen to counteract the deficit.

The same effects can be felt when acclimating to a high-altitude city such as Denver, Colorado, the "Mile-High City." When visiting sports teams from cities at lower altitudes play the home teams in Denver, they often travel to the city early to give their players a chance to acclimate; in addition, they supply oxygen on the sidelines. Others who experience high altitudes include fighter pilots, astronauts, and passengers and crew members on commercial flights; commercial aircraft and space vehicles must maintain pressurized cabins so that their passengers have sufficient oxygen, and fighter pilots wear oxygen masks.

In this chapter, you will learn how your body takes in oxygen and removes carbon dioxide via the respiratory system. Let's take a look at this amazing system that permits your body to make use of the oxygen from the air through mechanical ventilation and gas exchange.

NATIONAL GEOGRAPHIC

CHAPTER OUTLINE

CHAPTER PLANNER ✔

- ❏ Study the picture and read the opening story.
- ❏ Scan the Learning Objectives in each section:
  p. 372 ❏   p. 376 ❏   p. 382 ❏   p. 386 ❏   p. 390 ❏
- ❏ Read the text and study all visuals.
  Answer any questions.

Analyze key features

- ❏ InSight, pp. 372–373 ❏
- ❏ Process Diagram, p6. 376–377 ❏
- ❏ What a Health Provider Sees, p. 381 ❏
- ❏ Stop: Answer the Concept Checks before you go on:
  p. 375 ❏   p. 380 ❏   p. 386 ❏   p. 388 ❏   p. 392 ❏

End of chapter

- ❏ Review the Summary and Key Terms.
- ❏ Answer the Critical and Creative Thinking Questions.
- ❏ Answer What is happening in this picture?
- ❏ Complete the Self-Test and check your answers.

Respiratory Organs Move Air and Exchange Gases

LEARNING OBJECTIVES

1. **Identify** the structures of the respiratory system.

2. **Explain** the functions of the respiratory system.

3. **Compare** what happens during breathing, external respiration, and internal respiration.

The respiratory system consists of the nose, pharynx (throat), larynx (voice box), trachea (wind pipe), bronchi, bronchioles, and lungs (**Figure 13.1**). The major purpose of the respiratory system is gas exchange, which involves delivering oxygen from the air to the body's cells and tissues, removing carbon dioxide from them, and expelling that carbon diox-

---

InSight The respiratory system • Figure 13.1

The nose and pharynx make up the upper respiratory tract **a.**, which prepares the air for entry into the lower respiratory tract. The lower respiratory tract **b.** consists of the larynx, trachea, bronchi, and bronchioles.

a. Upper respiratory tract

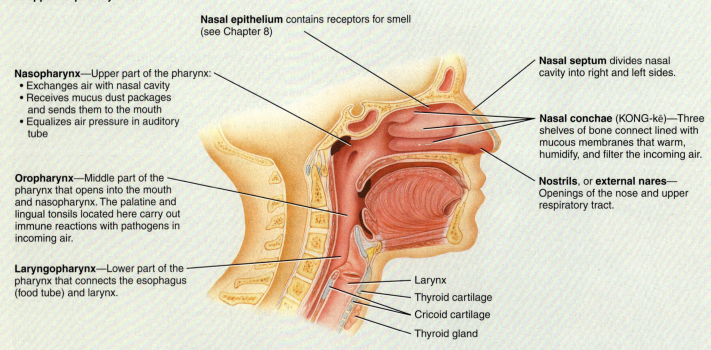

Nasal epithelium contains receptors for smell (see Chapter 8)

Nasal septum divides nasal cavity into right and left sides.

Nasopharynx—Upper part of the pharynx:
- Exchanges air with nasal cavity
- Receives mucus dust packages and sends them to the mouth
- Equalizes air pressure in auditory tube

Nasal conchae (KONG-kē)—Three shelves of bone connect lined with mucous membranes that warm, humidify, and filter the incoming air.

Oropharynx—Middle part of the pharynx that opens into the mouth and nasopharynx. The palatine and lingual tonsils located here carry out immune reactions with pathogens in incoming air.

Nostrils, or **external nares**—Openings of the nose and upper respiratory tract.

Laryngopharynx—Lower part of the pharynx that connects the esophagus (food tube) and larynx.

Larynx
Thyroid cartilage
Cricoid cartilage
Thyroid gland

Ask Yourself

Which of the following properly orders the flow of air from the environment to the alveolus?

a. Nasal cavity, larynx, pharynx, trachea, bronchus, bronchiole, alveolus

b. Nasal cavity, pharynx, larynx, trachea, bronchiole, bronchus, alveolus

c. Pharynx, nasal cavity, larynx, trachea, bronchus, bronchiole, alveolus

d. Nasal cavity, pharynx, larynx, trachea, bronchus, bronchiole, alveolus

e. Alveolus, bronchiole, bronchus, trachea, larynx, pharynx, nasal cavity

ide into the air. Without ventilation (breathing), however, the respiration process would soon be out of commission.

The Respiratory Organs Are Functionally Divided into Upper and Lower Respiratory Tracts

The **upper respiratory tract** consists of the **nose**, **nasal epithelium**, and **pharynx**; it warms, filters, and humidifies incoming air; detects smells in that air; and expels mucus from the respiratory tract. The **lower respiratory tract** starts with the **larynx**, which serves as the entrance into the trachea. The **epiglottis**, a portion of the larynx, covers the **glottis**, the vocal folds in the layrnx plus the space between them, during swallowing so that food does not enter the airways. The larynx is also involved in producing sounds. Within the larynx are several folds of tissues, the **vocal cords**, which aid in the creation of sound. As air passes through the larynx, the vocal cords vibrate, changing the flow of air through the passage.

The lower respiratory tract continues with the **trachea**, **bronchi**, **bronchioles**, and **lungs**. The trachea, bronchi, and bronchioles are a series of branching tubes that directs the air into and through the lung tissue. Because all of the airways resemble an upside-down tree with many branches, their arrangement is known as the bronchial tree.

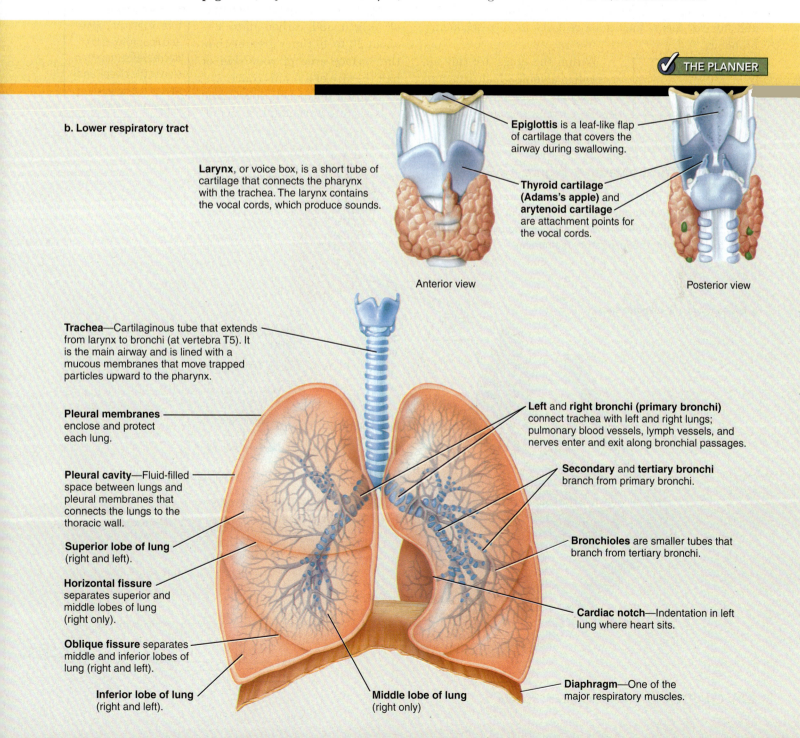

b. Lower respiratory tract

Epiglottis is a leaf-like flap of cartilage that covers the airway during swallowing.

Larynx, or voice box, is a short tube of cartilage that connects the pharynx with the trachea. The larynx contains the vocal cords, which produce sounds.

Thyroid cartilage (Adams's apple) and **arytenoid cartilage** are attachment points for the vocal cords.

Anterior view

Posterior view

Trachea—Cartilaginous tube that extends from larynx to bronchi (at vertebra T5). It is the main airway and is lined with a mucous membranes that move trapped particles upward to the pharynx.

Pleural membranes enclose and protect each lung.

Left and **right bronchi (primary bronchi)** connect trachea with left and right lungs; pulmonary blood vessels, lymph vessels, and nerves enter and exit along bronchial passages.

Pleural cavity—Fluid-filled space between lungs and pleural membranes that connects the lungs to the thoracic wall.

Secondary and **tertiary bronchi** branch from primary bronchi.

Superior lobe of lung (right and left).

Horizontal fissure separates superior and middle lobes of lung (right only).

Bronchioles are smaller tubes that branch from tertiary bronchi.

Oblique fissure separates middle and inferior lobes of lung (right and left).

Cardiac notch—Indentation in left lung where heart sits.

Inferior lobe of lung (right and left).

Middle lobe of lung (right only)

Diaphragm—One of the major respiratory muscles.

The lungs are enclosed in the **pleural cavity** and surrounded by a **pleural membrane**. The pleural membrane forms pleural fluid, which helps the lungs adhere to the thoracic wall, allowing the movement of the thoracic wall to draw air in and pushing it out of the lungs.

The lungs extend from the diaphragm to slightly above the clavicles and lie against the ribs. The broad bottom portion of each lung is its *base*; the narrow top portion is the *apex*. The left lung has an indentation in which the heart lies (cardiac notch). Because of the space occupied by the heart, the left lung is about 10% smaller than the right lung.

Deep grooves called *fissures* divide each lung into lobes. The left lung has only two lobes, the superior and inferior lobes. The right lung has three lobes: the superior, middle, and inferior lobes. Each lobe receives its own secondary bronchus.

> **alveolus** (al-VĒ-ō-lus; singular is *alveolus*) An air sac in the lungs.

Within the lungs are tiny air sacs called **alveoli**, which attach to the alveolar ducts. Gas exchange occurs between the lungs and the blood at the alveoli. Around the alveoli, the pulmonary arteriole and venule form lush networks of blood capillaries. The alveoli have thin walls for gas exchange with the red blood cells and plasma that carry oxygen and carbon dioxide (**Figure 13.2**). Gas exchange between the air spaces in the lungs and the blood takes place by diffusion across the alveolar and capillary walls. This exchange provides the oxygen needed by our cells, eliminates carbon dioxide produced by the body cells, and plays a crucial role in regulating the pH of the blood (a topic we will examine more closely in Chapter 15).

The lungs contain roughly 300 million alveoli. They provide a huge surface area for the exchange of oxygen and carbon dioxide—about 30 to 40 times greater than the surface area of your skin or half the size of a tennis court.

Within the alveoli, specialized cells secrete **surfactant**, a chemi-

> **surfactant** (sur-FAK-tant) A complex mixture of phospholipids and lipoproteins that decreases surface tension in the alveoli and makes the lungs more compliant (flexible).

Bronchioles and alveoli • Figure 13.2

a. Alveolar sacs are two or more alveoli that share a common opening into an alveolar duct. **b.** The exchange of respiratory gases occurs by diffusion across the alveolar and capillary walls, which together form the respiratory membrane.

a. Bronchioles and alveolar sacs

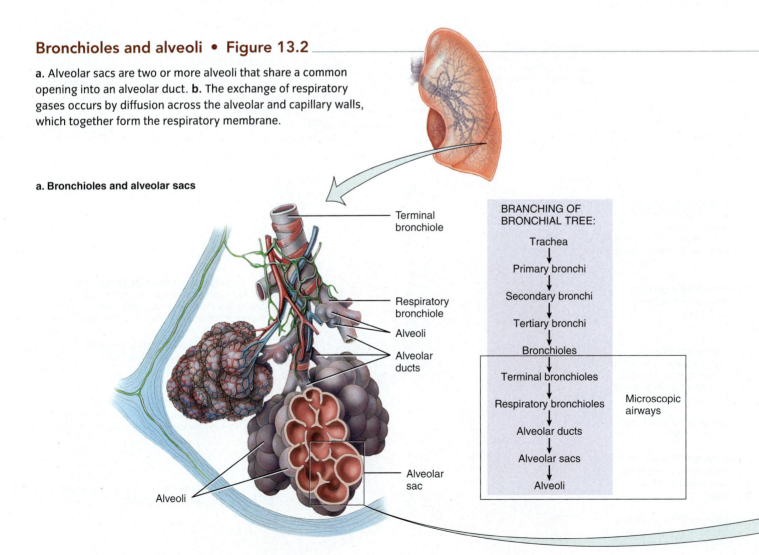

Terminal bronchiole

Respiratory bronchiole

Alveoli

Alveolar ducts

Alveoli

Alveolar sac

BRANCHING OF BRONCHIAL TREE:

Trachea
↓
Primary bronchi
↓
Secondary bronchi
↓
Tertiary bronchi
↓
Bronchioles
↓
Terminal bronchioles
↓
Respiratory bronchioles
↓
Alveolar ducts
↓
Alveolar sacs
↓
Alveoli

Microscopic airways

cal that improves the flexibility of the lung tissue, allowing the lungs to expand more easily. In addition, macrophages remove dust and debris from the alveolar space.

The Respiratory System Performs Two Important Processes: Breathing and Respiration

In order to have gas exchange, the alveoli must be filled with fresh air. Several processes must occur to get adequate amounts of oxygen from the air to the body cells and to get carbon dioxide waste products from the body cells back into the air:

- **Breathing** (*pulmonary ventilation*). Flow of air into and out of the lungs.
- **Respiration**. Exchange of gases across membranes:
  - **External respiration** is gas exchange between the alveoli and the blood in pulmonary capillaries—an exchange with the "external" environment. Here, the blood picks up oxygen from the air and releases carbon dioxide into the air.
  - **Internal respiration** is gas exchange between the tissue cells and the blood in the systemic capillaries—an exchange with the "internal" environment of the body. Here the blood releases oxygen into the tissues and picks up carbon dioxide from them.

CONCEPT CHECK 🛑 STOP

1. **Which** structures direct air from the upper respiratory tract into the alveoli of the lungs?
2. **What** is the major function of the respiratory system?
3. **What** is the difference between external and internal respiration?

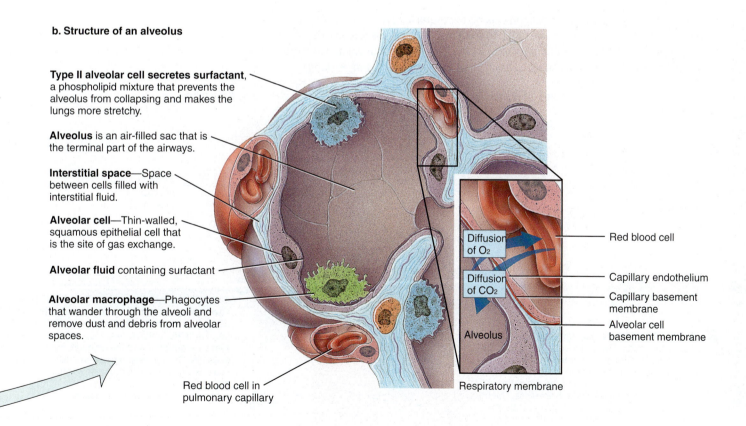

b. Structure of an alveolus

Type II alveolar cell secretes surfactant, a phospholipid mixture that prevents the alveolus from collapsing and makes the lungs more stretchy.

Alveolus is an air-filled sac that is the terminal part of the airways.

Interstitial space—Space between cells filled with interstitial fluid.

Alveolar cell—Thin-walled, squamous epithelial cell that is the site of gas exchange.

Alveolar fluid containing surfactant

Alveolar macrophage—Phagocytes that wander through the alveoli and remove dust and debris from alveolar spaces.

Red blood cell in pulmonary capillary

Diffusion of O₂
Diffusion of CO₂
Alveolus
Respiratory membrane

Red blood cell
Capillary endothelium
Capillary basement membrane
Alveolar cell basement membrane

Breathing Involves Changes in Pressures and Volumes

LEARNING OBJECTIVES

1. **Describe** the steps in the process of pulmonary ventilation.
2. **Identify** the various lung volumes and capacities.
3. **Explain** the different breathing patterns.

Pulmonary ventilation, which as we have seen is the flow of air between the atmosphere and the lungs, occurs due to differences in air pressure.

To understand the mechanics of breathing, as shown in **Figure 13.3**, you need to know some physical and anatomical principles:

1. Like fluid, air moves in response to changes in pressure; specifically, air moves from regions of higher pressure to regions of lower pressure. We inhale, or breathe in, when the pressure inside the lungs is less than the atmospheric air pressure. We exhale, or breathe out, when the pressure in the lungs is greater than the atmospheric air pressure.

2. According to Boyle's law, at a constant temperature, the pressure of a gas is inversely related to its volume. So when the volume of a gas increases the pressure decreases, and vice versa.

3. The lungs and chest wall are elastic. The elastic properties of the chest wall and lungs oppose each other: The chest tends to pull outward, while the lungs

PROCESS DIAGRAM

The breathing process • Figure 13.3

THE PLANNER

Various respiratory muscles contract and expand the thoracic cavity during normal breathing. Additional muscles are recruited for deep inhalations and forced exhalations. The contraction and relaxation of the respiratory muscles change the volume and pressure in the thoracic cavity and alveoli, causing air to flow into and out of the lungs.

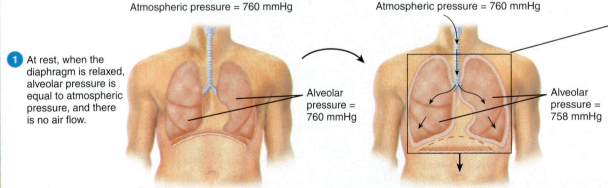

Atmospheric pressure = 760 mmHg

Atmospheric pressure = 760 mmHg

1 At rest, when the diaphragm is relaxed, alveolar pressure is equal to atmospheric pressure, and there is no air flow.

Alveolar pressure = 760 mmHg

Alveolar pressure = 758 mmHg

2 During inhalation, the diaphragm and external intercostals contract. The chest cavity expands, and the alveolar pressure drops below atmospheric pressure. Air flows into the lungs in response to the pressure gradient and the lung volume expands. During deep inhalation, the scalene and sternocleidomastoid muscles expand the chest further, thereby creating a greater drop in alveolar pressure.

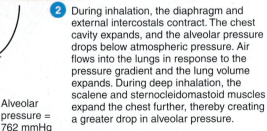

Atmospheric pressure = 760 mmHg

3 During exhalation, the diaphragm and external intercostals relax. The chest and lungs recoil, the chest cavity contracts, and the alveolar pressure increases above atmospheric pressure. Air flows out of the lungs in response to the pressure gradient, and the lung volume decreases. During forced exhalations, the internal intercostals and abdominal muscles contract, thereby reducing the size of the chest cavity further and creating a greater increase in alveolar pressure.

Alveolar pressure = 762 mmHg

tend to collapse inward. However, they adhere to one another through the negative pressure of the pleural space and suction created by the adhesive properties of pleural fluid, so they act in unison, much like a spring.

Muscles Contract and Relax to Move Air into and out of the Lungs During Ventilation

Breathing in is called **inhalation** or inspiration. When the respiratory muscles (such as the **diaphragm** and external intercostals; Figure 13.3) contract during inhalation, they expand the thoracic cavity (like pulling on a spring) and, subsequently, the lungs. The increased volume of the thoracic cavity and lungs results in decreased pressure in the alveoli. The alveolar pressure is now less than atmospheric pressure, so air flows into the lungs. The more muscles involved in the process, the more air that can be moved into the lungs.

Breathing out is called **exhalation** or expiration and begins when the respiratory muscles relax. During normal exhalation, the elastic properties of the lungs bring the thoracic cavity back to its original volume (like letting go of a spring). The decreased volume of the thoracic cavity increases the pressure in this space. The pressure in the alveolus is now greater than the atmospheric pressure, so air flows out of the lungs (see Figure 13.3).

Unlike quiet inhalation, there are no muscular contractions involved in quiet exhalation; it is a passive process. Exhalation becomes active only during forceful breathing, such as in playing a flute or during exercise. Forced exhalation involves contraction of muscles of the thorax and abdomen (for example, internal intercostals, external and internal obliques, transversus and rectus abdominis), causing a further decrease in the size of the thorax. This additional decrease in thoracic size increases the pressure, and more air leaves the lungs.

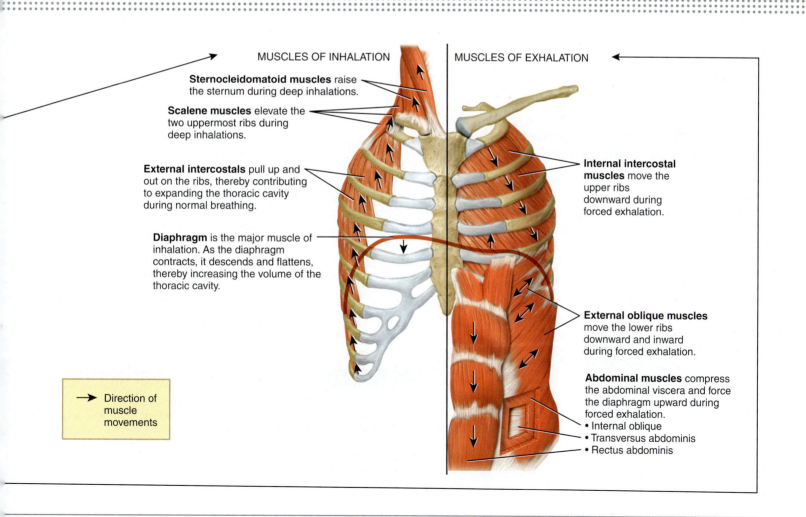

MUSCLES OF INHALATION   MUSCLES OF EXHALATION

Sternocleidomastoid muscles raise the sternum during deep inhalations.

Scalene muscles elevate the two uppermost ribs during deep inhalations.

External intercostals pull up and out on the ribs, thereby contributing to expanding the thoracic cavity during normal breathing.

Diaphragm is the major muscle of inhalation. As the diaphragm contracts, it descends and flattens, thereby increasing the volume of the thoracic cavity.

Internal intercostal muscles move the upper ribs downward during forced exhalation.

External oblique muscles move the lower ribs downward and inward during forced exhalation.

Abdominal muscles compress the abdominal viscera and force the diaphragm upward during forced exhalation.
• Internal oblique
• Transversus abdominis
• Rectus abdominis

→ Direction of muscle movements

As the lungs expand, the air molecules inside occupy a larger volume, which causes the air pressure inside to decrease. Because atmospheroic air pressure is now higher than the pressure inside the lungs, air moves into the lung. By contrast, when the lung volume decreases, the pressure inside the lung increases. Air then flows from an area of higher pressure in the alveoli to the area of lower pressure in the atmosphere.

The pressure differences during breathing are not huge. In fact, they are only about 2 mm of mercury (2 mm Hg), which is equal to 0.26% of atmospheric pressure (760 mm Hg). Yet these small pressure differences are sufficient to move about 500 mL of air into and out of the lungs with each breath (**Figure 13.4**). The greater the pressures created, the more air that will be moved.

The pressure in the pleural space is slightly lower than in the lungs. This pressure difference, along with the stickiness of the pleural fluid, keeps the lungs "adhered" to the chest wall so that the lungs and chest wall can move together. If air is introduced into the pleural space (as might occur with a puncture wound to the chest), it is called a **pneumothorax**. A condition known as *pleural effusion* occurs if fluid (for example, blood, pus, serous fluid) collects in the pleural space. In either case, the connection between the thoracic wall and the lung may be disrupted, and the lung can collapse. Without this connection to the wall of the chest, the lungs cannot inflate, so ventilation is not possible.

Respiratory Health Is Sometimes Tested Using a Spirometer

Like a balloon, your lungs are capable of expanding and filling with various volumes of air. While at rest, a healthy adult breathes about 12 times per minute, with each inhalation and exhalation moves about 500 mL of air into and out of the lungs. As illustrated in Figure 13.4a, the **tidal volume** is the volume of air that moves in and out during a normal quiet breath.

Tidal volume varies considerably from one person to the next and even in the same person at different times.

About 350 mL of the tidal volume, only 70%, actually reaches the respiratory bronchioles and alveolar sacs to participate in gas exchange. The other 150 mL (30%) does not participate in gas exchange because it remains in the airways of the nose, pharynx, larynx, trachea, bronchi, and bronchioles. Collectively, these conducting airways are known as **anatomic dead space**.

If you consciously take a very deep breath, your lungs can take in a good deal more air than 500mL. This additional amount of air is referred to as the **inspiratory reserve volume**. Similarly, you can expel more air beyond what exits your body during a normal quiet exhalation; this additional volume of air is called the **expiratory reserve volume**. Finally, there is a certain volume of air that remains in your lungs and airways that cannot be expelled because you cannot contract your airways to total collapse; this remaining volume of air is called the **residual volume**.

Lung capacities are combinations of specific lung volumes (the percentage of air that lungs can hold at any given period of time). The volume of air that is moved from a maximum inhalation to a maximum exhalation, for example, is called the **vital capacity**.

Total lung capacity is the sum of the vital capacity and residual volume, generally about 6000 mL in males and 4200 mL in females. Lung volumes and capacities vary with age, gender, and body size. For example, lung volume and capacities may be smaller in older people, females, and shorter people.

Health providers measure all of these volumes, except residual volume, using a device called a **spirometer**. The record produced by a spirometer is called a **spirogram** and is shown in Figure 13.4a. The vital capacity measurement is especially important because it can give the doctor an idea of lung capacity, lung flexibility, and levels of air flow to the alveolus. For example, people with obstructive lung disease, which affects the airways, will have normal or higher than normal lung capacities but lower than normal airflow. In contrast, people with restrictive pulmonary disease, which affects the structure of the lungs and their ability to expand, will have lower than normal lung volumes and capacities.

Put It Together

(For Figure 13.4) If a patient develops pneumothorax:
a. There will be no effect on the ventilation process.
b. Exhalation would be difficult.
c. Abdominal muscles would need to contract to inhale properly.
d. Lung pressures would not be able to decrease to admit air.
e. Vital capacity would increase.

Lung volumes and capacities • Figure 13.4

The volumes of air and lung capacities associated with breathing, deep inhalation, and deep exhalation can be measured with a spirometer and seen in a spirogram.

a. Pulmonary function tests use a spirometer to generate a spirogram. Note that spirograms are read from right to left.

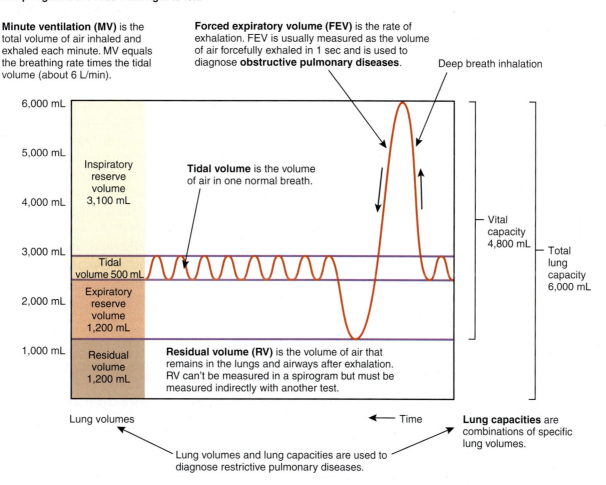

Minute ventilation (MV) is the total volume of air inhaled and exhaled each minute. MV equals the breathing rate times the tidal volume (about 6 L/min).

Forced expiratory volume (FEV) is the rate of exhalation. FEV is usually measured as the volume of air forcefully exhaled in 1 sec and is used to diagnose **obstructive pulmonary diseases**.

Deep breath inhalation

Tidal volume is the volume of air in one normal breath.

Inspiratory reserve volume 3,100 mL

Tidal volume 500 mL

Expiratory reserve volume 1,200 mL

Residual volume 1,200 mL

Residual volume (RV) is the volume of air that remains in the lungs and airways after exhalation. RV can't be measured in a spirogram but must be measured indirectly with another test.

Vital capacity 4,800 mL

Total lung capacity 6,000 mL

Lung volumes

← Time

Lung capacities are combinations of specific lung volumes.

Lung volumes and lung capacities are used to diagnose restrictive pulmonary diseases.

b. Changes in lung volumes/capacities and air flow rates can be used for diagnosing various pulmonary diseases.

	Disease	Lung volumes and capacities	Rates of air flow (e.g., FEV)
Obstructive pulmonary diseases affect the airways and hinder the flow of air.	• Asthma • Chronic pulmonary obstructive disease (COPD) • Emphysema • Chronic bronchitis	Normal or higher than normal	Lower than normal
Restrictive pulmonary diseases affect the structure of the lungs and limit the capacity of the lungs to expand.	• Pulmonary fibrosis • Sarcoidosis • Pulmonary edema	Lower than normal	Normal

Special Terms Are Used to Describe Breathing Patterns

WILEY PLUS Video

Several different breathing patterns can occur. Some of these patterns are normal, and some are not. Some are used to expel debris, and some occur in response to emotions. The following are some common terms used to describe certain breathing patterns:

- **Eupnea** (ŪP-nē-a). Normal, quiet breathing, which consists of shallow, deep, or combined shallow and deep breathing

- **Costal breathing** (chest breathing). Shallow breathing involving upward and downward movements of the chest due to contractions of the external intercostals

- **Diaphragmatic breathing** (abdominal breathing). Deep breathing involving outward movements of the abdomen due to contractions of the diaphragm.

- **Apnea**. Absence of breathing that can occur in premature infants who have incomplete brain development, in heavy snorers (**sleep apnea**), as a result of severe hypothermia

apnea (AP-nē-a) Temporary cessation of breathing.

(large drop in body temperature), and in some individuals who have suffered head trauma. With sleep apnea, some people periodically stop breathing and start only when the partial pressure of carbon dioxide (P_{CO_2}; see next section) builds up and stimulates breathing again.

- **Dyspnea**. Difficult breathing that can be due to airway obstructions (from foreign bodies, asthma attack, and so on), lungs that are less flexible because they contain excess fluid (as occurs in pulmonary edema or pneumonia), and thoracic injuries (such as rib fractures).

Table 13.1 shows some additional breathing patterns. Coughing and sneezing are protective reflexes that help maintain a clear airway. The other breathing patterns shown in the table are often used to express emotions. All of these movements are reflexes, but some can be initiated voluntarily.

Respiratory distress syndrome (RDS), a disease caused by very high surface tension in the alveolus that often leads to death among premature infants, was rather common in the 1950s. Since then, treatments for RDS have improved greatly (see *What a Health Provider Sees*).

Modified respiratory movements   Table 13.1

Movement	Description	Stimulus for Reflex
Coughing	A long-drawn and deep inhalation followed by a strong exhalation that suddenly sends a blast of air through the upper respiratory passages	A foreign body lodged in the larynx, pharynx, or epiglottis
Sneezing	Spasmodic contraction of muscles or exhalation that forcefully expels air through the nose and mouth	An irritation of the nasal mucosa
Hiccupping	Spasmodic contractions of the diaphragm followed by a spasmodic closure of the larynx, which produces a sharp sound on inhalation	Irritation of the sensory nerve endings of the gastrointestinal tract
Yawning	A deep inhalation through the widely opened mouth producing an exaggerated depression of the mandible	Drowsiness, fatigue, or someone else's yawning; precise cause is unknown
Sighing	A long-drawn and deep inhalation immediately followed by a shorter but forceful exhalation	Emotional
Sobbing	A series of convulsive inhalations followed by a single prolonged exhalation	Emotional
Crying	An inhalation followed by many short convulsive exhalations, during which the vocal cords vibrate; accompanied by characteristic facial expressions and tears	Emotional
Laughing	The same basic movements as crying, but the rhythm of the movements and the facial expressions usually differ from those of crying	Emotional

CONCEPT CHECK   STOP

1. **What** steps are involved in a normal inhalation?

2. **What** is the difference between inspiratory capacity and inspiratory reserve volume?

3. **What** is the difference between diaphragmatic breathing and costal breathing?

WHAT A HEALTH PROVIDER SEES

Respiratory Distress Syndrome (RDS)

In the 1950s, approximately 10,000 premature infants per year died of **respiratory distress syndrome (RDS)**. This disease is caused by very high surface tension in the alveoli. This surface tension is caused mostly by the thin film of water that coats the epithelial surface of the alveolis and helps prevent the tissues from drying out. Water contains numerous hydrogen bonds between the molecules that tend to hold the water molecules close together and resist separation. This is what causes water to bead up on a surface rather than spread out. During breathing, the lung tissue is stretched by the wall of the thorax. The lung tissue resists the stretching because it makes water molecules separate from one another. This is part of the mechanism that provides elasticity to the lung tissue and allows lungs to recoil when the muscles stop their contraction.

This high surface tension also requires the expenditure of a great deal of energy just to inflate the lungs. To overcome this situation, **Type II alveolar cells** produce surfactant, a phospholipid that mixes with the water layer in the alveolus, reduces the surface tension, and makes the lungs more flexible (or *compliant*). Surfactant also prevents the alveoli from collapsing, reducing the need to completely reinflate each alveolus with every breath and ensuring that the lungs remain in contact with the walls of the thorax.

Unfortunately, a fetus does not begin to secrete surfactant until 20 weeks gestation and does not secrete sufficient quantities to survive outside the womb until 26–28 weeks. Even babies born before the eighth month of gestation need to use more energy to ventilate than babies born closer to the normal end of pregnancy because of their lower levels of surfactant. RDS is more prominent in infants born to European American parents than those in other ethnic groups.

Symptoms of RDS include labored or irregular breathing, flaring of the **nostrils** upon inhalation, grunting during exhalation, and perhaps even a blue skin color (cyanosis). RDS can be detected in chest X-rays and with a blood test. Treatment depends on the severity of the symptoms and may include administration of supplemental oxygen and/or surfactant aerosols. The baby may also be put on a ventilator temporarily to help mechanically assist breathing until he or she starts to manufacture enough surfactant to make the lungs flexible and easy to expand. If the baby has to expend a tremendous amount of energy just to breathe, there will be little energy left for growth.

Pregnant women at risk for premature delivery may be given steroids to promote surfactant production by the fetus prior to delivery. Such treatments have lowered the RDS death rate dramatically, from 10,000 per year to about 1,000 per year.

Think Critically **1.** What does a blue color indicate about external respiration in a baby with RDS? **2.** Explain the benefits of surfactant administration to a baby with RDS.

Gases Are Exchanged at the Blood Capillaries

LEARNING OBJECTIVES

1. **Describe** the steps in the process of external respiration.

2. **Outline** the steps in the process of internal respiration.

3. **Identify** the routes of oxygen and carbon dioxide transport in the blood.

Air is a mixture of gases—nitrogen, oxygen, water vapor, carbon dioxide, and others. To understand gas exchange in the body, you must first become familiar with three physical principles of gases:

- Gases travel by diffusion from areas of higher concentration to areas of lower concentration.

- In a mixture of gases such as air, each gas exerts pressure independently of the other gases in the mixture. This pressure is called a **partial pressure** (abbreviated by P_x, where x is a particular gas). The sum of the partial pressures of all the gases in a mixture equals the total pressure. Partial pressure is a way to express the concentration of a gas. A gas with a higher partial pressure is more concentrated than one with a lower partial pressure. Therefore, gases diffuse from areas of higher partial pressures to areas of lower partial pressures.

- While partial pressures force gases across membranes, the solubility of the gases determines whether the gases will stay where they have been moved. **Solubility** indicates whether the material will remain dissolved in a liquid. For example, neither oxygen nor carbon dioxide is very soluble in the blood, so the body needs to overcome this issue to transport these important gases.

Diffusion Moves Gases Across the Capillary Membranes

By controlling the rate and depth of breathing, pulmonary ventilation can control the concentrations of oxygen (P_{O_2}) and carbon dioxide (P_{CO_2}) in the alveoli (**Figure 13.5**). In contrast, the P_{O_2} and P_{CO_2} of the blood are determined by the rate of metabolism of the cells that consume oxygen and produce carbon dioxide. As blood moves around the body, it encounters regions of different P_{O_2} and P_{CO_2} at the pulmonary and systemic capillaries.

In the pulmonary capillaries, where gaseous exchange with the air of the alveoli (external respiration) occurs:

- P_{O_2} is lower in the blood than in the air when the blood enters the pulmonary capillaries. Therefore, oxygen diffuses from the alveoli into the pulmonary capillaries.

- P_{CO_2} is higher in the blood than in the air when the blood enters the pulmonary capillaries. Therefore, carbon dioxide diffuses from the pulmonary capillaries into the alveoli.

In other words, external respiration in the lungs converts low-oxygen blood (deoxygenated) into high-oxygen blood (oxygenated). As the blood flows through the pulmonary capillaries, it picks up O_2 from alveolar air and unloads CO_2 into alveolar air.

In the systemic capillaries, where gaseous exchange with tissue cells (internal respiration) occurs:

- P_{O_2} is higher in the blood than in the interstitial fluid when it enters the systemic capillaries. Therefore, oxygen diffuses from the systemic capillaries into the interstitial fluid and on into the body cells.

- P_{CO_2} is lower in the blood than in the interstitial fluid when it enters the systemic capillaries. Therefore, carbon dioxide diffuses from the cells to the interstitial fluid to the systemic capillaries.

Because cells constantly use O_2 to produce ATP and constantly produce CO_2 as a by-product, internal respiration converts high-oxygen (oxygenated) blood into low-oxygen blood (deoxygenated). The low-oxygen blood returns to the heart and is pumped to the lungs for another cycle of external respiration.

Oxygen Is Transported Through the Blood Attached to Hemoglobin

Gases don't dissolve well in liquids—that is, they are not very soluble—except under pressure. Therefore, only a small percentage of oxygen (1.5%) or carbon dioxide (7%) gets transported in solution in the blood plasma. The rest gets transported bound to proteins or in other forms.

Partial pressures of oxygen and carbon dioxide during external and internal respiration • Figure 13.5

The changes in partial pressures of oxygen and carbon dioxide drive the diffusion of these gases in external and internal respiration. The partial pressures of various gases in the atmosphere are listed for comparison.

Partial pressures in atmosphere:

Nitrogen (P_{N_2})	597.4 mmHg (78.6%)
Oxygen (P_{O_2})	158.8 mmHg (20.9%)
Carbon dioxide (P_{CO_2})	0.3 mmHg (0.04%)
Other gases ($P_{other\ gases}$)	3.5 mmHg (0.46%)
Total:	**760.0 mmHg (100%)**

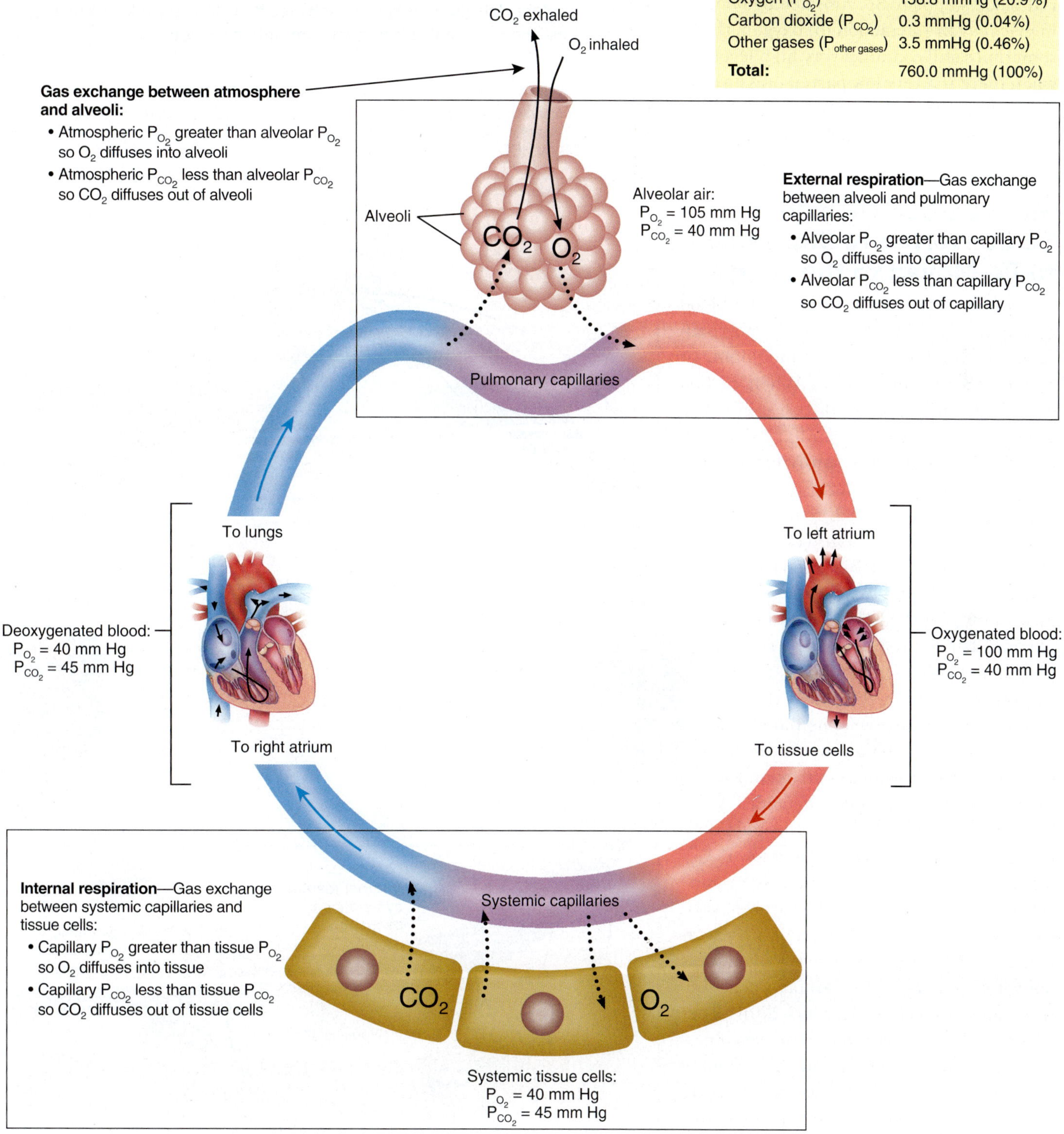

CO_2 exhaled

O_2 inhaled

Gas exchange between atmosphere and alveoli:
- Atmospheric P_{O_2} greater than alveolar P_{O_2} so O_2 diffuses into alveoli
- Atmospheric P_{CO_2} less than alveolar P_{CO_2} so CO_2 diffuses out of alveoli

Alveoli

CO_2 O_2

Alveolar air:
P_{O_2} = 105 mm Hg
P_{CO_2} = 40 mm Hg

External respiration—Gas exchange between alveoli and pulmonary capillaries:
- Alveolar P_{O_2} greater than capillary P_{O_2} so O_2 diffuses into capillary
- Alveolar P_{CO_2} less than capillary P_{CO_2} so CO_2 diffuses out of capillary

Pulmonary capillaries

To lungs

To left atrium

Deoxygenated blood:
P_{O_2} = 40 mm Hg
P_{CO_2} = 45 mm Hg

Oxygenated blood:
P_{O_2} = 100 mm Hg
P_{CO_2} = 40 mm Hg

To right atrium

To tissue cells

Internal respiration—Gas exchange between systemic capillaries and tissue cells:
- Capillary P_{O_2} greater than tissue P_{O_2} so O_2 diffuses into tissue
- Capillary P_{CO_2} less than tissue P_{CO_2} so CO_2 diffuses out of tissue cells

Systemic capillaries

CO_2 O_2

Systemic tissue cells:
P_{O_2} = 40 mm Hg
P_{CO_2} = 45 mm Hg

Most oxygen (98.5%) binds to hemoglobin, a protein in red blood cells; one hemoglobin molecule can bind four oxygen molecules. The resulting molecule, oxyhemoglobin, has a bright red color, leading to the characteristic appearance of arterial blood samples. Because hemoglobin has a very high *affinity for* (attraction to) oxygen, the blood can sometimes still contain a very high level of O_2, even when there is a very low partial pressure of O_2.

Figure 13.6 illustrates oxygen transport as well as our next topic, carbon dioxide transport.

Most Carbon Dioxide Must Be Converted to Bicarbonate to Be Moved to the Lungs

Carbon dioxide travels through the blood in the plasma (7%), bound to hemoglobin (23%) to form carbaminohemoglobin, or as bicarbonate ions (70%). When carbon dioxide from the cells diffuses into red blood cells in systemic capillaries, an enzyme called **carbonic anhydrase** combines it with water to form carbonic acid (H_2CO_3). Carbonic acid readily breaks apart into bicarbonate ions (HCO_3^-) and hydrogen ions (H^+). Bicarbonate ions, which are extremely soluble, quickly move out of red blood cells and into plasma. The hydrogen ions are rapidly picked up by the hemoglobin molecule to prevent shifts in the pH of the red blood cells. Because venous blood has more hemoglobin bound to CO_2 or hydrogen and less that is still bound to O_2, venous blood is a maroon color.

When the blood enters the pulmonary capillaries, the high P_{O_2} and low P_{CO_2} cause hemoglobin to pick up oxygen and release carbon dioxide and hydrogen ions. The hydrogen ions recombine with bicarbonate ions to form carbonic acid. Carbonic anhydrase changes the carbonic acid into water and carbon dioxide, which diffuses out of the red blood cells and into the alveoli.

The transformation of bicarbonate back into carbon dioxide is necessary to dispose of the carbon dioxide, as the highly soluble bicarbonate would never leave the blood. Carbon dioxide readily diffuses across the membrane and into the air.

Transport of oxygen and carbon dioxide in the blood • Figure 13.6

In external respiration, carbon dioxide is released from hemoglobin in exchange for oxygen, and bicarbonate is transformed back into carbon dioxide. In internal respiration, the opposite chemical changes occur: Hemoglobin lets go of oxygen in exchange for carbon dioxide and hydrogen ions, and bicarbonate is formed from carbon dioxide.

Transport of CO_2
7% dissolved in plasma
23% as Hb–CO_2 (carbaminohemoglobin)
70% as HCO_3^- (bicarbonate ions)

Transport of O_2
1.5% dissolved in plasma
98.5% as Hb–O_2 (oxyhemoglobin)

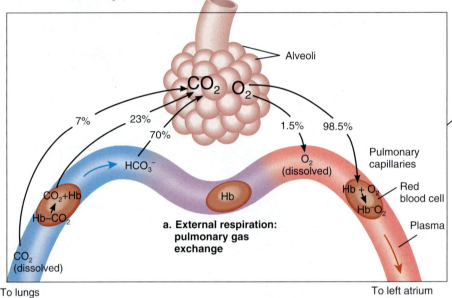

a. External respiration: pulmonary gas exchange

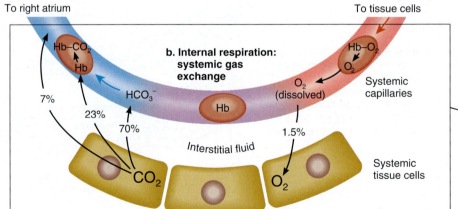

b. Internal respiration: systemic gas exchange

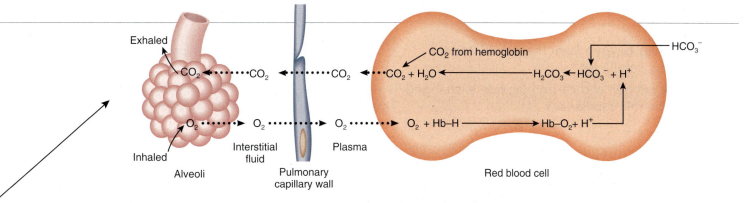

During external respiration, oxygen diffuses into red blood cells, and carbon dioxide diffuses out.

Alveolus and plasma	Direction of diffusion	Red blood cell (RBC)
The CO_2 that diffused out of the RBC into an alveolus is exhaled.	◄••• CO_2	Carbaminohemoglobin (Hb-CO_2) unbinds, yielding CO_2 and hemoglobin (Hb). The CO_2 diffuses out of the RBC into an alveolus.
O_2 diffuses from alveolus into plasma and RBC.	O_2 •••►	O_2 binds to Hb to become oxyhemoglobin (Hb-O_2) and displaces H^+.
Bicarbonate (HCO_3^-) diffuses from plasma into RBC.	HCO_3^- •••►	HCO_3^- combines with hydrogen ions (H^+) released from Hb to form carbonic acid (H_2CO_3).
The CO_2 that diffused out of the RBC into an alveolus is exhaled.	◄••• CO_2	The enzyme carbonic anhydrase converts H_2CO_3 into CO_2 and H_2O. The CO_2 diffuses out of the RBC into an alveolus.
	◄••• Cl^-	To maintain ionic balance in the wake of the HCO_3^- influx, chloride ions (Cl^-) move from the RBC into the plasma.

During internal respiration, oxygen diffuses from red blood cells into tissues, and carbon dioxide diffuses out.

Tissue cells and plasma	Direction of diffusion	Red blood cell
CO_2 diffuses from tissue cells into RBC.	CO_2 •••►	CO_2 reacts with Hb-O_2 to form Hb-CO_2, displacing O_2.
	◄••• O_2	The displaced O_2 from the RBC diffuses into the plasma and across the capillary membrane into the tissue cells.
	◄••• HCO_3^-	The enzyme carbonic anhydrase combines H_2O and CO_2 to form H_2CO_3, which breaks down spontaneously into HCO_3^- and H^+. H^+ binds to Hb in the RBC. HCO_3^- diffuses from RBC into plasma.
To maintain ionic balance in the wake of the HCO_3^- efflux, Cl^- moves from the plasma into the RBC.	Cl^- •••►	

1. **What** drives oxygen from the alveoli into the pulmonary capillaries?

2. **How** are oxygen and carbon dioxide exchanged in the systemic capillaries?

3. **How** are oxygen and carbon dioxide transported in the blood?

The Brain Controls Breathing

LEARNING OBJECTIVES

1. **Associate** the various respiratory areas of the brain with their roles in controlling respiration.

2. **Explain** the roles of chemoreceptors in influencing breathing.

3. **Outline** the negative feedback control of breathing by central and peripheral chemoreceptors.

A t rest, about 200 mL of O_2 are used each minute by body cells. During strenuous exercise, however, O_2 use typically increases 15- to 20-fold in normal healthy adults, and as much as 30-fold in elite endurance-trained athletes. Several mechanisms help match respiratory efforts to metabolic demand.

As you learned in Chapter 11, your heart rate is set internally in the heart muscle itself. In contrast, your breathing rate is set externally by the respiratory center, which consists of groups of neurons in the medulla oblongata and pons in the brain.

Structures of the Brainstem Regulate Breathing

The **respiratory center** consists of four specific areas that regulate breathing:

1. The inspiratory area.
2. The expiratory area.
3. The **pneumotaxic area** (sometimes called the *pontine center*) of the pons.
4. The **apneustic area**.

The basic rate and depth of breathing is set by the **inspiratory area** in the medulla oblongata, which sends nerve impulses to the muscles of inhalation (diaphragm, external intercostals) to contract for two seconds. Impulses then cease for three seconds, during which

pneumotaxic area
(noo-mō-TAK-sik) A part of the respiratory center in the pons that cyclically sends inhibitory nerve impulses to the inspiratory area, limiting inhalation and facilitating exhalation.

apneustic area
(ap-NOO-stik) A part of the respiratory center in the pons that sends stimulatory nerve impulses to the inspiratory area that activate and prolong inhalation and inhibit exhalation.

time the muscles relax and you exhale. This process occurs continuously to set the rhythm of normal breathing.

The neurons of the **expiratory area** in the medulla oblongata remain inactive during quiet breathing. However, the inspiratory area can adjust the depth and rate of breathing, based on input received from various body receptors about gas and activity levels. When you need to breathe harder, as during exercise, the inspiratory area sends nerve impulses to additional muscles of inhalation and also sends impulses to the expiratory muscles (internal intercostals, abdominal muscles) for forceful exhalation. **Figure 13.7** shows the relationships of the inspiratory and expiratory areas during normal quiet breathing and forceful breathing.

The pneumotaxic and apneustic areas influence the inspiratory area to modify the basic rhythm of breathing. The pneumotaxic area can shorten inspiration to create shorter, shallower breaths (as in panting) or prolong inhalation to allow slow, deep breaths (as with vocalization).

The cerebral cortex can also influence the respiratory centers. Cortical impulses allow you to voluntarily change the rate and depth of breathing. You can even voluntarily hold your breath, to prevent water or irritating gases from entering the lungs. However, you cannot hold your breath for very long. Even if you hold your breath long enough to cause fainting, breathing resumes when consciousness is lost. The brainstem structures immediately take

Control of the rate and depth of breathing by the respiratory center • Figure 13.7

The respiratory center consists of four groups of neurons in the medulla oblongata and pons.

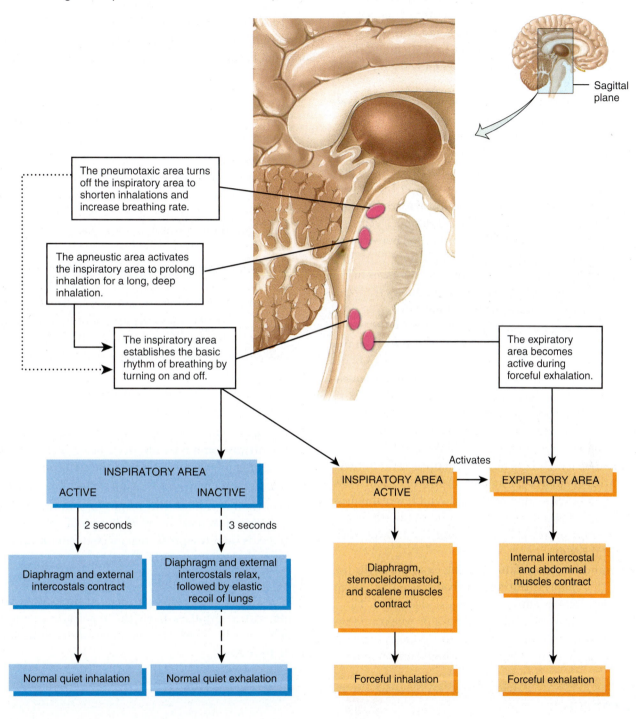

The pneumotaxic area turns off the inspiratory area to shorten inhalations and increase breathing rate.

The apneustic area activates the inspiratory area to prolong inhalation for a long, deep inhalation.

The inspiratory area establishes the basic rhythm of breathing by turning on and off.

The expiratory area becomes active during forceful exhalation.

Sagittal plane

Activates

INSPIRATORY AREA

ACTIVE — 2 seconds → Diaphragm and external intercostals contract → Normal quiet inhalation

INACTIVE — 3 seconds → Diaphragm and external intercostals relax, followed by elastic recoil of lungs → Normal quiet exhalation

INSPIRATORY AREA ACTIVE → Diaphragm, sternocleidomastoid, and scalene muscles contract → Forceful inhalation

EXPIRATORY AREA → Internal intercostal and abdominal muscles contract → Forceful exhalation

over the breathing process if the cerebrum attempts to do anything that becomes potentially life-threatening. As soon as the buildup of CO_2 and H^+ in your blood and cerebrospinal fluid (CSF) gets to a certain point, the brainstem sends a powerful stimulus to breathe.

When you laugh or cry, your respiration is also significantly different from normal, quiet breathing. Nerve impulses from the hypothalamus and limbic system stimulate the respiratory center, allowing emotional stimuli such as laughing or crying to alter respirations.

The Breathing Control Centers Are Influenced by Many Factors

Certain chemical stimuli determine how quickly and how deeply we breathe. The respiratory system functions to maintain proper levels of CO_2 and O_2 and is very responsive to changes in the level of either in body fluids. Sensory neurons that are responsive to chemicals are called **chemoreceptors**.

Chemoreceptors in the medulla, in the carotid bodies of the carotid artery, and in the aortic arch influence the respiratory centers through reflex pathways (**Figure 13.8**). The chemoreceptors in the medulla, called **central chemoreceptors**, monitor the P_{CO_2} and pH of the cerebrospinal fluid (CSF). The chemoreceptors in the carotid bodies and aortic arch, known as **peripheral chemoreceptors**, monitor the P_{O_2}, P_{CO_2}, and pH of the blood. Decreases in P_{O_2} and pH or increases in P_{CO_2} cause the chemoreceptors to send nerve impulses to the respiratory center to increase the rate of breathing. Conversely, increases in P_{O_2} and pH or decreases in P_{CO_2} reduce the impulses from the chemoreceptors and reduce the breathing rate.

The chemoreceptors participate in a negative feedback system that regulates the levels of CO_2, O_2, and H^+ in the blood. If these levels change, as shown in Figure 13.8, input from the central and peripheral chemoreceptors cause the inspiratory area to become highly active. The rate and depth of breathing increase. Rapid and deep breathing, called hyperventilation, allows the exhalation of more CO_2 until P_{CO_2} and H^+ are lowered to normal levels.

Severe deficiency of O_2, however, depresses the activity of the central chemoreceptors and inspiratory area. They do not respond well to any inputs and send fewer impulses to the muscles of respiration. As the rate of breathing decreases or stops entirely, P_{O_2} falls lower and lower, with possibly fatal results.

Remember the mountain-climbing example at the beginning of this chapter? Initially, the body responds to high altitudes with hyperventilation, as just described. However, full acclimatization may take days or weeks and involves other body systems, including the kidneys. Even so, humans cannot live for a long time without supplementary oxygen sources at altitudes above 5,950 meters (19,520 feet), and cannot live at all at altitudes above 8.000 meters (26,000 feet). This range, above 8,000 meters, is called the "death zone" in mountaineering.

In addition to the chemical influences just described, numerous other factors regularly impact the rate and depth of breathing:

- *Limbic system*. In response to emotional stresses, such as anxiety and fear, the limbic system stimulates the respiratory center to increase the rate and depth of breathing.

- *Proprioceptor stimulation*. As soon as you start exercising, your rate and depth of breathing increase—even before changes in P_{O_2}, P_{CO_2}, or H^+ level occur. When joints and muscles move (as in exercise), proprioceptors stimulate the inspiratory area to increase the rate and depth of breathing.

- *Temperature*. An increase in body temperature, which can be caused by a fever or by exercise, increases the rate of breathing. In contrast, a decrease in body temperature decreases the rate of breathing. A sudden cold stimulus (such as plunging into cold water) can slow or even temporarily stop breathing (apnea).

- *Pain*. Sudden pain may induce apnea, but prolonged somatic pain may increase breathing rate. Visceral pain may slow breathing.

- *Airway irritation*. Physical or chemical irritation of the airways immediately stops breathing, followed by coughing or sneezing to expel the irritants.

- *Inflation reflex*. Stretch receptors in the bronchi and bronchioles inhibit the inspiratory area and induce exhalation to prevent overinflation of the lungs. This reflex is mainly a protective mechanism for preventing excessive inflation of the lungs.

Defects or random impulses within the respiratory center can also change some patterns of breathing. For example, extra impulses from the respiratory center can asynchronously stimulate the respiratory muscles and cause **hiccups**.

CONCEPT CHECK

1. **Which** respiratory areas are involved in increasing the rate of breathing?

2. **How** is a change in pH of the cerebrospinal fluid detected?

3. **How** does an increase in P_{CO_2} of the blood stimulate breathing rate?

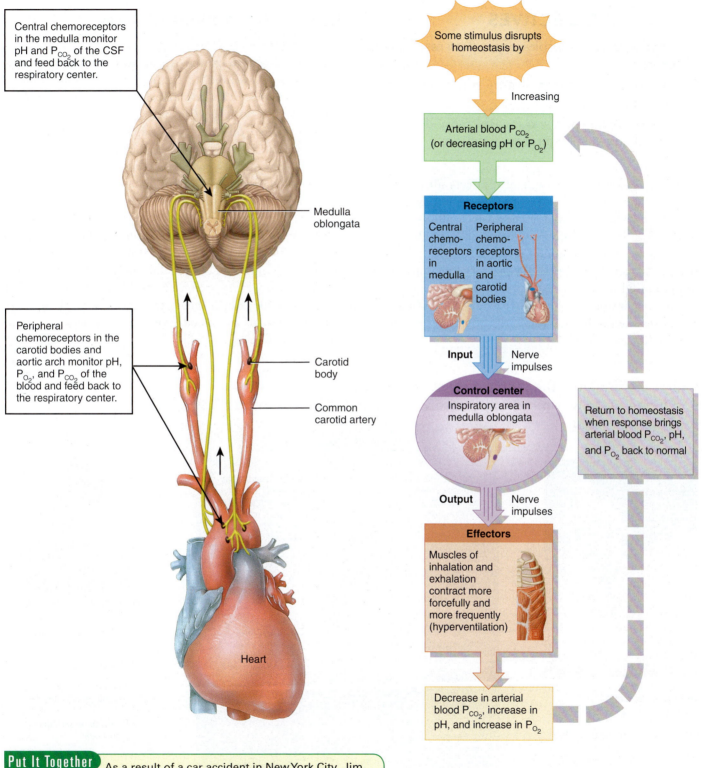

Central chemoreceptors in the medulla monitor pH and P_{CO_2} of the CSF and feed back to the respiratory center.

Peripheral chemoreceptors in the carotid bodies and aortic arch monitor pH, P_{O_2}, and P_{CO_2} of the blood and feed back to the respiratory center.

Medulla oblongata

Carotid body

Common carotid artery

Heart

Some stimulus disrupts homeostasis by

Increasing

Arterial blood P_{CO_2} (or decreasing pH or P_{O_2})

Receptors

Central chemoreceptors in medulla

Peripheral chemoreceptors in aortic and carotid bodies

Input   Nerve impulses

Control center

Inspiratory area in medulla oblongata

Output   Nerve impulses

Effectors

Muscles of inhalation and exhalation contract more forcefully and more frequently (hyperventilation)

Decrease in arterial blood P_{CO_2}, increase in pH, and increase in P_{O_2}

Return to homeostasis when response brings arterial blood P_{CO_2}, pH, and P_{O_2} back to normal

Put It Together As a result of a car accident in New York City, Jim experienced head trauma. His cranial nerve X was damaged. Which of the following would describe his response to a flight to Denver?
a. Jim would experience higher oxygen intake.
b. Jim would likely faint due to insufficient oxygen
c. Jim would not be affected in any way.

Good Respiratory Health Is Essential

LEARNING OBJECTIVES

1. **Explain** how the respiratory system responds to exercise.

2. **Describe** how the respiratory system changes with age.

3. **Link** smoking and various respiratory diseases with changes in respiratory function.

Good respiratory health affects all aspects of your life. The respiratory system oxygenates your cells and tissues, which allows them to function properly, helps cells produce ATP, and enables you to endure physical activities such as strenuous exercise and play. Respiratory functions can be compromised by aging, disease, and the effects of smoking and respiratory disorders. Let's start our discussion of respiratory health by examining the effects of exercise on the respiratory system.

Respiratory Actions Change with Physical Activity

When you begin to exercise, proprioceptors in your muscles and joints stimulate the respiratory center of the brain to increase the rate and depth of breathing (**Figure 13.9**). Because these proprioceptors are components of neurological reflexes, this change of breathing kicks in during the first few minutes of activity. ATP is needed for muscular activity, so as you continue to exercise, the cells rapidly begin to use up oxygen, generate carbon dioxide, and form hydrogen ions in the cellular respiration process. These changes alter the levels of the gases in the blood supply, stimulating peripheral chemoreceptors to initiate feedback to the respiratory center, resulting in increased breathing rate and depth. This altered respiration increases oxygenation of the blood and removal of carbon dioxide, which meets the needs of working muscles and returns the blood levels of the various gases to more normal levels.

Exercise stimulates the respiratory center, heart, and lungs to deliver more oxygen to working muscles • Figure 13.9

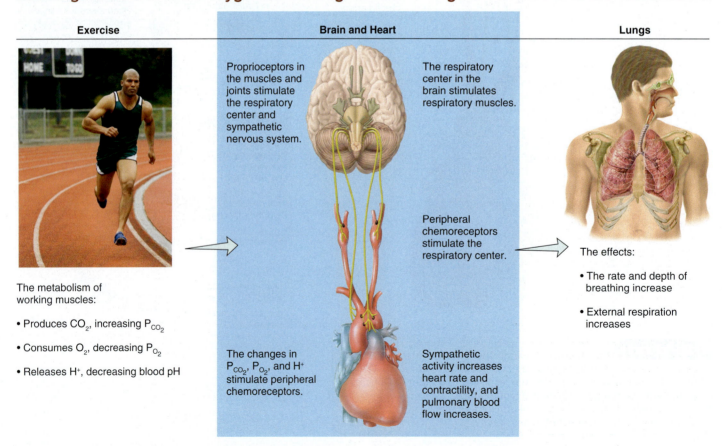

Exercise

The metabolism of working muscles:

- Produces CO_2, increasing P_{CO_2}
- Consumes O_2, decreasing P_{O_2}
- Releases H^+, decreasing blood pH

Brain and Heart

Proprioceptors in the muscles and joints stimulate the respiratory center and sympathetic nervous system.

The respiratory center in the brain stimulates respiratory muscles.

Peripheral chemoreceptors stimulate the respiratory center.

The changes in P_{CO_2}, P_{O_2}, and H^+ stimulate peripheral chemoreceptors.

Sympathetic activity increases heart rate and contractility, and pulmonary blood flow increases.

Lungs

The effects:

- The rate and depth of breathing increase
- External respiration increases

Aging

Airways and respiratory tissues become more rigid.

Ciliary action of the epithelial linings of the respiratory tracts decrease.

The chest wall becomes less pliable.

Aging decreases alveolar macrophage activities, which increases susceptibility to various diseases.

Age-related changes of the respiratory tissues also decrease the blood oxygen levels.

Smoking

The linings of the airways are irritated, causing inflammation and obstructive pulmonary diseases.

Substances in cigarette smoke stimulate uncontrolled cellular growth in the lungs.

Smoking destroys the alveoli.

The carbon monoxide in cigarette smoke displaces oxygen from hemoglobin and reduces the amount of oxygen that can be transported to cells and tissues.

Results

Airway flow decreases.

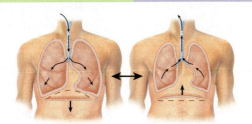

Lung capacity decreases as the elasticity of chest and lung tissues decreases.

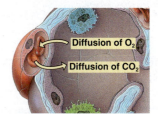

Diffusion of O_2

Diffusion of CO_2

Pulmonary gas exchange is affected, decreasing the level of O_2 and increasing the level of CO_2 in the blood.

Diseases and Behavioral Activities Can Also Affect the Breathing Process

Exercise stimulates the respiratory system, but aging and disease can compromise it. Airways and respiratory tissues become more rigid and less elastic with age. The chest wall also becomes less pliable. This loss of flexibility results in decreased lung capacity (**Figure 13.10**); vital capacity can decrease by as much as 35% by age 70. Aging also decreases **alveolar macrophage** activities and ciliary action of the epithelial linings of the respiratory tracts, which increases susceptibility to various diseases. Age-related changes of the respiratory tissues also decrease the blood oxygen levels.

Diseases such as bronchitis, pneumonia, and emphysema reduce the flow of air through the airways (*obstructive pulmonary diseases*) and decrease lung volumes and capacities (*restrictive pulmonary diseases*), as well as the diffusion of gases between the alveoli and the pulmonary capillaries. Pulmonary infections (such as pneumonia or tuberculosis) can destroy the alveolar membranes and cause multiple respiratory problems.

Infections (bacterial or viral) can also cause airway inflammation, mucus accumulation, and damage to alveoli. **Pulmonary edema**, a buildup of fluid in the interstitial spaces and alveoli, can occur as the pulmonary capillary membrane breaks down from disease, infection, or congestive heart failure. Such edema is generally caused by increases in filtration (see Chapter 11). The fluid buildup produces barriers to the diffusion of gases during external respiration. This is primarily because of the solubility issue discussed earlier. Oxygen must be forced to remain in solution; with the extra fluid that must be traversed before the oxygen arrives at the hemoglobin, much of the oxygen goes back into the air before it can be trapped in the blood supply. The excess fluid can also make the lung tissue less flexible and decrease pulmonary ventilation.

Smoking Can Damage Lung Tissue

Smoking is a major contributing factor in the development of pulmonary disease, as shown in Figure 13.10. Smoke irritates the linings of the airways, causing inflammation and obstructive diseases such as chronic bronchitis. Smoke contains chemicals that can paralyze the cilia of the tracheal lining, allowing various contaminants (such as bacteria, viruses, and toxic chemicals) to reach the alveoli. It also destroys the alveoli, which interferes with gas exchange and makes the lung tissues less elastic, causing emphysema. Cigarette smoke also contains carbon monoxide, which displaces oxygen from hemoglobin and reduces the amount of oxygen that can be transported to cells and tissues. Finally, substances in cigarette smoke stimulate uncontrolled cellular growth in the lungs; lung cancer is the leading cause of cancer deaths in the United States among both males and females.

1. **What** role do peripheral chemoreceptors have in the respiratory system's response to exercise?

2. **What** is the primary change in the respiratory system with age?

3. **What** are the consequences of COPD?

 THE PLANNER ✓

Summary

1 Respiratory Organs Move Air and Exchange Gases 372

- As shown, the upper respiratory tract consists of the **nose**, **nasal epithelium**, and **pharynx**; it warms, filters, and humidifies incoming air, detects smells in that air, and expels mucus from the respiratory tract.

The respiratory system • Figure 13.1

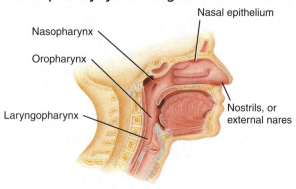

- The lower respiratory tract consists of a series of branching tubes that direct air into the lungs. The tubing includes the **larynx**, **trachea**, **bronchi**, and **bronchioles**—ending at the alveoli. The larynx serves as the entrance to the airways and produces sounds. The **epiglottis** is part of the larynx that covers the trachea during swallowing so that food does not enter the airways. The **lungs** are enclosed in the **pleural cavity** and surrounded by a **pleural membrane**. They adhere to the thoracic wall, which moves to draw air in and push it out. Within the lungs are tiny air sacs called **alveoli**, where gas exchange occurs between the lungs and the blood.

- The functions of the respiratory system include gas exchange, regulation of blood pH, production of sounds, and excretion of water vapor and heat. The respiratory system also houses the receptors for smell.

2 Breathing Involves Changes in Pressures and Volumes 376

- When the respiratory muscles contract during **inhalation**, as shown, they expand the thoracic cavity and stretch the lungs, increasing the volume of the lungs and decreasing the pressure in the lungs relative to atmospheric pressure. Air flows into the lungs in response to this change in pressure. In contrast, when the respiratory muscles relax during **exhalation**, the opposite events occur, and air flows out of the lungs. Deep breaths and forced exhalations use additional muscles.

The breathing process • Figure 13.3

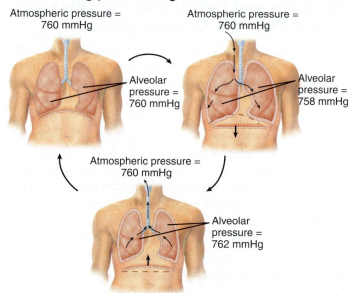

- Beyond the volume of air flowing into and out of the lungs during normal quiet breathing (**tidal volume**), the lungs are capable of expanding or collapsing to various volumes. These volumes and capacities are measured with a **spirometer** during a pulmonary function test and are used to diagnose various respiratory diseases.

- There are two major types of breathing: **costal breathing** (shallow chest breathing) and **diaphragmatic breathing** (deep abdominal breathing). Several modified respiratory movements force foreign objects from the respiratory system (coughing, sneezing) or express emotions (laughing, crying, sobbing).

- Oxygen is transported throughout the body bound mainly to hemoglobin in red blood cells (98.5%); only a small percentage travels in solution. Local changes in P_{O_2}, P_{CO_2}, pH, and temperature affect the capacity of hemoglobin to carry oxygen and may facilitate the loading or unloading of oxygen in various places.

- Carbon dioxide is transported in the body in three forms: bound to hemoglobin (23%) in red blood cells, as bicarbonate ions in the plasma (70%), and dissolved in solution (7%). The enzyme carbonic anhydrase within the red blood cell plays an important role in the conversion between carbon dioxide and bicarbonate.

3 **Gases Are Exchanged at the Blood Capillaries 382**

- As shown, **external respiration** is the exchange of gases between the air in the alveoli and the blood in the pulmonary capillaries. Also shown is **internal respiration**, the exchange of gases between blood in the systemic capillaries and the tissue cells. The diffusion of gases (O_2, CO_2) is driven by differences in the partial pressures of a given gas. (Gases move from high partial pressure to low partial pressure.) Oxygen partial pressure (P_{O_2}) is highest in the alveoli and lowest in the tissues, while carbon dioxide partial pressure (P_{CO_2}) is highest in the tissues and lowest in the alveoli.

Transport of oxygen and carbon dioxide in the blood • Figure 13.6

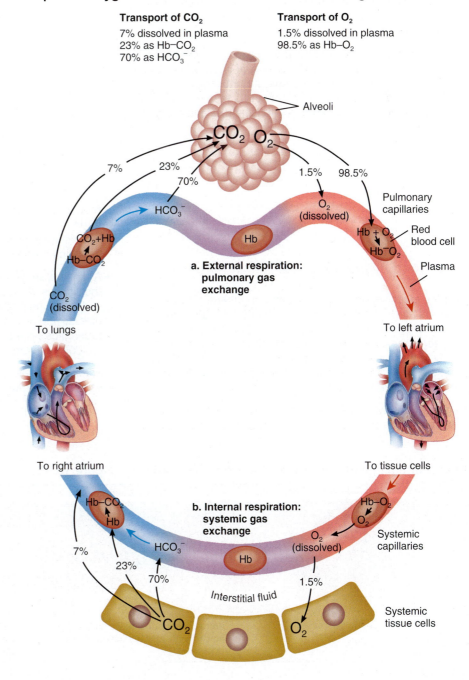

Transport of CO_2
7% dissolved in plasma
23% as Hb–CO_2
70% as HCO_3^-

Transport of O_2
1.5% dissolved in plasma
98.5% as Hb–O_2

Alveoli

a. External respiration: pulmonary gas exchange

Pulmonary capillaries
Red blood cell
Plasma
O_2 (dissolved)
CO_2 (dissolved)
HCO_3^-
Hb
Hb + O_2
Hb–O_2
CO_2+Hb
Hb–CO_2

To lungs
To left atrium

To right atrium
To tissue cells

b. Internal respiration: systemic gas exchange

Systemic capillaries
Interstitial fluid
Systemic tissue cells

Hb–CO_2
Hb
HCO_3^-
Hb
Hb–O_2
O_2
O_2 (dissolved)
CO_2
O_2

4 The Brain Controls Breathing 386

- As shown, the respiratory center in the medulla and pons consists of four areas. The inspiratory area controls the rate of normal quiet breathing. The expiratory area assists in forceful exhalation. The **pneumotaxic area** influences the inspiratory area to shorten the breathing rate. The **apneustic area** influences the inspiratory area to slow the breathing rate.

- The respiratory center is influenced by the cerebral cortex for voluntary control of breathing, by proprioceptors in muscles and joints activated by physical activity, and by central and peripheral **chemoreceptors** stimulated by metabolic activity.

- Central chemoreceptors in the medulla monitor the pH and P_{CO_2} of the cerebrospinal fluid. **Peripheral chemoreceptors** in the aortic arch and carotid bodies monitor the P_{O_2}, P_{CO_2}, and pH of the blood. The chemoreceptors feed information back to the respiratory center to alter the rate and depth of breathing.

Control of the rate and depth of breathing by the respiratory center • Figure 13.7

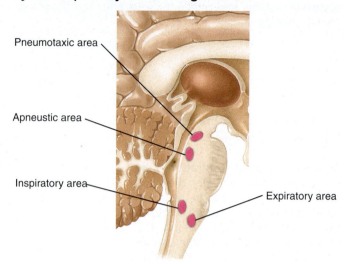

Pneumotaxic area

Apneustic area

Inspiratory area

Expiratory area

5 Good Respiratory Health Is Essential 390

- During exercise, proprioceptors and peripheral chemoreceptors modify the respiratory center of the brain, causing an increase in the rate and depth of breathing to increase oxygen supply to working muscles. The proprioceptors act early in this process, when muscles first begin the activity. The chemoreceptors act later, as levels of metabolic wastes in the blood begin to increase.

- Aging primarily influences the elasticity of the lungs and chest wall, which in turn reduces vital capacity. This also reduces pulmonary gas exchange and may reduce the ability to do prolonged or strenuous exercise. In addition, elderly people are susceptible to various respiratory diseases, including chronic bronchitis, emphysema, and pneumonia.

- As shown, smoking, and aging can reduce the flow of air in the airways, decrease the elasticity of the lungs and chest, and lessen pulmonary gas exchange. In addition, respiratory disease and pulmonary infections such as pneumonia and tuberculosis can destroy the alveolar membranes and cause multiple respiratory problems.

- Smoking increases the risk of developing pulmonary diseases such as chronic obstructive pulmonary disease and lung cancer.

Effects of aging and smoking on the respiratory system • Figure 13.10

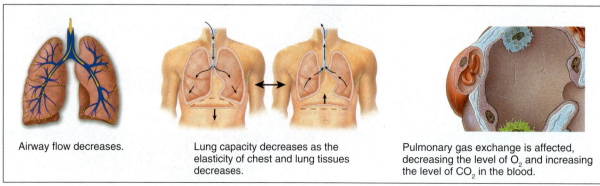

Airway flow decreases.

Lung capacity decreases as the elasticity of chest and lung tissues decreases.

Pulmonary gas exchange is affected, decreasing the level of O_2 and increasing the level of CO_2 in the blood.

Key Terms

- alveolar macrophage 391
- alveolus 374
- anatomic dead space 378
- apnea 380
- apneustic area 386
- breathing 375
- bronchiole 373
- bronchus 373
- carbonic anhydrase 384
- central chemoreceptor 388
- chemoreceptor 388
- costal breathing 380
- diaphragm 377
- diaphragmatic breathing 380
- dyspnea 380
- epiglottis 373

- eupnea 380
- exhalation 377
- expiratory area 386
- expiratory reserve volume 378
- external respiration 375
- glottis 373
- hiccup 388
- inhalation 377
- inspiratory area 386
- inspiratory reserve volume 378
- internal respiration 375
- interstitial space 375
- larynx 373
- lower respiratory tract 373

- lung 373
- lung capacity 378
- nasal epithelium 373
- nose 373
- nostril 381
- obstructive pulmonary disease 379
- partial pressure 382
- peripheral chemoreceptor 388
- pharynx 373
- pleural cavity 374
- pleural membrane 374
- pneumotaxic area 386
- pneumothorax 378
- pulmonary edema 391
- residual volume 378

- respiration 375
- respiratory distress syndrome (RDS) 381
- restrictive pulmonary disease 379
- sleep apnea 380
- solubility 382
- spirogram 378
- spirometer 378
- surfactant 374
- tidal volume 378
- total lung capacity 378
- trachea 373
- type II alveolar cell 381
- upper respiratory tract 373
- vital capacity 378
- vocal cord 373

Critical and Creative Thinking Questions

1. Poorly ventilated furnaces often emit dangerous levels of carbon monoxide. Carbon monoxide binds with greater affinity to hemoglobin than does oxygen. What would be the consequences to gas exchange of carbon monoxide inhalation?

2. According to Jim's wife, periodically when he is sleeping Jim stops breathing for long periods of time, takes a long, noisy breath, and then begins breathing normally again. From what disorder does Jim suffer, and what stimulates him to breathe?

3. A quarterback must play in Denver, where the oxygen concentration in the atmosphere is approximately 59% than at sea level, where he lives. How will this affect his rate of breathing? What physiological mechanisms account for your answer?

4. Through prolonged exposure to coal dust, a miner's lungs become *fibrotic* (that is, the elastic smooth muscle in the lungs and airways is replaced with rigid scar tissue). He often complains of fatigue and lack of energy. What effects would you expect this fibrotic tissue to have on his breathing, and what diagnostic tests would you use to confirm your diagnosis?

What is happening in this picture?

Symptoms of asthma include difficulty breathing, coughing, wheezing, chest tightness, and anxiety. Rescue inhalers contain albuterol, which relaxes airway smooth muscles.

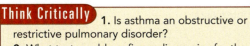
Think Critically
1. Is asthma an obstructive or restrictive pulmonary disorder?
2. What test would confirm a diagnosis of asthma?
3. How would albuterol stop an asthma attack?

Self-Test

(Check your answers in Appendix C.)

1. In general, carbon dioxide is transported in the blood _____.
   a. dissolved in solution
   b. bound to hemoglobin
   c. bound to other proteins
   d. as bicarbonate ions

Use this figure to answer questions 2–4.

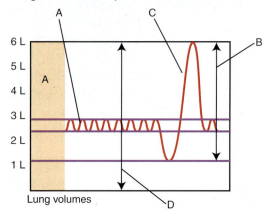

Lung volumes

2. The lung capacity indicated by D is known as _____.
   a. vital capacity
   b. inspiratory capacity
   c. total lung capacity
   d. expiratory capacity

3. The rate indicated by C would be used to diagnose _____.
   a. restrictive pulmonary disease
   b. obstructive pulmonary disease
   c. defect in pulmonary gas exchange
   d. defect in respiratory center activity

4. The lung volume indicated by A is _____.
   a. tidal volume
   b. inspiratory reserve volume
   c. expiratory reserve volume
   d. residual volume

5. Which lung volume cannot be measured using a spirometer?
   a. tidal volume
   b. inspiratory reserve volume
   c. expiratory reserve volume
   d. residual volume

6. Which area of the respiratory center is activated during forced exhalation?
   a. inspiratory area   c. expiratory area
   b. apneustic area   d. pneumotaxic area

Use this figure to answer questions 7–9.

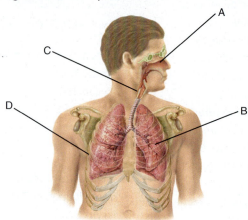

7. Which of the labeled areas produces sounds?
   a. A   c. C
   b. B   d. D

8. To inhale, the respiratory muscles pull on which of the following structures?
   a. A   c. C
   b. B   d. D

9. In which labeled structure would you find the olfactory epithelia?
   a. A   c. C
   b. B   d. D

10. Which of the following areas has the highest P_{CO_2}?
   a. tissue cells
   b. alveoli
   c. pulmonary capillaries
   d. left atrium

11. Smokers often have _____.
   a. asthma
   b. pulmonary edema
   c. chronic obstructive pulmonary disease
   d. pneumonia

12. The _____ muscle is used exclusively for forced exhalation.
   a. diaphragm
   b. sternocleidomastoid
   c. external intercostal
   d. internal intercostal

Use this figure for questions 13–15.

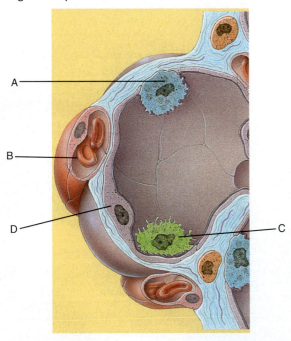

13. Which of the following cells would be involved in a baby with RDS?
   a. A
   b. B
   c. C
   d. D

14. What is the function of the cell labeled C?
   a. secretes surfactant
   b. clears debris
   c. carries oxygen
   d. secretes mucus

15. Which of the labeled cells is most affected by carbon monoxide?
   a. A
   b. B
   c. C
   d. D

Use this figure to answer questions 16-17.

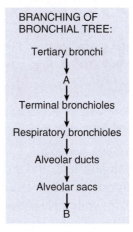

BRANCHING OF BRONCHIAL TREE:

Tertiary bronchi
↓
A
↓
Terminal bronchioles
↓
Respiratory bronchioles
↓
Alveolar ducts
↓
Alveolar sacs
↓
B

16. What is the name of area A in the bronchial tree, after the tertiary bronchi?
   a. the secondary bronchioles
   b. the alveolar sacs
   c. the alveoli
   d. the bronchioles

17. What is the name of area B in the bronchial tree, after the alveolar sacs?
   a. the secondary bronchioles
   b. the alveolar sacs
   c. the alveoli
   d. the bronchioles

18. Which of the following is *not* correct regarding the respiration process?
   a. In external respiration, hemoglobin must be used to help O_2 remain soluble in the blood.
   b. Oxygen is released in the systemic capillaries because of changes in pH, CO_2, and temperature.
   c. In internal respiration, CO_2 forms inside the red blood cells.
   d. During both internal and external respiration, gases diffuse across the membranes because of gradients of partial pressure.
   e. Carbonic anhydrase helps convert CO_2 to a more soluble form that can be transported.

THE PLANNER ✓

Review your Chapter Planner on the chapter opener and check off your completed work.

14 The Digestive System, Nutrition, and Metabolism

Eating is something you do every day without thinking much about it. After you eat that delicious breakfast of waffles, sausages, and eggs, its nutrients (including sugars, fats, and proteins) are absorbed by your digestive system and whisked away to provide fuel for your daily activities. However, the nutrient molecules must be broken down to their most basic form prior to their absorption in the digestive system. This process requires mechanical actions to physically break down the food particles and numerous secretions from the digestive organs to chemically break down the food molecules. After absorption, nutrients are further processed by a variety of cells in the body—converted into cellular components and secretions or used to form ATP (the cell's energy-storing molecule).

The digestive system is composed of a number of organs, many of which are joined as a large tube. The remaining accessory organs produce the majority of the secretions that are used to process the foods we eat. In this chapter, we examine what happens as you eat a meal and how the nutrients are processed after they get to your cells. By the end of this chapter, you should be able to answer a wide variety of questions related to the seemingly simple process of eating your breakfast: How is food digested? How is it absorbed? What is the fate of each type of nutrient? How are those nutrients stored or broken down? How do problems with metabolism lead to diseases and, on the flip side, how do diseases lead to problems with metabolism?

CHAPTER OUTLINE

CHAPTER PLANNER ✔

- ❏ Study the picture and read the opening story.
- ❏ Scan the Learning Objectives in each section:
  p. 400 ❏   p. 416 ❏   p. 418 ❏   p. 422 ❏   p. 427 ❏
- ❏ Read the text and study all visuals. Answer any questions.

Analyze key features

- ❏ InSight, pp. 400–401 ❏   pp. 406–407 ❏   pp. 408–409 ❏ p. 413 ❏
- ❏ Process Diagram, p. 405 ❏   p. 419 ❏
- ❏ What a Health Provider Sees, p. 426 ❏
- ❏ Stop: Answer the Concept Checks before you go on:
  p. 415 ❏   p. 416 ❏   p. 422 ❏   p. 426 ❏   p. 430 ❏

End of chapter

- ❏ Review the Summary and Key Terms.
- ❏ Answer the Critical and Creative Thinking Questions.
- ❏ Answer What is happening in this picture?
- ❏ Complete the Self-Test and check your answers.

Let's Journey Through the Digestive System

LEARNING OBJECTIVES

1. **List** the organs of the gastrointestinal tract and their functions.

2. **Identify** the accessory organs of the digestive system and the function of their secretions.

3. **Describe** the tissue layers that form the walls of the digestive tract.

4. **Explain** the processes associated with digestion.

5. **Compare** the absorption of water-soluble nutrients and the absorption of fat-soluble nutrients.

T he cells in your body require many small molecules to function properly, such as carbohydrates, lipids, and amino acids (see Chapter 2). However, food is most commonly available in much larger forms, such as your waffle, sausage, and egg breakfast. Your cells cannot use the food in such a meal directly, so the body must break down food into smaller, usable components. This is the job of the **digestive system**.

The digestive system basically consists of a long, continuous tube called the **gastrointestinal (GI)**

InSight   The digestive system   •   Figure 14.1

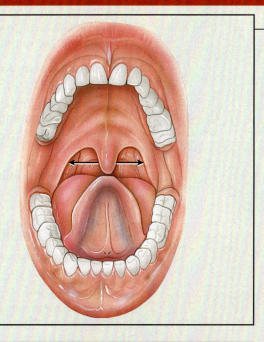

Mouth (oral cavity) forms the chamber where food is chewed.
- **Teeth** mechanically digest food by chewing.
- Muscular **tongue** manipulates food and holds taste sensors.
- **Salivary glands** (outside mouth) secrete **saliva**, which lubricates food and begins chemical digestion with the enzyme **salivary amylase**.

Pharynx and **esophagus** are muscular conduits that propel swallowed food from the mouth to the stomach.

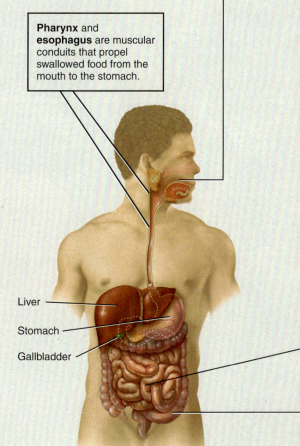

Liver

Stomach

Gallbladder

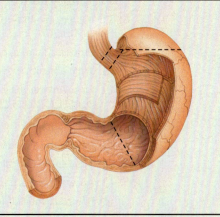

Stomach is a muscular mixing chamber and storage container for food.
- Contractions of stomach muscle mechanically mix the food.
- Secretions of hydrochloric acid, pepsin, and lipase chemically break down food.
- Intrinsic factor secretion assists absorption of vitamin B$_{12}$.
- Some absorption of water, ions, short-chain fatty acids, and drugs (alcohol, aspirin, ibuprofen) occurs in the stomach.
- Stomach controls rate of food expulsion into the small intestine (gastric emptying).

tract, or *alimentary canal*. The GI tract contains food from the time it is eaten until it is digested and absorbed or eliminated from the body. Extending from the mouth to the anus, the GI tract is about 5–7 meters long (16.5–23 ft). It is divided into specialized sections referred to as digestive organs, which include the mouth, pharynx, esophagus, stomach, small intestine, and large intestine (**Figure 14.1**).

In addition to the digestive organs of the GI tract, there are several **accessory digestive organs**, including the salivary glands, liver, pancreas, and gall bladder. These accessory organs secrete substances into the GI tract or perform other functions.

In sum, the digestive system performs several vital functions:

- *Ingestion*. Taking food into the mouth
- *Secretion*. Releasing water, acid, buffers, and enzymes into the lumen of the GI tract
- *Mixing and propulsion*. Churning and pushing food through the GI tract
- *Digestion*. Physically and chemically breaking down food
- *Absorption*. Passing digested products from the GI tract into the blood and lymph
- *Defecation*. Eliminating feces from the GI tract

THE PLANNER

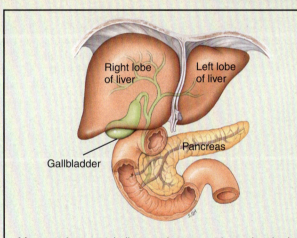

Liver receives, metabolizes, and stores nutrients absorbed from the small intestine. It secretes bile that breaks down large lipid globules. **Gallbladder** stores bile secreted by the liver. **Pancreas** (behind stomach) secretes sodium bicarbonate and enzymes that break down food into amino acids, lipids, and simple carbohydrates.

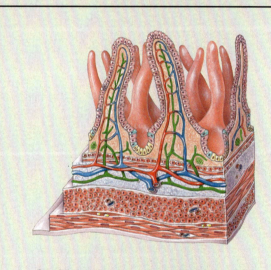

Small intestine absorbs sugars, lipids, peptides, water, ions, and vitamins from food. It also secretes hormones that acts on other digestive organs.

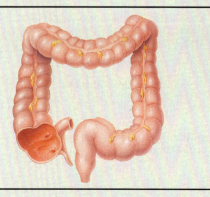

Large intestine absorbs water, ions, and vitamins. Resident bacteria break down remaining carbohydrates and proteins (bilirubin). Chyme becomes semi-solid **feces**, which get expelled through the **anus**.

The GI Tract Is Supported by the Peritoneum, and Each of Its Sections Has Four Layers

The **peritoneum** (per′-i-tō-NE-um) is the membrane that covers the abdominal organs and lines the abdominal cavity; it contains large folds that bind the organs to one another and to the walls of the abdominal cavity. Two of the major folds are the greater omentum and the mesentery. The *greater omentum* drapes over the intestines like a fatty apron, and the *mesentery* binds the small intestine to the abdominal wall. **Figure 14.2** shows the relationship between the peritoneum and the organs of the digestive system.

Each section of the GI tract has four layers. From the inside out, they are the mucosa, submucosa, muscularis,

The peritoneum and the four layers of the gastrointestinal tract • Figure 14.2

a. The peritoneum

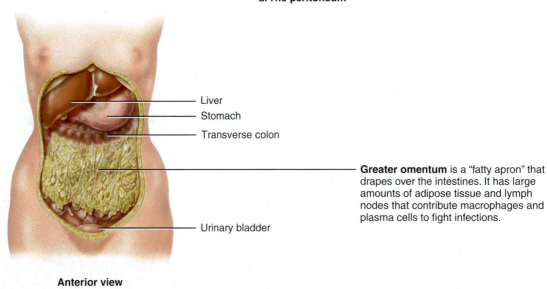

- Liver
- Stomach
- Transverse colon

Greater omentum is a "fatty apron" that drapes over the intestines. It has large amounts of adipose tissue and lymph nodes that contribute macrophages and plasma cells to fight infections.

- Urinary bladder

Anterior view

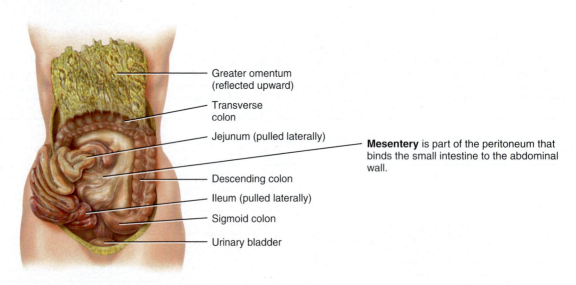

- Greater omentum (reflected upward)
- Transverse colon
- Jejunum (pulled laterally)
- Descending colon
- Ileum (pulled laterally)
- Sigmoid colon
- Urinary bladder

Mesentery is part of the peritoneum that binds the small intestine to the abdominal wall.

Anterior view (greater omentum lifted and small intestine pulled laterally to right side)

and serosa (Figure 14.2b). These layers include epithelial tissue (to help absorb food), connective tissue (for support), and muscle (to propel the food on its way).

The GI tract has a rich blood supply for absorbing nutrients and networks of neurons. The **enteric nervous system (ENS)**, the "brain of the gut," coordinates its movements and secretions. ENS neurons within the submucosa control the secretions of the organs of the GI tract, and ENS neurons within the muscularis control the frequency and strength of its contractions. The ENS is subject to regulation by the autonomic nervous system (ANS). You will learn more about the neural control of digestion later in this chapter.

Let's take a closer look at the digestive system by following your breakfast of waffles, sausages, and eggs on its journey.

b. The four layers of the gastrointestinal tract: serosa, muscularis, submucosa, and mucosa

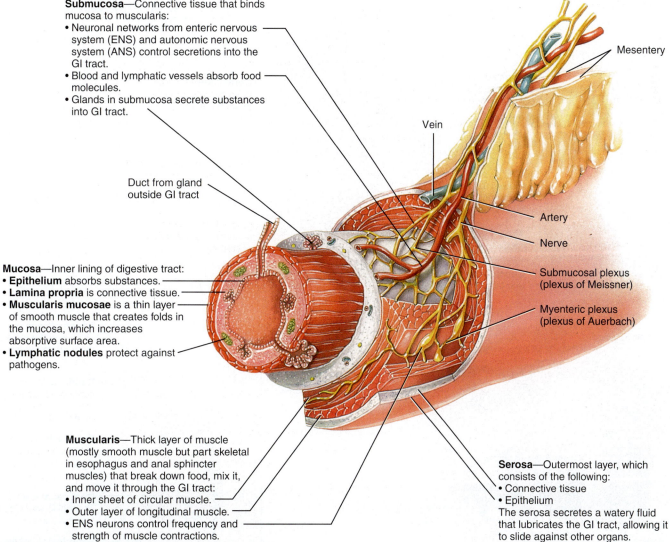

Submucosa—Connective tissue that binds mucosa to muscularis:
- Neuronal networks from enteric nervous system (ENS) and autonomic nervous system (ANS) control secretions into the GI tract.
- Blood and lymphatic vessels absorb food molecules.
- Glands in submucosa secrete substances into GI tract.

Duct from gland outside GI tract

Mucosa—Inner lining of digestive tract:
- **Epithelium** absorbs substances.
- **Lamina propria** is connective tissue.
- **Muscularis mucosae** is a thin layer of smooth muscle that creates folds in the mucosa, which increases absorptive surface area.
- **Lymphatic nodules** protect against pathogens.

Muscularis—Thick layer of muscle (mostly smooth muscle but part skeletal in esophagus and anal sphincter muscles) that break down food, mix it, and move it through the GI tract:
- Inner sheet of circular muscle.
- Outer layer of longitudinal muscle.
- ENS neurons control frequency and strength of muscle contractions.

Mesentery

Vein

Artery

Nerve

Submucosal plexus (plexus of Meissner)

Myenteric plexus (plexus of Auerbach)

Serosa—Outermost layer, which consists of the following:
- Connective tissue
- Epithelium
The serosa secretes a watery fluid that lubricates the GI tract, allowing it to slide against other organs.

Ask Yourself

Abdominal fat is located in which of the following?
a. Submucosa
b. Greater omentum
c. Mesentery
d. Mucosa

Digestion Begins in the Mouth

Digestion begins when you bring food into your mouth (**ingestion**). The jaw muscles and teeth help you chew (masticate) your food, as shown in **Figure 14.3**. Chewing physically breaks the food into smaller pieces, which increases the surface area of your waffles, sausages, and eggs that is available for subsequent chemical digestion. Various teeth are specialized for the different ways in which they break apart food (tearing off chunks of sausage, cutting up eggs and waffles, grinding the sausage into bits that can be swallowed). Salivary glands secrete fluid called *saliva* through ducts that lead into the mouth. Saliva mixes with the food pieces, lubricates and moistens the food, kills bacteria, and begins to digest starches in your meal. Movements of your tongue move the food to contact the teeth and help shape the chewed food into a soft, rounded, flexible mass called a **bolus**.

Ingestion of food • Figure 14.3

The mouth receives the food. The mouth is defined by the hard and soft palates, the tongue, and the cheeks. The teeth and salivary glands are also associated with the mouth.

WILEY PLUS | NATIONAL GEOGRAPHIC | Video

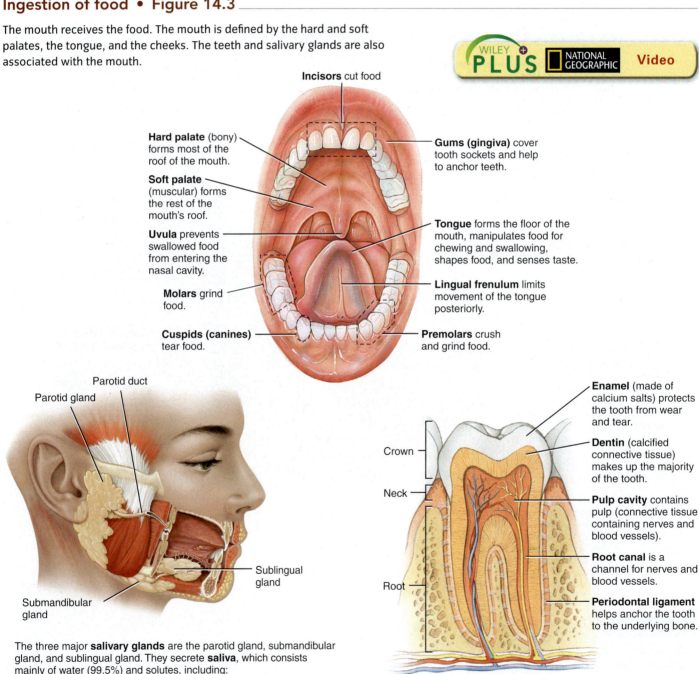

Incisors cut food

Hard palate (bony) forms most of the roof of the mouth.

Soft palate (muscular) forms the rest of the mouth's roof.

Uvula prevents swallowed food from entering the nasal cavity.

Molars grind food.

Cuspids (canines) tear food.

Gums (gingiva) cover tooth sockets and help to anchor teeth.

Tongue forms the floor of the mouth, manipulates food for chewing and swallowing, shapes food, and senses taste.

Lingual frenulum limits movement of the tongue posteriorly.

Premolars crush and grind food.

Parotid duct

Parotid gland

Sublingual gland

Submandibular gland

Enamel (made of calcium salts) protects the tooth from wear and tear.

Dentin (calcified connective tissue) makes up the majority of the tooth.

Crown

Neck

Pulp cavity contains pulp (connective tissue containing nerves and blood vessels).

Root canal is a channel for nerves and blood vessels.

Root

Periodontal ligament helps anchor the tooth to the underlying bone.

Sagittal section of a molar

The three major **salivary glands** are the parotid gland, submandibular gland, and sublingual gland. They secrete **saliva**, which consists mainly of water (99.5%) and solutes, including:
• **Salivary amylase**—Enzyme that begins digestion of starch
• **Mucus**—Lubricates food
• **Lysozyme**—Enzyme that kills bacteria

Deglutition • Figure 14.4

THE PLANNER ✓

The process of deglutition (swallowing) involves the mouth, pharynx, and esophagus in voluntary and involuntary phases.

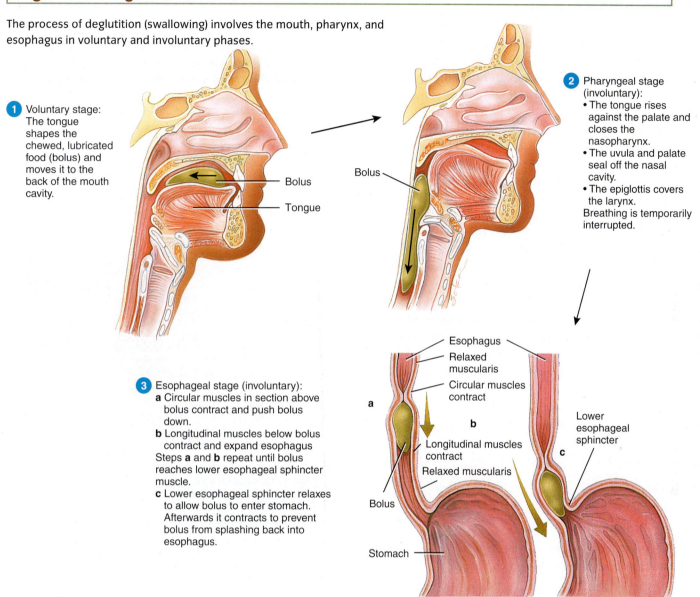

1 Voluntary stage: The tongue shapes the chewed, lubricated food (bolus) and moves it to the back of the mouth cavity.

Bolus

Tongue

2 Pharyngeal stage (involuntary):
• The tongue rises against the palate and closes the nasopharynx.
• The uvula and palate seal off the nasal cavity.
• The epiglottis covers the larynx. Breathing is temporarily interrupted.

Bolus

3 Esophageal stage (involuntary):
a Circular muscles in section above bolus contract and push bolus down.
b Longitudinal muscles below bolus contract and expand esophagus Steps **a** and **b** repeat until bolus reaches lower esophageal sphincter muscle.
c Lower esophageal sphincter relaxes to allow bolus to enter stomach. Afterwards it contracts to prevent bolus from splashing back into esophagus.

Esophagus
Relaxed muscularis
Circular muscles contract
Lower esophageal sphincter
a
b
c
Longitudinal muscles contract
Relaxed muscularis
Bolus
Stomach

Swallowing Involves Voluntary and Involuntary Stages

Once chewed, the bolus of food is swallowed, a process called **deglutition** (dē′-glū-TISH-un). Deglutition involves several steps, including a voluntary stage, a pharyngeal stage (involuntary), and an esophageal stage (involuntary)

> **peristalsis** (per′-i-STAL-sis) Successive muscular contractions along the wall of a hollow muscular structure.

(**Figure 14.4**). This process, which involves the mouth, pharynx, and esophagus, moves the bolus from the mouth through the esophagus to the stomach by waves of muscular movements called **peristalsis**.

If the lower esophageal sphincter fails to close adequately, the acidic contents of the stomach will back up into the lower esophagus. This condition is called **gastroesophageal reflux disease (GERD)**. The stomach acid irritates the mucosal lining of the esophagus and gives the feeling of heartburn. If the condition is chronic, the acid can permanently damage the lining of the esophagus. GERD can be controlled by avoiding foods that stimulate stomach acid secretion (such as chocolate, coffee, tomatoes, fatty foods, and the orange juice that goes so well with your breakfast), neutralizing the acid with antacids, or taking drugs that inhibit stomach acid secretion.

Once a bolus of food is inside the stomach, digestion begins in earnest. Let's take a closer look at the work of this important digestive organ.

Let's Journey Through the Digestive System **405**

The Stomach Begins Digestion in Earnest

The stomach receives multiple boluses of food swallowed from the mouth during a meal. Think about how many bites you would take to consume that generous breakfast. Because you eat and swallow faster than you can digest, the stomach serves as a storage site for ingested food. The *fundus* and *body* regions contain folds of mucosa (rugae) that can expand to accommodate the volume of your meal. The *pylorus* region is the site of most of the gastric digestion.

The lining of the stomach is composed of a thick mucosa embedded with numerous **gastric pits**, where various secretory cells lie (**Figure 14.5**). The presence of food inside the stomach stimulates the secretion of **gastric juices**, which are composed of mucus, hydrochloric acid (HCl), and digestive enzymes. Numerous peristaltic contractions from the fundus to the pylorus of the stomach mix the gastric juices with your breakfast and reduce it to a thick liquid called **chyme** (KĪM). The hydrochloric acid secreted into the chyme kills bacteria and activates the enzyme **pepsin**, which breaks down those sausage and egg proteins into peptides (see Chapter 2). Gastric juice contains another enzyme called **gastric lipase**, which breaks down the generous quantity of lipids included in your morning meal.

Why doesn't hydrochloric acid burn a hole in your stomach? There are three reasons:

- Hydrochloric acid (HCl) secretion occurs only when there is food present in the stomach.
- The chyme dilutes the acid.
- The thick layer of mucus covering the mucosa protects the underlying stomach tissues.

In addition to mixing, secreting, and storing, the stomach can also absorb some substances. For example, mucous cells within the gastric pits absorb some water, ions, short-chain fatty acids, and some drugs (for example, aspirin, alcohol, ibuprofen) from the stomach's lumen.

Finally, the stomach empties its contents into the small intestine in a process called **gastric emptying**. The peristaltic motions of the stomach muscles propel small bursts of chyme through the partially closed pyloric sphincter into the duodenum (first section) of the small intestine. Once a burst of chyme passes through it, the pyloric sphincter closes, and reflexes

The stomach is a muscular pouch where food is stored and digestion begins. The mucosa is thicker in the stomach than in other areas of the digestive system. Numerous gastric pits compose the stomach lining. Within these gastric pits, various cells produce secretions to assist the digestion process.

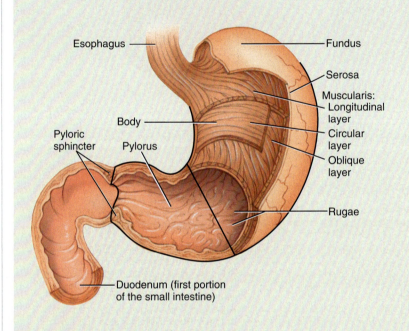

Esophagus — Fundus — Serosa — Muscularis: Longitudinal layer — Body — Circular layer — Pyloric sphincter — Pylorus — Oblique layer — Rugae — Duodenum (first portion of the small intestine)

Anterior view of the stomach

may slow the peristaltic movements and the exit of chyme. The process continues until the stomach is empty (about 1–2 hours after you finish a meal). Various foods spend different amounts of time in the stomach. Fat-rich meals spend more time in your stomach than protein-rich ones, which in turn hang out longer in this part of the digestive system than carbohydrate-rich meals.

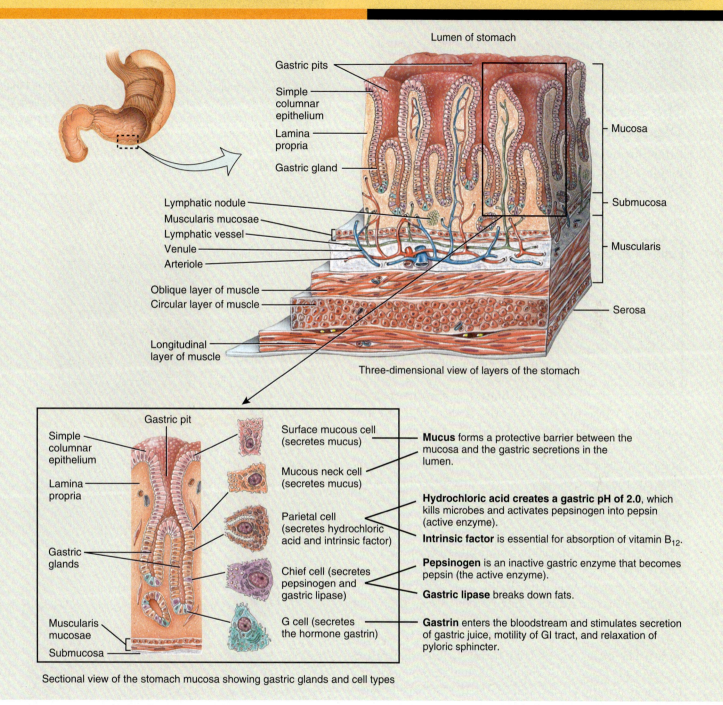

Three-dimensional view of layers of the stomach

Lumen of stomach

Gastric pits
Simple columnar epithelium
Lamina propria
Gastric gland

Mucosa

Lymphatic nodule
Muscularis mucosae
Lymphatic vessel
Venule
Arteriole

Submucosa

Oblique layer of muscle
Circular layer of muscle

Muscularis

Longitudinal layer of muscle

Serosa

Gastric pit

Simple columnar epithelium

Lamina propria

Gastric glands

Muscularis mucosae

Submucosa

Surface mucous cell (secretes mucus)

Mucous neck cell (secretes mucus)

Parietal cell (secretes hydrochloric acid and intrinsic factor)

Chief cell (secretes pepsinogen and gastric lipase)

G cell (secretes the hormone gastrin)

Mucus forms a protective barrier between the mucosa and the gastric secretions in the lumen.

Hydrochloric acid creates a gastric pH of 2.0, which kills microbes and activates pepsinogen into pepsin (active enzyme).

Intrinsic factor is essential for absorption of vitamin B_{12}.

Pepsinogen is an inactive gastric enzyme that becomes pepsin (the active enzyme).

Gastric lipase breaks down fats.

Gastrin enters the bloodstream and stimulates secretion of gastric juice, motility of GI tract, and relaxation of pyloric sphincter.

Sectional view of the stomach mucosa showing gastric glands and cell types

To summarize, the main functions of the stomach include the following:

- Mixes saliva, food, and gastric juice to form chyme.
- Serves as a reservoir for food before release into the small intestine.

- Secretes gastric juice, which contains HCl (kills bacteria and denatures protein), pepsin (begins the digestion of proteins), intrinsic factor (aids absorption of vitamin B_{12}), and gastric lipase (aids digestion of triglycerides).
- Secretes gastrin into blood.

While the stomach digests proteins into peptides through the action of pepsin, the majority of food is digested and absorbed in the small intestine. Let's take a closer look at the role of the small intestine in transforming your breakfast into fuel.

The Small Intestine Is the Site of Most Digestion and Absorption

After the stomach turns your breakfast into chyme and empties it into the small intestine, the real work of digestion begins. The small intestine consists of three parts—duodenum, jejunum, and ileum (**Figure 14.6**)—and is about 5 m (16 ft) long in an adult.

> **villi** (VIL-T) Projections of the intestinal mucosal cells containing connective tissue, blood vessels, and a lymphatic vessel that function in the absorption of the end products of digestion.

The small intestine has the same layers as other portions of the GI tract, but the inner surface is highly folded, both macroscopically into circular folds and microscopically into **villi** and microvilli. This folded surface is essential for the functioning of the small intestine, where the vast majority of materials are digested and absorbed. These internal features of the small intestine increase the surface area by 300–400% while maintaining a relatively small outer surface for the organ. The villi are lined with absorptive cells, secretory cells, and endocrine cells. In the crevices between villi are intestinal glands, which secrete intestinal juices that contain water, mucus, and enzymes.

Digestion in the intestine starts at the **duodenum**, which is the first part of the small intestine and is about 25 cm (10 in) long. The **jejunum** and **ileum** are each about 2–2.5 m (6–8 ft) long. Acidic chyme enters the duodenum and stimulates S cells to produce secretin, which in turn stimulates the pancreas to secrete pancreatic bicarbonate into the duodenum (Figure 14.7 on the next page). The bicarbonate-rich pancreatic juice neutralizes the acid in the chyme. The presence of amino acids and fatty acids in the chyme stimulates the CCK cells to secrete **cholecystokinin (CCK)**, which increases pancreatic enzyme secretion. CCK also causes bile release from the gallbladder and inhibits gastric emptying. Localized muscle movements called **segmentations** mix the various secretions with the chyme. Peristalsis moves the chyme along the length of the small intestine as digestion and absorption occur.

The small intestine consists of three regions (duodenum, jejunum, ileum) with highly folded layers. Water-soluble substances absorbed through the epithelial lining go into blood vessels. Fat-soluble substances proceed into lymph vessels called lacteals and eventually make their way to the blood circulation.

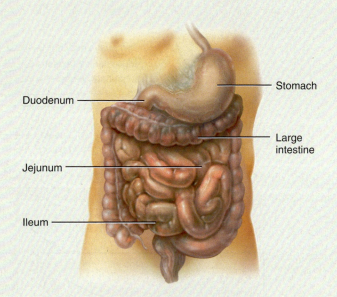

Duodenum

Stomach

Jejunum

Large intestine

Ileum

Anterior view of external anatomy

Circular folds of mucosa and submucosa increase the surface area for absorption of nutrients.

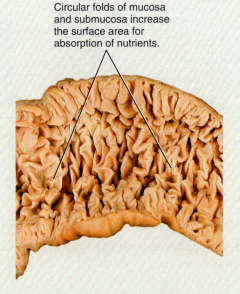

Internal anatomy of the jejunum

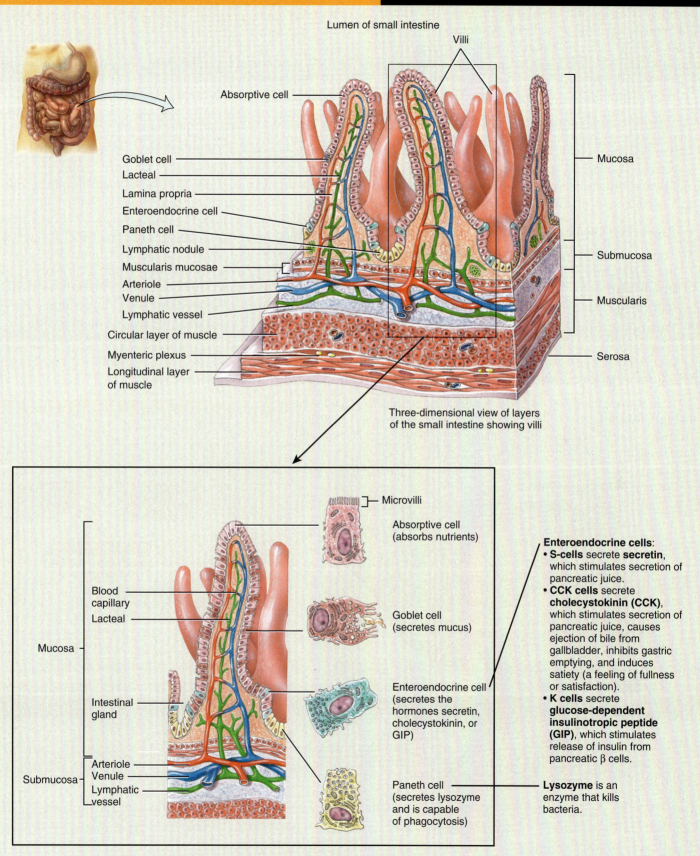

Lumen of small intestine

Villi

Absorptive cell

Goblet cell
Lacteal
Lamina propria
Enteroendocrine cell
Paneth cell
Lymphatic nodule
Muscularis mucosae
Arteriole
Venule
Lymphatic vessel
Circular layer of muscle
Myenteric plexus
Longitudinal layer
of muscle

Mucosa

Submucosa

Muscularis

Serosa

Three-dimensional view of layers
of the small intestine showing villi

Microvilli

Absorptive cell
(absorbs nutrients)

Blood
capillary
Lacteal

Mucosa

Goblet cell
(secretes mucus)

Intestinal
gland

Enteroendocrine cell
(secretes the
hormones secretin,
cholecystokinin, or
GIP)

Arteriole
Venule
Lymphatic
vessel

Submucosa

Paneth cell
(secretes lysozyme
and is capable
of phagocytosis)

Enteroendocrine cells:
• **S-cells** secrete **secretin**,
which stimulates secretion of
pancreatic juice.
• **CCK cells** secrete
cholecystokinin (CCK),
which stimulates secretion of
pancreatic juice, causes
ejection of bile from
gallbladder, inhibits gastric
emptying, and induces
satiety (a feeling of fullness
or satisfaction).
• **K cells** secrete
**glucose-dependent
insulinotropic peptide
(GIP)**, which stimulates
release of insulin from
pancreatic β cells.

Lysozyme is an
enzyme that kills
bacteria.

Enlarged villus showing lacteal, capillaries, intestinal glands, and cell types

Let's Journey Through the Digestive System **409**

Figure 14.7 shows the liver, pancreas, and gallbladder. Liver cells (hepatocytes) make bile. **Bile salts** break down large lipid globules into small ones through emulsification, the breakdown of large lipid globules into a suspension of small lipid globules. **Bile pigment** consists mainly of bilirubin, which is derived from the destruction of hemoglobin from worn-out red blood cells (see Chapter 10).

The **gallbladder** stores bile and releases it into the duodenum of the small intestine through the common bile duct when stimulated by CCK.

Pancreatic acinar cells secrete **pancreatic juice**, which is clear, colorless, and consists of water, sodium bicarbonate, and various enzymes. The sodium bicarbonate neutralizes stomach acid. The enzymes include the following:

- **Pancreatic amylase**, which breaks down starches.
- **Trypsin**, **chymotrypsin**, and **carboxypeptidase**, which break proteins into amino acids. (Carboxypeptidase is inactive in the pancreas but is activated by enterokinase in the duodenum.)
- **Pancreatic lipase**, which breaks triglycerideas into fatty acids.

- **Ribonuclease** and **deoxyribonuclease**, which break down nucleic acids.
- Pancreatic juice flows into the duodenum through the pancreatic duct.

The actions of several digestive enzymes from the stomach, pancreas, and small intestine completely break down carbohydrates, fats, proteins, and nucleic acids into their simplest monomers (small building-block molecules). Bile salts from the liver and gallbladder break down fat droplets into smaller ones called **micelles**, much as dishwashing detergent breaks up grease in the dishwater.

As the chyme passes the absorptive cells on the villi, sugars, amino acids, and lipids are absorbed into the epithelial cells by diffusion and active and passive transport processes (**Figure 14.8**). The removal of these materials from the lumen creates a gradient that allows water to be absorbed by osmosis. Vitamins are absorbed by simple diffusion. This can happen directly, as is the case with water-soluble vitamins (B, C), or indirectly—along with lipids—for the fat-soluble vitamins (A, D, E, and K).

The liver and pancreas secrete substances into the duodenum • Figure 14.7 _____

Bile is made in the liver, stored in the gallbladder, and subsequently released into the duodenum through the common duct. Pancreatic juice gets secreted from exocrine cells of the pancreas called acinar cells. Pancreatic juice flows into the duodenum through the pancreatic duct.

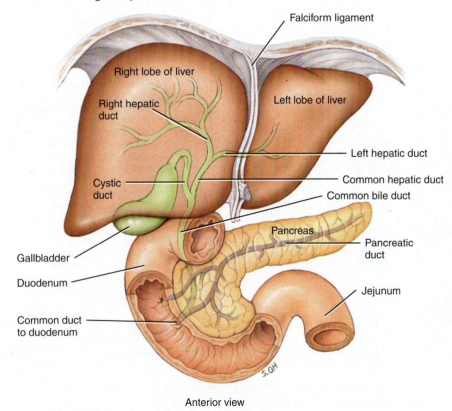

Anterior view

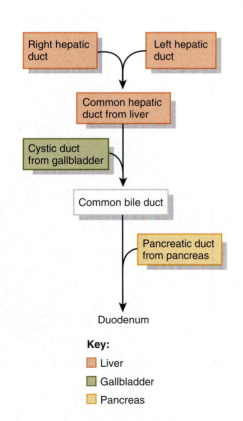

Key:
■ Liver
■ Gallbladder
■ Pancreas

The small intestine absorbs nutrients • Figure 14.8

The absorptive cells on the villi of the small intestine absorb various molecules through simple or facilitated diffusion as well as active transport. Depending on the nature of the absorbed substance, it passes from the absorptive cell either into the blood and hepatic portal circulation (amino acids, sugars) or into the lacteal and lymphatic system (lipids).

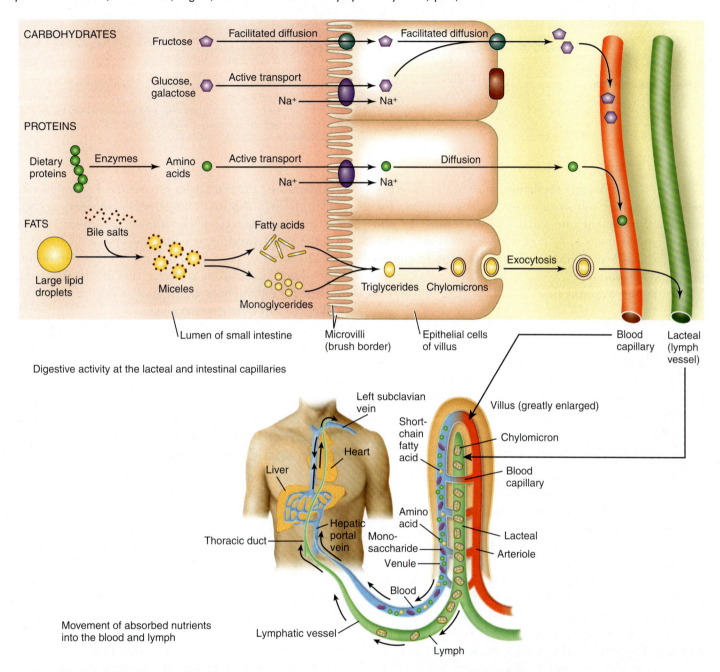

Digestive activity at the lacteal and intestinal capillaries

Movement of absorbed nutrients into the blood and lymph

Absorbed substances must then pass from the intestinal epithelial cells in the villi into the circulation. Amino acids, sugars, and short-chain fatty acids pass from the epithelial cells into the blood by diffusion. In the blood, these substances travel through the hepatic portal circulation to the liver for processing and then subsequently into the general circulation. Fats are handled differently. Inside the absorptive cells, various lipids (for example, long-chain fatty acids, monoglycerides, cholesterol, triglycerides) are packaged into protein-coated spheres called **chylomicrons**. Chylomicrons exit the absorptive cells by exocytosis into lymph vessels within villi called **lacteals**. Chylomicrons travel through the lymphatic system and into the general circulation to adipose tissue and the liver. There, the chylomicrons are removed, and lipids are stored for future use.

Let's Journey Through the Digestive System   411

Major hormones that control digestion Table 14.1

Hormone	Where produced	Stimulant	Action
Gastrin	Stomach mucosa (pyloric region)	Stretching of stomach, partially digested proteins and caffeine in stomach, and low pH of stomach chyme.	Stimulates secretion of gastric juice, increases motility of GI tract, and relaxes pyloric sphincter.
Secretin	Intestinal mucosa	Acidic chyme that enters the small intestine.	Stimulates secretion of pancreatic juice rich in bicarbonate ions.
Cholecystokinin (CCK)	Intestinal mucosa	Amino acids and fatty acids in chyme in small intestine.	Inhibits gastric emptying, stimulates secretion of pancreatic juice rich in digestive enzymes, causes ejection of bile from the gallbladder, and induces a feeling of satiety (feeling full to satisfaction).

When chyme reaches the ileum, most of the bile salts are reabsorbed and returned by the blood to the liver for recycling. Insufficient bile salts, whether caused by obstruction of bile ducts or by liver disease, can result in the loss of up to 40% of dietary lipids in feces due to diminished lipid absorption.

In sum, these are the functions of the small intestine:

- Segmentations mix chyme with digestive juices and bring food into contact with the mucosa for absorption. Peristalsis propels chyme through the small intestine.

- The small intestine completes the digestion of carbohydrates, proteins, and lipids and begins and completes the digestion of nucleic acids.

- The small intestine absorbs about 90% of nutrients and water that pass through the digestive system.

Digestion also involves the activity of a number of hormones (**Table 14.1**). Many of these substances are released into the GI tract and are thought to act locally, while others are released into the bloodstream. The exact physiological roles of these "gut hormones" are not known but are being investigated.

Many of the absorbed nutrients are processed by the liver. The hepatic portal vein transports nutrient-rich venous blood to the liver, where it is distributed between rows of hepatocytes (liver cells) through hepatic sinusoids. The hepatocytes extract nutrients and process them. As you will see later in this chapter, hepatocytes combine simple carbohydrates to form glycogen, make lipids from glucose and/or amino acids, distribute amino acids to the blood, and break down excessive amino acids and worn-out proteins into glucose or triglycerides. After extraction and processing, the blood drains into the central veins to the hepatic vein and ultimately into the inferior vena cava.

Meanwhile, the remainder of your meal completes its journey through the small intestine. Segmentation stops and the peristaltic wave pushes the chyme forward slowly down the small intestine, reaching the end of the ileum in 90 to 120 minutes. Then another wave of peristalsis begins in the stomach. Altogether, chyme remains in the small intestine for 3 to 5 hours. By the time the chyme reaches the ileum, most nutrients have been absorbed. But water and some vitamins, carbohydrates, and ions remain. The final steps of digestion occur in the large intestine. Let's turn our attention next to this last organ in the gastrointestinal tract.

The Large Intestine Absorbs Water and Eliminates Wastes

The large intestine is the last part of the GI tract that your breakfast visits on its travels through the digestive system. It is about 6.5 cm (2.5 in) in diameter and 1.5 m (5 ft) long. It extends from the ileum to the anus and is attached to the posterior abdominal wall by a double layer of peritoneum called the mesocolon. It has four major segments: cecum, colon, rectum, and anal canal (**Figure 14.9**). The ileocecal valve regulates the movement of material into the large intestine, while the internal and external anal sphincters regulate the movement of material out of the large intestine.

Like the small intestine, the large intestine needs lots of surface area to complete its work, which includes absorption of water, certain ions (Na^+ and Cl^-), and some vitamins (B and K). Unlike the small intestine, however, the large intestine's surface changes are visible from the outside of the tubing. Teniae coli, bands of muscle fibers, pucker up the large intestine to form pouch-like sacs called haustra.

The large intestine consists of four regions (cecum, colon, rectum, anal canal). The layers of the walls of the large intestine are like those of organs elsewhere in the GI tract. The lumen contains bacteria that have important digestive functions. The mucosa contains absorptive cells and mucus-secreting cells. The large intestine has several functions, including digestion; absorption of ions, vitamins, and water; and elimination of feces.

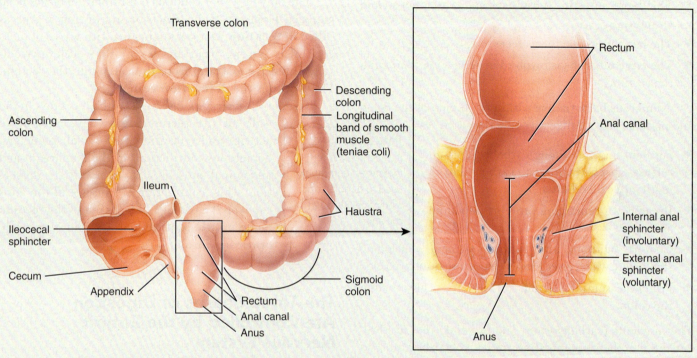

Regions of the large intestine

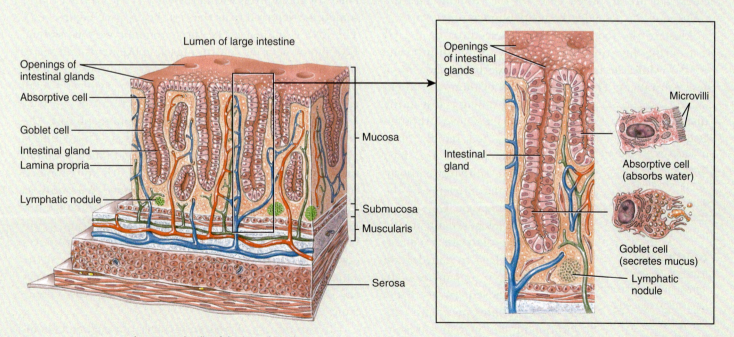

Layers and cells of the large intestine

To summarize, these are the large intestine's main functions:

- Peristalsis drives the contents of the colon into the rectum.
- Bacteria in the large intestine convert proteins to amino acids, break down amino acids, and produce some B vitamins and vitamin K.
- The large intestine absorbs some water, ions, and vitamins.
- The large intestine forms feces.
- The large intestine is involved in defecation (emptying the rectum).

Numerous bacteria live within the lumen of the colon and perform several digestive functions:

- Ferment any remaining carbohydrates and release gases (hydrogen, carbon dioxide, methane) called **flatus**.
- Break down remaining proteins into amino acids.
- Convert bilirubin into simpler pigments, including stercobilin, which gives feces a characteristic brown color.
- Produce essential vitamins (B, K) that are absorbed in the colon.

> **flatus** (FLĀ-tus) Gas in the stomach or intestines; commonly used to denote expulsion of gas through the anus.

The nutrients (carbohydrates and proteins) that are broken down in the large intestine are not able to be absorbed and are passed out with the feces. The only nutrient molecules that are absorbed here are the vitamins that are produced in the colon by the resident microorganisms.

Most of the food and materials secreted by the GI tract during digestion is absorbed by the small intestine (~90%), while the large intestine absorbs another 9% of these substances. Therefore, only 1% of the material entering the GI tubing is lost in the feces. If chyme moves through the intestines too rapidly or if enzymes that process the food materials are missing, absorption of fluid and nutrients by the intestines decreases. Excess water and nutrients remain in the feces, causing **diarrhea**. Frequent diarrhea can lead to dehydration and electrolyte imbalances. In contrast, if movement of chyme through the intestines slows, absorption of fluid increases, resulting in **constipation**.

> **diarrhea** (dī-a-RĒ-a) Frequent defecation of liquid feces, caused by increased motility of the intestines or absence of enzymes to process the food nutrients.
>
> **constipation** (kon'-sti-PĀ-shun) Infrequent or difficult defecation caused by decreased motility of the intestines.

The remains of your breakfast will take about 3 to 10 hours to journey through the large intestine. As water gets absorbed from the chyme, it changes from a liquid to a semi-solid or solid mass called **feces**. Feces consists mainly of water, inorganic salts, epithelial cells that have sloughed off the mucosa of the GI tract, bacteria, products of bacterial decomposition, unabsorbed digested material, and indigestible parts of food.

The only task that remains is elimination of the fecal material from the body, which occurs during a process called **defecation**. Defecation begins with waves of mass peristalsis (strong peristalic wave) that occur periodically and drive material from the colon into the rectum. This initiates the **defecation reflex**.

The defecation reflex involves stretch receptors in the walls of the rectum, parasympathetic nerve activity, and relaxation of the anal sphincters. Sometimes contractions of the diaphragm and abdominal muscles can help initiate defecation, while contraction of muscles in the anus can stop the passage of feces out of the rectum. Some muscle contractions are voluntary, while others are involuntary.

The Three Phases of Digestion Are Controlled by the Enteric Nervous System

Digestion is controlled by the enteric nervous system (ENS), a network of neurons within the walls of the GI tract that is sometimes referred to as the "gut brain" (see Figure 14.2). The ENS is influenced by the autonomic nervous system. ENS motor neurons control muscle movements and secretions within the GI tract, and sensory neurons detect the presence of food, using chemical and mechanical stretch receptors within the gastrointestinal organs.

Digestion occurs in three overlapping phases: the cephalic, gastric, and intestinal phases. The cephalic phase responds to the sensations of food (such as smell, taste, and sight) and prepares the mouth and stomach to receive it. This relatively short phase lasts only as long as eating the meal. The gastric phase continues gastric secretions and motility, mixing food and regulating gastric emptying. The intestinal phase promotes continued digestion through the small intestine. At the same time, it inhibits gastric emptying, which prevents the duodenum from becoming overloaded with chyme from the stomach.

To summarize, digestion involves physical and chemical processes in several organs. The physical processes include chewing, mixing, and peristalsis, all of which involve muscle contractions. The chemical processes include secretions of fluids (saliva, gastric juices, pancreatic juices, bile) that consist mostly of water, mucus, ions, bile salts, and digestive enzymes. The chemical secretions and mechanical mixing work together to break down food into simple carbohydrates, lipids, amino acids, nucleic acids, and vitamins. These substances are absorbed mainly in the small intestine, but some are also absorbed in the stomach and large intestine. What remains of your breakfast after it passes through the large intestine moves into the rectum and is excreted from the body. For a summary of the functions of the digestive organs, see **Table 14.2**.

Summary of digestive organs and their functions	Table 14.2	
Organ	**Functions**
Mouth	See other listings in this table for the functions of the tongue, salivary glands, and teeth, all of which are in the mouth. Additionally, the lips and cheeks keep food between the teeth during mastication, and buccal glands lining the mouth produce saliva.
Tongue	Maneuvers food for mastication, shapes food into a bolus, maneuvers food for deglutition, detects taste and touch sensations, and initiates digestion of triglycerides.
Salivary glands	Produce saliva, which softens, moistens, and dissolves foods; cleanses mouth and teeth; and initiates the digestion of starch.
Teeth	Cut, tear, and pulverize food to reduce solids to smaller particles for swallowing.
Pharynx	Receives a bolus from the mouth and passes it into the esophagus.
Esophagus	Receives a bolus from the pharynx and moves it into the stomach. This requires relaxation of the upper esophageal sphincter and secretion of mucus.
Stomach	Mixing waves soak food, mix it with the secretions of gastric glands (gastric juice), and reduce food to chyme. Gastric juice activates pepsin and kills many microbes in food. Intrinsic factor aids absorption of vitamin B_{12} from the colon. The stomach serves as a reservoir for food before releasing it into the small intestine.
Pancreas	Pancreatic juice buffers acidic gastric juice in chyme (creating the proper pH for digestion in the small intestine), stops the action of pepsin from the stomach, and contains enzymes that digest carbohydrates, proteins, triglycerides, and nucleic acids.
Liver	Produces bile, which is needed for the emulsification and absorption of lipids in the small intestine.
Gallbladder	Stores and concentrates bile and releases it into the small intestine.
Small intestine	Segmentations mix chyme with digestive juices; peristatic contractions propel chyme toward the ileocecal sphincter; digestive secretions from the small intestine, panceas, and liver complete the digestion of carbohydrates, proteins, lipids, and nucleic acids; circular folds, villi, and microvilli increase surface area for absorption; site where about 90% of nutrients and water are absorbed.
Large intestine	Haustral churning, peristalsis, and mass peristalsis drive the contents of the colon into the rectum; bacteria produce some B vitamins and vitamin K; absorption of some water, ions, and vitamins; defecation.

CONCEPT CHECK

1. **What** process begins physical digestion?
2. **Which** layer of the GI tract is responsible for absorption?
3. **What** are the conditions in the stomach that help break down proteins?
4. **How** is the surface of the small intestine specialized for absorption?
5. **Where** is bile made, and what does it do?
6. **What** is the function of pancreatic juice?
7. **Which** substances are absorbed in the large intestine?
8. **How** is the defecation reflex controlled?
9. **What** types of materials are absorbed into the lacteals?

Your Diet Contains Many Nutrients

LEARNING OBJECTIVES

1. **Define** the term *nutrient* and identify the six main types of nutrients.

2. **Outline** the guidelines for healthy eating.

3. **Identify** minerals in your diet and their importance.

4. **List** the principal vitamins and explain their functions.

You have learned that the purpose of the digestive system is to break down food and absorb nutrients. But what are nutrients? **Nutrients** are chemical substances in food that the cells of your body need for growth, maintenance, and repair. Nutrients include carbohydrates, lipids, proteins, water, minerals, and vitamins. **Essential nutrients** are specific substances that your body cannot make on its own in sufficient quantities to meet its needs. Essential nutrients include some amino acids, fatty acids, vitamins, and minerals.

When planning your diet, there are two things to consider:

- How many **Calories** your food contains. Do not confuse a Calorie (with a capital C) with a calorie (cal), the amount of heat needed to raise the temperature of 1 g of water from 14°C to 15°C; 1 Calorie equals 1,000 calories.

- The distribution of the food types in the diet (such as, carbohydrates, fats, and proteins). Diets vary worldwide with food availability and cultures. Most experts recommend a diet with the following distribution of calories: 50–60% carbohydrates, 30% fats, and 12–15% proteins.

Suggestions for healthy eating include the following:

- Eat a variety of foods.
- Maintain a healthy weight.
- Choose foods low in fat, saturated fat, and cholesterol.
- Eat plenty of vegetables, fruits, and grain products.
- Use sugars in moderation only.

In 2005, the U.S. Department of Agriculture (USDA) adopted a new, personalized approach to nutritional guidelines called *MyPyramid* (**Figure 14.10**).

Your diet should include **minerals**, inorganic elements that constitute about 4% of your body weight. Most minerals are found in your bones and teeth. Minerals have important functions in the structure of bone, regulation of enzymatic functions, maintenance of pH of body fluids, osmosis, and generation of action potentials in nerves and muscles. For example, the mineral magnesium is an important cofactor for enzymes that convert ADP to ATP (see Chapter 2). Minerals that are vital to the body include calcium, phosphorus, potassium, sulfur, sodium, chloride, and magnesium.

Vitamins are organic nutrients that are required in small amounts to maintain normal growth and metabolism. Most vitamins serve as *coenzymes* (molecules that help enzymes do their jobs) in chemical reactions. They are classified as either **fat-soluble vitamins** (for example, vitamins A, D, E, and K) or **water-soluble vitamins** (for example, vitamins B and C). Fat-soluble vitamins are absorbed from the digestive tract and transported in chylomicrons. In contrast, water-soluble vitamins are dissolved in body fluids. While excess quantities of fat-soluble vitamins can be stored in cells, excess quantities of water-soluble vitamins cannot and are instead excreted in the urine.

Two vitamins (C, E) and a **pro-vitamin** (beta-carotene) inactivate damaging free radicals of oxygen. These vitamins are also called **antioxidant vitamins**.

> **Calorie** (KAL-ō-rē) An expression of the amount of energy in various foods and a measurement of metabolic rate that is equal to 1,000 calories.

> **pro-vitamin** A chemical precursor to a vitamin.

CONCEPT CHECK

1. **What** is an essential nutrient?
2. **What** are the recommendations for calorie distribution for protein, fat, and carbohydrate intake?
3. **Why** is calcium important for your body?
4. **What** is the function of vitamin D?

The USDA MyPyramid nutritional guidelines • Figure 14.10

The five color bands represent the five basic food groups, and the size of each band represents the proportion of food a person should choose on a daily basis. The wide base represents foods with little or no solid fats or added sugars. The narrow top represents foods with more solid fats and added sugars. The person climbing the stairs is a reminder that daily physical activity is needed for healthy living.

MyPyramid
STEPS TO A HEALTHIER YOU
MyPyramid.gov

GRAINS Make half your grains whole	VEGETABLES Vary your veggies	FRUITS Focus on fruits	MILK Get your calcium-rich foods	MEAT & BEANS Go lean with protein
Eat at least 3 oz. of whole-grain cereals, breads, crackers, rice, or pasta every day 1 oz. is about 1 slice of bread, about 1 cup of breakfast cereal, or ½ cup of cooked rice, cereal, or pasta	Eat more dark-green veggies like broccoli, spinach, and other dark leafy greens Eat more orange vegetables like carrots and sweetpotatoes Eat more dry beans and peas like pinto beans, kidney beans, and lentils	Eat a variety of fruit Choose fresh, frozen, canned, or dried fruit Go easy on fruit juices	Go low-fat or fat-free when you choose milk, yogurt, and other milk products If you don't or can't consume milk, choose lactose-free products or other calcium sources such as fortified foods and beverages	Choose low-fat or lean meats and poultry Bake it, broil it, or grill it Vary your protein routine — choose more fish, beans, peas, nuts, and seeds

For a 2,000-calorie diet, you need the amounts below from each food group. To find the amounts that are right for you, go to MyPyramid.gov.

Eat 6 oz. every day	Eat 2½ cups every day	Eat 2 cups every day	Get 3 cups every day; for kids aged 2 to 8, it's 2	Eat 5½ oz. every day

Find your balance between food and physical activity
- Be sure to stay within your daily calorie needs.
- Be physically active for at least 30 minutes most days of the week.
- About 60 minutes a day of physical activity may be needed to prevent weight gain.
- For sustaining weight loss, at least 60 to 90 minutes a day of physical activity may be required.
- Children and teenagers should be physically active for 60 minutes every day, or most days.

Know the limits on fats, sugars, and salt (sodium)
- Make most of your fat sources from fish, nuts, and vegetable oils.
- Limit solid fats like butter, stick margarine, shortening, and lard, as well as foods that contain these.
- Check the Nutrition Facts label to keep saturated fats, *trans* fats, and sodium low.
- Choose food and beverages low in added sugars. Added sugars contribute calories with few, if any, nutrients.

Nutrients Are Metabolized in a Number of Ways

LEARNING OBJECTIVES

1. **Outline** the pathways of glycolysis, the Krebs cycle, and oxidative phosphorylation.
2. **Compare** the metabolism of carbohydrates, proteins, and lipids.
3. **Explain** how carbohydrates are made by gluconeogenesis.

Metabolism (me-TAB-ō-lizm) refers to all the chemical reactions in the body. Most of the chemical reactions are catalyzed (sped up) by enzymes (see Chapter 2). Some enzymes work together with coenzymes, which are organic molecules that temporarily carry atoms (or their components) during a reaction. Two important examples are the electron carriers **nicotinamide adenine dinucleotide (NAD⁺)** (referred to as NADH when carrying electrons) and **flavin adenine dinucleotide (FAD)** (referred to as FADH₂ when carrying electrons).

Metabolic reactions include synthesis reactions (**anabolic reactions**, or *anabolism*) and decomposition reactions (**catabolic reactions**, or *catabolism*). In anabolic reactions, simple organic molecules are combined to make more complex ones; the process usually requires energy, generated by splitting ATP into ADP and phosphate. Conversely, in catabolic reactions, complex organic molecules are broken down into simpler ones; this process usually releases energy, which is stored in the form of ATP, a molecule created from ADP and phosphate (**Figure 14.11**). Of the energy released in catabolic reactions, only 40% is captured as ATP. The rest of the energy is released as heat, which warms your body.

To provide ATP for anabolic reactions, your body breaks down some of the carbohydrates, fats, and proteins absorbed by the digestive system. Most cells make ATP from glucose. Some cells, such as neurons and red blood cells, typically use only glucose to make ATP. Most other cells can use lipids (fats) and amino acids (proteins) when glucose is unavailable.

Carbohydrates Are Converted to Glucose

Let's take a look at the catabolism of glucose as an example of the basic process of making ATP. During digestion, car-

Role of ATP in linking catabolic and anabolic reactions • Figure 14.11

When complex molecules are split (catabolism), some energy is transferred to make ATP. Conversely, when simple molecules combine to form complex ones (anabolism), energy is usually supplied by splitting ATP.

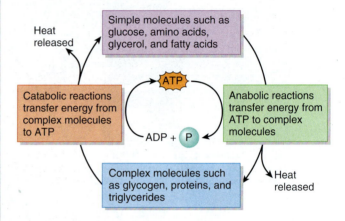

bohydrates are broken down into many monosaccharides (for example, glucose, fructose, galactose), all of which are absorbed by the small intestine. Shortly after absorption, the monosaccharides are converted to glucose. The fate of glucose depends on the needs of your body's cells:

- If cells require ATP, then glucose is broken down via **oxidation**.
- Excess glucose is stored as glycogen in liver and muscle cells.
- If the glycogen stores are full, liver cells convert excess glucose into triglycerides for storage in adipose tissue. Triglycerides can be converted back to glucose when needed.

> **oxidation** (ok-si-DĀ-shun) The removal of electrons from a molecule or, less commonly, the addition of oxygen to a molecule that results in a decrease in the energy content of the molecule.

Let's see how your cells break down glucose to make ATP.

Cellular Respiration Creates ATP

Like a fire combines fuel and oxygen to release energy, your cells "burn glucose" by combining it with oxygen in a process called **cellular respiration**. The overall process can be summarized as follows:

$$1 \text{ glucose} + 6\,O_2 \rightarrow 6\,CO_2 + 6\,H_2O + 36 \text{ ATP}$$

Cellular respiration • Figure 14.12

During cellular respiration, glucose molecules are broken down into carbon dioxide, ATP is made, oxygen is consumed, and water is made. The process involves glycolysis, formation of acetyl coenzyme A, the Krebs cycle, and oxidative phosphorylation. Glycolysis occurs in the cytosol and requires no oxygen; the subsequent processes occur in the mitochondrion, and use oxygen.

1 **Glycolysis.** In the cytosol, glucose (6 carbons) is broken down into 2 pyruvic acid molecules (3 carbons each) in a 10-step pathway. ATP must be used to begin the process.

2 **Formation of acetyl coenzyme A** (2 carbons). Each pyruvic acid from glycolysis enters the mitochondrion, loses a carbon dioxide, and is attached to coenzyme A. Electrons are transferred to NADH.

3 **Krebs cycle.** In the matrix of the mitochondrion, the acetyl groups are broken down. In addition:
• More CO_2 is released.
• A small amount of ATP is formed.
• Many electrons are transferred to NADH and $FADH_2$ (electron carriers).

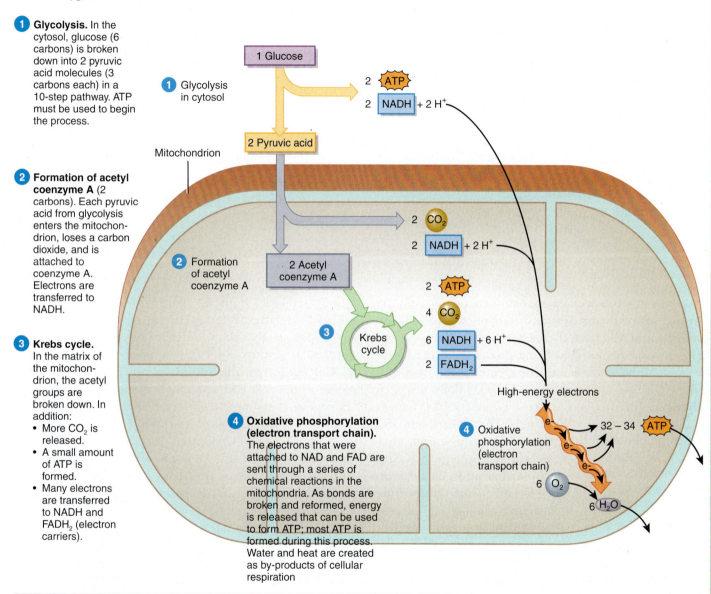

4 **Oxidative phosphorylation (electron transport chain).** The electrons that were attached to NAD and FAD are sent through a series of chemical reactions in the mitochondria. As bonds are broken and reformed, energy is released that can be used to form ATP; most ATP is formed during this process. Water and heat are created as by-products of cellular respiration

This seemingly simple equation involves four complex biochemical pathways: **1** **glycolysis**, **2** the formation of **acetyl coenzyme A (acetyl CoA)**, **3** the **Krebs cycle**, and **4** **oxidative phosphorylation** (or the **electron transport chain [ETC]**).

As you can see from **Figure 14.12** and the formula just shown, each glucose molecule that goes through cellular respiration produces 36 ATP molecules (2 from glycolysis, 2 from the Krebs cycle, and 32 from oxidative phosphorylation). Glycolysis is an *anaerobic* process—that is, it requires no oxygen. (If there is no oxygen available after the process finishes, however, pyruvic acid will be converted into lactic acid and very little ATP will be made from the glucose.) Because the formation of acetyl CoA, the Krebs cycle, and oxidative phosphorylation use oxygen, this part of the process of cellular respiration is also known as **aerobic respiration.**

Pathways for making glucose • Figure 14.13

The liver and skeletal muscles store about 500 g (1.1 lb) of glycogen, which is made from glucose following stimulation by insulin. When stimulated by glucagon and epinephrine, liver and skeletal muscle cells can break glycogen into glucose (glycogenolysis) for release into the blood and subsequent use by other cells. When these same cells are stimulated by glucagon and cortisol, glucose can be made from other molecules in a process called gluconeogenesis.

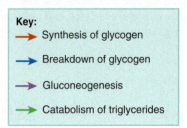

Key:
→ Synthesis of glycogen
→ Breakdown of glycogen
→ Gluconeogenesis
→ Catabolism of triglycerides

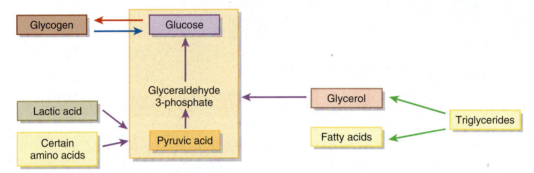

Glucose is the primary source of energy for cells. As a result, glucose must always be supplied. This glucose is mostly acquired from the food we eat. If glucose is not being absorbed from food being processed in the digestive system, however, it must be created from other materials. Glucose can be made by breaking down glycogen (**glycogenolysis**) stored in the liver and skeletal muscle (**Figure 14.13**). Glucose can also be made from pyruvic acid, glyceraldehyde-3-phosphate (G3P), certain amino acids, lactic acid, and glycerol (from triglyceride breakdown) through a process called **gluconeogenesis**. The liver is capable of converting these molecules into glucose.

> **gluconeogenesis**
> (gloo´-kō-nē-ō-JEN-e-sis) The synthesis of glucose from certain amino acids, glycerol, pyruvic acid, or lactic acid; essentially, the reverse of glycolysis.

If glucose availability and glycogen stores become depleted, your body breaks down fats to make glucose. Once fat reserves have been exhausted, your body resorts to breaking down proteins to meet its energy demands. Let's take a closer look at how fat and proteins can be broken down to make ATP.

Lipids Are Broken Down into Intermediates of Cellular Respiration

Like carbohydrates, lipids can be broken down to make ATP, using some of the same pathways that are used for glucose oxidation. Lipid catabolism involves a type of lipid called a triglyceride and occurs in muscle, liver, and fat cells. First, triglycerides are split into glycerol and three fatty acids (**Figure 14.14**). The glycerol gets converted to G3P, an intermediate in glycolysis. Depending on the needs of the cell, G3P can be (1) made into glucose (via gluconeogenesis) or (2) converted to carbon dioxide, water, and ATP (via cellular respiration). The fatty acids are converted to acetyl CoA (used in cellular respiration) or **ketones** (or *ketone bodies*) for use as an energy source in other cells. Ketones are transported from liver cells to other cells, where they can be easily changed back to acetyl CoA and metabolized through cellular respiration.

If the body has no immediate need to use lipids to make ATP, the triglycerides are stored for future use. Triglycerides can also be created from excess glucose, using these same biochemical pathways. Whether lipids are synthesized via **lipogenesis** or broken down via **lipolysis** depends on the cell's energy needs and stimulation by various hormones. Insulin stimulates lipid formation, while epinephrine, norepinephrine, and cortisol stimulate lipid breakdown. Because most lipids do not dissolve in water, they must be shuttled among various cells and tissues, wrapped in water-soluble packages called **lipoproteins**.

> **lipogenesis** (li-pō-GEN-e-sis) The synthesis of triglycerides.
>
> **lipolysis** (lip-OL-i-sis) The splitting of fatty acids from a triglyceride or phospholipids.

There are several types of lipoproteins. Chylomicrons are made in the small intestine and transport dietary lipids to adipose cells for storage (see Figure 14.8). **Very-low density lipoproteins** (VLDLs) transport triglycerides from liver cells to adipose cells for storage. After After depositing some triglycerides, VLDLs become LDLs. **Low-density lipoproteins** (LDLs) carry 75% of the cholesterol in blood and deliver it to cells throughout the body. LDL is known as "bad" cholesterol because it can lead to fatty plaques in blood vessels. Blood LDL levels below 130mg/dL are desirable. **High-density lipoprotein** (HDL) carries excess cholesterol from body cells to liver cells for disposal. HDL is considered "good" cholesterol and blood levels over 40 mg/dL are desirable.

Now that we have seen what happens to lipids, let's take a closer look at protein catabolism.

Proteins Can Also Be Metabolized to Create ATP

During digestion, proteins are broken down into amino acids, which are absorbed in the small intestine. Unlike carbohydrates and lipids, amino acids cannot be stored. Instead, they are used to make new proteins (see Chapter 3), converted to glucose or triglycerides, or catabolized to make ATP. Furthermore, when the new proteins wear out, cells degrade them to amino acids, which are either metabolized or recycled to make new proteins. Cells take up amino acids when stimulated by insulin-like growth factor (IGF) and insulin. During amino acid catabolism, the amino group is removed first (via deamination), yielding highly toxic ammonia; the liver converts ammonia to urea, which is excreted into the urine. The remaining organic component is some intermediate in cellular respiration, mostly Krebs cycle components . These Krebs cycle intermediates are then metabolized completely to make ATP. Some amino acids can be converted into glucose, fatty acids, or ketone bodies (see Figures 14.13 and 14.14).

Hormones and Chemical Levels Regulate Metabolic Activities

If carbohydrates, lipids, and proteins can be broken down, stored, or interconverted, how does the cell know what to do? Regulation of metabolism is based on the levels of ATP in the cell and is controlled by the enzymatic reactions of certain key metabolic products:

- **Glucose-6-phosphate (G6P)**. The cell traps glucose inside by transferring a phosphate from ATP to glucose. G6P can be metabolized to form ATP or converted into other types of molecules needed by the body, depending on cell ATP level and blood glucose levels.

Pathways for making and breaking down lipids • Figure 14.14

Triglycerides can be made from or split into glycerol and fatty acids. These substances can enter cellular respiration pathways at different points to make ATP. These same pathways can be used to make triglycerides from glucose or amino acids.

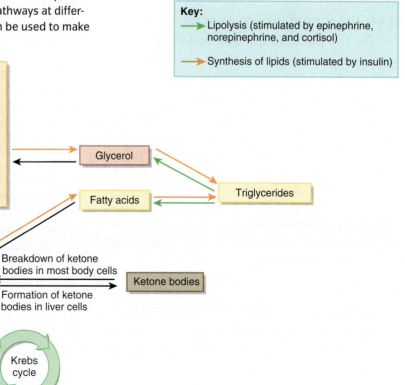

Key:
→ Lipolysis (stimulated by epinephrine, norepinephrine, and cortisol)
→ Synthesis of lipids (stimulated by insulin)

- **Pyruvic acid**. This end product of glycolysis can be metabolized further to create ATP or can be converted to make amino acids or glucose, depending on the ATP level, blood glucose level, and amino acid requirements. If oxygen is unavailable, pyruvic acid is converted into lactic acid and then must be transformed back into a more functional material once oxygen become available.

- **Acetyl CoA**. This product of pyruvic breakdown, fatty acid oxidation, and deamination of certain amino acids can be metabolized in the Krebs cycle for ATP production, depending on the cell's ATP levels and oxygen availability. As already noted, under certain conditions, acetyl CoA can also be converted to ketones for transport to other cells.

The fate of these molecules can also be influenced by various hormones, including insulin, glucagon, cortisol, epinephrine, and norepinephrine.

CONCEPT CHECK

1. **Which** metabolic pathway occurs exclusively in the cytosol?
2. **During** which part of cellular respiration is most ATP created?
3. **Where** does aerobic respiration take place in the cell?
4. **What** is gluconeogenesis, and what nutrients feed into its pathways?
5. **Where** do the products of lipolysis feed into the pathways of cellular respiration?
6. **What** is the first step in catabolizing amino acids?

Diabetes and Obesity Are Metabolic Disorders

LEARNING OBJECTIVES

1. **Distinguish** between type 1 and type 2 diabetes.
2. **Explain** the physiological consequences of diabetes.
3. **Define** *obesity* and explain its consequences.
4. **Describe** the possible treatments for obesity.

Diabetes mellitus is one of many disorders that can affect the endocrine system (see Chapter 9). This disorder involves the hormone insulin. Although its causes are endocrine in nature, diabetes may be best understood when you consider its effects on metabolism. Diabetes is also closely linked with obesity, which has reached epidemic proportions in the United States. Obesity is a complex metabolic disease involving the balance between food intake and energy expenditure. Let's begin our discussion of metabolic disorders by examining diabetes.

Diabetes Is Like Prolonged Starvation

In most people, blood sugar levels do not change much after eating a meal. However, eating can be a dangerous undertaking for the millions of Americans affected by diabetes mellitus. As mentioned in Chapter 9, there are two major types of diabetes mellitus. **Type 1 diabetes** involves a lack of insulin because the patient's immune system has destroyed the pancreatic β cells. **Type 2 diabetes** involves decreased sensitivity of target cells to insulin and is often associated with obesity. Patients with type 1 diabetes have no insulin in their bloodstream, while those with type 2 diabetes may have normal, or even elevated, levels of insulin in their blood. The major symptoms of both types of diabetes include elevated blood glucose, increased urine flow, increased appetite, and poor circulation (especially in the hands and feet). How does diabetes produce these symptoms?

Diabetes is a disease that essentially involves the ineffectiveness of insulin, either due to lack of the hormone or insensitivity to it. As a diabetic patient eats, glucose, triglycerides, fatty acids, and amino acids are absorbed into the blood but are not taken up by cells. (Remember that cellular uptake of these substances requires insulin.) This increases the concentrations of these substances; elevated concentrations of glucose, triglycerides, fatty acids, and amino acids have important consequences for kidney function, peripheral circulation, respiration, nervous system activity, and vision (**Figure 14.15**).

The metabolic events in diabetes • Figure 14.15

The primary causes of diabetes are lack of insulin (Type I diabetes) and insulin resistance (Type II diabetes).

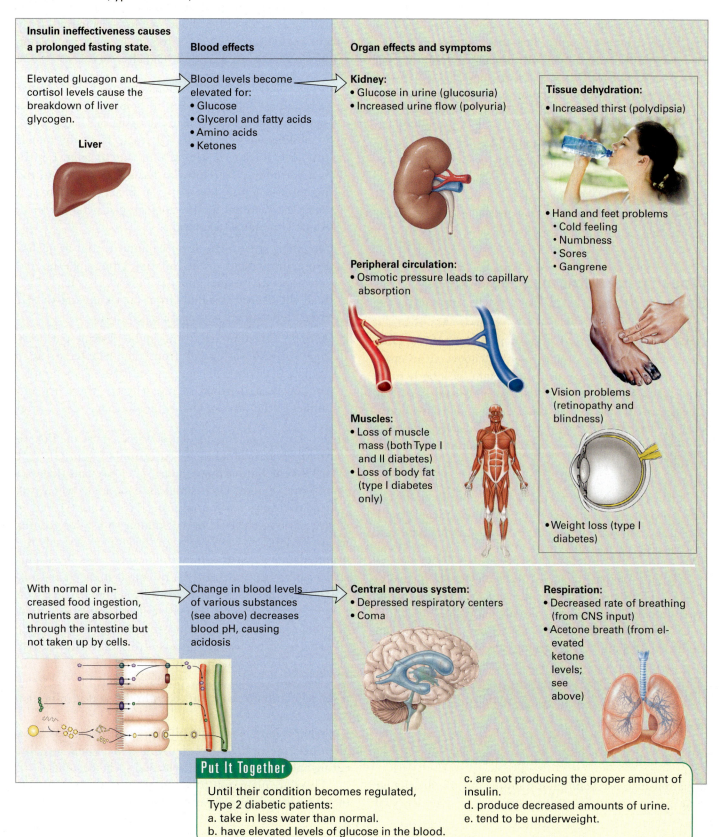

Insulin ineffectiveness causes a prolonged fasting state.

Elevated glucagon and cortisol levels cause the breakdown of liver glycogen.

Liver

Blood effects

Blood levels become elevated for:
• Glucose
• Glycerol and fatty acids
• Amino acids
• Ketones

Organ effects and symptoms

Kidney:
• Glucose in urine (glucosuria)
• Increased urine flow (polyuria)

Peripheral circulation:
• Osmotic pressure leads to capillary absorption

Muscles:
• Loss of muscle mass (both Type I and II diabetes)
• Loss of body fat (type I diabetes only)

Tissue dehydration:
• Increased thirst (polydipsia)

• Hand and feet problems
  • Cold feeling
  • Numbness
  • Sores
  • Gangrene

• Vision problems (retinopathy and blindness)

• Weight loss (type I diabetes)

With normal or increased food ingestion, nutrients are absorbed through the intestine but not taken up by cells.

Change in blood levels of various substances (see above) decreases blood pH, causing acidosis

Central nervous system:
• Depressed respiratory centers
• Coma

Respiration:
• Decreased rate of breathing (from CNS input)
• Acetone breath (from elevated ketone levels; see above)

Put It Together

Until their condition becomes regulated, Type 2 diabetic patients:
a. take in less water than normal.
b. have elevated levels of glucose in the blood.
c. are not producing the proper amount of insulin.
d. produce decreased amounts of urine.
e. tend to be underweight.

Although there is some overlap, treatments for diabetes depend on the type. Type 1 diabetes patients need insulin. They must take multiple injections of insulin daily, monitor their blood glucose often, watch their diet (especially for foods high in sugar), and exercise. Patients with type 2 diabetes must diet and exercise to control their weight. Like type 1 patients, they must monitor their blood sugar and watch their diet, being especially aware of sugar content. Depending on severity, Type 2 diabetics may also need supplemental injections of insulin to help control their blood sugar.

The key to living with diabetes is control of blood glucose. This can be accomplished through frequent monitoring of blood glucose levels, careful dietary planning, exercise, and weight management. As noted earlier, many diabetics may also need daily insulin injections. Let's take a closer look at this increasingly prevalent disorder.

Obesity Is an Imbalance Between Energy Intake and Energy Expenditure

The percentage of overweight and obese Americans is about 65%, an increase of 40% from the late 1970s. **Obesity** is defined as body weight over 20% above an accepted standard for height, age, sex, and weight, or a **body mass index (BMI)** greater than 30 kg/m^2 (**Figure 14.16**). Obese patients are at increased risk for high blood pressure, heart attack, stroke, pulmonary disease, diabetes mellitus, arthritis, varicose veins, gallbladder disease, and certain cancers (for example, breast, uterus, and colon cancers).

Obesity is basically an imbalance between energy intake and energy expenditure. Energy intake includes both eating and absorption of nutrients. Energy expenditure includes three aspects:

- The rate of basal metabolism accounts for 65% of energy expenditure. Basal metabolism is influenced by thyroid hormones and results in production of the amount of ATP necessary to keep you alive (that is, running your heart, keeping you breathing, circulating your blood, producing urine) when you are quietly resting.

- Physical activity accounts for 30–35% of energy expenditure (lower in sedentary lifestyles)

- The **thermic effect of food**, the heat produced while digesting a meal, accounts for 5–10% of energy expenditure.

So if you increase your caloric intake but do not change your energy expenditure (or, worse yet, reduce it), your weight will increase. Conversely, if you reduce your food intake and increase your energy expenditure (such as by exercising), your weight will decrease.

From a physiological standpoint, scientists are just beginning to unravel how body weight is regulated. Body weight regulation includes a feedback system involving sensors, an integrating system, and effectors:

- *Sensors*. Fat tissues and various organs of the GI tract use nerves and hormones to send to the brain a signal about the amount of fat tissue or the presence or absence of food in the GI tract. These hormones primarily include leptin, insulin, and ghrelin. **Leptin** is a hormone secreted by fat tissue; the more fat, the higher the leptin levels in the blood. **Ghrelin** is a hormone secreted by the stomach just before a meal, which stimulates food intake.

- *Integrating system*. The hypothalamus in the brain has two neural circuits—one for stimulating appetite and one for suppressing appetite. These circuits, which read the chemical signals from fat tissue and the GI tract, determine whether you need to eat.

- *Effectors*. Effectors are neural and endocrine pathways that respond to signals from the medulla to alter feeding behavior and energy metabolism. For example, the levels of thyroxin secreted by the thyroid gland influence the rates of energy metabolism.

In the short term, this feedback system must operate to control each individual meal. For example, your GI tract must let your brain know when it is empty and that you should start eating. Or it should let you know when it is full and that you should stop eating. When your GI tract is empty, parasympathetic nerves inhibit the satiety center in the medulla, and ghrelin from the stomach signals the hypothalamic appetite-stimulating center, thereby stimulating you to eat. When the GI tract and stomach are full, sympathetic nerves signal the satiety center to cause you to stop eating; ghrelin levels also fall and no longer stimulate the hypothalamic appetite-stimulating center.

The feedback system also controls the body's stores of fat, thereby regulating body weight in the longer term. The hormone leptin is the primary signal for the long-term regulation of body fat, but ghrelin and insulin also exert effects.

What causes obesity? While no single cause has been identified, multiple factors contribute, including heredity, eating habits, sedentary lifestyle, and social customs.

There are many approaches to the treatment of obesity, targeting both energy inputs and energy expenditures:

Body mass index • Figure 14.16

Body mass index (BMI) is a ratio of height to weight that is used to track overweight and obesity. BMI is calculated by dividing your weight (in kilograms) by the square of your height (in meters). Those numbers are shown in the body of the table, with the more familiar U.S. units at the top and sides for your reference. Overweight is defined as a BMI = 25–30, and obesity is defined as a BMI greater than 30. Where your height and weight intersect, your BMI tells you whether you are underweight, at a healthy weight, overweight, or obese. BMI is only a rough estimate of the amount of body fat; it does not take into account muscular build or bone density.

Weight in Pounds

Height	120	130	140	150	160	170	180	190	200	210	220	230	240	250
4'6"	29	31	34	36	39	41	43	46	48	51	53	56	58	60
4'8"	27	29	31	34	36	38	40	43	45	47	49	52	54	56
4'10"	25	27	29	31	34	36	38	40	42	44	46	48	50	52
5'0"	23	25	27	29	31	33	35	37	39	41	43	45	47	49
5'2"	22	24	26	27	29	31	33	35	37	38	40	42	44	46
5'4"	21	22	24	26	28	29	31	33	34	36	38	40	41	43
5'6"	19	21	23	24	26	27	29	31	32	34	36	37	39	40
5'8"	18	20	21	23	24	26	27	29	30	32	34	35	37	38
5'10"	17	19	20	22	23	24	26	27	29	30	32	33	35	36
6'0"	16	18	19	20	22	23	24	26	27	28	30	31	33	34
6'2"	15	17	18	19	21	22	23	24	26	27	28	30	31	32
6'4"	15	16	17	18	20	21	22	23	24	26	27	28	29	30
6'6"	14	15	16	17	19	20	21	22	23	24	25	27	28	29
6'8"	13	14	15	17	18	19	20	21	22	23	24	25	26	28

Height in Feet and Inches

☐ Underweight ☐ Healthy weight ☐ Overweight ☐ Obese

Ask Yourself

If you are 6'0" tall and weigh 210 pounds, you would be considered:
a. Underweight
b. Healthy weight
c. Overweight
d. Obese

   Energy inputs
− Energy expenditures
= Change in body weight

Behavior therapy, which is used to change lifestyle, involves increasing physical activity, eliminating binge or emotional eating, and watching your diet (monitoring your caloric intake, frequency of meals and snacks, and types of foods that you eat—see *What a Health Provider Sees* on the next page). If a health care provider assesses obesity as severe, then drug therapies may be instituted; most Food and Drug Administration (FDA)-approved obesity drugs target appetite suppression, while one targets absorption of fat in the small intestine. Extreme obesity may warrant surgical interventions, such as gastric bypass or gastric banding. These surgeries limit the capacity of the stomach to store food. *Gastric bypass* is a complex, irreversible surgery in which a portion of the stomach is removed. *Gastric banding* is a reversible procedure in which a band is placed around a portion of the stomach to reduce its size; this procedure has become a more popular technique. Some bands can be adjusted to vary the available size of the stomach as weight loss progresses.

WHAT A HEALTH PROVIDER SEES

Dieting and Weight Loss

There are so many diets available. Patients often ask their health providers which one is best. Diet strategies include three main approaches:

- Reduce total caloric intake but keep the distribution between food types the same (carbohydrates vs. fats vs. proteins).

- Alter the distribution of the diet to restrict or eliminate certain food types. Some diet plans severely restrict fats in favor of carbohydrates. Other plans eliminate carbohydrates in favor of high-fat, high-protein foods.

- Reduce calories while encouraging consumption of foods that have a low **glycemic index** or **glycemic load**. Glycemic index and glycemic load are measures of the degree to which eating a particular food increases blood glucose. Foods with low glycemic index or glycemic load cause blood glucose to rise only slightly, and the changes occur slowly. For example, a baked sweet potato has a lower glycemic index than a baked white potato (63 vs. 158).

A diet plan should be effective and relatively easy to follow, and should always be accompanied by exercise. Patients should understand that dieting is not just a temporary solution but a lifestyle change. Otherwise, many patients will gain the weight back within about two years.

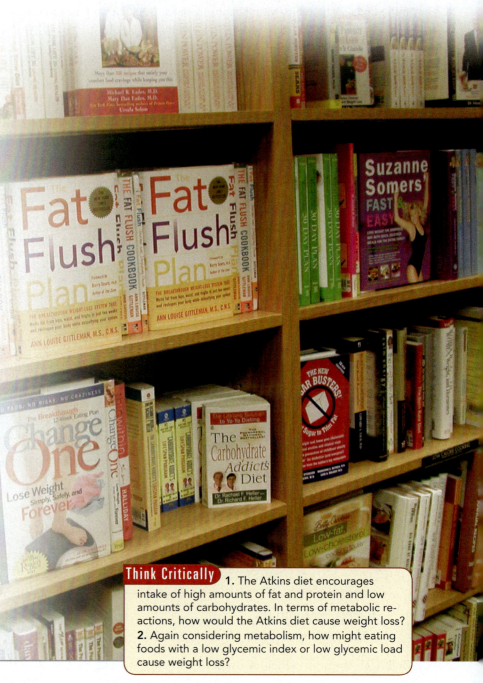

Think Critically 1. The Atkins diet encourages intake of high amounts of fat and protein and low amounts of carbohydrates. In terms of metabolic reactions, how would the Atkins diet cause weight loss? 2. Again considering metabolism, how might eating foods with a low glycemic index or low glycemic load cause weight loss?

CONCEPT CHECK STOP

1. **How** do the insulin signaling disruptions differ in type 1 versus type 2 diabetes?

2. **How** can chronically elevated blood glucose cause poor peripheral circulation in a diabetic?

3. **What** is body mass index?

4. **How** would inhibition of lipase help to treat obesity?

The Heat from Metabolism Must Be Regulated

LEARNING OBJECTIVES

1. **Explain** how body heat is produced and how it is lost.

2. **Describe** how body temperature is regulated.

3. **Outline** the responses of your body to fever and hypothermia.

As you learned earlier in the chapter, catabolism of food transfers only about 40% of the energy to ATP for cellular functions. The rest of the energy is lost as heat, which is what makes us warm-blooded animals. The rate at which this heat is produced is called the **basal metabolic rate (BMR)**. BMR is about 1,200–1,800 Calories/day for adults and can be affected by many factors, including physical activity, hormones, nervous activity, age, and gender (**Figure 14.17**).

Because heat is continuously produced, our bodies must have a way to remove heat so that body temperature does not rise constantly. The processes of heat loss include radiation, conduction, convection, and evaporation (see Figure 14.17). So, overall body temperature is a balance between heat production and heat loss. Your body has a system of temperature regulation in place to maintain this balance, especially in your head and trunk. Let's take a closer look.

Body temperature is a balance between heat production and heat loss • Figure 14.17

Cellular metabolism constantly generates heat. The rate of heat production, called the basal metabolic rate (BMR), is influenced by many factors. The heat produced must be removed by several processes to maintain a constant body temperature.

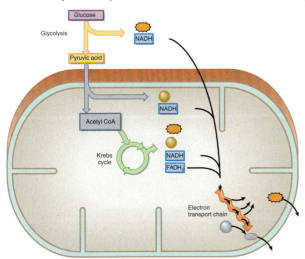

Factors in heat production

Gender: Males have higher BMR than females.

Nervous system: The sympathetic nervous system secretes epinephrine and norepinephrine, which increases BMR.

Hormones: Thyroid hormones, testosterone, insulin, and hGH increase BMR.

Activity: Physical activity increases BMR.

Eating: The thermic effect of food increases BMR.

Body temperature: BMR is proportional to body temperature.

Age: BMR decreases with age.

Sleep: BMR is lower during sleep.

Malnutrition: Malnutrition lowers BMR.

Climate: BMR is lower in the tropics.

Methods of heat loss

Evaporation: Conversion of liquid to vapor as in sweating; removes about 22% of heat at rest.

Conduction: Heat loss between two materials in contact, like water poured on a runner or a swimmer in a pool.

Radiation: Transfer of heat (infrared rays) between warm objects and cooler ones without contact. Your body constantly radiates heat to its surroundings.

Convection: Transfer of heat by movement of a gas or liquid between areas of different temperatures. A fan blowing air over you removes heat by convection.

Like a heating/air conditioning unit in your house, the regulatory system in charge of your body's temperature has sensors (thermoreceptors in skin and hypothalamus), a thermostat (hypothalamus) that sets the temperature, a furnace (BMR, shivering), and a cooling system (blood flow through skin, sweat). When you are exposed to the cold (hypothermia), the system senses the drop in temperature and responds by increasing heat production and minimizing heat loss so that the body temperature increases and returns to its normal 37°C (**Figure 14.18**).

Negative-feedback mechanisms that increase heat production and decrease heat loss in response to decreasing body temperature • Figure 14.18

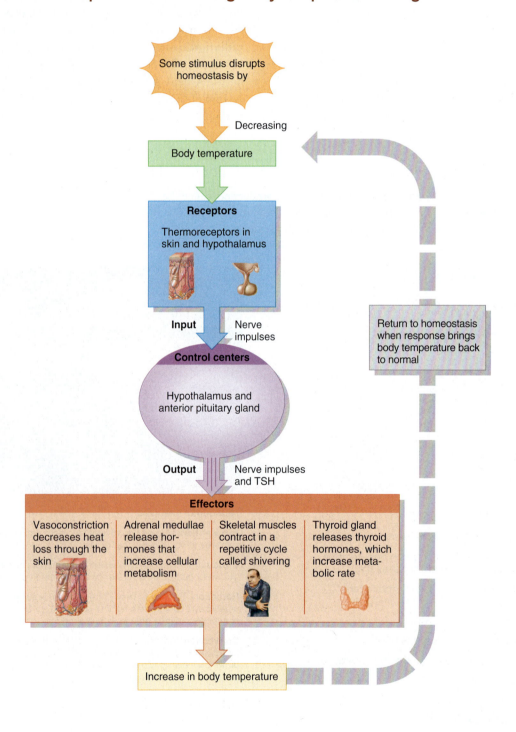

The situation is quite different when you are hot (exercising vigorously, have a fever, or are outside on a sunny summer day): the system senses the increase in temperature. The response of the body is to decrease heat production and increase heat loss. As a result, body temperature decreases to normal (**Figure 14.19**).

Negative-feedback mechanisms that decrease heat production and increase heat loss in response to increasing body temperature • Figure 14.19

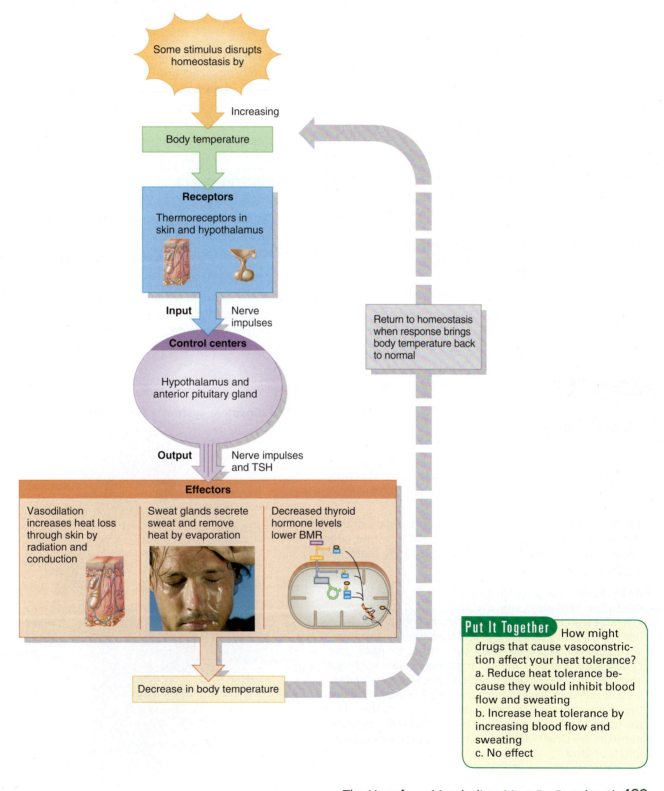

Some stimulus disrupts homeostasis by

Increasing

Body temperature

Receptors
Thermoreceptors in skin and hypothalamus

Input   Nerve impulses

Control centers
Hypothalamus and anterior pituitary gland

Output   Nerve impulses and TSH

Effectors

Vasodilation increases heat loss through skin by radiation and conduction

Sweat glands secrete sweat and remove heat by evaporation

Decreased thyroid hormone levels lower BMR

Decrease in body temperature

Return to homeostasis when response brings body temperature back to normal

Put It Together How might drugs that cause vasoconstriction affect your heat tolerance?
a. Reduce heat tolerance because they would inhibit blood flow and sweating
b. Increase heat tolerance by increasing blood flow and sweating
c. No effect

The Heat from Metabolism Must Be Regulated **429**

Your temperature regulatory mechanism works fairly well for moderate changes in temperature. However, in extreme cases of **hypothermia** (in which core body temperature is lower than 35°C /94°F), the system may be active but unable to compensate for the heat losses. Some conditions that cause hypothermia include extreme cold stress (for example, immersion in icy water or exposure to very cold air), metabolic diseases (hypoglycemia, hypothyroidism, adrenal insufficiency), drugs (alcohol, antidepressants, sedatives, tranquilizers), burns, and malnutrition. Symptoms of hypothermia include cold sensation, shivering, vasoconstriction, slow heart rate, loss of spontaneous movement, and coma. Death from hypothermia is usually due to cardiac arrhythmias and cardiac arrest. Generally, elderly people are at increased risk for hypothermia because their BMR is already reduced.

The hypothalamic thermostat can be reset to higher temperatures by infections (viruses, bacteria), bacterial toxins, hyperthyroidism, tumors, and immune reactions to vaccinations. The result is a fever. When phagocytes fight infections, they secrete chemicals called **pyrogens** that reset the hypothalamic thermostat to a higher temperature by causing it to secrete **prostaglandins**. Often, this higher body temperature helps fight the infection. However, at the higher

> **prostaglandin**
> (pros′-ta-GLAN-din) A membrane-associated lipid that is released in small quantities and act as a local hormone.

set point (higher temperature), your body senses the surrounding temperature as colder and adapts to the higher set point (see Figure 14.17); this is why a feverish person may feel chilled and shiver. The fever may "break" when the original cause of the change in the set point is eliminated. This "breaking" of the fever is usually accompanied by sweating.

While some fever may be beneficial, it is important to prevent the core body temperature from rising to a life-threatening level of 40–42°C or 104–106°F. Temperatures above 42°C are generally fatal. To reduce fevers, you can take pain relievers that reduce prostaglandin production, such as aspirin, acetaminophen (Tylenol™), and ibuprofen (Advil™). Aspirin should not be given to feverish children and teenagers because it has been associated with Reye's syndrome, a disorder characterized by vomiting and brain dysfunction that often progresses to coma and death.

CONCEPT CHECK

1. **What** is the main source of heat production in your body?
2. **Where** is the thermostat of your temperature regulation system located?
3. **How** does your body respond to control your temperature when you are working outside on a hot summer day?

Summary

1 Let's Journey Through the Digestive System 400

- The digestive system, as shown, consists of organs of the **GI tract** (mouth, esophagus, stomach, small intestine, large intestine) and accessory organs (salivary glands, teeth, liver, pancreas, gallbladder).

- As food passes through the GI tract, it is digested into smaller nutrient molecules, including simple sugars, triglycerides, and amino acids. This occurs both physically (through chewing and mixing) and chemically (via various secretions containing enzymes, such as **gastric juice**, pancreatic juice, and bile). Most of the nutrients and water get absorbed in the small intestine and large intestine. Water-soluble nutrients pass through the blood to the liver, where they are processed and then make their way to other cells in the body where they are stored or metabolized. Fat-soluble nutrients are directed into the lymphatic system and then into the blood.

The digestive system • Figure 14.1

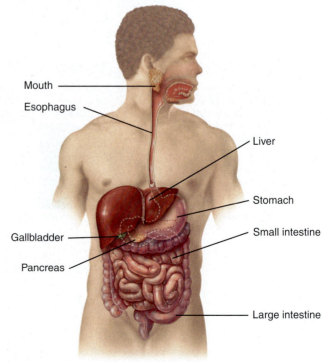

Mouth
Esophagus
Liver
Gallbladder
Pancreas
Stomach
Small intestine
Large intestine

- The remnants of digestion (**feces**) contain dead cells, undigested/unabsorbed material, the breakdown products of bilirubin, and gases (carbon dioxide, methane). Feces are excreted from the body through the anus in a coordinated series of muscle contractions involving the rectum, rectal sphincters, diaphragm, and abdominal muscles.

2 Your Diet Contains Many Nutrients 416

- Your diet consists of water and five types of nutrients: carbohydrates, lipids, proteins, minerals, and vitamins. Many experts recommend that dietary calories be distributed as follows: 50–60% from carbohydrates, 30% or less from fats, and 12–15% from proteins. As shown, the MyPyramid guide from the USDA represents a personal approach to diet and encourages exercise.

The USDA MyPyramid nutritional guidelines • Figure 14.10

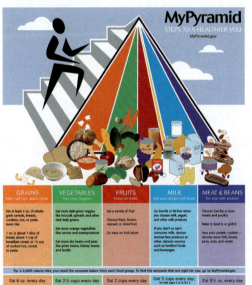

- Minerals have many important functions in the structure of bone, regulation of enzymatic functions, maintenance of body fluids pH, regulation of osmosis, and generation of action potentials in nerves and muscles. Important minerals include calcium, magnesium, phosphorus, potassium, sodium, iron, manganese, copper, and zinc.

- Vitamins are organic nutrients that are required in small amounts to maintain normal growth and metabolism. Most vitamins are cofactors or coenzymes. They are classified as **fat-soluble vitamins** (vitamins A, D, E, K) or **water-soluble vitamins** (vitamins B and C).

3 Nutrients Are Metabolized in a Number of Ways 418

- **Metabolism** refers to all of the chemical reactions in the body. Most chemical reactions are catalyzed by enzymes, many of which work together with coenzymes. Metabolic reactions include synthesis reactions (**anabolic reactions**, or anabolism) and decomposition reactions (**catabolic reactions**, or catabolism). In anabolic reactions, which usually require energy, simple organic molecules are combined to make more complex ones. In catabolic reactions, which usually release energy, complex organic molecules are broken down into simpler ones. Of the energy released in catabolic reactions, only 40% is captured as ATP. The rest is released as heat.

- The main carbohydrate is glucose, which is oxidized into carbon dioxide and water by **glycolysis**, the formation of **acetyl CoA**, the **Krebs cycle**, and **oxidative phosphorylation**, as shown below. These processes take place in the cytosol and mitochondria and produce 36 ATPs per glucose molecule. Glucose can also be stored as glycogen in the liver and skeletal muscle or converted to triglycerides in adipose tissue. When blood glucose is low, glucose can be mobilized from glycogen (**glycogenolysis**) or made from other molecules by gluconeogenesis.

- Absorbed amino acids are either made into proteins or metabolized. The amino groups are removed (deamination), and many become intermediates of the Krebs cycle and are subsequently metabolized to make ATP.

Cellular respiration • Figure 14.12

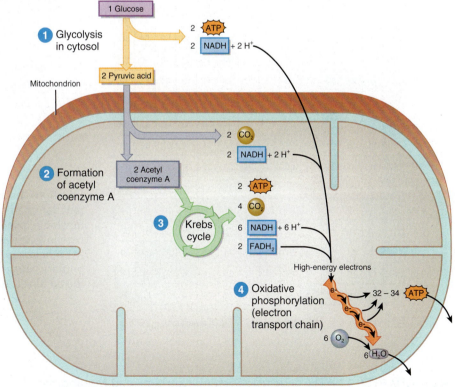

4 Diabetes and Obesity Are Metabolic Disorders 422

- **Diabetes mellitus** has two main forms: **Type 1** (lack of insulin) and **Type 2** (insulin resistance or insensitivity). Regardless of the type, the body reacts as if it is in a fasting state and strives to increase blood glucose by metabolizing body fuel stores. The increased blood glucose causes most diabetic symptoms, including glucosuria, increased urine flow, decreased capillary filtration, tissue dehydration, and increased appetite. Treatment depends on the type of diabetes but can include monitoring of diet, exercise, and insulin injection.

- Body weight regulation is a balance between energy intake (food intake, absorption) and energy expenditure (basal metabolism, physical activity, **thermic effects of food**). Body fat stores are regulated by a system involving the hormone leptin, appetite control centers in the hypothalamus, and activity of the sympathetic nervous system and thyroid hormones, which govern food intake and energy metabolism.

- As shown, obesity is defined as body weight greater than 20% of accepted standards for age, height, and gender or BMI of 30 or higher, as shown. Many factors contribute to **obesity**, including genetic factors, eating habits, sedentary lifestyle, and social customs. Obesity treatments include behavior therapy (dieting, exercise), drugs, and possibly surgery (gastric bypass, gastric banding).

Body mass index • Figure 14.16

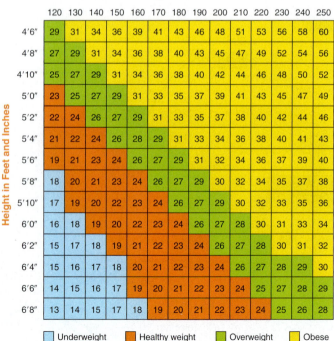

Weight in Pounds

	120	130	140	150	160	170	180	190	200	210	220	230	240	250
4'6"	29	31	34	36	39	41	43	46	48	51	53	56	58	60
4'8"	27	29	31	34	36	38	40	43	45	47	49	52	54	56
4'10"	25	27	29	31	34	36	38	40	42	44	46	48	50	52
5'0"	23	25	27	29	31	33	35	37	39	41	43	45	47	49
5'2"	22	24	26	27	29	31	33	35	37	38	40	42	44	46
5'4"	21	22	24	26	28	29	31	33	34	36	38	40	41	43
5'6"	19	21	23	24	26	27	29	31	32	34	36	37	39	40
5'8"	18	20	21	23	24	26	27	29	30	32	34	35	37	38
5'10"	17	19	20	22	23	24	26	27	29	30	32	33	35	36
6'0"	16	18	19	20	22	23	24	26	27	28	30	31	33	34
6'2"	15	17	18	19	21	22	23	24	26	27	28	30	31	32
6'4"	15	16	17	18	20	21	22	23	24	26	27	28	29	30
6'6"	14	15	16	17	19	20	21	22	23	24	25	27	28	29
6'8"	13	14	15	17	18	19	20	21	22	23	24	25	26	28

Height in Feet and Inches

☐ Underweight ☐ Healthy weight ☐ Overweight ☐ Obese

5 The Heat from Metabolism Must Be Regulated 427

- Body temperature is a balance between heat production and heat loss. Heat production is mainly due to the basal **metabolic rate** (BMR), which can be influenced by physical activity, hormones, nervous activity, food intake, age, and other factors. Heat loss is mainly due to radiation, conduction, convection, and evaporation.

- The temperature regulation system includes thermoreceptors in the skin and hypothalamus, a "thermostat" in the hypothalamus, and effectors including sympathetic nerve activity and thyroid hormones. If body temperature drops, it sets off a negative-feedback system in which vasoconstriction conserves body heat, muscle contractions generate heat, and adrenal and thyroid hormones increase BMR. These processes work to restore normal body temperature.

- Increases in body temperature trigger a negative-feedback system, as shown, that works to increase radiative heat loss by vasodilation and evaporative heat loss by sweating. Decreases in thyroid hormones reduce BMR, thereby reducing heat production. All these processes work to restore normal body temperature.

Negative-feedback mechanisms that decrease heat production and increase heat loss in response to increasing body temperature • Figure 14.19

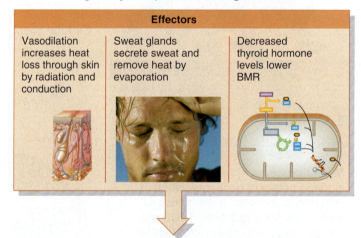

Effectors		
Vasodilation increases heat loss through skin by radiation and conduction	Sweat glands secrete sweat and remove heat by evaporation	Decreased thyroid hormone levels lower BMR

- Infections, bacterial toxins, hyperthyroidism, and immune reactions can cause fevers in which the hypothalamic thermostat is reset to higher temperatures. The resetting process involves pyrogens and prostaglandins that work in the hypothalamus. The elevated temperatures continue until the primary cause is eliminated. Fevers can be reduced through the administration of aspirin, acetaminophen, or ibuprofen.

Key Terms

- accessory digestive organ 401
- acetyl coenzyme A (acetyl CoA) 419
- aerobic respiration 419
- anabolic reaction 418
- antioxidant vitamin 416
- basal metabolic rate (BMR) 427
- bile pigment 410
- bile salt 410
- body mass index (BMI) 424
- bolus 404
- Calorie 416
- carboxypeptidase 410
- catabolic reaction 418
- cellular respiration 418
- cholecystokinin (CCK) 408
- chylomicron 411
- chyme 406
- chymotrypsin 410
- constipation 414
- defecation 414
- defecation reflex 414
- deglutition 405
- deoxyribonuclease 410
- diabetes mellitus 422
- diarrhea 414
- digestive system 400
- duodenum 408
- electron transport chain (ETC) 419
- enteric nervous system (ENS) 403
- essential nutrient 416

- fat-soluble vitamin 416
- feces 414
- flatus 414
- flavin adenine dinucleotide (FAD) 418
- gallbladder 410
- gastric emptying 406
- gastric juice 406
- gastric lipase 406
- gastric pit 406
- gastroesophageal reflux disease (GERD) 405
- gastrointestinal (GI) tract 400
- ghrelin 424
- gluconeogenesis 420
- glucose-6-phosphate 421
- glycemic index 426
- glycemic load 426
- glycogenolysis 420
- glycolysis 419
- high-density lipoprotein 421
- hypothermia 430
- ileum 408
- ingestion 404
- jejunum 408
- ketone 420
- Krebs cycle 419
- lacteal 411
- leptin 424
- lipogenesis 420
- lipolysis 420

- lipoprotein 420
- low-density lipoprotein 421
- metabolism 418
- micelle 410
- mineral 416
- nicotinamide adenine dinucleotide (NAD^+) 418
- nutrient 416
- obesity 424
- oxidation 418
- oxidative phosphorylation 419
- pancreatic acinar cell 410
- pancreatic amylase 410
- pancreatic juice 410
- pancreatic lipase 410
- pepsin 406
- peristalsis 405
- peritonium 402
- prostaglandin 430
- pro-vitamin 416
- pyrogen 430
- pyruvic acid 422
- ribonuclease 410
- segmentation 408
- thermic effect of food 424
- trypsin 410
- type 1 diabetes 422
- type 2 diabetes 422
- villi 408
- very-low density lipoprotein 421
- vitamin 416
- water-soluble vitamin 416

Critical and Creative Thinking Questions

1. Jane loves chocolate but wants to lose weight. So she eats large amounts of sugar-free chocolate, which contains an artificial sweetener that cannot be absorbed by the small intestine. She complains that she has diarrhea, gas, and intestinal cramps after eating the sugar-free chocolate. How would eating the chocolate lead to these symptoms?

2. A popular drug for treating heartburn inhibits gastrin receptors in the stomach. How does this action treat heartburn?

3. Because Roberto has a pancreatic tumor that secretes high levels of insulin, he often has episodes of extremely low blood sugar, or hypoglycemia. During these episodes, he feels faint and has even collapsed into a coma on occasion. How does the tumor cause these symptoms?

4. To treat his high blood pressure, Mike takes a vasodilator daily. He finds that he is not able to tolerate cold temperatures as well as he did before he was on the medication. Explain how taking a vasodilator would reduce Mike's cold tolerance.

5. Jerry is extremely obese and agrees to have gastric banding surgery, in which about half of his stomach will be banded so that it receives no food. What will be the consequences with respect to the gastric phase of digestion?

6. Susan is being treated for an infection with a broad-spectrum antibiotic. She complains that the medicine is giving her diarrhea. Her doctor suggests that she eat yogurt (containing live organism cultures) during the course of the antibiotic treatment. How does the antibiotic cause Susan's diarrhea, and how might the yogurt help?

Critical and Creative Thinking Questions **433**

What is happening in this picture?

Bulimia is an eating disorder in which people consume large amounts of food in short periods to satisfy emotional needs, and then induce vomiting.

Think Critically 1. Why would a diet like the one shown here be typical of an individual with bulimia?
2. If the person in this picture "purges" regularly, what will happen to her mouth and esophagus?

Self-Test

(Check your answers in Appendix C.)

1. Where does physical digestion begin?
   a. stomach
   b. small intestine
   c. mouth
   d. large intestine

Use this diagram to answer questions 2–5.

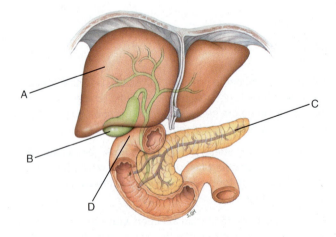

2. Which labeled organ secretes CCK?
   a. A   c. C
   b. B   d. D

3. Nutrients are extracted and processed in which organ?
   a. A   c. C
   b. B   d. D

4. In which labeled organ is bile stored?
   a. A   c. C
   b. B   d. D

5. Which organ in the figure secretes insulin and glucagon?
   a. A   c. C
   b. B   d. D

6. Lack of insulin characterizes _____.
   a. type 1 diabetes
   b. gestational diabetes
   c. type 2 diabetes
   d. hypoglycemia

Use this diagram to answer questions 7–10.

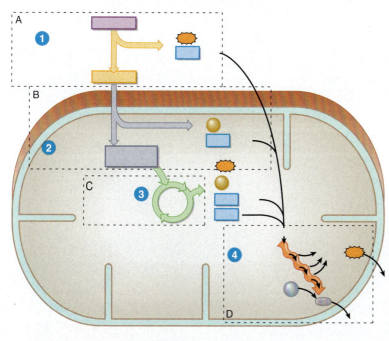

7. Which process is oxidative phosphorylation?

  a. A   c. C

  b. B   d. D

8. Lipids get broken down into glycerol and fatty acids. In which process do fatty acids get broken down?

  a. A   c. C

  b. B   d. D

9. Which label shows the process of glycolysis?

  a. A   c. C

  b. B   d. D

10. Which step involves the movements of H^+ from one side of a membrane to another?

  a. A   c. C

  b. B   d. D

11. Which part of the central nervous system is the thermostat for temperature regulation?

  a. medulla

  b. cerebrum

  c. hypothalamus

  d. thalamus

Use this figure to answer questions 12–15.

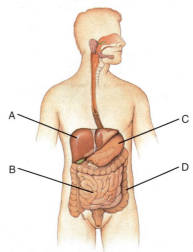

12. Bacteria play an important role in digestion in which organ?

  a. A   c. C

  b. B   d. D

13. Which organ is the target in the surgical treatment of obesity?

  a. A   c. C

  b. B   d. D

14. In which organ does gluconeogenesis occur?

  a. A   c. C

  b. B   d. D

15. The organ labeled in B is primarily responsible for _____.

  a. production of B and K vitamins

  b. absorption of nutrients

  c. extraction and processing of nutrients

  d. elimination of wastes

16. What is the function of the circular folds in the mucosa and submucosa of the small intestine?

  a. They increase the surface area for the absorption of nutrients.

  b. They secrete bilirubin to help absorb nutrients.

  c. They decrease the surface area for the absorption of nutrients

  d. They increase motility in the small intestine.

17. Which of the following is an irreversible approach to treating severe obesity?

  a. drug therapy

  b. gastric banding

  c. gastric bypass

  d. behavior therapy

THE PLANNER ✔

Review your Chapter Planner on the chapter opener and check off your completed work.

15 The Urinary System and Fluid, Electrolyte, and Acid–Base Balance

Sports drinks are becoming increasingly popular. They are available in a number of different brands and formulations. These products have been developed because exercise is one of the many factors that can change the makeup of your body fluids. As sweat trickles down your face and back, you lose fluids to evaporation, which allows your body to cool. You lose salts along with these fluids, which is sometimes evident on dark-colored workout clothes after they dry. Your muscles produce metabolites that get carried away by increased blood flow. These metabolites can change the osmolarity (solute concentration) and pH of your blood. If you are not careful to take in sufficient quantities of water, you can become severely dehydrated, which can affect your blood pressure and even cause you to collapse.

The composition of your blood and body fluids is regulated by your kidneys, the primary organs of the urinary system. The intricate anatomy and physiology of the kidneys keep your body fluid composition in balance, eliminating wastes and adjusting to various physiological conditions. Let's take a closer look at the kidneys and the other components of this remarkable system.

CHAPTER PLANNER ✓

- ❑ Study the picture and read the opening story.
- ❑ Scan the Learning Objectives in each section:
  p. 438 ❑   p. 450 ❑   p. 453 ❑   p. 458 ❑
- ❑ Read the text and study all visuals. Answer any questions.

Analyze key features

- ❑ InSight, pp. 440–441 ❑
- ❑ Process Diagram, p. 445 ❑   p. 448 ❑   p. 452 ❑
- ❑ What a Health Provider Sees, p. 462 ❑
- ❑ Stop: Answer the Concept Checks before you go on:
  p. 449 ❑   p. 452 ❑   p. 458 ❑   p. 463 ❑

End of chapter

- ❑ Review the Summary and Key Terms.
- ❑ Answer the Critical and Creative Thinking Questions.
- ❑ Answer What is happening in this picture?
- ❑ Complete the Self-Test and check your answers.

The Urinary System Plays a Vital Role in Maintaining Homeostasis

LEARNING OBJECTIVES

1. **Outline** the components of the urinary system and their functions.

2. **Identify** the macroscopic and microscopic structures of the kidneys.

3. **Describe** the processes involved in urine formation.

4. **Identify** the role of each portion of the nephron in urine formation.

Maintaining homeostasis—keeping a stable internal environment—is critical for survival, and the urinary system plays a vital role in this process. As body cells carry out their metabolic functions, they consume oxygen and nutrients and produce substances, such as carbon dioxide, that have no useful functions and need to be eliminated from the body.

As we saw in the chapter opener, these substances or metabolites can change the composition and pH of your blood, not only during exercise but in the course of your normal day-to-day life. While the respiratory system rids the body of carbon dioxide, the urinary system disposes of most other unneeded substances, including the nitrogen-containing wastes, toxins, and drugs. As you will see later in this chapter, the urinary system also helps maintain fluid and electrolyte balance.

The **urinary system** consists of the **kidneys**, **ureters**, **urinary bladder**, and **urethra** (**Figure 15.1**). The ureters connect the kidneys to the urinary bladder, and the urethra connects the urinary bladder to the outside. Together these organs make up an elaborate system that carry out the following functions:

- Filtration of blood plasma.
- Reabsorption of essential substances.
- Secretion of nonessential substances.

The kidneys do the major work of the urinary system, filtering gallons of fluid from the bloodstream every day. The other parts of the system are primarily passageways and temporary storage areas. After the kidneys filter blood, they return most of the water and many of the solutes to the bloodstream. The remaining water and solutes make up **urine**, which passes from the kidneys through the ureters to the urinary bladder. Urine is stored in the urinary bladder until it is expelled from the body through the urethra.

Let's begin our journey through the urinary system by examining the kidneys. Along the way, we will see how your kidneys help you maintain homeostasis, even during strenuous exercise.

The Kidneys Do the Major Work of the Urinary System

The kidneys are a pair of reddish organs shaped like kidney beans. They lie on either side of the vertebral column between the peritoneum and the back wall of the abdominal cavity, at the level of the 12th thoracic and first three lumbar vertebrae. The 11th and 12th pairs of ribs provide some protection for the superior parts of the kidneys. The right kidney is slightly lower than the left because the liver occupies a large area above the kidney on the right side.

An adult kidney is about the size of a bar of bath soap. Surrounding each kidney is the smooth, transparent renal capsule, a connective tissue sheath that helps maintain the shape of the kidney and acts as a barrier against trauma. Adipose (fatty) tissue surrounds the renal capsule and cushions the kidney. Along with a thin layer of dense irregular connective tissue, the adipose tissue anchors the kidney to the posterior abdominal wall.

The major processes of filtration, reabsorption, and secretion occur in the kidneys. Specifically, the functions of the kidneys include the following.

- **Regulation of blood volume and blood pressure.** The kidneys adjust the volume of water in the blood by returning water to the blood or eliminating it in the urine. They help regulate blood pressure by secreting the enzyme renin.

- **Regulation of blood composition and blood pH.** The kidneys affect blood composition by regulating the blood levels of several ions. They also help regulate blood pH through regulation of the concentrations of H^+ and blood bicarbonate ions (HCO_3^-).

- **Production of two hormones.** The kidneys produce calcitriol, the active form of vitamin D, which helps regulate calcium homeostasis. They also produce erythropoietin, which stimulates the production of red blood cells.

- **Excretion of wastes.** By forming urine, the kidneys help excrete wastes—substances that have no useful function in the body. Some of these wastes result from metabolic processes in the body, as we saw with the exercise example that began this chapter. Such metabolic wastes include ammonia, urea, bilirubin, creatinine, and uric acid. Other wastes are foreign substances from a person's diet, such as drugs and environmental toxins.

The urinary system • Figure 15.1

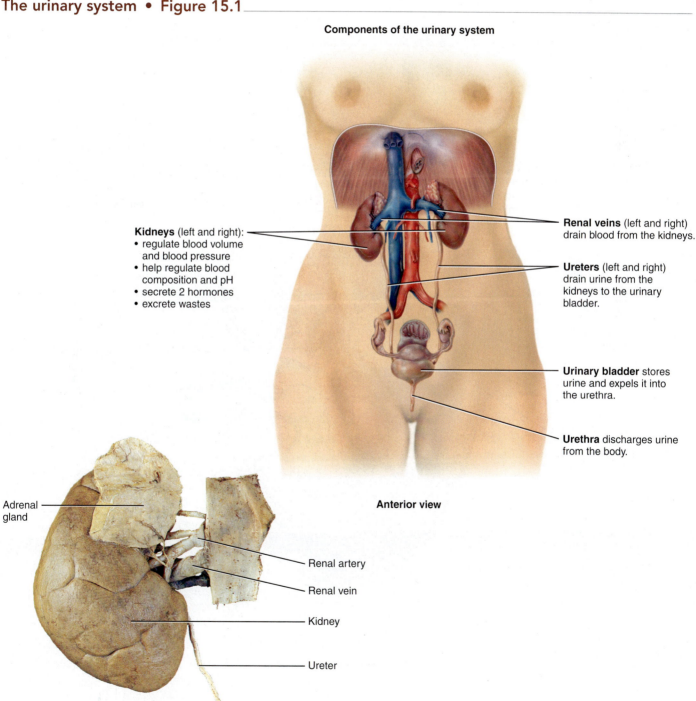

Components of the urinary system

Kidneys (left and right):
- regulate blood volume and blood pressure
- help regulate blood composition and pH
- secrete 2 hormones
- excrete wastes

Renal veins (left and right) drain blood from the kidneys.

Ureters (left and right) drain urine from the kidneys to the urinary bladder.

Urinary bladder stores urine and expels it into the urethra.

Urethra discharges urine from the body.

Anterior view

Adrenal gland

Renal artery

Renal vein

Kidney

Ureter

Anterior view

The Kidney Is a Complex Filter

Internally, each kidney itself has two main regions: the renal cortex and the renal medulla (**Figure 15.2**). The cortex receives most of the blood supply, and the medulla contains tubules that collect and concentrate the fluid that becomes the urine.

Urine formed in the kidney drains into a large, funnel-shaped cavity called the **renal pelvis**. The rim of the renal pelvis contains cuplike structures called **major** and **minor calyces** (KĀL-i-sēz; the singular form is calyx). Urine flows from several ducts within the kidney into a minor calyx. From there, it moves through a major calyx into the renal pelvis, which connects to a ureter. Water and solutes in the fluid that drains into the renal pelvis are excreted, or eliminated from the body.

InSight   Internal structure of the kidney • Figure 15.2

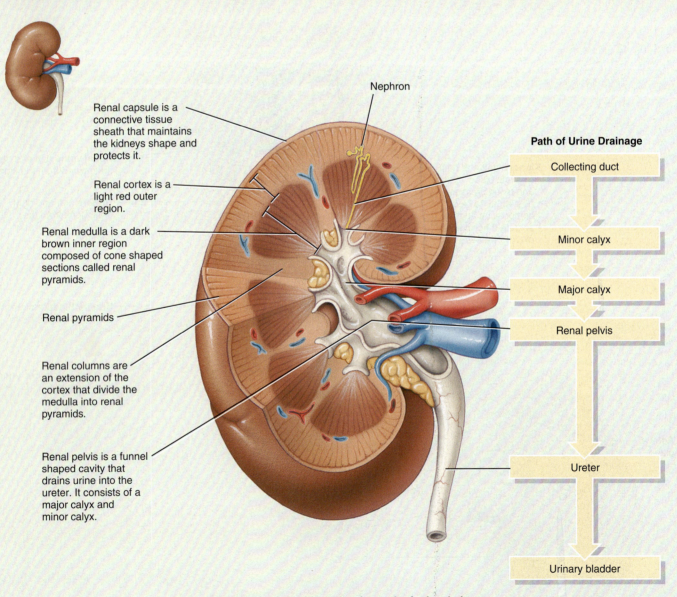

Nephron

Renal capsule is a connective tissue sheath that maintains the kidneys shape and protects it.

Renal cortex is a light red outer region.

Renal medulla is a dark brown inner region composed of cone shaped sections called renal pyramids.

Renal pyramids

Renal columns are an extension of the cortex that divide the medulla into renal pyramids.

Renal pelvis is a funnel shaped cavity that drains urine into the ureter. It consists of a major calyx and minor calyx.

Path of Urine Drainage

Collecting duct

Minor calyx

Major calyx

Renal pelvis

Ureter

Urinary bladder

a. Frontal section of right kidney showing path of urine drainage

About 20–25% of resting cardiac output (1,200 mL of blood each minute) flows through the kidneys via the renal arteries and veins. Within each kidney, the renal artery divides into smaller and smaller vessels, shown in the figure, that eventually deliver blood to the afferent arterioles. Each **afferent arteriole** divides into a tangled capillary network called a **glomerulus** (glō-MER-ū-lus).

The capillaries of the glomerulus reunite to form an efferent arteriole. Upon leaving the glomeruluis, each **efferent arteriole** divides to form a network of capillaries around the kidney tubules (described next). These capillaries eventually reunite to form veins, which merge into larger and larger veins. Eventually, all these smaller veins drain into the **renal vein**.

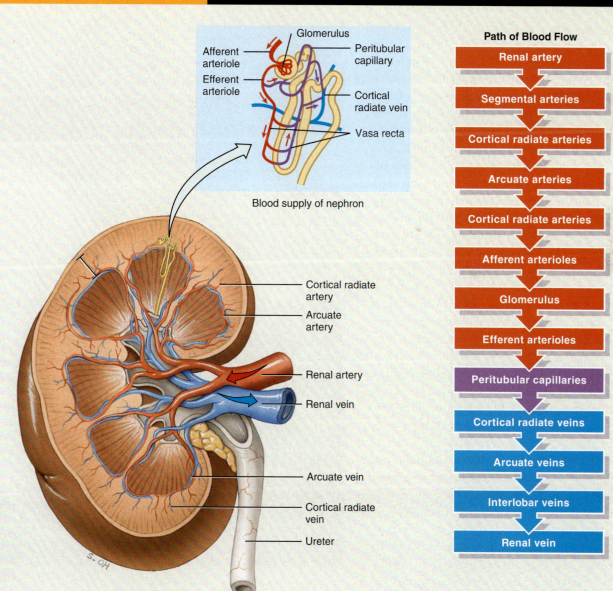

b. Frontal section of right kidney showing path of blood flow

Within each kidney, there are approximately 1 million functional units called **nephrons** (NEF-ronz). Each nephron consists of two parts: a collecting cup-like structure called the **renal corpuscle** and a long tube called the **renal tubule** (**Figure 15.3**).

Blood plasma is filtered in the renal corpuscle. The two parts that make up a renal corpuscle are the glomerulus and the glomerular (Bowman's) capsule, a double-walled cup of epithelial cells that surrounds the glomerular capillaries. The filtered fluid, also called the *filtrate*, first enters the glomerular capsule and then passes into the renal tubule.

The renal tubule is twisted and divided into three main segments. Taking them in the order that fluid passes through them, they are the **proximal convoluted tubule**, a long hairpin loop called the **nephron loop**, and the **distal convoluted tubule**. The distal convoluted tubules of several nephrons empty into the common **collecting duct**. Each nephron extends from the cortex into the medulla. Some nephrons have short nephron loops that extend only minimally into the medulla (**cortical nephrons**), while others have long nephron loops that extend deep into the medulla (**juxtamedullary nephrons**).

The kidney works by filtering almost everything from the blood plasma into the renal corpuscle of the nephron, by selectively reabsorbing only those substances that the body requires from the filtrate as it flows along the length of the renal tubule, and by secreting additional nonessential substances into the filtrate. Substances required by the body include water, ions, and glucose; nonessential substances include urea, drugs, and toxins. What remains in the lumen of the tubules becomes urine and gets drained by the collecting ducts into the renal pelvis, ureter, and urinary bladder.

For example, let's look back at the exercise example at the beginning of this chapter. During and after strenuous exercise, you lose some body fluids through sweating and evaporation. The kidney works to return fluid to the tissues by secreting less water into the urine. The kidney also reabsorbs some salts that your body may need because they were lost along with the sweat. However, the kidney also works to excrete the metabolites produced by your working cells.

How do all these processes happen? Let's examine the processes of filtration, reabsorption, and secretion in more detail.

The structure of the nephron, showing vascular supply and direction of blood flow • Figure 15.3

Renal corpuscle is where blood plasma is filtered:
• Located in cortex
• Consists of **glomerulus** (capillary network) and **glomerular capsule** or **Bowman's capsule** (double-walled cup of epithelial cells)
• Inner wall of glomerular capsule consists of modified simple squamos epithelial cells called **podocytes**, which have fingerlike projections called **pedicels** that wrap around capillary endothelial cells
• Outer wall of glomerular capsule consists of simple squamous epithelial cells

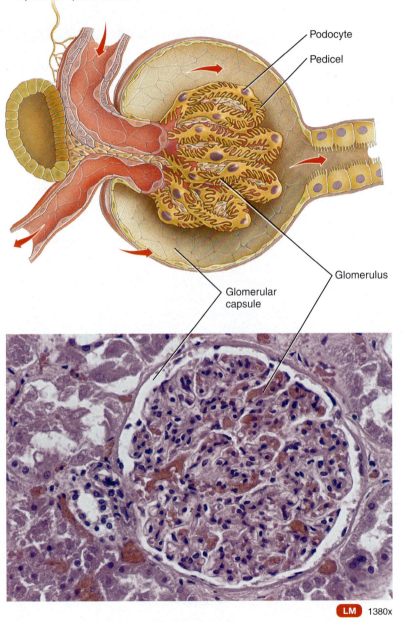

Podocyte
Pedicel
Glomerulus
Glomerular capsule

LM 1380x

Renal corpuscle

Structure of nephron showing vascular supply and direction of fluid flow

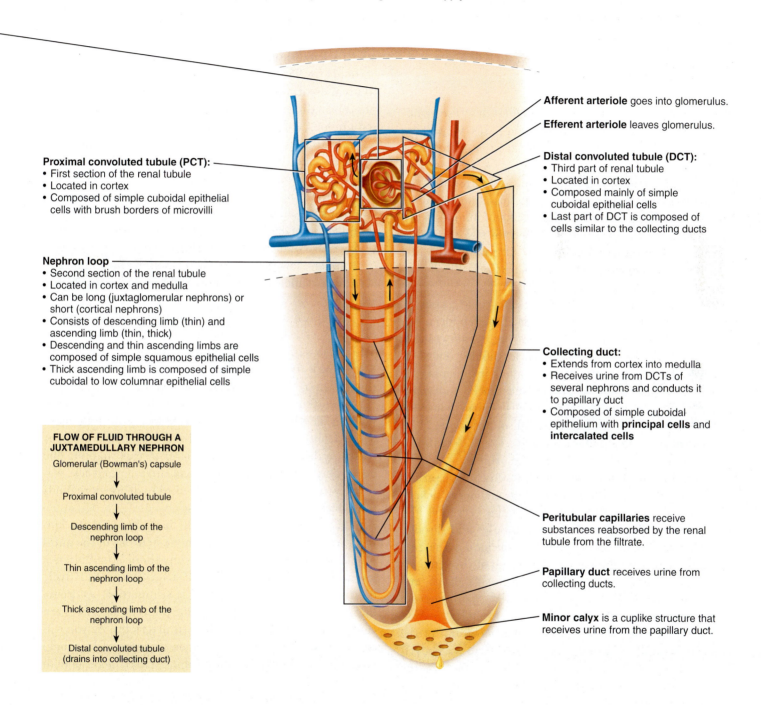

Proximal convoluted tubule (PCT):
- First section of the renal tubule
- Located in cortex
- Composed of simple cuboidal epithelial cells with brush borders of microvilli

Nephron loop
- Second section of the renal tubule
- Located in cortex and medulla
- Can be long (juxtaglomerular nephrons) or short (cortical nephrons)
- Consists of descending limb (thin) and ascending limb (thin, thick)
- Descending and thin ascending limbs are composed of simple squamous epithelial cells
- Thick ascending limb is composed of simple cuboidal to low columnar epithelial cells

FLOW OF FLUID THROUGH A JUXTAMEDULLARY NEPHRON

Glomerular (Bowman's) capsule
↓
Proximal convoluted tubule
↓
Descending limb of the nephron loop
↓
Thin ascending limb of the nephron loop
↓
Thick ascending limb of the nephron loop
↓
Distal convoluted tubule (drains into collecting duct)

Afferent arteriole goes into glomerulus.

Efferent arteriole leaves glomerulus.

Distal convoluted tubule (DCT):
- Third part of renal tubule
- Located in cortex
- Composed mainly of simple cuboidal epithelial cells
- Last part of DCT is composed of cells similar to the collecting ducts

Collecting duct:
- Extends from cortex into medulla
- Receives urine from DCTs of several nephrons and conducts it to papillary duct
- Composed of simple cuboidal epithelium with **principal cells** and **intercalated cells**

Peritubular capillaries receive substances reabsorbed by the renal tubule from the filtrate.

Papillary duct receives urine from collecting ducts.

Minor calyx is a cuplike structure that receives urine from the papillary duct.

Urine Formation Involves Three Processes and Helps Maintain the Blood's Volume and Composition

To produce urine, nephrons and collecting ducts perform three basic processes, summarized in **Figure 15.4**:

- Glomerular filtration
- Tubular reabsorption
- Tubular secretion

These processes are much like the children's "fish pond" game that you find in many amusement parks. We will use that game as an analogy as we look at each of the three processes.

The game starts by letting water and differently colored plastic fish flow into a stream; this is analogous to filtration, which is defined as the forcing of fluids and dissolved substances smaller than a certain size through a membrane by pressure. Glomerular filtration is the first step of urine production. Blood pressure forces blood plasma (consisting of water and dissolved substances) across the wall of glomerular capillaries into the renal corpuscle.

Along the length of the stream, little children use poles to remove the plastic fish from the stream (usually getting a little wet in the process). In tubular reabsorption, specific transport mechanisms allow cell membranes along the renal tubule to absorb substances and water from the filtrate. Tubule and duct cells return about 99% of the filtered water and many useful solutes to the blood flowing through peritubular capillaries. Only 1% of the filtered water actually leaves the body in urine.

Occasionally, children lean too far over the stream and drop items that they are carrying (loose change, tickets, toys, and so on) into the stream. Tubular secretion also takes place as fluid flows along the tubules and through the collecting duct. Specific transport mechanisms secrete substances into the fluid along the length of the renal tubule. They remove substances such as wastes, drugs, and excess ions from blood in the peritubular capillaries and transport them into the fluid in the renal tubules.

What remains in the fishing stream flows out of the game area; similarly, urine gets excreted from the body. It is the balance of these processes that determine what the children keep (composition of the blood) and what gets thrown away (composition of the urine).

Figure 15.4 also compares the substances that are filtered, reabsorbed, and excreted in urine per day in an adult male. Although the values shown are typical, they vary considerably according to diet and gender. For example, the kidneys filter about 180 liters (about 48 gallons) daily in adult males and about 150 liters daily (about 40 gallons) in adult females.

The volume of urine eliminated per day in a normal adult is 1 to 2 liters per day (about 1 to 2 quarts). Water accounts for 95% of the total volume of urine. In addition to urea, creatinine, potassium, and ammonia, typical solutes normally present in urine include uric acid as well as sodium ions, chloride ions, and other ions.

Filtration is driven by net pressures within the renal corpuscle, and tubular reabsorption and tubular secretion are driven by membrane transport processes across the cells of the renal tubule, such as diffusion, osmosis, passive transport, and active transport. (For a review of these processes, see Chapter 3.) Tubular reabsorption and secretion are also influenced by the rate of filtrate flow through the tubule; slow flow rates yield greater absorption and secretion, while fast flow rates yield less absorption and secretion.

As nephrons perform their functions, they help maintain homeostasis of the blood's volume and composition. They also adapt to changing physiological conditions. Think back to the exercise example at the beginning of this chapter. As you sweat from exercise, evaporation causes you to lose body fluids. If you do not increase your fluid intake, the kidneys compensate by reabsorbing more water from the fluid. We will discuss how that happens when we discuss water balance later in this chapter. The urine may appear to be more concentrated and less watery. Even so, you will most likely need to drink more fluids to replace the water you have lost.

Like most blood vessels of the body, those of the kidneys are supplied by the sympathetic neurons of the autonomic nervous system. At rest, sympathetic stimulation is low and the afferent and efferent arterioles are relatively dilated. With greater sympathetic stimulation, as occurs during exercise or hemorrhage, the afferent arterioles are constricted more than the efferent arterioles. As you will see shortly, this constriction leads to changes that help conserve blood volume and permit greater blood flow to other body tissues.

Let's take a closer look at glomerular filtration.

Functional processes in the nephron of the kidney • Figure 15.4

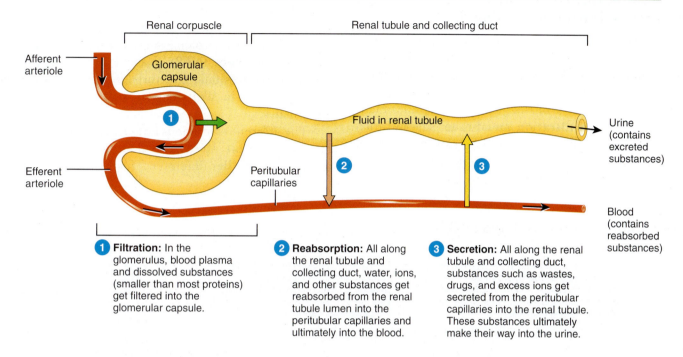

1 Filtration: In the glomerulus, blood plasma and dissolved substances (smaller than most proteins) get filtered into the glomerular capsule.

2 Reabsorption: All along the renal tubule and collecting duct, water, ions, and other substances get reabsorbed from the renal tubule lumen into the peritubular capillaries and ultimately into the blood.

3 Secretion: All along the renal tubule and collecting duct, substances such as wastes, drugs, and excess ions get secreted from the peritubular capillaries into the renal tubule. These substances ultimately make their way into the urine.

Substance	Filtered* (enters glomerular capsule per day)	Reabsorbed (returned to blood per day)	Urine (excreted per day)
Water	180 liters	178–179 liters	1–2 liters
Proteins	2.0 g	1.9 g	0.1 g
Sodium ions (Na$^+$)	579 g	575 g	4 g
Chloride ions (Cl$^-$)	640 g	633.7 g	6.3 g
Bicarbonate ions (HCO$_3^-$)	275 g	274.97 g	0.03 g
Glucose	162 g	162 g	0 g
Urea	54 g	24 g	30 g[†]
Potassium ions (K$^+$)	29.6 g	29.6 g	2.0 g[‡]
Uric acid	8.5 g	7.7 g	0.8 g
Creatinine	1.6 g	0 g	1.6 g

*Assuming GFR is 180 liters per day.

[†]In addition to being filtered and reabsorbed, urea is secreted.

[‡]After virtually all filtered K$^+$ is reabsorbed in the convoluted tubules and nephron loop, a variable amount of K$^+$ is secreted by principal cells in the collecting duct.

Glomerular Filtration Moves a Large Amount of Fluid into the Glomerular Capsule

Glomerular filtration occurs in the renal corpuscle and involves a movement of fluid from the glomerulus into the glomerular capsule. The endothelial cells of the glomerular capillaries, the basal lamina, and the *podocytes* (the cells of the inner layer of the glomerular capsule) make up a three-layered **filtration membrane**. The filtration membrane allows the passage of blood plasma and small proteins but prevents the passage of blood cells and medium- or large-sized proteins.

The pressure that causes filtration is the blood pressure in the glomerular capillaries. Two other pressures oppose glomerular filtration: Blood colloid osmotic pressure and glomerular capsule pressure. Normally, bood pressure is greater than the two opposing forces, producing a **net filtration pressure (NFP)** of about 10 mm Hg (**Figure 15.5**). NFP drives fluid flow through the nephron, about 150 liters daily in adult females and 180 liters daily in adult males.

The afferent arteriole is larger in diameter than the efferent arteriole, so blood is flowing from a larger diameter to a smaller diameter. This helps to raise blood

Glomerular filtration • Figure 15.5

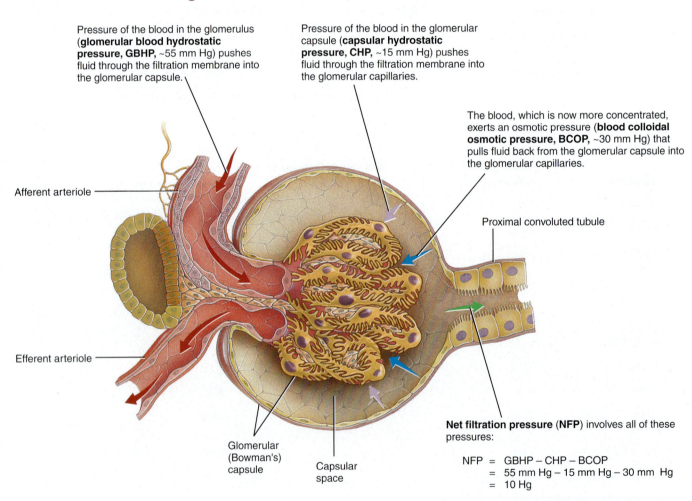

Pressure of the blood in the glomerulus (**glomerular blood hydrostatic pressure, GBHP**, ~55 mm Hg) pushes fluid through the filtration membrane into the glomerular capsule.

Pressure of the blood in the glomerular capsule (**capsular hydrostatic pressure, CHP**, ~15 mm Hg) pushes fluid through the filtration membrane into the glomerular capillaries.

The blood, which is now more concentrated, exerts an osmotic pressure (**blood colloidal osmotic pressure, BCOP**, ~30 mm Hg) that pulls fluid back from the glomerular capsule into the glomerular capillaries.

Afferent arteriole

Efferent arteriole

Proximal convoluted tubule

Glomerular (Bowman's) capsule

Capsular space

Net filtration pressure (NFP) involves all of these pressures:

$$\text{NFP} = \text{GBHP} - \text{CHP} - \text{BCOP}$$
$$= 55 \text{ mm Hg} - 15 \text{ mm Hg} - 30 \text{ mm Hg}$$
$$= 10 \text{ Hg}$$

NFP drives fluid flow from glomerular capillaries into glomerular capsule and through the renal tubule.

pressure in the glomerular capillaries. When blood pressure increases or decreases slightly, changes in the diameters of the afferent and efferent arterioles can actually keep net filtration pressure steady and maintain normal glomerular filtration.

The rate at which fluid is filtered in the renal corpuscle is called the **glomerular filtration rate (GFR)**. GFR is about 105 mL/minute in females and 125 mL/minute in males. As mentioned earlier, reabsorption of substances from the filtrate depends on the concentrations and rates of flow. If GFR is too high, substances are not adequately reabsorbed and leave the body in the urine (because it's difficult to collect materials that are flowing by too quickly). If GFR is too low, substances get reabsorbed, and waste products are not adequately excreted (because it's easier to collect materials that are moving slowly through the tubing).

Remember what you read about the afferent arterioles being more constricted than the efferent arterioles during exercise? Now you can follow the steps more closely: as a result of the constriction, blood flow into the glomerular capillaries is greatly decreased, net filtration pressure decreases, and GFR drops. These changes reduce urine output, which helps conserve blood volume and permits greater blood flow to other body tissues needing it for exercise.

Because the filtration membrane normally prevents the passage of medium- to large-sized proteins, excessive amounts of the blood protein albumin in the urine (**albuminuria**, al′-bū-mi-NOO-rē-a) indicate that the filtration membrane is damaged due to injury, disease, high blood pressure, or kidney cell damage.

Now let's take a closer look at what happens during tubular reabsorption and tubular secretion.

Reabsorption and Secretion Occur Along the Length of the Renal Tubule

The filtrate gets modified as it passes through the renal tubule. Most of the water and dissolved solutes get reabsorbed. The reabsorbed solutes include large quantities of ions (such as Na^+, K^+, Cl^-, Ca^{2+}, Mg^{2+}, HCO_3^-, PO_4^{2-}, and SO_4^{2-}), all of the glucose, and all of the amino acids. Among the many waste substances secreted into the filtrate are ammonia and urea (by-products of protein catabolism), creatinine (a waste product from muscle creatine—see Chapter 5), drugs (such as penicillin), and some ions (including H^+ and K^+).

As a result of tubular secretion, some drugs pass from blood into the urine and can be detected by urine tests. For example, urine tests can detect the presence of performance-enhancing drugs such as anabolic steroids, human growth hormone, and amphetamines in the urine of athletes. Urine tests can also be used to detect the presence of alcohol or illegal drugs such as marijuana, cocaine, and heroin.

How does the kidney know which substances to absorb and which to secrete? The presence of specific transporters on the lumen side of the tubular membrane cells and others on the interstitial fluid side of the tubular cells determines which substances are absorbed and which are secreted. For example, as a sodium ion (Na^+) in the filtrate passes a proximal tubule cell with an Na^+ channel, the Na^+ enters the cell because its concentration is higher in the filtrate than inside the cell. The Na^+ subsequently gets pumped out the other side of the cell by the sodium-potassium pump and gets absorbed from the interstitial fluid into the blood.

Because it proceeds from one side of a cell to the other, this pathway for sodium reabsorption through a cell is a **transcellular route** of reabsorption (*trans* = across); think of the ion as moving across the cell. Transcellular routes involve many different active and passive transport mechanisms, including *symporters* (cotransporters that move substances in the same direction; *sym* = same), channels, *antiporters* (cotransporters that move substances in opposite direction; *anti* = against), and pumps (see Chapter 3).

It is also possible for the Na^+ to pass between the cells through tight junctions (**paracellular route**; *para* = beside); think of the ion as moving between or beside cells. Paracellular routes usually involve passive transport in response to osmotic and/or electrochemical gradients. Most transcellular reabsorption is linked to the reabsorption of Na^+.

The movement of some ions (such as H^+ and HCO_3^-) depends on the activity of enzymes, such as **carbonic anhydrase**, found on the surfaces and inside renal tubule cells. This enzyme assists the kidney in the regulation of the blood pH—a topic that is discussed further at the end of this chapter.

The rate and amount of reabsorption of any substance depends on the concentration of the substance in the filtrate and the flow rate of the filtrate. Generally, high concentrations of any substance in the filtrate lead to greater reabsorption, but there are limits. If the concentration of the substance exceeds the capacity of the transporters, the excess amount is excreted in the urine. Likewise, if the flow rate of the filtrate past the transporters is too fast, the transporters cannot "snag" the substance, and it is excreted.

Reabsorption and secretion in the nephron • Figure 15.6

The reabsorption of water and solutes, along with the secretion of solutes, are shown as they proceed through the various segments of the renal tubule.

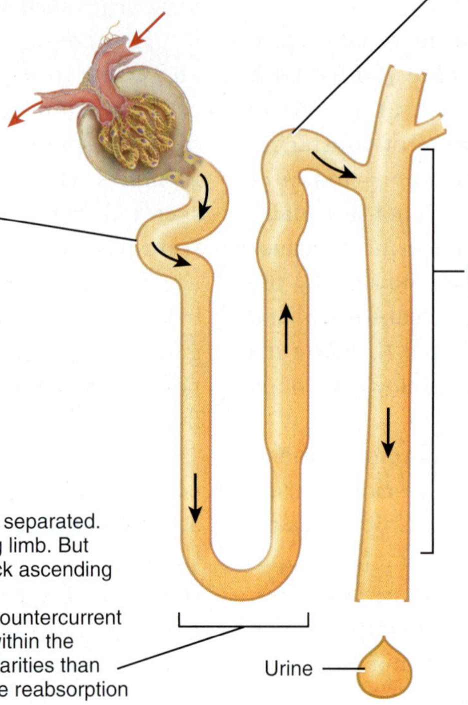

1 In the PCT, Na$^+$, K$^+$, and water are absorbed equally. So the osmolarity of the filtrate remains the same as that of blood plasma. H$^+$, NH$_4^+$, urea, and creatinine are secreted.

2 In the nephron loop, the following occurs:
- The absorption of water, Na$^+$, and K$^+$ are separated. Water is absorbed only in the descending limb. But more Na$^+$ and K$^+$ are absorbed in the thick ascending limb.
- The differences in reabsorption and the countercurrent flow set up a standing osmotic gradient within the medulla. (Deep layers have higher osmolarities than superficial ones, which is important for the reabsorption of water from the collecting ducts.)
- When the filtrate leaves the nephron loop, it is hypotonic compared to the blood plasma.
- Urea is secreted here as well as in the PCT.

3 In the early DCT, Na$^+$, Cl$^-$, and water are absorbed. Also, Ca^{2+} is reabsorbed here, under regulation by parathyroid hormone (PTH).

4 In the late DCT and collecting ducts, the following occurs :
- Principal cells reabsorb Na$^+$ and secrete K$^+$. This reabsorption is influenced by the hormone **aldosterone**.
- Intercalated cells reabsorb K$^+$ and HCO$_3^-$ while secreting H$^+$.
- The permeability of the cells to water is regulated by the hormone **ADH**. When stimulated by ADH, water channels (aquaporin-2 channels) are inserted into the apical membranes, water is reabsorbed according to the osmotic gradient set up in the medulla by the nephron loop, and urine is concentrated. In the absence of ADH, the remaining water does not get reabsorbed, resulting in dilute urine.

Urine

The Nephron Loop Contributes to Water Conservation in the Body

The amounts of solutes and water reabsorbed by the renal tubule are not equal in all areas of the tubule. Approximately two-thirds of the water and solutes are reabsorbed in the proximal convoluted tubule (PCT), and the tubular fluid is isosmotic (that is, has the same osmolarity) compared to the blood plasma (**Figure 15.6**). Secretion of certain materials (for example, urea, creatinine, ammonium ion, and hydrogen ion) can also occur in the PCT.

However, as the fluid moves into the nephron loop, it passes through an interstitial fluid gradient composed of sodium chloride. This gradient is more and more concentrated deeper in the medulla of the kidney and creates an osmotic gradient to promote water movement. In the descending limb of the nephron loop, which dips deep into the medulla, the medullary gradient promotes reabsorption of water, returning it to the blood supply of the vasa recta.

As the filtrate moves back toward the cortex through the ascending limb of the nephron loop, the interstitial fluid gradient becomes more dilute. To prevent water from reentering the tube, the ascending limb is not permeable to water. Sodium chloride is pulled back out of the filtrate in the ascending limb via an active transport process to maintain the medullary gradient. This process, known as the **countercurrent multiplier**, concentrates the urine to minimize water loss from the body and helps to maintain proper blood volume.

As the NaCl is removed from the filtrate in the ascending limb of the nephron loop, the tubular fluid becomes hypotonic (that is, has less osmolarity). In the distal convoluted tubule (DCT) and collecting ducts, much of the remaining sodium is reabsorbed, and some potassium is secreted, but this process depends on stimulation by the hormone **aldosterone**. The amount of calcium ions reabsorbed in the DCT depends on stimulation by **parathyroid hormone (PTH)**. Finally, the amount of water reabsorbed in these latter segments depends on stimulation by **antidiuretic hormone (ADH)**, which we discuss later in the chapter. Additional materials (such as calcium, urea, bicarbonate ion, and hydrogen ion) are also reabsorbed and secreted in the distal convoluted tubule and collecting ducts.

When the filtrate leaves the collecting ducts, it becomes urine. The urine passes into the calyx, the renal pelvis, and then the ureter. It is temporarily stored in the urinary bladder and is eventually excreted from the body via the urethra. Urine has distinct characteristics of color, turbidity, odor, pH, and specific gravity, as described in **Table 15.1**.

The composition of urine can sometimes indicate metabolic imbalances and disease conditions. Urine is often tested to determine whether abnormal components are present and to help diagnose disease (**Table 15.2**).

CONCEPT CHECK	

1. **Which** organ of the urinary system stores urine?
2. In **which** layer of the kidney is a renal column found?
3. In **which** part of the nephron is sodium reabsorption controlled by hormones?
4. In **which** urine formation process does blood pressure determine how much material crosses the membrane?
5. **Where** does most reabsorption occur in the nephron?

Physical characteristics of normal urine   Table 15.1

Characteristic	Description
Volume	One to two liters (about 1 to 2 quarts) in 24 hours but varies considerably.
Color	Yellow or amber, normally, is due to urochrome (pigment produced from breakdown of bile) and urobilin (from breakdown of hemoglobin). Concentrated urine is darker in color. Diet, medications, and certain diseases affect color. Kidney stones may produce blood in urine.
Turbidity	Transparent when freshly voided, but becomes turbid (cloudy) after awhile.
Odor	Mildly aromatic but becomes ammonia-like after a time. Some people inherit the ability to form methylmercaptan from digested asparagus, which gives urine a characteristic odor.
pH	Ranges between pH 4.6 and 8.0; average 6.0; varies considerably with diet. High-protein diets increase acidity; vegetarian diets increase alkalinity.
Specific gravity (density)	The ratio of the weight of a volume of a substance to the weight of an equal volume of distilled water. Urine specific gravity ranges from 1.001 to 1.035. The higher the concentration of solutes, the higher the specific gravity.

Summary of abnormal constituents in urine   Table 15.2

Abnormal Constituent	Comments
Albumin	A normal constituent of blood plasma that usually appears in only very small amounts in urine because it is too large to be filtered. The presence of excessive albumin in the urine, *albuminuria* (al′-bū-mi-NOO-rē-a), indicates an increase in the permeability of filtering membranes due to injury or disease, increased blood pressure, or damage to kidney cells.
Glucose	*Glucosuria*, the presence of glucose in the urine, usually indicates diabetes mellitus.
Red blood cells (erythrocytes)	*Hematuria* (hēm-a-TOO-rē-a), the presence of hemoglobin from ruptured red blood cells in the urine, can occur with acute inflammation of the urinary organs as a result of disease or irritation from kidney stones, tumors, trauma, and kidney disease.
White blood cells (leukocytes)	The presence of white blood cells and other components of pus in the urine, referred to as *pyuria* (pī-Ū-rē-a), indicates infection in the kidneys or other urinary organs.
Ketone bodies	High levels of ketone bodies in the urine, called *ketonuria* (kē-tō-NOO-rē-a), may indicate diabetes mellitus, anorexia, starvation, or too little carbohydrate in the diet.
Bilirubin	When red blood cells are destroyed by macrophages, the globin portion of hemoglobin is split off and the heme is converted to biliverdin. Most of the biliverdin is converted to bilirubin. An above-normal level of bilirubin in urine is called *bilirubinuria* (bil′-ē-roo-bi-NOO-rē-a).
Urobilinogen	The presence of urobilinogen (breakdown product of hemoglobin) in urine is called *urobilinogenuria* (ū′-rō-bi-lin′-ō-jē-NOO-rē-a). Trace amounts are normal, but elevated urobilinogen may be due to hemolytic or pernicious anemia, infectious hepatitis, obstruction of bile ducts, jaundice, cirrhosis, congestive heart failure, or infectious mononucleosis.
Microbes	The number and type of bacteria vary with specific infections in the urinary tract. One of the most common is *E. coli*. The most common fungus to appear in urine is *Candida albicans*, a cause of vaginitis. The most frequent protozoan seen is *Trichomonas vaginalis*, a cause of vaginitis in females and urethritis in males.

The Urinary Bladder Stores Urine and Expels It from the Body

LEARNING OBJECTIVES

1. **Describe** the function of the ureters, urinary bladder, and urethra.
2. **Compare** the structure of the urethra in males and females.
3. **Explain** the process of micturition.

The fluid leaving the collecting ducts of the nephrons is now called urine. Urine passes from the kidneys to the urinary bladder via the ureters (**Figure 15.7a**). The ureters pass under the urinary bladder for several centimeters, causing the bladder to compress the ureters and thus precent backflow of urine when pressure builds up in the urine during urination.

The Urinary Bladder Is a Temporary Storage Area

The urinary bladder is a hollow muscular organ situated in the pelvic cavity behind the pubic symphysis. In males, it is directly in front of the rectum. In females, it is in front of the vagina and below the uterus. Folds of the peritoneum hold the urinary bladder in position.

The muscular layer of the urinary bladder consists of three layers of smooth muscle called the **detrusor muscle** (de-TROO-ser). The urinary bladder's smooth muscle al-

The structures of the ureters, urinary bladder, and urethra and their functions • Figure 15.7

a. The ureters, urinary bladder, and urethra

Ureters transport urine from the kidneys to the urinary bladder. The ureters tunnel a short distance within the urinary bladder wall. As the urinary bladder fills, it expands and compresses the ureters, thereby preventing the backflow of urine.

When empty, the urinary bladder looks like a deflated balloon. As it fills, it becomes round and then pear-shaped. The bladder holds an average of 700–800 mL of urine.

Ureteral openings into the urinary bladder

Frontal plane

Rugae in the mucosa of the urinary bladder and the lining transitional epithelium allow the urinary bladder to expand as it fills.

Detrusor muscle consists of three layers of smooth muscle that stretch when the urinary bladder fills and contract to push out urine.

Peritoneum helps holds the urinary bladder in place.

Internal urethral sphincter is a smooth muscle that opens and closes the urethra involuntarily.

Urethra is a small tube that leads from the urinary bladder to the outside.

External urethral sphincter is a skeletal muscle that opens and closes voluntarily.

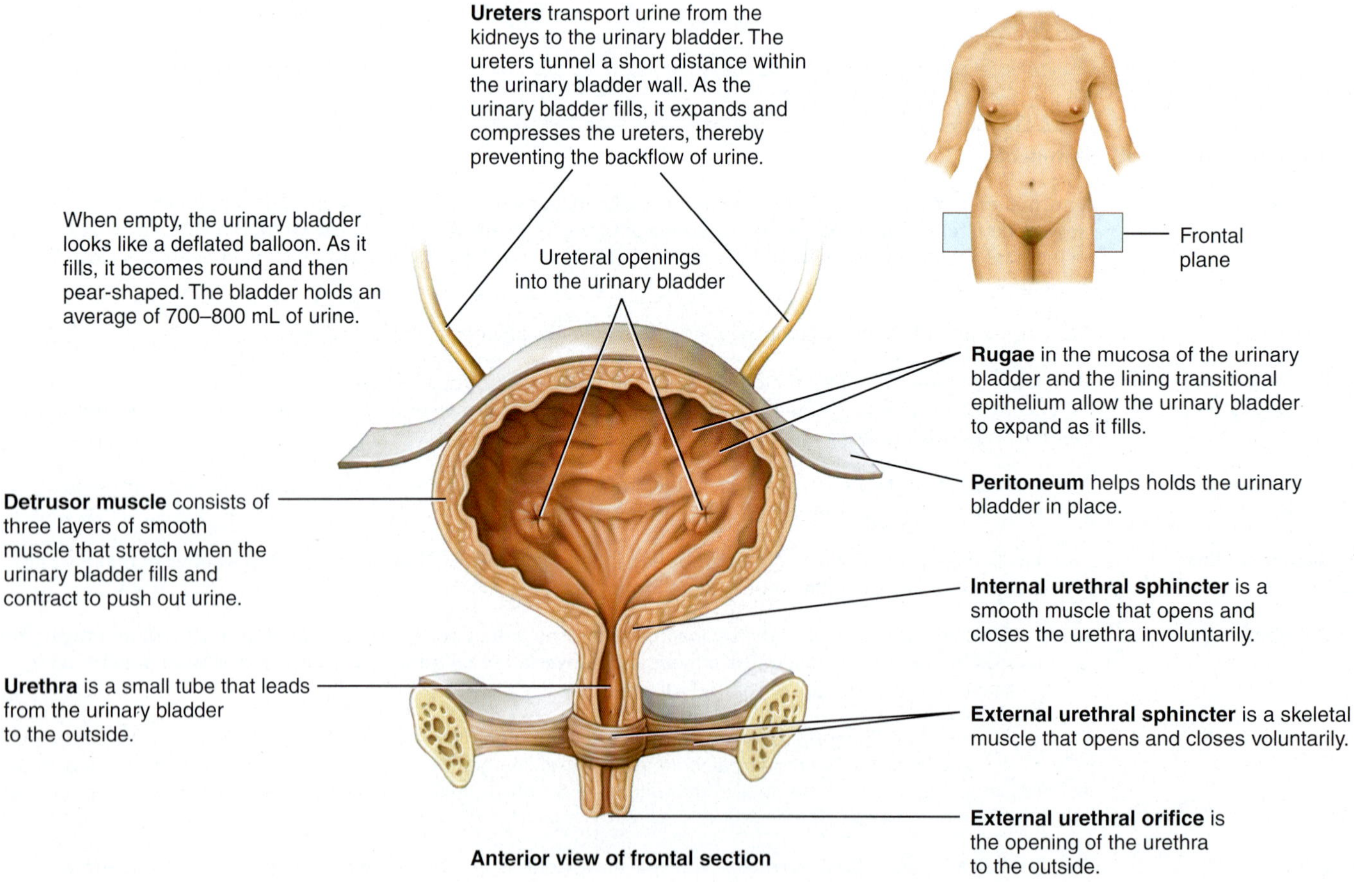

Anterior view of frontal section

External urethral orifice is the opening of the urethra to the outside.

lows the urinary bladder to expand as it fills, and contractions of the smooth muscle generate the pressure to expel urine from the body through the urethra.

When empty, the urinary bladder looks like a deflated balloon. As it fills, it becomes round and then pear-shaped. The urinary bladder holds an average of 700 to 800 mL of urine. It is smaller in females because the uterus occupies the space just superior to the urinary bladder.

Urethral Structure Varies Between the Sexes

The urethra (u-RĒ-thra) is a small tube leading from the floor of the urinary bladder to the exterior of the body. Around the opening to the urethra is an internal urethral sphincter composed of smooth muscle. The opening and closing of the internal sphincter is involuntary. Below the internal sphincter is the external urethral sphincter, which is composed of skeletal muscle and is under voluntary control.

The structure of the urethra is different in males and females (**Figure 15.7b**). In females, it lies directly behind the pubic symphysis and is embedded in the front wall of the vagina. The opening of the urethra to the exterior, the *external urethral orifice*, lies between the clitoris and the vaginal opening. In males, the urethra passes vertically through the prostate, the deep perineal muscles, and finally the penis. The male urethra also serves as the duct through which semen is ejaculated.

The shortened length and slightly larger diameter of the female urethra makes females susceptible to urinary tract infections (UTIs) from bacteria outside the body. The most common type of urinary tract infection occurs in the urinary bladder—a condition known as *cystitis*. This is nearly always caused by the entry of microorganisms via the urethra. Symptoms of urinary tract infections include painful urination, often accompanied by bleeding. Most UTIs can be treated easily with antibiotics.

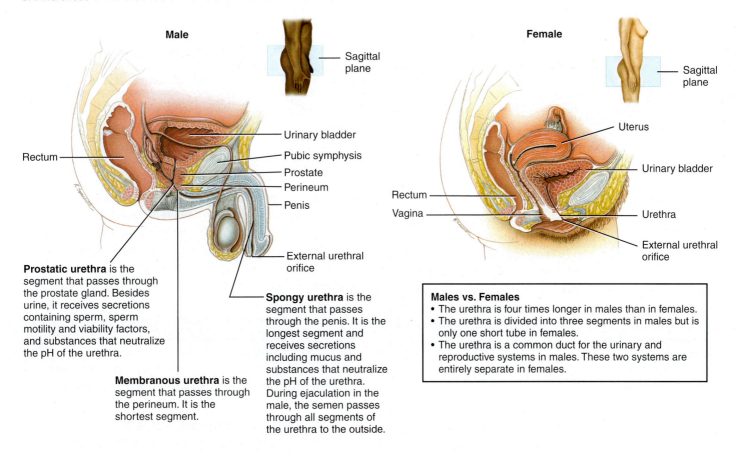

b. Differences in the urethras of males and females

Male

Sagittal plane

Rectum

Urinary bladder
Pubic symphysis
Prostate
Perineum
Penis

External urethral orifice

Prostatic urethra is the segment that passes through the prostate gland. Besides urine, it receives secretions containing sperm, sperm motility and viability factors, and substances that neutralize the pH of the urethra.

Membranous urethra is the segment that passes through the perineum. It is the shortest segment.

Spongy urethra is the segment that passes through the penis. It is the longest segment and receives secretions including mucus and substances that neutralize the pH of the urethra. During ejaculation in the male, the semen passes through all segments of the urethra to the outside.

Female

Sagittal plane

Uterus

Urinary bladder

Urethra

External urethral orifice

Rectum
Vagina

Males vs. Females
- The urethra is four times longer in males than in females.
- The urethra is divided into three segments in males but is only one short tube in females.
- The urethra is a common duct for the urinary and reproductive systems in males. These two systems are entirely separate in females.

The micturition reflex • Figure 15.8

The micturition reflex involves the urinary bladder and parasympathetic nervous system. It is also influenced by the cerebral cortex.

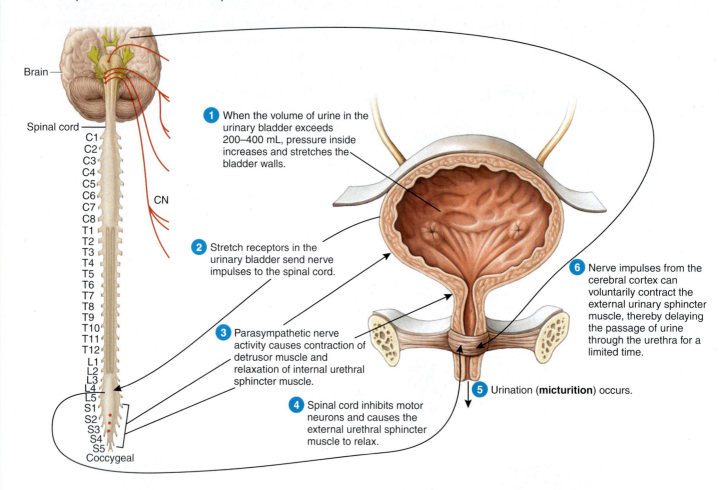

Brain

Spinal cord
C1
C2
C3
C4
C5
C6
C7
C8
T1
T2
T3
T4
T5
T6
T7
T8
T9
T10
T11
T12
L1
L2
L3
L4
L5
S1
S2
S3
S4
S5
Coccygeal

CN

1 When the volume of urine in the urinary bladder exceeds 200–400 mL, pressure inside increases and stretches the bladder walls.

2 Stretch receptors in the urinary bladder send nerve impulses to the spinal cord.

3 Parasympathetic nerve activity causes contraction of detrusor muscle and relaxation of internal urethral sphincter muscle.

4 Spinal cord inhibits motor neurons and causes the external urethral sphincter muscle to relax.

5 Urination (**micturition**) occurs.

6 Nerve impulses from the cerebral cortex can voluntarily contract the external urinary sphincter muscle, thereby delaying the passage of urine through the urethra for a limited time.

Urination Is a Spinal Reflex

micturition (mik'-choo-RISH-un) The act of expelling urine from the urinary bladder.

As urine accumulates in the bladder, it stimulates a reflex that initiates the process of urination, or **micturition**. The **micturition reflex** involves the urinary bladder's smooth muscles and sphincter muscles and is regulated by the parasympathetic nervous system (**Figure 15.8**). The cerebral cortex can temporarily suppress the reflex; young children learn how to control this reflex when toilet training.

incontinence (in-KON-ti-nens) Inability to retain urine, semen, or feces through loss of sphincter control.

Urinary **incontinence** (in-KON-ti-nens) occurs when the urinary bladder's sphincter muscles weaken. Causes of urinary incontinence include frequent urinary tract infections, side effects of medications, age, trauma, neuromuscular disease,

constipation, and enlarged prostate (in males). Childbirth is a common cause of incontinence in females due to trauma to nerves and muscles in the perineal and pelvic areas as the baby passes through the birth canal. Treatment for urinary incontinence depends on the cause of the condition and may include medication, muscle strengthening exercises, or behavior modifications. In some situations, surgery may be part of the treatment regimen.

CONCEPT CHECK   STOP

1. **Which** tube transports urine into the urinary bladder?

2. **Which** muscle group generates the force necessary to expel urine from the urinary bladder?

3. **What** types of materials enter the male prostatic urethra?

The Kidneys Regulate the Composition of Body Fluids

LEARNING OBJECTIVES

1. **Describe** the electrolyte composition of various fluid compartments in the body.

2. **Identify** the sources of water gain and loss and the mechanisms involved in controlling blood water levels.

3. **Outline** the role of the renin–angiotensin–aldosterone system in NaCl reabsorption, blood volume, and blood pressure regulation.

4. **Explain** how parathyroid hormone affects the kidneys to regulate Ca^{2+} levels in the blood.

Water is by far the largest single component of the body, making up 45–75% of total body mass, depending on age and gender. In lean adults, body fluids make up between 55% and 60% of total body mass. Babies are about 75% water, as they have low bone mass and little fat. Water content declines with age and may be only 45% of body weight in old age.

Body fluids are distributed in three main compartments within the body, called fluid compartments. About two-thirds of body fluid is **intracellular fluid (ICF)**, the fluid within cells. The other third, called **extracellular fluid (ECF)** is outside cells and includes all remaining body fluids, notably interstitial fluid and blood plasma, along with other extracellular fluids (**Figure 15.9**).

Fluid can travel among the three major fluid compartments in two ways, as shown by the black arrows in Figure 15.9. First, fluid can be exchanged between cells and their surrounding interstitial fluid because the plasma membrane of each cell is a selectively permeable barrier. Second, fluid can be exchanged between blood plasma in the capillaries and interstitial fluid. Exchanges take place constantly at both levels throughout the body to maintain fluid balance.

The body is in fluid balance when the required amounts of water and solutes are present, in the right proportions, in the various fluid compartments. We'll take a look at those solutes next.

Distribution of water in various compartments of the body • Figure 15.9

Distribution of body fluids in average adult

Total body mass (female)
- 45% Solids
- 55% Fluids

Total body mass (male)
- 40% Solids
- 60% Fluids

Total body fluid
- 2/3 Intracellular fluid (ICF)
- 1/3 Extracellular fluid (ECF)

Extracellular fluid
- 80% Interstitial fluid
- 20% Plasma

- Interstitial fluid also includes lymph, CSF, synovial fluid, aqueous and vitreous humor (eyes), pleural, peritoneal, and pericardial fluids
- Water continuously exchanges between fluid compartments.

Tissue cells

Blood capillary

Electrolyte and protein anion concentrations in body fluids • Figure 15.10

Sodium (Na⁺), and chloride (Cl⁻) are the most abundant ions in extracellular fluid. Potassium ions (K⁺) are the most abundant ions in intracellular fluid.

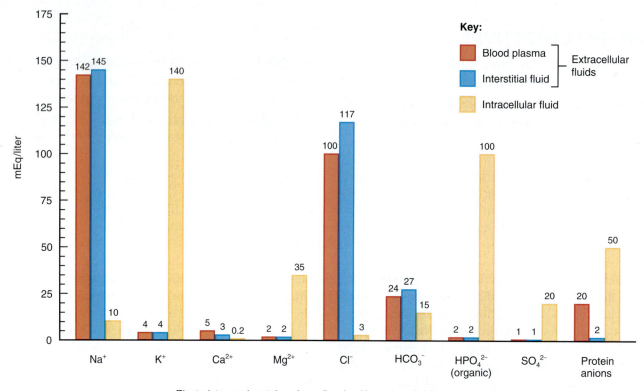

Key:
- Blood plasma ⎫ Extracellular
- Interstitial fluid ⎭ fluids
- Intracellular fluid

Electrolytes and protein anions dissolved in various fluid compartments

Fluid Balance Depends Primarily on Electrolyte Balance

The solutes or dissolved substances in body fluids consist of various ions and proteins (**Figure 15.10**). While water moves freely among the body's fluid compartments, its flow is driven by the concentrations of these ions, also called **electrolytes**.

> **electrolyte** (ē-LEK-trō-lītz) A compound that separates into ions when dissolved in water and that conducts electricity.

The ions formed when electrolytes break apart serve four general functions in the body:

1. Certain ions control the osmosis of water between fluid compartments, for example, sodium ions (Na⁺), chloride ions (Cl⁻), and potassium ions (K⁺).

2. Ions maintain acid–base balance required for normal cellular activities: for example, H⁺ (acid) and the bicarbonate ion, HCO₃⁻ (base).

3. Ions such as Na⁺, K⁺, and calcium (Ca²⁺) carry electrical current for action potentials.

4. Several ions, such as Ca²⁺ and magnesium (Mg²⁺), are cofactors (nonprotein substances) for enzymes.

Here we'll focus on the first function just listed, control of osmosis between fluid compartments. The distribution of electrolytes in plasma, interstitial fluid, and intracellular fluid regulates the osmotic gradient and therefore has an impact on movement of water between fluid compartments. Na⁺ plays a pivotal role in fluid and electrolyte balance because it accounts for almost half of the osmotic pressure of extracellular fluid. Cl⁻ moves easily between the intracellular and extracellular compartments and helps to balance the level of anions in each compartment. Very small changes in the concentration of any of these solutes can cause water to move from one compartment to another.

What happens if we lose electrolytes through sweating or if we do not take in enough? In fact, intake of water and electrolytes rarely occurs in exactly the same proportion as their presence in body fluids; the kidneys help the body to achieve the right balance and maintain homeostasis in two ways. The kidneys can excrete excess water by producing dilute urine, and they can excrete excess electrolytes by producing concentrated urine.

Because water is a primary component of blood, a change in the water level changes blood volume and, therefore, blood pressure. Because of its influence on the reabsorption of water and NaCl from the filtrate, the kidney plays a major role in blood pressure homeostasis. Water and NaCl reabsorption is regulated by several hormones, including antidiuretic hormone (ADH), atrial natriuretic peptide (ANP), and the renin–angiotensin–aldosterone system (see Chapter 9). Let's take a closer look at water regulation in the body.

Water Regulation Involves a Balance of Intake and Loss

On a daily basis, the amount of water that you take in must be kept equal to the amount that you lose (**Figure 15.11**). You take in water in three ways: by drinking, by eating foods that contain water, and through chemical reactions such as respiration inside cells. However, you also lose water in four ways: through excretion by the kidneys, through evaporation from the skin surface, through exhalation from the lungs, and through the gastrointestinal tract, in feces. In women of reproductive age, additional water is lost through menstruation. Normally, daily water loss and gain are both equal to about 2500 mL.

However, activities such as exercise, working in a hot environment, or not drinking enough water can upset the water balance and change the osmolarity (or solute concentration) of your blood. That was the case with the exercise example we saw in the chapter opener. Water may also be lost in vomit or diarrhea during a GI tract infection. The condition in which water loss is greater than water gain is called **dehydration**.

Daily water balance • Figure 15.11 _____

The sources and volumes of water produced and lost in a day. Normally, daily water loss and gain are both equal to 2500 mL. The numbers are averages for adults.

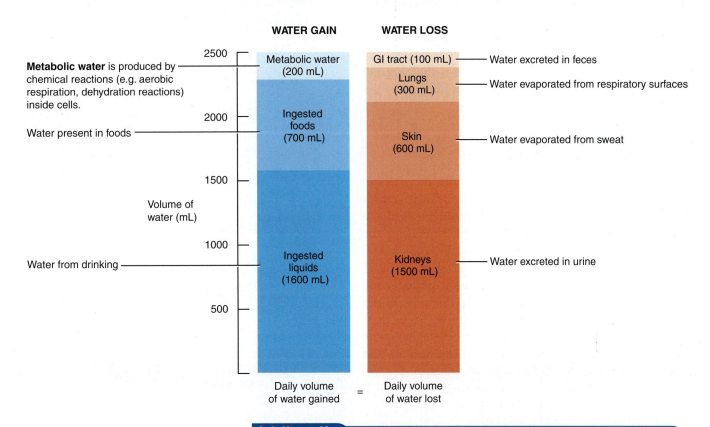

WATER GAIN

WATER LOSS

Metabolic water is produced by chemical reactions (e.g. aerobic respiration, dehydration reactions) inside cells.

Water present in foods

Volume of water (mL)

Water from drinking

Metabolic water (200 mL)

Ingested foods (700 mL)

Ingested liquids (1600 mL)

GI tract (100 mL) —— Water excreted in feces

Lungs (300 mL) —— Water evaporated from respiratory surfaces

Skin (600 mL) —— Water evaporated from sweat

Kidneys (1500 mL) —— Water excreted in urine

Daily volume of water gained = Daily volume of water lost

Ask Yourself

The caffeine in coffee, tea, and cola sodas is a naturally occurring *diuretic*, a substance that slows reabsorption of water by the kidneys and therefore causes an elevated urine flow rate. What is a likely effect of drinking these beverages?
a. More substances and water will be excreted than normal.
b. Fewer substances and water will be excreted than normal.

The Kidneys Regulate the Composition of Body Fluids **455**

Osmoreceptors in the hypothalamus sense these changes and stimulate the posterior pituitary to secrete ADH. ADH stimulates water reabsorption in the kidneys, which produce concentrated urine and help to restore water balance (**Figure 15.12**). The osmoreceptors also stimulate thirst and cause you to drink more water.

Exactly how does ADH work? As discussed with the countercurrent multiplier earlier in this chapter, the kidneys set up an osmotic gradient from the cortex through the medulla due to increasing levels of NaCl in the interstitial fluid. By the time the filtrate reaches the distal tubules and collecting ducts, the fluid is hypotonic compared with the interstitial fluid that surrounds the tubing. When ADH is present, the apical membranes of the principal cells of the distal tubule and collecting ducts become permeable to water. As the hypotonic filtrate descends through the collecting ducts, water flows out of the collecting ducts in response to this osmotic gradient. Increased water reabsorption produces concentrated urine and reduces water loss from the body. If no ADH is present, the membranes are impermeable to water, so no water is reabsorbed as the filtrate passes through the DCT and collecting duct, resulting in dilute urine.

Because water follows the osmotic gradients set up by NaCl movements, its flow can also be influenced by the regulation of reabsorption of NaCl, which is the responsibility of the renin–angiotensin–aldosterone system.

Water Levels Also Depend on Changes in NaCl Levels

You just learned that the blood water concentration changes when you become dehydrated. Blood volume also decreases with dehydration. These changes in blood volume lead to a decrease in blood pressure. The juxtaglomerular cells in the afferent arteriole of the nephron sense this decrease and respond by releasing the enzyme renin. Through a cascade of events referred to as the **renin–angiotensin–aldosterone system**, renin yields angiotensin II (AII), which causes the adrenal glands to secrete aldosterone. Aldosterone increases sodium reabsorption in the kidneys, which indirectly increases water reabsorption to restore blood volume and blood pressure (**Figure 15.13**). Aldosterone regulates reabsorption of only about 1–4% of the filtered sodium, which amounts to about 5.8 g, on a daily basis.

If blood volume and blood pressure were to increase (due to excessive fluid intake, for example), the kidneys would need to excrete sodium and water to restore blood volume. To accomplish this, atrial cells of the heart release the hormone atrial natriuretic peptide (ANP), which acts on the kidney (directly and indirectly).

Negative feedback regulation of water reabsorption in the kidney • Figure 15.12 _____

The hypothalamus is responsible for monitoring blood water concentrations. ADH can help restore water homeostasis when blood becomes too concentrated.

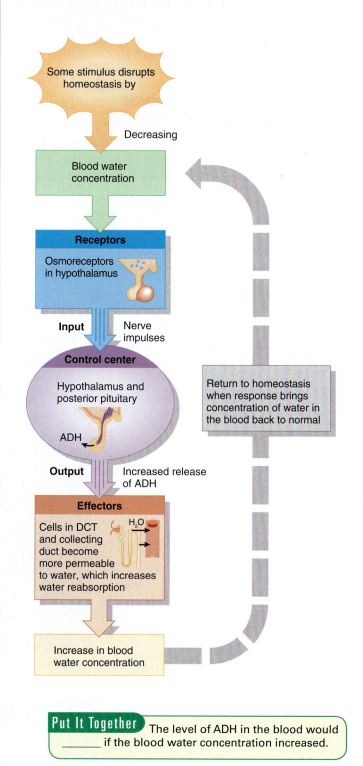

Put It Together The level of ADH in the blood would _____ if the blood water concentration increased.

Renin–angiotensin–aldosterone system • Figure 15.13

In a complex series of events involving several organs, the renin–angiotensin–aldosterone system increases sodium and water reabsorption in the kidney and restores blood volume and blood pressure.

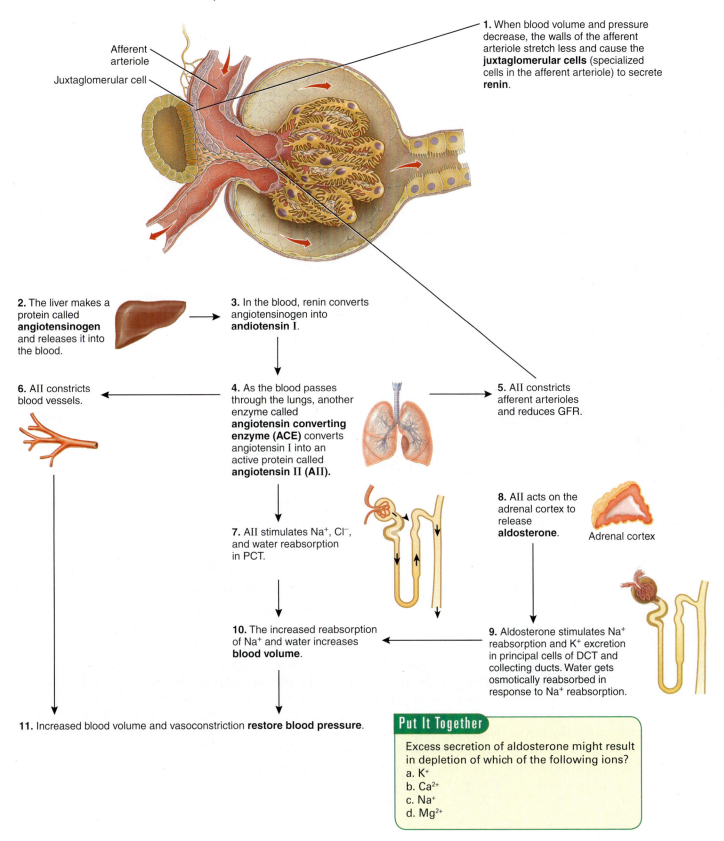

Afferent arteriole

Juxtaglomerular cell

1. When blood volume and pressure decrease, the walls of the afferent arteriole stretch less and cause the **juxtaglomerular cells** (specialized cells in the afferent arteriole) to secrete **renin**.

2. The liver makes a protein called **angiotensinogen** and releases it into the blood.

3. In the blood, renin converts angiotensinogen into **andiotensin I**.

6. AII constricts blood vessels.

4. As the blood passes through the lungs, another enzyme called **angiotensin converting enzyme (ACE)** converts angiotensin I into an active protein called **angiotensin II (AII).**

5. AII constricts afferent arterioles and reduces GFR.

7. AII stimulates Na^+, Cl^-, and water reabsorption in PCT.

8. AII acts on the adrenal cortex to release **aldosterone**.

Adrenal cortex

10. The increased reabsorption of Na^+ and water increases **blood volume**.

9. Aldosterone stimulates Na^+ reabsorption and K^+ excretion in principal cells of DCT and collecting ducts. Water gets osmotically reabsorbed in response to Na^+ reabsorption.

11. Increased blood volume and vasoconstriction **restore blood pressure**.

Put It Together

Excess secretion of aldosterone might result in depletion of which of the following ions?
a. K^+
b. Ca^{2+}
c. Na^+
d. Mg^{2+}

Finally, when you ingest a food that is high in salt content, such as theater popcorn or several strips of bacon, the renin–angiotensin–aldosterone system is suppressed, and the release of ANP is increased to remove excess sodium chloride and water.

How Do Kidney Stones Form?

Although regulation of blood water levels is critical for survival, when very concentrated urine is produced, minerals such as calcium, phosphorus, and oxalates may begin to crystallize out of solution. This can lead to formation of urinary **calculi** (or *stones*; singular is calculus). Alkaline urine pH and urinary tract infections also increase the chance of calculus formation. These stones can form in either the kidneys or the urinary bladder.

Treatments for urinary stones can involve dissolving them by using urinary acidifier medications; breaking them up into smaller pieces by using sound waves, ultrasound, or lasers; or letting them pass naturally through the tubing. Stone formation can often be prevented by consuming adequate amounts of fluids and reducing the intake of certain foods that promote release of excess amounts of mineral substances into the urine.

CONCEPT CHECK STOP

1. **Which** electrolytes are the most important for driving the movements of water?
2. By **what** routes does your body lose water?
3. **What** does aldosterone do to regulate blood pressure in the body?
4. **How** does the kidney regulate Ca^{2+} levels in the blood?

The Kidneys Help Maintain the Acid–Base Balance of Body Fluids

LEARNING OBJECTIVES

1. **List** the various pH buffer systems in the body.
2. **Describe** how the lungs maintain normal blood pH.
3. **Explain** how the kidneys maintain normal blood pH.
4. **Identify** the physiologic changes and compensations in acidosis and alkalosis.

Your body continuously produces acid in the form of hydrogen ions (H^+) as a result of metabolism. Your diet also contributes to H^+ production; diets rich in meat lead to more acid production than those rich in fruits and vegetables. Yet the H^+ concentration (pH) of the blood must be maintained within a narrow range, between 7.35 and 7.45, for the cells of the body to function properly. The H^+ produced when you eat a bucket of chicken is released into the blood, and the body needs a way to neutralize it or remove it from the body to maintain a normal blood pH. Your body has three major mechanisms that help to keep the blood pH within the normal range. We will discuss each one in turn.

- Buffer systems in the blood and intracellular compartments.
- Exhalation of carbon dioxide.
- Kidney excretion.

As you will see, kidney excretion is the slowest mechanism, but the only one that eliminates acids.

Buffers Help to Maintain pH Levels Within a Narrow Range

Recall from Chapter 2 that a buffer can rapidly absorb or release H^+ in response to changes in the H^+ levels in order to maintain normal or nearly normal pH. Only the H^+ that is "free" in solution can contribute to pH, so binding H^+ raises the pH, and releasing H^+ helps lower the pH. The body's major pH buffer systems consist of protein, carbonic acid–bicarbonate, and phosphate buffers (**Figure 15.14**). All three buffer systems work together, as a change in H^+ concentration in one fluid compartment causes a change in other compartments.

The protein buffer system is the most abundant buffer in intracellular fluid and blood plasma. Proteins such

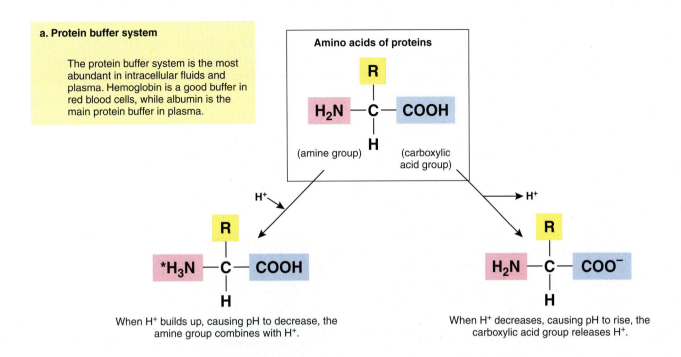

a. Protein buffer system

The protein buffer system is the most abundant in intracellular fluids and plasma. Hemoglobin is a good buffer in red blood cells, while albumin is the main protein buffer in plasma.

Amino acids of proteins

(amine group)   (carboxylic acid group)

When H⁺ builds up, causing pH to decrease, the amine group combines with H⁺.

When H⁺ decreases, causing pH to rise, the carboxylic acid group releases H⁺.

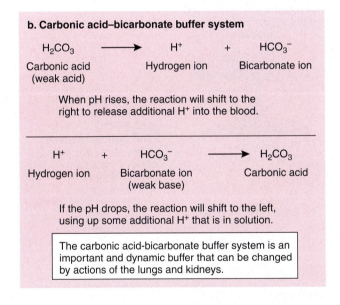

b. Carbonic acid–bicarbonate buffer system

$$H_2CO_3 \longrightarrow H^+ + HCO_3^-$$

Carbonic acid   Hydrogen ion   Bicarbonate ion
(weak acid)

When pH rises, the reaction will shift to the right to release additional H⁺ into the blood.

$$H^+ + HCO_3^- \longrightarrow H_2CO_3$$

Hydrogen ion   Bicarbonate ion   Carbonic acid
(weak base)

If the pH drops, the reaction will shift to the left, using up some additional H⁺ that is in solution.

The carbonic acid-bicarbonate buffer system is an important and dynamic buffer that can be changed by actions of the lungs and kidneys.

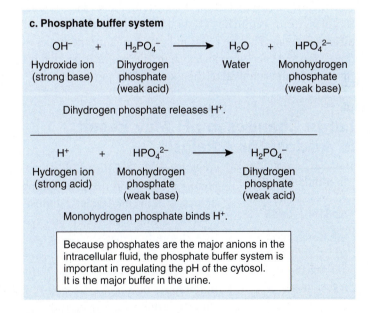

c. Phosphate buffer system

$$OH^- + H_2PO_4^- \longrightarrow H_2O + HPO_4^{2-}$$

Hydroxide ion   Dihydrogen   Water   Monohydrogen
(strong base)   phosphate   phosphate
                 (weak acid)   (weak base)

Dihydrogen phosphate releases H⁺.

$$H^+ + HPO_4^{2-} \longrightarrow H_2PO_4^-$$

Hydrogen ion   Monohydrogen   Dihydrogen
(strong acid)   phosphate   phosphate
                (weak base)   (weak acid)

Monohydrogen phosphate binds H⁺.

Because phosphates are the major anions in the intracellular fluid, the phosphate buffer system is important in regulating the pH of the cytosol. It is the major buffer in the urine.

as hemoglobin in red blood cells and albumin in blood plasma either release or combine with H⁺ to maintain normal pH.

The carbonic acid–bicarbonate buffer system is an important system for the blood. It is based on the bicarbonate ion (HCO_3^-), which can act as a weak base, and carbonic acid (H_2CO_3), which can act as a weak acid. This buffer sys-tem works with mechanisms of the kidneys and lungs, as we will see next. Carbonic acid can dissociate or form, depending on whether there is an excess or shortage of H⁺.

The phosphate system regulates the cytosol and is the most important buffer in the urine. Its mechanism is simi-lar to the mechanism for the carbonic acid–bicarbonate buffer system.

The Lungs and Kidneys Can Help Compensate for Changes in pH

The simple act of breathing also plays a role in maintaining the pH of body fluids. As we have seen, the carbonic acid–bicarbonate system is a dynamic mechanism for buffering the pH of the blood and body fluids. Carbonic acid (H_2CO_3) can be eliminated by exhaling CO_2, and the concentration of CO_2 can be rapidly regulated by breathing (**Figure 15.15**). Increasing the rate and depth of breathing lowers the CO_2 concentration and shifts the equilibrium to the left, thereby reducing the H^+ concentration and increasing blood pH. Conversely, decreasing the rate and depth of breathing shifts the equilibrium to the right, thereby increasing the H^+ concentration and decreasing blood pH. The rate of breathing is influenced by chemoreceptors in the carotid artery, aortic arch, and medulla. This respiratory mechanism is powerful and can respond rapidly, within minutes, to changes in blood pH, but it can only alter the carbonic acid concentration. It cannot fully compensate for any dramatic changes in pH.

The kidneys can respond to changes in blood pH by permanently removing excess H^+ or HCO_3^- from the body. The kidneys can alter the reabsorption of H^+ or HCO_3^-, thereby shifting the equilibrium of the carbonic acid–bicarbonate buffer system, removing H^+ or HCO_3^- from the body and altering blood pH. The renal mechanism can take hours or days, and is therefore slower than the respiratory mechanism—but the removal of the excess H^+ is permanent.

Large Changes in pH May Result in Acidosis or Alkalosis

> **acidosis** (as-i-DŌ-sis) A condition in which blood pH is below 7.35.
>
> **alkalosis** (al-ka-LŌ-sis) A condition in which blood pH is higher than 7.45.

Acid–base imbalances lead to changes in blood pH, including **acidosis** and **alkalosis**. Acidosis depresses the central nervous system through depression of synaptic transmission. If the systemic arterial blood pH falls below 7, the individual becomes disoriented, then becomes comatose, and may die. Alkalosis causes overexcitability in both the central nervous system and peripheral nerves. Neurons conduct impulses repetitively, even when not stimulated; the results are nervousness, muscle spasms, and even convulsions and death.

There are many causes and two forms of acidosis and alkalosis: respiratory forms, which result from an alteration of CO_2 levels, and metabolic forms, which are due to changes in the levels of H^+ or HCO_3^-. Your body compensates for these imbalances in various ways.

Regulation of the carbonic acid–bicarbonate buffer system by the respiratory and renal systems • Figure 15.15

The respiratory system can rapidly alter the bicarbonate buffer system in the blood to compensate for changes in pH, but only the kidneys are capable of permanent compensation by excreting H^+ or HCO_3^-.

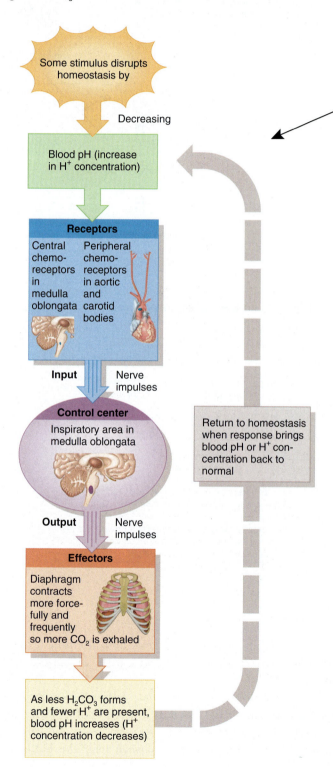

Some stimulus disrupts homeostasis by

Decreasing

Blood pH (increase in H^+ concentration)

Receptors
Central chemoreceptors in medulla oblongata | Peripheral chemoreceptors in aortic and carotid bodies

Input — Nerve impulses

Control center
Inspiratory area in medulla oblongata

Output — Nerve impulses

Effectors
Diaphragm contracts more forcefully and frequently so more CO_2 is exhaled

As less H_2CO_3 forms and fewer H^+ are present, blood pH increases (H^+ concentration decreases)

Return to homeostasis when response brings blood pH or H^+ concentration back to normal

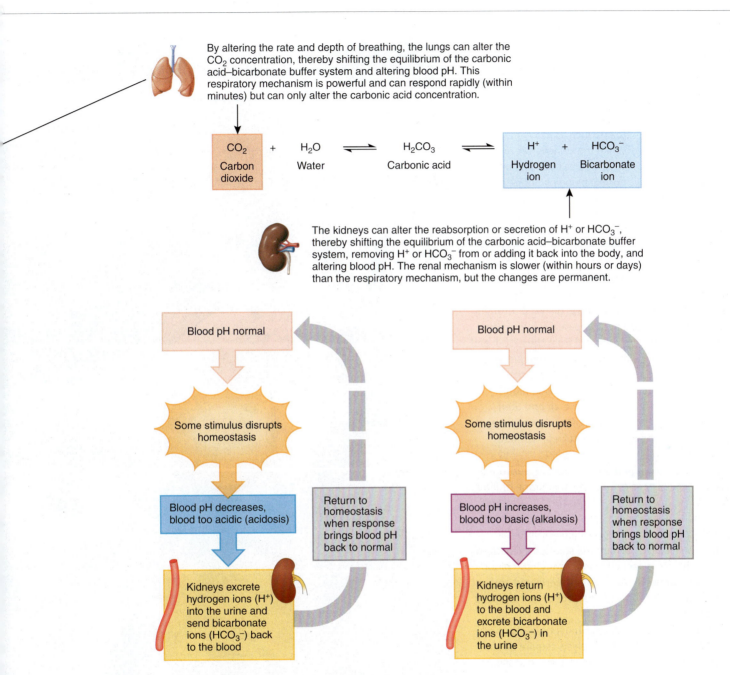

By altering the rate and depth of breathing, the lungs can alter the CO_2 concentration, thereby shifting the equilibrium of the carbonic acid–bicarbonate buffer system and altering blood pH. This respiratory mechanism is powerful and can respond rapidly (within minutes) but can only alter the carbonic acid concentration.

| CO_2 Carbon dioxide | + | H_2O Water | $\rightleftharpoons$ | H_2CO_3 Carbonic acid | $\rightleftharpoons$ | H^+ Hydrogen ion | + | HCO_3^- Bicarbonate ion |

The kidneys can alter the reabsorption or secretion of H^+ or HCO_3^-, thereby shifting the equilibrium of the carbonic acid–bicarbonate buffer system, removing H^+ or HCO_3^- from or adding it back into the body, and altering blood pH. The renal mechanism is slower (within hours or days) than the respiratory mechanism, but the changes are permanent.

Blood pH normal

Some stimulus disrupts homeostasis

Blood pH decreases, blood too acidic (acidosis)

Return to homeostasis when response brings blood pH back to normal

Kidneys excrete hydrogen ions (H^+) into the urine and send bicarbonate ions (HCO_3^-) back to the blood

Blood pH normal

Some stimulus disrupts homeostasis

Blood pH increases, blood too basic (alkalosis)

Return to homeostasis when response brings blood pH back to normal

Kidneys return hydrogen ions (H^+) to the blood and excrete bicarbonate ions (HCO_3^-) in the urine

Put It Together

Patients with unregulated diabetes mellitus often develop a pH imbalance because they lack insulin that allows the cells to utilize glucose. As a result, the body is forced to use other materials, many of which are acids, as energy sources. This causes the H^+ level to increase in the blood. Which of the following would occur as the body tries to regulate blood pH?

a. Kidneys increase H^+ reabsorption.
b. Breathing rate decreases.
c. pH decreases.
d. Kidneys increase HCO_3^- secretion.
e. Urinary system increases excretion of H^+.

WHAT A HEALTH PROVIDER SEES

Renal Failure and Dialysis

Kidney diseases such as *nephritis* (inflammation of the kidneys) and *glomerulonephritis* (inflammation of the glomeruli) can interfere with the blood supply to the kidney and lead to renal failure. Renal failure may also occur as a complication of diabetes mellitus or from exposure to toxins such as heavy metals.

dialysis (dī-AL-i-sis) The removal of waste products from blood by diffusion through a selectively permeable membrane. The verb form of dialysis is *dialyze* (DĪ-a-līz).

When the kidneys fail, metabolic wastes build up, and the normal blood composition goes awry. Without treatment, renal failure can quickly cause death. To treat renal failure, patients must undergo dialysis. There are two types of dialysis available.

In **hemodialysis**, the patient's blood is pumped through a machine, where the blood is dialyzed through an artificial membrane against a prepared solution. During this process, wastes are removed from the blood, and plasma components are either prevented from passing through the membrane or are added back into the blood after the process has occurred. The cleansed blood is passed through a detector to remove air and then returned to the body. Hemodialysis often destroys fragile red blood cells as they pass through tubes and membranes under pressure.

In contrast, **peritoneal dialysis** uses the peritoneum of the abdominal cavity as a dialyzing membrane to filter the blood. The peritoneum has a large surface area and numerous blood vessels and is a very effective filter. An artificial dialyzing solution is infused directly into the person's peritoneal cavity, where the solution exchanges components with the plasma flowing through the blood vessels in the peritoneum. After a predetermined amount of time, the solution is removed, discarded, and replaced with fresh solution.

Regardless of the type, dialysis must be performed frequently. Most people on hemodialysis require about 6–12 hours a week, typically divided into three sessions. Peritoneal dialysis is done multiple times per day. Dialysis is only a temporary solution; ultimately, a kidney transplant is the only permanent treatment for renal failure.

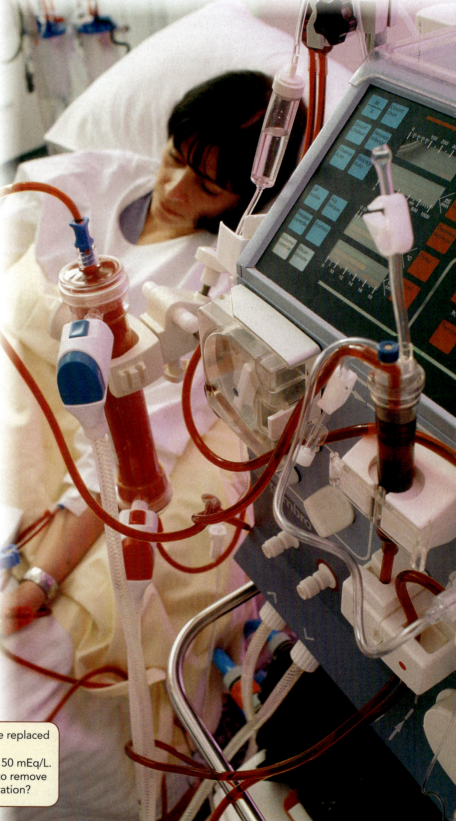

Think Critically
1. What functions of the kidney cannot be replaced by dialysis?
2. A dialysis patient's blood has an NaCl concentration of 150 mEq/L. What should the NaCl composition of the dialysis fluid be to remove excess NaCl and return the blood to normal NaCl concentration?

For example, moving to a high altitude (low oxygen) causes rapid, shallow breathing. The rapid breathing reduces CO_2 concentration and shifts the bicarbonate system to the left, using up some of the H^+, thereby causing alkalosis (see Figure 15.14). Within a few days, the kidneys excrete the excess HCO_3^-, and blood pH returns to normal. If your body cannot compensate for large swings in pH, medical treatment may be required (see *What a Health Provider Sees*).

CONCEPT CHECK STOP

1. **Which** pH buffer system is important in regulating the pH of the cytosol?

2. **How** does a decrease in the rate and depth of breathing change blood pH?

3. **How** do the kidneys respond to an increase in blood pH?

4. **What** is metabolic acidosis, and how does the body compensate for it?

THE PLANNER ✔

Summary

1 The Urinary System Plays a Vital Role in Maintaining Homeostasis 438

• The **kidneys** regulate the composition of the blood. The basic unit of the kidney, the **nephron**, is composed of a **renal corpuscle** and a **renal tubule**. The tubule is divided into four segments: the **proximal convoluted tubule (PCT)**, **nephron loop**, **distal convoluted tubule (DCT)**, and **collecting duct**.

• Blood plasma is filtered in the renal corpuscle. As the filtrate passes through the renal tubule, water and various solutes (sodium, potassium, glucose, amino acids, chloride) are reabsorbed in different segments. As shown, reabsorption returns materials to the blood and occurs via various passive and active transport mechanisms on the membranes of tubule cells.

Functional processes in the nephron of the kidney • Figure 15.4

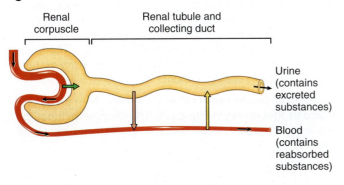

• As the filtrate passes through the tubule, various substances (such as H^+, urea, and drugs) are secreted into the filtrate. Secretion eliminates extra waste materials from the blood. What remains of the filtrate at the end of the nephron becomes **urine**.

2 The Urinary Bladder Stores Urine and Expels It from the Body 450

• Urine passes from the kidneys through the ureters to the urinary bladder. As shown, the urinary **bladder** is a hollow organ made of mucosa and smooth muscle. The urinary bladder can stretch to store urine.

The ureters, urinary bladder, and urethra • Figure 15.7

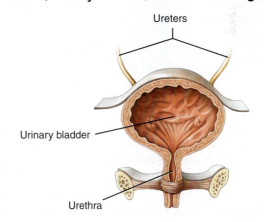

• The micturition reflex involves the urinary bladder, parasympathetic nervous system, and spinal cord; it causes the bladder to expel urine (**micturition**). The flow of urine into the urethra can be suppressed for a limited amount of time by nerve impulses from the cerebral cortex.

• Urine passes from the urinary bladder to the outside through the **urethra**. In females, the urethra is short and separate from the reproductive system. In males, the urethra is longer, separated into segments, and shared with the reproductive system; both urine and semen pass through the male urethra.

3 The Kidneys Regulate the Composition of Body Fluids 453

- Water is distributed among **intracellular** and **extracellular fluid compartments**. The extracellular compartment includes interstitial fluid and blood plasma, as shown. Within each fluid compartment, **electrolytes** are dissolved. Sodium, chloride, and bicarbonate are the main electrolytes in extracellular compartments. Potassium, magnesium, bicarbonate, and phosphate are the primary electrolytes in intracellular fluid.

Distribution of water in various compartments of the body • Figure 15.9

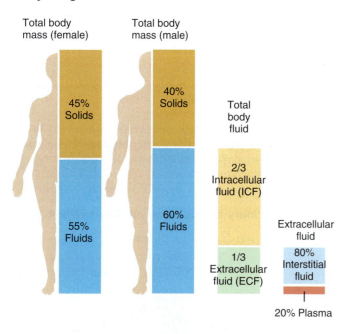

- Water intake and water loss are balanced on a daily basis. The kidneys regulate water balance by altering reabsorption in the distal tubule and **collecting duct** under stimulation by **antidiuretic hormone (ADH)**.

- The balance of water, sodium, and chloride are linked with blood volume and blood pressure. If blood volume decreases, the kidneys maintain sodium and chloride balance, blood volume, and blood pressure by reabsorbing these ions in the distal tubules under stimulation by the **renin–angiotensin–aldosterone system**. When blood volume is increased, the heart releases atrial natriuretic peptide (ANP), which stimulates loss of sodium, chloride, and water in the distal convoluted tubule.

- The kidneys regulate blood plasma calcium levels by reabsorbing calcium in the **distal convoluted tubule (DCT)** when stimulated by **parathyroid hormone (PTH)**. The kidneys also secrete calcitriol (vitamin D), which increases calcium absorption in the intestine.

4 The Kidneys Help Maintain the Acid–Base Balance of Body Fluids 458

- Your body produces acids (H^+) through metabolism. More acid is produced when you ingest a meat-rich diet than when you ingest a diet rich in fruits and vegetables. To deal with the acid produced, your fluids have three major buffer systems: protein, carbonic acid–bicarbonate (shown below), and phosphate buffers.

Carbonic acid–bicarbonate buffer system • Figure 15.14b

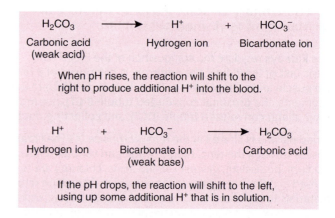

- The carbonic acid–bicarbonate system is a dynamic buffer that keeps the blood pH in the range 7.35–7.45. The respiratory system can rapidly alter blood pH through changes in the rate and depth of breathing. This changes the carbon dioxide concentration of the blood and alters the equilibrium of the bicarbonate system. The kidneys can alter blood pH by changing the reabsorption of H^+ or HCO_3^-. The kidneys change blood pH more slowly than the respiratory system, but the changes are permanent.

- Several conditions cause **alkalosis** (increases in blood pH) or **acidosis** (decreases in blood pH). The changes in blood pH are compensated by actions of the respiratory system and kidneys.

Key Terms

- acidosis 460
- afferent arteriole 441
- albuminuria 447
- aldosterone 448
- alkalosis 460
- antidiuretic hormone (ADH) 448
- carbonic anhydrase 447
- calculus 458
- collecting duct 442
- cortical nephron 442
- countercurrent multiplier 448
- dehydration 455
- detrusor muscle 450
- dialysis 462
- distal convoluted tubule 442
- efferent arteriole 441
- electrolyte 454
- extracellular fluid (ECF) 453
- filtration membrane 446
- glomerular filtration rate (GFR) 447
- glomerulus 441
- hemodialysis 462
- incontinence 452
- intracellular fluid (ICF) 453
- juxtamedullary nephron 442
- kidney 438
- major calyx 440
- micturition 452
- micturition reflex 452
- minor calyx 440
- nephron 442
- nephron loop 442
- net filtration pressure (NFP) 446
- paracellular route 447
- parathyroid hormone (PTH) 448
- peritoneal dialysis 462
- proximal convoluted tubule 442
- renal corpuscle 442
- renal pelvis 440
- renal tubule 442
- renal vein 441
- renin–angiotensin–aldosterone system 456
- transcellular route 447
- ureter 438
- urethra 438
- urinary bladder 438
- urinary system 438
- urine 438

Critical and Creative Thinking Questions

1. Jimmy is an active 10-year-old boy. He notices that his first-morning urine is a dark yellow color, much darker than urine later in the day. He is worried that something may be wrong and tells his mother. Should Jimmy be worried? Explain why or why not.

2. A man is trapped in a raft at sea for days, with no fresh water. Why can't he drink seawater to quench his thirst? Explain what drinking seawater would do to the man's kidney functions.

3. A diabetic develops acidosis, and his blood pH drops. What compensatory mechanisms are going on in his body to try to restore normal blood pH?

4. Bob notices dark red blood in his urine. What part of the nephron might be affected? Explain how the kidney normally prevents blood from entering the urine.

5. Selena has high blood pressure and takes a medication that inhibits angiotensin-converting enzyme. Her doctor recommends that she watch her diet to keep her intake of sodium chloride low while she's on the medication. If Selena has a high salt intake, what will happen? Explain the mechanism that would be affected if she indulges in French fries.

What is happening in this picture?

Kidney transplants are necessary to treat kidney disease and renal failure. The most common cause of kidney disease is nephritis.

Think Critically 1. What treatment would this patient undergo while awaiting a transplant? 2. Explain the pathway of blood from the aorta through the kidney and back to the inferior vena cava.

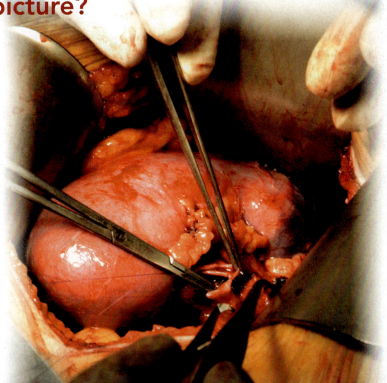

Self-Test

(Check your answers in Appendix C.)

1. A patient excretes bicarbonate in his urine. What does this tell you about conditions in the nephron?

   a. The amount of bicarbonate exceeds the amount of hydrogen ion secreted by the tubule.

   b. The amount of bicarbonate is smaller than the amount of hydrogen ion secreted by the tubule.

   c. The amount of bicarbonate equals the amount of hydrogen ion secreted by the tubule.

   d. The amount of bicarbonate reabsorbed exceeds the amount of hydrogen ion secreted by the tubule.

Use this figure to answer questions 2–4.

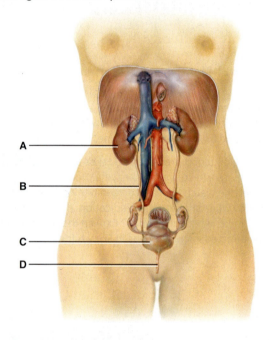

2. The function of the structure labeled B is to _____.

   a. transport urine from the body

   b. store urine

   c. transport urine to the bladder

   d. filter the blood

3. During a urinary tract infection, the blood most commonly comes from which of the labeled structures?

   a. A   c. C

   b. B   d. D

4. Which of the labeled structures is susceptible to nephritis?

   a. A   c. C

   b. B   d. D

5. Aldosterone acts on the _____.

   a. glomerulus

   b. proximal convoluted tubule

   c. distal convoluted tubule

   d. nephron loop

Use this figure to answer questions 6–9.

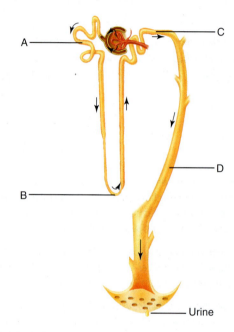

6. Which labeled segment sets up an osmotic gradient in the medulla?

   a. A   c. C

   b. B   d. D

7. Which segment absorbs approximately two-thirds of all filtered water and solutes?

   a. A   c. C

   b. B   d. D

8. The structure shown in D is the target of _____.

   a. ANP

   b. PTH

   c. AII

   d. ADH

9. PTH acts on which segment to control calcium reabsorption?

   a. A   c. C

   b. B   d. D

10. The major electrolyte in the intracellular fluid is _____.

    a. sodium

    b. potassium

    c. chloride

    d. bicarbonate

11. The _____ stores urine.

    a. kidney

    b. ureter

    c. urethra

    d. urinary bladder

12. Which of the following best describes how the kidneys respond to changes in blood pH?

    a. They increase the CO_2 concentration.

    b. They conserve more fluid.

    c. They secrete H^+.

    d. They secrete more fluid.

13. In which part of the nephron would you find a hypotonic filtrate?

    a. glomerular capsule

    b. nephron loop

    c. distal convoluted tubule

    d. proximal convoluted tubule

Use this figure to answer questions 14–15.

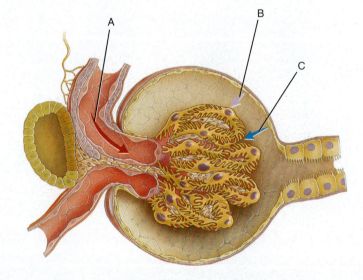

14. Which of the labeled forces would be most affected by AII?

    a. A

    b. B

    c. C

    d. All would be equally affected.

15. The net filtration pressure is calculated by which of the following?

    a. A − B − C

    b. B − A + C

    c. C − A − B

    d. C + B − A

Use this figure to answer questions 16–18.

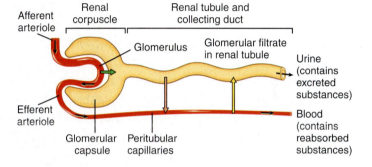

16. Filtration occurs in which part of the nephron?

    a. Afferent arteriole

    b. Renal corpuscle

    c. Peritubular capillaries

    d. Renal tubule and collecting duct

17. Tubular secretion occurs in which part of the nephron?

    a. Afferent arteriole

    b. Renal corpuscle

    c. Peritubular capillaries

    d. Renal tubule and collecting duct

18. Reabsorption into the bloodstream occurs in which part of the nephron?

    a. Afferent arteriole

    b. Renal corpuscle

    c. Peritubular capillaries

    d. Renal tubule and collecting duct

THE PLANNER

Review your Chapter Planner on the chapter opener and check off your completed work.

The Reproductive Systems

Jason and Bonnie have been unsuccessful in starting a family and have decided to consult a fertility specialist to determine what is wrong. One out of every six couples experience **infertility**, defined as being unable to get pregnant for at least 12 consecutive months. There are a number of possible causes: About 40% of the cases are related to an issue with the male; 40% are related to an issue with the female; and the remaining 20% are due to a combination of male and female issues.

Infertility in the male is often associated with low sperm counts or high numbers of defective sperm. This can be due to harmful environmental chemicals (cigarette smoke, lead, pesticides, and so on), high testicular temperatures (due to tight clothing or certain occupations), or abnormal hormone levels. In the female, the most common causes are related to a lack of egg release and abnormal pathways for the sperm and egg. Abnormal hormone levels and infections (especially sexually transmitted diseases) are most often involved. For both the males and females, fertility rates continue to decrease after age 25. For some couples, the problem lies in compatibility between the male and female, even though both are "functioning normally."

Many infertile couples turn to reproductive specialists and assisted reproductive technologies to improve their chances of a successful pregnancy. In one technique, *in vitro* **fertilization (IVF)**, the woman is given drugs to enhance her egg production, and the eggs are harvested. Following injection with sperm in the laboratory, the resulting embryos are allowed to develop briefly and are then transferred into the woman's uterus. If all goes well, she will deliver a healthy baby. Often, more than one embryo is transferred for better chances of success; if more than one implants, multiple births may occur. This technology has enabled many infertile couples to have children. However, the fact that it can also result in multiple births has become controversial in recent cases and has led to medical board review of IVF candidate selection.

In this chapter, we examine the male and female reproductive systems, as well as the growth and development of a new human being from fertilization to birth.

CHAPTER OUTLINE

CHAPTER PLANNER ✔

- ❏ Study the picture and read the opening story.
- ❏ Scan the Learning Objectives in each section:
  p. 470 ❏ p. 480 ❏ p. 484 ❏ p. 488 ❏ p. 498 ❏ p. 502 ❏
- ❏ Read the text and study all visuals. Answer any questions.

Analyze key features

- ❏ InSight, p. 471 ❏   p. 477 ❏
- ❏ Process Diagram, p. 474 ❏   p. 480   ❏ p. 483 ❏   p. 489 ❏
- ❏ What a Health Provider Sees, p. 487 ❏
- ❏ Stop: Answer the Concept Checks before you go on:
  p. 479 ❏ p. 484 ❏ p. 487 ❏ p. 497 ❏ p. 500 ❏ p. 503 ❏

End of chapter

- ❏ Review the Summary and Key Terms.
- ❏ Answer the Critical and Creative Thinking Questions.
- ❏ Answer What is happening in this picture?
- ❏ Complete the Self-Test and check your answers.

The Reproductive Organs Make, Deliver, and Receive the Sex Cells

LEARNING OBJECTIVES

1. **Identify** the organs of the male reproductive system and their functions.

2. **Describe** how sperm are made.

3. **Explain** the functions of hormones in the male reproductive system.

4. **Identify** the organs of the female reproductive system and their functions.

 e reproduce by **sexual reproduction**, a process in which sex cells (**gametes**) unite to form offspring. The gametes are formed in the **gonads**. The gametes unite inside the female in a process called **fertilization**. The female then nurtures the growing embryo until birth. So the reproductive organs are specialized for making gametes and bringing them together for fertilization; in the case of the female, they also provide a "home" for nurturing the growing embryo.

> **gonad** (GŌ-nad) A gland that produces gametes and secretes hormones; the ovaries in the female and the testes in the male.

Let's start by examining the male reproductive organs.

Male Reproductive Organs Make and Deliver Sperm

The organs of the male reproductive system are the following. (**Figure 16.1**):

- The testes.
- A system of ducts: the epididymis, vas deferens, ejaculatory ducts, and urethra.
- Accessory sex glands: seminal vesicles, prostate, and bulbourethral glands.
- Supporting structures: the scrotum and the penis.

The male's **testes** (singular: *testis*) make gametes called **sperm**. The testes are paired oval glands that lie outside the body's core, in a sac called the **scrotum**. They also secrete the hormone **testosterone**, which we will discuss soon.

The scrotum is a pouch that supports the testes. It consists of loose skin, and smooth muscle. Internally, the scrotum is divided into two sacs, each containing a single testis. The scrotum keeps the sperm at an optimal temperature for their development, slightly less than the core body temperature. Higher temperatures, which may be caused by tight clothing or certain occupations, may harm the sperm and cause infertility in the male, as we noted in the chapter opener.

The sperm mature in the tightly coiled ducts of the **epididymis** and are stored in a tube called the **vas deferens** (or *ductus deferens*).

After they are made and matured, the sperm cells will ultimately be delivered to the female via the penis during **sexual intercourse**. When a male becomes sexually aroused, the spongy tissue of this accessory reproductive organ (**Figure 16.1b**) becomes engorged with blood until it is stiff and erect; this configuration is necessary for the penis to be inserted inside the female's vagina.

> **sexual intercourse** The insertion of the erect penis of a male into the vagina of a female. Also called *coitus* (KŌ-i-tus).

During sexual arousal, the sperm get swept (by peristaltic contractions) through the vas deferens and ejaculatory duct into the urethra, where they are mixed with secretions from other accessory reproductive organs. The accessory organs include the following:

- **Prostate**. The prostate secretes a milky, slightly acidic fluid that helps semen coagulate after ejaculation and subsequently breaks down the clot.
- **Seminal vesicles**. The seminal vesicles secrete alkaline, viscous fluid that helps neutralize acid in the female reproductive tract, provides fructose for ATP production by sperm, contributes to sperm motility and viability, and helps semen coagulate after ejaculation.
- **Bulbourethral gland**. The bulbourethral (Cowper's) glands secrete alkaline fluid that neutralizes the acidic environment of the urethra, and mucus that lubricates the lining of the urethra and the tip of the penis during sexual intercourse.

The mixture of sperm and accessory organ secretions is called **semen**. Semen gets expelled from the body through the urethra of the penis during **ejaculation**.

> **ejaculation** (e-jak-ū-LĀ-shun) The reflex ejection or expulsion of semen from the penis.

a. Sagittal section showing male reproductive organs

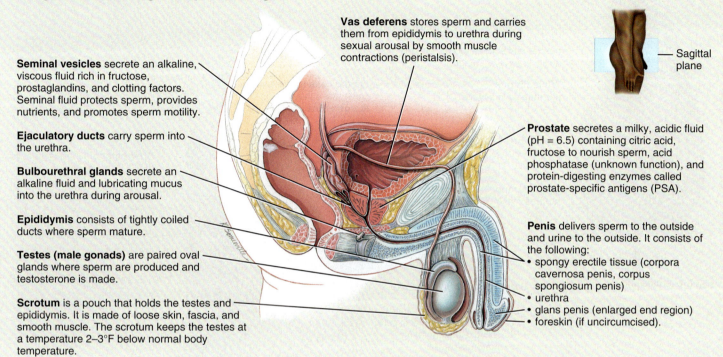

Seminal vesicles secrete an alkaline, viscous fluid rich in fructose, prostaglandins, and clotting factors. Seminal fluid protects sperm, provides nutrients, and promotes sperm motility.

Ejaculatory ducts carry sperm into the urethra.

Bulbourethral glands secrete an alkaline fluid and lubricating mucus into the urethra during arousal.

Epididymis consists of tightly coiled ducts where sperm mature.

Testes (male gonads) are paired oval glands where sperm are produced and testosterone is made.

Scrotum is a pouch that holds the testes and epididymis. It is made of loose skin, fascia, and smooth muscle. The scrotum keeps the testes at a temperature 2–3°F below normal body temperature.

Vas deferens stores sperm and carries them from epididymis to urethra during sexual arousal by smooth muscle contractions (peristalsis).

Sagittal plane

Prostate secretes a milky, acidic fluid (pH = 6.5) containing citric acid, fructose to nourish sperm, acid phosphatase (unknown function), and protein-digesting enzymes called prostate-specific antigens (PSA).

Penis delivers sperm to the outside and urine to the outside. It consists of the following:
- spongy erectile tissue (corpora cavernosa penis, corpus spongiosum penis)
- urethra
- glans penis (enlarged end region)
- foreskin (if uncircumcised).

b. Anterior view of scrotum and testes

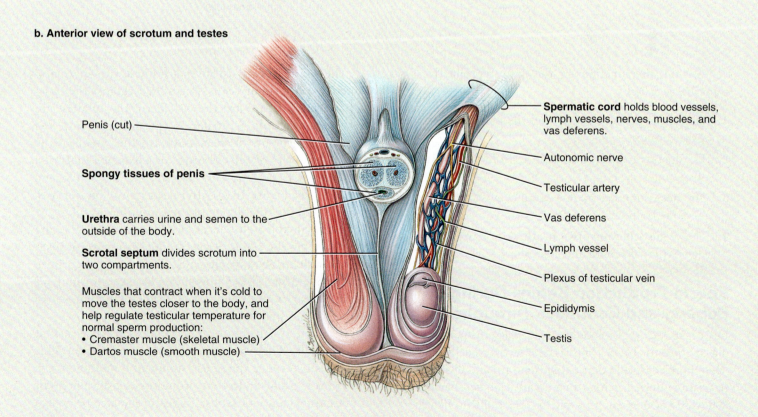

Penis (cut)

Spongy tissues of penis

Urethra carries urine and semen to the outside of the body.

Scrotal septum divides scrotum into two compartments.

Muscles that contract when it's cold to move the testes closer to the body, and help regulate testicular temperature for normal sperm production:
- Cremaster muscle (skeletal muscle)
- Dartos muscle (smooth muscle)

Spermatic cord holds blood vessels, lymph vessels, nerves, muscles, and vas deferens.

Autonomic nerve

Testicular artery

Vas deferens

Lymph vessel

Plexus of testicular vein

Epididymis

Testis

The Reproductive Organs Make, Deliver, and Receive the Sex Cells   **471**

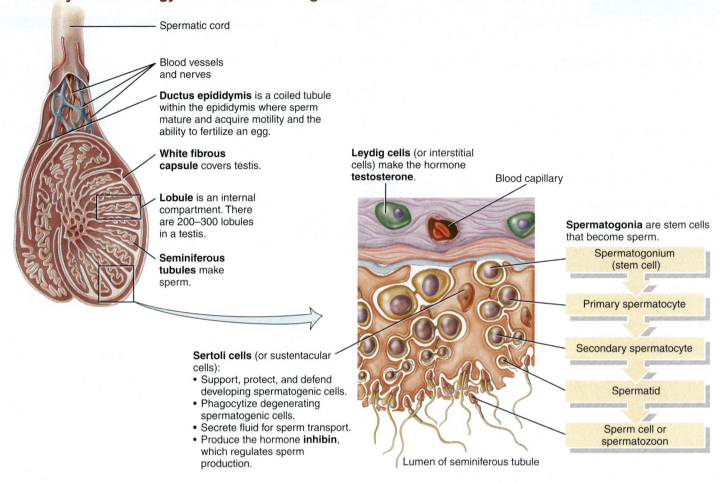

Spermatic cord

Blood vessels and nerves

Ductus epididymis is a coiled tubule within the epididymis where sperm mature and acquire motility and the ability to fertilize an egg.

White fibrous capsule covers testis.

Lobule is an internal compartment. There are 200–300 lobules in a testis.

Seminiferous tubules make sperm.

Sertoli cells (or sustentacular cells):
• Support, protect, and defend developing spermatogenic cells.
• Phagocytize degenerating spermatogenic cells.
• Secrete fluid for sperm transport.
• Produce the hormone **inhibin**, which regulates sperm production.

Leydig cells (or interstitial cells) make the hormone **testosterone**.

Blood capillary

Spermatogonia are stem cells that become sperm.

Spermatogonium (stem cell)

Primary spermatocyte

Secondary spermatocyte

Spermatid

Sperm cell or spermatozoon

Lumen of seminiferous tubule

In sum, the functions of the male reproductive systems are as follows.

1. The testes produce sperm and the male sex hormone testosterone.

2. The ducts transport, store, and assist in the maturation of sperm.

3. The accessory sex glands secrete most of the liquid portion of semen.

4. The penis contains the urethra, a passageway for ejaculation of semen and excretion of urine.

Sperm Production Begins During Puberty and Continues Throughout Life

Let's take a closer look at the testes, where sperm is produced (**Figure 16.2**). The testes are covered by a dense white fibrous capsule that extends inward and divides each testis into internal compartments called lobules. Each of the 200 to 300 lobules contains one to three tightly coiled seminiferous tubules that produce sperm, as we'll discuss soon.

The **seminiferous tubules** are lined with spermatogenic (sperm-forming) cells. Positioned against the basement membrane, toward the outside of the tubules, are the **spermatogonia** (singular is spermatogonium), the stem cell precursors of sperm. Toward the lumen of the tubule are layers of cells in order of advancing maturity: *primary spermatocytes, secondary spermatocytes, spermatids,* and *sperm cells* (see Figure 16.3). After a sperm cell has formed, it is released into the lumen of the seminiferous tubule.

Large **Sertoli cells**, located between the developing sperm cells in the seminiferous tubules, support, protect, and nourish spermatogenic cells. They also phagocytize any degenerating spermatogenic cells, secrete fluid for sperm transport, and release the hormone inhibin, which helps regulate sperm production.

Between the seminiferous tubules are clusters of **Leydig cells**. These are the cells that secrete the hormone testosterone, the most important androgen. An androgen is a hormone that promotes the development of masculine characteristics. Testosterone also promotes the male's sex drive.

Males make sperm continuously from puberty onward, in a process called **spermatogenesis** (**Figure 16.3**). Spermatogenesis involves a process first introduced in Chapter 3 called **meiosis**. In this process of gamete production, gametes receive a single set of 23 chromosomes, symbolized as *n*. In this way, the new individual receives the proper number of chromosomes when the male and female gametes join.

Meiosis is followed by **spermiogenesis**, in which the sperm are reshaped to form the classical ovoid cell with the tail that we refer to as the sperm.

Sperm are produced at the rate of 300 million per day. Once ejaculated, they do not survive more than 48

meiosis (mē-Ō-sis)
A type of cell division that occurs during production of gametes, involving two successive nuclear divisions that result in cells with a haploid (*n*)—one half the normal—number of chromosomes.

hours in the female reproductive tract. The major parts of a sperm cell are the head and tail. The head contains the nuclear material (DNA) and an acrosome. An **acrosome** is a vesicle containing enzymes that aid penetration into the egg (or oocyte) by the sperm cell. The tail of a sperm cell is divided into four parts: neck, middle piece, principal piece, and end piece. The *neck* is the constricted region just behind the head. The *middle piece* contains mitochondria that provide ATP for locomotion. The *principal piece* is the longest portion of the tail, and the *end piece* is the terminal, tapering portion of the tail.

Spermatogenesis and the structure of sperm • Figure 16.3

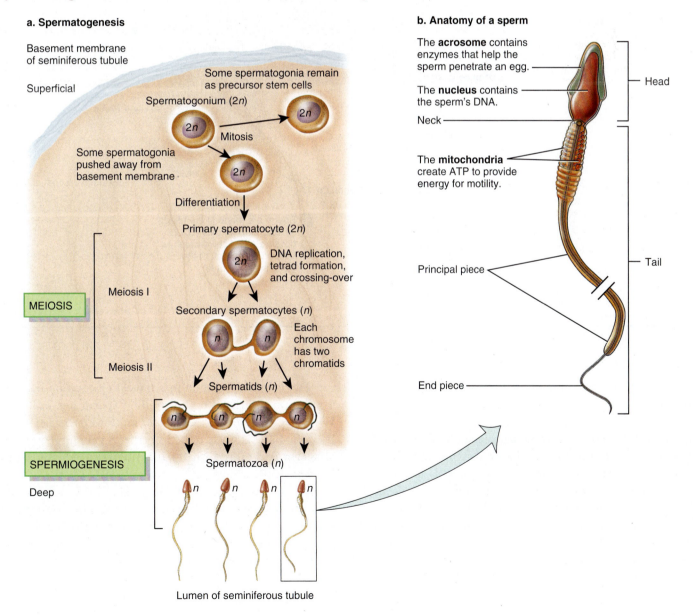

a. Spermatogenesis

Basement membrane of seminiferous tubule

Superficial

Some spermatogonia remain as precursor stem cells

Spermatogonium (2*n*)

2*n*

2*n*

Mitosis

Some spermatogonia pushed away from basement membrane

2*n*

Differentiation

Primary spermatocyte (2*n*)

2*n*

DNA replication, tetrad formation, and crossing-over

MEIOSIS

Meiosis I

Secondary spermatocytes (*n*)

n   *n*

Each chromosome has two chromatids

Meiosis II

Spermatids (*n*)

n   *n*   *n*   *n*

SPERMIOGENESIS

Spermatozoa (*n*)

Deep

n   *n*   *n*   *n*

Lumen of seminiferous tubule

b. Anatomy of a sperm

The **acrosome** contains enzymes that help the sperm penetrate an egg.

The **nucleus** contains the sperm's DNA.

Neck

The **mitochondria** create ATP to provide energy for motility.

Principal piece

End piece

Head

Tail

Path of sperm through the reproductive tract • Figure 16.4

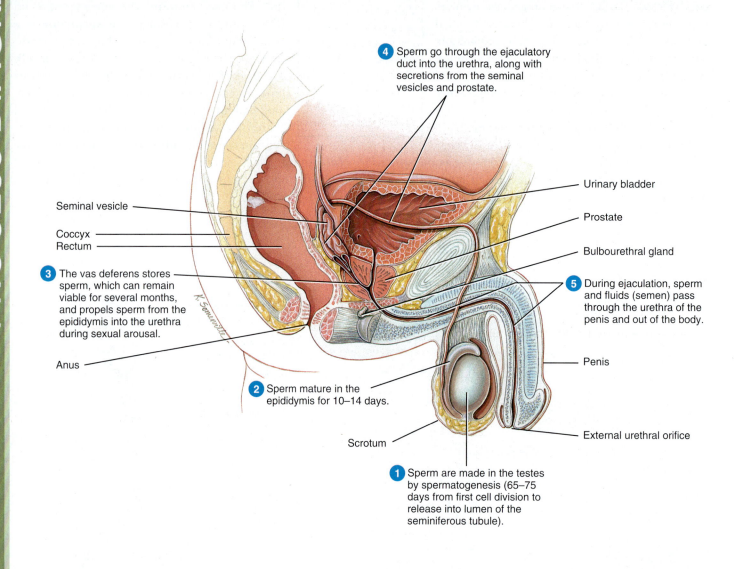

4 Sperm go through the ejaculatory duct into the urethra, along with secretions from the seminal vesicles and prostate.

Seminal vesicle

Coccyx

Rectum

3 The vas deferens stores sperm, which can remain viable for several months, and propels sperm from the epididymis into the urethra during sexual arousal.

Anus

2 Sperm mature in the epididymis for 10–14 days.

Scrotum

Urinary bladder

Prostate

Bulbourethral gland

5 During ejaculation, sperm and fluids (semen) pass through the urethra of the penis and out of the body.

Penis

External urethral orifice

1 Sperm are made in the testes by spermatogenesis (65–75 days from first cell division to release into lumen of the seminiferous tubule).

We can now trace the path of sperm through the male reproductive tract (**Figure 16.4**). Sperm are made in the testes and released into the lumen of the seminiferous tubule. They mature in the epididymis for 10 to 14 days. Sperm are stored in the vas deferens and can remain viable for months. The vas deferens also propels sperm from the epididymis into the urethra during sexual arousal. Upon ejaculation, sperm go through the ejaculatory duct into the urethra and mix with secretions from the seminal vesicles, prostate, and bulbourethral glands to form semen, as we saw earlier.

The volume of semen in a typical ejaculation is 2.5 to 5 milliliters, with 50 to 150 million sperm per milliliter. When the number falls below 20 million sperm per milliliter, the male is likely to be infertile. A very large number of sperm is required for fertilization because only a tiny fraction ever reaches the secondary oocyte (egg).

Semen has a slightly alkaline pH of 7.2 to 7.7, a milky white color, and a sticky texture. Semen also contains an antibiotic that can destroy certain bacteria and may help control naturally occurring bacteria in the semen and in the lower female reproductive tract.

Spermatogenesis and the production of the male hormones testosterone and **dihydrotestosterone (DHT)** are regulated by the anterior pituitary gland through the secretions of **luteinizing hormone (LH)** and **follicle-stimulating hormone (FSH)**; the latter is also called interstitial cell–stimulating hormone (ICSH).

Testosterone and dihydrotestosterone both bind to the same androgen receptors, producing several effects:

- *Prenatal development.* Before birth, testosterone stimulates the male pattern of development of reproductive system ducts and the descent of the testes. DHT stimulates the development of the external genitals.

- *Development of male sexual characteristics.* During puberty, the hormone testosterone stimulates further development of the sex organs. In addition, testosterone stimulates development of secondary sex characteristics, or changes in nonreproductive organs. These changes include increased hair growth (facial, axillary, chest, and pubic), enlargement of the larynx, increased oil gland secretion, and the skeletal and muscular development that leads to wide shoulders and narrow hips.

- *Development of sexual function.* Androgens contribute to male sexual behavior and spermatogenesis, as well as to sex drive in both males and females. Recall that the adrenal cortex is the main source of androgens in females.

- *Stimulation of anabolism.* Androgens are anabolic hormones; that is, they stimulate protein synthesis. This effect is obvious in the heavier muscle and bone mass of most men as compared to women.

FSH and testosterone act together to stimulate spermatogenesis. Once the degree of spermatogenesis required for male reproductive function has been achieved, Sertoli cells release **inhibin**, a hormone named for its inhibition of FSH secretion by the anterior pituitary. Testosterone and inhibin, which are both produced in the testes, affect the anterior pituitary gland via negative feedback (**Figure 16.5**).

As we look back to the couple discussed in the chapter opener, we can see that we've discussed topics that help explain the possible causes of infertility for the man. As we've seen, counts below 20 million sperm per millimeter in the semen would be considered low. High testicular temperature can also harm sperm. Finally, we've seen that sperm production is regulated by hormones, so abnormal hormone levels might lead to male infertility.

Now let's take a look at the female reproductive system.

Hormonal control in the male • Figure 16.5

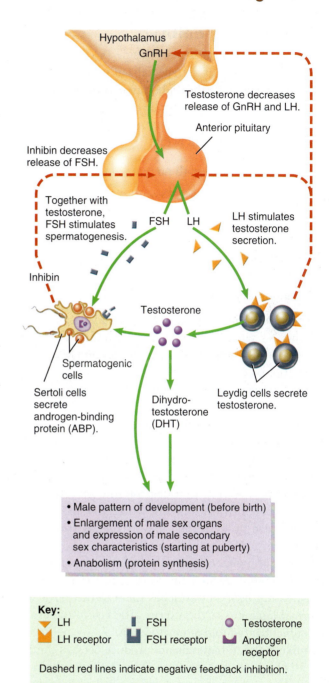

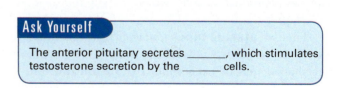

Ask Yourself

The anterior pituitary secretes _____, which stimulates testosterone secretion by the _____ cells.

Female Reproductive Organs Provide the Site of Fertilization and Nurture the Developing Embryo

Because fertilization takes place inside the female's body, her reproductive organs are specialized to make gametes, receive the sperm from the male, serve as the site of fertilization, and house and nurture the developing embryo and later fetus until birth. The organs of the female reproductive system include the following.

- The ovaries, which are the female gonads.
- The uterine (fallopian) tubes, or oviducts.
- The uterus.
- The vagina.
- External organs, collectively called the vulva or pudendum.
- The mammary glands.

Around the time of puberty, the female begins to exhibit female secondary sex characteristics, changes regulated by organs to nonreproductive organs and features. In females, the secondary sex characteristics include development of breasts, growth of pubic and axillary hair, and a rotation of the pelvis to open the birth canal.

Females make gametes called egg cells, or *oocytes*, in the **ovaries** (**Figure 16.6a**) through a process called *oogenesis*. The ovaries are paired organs that produce secondary oocytes and hormones. As we will see soon, the secondary oocytes are cells that develop into mature ova, or eggs, following fertilization. The hormones produced by the ovaries are progesterone and estrogens (the female sex hormones), inhibin, and relaxin.

A series of ligaments hold the ovaries in position. The ovarian ligament anchors the ovaries to the uterus, and the suspensory ligament attaches them to the pelvic wall. The broad ligament of the uterus attaches to the ovaries by a double-layered fold of peritoneum.

Unlike males, females do not produce oocytes continuously; all of her oocytes are formed by the time a baby girl is born. The oocytes do not mature until after puberty; then they further develop—usually one at a time—in a monthly reproductive cycle, which we will discuss later in the chapter.

The two **uterine tubes** (*fallopian tubes*) extend laterally from the uterus toward the ovaries. The open, funnel shaped end of each tube, the **infundibulum**, lies close to the ovary but is open to the pelvic cavity. The ovaries release the mature, or secondary, oocytes into the uterine tubes where fertilization usually takes place. Fingerlike projections called **fimbriae** (FIM-brē-ē) catch the egg released from the ovary and direct it into the tube. The uterine tubes also transport the secondary oocyte to the uterus.

The **uterus** serves as part of the pathway for sperm deposited in the vagina to reach the uterine tubes. It is also the site of implantation of a fertilized ovum, development of a fetus during pregnancy, and labor. During reproductive cycles when implantation does not occur, the uterus is the source of menstrual flow. The uterus is situated between the urinary bladder and the rectum and is shaped like an inverted pear.

Part of the uterus includes the dome-shaped portion superior to the uterine tubules called the **fundus**, the tapering central portion called the **body**, and the narrow portion opening into the vagina called the **cervix**. The interior of the body of the uterus is called the **uterine cavity**. (**Figure 16.6b**).

The middle muscular layer of the uterus, the **myometrium**, consists of smooth muscle and forms the bulk of the uterine wall. During childbirth, coordinated contractions of uterine muscles help expel the fetus.

The innermost part of the uterine wall, the **endometrium**, is a mucous membrane. It nourishes a growing fetus or is shed each month during menstruation if fertilization does not occur. The endometrium contains many endometrial glands whose secretions nourish sperm and the zygote.

The **vagina** is a tubular canal that extends from the exterior of the body to the uterine cervix. It is the receptacle for the penis and semen during sexual intercourse, the outlet for menstrual flow, and the passageway for childbirth. The vagina is situated between the urinary bladder and the rectum.

The mucosa of the vagina contain large stores of glycogen. When glygogen decomposes, it produces organic acids, which retard microbial growth but are also harmful to sperm. Alkaline components of semen, mostly from the seminal vesicles, neutralize the acidity of the vagina and increase the viability of sperm. The muscular layer of the vagina is composed of smooth muscle that can stretch to receive the penis during intercourse and allow for childbirth.

A thin fold of mucous membrane called the **hymen** may partially cover the vaginal orifice, the vaginal opening. The hymen usually breaks and bleeds during the first sexual intercourse, but it can also break when participating in sports or other activities.

a. Sagittal section showing female reproductive organs

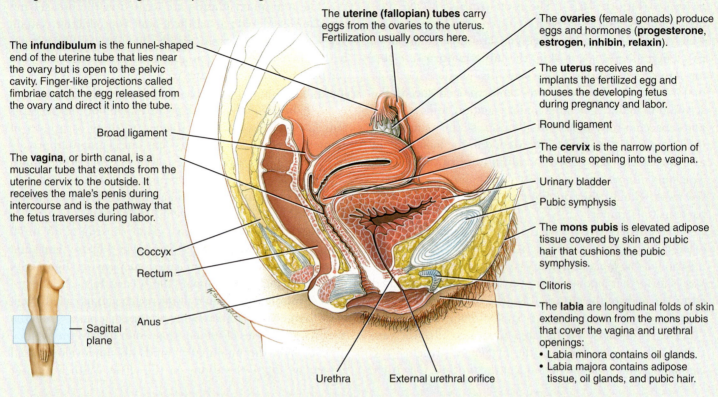

The **infundibulum** is the funnel-shaped end of the uterine tube that lies near the ovary but is open to the pelvic cavity. Finger-like projections called fimbriae catch the egg released from the ovary and direct it into the tube.

Broad ligament

The **vagina**, or birth canal, is a muscular tube that extends from the uterine cervix to the outside. It receives the male's penis during intercourse and is the pathway that the fetus traverses during labor.

Coccyx

Rectum

Sagittal plane

Anus

The **uterine (fallopian) tubes** carry eggs from the ovaries to the uterus. Fertilization usually occurs here.

Urethra

External urethral orifice

The **ovaries** (female gonads) produce eggs and hormones (**progesterone, estrogen, inhibin, relaxin**).

The **uterus** receives and implants the fertilized egg and houses the developing fetus during pregnancy and labor.

Round ligament

The **cervix** is the narrow portion of the uterus opening into the vagina.

Urinary bladder

Pubic symphysis

The **mons pubis** is elevated adipose tissue covered by skin and pubic hair that cushions the pubic symphysis.

Clitoris

The **labia** are longitudinal folds of skin extending down from the mons pubis that cover the vagina and urethral openings:
• Labia minora contains oil glands.
• Labia majora contains adipose tissue, oil glands, and pubic hair.

b. Posterior view of the female reproductive organs

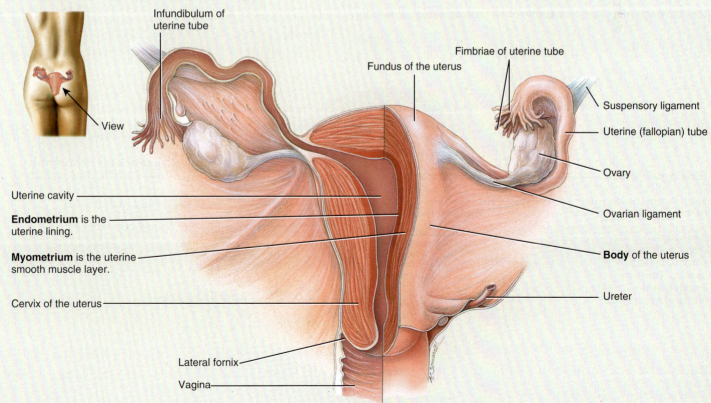

Infundibulum of uterine tube

Fundus of the uterus

Fimbriae of uterine tube

View

Uterine cavity

Endometrium is the uterine lining.

Myometrium is the uterine smooth muscle layer.

Cervix of the uterus

Suspensory ligament

Uterine (fallopian) tube

Ovary

Ovarian ligament

Body of the uterus

Ureter

Lateral fornix

Vagina

The *perineum* is the diamond-shaped area between the thighs and buttocks of both males and females that contains the external genitals and anus.

The female external genitals are called the *vulva* or *pudendum* (**Figure 16.7**). The **mons pubis** (MONZ PŪ-bis) is an elevation of adipose tissue covered by pubic hair, which cushions the pubic symphysis. From the mons pubis, two longitudinal folds of skin, the **labia majora** (LĀ-be-a ma-JO-ra), extend down and back. In females, the labia majora develop from the same embryonic tissue as the scrotum in males. The labia majora contain adipose tissue and sebaceous (oil) and sweat glands. Like the mons pubis, they are covered by pubic hair. Medial to the labia majora are two folds of skin called the **labia minora** (mi-NŌ-ra). The labia minora do not contain pubic hair or fat and have few sweat glands; they do, however, contain numerous sebaceous (oil) glands.

The **clitoris** (KLIT-o-ris) is a small, cylindrical mass of erectile tissue and nerves. It is located at the interior junction of the labia minora. A layer of skin called the *prepuce* is formed at a point where the labia minora unite and cover the body of the clitoris. Like the penis, the clitoris is capable of enlargement upon sexual stimulation.

The region between the labia minora is called the **vestibule**. This region contains the hymen (if present), the vaginal opening, and the opening of the urethra. On either side of the vaginal orifice are the *greater vestibular glands*, which produce a small quantity of mucus during sexual arousal and intercourse that adds to cervical mucus and provides lubrication.

The **mammary glands** are located in the breasts and are also considered part of the female reproductive system. They are modified sweat glands that develop during puberty and change during pregnancy to produce milk to feed the baby after birth (**Figure 16.8**). We will discuss development, birth, delivery, and lactation later in the chapter.

The breasts lie over the pectoralis major and serratus anterior muscles and are attached to them by a layer of connective tissue. Each breast has one pigmented projection, the nipple, with a series of closely spaced openings where milk emerges. The circular pigmented area surrounding the skin is called the areola (a-RE-o-la). This region contains modified sebaceous (oil) glands.

Internally, each mammary gland consists of 15 to 20 lobes arranged radially and separated by adipose tissue and strands of connective tissue called **suspensory ligaments of the breast**, which support the breast. In each lobe are smaller **lobules**, in which milk-secreting glands called alveoli are found. When milk is being produced, it passes from the alveoli into a series of tubules that drain toward the nipple.

At birth, the mammary glands are undeveloped and appear as slight elevations on the chest. With the onset

Components of the vulva • Figure 16.7

Like the penis, the clitoris is capable of erection on sexual stimulation.

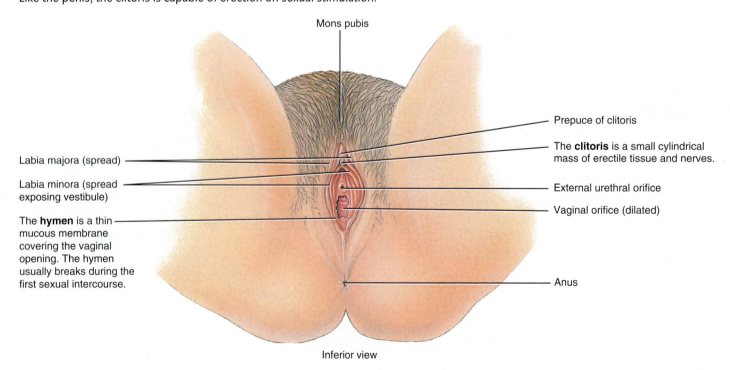

Mons pubis

Prepuce of clitoris

The **clitoris** is a small cylindrical mass of erectile tissue and nerves.

Labia majora (spread)

Labia minora (spread exposing vestibule)

External urethral orifice

Vaginal orifice (dilated)

The **hymen** is a thin mucous membrane covering the vaginal opening. The hymen usually breaks during the first sexual intercourse.

Anus

Inferior view

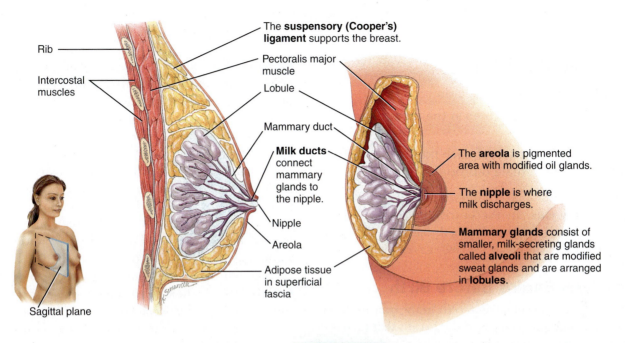

Rib

Intercostal muscles

The **suspensory (Cooper's) ligament** supports the breast.

Pectoralis major muscle

Lobule

Mammary duct

Milk ducts connect mammary glands to the nipple.

Nipple

Areola

Adipose tissue in superficial fascia

Sagittal plane

The **areola** is pigmented area with modified oil glands.

The **nipple** is where milk discharges.

Mammary glands consist of smaller, milk-secreting glands called **alveoli** that are modified sweat glands and are arranged in **lobules**.

Put It Together

Which of the following parts of the breast contains pigments and oil glands?

a. nipple
b. areola
c. mammary glands
d. adipose tissue

of puberty, under the influence of estrogens and progesterone, the female breasts begin to develop. The duct system matures and fat is deposited, which increases breast size. The areola and nipple also enlarge and become more darkly pigmented.

In sum, the functions of the female reproductive system are as follows.

- The ovaries produce secondary oocytes and hormones, including estrogens, progesterone, inhibin, and relaxin.
- The uterine tubes transport a secondary oocyte to the uterus, and these tubes are normally the site where fertilization occurs.

- The uterus is the site of implantation of a fertilized ovum, development of the fetus during pregnancy, and labor.
- The vagina receives the penis during sexual intercourse and is a passageway for childbirth.
- The mammary glands synthesize, secrete, and eject milk for nourishment of the newborn.

Now that we have examined the female reproductive system, we can understand the common causes of female infertility discussed in the chapter opener: a lack of egg release, abnormal pathways for the sperm and egg, abnormal hormone levels, and infections (especially sexually transmitted diseases).

CONCEPT CHECK

1. **What** are the male's accessory sex organs, and what do they do?

2. **How** do Sertoli cells in the testes nurture and protect the developing sperm cells?

3. **What** does the hormone testosterone do?

4. **What** role does the uterus play in reproduction?

The Female Reproductive Cycle Shows That Timing Is Everything

LEARNING OBJECTIVES

1. **Describe** how eggs are made.
2. **Explain** the major events in the female reproductive cycle.
3. **Identify** the function of each hormone associated with the female reproductive cycle.

 emales make gametes called egg cells, or **oocytes**, in the ovaries through a process called **oogenesis**. Let's look at the ovary more closely to see how eggs develop.

Oogenesis Begins Before Birth

During early fetal development, cells in the ovaries differentiate into **oogonia**, which give rise to cells that turn into **primary oocytes**. These cells begin meiosis during fetal development but do not complete it until puberty. At birth, 200,000 to 2,000,000 primary oocytes remain in each ovary. Of these, about 40,000 remain at puberty, but only 400 go on to mature and ovulate during a woman's reproductive lifetime. The others degenerate. In other words, females have all the eggs they will ever have by birth.

The surface of the ovary is covered by the **germinal epithelium**. Within the ovarian cortex, ovaries contain many small sac-like structures called **ovarian follicles**. Each follicle consists of an oocyte and a number of supporting cells that nourish the developing oocyte and begin to secrete estrogens as the follicle grows larger. The primordial follicles grow and differentiate in various stages within the ovary (**Figure 16.9**).

The follicle enlarges until it is a **mature (graafian) follicle**, a large, fluid-filled follicle that is preparing to rupture and expel a secondary oocyte. At this stage, the developing egg is ready to be ejected from the ovary, a process called **ovulation**.

The remnants of an ovulated follicle develop into a **corpus luteum** (plural is *lutea*). If no pregnancy occurs, the cor-

PROCESS DIAGRAM

Development of ovarian follicles • Figure 16.9

THE PLANNER ✓

The arrows within the ovary indicate the developmental stages that occur during maturation of an ovum during the ovarian cycle. (The stages shown occur sequentially in the same place on the ovary during a single monthly cycle but are drawn in different areas on the ovary for demonstration purposes.)

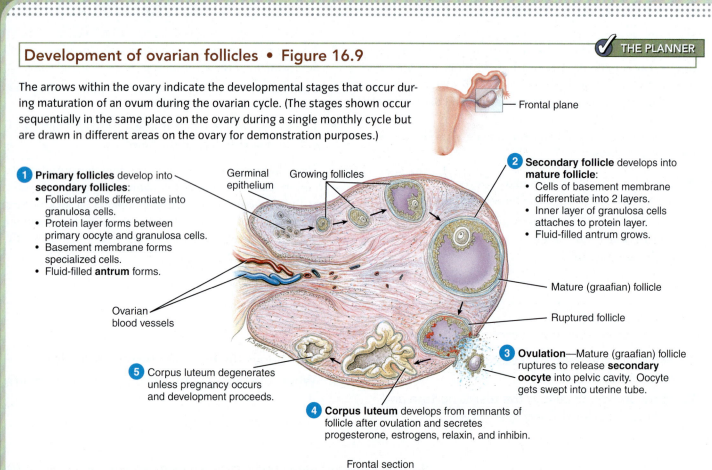

1 **Primary follicles** develop into **secondary follicles**:
- Follicular cells differentiate into granulosa cells.
- Protein layer forms between primary oocyte and granulosa cells.
- Basement membrane forms specialized cells.
- Fluid-filled **antrum** forms.

Germinal epithelium

Growing follicles

Frontal plane

2 **Secondary follicle** develops into **mature follicle**:
- Cells of basement membrane differentiate into 2 layers.
- Inner layer of granulosa cells attaches to protein layer.
- Fluid-filled antrum grows.

Mature (graafian) follicle

Ruptured follicle

3 **Ovulation**—Mature (graafian) follicle ruptures to release **secondary oocyte** into pelvic cavity. Oocyte gets swept into uterine tube.

Ovarian blood vessels

5 Corpus luteum degenerates unless pregnancy occurs and development proceeds.

4 **Corpus luteum** develops from remnants of follicle after ovulation and secretes progesterone, estrogens, relaxin, and inhibin.

Frontal section

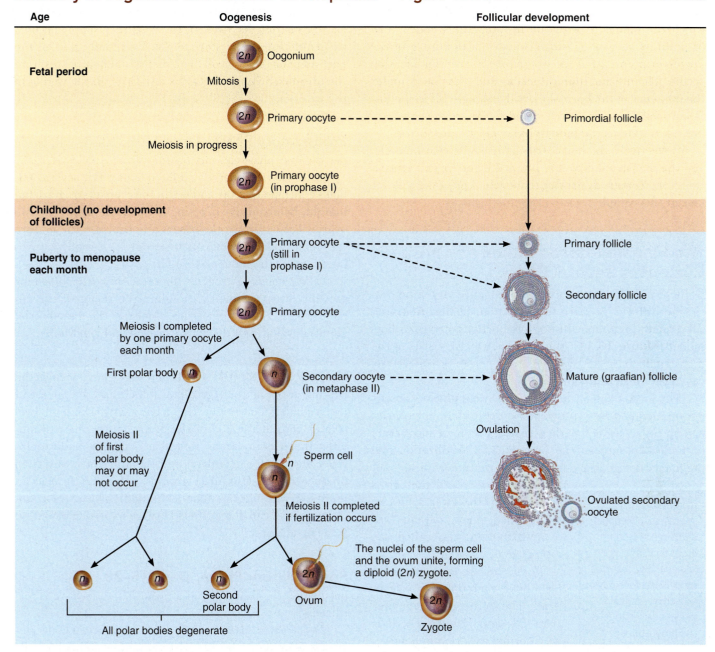

Age	Oogenesis	Follicular development

Fetal period

2n Oogonium

Mitosis

2n Primary oocyte - - - - - - - - - - -> Primordial follicle

Meiosis in progress

2n Primary oocyte (in prophase I)

Childhood (no development of follicles)

Puberty to menopause each month

2n Primary oocyte (still in prophase I) - - - -> Primary follicle

2n Primary oocyte

Secondary follicle

Meiosis I completed by one primary oocyte each month

First polar body n

n Secondary oocyte (in metaphase II) - - - -> Mature (graafian) follicle

Meiosis II of first polar body may or may not occur

Ovulation

n Sperm cell

Meiosis II completed if fertilization occurs

Ovulated secondary oocyte

n   n   n   The nuclei of the sperm cell and the ovum unite, forming a diploid (2n) zygote.

Second polar body   2n Ovum

All polar bodies degenerate

2n Zygote

pus luteum produces progesterone, estrogens, relaxin, and inhibin until it degenerates and turns into fibrous tissue.

The series of events in which a follicle develops in the ovary, releases a secondary oocyte during ovulation, and becomes a corpus luteum is called the **ovarian cycle**. This ovarian cycle occurs about once per month during the female's reproductive years. Hormones from the hypothalamus (gonadotropin-releasing hormone, GnRH) and the anterior pituitary gland (luteinizing hormone, LH) and follicle-stimulating hormone, FSH) drive the cycle.

Like spermatogenesis, oogenesis involves meiosis (**Figure 16.10**), a form of sexual reproduction in which each gamete receives half the normal number of chromosomes

(n). One of the cells produced from meiosis is the **secondary oocyte**; the other is called a **polar body** and is essentially a packet of discarded nuclear material. Once a secondary oocyte is formed, it begins meiosis II and then stops. These events take place in the mature (graafian) follicle. At ovulation, the follicle will rupture and release the secondary oocyte, which has undergone only the first half of meiosis.

If the secondary oocyte is not fertilized, it deteriorates and never completes meiosis. If the secondary oocyte is penetrated by a sperm in the uterine tubes, however, it will complete meiosis II and produce another polar body and the **ovum**. The ovum and sperm combine their chromosomes in the fertilized egg, which is the first cell of the new offspring.

The developing follicles and corpus lutea also secrete hormones (estrogens, progesterone, relaxin, inhibin), which communicate with the hypothalamus and anterior pituitary, as well as evoke changes in the lining of the uterus in preparation for pregnancy in case fertilization occurs. The uterine lining thickens and transforms under the influence of the hormones, and it sloughs off when hormone levels dwindle. These cyclical changes in the uterus are referred to as the **menstrual cycle**, or *uterine cycle*.

Collectively, the hormonal changes, the ovarian cycle, and the menstrual cycle are lumped together in a process called the **female reproductive cycle**. Let's take a closer look at this complex series of events.

The Female Reproductive Cycle Has Several Phases

In the first half of the female reproductive cycle, the ovaries and uterus prepare for ovulation, fertilization, and subsequent pregnancy through the actions of various hormones (**Figure 16.11**); this portion, which consists of the menstrual phase and preovulatory phase, is collectively called the **follicular phase**.

The second half of the cycle (the **luteal phase**) consists of ovulation and the postovulatory phase. Here, the ovaries and uterus act as though the female is pregnant and secrete hormones to support the pregnancy. If fertilization and pregnancy do occur, the luteal phase continues throughout pregnancy. However, if the egg is not fertilized, the luteal phase is programmed to end and give way to **menstruation**, also called the *menstrual phase*, or *menses*.

Let's look at each phase of the female reproductive cycle in detail. The entire cycle can take about 28 days, though there is some normal variation in length.

> **menstruation** (men'-stroo-Ā-shun) Periodic discharge of blood, tissue fluid, mucus, and epithelial cells that usually lasts for 5 days; caused by a sudden reduction in estrogens and progesterone.

Menstrual Phase, Days 1–5. By convention, the first day of menstruation begins the cycle because it is the most outwardly visible sign.

In the ovaries: several ovarian follicles grow and enlarge.

In the uterus: menstrual flow from the uterus consists of 50 to 150 mL of blood and tissue cells from the endome-trium. This discharge occurs because the declining level of progesterones and estrogens causes the uterine arteries to constrict. As a result, the cells they supply become oxygen-deprived and start to die. Eventually, part of the endometrium sloughs off. The menstrual flow passes from the uterine cavity to the cervix and through the vagina to the exterior.

Preovulatory Phase, Days 6–13. The length of the preovulatory phase can vary the most from female to female.

In the ovary: under the influence of FSH, several follicles continue to grow and begin to secrete estrogens and inhibin. By about day 6, a single follicle in one of the two ovaries has outgrown the others and becomes the mature (graafian) follicle, which enlarges until it is ready for ovulation under the influence of LH.

In the uterus: estrogens liberated into the blood by growing ovarian follicles stimulate the repair of the endometrium. The endometrium thickens, the endometrial glands develop, and the arterioles coil and lengthen.

Ovulation, Day 14. Ovulation, as we have seen, is the rupture of the mature (graafian) follicle and the release of the secondary oocyte into the pelvic cavity. The high level of estrogens at the end of the preovulatory phase stimulates the hypothalamus to release more gonadotropin-releasing hormone (GnRH) and the anterior pituitary to produce more LH. GnRH promotes the release of even more LH, which brings about ovulation. An over-the-counter home test can detect this LH surge to predict ovulation a day in advance.

Postovulatory Phase, Days 15–28. Events in the postovulatory phase depend on whether the oocyte is fertilized.

In one ovary: after ovulation, the mature follicle collapses. Stimulated by LH, the remaining follicular cells enlarge and form the corpus luteum, which secretes progesterone, estrogens, relaxin, and inhibin.

If the oocyte is not fertilized, the corpus luteum lasts for two weeks, after which it degenerates. As the hormone levels decrease, release of GnRH, FSH, and LH rise due to loss of negative feedback. Follicular growth resumes and a new ovarian cycle begins.

> **Put It Together**
>
> (for Figure 16.11) One method of contraception involves administration of hormones to "trick" the body into thinking that a pregnancy has already occurred. As a result, the ovary will not produce additional follicles for ovulation. Which of the following hormones would you need to include in a pill in order to "fool" the female body into thinking that it is pregnant?
> a. LH
> b. progesterone
> c. GnRH
> d. FSH

The female reproductive cycle • Figure 16.11

The female reproductive cycle can last from 24 to 36 days; the average is about 28 days. The preovulatory phase is more variable than other phases. In this diagram, fertilization and implantation of the embryo have not occurred. The inset shows the positive feedback that occurs to produce the midcycle LH surge.

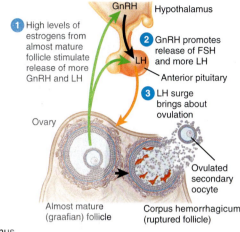

1 High levels of estrogens from almost mature follicle stimulate release of more GnRH and LH

GnRH Hypothalamus

2 GnRH promotes release of FSH and more LH

Anterior pituitary

3 LH surge brings about ovulation

Ovary

Ovulated secondary oocyte

Almost mature (graafian) follicle

Corpus hemorrhagicum (ruptured follicle)

a. Hormonal regulation of changes in the ovary and uterus

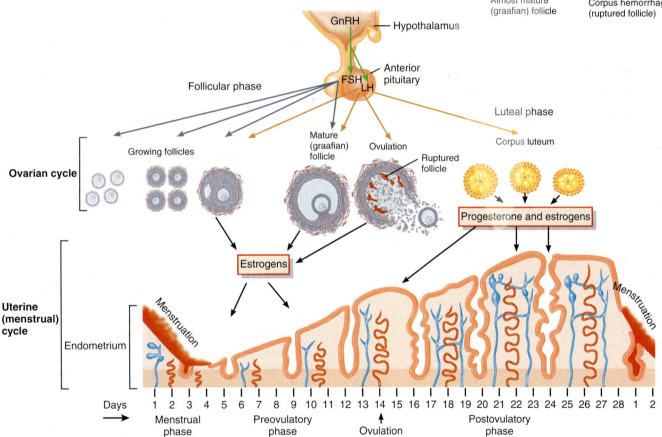

GnRH — Hypothalamus

Follicular phase

FSH LH — Anterior pituitary

Luteal phase

Mature (graafian) follicle

Ovulation

Corpus luteum

Ovarian cycle

Growing follicles

Ruptured follicle

Progesterone and estrogens

Estrogens

Uterine (menstrual) cycle

Menstruation

Endometrium

Menstruation

Days → 1 2 3 4 5 6 7 8 9 10 11 12 13 14 15 16 17 18 19 20 21 22 23 24 25 26 27 28 1 2

Menstrual phase

Preovulatory phase

Ovulation

Postovulatory phase

b. Changes in concentration of anterior pituitary and ovarian hormones

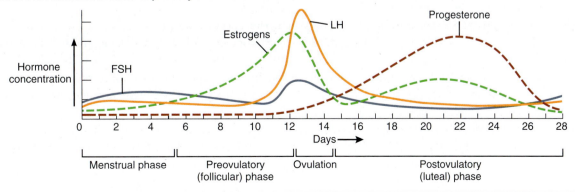

Hormone concentration

FSH

Estrogens

LH

Progesterone

0 2 4 6 8 10 12 14 16 18 20 22 24 26 28

Days →

Menstrual phase

Preovulatory (follicular) phase

Ovulation

Postovulatory (luteal) phase

If the secondary oocyte is fertilized and begins to divide, the corpus luteum does not degenerate; it is rescued by human chorionic gonadotropin (hCG), a hormone produced by the embryo beginning about eight days after fertilization. Like LH, hCG stimulates the secretory activity of the corpus luteum. The presence of hCG in maternal blood or urine is an indicator of pregnancy, and hCG is the hormone detected by home pregnancy tests.

In the uterus. Progesterone and estrogens produced by the corpus luteum promote growth of the endometrial glands, which begin to secrete glycogen. The hormones also promote thickening of the endometrium. These changes peak about one week after ovulation, at the time a fertilized ovum might arrive at the uterus.

CONCEPT CHECK

1. **What** is the difference between an oocyte and a follicle?
2. **What** happens during the preovulatory phase of the female reproductive cycle?
3. **What** is the function of follicle-stimulating hormone (FSH)?

Fertilization Requires the Egg and Sperm to Get Very Close to One Another

LEARNING OBJECTIVES

1. **Explain** why timing is so critical for successful fertilization.
2. **List** the different categories of contraceptives and explain how they work.

From a biological standpoint, the purpose of sexual intercourse is to deliver sperm in close proximity to the secondary oocyte so that fertilization can take place.

During sexual arousal or excitement, there is altered blood flow to the genitals and increased glandular secretions; these responses produce an erect penis in the male for insertion into the female and prepare the female's vagina to receive the penis. Once aroused, the male inserts his penis into the female's vagina.

Fertilization Must Occur While Both the Egg and Sperm Are Still Viable

After ovulation, the secondary oocyte remains functional for only about 24 hours; contrast this with sperm, which can remain active for about 48 hours after being delivered to the female. For a woman to become pregnant, the timing of sexual intercourse must occur within a narrow window of time during the female reproductive cycle. Sperm require nearly 6 hours to journey from their site of delivery to the egg that is waiting in the uterine tube. Therefore, sperm delivered up to 2 days before or as late as 18 hours after ovulation will have the best opportunity to fertilize the egg.

Some over-the-counter ovulation tests can detect high levels of LH during the midcycle surge and can predict ovulation about 1 day in advance. If sensitive thermometers called *basal body temperature thermometers* are used daily, a woman can also estimate when she is ovulating. Ovulation is usually associated with about a half-degree increase in body temperature that lasts throughout the luteal phase. If a woman wishes to become pregnant, she can increase her sexual activity at this time. If pregnancy is not desired, intercourse can be avoided during this period, or other contraceptive measures can be used.

Once semen is delivered into the vagina near the cervix of the uterus during sexual intercourse, it coagulates. After 10 to 20 minutes, prostate-specific antigen (PSA) and other enzymes in semen cause it to re-liquefy, and prostaglandins in semen increase sperm motility. Sperm must swim from the vagina through the cervix and uterus to reach the egg in the uterine tube. Depending on the timing of the female's reproductive cycle, the mucus within the cervix may present the first obstacle. During the ovulatory or early postovulatory phase, cervical mucus is thin and watery, making it relatively easy to penetrate. During the preovulatory or late postovulatory phase, the cervical mucus is thick, which makes it more difficult for sperm to navigate.

Once past the cervix, sperm swim along the walls of the uterus into the uterine tubes. Contractions of the walls of the uterus aid the sperm in their passage.

capacitation (ka'-pas-i-TĀ-shun) The functional changes that sperm undergo in the female reproductive tract that allow them to fertilize a secondary oocyte.

At the same time, secretions from the female reproductive tract cause sperm to become activated or undergo **capacitation**. Sperm motility increases further, and the acrosome—a vesicle containing enzymes that aid penetration—is stripped of cholesterol and glycoproteins as the sperm rubs against the walls of the uterus and uterine tube; both of these events make it easier for the sperm to fuse with the secondary oocyte. The enzymes of the acrosome digest through the corona radiata cells that surround the egg, allowing physical contact between the membranes of the egg and sperm. Despite the fact that as many as 500,000,000 sperm could be present in a single ejaculate, a mere 100 of these sperm locate the egg, and only 1 gets to fertilize it.

To prevent fertilization, several contraceptive methods can be used during or before intercourse. Let's take a closer look at some of the options.

Contraceptive Methods Interrupt Different Stages of the Fertilization Process

contraception (kon'-tra-SEP-shun) The prevention of fertilization or impregnation without the destruction of fertility.

There are several strategies for **contraception**. These methods accomplish their common goal by interrupting different stages of the fertilization process. Some contraceptive methods are more successful than others, and we'll discuss their effectiveness rates based on typical use.

- *Abstinence.* Total abstinence, or the avoidance of sexual intercourse, is the only method of preventing pregnancy that is 100% reliable.
- *Sterilization.* Blocking sperm or eggs from moving through their respective tubes. Sterilization is a procedure that renders an individual incapable of reproduction. Sterilization methods are very effective at preventing pregnancy, with a failure rate of less than 1%. The most common means of sterilization in males is *vasectomy*, in which a portion of each vas deferens is removed. Even though sperm production continues in the testes, sperm can no longer reach the exterior. Instead, they degenerate and are destroyed by phagocytosis. Blood testosterone level is normal, so a vasectomy has no effect on sexual desire or performance. Sterilization in females most often is achieved by *tubal ligation*, in which both uterine tubes

are tied closed and then cut. As a result, the secondary oocyte cannot pass through the uterine tubes, and sperm cannot reach the oocyte.

- *Hormonal methods.* Preventing ovulation. Aside from total abstinence or surgical sterilization, hormonal methods are the most effective means of birth control. In typical use, they may be 97% effective. Used by 50 million women worldwide, *oral contraceptives* ("the pill") contain various mixtures of synthetic estrogens and progestins (chemicals with actions similar to those of progesterone). They prevent pregnancy mainly by negative feedback inhibition of anterior pituitary secretios of FSH and LH. The low levels of these hormones usually prevent development of a dominant follicle.
- *Barrier methods.* Preventing sperm from accessing the uterus and uterine tubes. Barrier methods are effective 80% to 85% of the time in typical use. Barrier methods include a condom, a vaginal pouch, or a diaphragm. A *condom* is a nonporous, latex covering placed over the penis that prevents deposition of sperm in the female reproductive tract. A *vaginal pouch* is made of two flexible rings connected to a polyurethane sheath. The pouch is fitted within the vagina, from the external genitals to the cervix. A *diaphragm* is a rubber, dome-shaped structure that fits over the cervix and is used in conjunction with a spermicide.
- *Spermicides.* Killing sperm. Spermicides are about 75% effective in typical use. Spermicides come in various foams, creams, suppositories, and douches that contain sperm-killing agents. A spermicide is more effective when used with a condom or a diaphragm.
- *Intrauterine devices.* Preventing implantation of a fertilized egg into the uterus. An IUD is a small object made of plastic, copper, or stainless steel that is inserted into the cavity of the uterus. IUDs cause changes in the uterine lining that prevent implantation of a fertilized ovum and are about 99% effective.

Of the various contraceptive methods, sterilizations (for example, vasectomy, tubal ligation) are generally irreversible; however, expensive surgical techniques are available to attempt to reverse these procedures. Hormonal methods carry risks of heart attack and stroke, especially in female smokers and those with histories of heart disease and hypertension.

While some contraceptive methods (condoms, total abstinence) help prevent the spread of **sexually transmitted diseases (STDs)**, most methods do not. STDs can be caused by bacteria, viruses, protozoa, and fungi.

Some of the STDs caused by bacteria include chlamydia, gonorrhea, and syphilis. Chlamydia usually does not have any symptoms and may cause urethritis in males and can lead to pelvic inflammatory disease in females. The symptoms of gonorrhea include urethritis with excess pus discharge; the disease may be asymptomatic in females, leading to sterility. Syphilis begins with a primary stage which results in a painless open sore. Symptoms of the secondary stage include rash, fever, and joint pain. In the tertiary stage, the organs begin to deteriorate. All three STDs are treatable with antibiotics, except for the third stage of syphilis.

STDs caused by viruses include genital herpes and genital warts. The symptoms of genital herpes include painful blisters on the external genitals of males and females, with possible internal blistering in females. The symptoms of genital warts include cauliflower growths on the external genital area and internal growths in females; they can also appear on or around the anus. Both genital herpes and genital warts are incurable. Outbreaks of genital herpes can be controlled with anti-inflammatory drugs, and genital warts can be removed cryogenically.

Table 16.1 summarizes some of the common contraceptive methods as well as their effectiveness. Research is ongoing for more effective contraceptives; see *What a Health Provider Sees* for one example.

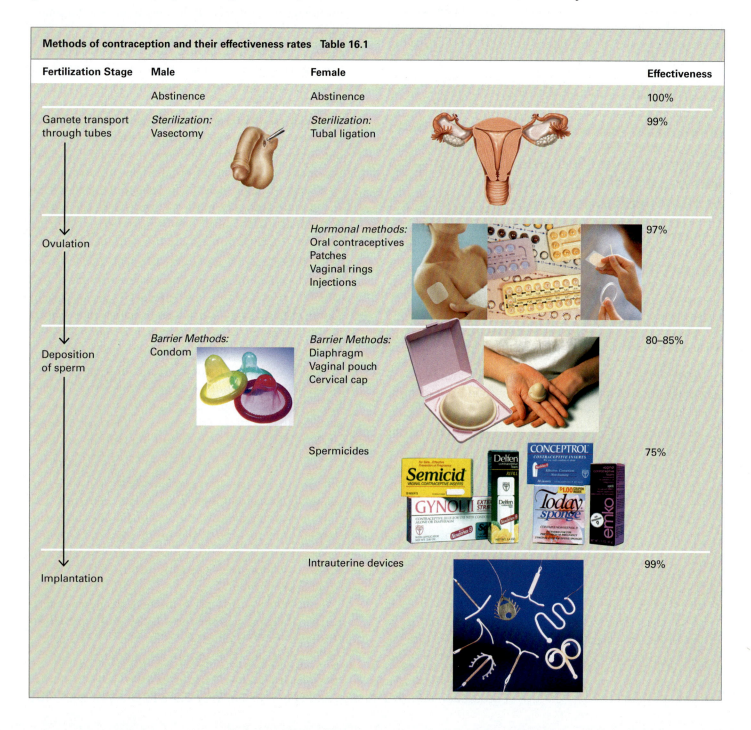

Methods of contraception and their effectiveness rates   Table 16.1

Fertilization Stage	Male	Female		Effectiveness
	Abstinence	Abstinence		100%
Gamete transport through tubes	*Sterilization:* Vasectomy	*Sterilization:* Tubal ligation		99%
Ovulation		*Hormonal methods:* Oral contraceptives Patches Vaginal rings Injections		97%
Deposition of sperm	*Barrier Methods:* Condom	*Barrier Methods:* Diaphragm Vaginal pouch Cervical cap		80–85%
		Spermicides		75%
Implantation		Intrauterine devices		99%

WHAT A HEALTH PROVIDER SEES

The Male Birth Control Pill

Birth control pills first became available for females in the 1960s. The basis of birth control pills was to use various combinations of estrogens and progesterone to "fool" the body into thinking that it is pregnant. The hormones in birth control pills suppress follicle development and ovulation without altering sexual desire. If it works so well for women, why is there no male birth control pill? There have been several attempts to produce such a form of male contraception, but none has achieved commercial success.

One strategy is to use high doses of testosterone to suppress spermatogenesis. While high testosterone levels can successfully suppress spermatogenesis and reduce secretion of GnRH, LH, and FSH through negative feedback, they can also suppress sexual desire and even cause sterility. Another strategy has been to use high concentrations of inhibin to suppress FSH secretion, with the goal of controlling spermatogenesis without affecting the LH and testosterone secretions that control sex drive. However, this approach has also met with little success.

In 2000, a new hormone called **gonadotropin-inhibiting hormone (GnIH)** was found in birds; it suppresses the secretion of GnRH. Recently, GnIH was found in the human hypothalamus and was shown to suppress GnRH in neurons within the laboratory. Research into this potential new male birth control method is currently under way.

Think Critically 1. How could the suppression of GnRH secretion suppress spermatogenesis?
2. Given what you have learned about male reproductive hormones, how could you invent a form of male birth control that inhibits spermatogenesis without altering the male sex drive?

CONCEPT CHECK STOP

1. **Why** would a couple have trouble conceiving if the sperm were delivered to the female 24 hours after ovulation?

2. **How** do intrauterine devices (IUDs) work?

Pregnancy Lasts from Fertilization to Delivery

LEARNING OBJECTIVES

1. **Explain** the major events in embryonic and fetal development.

2. **Describe** the differences between fetal circulation and normal human circulation.

3. **Identify** the hormones associated with pregnancy and their functions.

From the time fertilization occurs to delivery of a baby, a typical pregnancy lasts about 38 weeks. However, the due date is calculated from the first day of the mother's last menstrual period—a documentable time of the cycle—which occurred 2 weeks prior to fertilization. During this time of development, the fertilized egg divides and grows from 1 cell to approximately 1 trillion cells. The cells differentiate into tissues and organs as the embryo develops a human shape and appearance.

Most of the changes occur within the *embryonic period*, which lasts from fertilization through 8 weeks. After 8 weeks, the embryo is considered to be a fetus. By the early part (8 to 12 weeks) of the *fetal period*, the organs have formed and differentiated. Growth is the major event during the remainder of the fetal period; no new changes in shape or organs occur, but the length of the fetus increases 12-fold, with significantly greater changes in weight. By about 33 weeks, the fetus is *viable* (can live outside the mother), so the physiological events of labor can begin and the baby can be born.

Let's take a closer look at the events of embryonic development.

During the Embryonic Phase, the Embryo Changes from a Single Cell to a Differentiated Organism

> **zygote** (ZĪ-g-ōt) The single cell resulting from the union of male and female gametes; the fertilized ovum.

The first step in development is fertilization, which occurs in the uterine tubes (**Figure 16.12**). The fertilized egg is now called a **zygote** and contains 46 chromosomes: 23 from the ovum and 23 from the sperm.

Week 1 of development. In the first two days, the zygote immediately undergoes rapid mitotic cell divisions of the zygote called **cleavage**, forming numerous cells that resemble the original cell. (Refer to Chapter 3 for a discussion of mitosis.) The progressively smaller cells produced by cleavage are called **blastomeres**. Through successive divisions within the first week, it becomes a **morula**. The morula is surrounded by the zona pellucida, a clear glycoprotein layer. By the end of the fourth day, the number of cells in the morula increases as it continues to move through the uterine tube toward the uterine cavity.

> **morula** (MOR-ū-la) A solid sphere of cells produced by successive cleavages of a fertilized ovum about 4 days after fertilization.

When the morula enters the uterine cavity on day 4 or 5, glycogen-rich secretion from the glands of the endometrium penetrate the morula and reorganize the cells around a large, fluid-filled cavity called a **blastocyst cavity**. The developing mass is now called a **blastocyst**. Though it has hundreds of cells, the blastocyst is about the same size as the original zygote. The blastocyst remains free within the uterine cavity for 1–2 days before it attaches to the uterine wall.

> **blastocyst** (BLAS-tō-sist) In the development of an embryo, a hollow ball of cells that consists of a blastocyst cavity (the internal cavity), a trophoblast (outer cells), and an inner cell mass.

The blastocyst begin to rearrange into two distinct structures: the **inner cell mass** is located internally and eventually develops into the embryo while the **trophoblast** (TROF-o-blast) is an outer superficial layer of cells that forms the wall of the blastomere. It will ultimately develop into the fetal portion of the placenta, the site of exchange of nutrients and wastes between the mother and fetus.

However, the blastocyst is not safe until it nestles into the uterine lining, about day 6 after fertilization, during the second week, in a process called **implantation** (see Figure 16.12). As the blastocyst implants, the inner cell mass is oriented toward the endometrium.

Events associated with the first week of development • Figure 16.12

The first week of development is characterized by cleavage of the zygote into a morula and then into a more complex blastocyst by successive cell divisions. During this change, the embryo travels from the uterine tube into the uterus.

- Blastomeres
- Zona pellucida

- Nucleus
- Cytoplasm

3 Morula (3–4 days postfertilization)
- Successive cell divisions of the blastomeres produce a solid sphere of cells called a **morula**.
- The morula is surrounded by the zona pellucida and is the same size as the original zygote.

2 Cleavage (1–2 days postfertilization)
- The zygote divides into two smaller cells called blastomeres.
- Each blastomere divides again to produce a four-celled **embryo**.

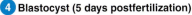

- Inner cell mass
- Blastocyst cavity
- Trophoblast

4 Blastocyst (5 days postfertilization)
- When the morula enters the uterine cavity, glycogen-rich fluid secretions from the endometrium cause the cells to shift around, forming a fluid-filled cavity (**blastocyst cavity**). This hollow sphere of cells is called a **blastocyst**.
- The **inner cell mass** becomes the growing embryo.
- The outer cell layer, called the **trophoblast**, will become the fetal portion of the placenta, where nutrients and wastes will be exchanged with the mother.

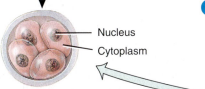

- Sperm cell

PATH OF SPERM CELL:

- Corona radiata
- Zona pellucida
- Plasma membrane of secondary oocyte
- Cytoplasm of secondary oocyte

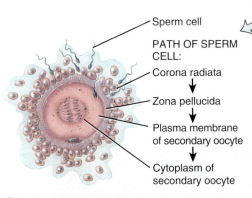

1 Fertilization (12–24 hours postovulation)
- Many sperm bind to the secondary oocyte.
- The enzymes of the sperm's acrosome digest through the **corona radiata** (the innermost layer of granulosa cells that is attached to the zona pellucida around a secondary oocyte).
- Once one sperm penetrates and touches the egg's membrane, the zona pellucida changes chemically and presents a barrier to other sperm cells.
- When the sperm enters the secondary oocyte, meiosis II is triggered.
- The nuclei of the sperm and oocyte fuse to produce a single **zygote**.

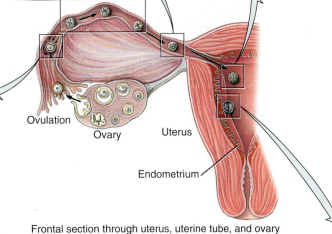

- Ovulation
- Ovary
- Uterus
- Endometrium

Frontal section through uterus, uterine tube, and ovary

WILEY PLUS | NATIONAL GEOGRAPHIC | Video

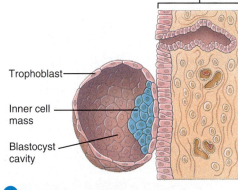

Endometrium of uterus

- Trophoblast
- Inner cell mass
- Blastocyst cavity

5 Implantation (6 days postfertilization)
- After floating free for 2 days, the blastocyst attaches to the uterine wall (implantation).
- The blastocyst's inner cell mass orients toward the endometrium.

Ask Yourself

1. The _____ is the embryonic stage that implants into the uterus.
2. At implantation, the inner cell mass of the blastocyst is oriented _____ the endometrium.

Week 2 of development. Successful implantation provides a safe environment for the embryo, which will spend the next week penetrating the endometrium to make connections to the uterine blood supply to obtain nutrients, exchange gases, and eliminate wastes. These connections are made through the formation of *chorionic villi*, which are finger-like projections from the blastocyst to the tissues of the mother's uterus (**Figure 16.13a**). Oxygen and nutrients in the mother's blood diffuse across the cell membranes into the capillaries of the chorionic villi. Waste products such as carbon dioxide diffuse in the opposite direction.

Also at this time, the embryo begins secreting the hormone **human chorionic gonadotropin (hCG)**; hCG prevents destruction of the corpus luteum in the ovary, allowing continued secretion of the estrogens and proges-

> **gastrulation** (gas'-troo-LĀ-shun) The migration of groups of cells from the epiblast (primitive ectoderm) that transform a bilaminar embryonic disc into a trilaminar embryonic disc consisting of the three primary germ layers; transformation of the blastula into the gastrula.
>
> **germ layers** The three major embryonic tissues (ectoderm, mesoderm, and endoderm) from which the various tissues and organs of the body develop.

terone that maintain the mother's body in a pregnant state (that is, maintaining the uterine lining and suppressing menses).

Embryonic Germ Layers Differentiate to Form the Various Body Components

Week 3 of development. Beginning in the third week, the embryo undergoes many changes in shape and complexity. Cells within the flat, two-layered embryonic disc migrate inward to become a three-layered structure in a process called **gastrulation**. The three layers of cells are called the **germ layers** and consist of the ectoderm, mesoderm, and endoderm. These layers later differentiate into various organs.

Connections between the embryo and the mother • Figure 16.13

a. Formation of chorionic villi

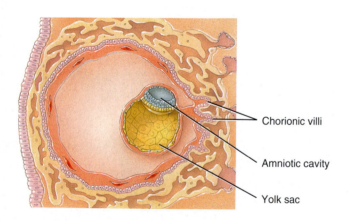

Chorionic villi

Amniotic cavity

Yolk sac

Frontal section through uterus showing blastocyst, about 13 days after fertilization

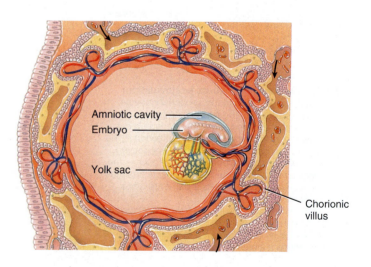

Amniotic cavity
Embryo

Yolk sac

Chorionic villus

Frontal section through uterus showing embryo and vascular supply, about 21 days after fertilization

- The **ectoderm** (EK-tō-derm) is the primary germ layer that gives rise to the nervous system and the epidermis of skin and its derivatives.

- The **mesoderm** (MĒS-ō-derm) is the middle primary germ layer that gives rise to connective tissues, blood and blood vessels, and muscles.

- The **endoderm** (EN-dō-derm) is the primary germ layer of the developing embryo, which gives rise to the gastrointestinal tract, urinary bladder, urethra, and respiratory tract.

Following gastrulation, at about 22 to 24 days, cells of the mesoderm form a cylinder called a **notochord**. The notochord induces the overlying ectoderm to form a neural tube that eventually becomes the nervous system; this process is called **neurulation**.

> **notochord** (NŌ-tō-cord) A flexible rod of mesodermal tissue that helps form part of the backbone and intervertebral discs.
>
> **neurulation** (noor-oo-LĀ-shun) The process by which the neural plate, neural folds, and neural tube form.

As the embryo develops and changes into a fetus, this maternal–fetal connection becomes more specialized as the **placenta** and **umbilical cord** (**Figure 16.13b**). The placenta is the site of the exchange of nutrients between the mother and fetus. As with the chorionic villi, nutrients and oxygen diffuse across the capillaries; the blood of the mother and fetus does not mix. The actual connection between the placenta and embryo (and later the fetus) is the *umbilical cord*.

Once the placenta has formed, the embryonic body is surrounded by a thin membrane called the *amnion*. Eventually, the amnion surrounds the entire embryo and is filled with *amniotic fluid*, which serves as a shock absorber for the fetus, helps regulate body temperature, helps prevent drying out, and prevents adhesions between the skin of the fetus and surrounding tissues.

b. Placenta and umbilical cord

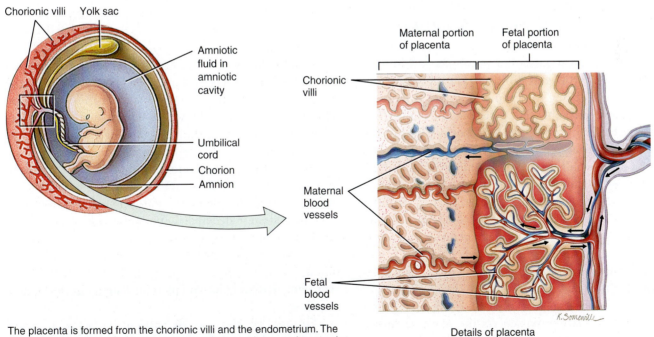

The placenta is formed from the chorionic villi and the endometrium. The chorion portion of the placenta exchanges gases, nutrients, and wastes by diffusion between the adjacent blood vessels of the mother and the fetus. When fully developed, the chorion is pancake-shaped and connects to the fetus through the umbilical cord. The amnion portion of the placenta forms a protective, fluid-filled sac around the baby. The placenta is shed as the afterbirth once the baby is born.

Details of placenta and umbilical cord

Weeks 4–8 of development. During the fourth week after fertilization, the embryo undergoes dramatic changes in shape and size, nearly tripling its size. All major organs appear from weeks 4 through 8 of embryonic development.

By 5 weeks of development, the embryo has expressed neither male nor female reproductive structures. Instead, it has precursors to both systems. Some of these precursor structures develop exclusively into male organs; some develop only into female organs; and others can become either male or female, depending on the hormonal influence. If the embryo is male (that is, has a Y chromosome), it will begin to secrete testosterone and form testes and male reproductive organs and external genitalia by 7 to 8 weeks; the female precursor organs degenerate. In the absence of testosterone (if the embryo has only X chromosomes) the high level of estrogens from the mother's body will stimulate the male precursor organs to degenerate and the female organs to develop.

By the end of the eighth week, all major body systems have begun to develop, although their functions are minimal. The embryo now has clearly human characteristics, with ears, eyes, arms and legs, even fingers and toes. The embryo also has a four-chambered heart, and the external genitals begin to differentiate. The embryonic period now ends, and the fetal period begins.

The Fetal Period Is Devoted to Growth and Refinement of Body Structures

During the fetal period, which lasts from week 9 until birth, tissues and organs that developed during the embryonic period grow and differentiate. Very few new structures appear during the fetal period, but the rate of body growth is remarkable. For example, during the last two-and-one-half months of the fetal period, half of the full-term weight is added. At the beginning of the fetal period, the head is half the length of the body. By the end of the fetal period, however, the head size is one-quarter the length of the body. During the same period, the fetal limbs increase in size from one-eighth to one-half the fetal length. By birth, both the external genitals and internal reproductive organs have developed. The fetus is also less vulnerable to the damaging effects of drugs, radiation, and microbes than it was as an embryo.

The fetus obtains its nourishment and oxygen and eliminates wastes and carbon dioxide via the placenta rather than through the exchange with the external environment. The lungs, kidneys, and gastrointestinal organs do not begin to function until after birth. Therefore, circulation of blood through the fetal vessels (**Figure 16.14**) is a bit different than in an individual after birth.

A significant portion of the fetal blood flows through vessels to and from the chorionic portion of the placenta. Minimal amounts of blood are sent through the pulmonary circulation as blood is routed away from the lungs through special vascular pathways: the foramen ovale and ductus arteriosus. This is because the lungs are not currently needed for ventilation (breathing) and external respiration.

The **foramen ovale** (fō-RĀ-men ō-val-ē) exists in the septum between the right and left atria. About one-third of the blood that enters the right atrium passes through the foramen ovale into the left atrium and joins the systemic circulation. The blood that does pass into the right ventricle is pumped into the pulmonary trunk, but little of this blood reaches the nonfunctioning fetal lungs. Instead, most is sent through the **ductus arteriosus** (ar-te-re-Ō-sus), a vessel that connects the pulmonary trunk with the aorta, so that most blood blood bypasses the fetal lungs. The blood in the aorta is carried to all fetal tissues through the systemic circulation. When the common iliac arteries branch into the external and internal iliacs, part of the blood flows into the internal iliacs, into the umbilical arteries, and back to the placenta for another exchange of materials.

At the time of birth, the umbilical cord is tied off and blood no longer flows through the umbilical arteries; they fill with connective tissue. The placental vessels seal off, along with the unusual vascular pathways that were in place to help bypass the pulmonary circulation. When an infant takes his or her first breath, the lungs expand and blood flow to the lungs increases. Blood returning from the lungs to the heart increases pressure in the left atrium. This closes the foramen ovale, and permanent closure occurs in about a year. Within a week after birth, the typical circulation pattern is usually established.

Many fetuses born prematurely at 26 to 29 weeks of development survive if given intensive care because their lungs can provide adequate ventilation and the central nervous system is developed enough to control breathing and body temperature. Fetuses 33 weeks and older usually survive if born prematurely. Even after birth, however, an infant is not completely developed. An additional year is required, especially for complete development of the nervous system.

Figures 16.15a and **16.15b** on the following pages show some photos from the embryonic and fetal periods and detail the changes taking place during these remarkable stages.

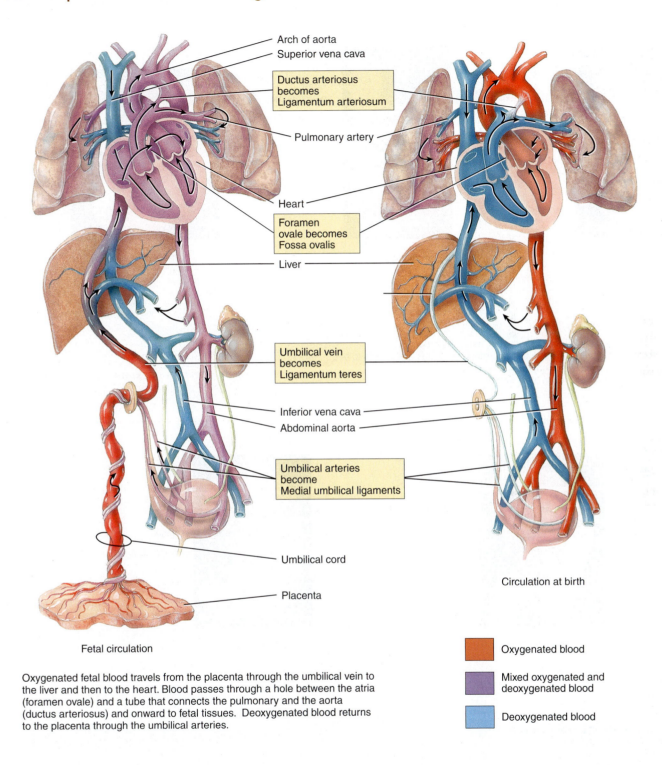

Arch of aorta
Superior vena cava

Ductus arteriosus
becomes
Ligamentum arteriosum

Pulmonary artery

Heart

Foramen
ovale becomes
Fossa ovalis

Liver

Umbilical vein
becomes
Ligamentum teres

Inferior vena cava
Abdominal aorta

Umbilical arteries
become
Medial umbilical ligaments

Umbilical cord

Placenta

Fetal circulation

Circulation at birth

Oxygenated blood

Mixed oxygenated and
deoxygenated blood

Deoxygenated blood

Oxygenated fetal blood travels from the placenta through the umbilical vein to the liver and then to the heart. Blood passes through a hole between the atria (foramen ovale) and a tube that connects the pulmonary and the aorta (ductus arteriosus) and onward to fetal tissues. Deoxygenated blood returns to the placenta through the umbilical arteries.

Summary of the changes during embryonic and fetal development • Figure 16.15

a. Embryonic period

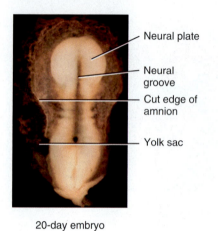

- Neural plate
- Neural groove
- Cut edge of amnion
- Yolk sac

20-day embryo

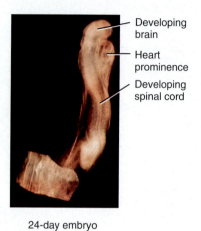

- Developing brain
- Heart prominence
- Developing spinal cord

24-day embryo

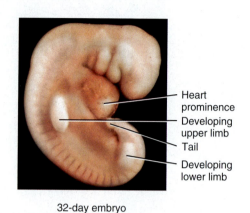

- Heart prominence
- Developing upper limb
- Tail
- Developing lower limb

32-day embryo

1–4 weeks

0.6 cm (3/16 in.),
weight minimal

Primary germ layers and notochord develop; neurulation occurs.
Blood vessel formation begins and blood forms; heart forms and begins to beat.
Chorionic villi develop and placental formation begins.
Eyes and ears begin to develop; body systems begin to form.

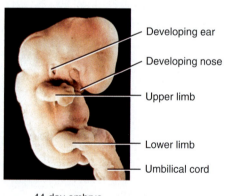

- Developing ear
- Developing nose
- Upper limb
- Lower limb
- Umbilical cord

44-day embryo

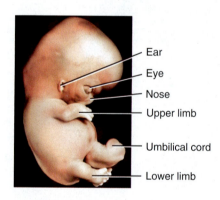

- Ear
- Eye
- Nose
- Upper limb
- Umbilical cord
- Lower limb

52-day embryo

5-8 weeks

3 cm (1.25 in.),
1 g (1/30 oz)

Limbs become distinct and digits appear.
Heart becomes four-chambered.
Face is more humanlike: Nose develops and is flat; the eyes are far apart and eyelids are fused.
Bone formation begins.
Blood cells start to form in liver.
External genitals begin to differentiate.
Tail disappears.
Major blood vessels form.
Internal organs continue to develop.

b. Fetal period

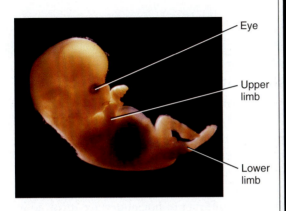

Ten-week fetus

Eye

Upper limb

Lower limb

9–12 weeks

7.5 cm (3 in.),
30 g (1 oz)

Head is about half the length of the body, and length nearly doubles.
Brain continues to enlarge.
Face is broad, with eyes fully developed, closed, and widely separated.
External ears develop and are low set.
Bone formation continues.
Upper limbs almost reach final relative length.
Gender is distinguishable from external genitals.
Red bone marrow, thymus, and spleen participate in blood cell formation.
Fetus begins to move, but movements cannot yet be felt by the mother.
Heartbeat can be detected.
Body systems continue to develop.

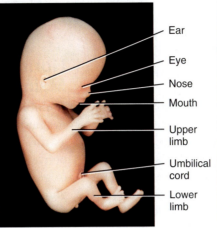

Thirteen-week fetus

Ear

Eye

Nose

Mouth

Upper limb

Umbilical cord

Lower limb

13–16 weeks

18 cm (6.5–7 in.),
100 g (4 oz)

Head is relatively smaller than rest of body.
Eyes and ears move to their final positions.
Lower limbs lengthen.
Fetus appears even more humanlike.
Rapid development of body systems occurs.

17–20 weeks

25–30 cm (10–12 in.),
200–450 g (0.5–1 lb)

Head is more proportionate to rest of body.
Eyebrows and head hair are visible.
Vernix caseosa (fatty secretions of oil glands and dead epithelial cells) and lanugo (delicate fetal hair) cover fetus.
Fetal movements are commonly felt by mother (quickening).
Growth slows but lower limbs continue to lengthen.

21–25 weeks

27–35 cm (11–14 in.),
550–800 g (1.25–1.5 lb)

Head becomes even more proportionate to rest of body.
Weight gain is substantial, and skin is pink and wrinkled.

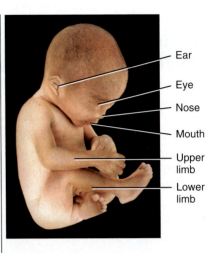

Twenty-six-week fetus

Ear

Eye

Nose

Mouth

Upper limb

Lower limb

26–29 weeks

32–42 cm (13–17 in.)
1110–1350 g (2.5–3 lb)

Head and body are more proportionate and eyes are open.
Toenails are visible.
Body fat is 3.5% of total body mass.
Testes begin to descend toward scrotum (28–32 weeks).
Red bone marrow is major site of blood cell production.

30–34 weeks

41–45 cm (16.5–18 in.),
2000–2300 g (4.5–5 lb)

Skin is pink and smooth.
Fetus assumes upside-down position.
Body fat is 8% of total body mass.

35–38 weeks

50 cm (20 in.),
3200–3400 g (7–7.5 lb)

Skin is usually bluish-pink, and growth slows as birth approaches.
Body fat is 16% of total body mass.
Testes are usually in scrotum in full-term male infants.

Birth

Even after birth, an infant is not completely developed; an additional year is required, especially for complete development of the nervous system.

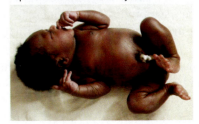

So far, we have focused our attention on the developmental changes in the embryo and fetus. Let's turn our attention to changes in the mother.

Pregnancy Changes the Mother's Physiology

Pregnancy brings about striking changes in the mother's anatomy as well as her physiology. The ability of the body to adapt to all these changes is truly remarkable. By about the end of the third month of pregnancy, the uterus occupies most of the pelvic cavity. As the fetus continues to grow, the uterus extends higher into the abdominal cavity.

By 38 weeks (that is, a full-term pregnancy), the fetus and uterus occupy most of the abdominal cavity (**Figure 16.16**). The fetus exerts pressure on the mother's diaphragm, liver, intestines, and stomach.

For some women, changes in the skin during pregnancy may include increased pigmentation around the eyes and cheekbones in a masklike pattern, in the areolae of the breasts, and in the lower abdomen. Stretch marks over the abdomen can occur as the uterus enlarges, and hair loss increases. Pregnancy-induced physiological changes include weight gain due to the fetus, amniotic fluid, the placenta, uterine enlargement, and increased total body water; increased storage of nutrients; marked breast enlargement in preparation for lactation; and lower back pain.

Several changes occur in the mother's cardiovascular system. Stroke volume increases by about 30% and cardiac output rises by 20–30% due to increased maternal blood flow to the placenta and increased metabolism. These increases are needed to meet the additional demands of the fetus for nutrients and oxygen.

Pulmonary function is also altered during pregnancy to meet the oxygen demands of the fetus. Total body oxygen consumption can increase by 10–20%. Difficult breathing also occurs as the expanding uterus pushes on the diaphragm.

Pregnant women also experience an increase in appetite. Pressure on the stomach may force the stomach contents into the esophagus, resulting in heartburn. A general decrease in gastrointestinal tract motility can cause constipation, delay gastric emptying time, and produce nausea, vomiting, and heartburn. Pressure on the urinary bladder by the enlarging uterus can produce urinary symptoms, such as increased frequency and urgency of urination, and stress incontinence.

Finally, changes in the reproductive system include edema and increased blood flow to the vagina. The uterus increases from its nonpregnant mass of 60–80 g (2–3 ounces) to 900–1200 g (2 pounds) at term because of increased numbers of muscle fibers in the uterus as well as their enlargement.

Let's turn our attention to the hormonal regulation of pregnancy.

Hormones Are Important for Maintaining the Pregnancy

A number of different hormones are released during pregnancy. The level of each of these hormones changes as the pregnancy progresses. There appears to be a link between some of these hormones and the development of gestational diabetes, which results in very high blood glucose levels in the mother (see *What a Health Provider Sees* in Chapter 9). A number of hormones are involved in pregnancy:

- **Human chorionic gonadotropin (hCG)** secreted by the chorion maintains the corpus luteum. hCG secretion rises and peaks by about the fourth month of pregnancy. By this time, the placenta is established, and the corpus luteum is no longer needed. This hormone is the material measured by a pregnancy test and is also the suspected cause of the morning sickness that many women experience during the first trimester of pregnancy.

- **Estrogens** and **progesterone** are secreted initially by the corpus luteum and later by the chorion, starting at 3 to 4 weeks, and continue to be secreted throughout pregnancy.

- **Relaxin** secreted by the corpus luteum and the placenta softens the pubic symphysis, relaxes sacroiliac ligaments, and dilates the cervix in preparation for labor.

- **Human chorionic somatomammotropin (hCS)**, (also called *human placental lactogen* [*hPL*]) is secreted by the chorion; secretion increases as the placenta grows, and peaks at about 32 weeks. hCS is thought to help prepare the breast tissue for milk production and lactation.

- **Corticotropin-releasing hormone (CRH)** is secreted by the placenta starting at 12 weeks, and secretion increases until the end of pregnancy. In nonpregnant women, this hormone is secreted only by the hypothalamus. CRH secretion by the placenta is thought to be important for timing birth.

Next, we'll look at what happens at the end of development, during the process of labor and delivery.

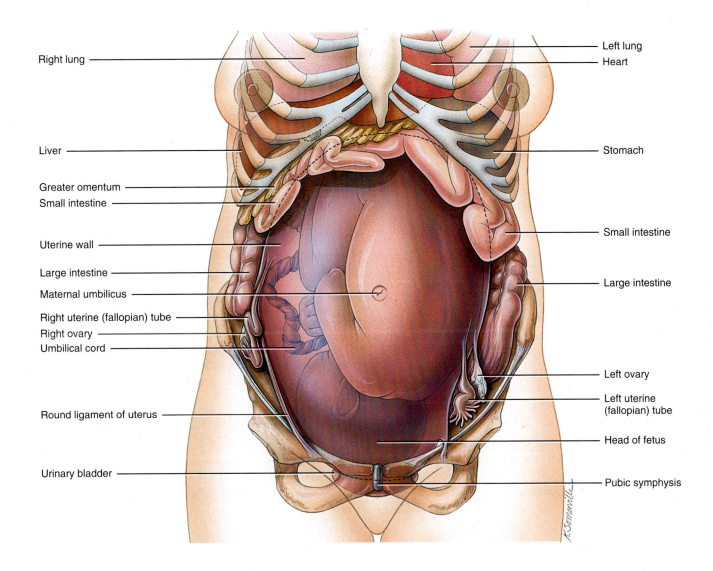

Right lung

Left lung

Heart

Liver

Stomach

Greater omentum

Small intestine

Uterine wall

Small intestine

Large intestine

Large intestine

Maternal umbilicus

Right uterine (fallopian) tube

Right ovary

Umbilical cord

Left ovary

Left uterine (fallopian) tube

Round ligament of uterus

Head of fetus

Urinary bladder

Pubic symphysis

Anterior view of position of organs at end of full-term pregnancy

---

CONCEPT CHECK **STOP**

1. **How** does gastrulation occur?

2. **How** is the fetal circulation different from the typical human circulation?

3. **What** is the function of the hormone relaxin?

Giving Birth Requires a Complex Series of Hormonal Changes

LEARNING OBJECTIVES

1. **Outline** the events of labor and delivery.
2. **Describe** the process of lactation.

After fertilization has taken place, the embryo has matured into a fetus, and the 38 weeks of pregnancy have passed, it is time for the baby to be born. Let's look at how the process of delivery of the baby is initiated and what brings it to an end.

Labor Has Several Stages

labor The process of giving birth in which a fetus is expelled from the uterus through the vagina.

prostaglandins (pros'-ta-GLAN-dinz) Membrane-associated lipids that are released in small quantities and act as a local hormone.

Near the end of pregnancy, several hormones come into play to prepare for and initiate **labor**. High levels of estrogens near the end of pregnancy override the inhibitory effects of progesterone on uterine contractions. Estrogens stimulate the placenta to secrete **prostaglandins**, which induce secretion of enzymes that digest collagen fibers in the cervix, causing it to soften. Estrogens also induce receptors for oxytocin in uterine muscle fibers. In addition, release of the hormone relaxin increases the flexibility of the pubic symphysis and helps dilate the cervix.

During labor, uterine contractions force the baby's head into the cervix, which stretches. The stretching induces more uterine contractions through a positive feedback mechanism involving secretion of oxytocin by the hypothalamus (**Figure 16.17a**).

Uterine contractions occur in waves that start at the top of the uterus and move downward, eventually expelling the fetus. True labor begins when uterine contractions occur at regular intervals, usually causing pain. As the interval between contractions shortens, the contractions intensify. Another symptom of true labor in some women is localization of pain in the back that is intensified by walking. The reliable indicator of true labor is dilation of the cervix and the "show," a discharge of blood-containing mucus that appears in the cervical canal during labor. In *false labor*, pain is felt in the abdomen at irregular intervals, but it does not intensify and walk-ing does not alter it significantly. There is no "show" and no cervical dilation.

Labor occurs in three stages:

1. *Stage of dilation.* The cervix dilates to provide room for the baby's head and body to pass. This stage typically lasts 6–12 hours, features regular contractions of the uterus, usually a rupturing of the amniotic sac, and complete dilation (to 10 cm or about 4 inches) of the cervix. If the amniotic sac does not rupture spontaneously, it is ruptured intentionally.

2. *Stage of expulsion.* The baby is expelled during this stage, which can take anywhere from 10 minutes to several hours from complete cervical dilation.

3. *Placental stage.* The placenta or "afterbirth" is expelled by powerful uterine contractions during this stage. These contractions also constrict blood vessels that were torn during delivery, thereby reducing the likelihood of hemorrhage. (**Figure 16.17b**). This stage may take 5 to 30 minutes or more after delivery.

The positive feedback mechanism ends at childbirth, or **parturition**.

parturition (par'-too-RISH-un) The act of giving birth to young; also called *childbirth* or *delivery*.

As a rule, labor lasts longer with first babies, typically about 14 hours. For women who have previously given **birth**, labor may last an average of 8 hours—although the time varies enormously among births.

Usually, the fetus is positioned with its head down toward the vagina. A heads-up or sideways orientation is called a **breech** position. If the fetus is in the breech position, the obstetrician may try to turn it manually prior to birth or may elect to deliver the baby surgically using an operation called a cesarean section (or C-section).

A premature baby is one who weighs less than 2500 g (5.5 lb) at birth. Causes include poor prenatal care, drug abuse, history of a previous premature delivery, and mother's age (below 16 or over 35). The body of a premature infant is not yet ready to sustain some critical functions, and most likely needs medical intervention to survive. The major problem after delivery of an infant under 36 weeks of gestation is respiratory distress syndrome (RDS), caused by insufficient surfactant. RDS can be eased by use of artificial surfactant and a ventilator that delivers oxygen until the infant's lungs can operate on their own.

a. Positive feedback mechanism that governs uterine contractions during labor

b. Stages of labor

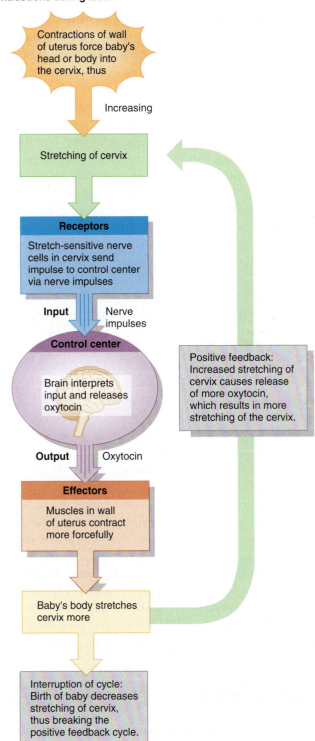

Contractions of wall of uterus force baby's head or body into the cervix, thus

Increasing

Stretching of cervix

Receptors
Stretch-sensitive nerve cells in cervix send impulse to control center via nerve impulses

Input Nerve impulses

Control center
Brain interprets input and releases oxytocin

Positive feedback: Increased stretching of cervix causes release of more oxytocin, which results in more stretching of the cervix.

Output Oxytocin

Effectors
Muscles in wall of uterus contract more forcefully

Baby's body stretches cervix more

Interruption of cycle: Birth of baby decreases stretching of cervix, thus breaking the positive feedback cycle.

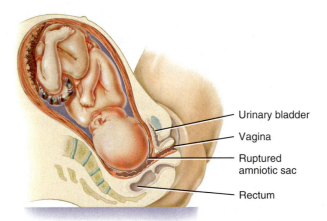

Urinary bladder

Vagina

Ruptured amniotic sac

Rectum

Stage 1. Dilation
• Extends from onset of labor to complete dilation of cervix (6-12 h)
• Regular uterine contractions
• Ruptured amniotic sac
• Completed dilation of cervix (10 cm)

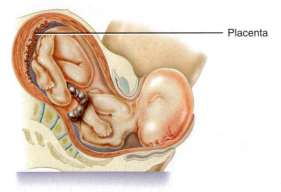

Placenta

Stage 2. Expulsion extends from complete dilation of cervix to delivery of the baby (10 min to several hours).

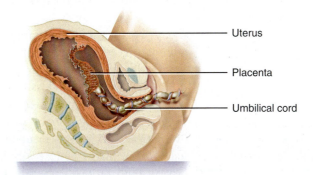

Uterus

Placenta

Umbilical cord

Stage 3. Placental stage extends from the time of delivery of the baby to the delivery of the placenta. (5-30 minutes or more). Uterine contractions constrict blood vessels torn during delivery and expel the placenta.

Giving Birth Requires a Complex Series of Hormonal Changes **499**

About 7% of pregnant women do not deliver by 2 weeks after their due date. Post-term pregnancies carry an increased risk of brain damage to the fetus, and even fetal death, due to inadequate supplies of oxygen and nutrients from an aging placenta. Delivery may be facilitated by inducing labor, initiated by administration of oxytocin, or by surgical delivery through a cesarean section.

After the delivery of the baby and placenta, the mother's reproductive organs and physiology return to their pre-pregnancy state. This process usually takes about 6 weeks. Following delivery, the baby can feed on its mother's milk. Let's take a closer look at this process.

Lactation Provides Food for the Newborn

Milk production by mammary glands is maintained by the hormone **prolactin (PRL)**, which is secreted by the anterior pituitary. Although prolactin stimulates milk production, the mammary glands do not produce milk during pregnancy because of inhibition by high levels of estrogens and progesterone. When levels of these hormones decrease after birth, the mammary glands can produce milk.

> **Lactation** (lak-TĀ-shun) The secretion and ejection of milk by the mammary glands.

Lactation occurs via a positive feedback mechanism involving the hormone oxytocin. The reflex involves the baby's suckling action on the mother's nipples, the hypothalamus, the anterior pituitary, and the mammary glands (**Figure 16.18**). Oxytocin causes the release of milk into the mammary ducts. Milk formed by the glandular cells of the breasts is stored until the baby begins active suckling. Stimulation of touch receptors in the nipple initiate sensory nerve impulses that are relayed to the hypothalamus. In response, secretion of oxytocin from the posterior pituitary increases.

During the first few days postpartum, the mammary glands secrete a nutrient-rich fluid called **colostrum**. Compared to milk, colostrum has less lactose and no fat; however, it is easily digested and helps the baby survive until true milk comes in about 4 days after birth. Colostrum has a very high protein content. The primary proteins are antibodies, made by the immune system and released into the colostrum, which the baby is able to absorb intact during the first few days of life.

> **colostrum** (kō-LOS-trum) A thin, cloudy fluid secreted by the mammary glands a few days and after delivery, before true milk is produced.

These antibodies are the baby's main source of adaptive immunity (see Chapter 12) until its own immune system begins to manufacture additional antibodies several weeks or months after birth.

Frequent suckling by the baby (8 to 10 times per day) can block ovarian cycles during the first few months after birth; however, this is not a reliable form of birth control. The effect is inconsistent, and ovulation commonly precedes the first menstrual period after the delivery of a baby. As a result, the mother can never be certain she is not fertile.

The primary benefit of breastfeeding is nutritional: Human milk is a sterile solution that contains fatty acids, lactose, amino acids, minerals, vitamins, and water that are ideal for the baby's digestion, brain development, and growth. As already noted, breast milk contains maternal antibodies, which help protect the baby from infection. Breast milk also contains some immune cells, including T lymphocytes and macrophages, which attack invading microbes directly. In fact, health care providers in both industrialized and developing countries have long observed that breast-fed babies contract fewer infections than babies fed with formula. Breast milk is more nutritious and beneficial to the baby than commercial formulas.

The mother also receives benefits from breastfeeding her baby. Breastfeeding helps her lose weight gained during the pregnancy much more quickly than if she does not breastfeed. In addition, she is likely to experience less postpartum bleeding, and her uterus returns to its pre-pregnancy size more rapidly. There is even some evidence that breastfeeding may reduce the risk of breast and reproductive organ cancers.

Years before oxytocin was discovered, it was common practice in midwifery to let a first-born twin nurse at the mother's breast to speed the birth of the second child. Now we know why this practice is helpful—it stimulates the release of oxytocin. Even after a single birth, nursing promotes expulsion of the placenta and helps the uterus return to its normal size. Synthetic oxytocin is often given not only to induce labor but also to increase uterine tone and control hemorrhage just after giving birth.

CONCEPT CHECK

1. **What** happens during the first stage of labor?
2. **How** does the baby's suckling stimulate milk ejection?

A positive feedback mechanism governs lactation.

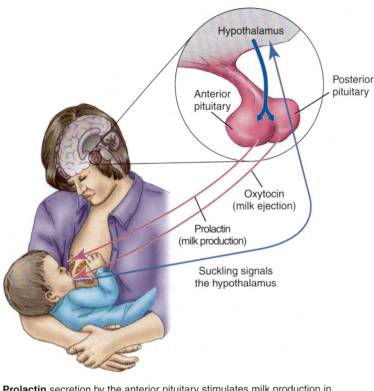

Prolactin secretion by the anterior pituitary stimulates milk production in the mammary glands. Milk ejection or let-down begins when the baby suckles on the mother's nipple. The stretch receptors stimulate the hypothalamus and posterior pituitary to release **oxytocin**, which stimulates milk ejection by the mammary gland.

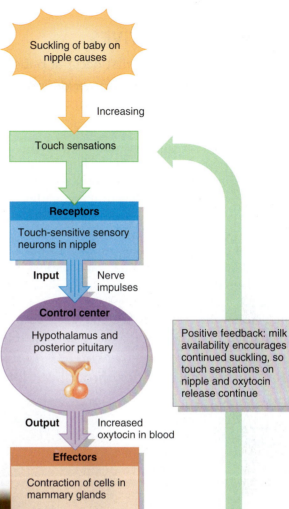

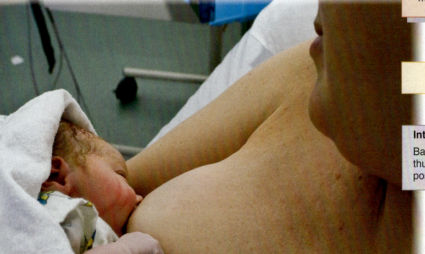

Put It Together A non-pregnant woman begins secreting breast milk. Which of the following might be the most likely cause of this symptom?
a. Over-stimulation of the nipple area by tight clothing
b. Pituitary tumor that causes decreased oxytocin levels
c. Pituitary tumor that causes increased prolactin secretion
d. Trauma to the posterior pituitary gland

Aging Alters Reproductive Capacity

LEARNING OBJECTIVES

1. **Describe** the changes that occur at puberty.
2. **Identify** the changes that occur during menopause.
3. **Explain** the changes that occur in the reproductive systems of older men.

At birth, humans have all their reproductive organs, but these organs are inactive because the hypothalamic–pituitary–gonadal axis does not yet function (see Chapter 9). When we reach about 10 years of age, hormonal changes occur that cause us to start **puberty** and become sexually functional (**Figure 16.19**). Females begin going through their reproductive cycle and have their first menses, called **menarche** (me-NAR-ke), and males begin producing sperm. Both genders develop secondary

> **puberty** (PŪ-ber-tē) The time of life during which the secondary sex characteristics begin to appear and the capability for sexual reproduction is possible; usually occurs between ages 10 and 17.

sex characteristics and are capable of reproduction.

The onset of puberty is marked by bursts of LH and FSH secretion, each triggered by a burst of GnRH. The cause of the GnRH burst is not yet clear, but a role for the hormone leptin is starting to unfold. Just before puberty, leptin levels rise in proportion to adipose tissue mass. Leptin may signal the hypothalamus that long-term energy stores (triglycerides in adipose tissue) are adequate for reproductive functions to begin.

Reproductive capabilities throughout life • Figure 16.19

Childhood

(Birth to age 10)

- The reproductive system is in a juvenile state and does not function.
- Neither male spermatogenesis nor female reproductive cycles occur.

Puberty

(about ages 10-14)

- Bursts of GnRH from the hypothalamus stimulate bursts of LH and FSH from the pituitary.
- Secondary sex characteristics appear in both genders.
- Female reproductive cycles begin and girls have their first menses (menarche).
- Male spermatogenesis begins.
- Both genders are capable of reproduction.

Young and early adulthood

(ages 15-40)

- Both genders are fully capable of reproduction.

Females capable of reproduction

Males capable of reproduction

From puberty through middle age, both genders are fully capable of reproduction. Upon reaching middle age, females undergo **menopause**; the female reproductive cycles diminish, and the women can no longer reproduce. Between the ages of 40 and 50 the pool of remaining ovarian follicles becomes exhausted. As a result, the ovaries become less responsive to hormonal stimulation. The production of estrogens declines, despite copious secretions of FSH and LH by the anterior pituitary. Many women experience hot flashes and heavy sweating, which coincide with bursts of GnRH release. Other symptoms of menopause are headache, hair loss, muscular pains, vaginal dryness, insomnia, depression, weight gain, and mood swings. Some atrophy of the reproductive organs occurs in postmenopausal women. Because of the loss of estrogens, most women also experience a decline in bone mineral density after menopause. Sexual desire does not decline, however; it may be maintained by adrenal androgens.

menopause (MEN-ō-pawz) The termination of the menstrual cycle.

In contrast, males can continue to reproduce well into old age (80s or 90s), but the number of viable sperm diminishes. At about age 55, a decline in testosterone synthesis leads to reduced muscle strength, fewer viable sperm, and decreased sexual desire. Although males are capable of reproduction at older ages, many older males develop reproductive system issues because of the age-related changes of their cardiovascular systems. A poorly functioning heart or blocked vascular flow will not allow adequate blood to be routed to the reproductive system for erection.

As for diseases of the reproductive systems, the risk of uterine cancer peaks at about 65 years of age for women. Cervical cancer is more common in younger women. For men, enlargement of the prostate occurs in approximately one-third of all males over age 60.

CONCEPT CHECK — STOP

1. **What** signals the occurrence of puberty in the female?
2. **What** is menopause?
3. **What** reproductive changes occur in men over age 55?

Middle age

(ages 40–late 60s)

- In females, the number of viable follicles becomes exhausted, and the production of estrogens declines (FSH, LH, and GnRH levels are high). Menopause (the cessation of menses) begins.
- Most healthy men are still capable of reproduction, but some men may have a decline in testosterone production and decreased sexual desire.

Elderly

(ages 69 and older)

- Males may be reproductively capable into their 80s or 90s. However, many have reduced muscle strength, fewer viable sperm, and decreased sexual desire.
- Females are in menopause and are no longer capable of reproduction.

Summary

1 The Reproductive Organs Make, Deliver, and Receive the Sex Cells 470

• As shown, the male reproductive organs include the testes, **scrotum, epididymis, ductus (vas) deferens, prostate, seminal vesicles, bulbourethral gland,** and penis. Sperm are produced in the testes, mature in the epididymis, and are stored in the vas deferens. Upon **ejaculation,** sperm are mixed with secretions from the prostate, seminal vesicles, and bulbourethral gland, forming a fluid called **semen.** Pelvic muscle contractions forcefully expel the semen through the urethra of the penis.

The male reproductive system • Figure 16.1a

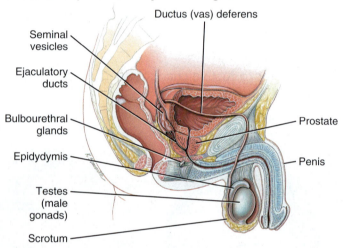

Ductus (vas) deferens

Seminal vesicles

Ejaculatory ducts

Bulbourethral glands

Epidydmis

Prostate

Penis

Testes (male gonads)

Scrotum

• **Spermatogenesis** and male secondary sex characteristics are maintained by secretions of GnRH from the hypothalamus, LH and FSH from the anterior pituitary, and testosterone and **dihydrotestosterone** from the testes via a negative feedback mechanism.

The female reproductive system • Figure 16.6

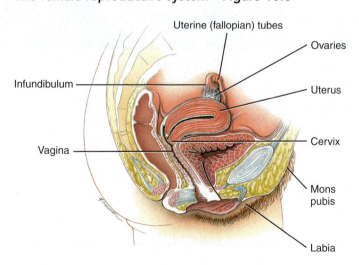

Uterine (fallopian) tubes

Ovaries

Infundibulum

Uterus

Vagina

Cervix

Mons pubis

Labia

• As shown, the female reproductive organs include the **ovaries, uterine tubes, uterus,** and **vagina. Oocytes** develop within follicles in the ovaries and are released into the uterine tubes. The uterus protects and nurtures the growing embryo/fetus. The **vagina** receives the male's penis during **sexual intercourse** and is the passageway through which the baby will pass during labor and delivery.

2 The Female Reproductive Cycle Shows That Timing Is Everything 480

• **Oogenesis,** follicle development, ovulation, and female secondary sex characteristics are maintained by secretions of GnRH from the hypothalamus, LH and FSH from the anterior pituitary, and **estrogens** and **progesterone** from the ovaries through a negative feedback mechanism.

• In the **ovarian cycle,** follicles develop in the ovary, as shown. One follicle (or occasionally more) completes development, releasing a secondary oocyte from the ovary at ovulation when stimulated by a surge of LH and FSH from the anterior pituitary. The remnants of the follicle remain in the ovary as the **corpus luteum.** If fertilization does not take place, the corpus luteum degenerates within 2 weeks, and a new cycle begins.

Development of ovarian follicles • Figure 16.9

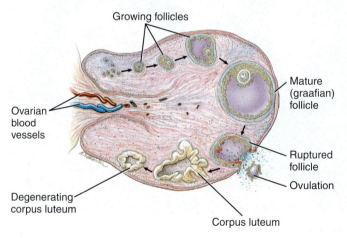

Growing follicles

Mature (graafian) follicle

Ovarian blood vessels

Ruptured follicle

Ovulation

Degenerating corpus luteum

Corpus luteum

- In the **menstrual cycle**, estrogens produced by the developing follicle stimulate the growth of the uterine lining, and progesterone converts the **endometrium** into a form that is ready to receive the fertilized egg. Progesterone produced by the corpus luteum maintains the lining after ovulation. When the corpus luteum degrades, the uterine lining is shed as the menses, and the cycle repeats. The female reproductive cycle—which includes coordinated hormonal changes, the **ovarian cycle**, and the menstrual cycle—lasts approximately 28 days.

3 Fertilization Requires the Egg and Sperm to Get Very Close to One Another 484

- **Sperm** can survive for only about 48 hours after ejaculation, and the egg remains viable for only about 24 hours. Sperm require about 6 hours of transit time from their delivery site. Therefore, timing of sperm delivery is important for successful fertilization.

- Numerous contraceptive products are available to help prevent pregnancy. These products are designed to disrupt the fertilization or implantation processes. Sterilization prevents the egg and sperm from passing through the tubing in their respective reproductive systems. Barrier methods, as shown, and chemical contraceptives prevent sperm from reaching the egg after ejaculation. Hormonal contraceptives disrupt the female reproductive cycle. Implantation-inhibiting contraceptives alter the endometrium and prevent the embryo from attaching to the uterus.

Methods of contraception and their effectiveness rates —barrier methods • Table 16.1

4 Pregnancy Lasts from Fertilization to Delivery 488

- Once deposited in the vagina, sperm travel through the cervical mucus, along the walls of the uterus, and into the uterine tubes. If they encounter a secondary oocyte, they will attempt to penetrate its outer layers and plasma membrane. Once a single sperm penetrates the zona pellucida, this region swells outward to block the activity of other sperm. Fertilization occurs when the nuclei of the sperm and oocyte fuse and the oocyte becomes a **zygote**.

- During the first week of development, the zygote divides many times and changes to a hollow sphere of cells called a **blastocyst**. During the second week, the blastocyst implants into the uterine wall, where it develops connections to the maternal blood supply to receive nutrients, exchange gases, and eliminate wastes via the formation of chorionic villi. Later, as shown, the placenta and **umbilical cord** develop from the chorionic villi and endometrium to support the growing embryo.

Connections between the embryo and the mother • Figure 16.13b

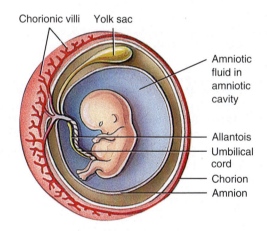

- During the first 8 weeks of development, the blastocyst changes size and shape into an embryo. Some key changes include **gastrulation**, the formation of three layers of cells (germ layers); all organs will form from various germ layers. Another key event is **neurulation**, in which the nervous system develops from the **ectoderm** (one of the three germ layers).

- From 8 weeks onward, the developing organism is a fetus. By the 12th week, all organs have developed. The fetus continues to grow; it is considered full term at 38 weeks.

5 Giving Birth Requires a Complex Series of Hormonal Changes 498

• At some point near the end of pregnancy, **labor** is initiated. A positive feedback mechanism involving the uterus, hypothalamus, and posterior pituitary using the hormone **oxytocin** produces wave-like uterine contractions. The contractions first dilate the cervix to about 10 cm (stage 1), expel the baby (stage 2 as shown), and deliver the **placenta** following birth (stage 3).

Labor • Figure 16.17b

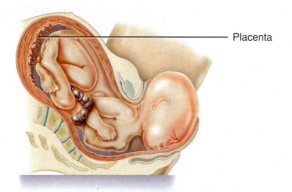

Placenta

Stage 2. Expulsion extends from complete dilation of cervix to delivery of the baby (10 min to several hours).

• During the process of **lactation**, milk is produced and ejected to feed the baby. The hormone **prolactin** maintains milk production in the mammary glands. Milk ejection from the glands occurs through a positive feedback mechanism among the nipples, hypothalamus, anterior pituitary, and mammary glands involving the hormone oxytocin.

6 Aging Alters Reproductive Capacity 502

• The reproductive systems of both genders are intact but inactive from birth to about 10 years of age. At **puberty**, rhythmic secretions of GnRH, LH, and FSH stimulate the ovarian and menstrual cycles in the female and spermatogenesis in the male, along with sex hormone production and development of secondary sex characteristics in both genders. Both genders become capable of sexual reproduction. Between ages 15 and 40 (as shown), when couples often marry, both genders are fully capable of reproduction.

Reproductive capabilities throughout life • Figure 16.19

• Between ages 40 and 50, females undergo **menopause**, the cessation of menses. The number of viable follicles is exhausted, and the ovarian and menstrual cycles cease. The female is no longer capable of reproduction.

• The reproductive capacity of healthy men extends beyond middle age and into the 80s and 90s. However, after age 55, men may experience decreases in the number of viable sperm, reduced muscle strength, and decreased sexual desire.

Key Terms

- acrosome 473
- birth 498
- blastocyst 488
- blastocyst cavity 488
- blastomere 488
- body 476
- breech 498
- bulbourethral gland 470
- capacitation 485
- cervix 476
- cleavage 488
- clitoris 478
- colostrum 500
- contraception 485
- corpus luteum 480
- corticotrophin-releasing hormone (CRH) 496
- dihydrotestosterone (DHT) 475
- ductus arteriosus 492
- ectoderm 491
- ejaculation 470
- endoderm 491
- endometrium 476
- epididymis 470
- estrogen 496
- female reproductive cycle 482
- fertilization 470
- fimbriae 476
- follicle-stimulating hormone (FSH) 475
- follicular phase 482
- foramen ovale 492
- fundus 476
- gamete 470
- gastrulation 490
- gonad 470
- germ layer 490
- germinal epithelium 480
- gonadotropin-inhibiting hormone (GnIH) 487
- human chorionic gonadotropin (hCG) 490
- human chorionic somato-mammotropin (hCS) 496
- hymen 476
- implantation 488
- infertility 468
- infundibulum 476
- inhibin 475
- inner cell mass 488
- *in vitro* fertilization (IVF) 468
- labia majora 478
- labia minora 478
- labor 498
- lactation 500
- Leydig cells 472
- lobule 478
- luteal phase 482
- luteinizing hormone (LH) 475
- mammary gland 478
- mature (graafian) follicle 480
- meiosis 473
- menarche 502
- menopause 503
- menstrual cycle 482
- menstruation 482
- mesoderm 491
- mons pubis 478
- morula 488
- myometrium 476
- neurulation 491
- notochord 491
- oocyte 480
- oogenesis 480
- oogonia 480
- ovarian cycle 481
- ovarian follicle 480
- ovary 476
- ovulation 480
- ovum 481
- parturition 498
- placenta 491
- polar body 481
- primary oocytes 480
- progesterone 496
- prolactin (PRL) 500
- prostaglandins 498
- prostate 470
- puberty 502
- relaxin 496
- scrotum 470
- secondary oocyte 481
- semen 470
- seminal vesicle 470
- seminiferous tubules 472
- Sertoli cells 472
- sexual intercourse 470
- sexual reproduction 470
- sexually transmitted disease (STD) 485
- sperm 470
- spermatogenesis 473
- spermatogonia 472
- spermiogenesis 473
- suspensory ligament of the breast 478
- testis 470
- testosterone 470
- trophoblast 488
- umbilical cord 491
- uterine cavity 476
- uterine tube 476
- uterus 476
- vagina 476
- vas deferens 470
- vestibule 478
- zygote 488

Critical and Creative Thinking Questions

1. Elaine was in a car accident in which she experienced head trauma. Before her accident, Elaine had regular menstrual cycles, but after the accident, her cycles were irregular and then stopped completely. What might have happened during the accident? What physiological mechanism explains these results?

2. To treat some child molesters, high doses of testosterone can be administered as a form of chemical castration. How does this affect male reproductive function and sex drive?

3. Alcohol causes vasodilation; excessive alcohol intake can render a man incapable of having sexual intercourse. What physiological mechanisms are important for the male portion of sexual intercourse, and how could they be affected by excessive alcohol levels?

4. Julia is breast-feeding her newborn baby. Between feedings, she wears her favorite T-shirt, which has become a little tight across her breasts, and notices that the T-shirt becomes wet. Explain why this would happen.

5. Spina bifida is a birth defect in which the baby is born with an open spine, accompanied by brain and spinal cord defects. Explain the process by which the central nervous system develops and how a problem in this process can lead to spina bifida.

6. Obstetricians are careful when administering pain medications to women in labor because they might act as anesthetics and slow the progression of labor. Pain killers and anesthetics interfere with nerve transmissions. What is the physiological mechanism involved in labor, and how might it be affected by an anesthetic?

What is happening in this picture?

Male ejaculate contains millions of sperm; the ovary releases a single secondary oocyte. For fertilization to occur,
one sperm must penetrate the oocyte plasma membrane
and fuse with its nucleus.

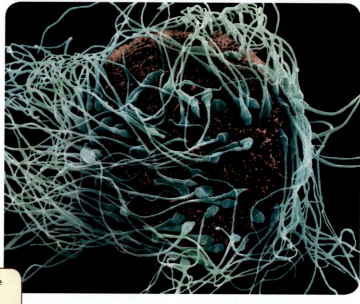

Think Critically 1. How does the oocyte
stop fertilization by multiple sperm?
2. What role does the sperm play in oocyte
meiosis?

Self-Test

(Check your answers in Appendix C.)

1. Which cell nurtures and protects the developing sperm?

   a. spermatogonium

   b. Sertoli cell

   c. Leydig cell

   d. epididymal cell

Use this figure to answer questions 2–4.

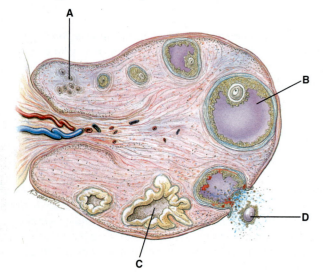

2. Which of the labeled structures secretes progesterone?

   a. A

   b. B

   c. C

   d. D

3. Which of the labeled structures can be fertilized by a sperm?

   a. A

   b. B

   c. C

   d. D

4. The process shown in D is triggered by _____.

   a. continuous periodic secretion of FSH and LH

   b. continuous secretion of estrogen

   c. a surge of LH and FSH

   d. continued secretion of progesterone

5. _____ induces labor if injected into a pregnant woman.

   a. Oxytocin

   b. Progesterone

   c. Inhibin

   d. Testosterone

6. _____ gives rise to three germ layers in the embryo.

   a. Fertilization

   b. Neurulation

   c. Gastrulation

   d. Morulation

7. The _____ phase of the male sexual response is associated with ejaculation.

   a. excitement

   b. plateau

   c. orgasm

   d. resolution

Use this figure to answer questions 8–10.

8. In which of the labeled organs do sperm mature?

   a. A   c. C

   b. B   d. D

9. The organ labeled B produces _____.

   a. LH

   b. testosterone

   c. FSH

   d. GnRH

10. Wave-like contractions propel the sperm through _____.

   a. A   c. C

   b. B   d. D

11. Which of the following events releases a hormone that rescues the corpus luteum?

   a. capacitation

   b. fertilization

   c. gastrulation

   d. implantation

12. Which of the following is part of a positive feedback mechanism?

   a. testosterone secretion

   b. milk ejection

   c. secretion of estrogens

   d. inhibin secretion

Use this figure to answer questions 13–15.

13. In which of the labeled stages would you find a blastocyst?

   a. A   c. C

   b. B   d. D

14. The cells pictured in part C are most likely _____.

   a. oocytes

   b. zygotes

   c. morulas

   d. blastocysts

15. In which labeled event do you find a zygote?

   a. A   c. C

   b. B   d. D

THE PLANNER ✓

Review your Chapter Planner on the chapter opener and check off your completed work.

Appendix A   Periodic Table

The periodic table lists the known **chemical elements,** the basic units of matter. The elements in the table are arranged left-to-right in rows in order of their **atomic number,** the number of protons in the nucleus. Each horizontal row, numbered from 1 to 7, is a **period.** All elements in a given period have the same number of electron shells as their period number. For example, an atom of hydrogen or helium each has one electron shell, while an atom of potassium or calcium each has four electron shells. The elements in each column, or **group,** share chemical properties. For example, the elements in column IA are very chemically reactive, whereas the elements in column VIIIA have full electron shells and thus are chemically inert.

Scientists now recognize 117 different elements; 92 occur naturally on Earth, and the rest are produced from the natural elements using particle accelerators or nuclear reactors. Elements are designated by **chemical symbols,** which are the first one or two letters of the element's name in English, Latin, or another language.

Twenty-six of the 92 naturally occurring elements normally are present in your body. Of these, just four elements—oxygen (O), carbon (C), hydrogen (H), and nitrogen (N) (coded blue)—constitute about 96% of the body's mass. Eight others—calcium (Ca), phosphorus (P), potassium (K), sulfur (S), sodium (Na), chlorine (Cl), magnesium (Mg), and iron (Fe) (coded pink)— contribute 3.8% of the body's mass. An additional 14 elements, called **trace elements** because they are present in tiny amounts, account for the remaining 0.2% of the body's mass. The trace elements are aluminum, boron, chromium, cobalt, copper, fluorine, iodine, manganese, molybdenum, selenium, silicon, tin, vanadium, and zinc (coded yellow). Table 2.1 on page 29 provides information about the main chemical elements in the body.

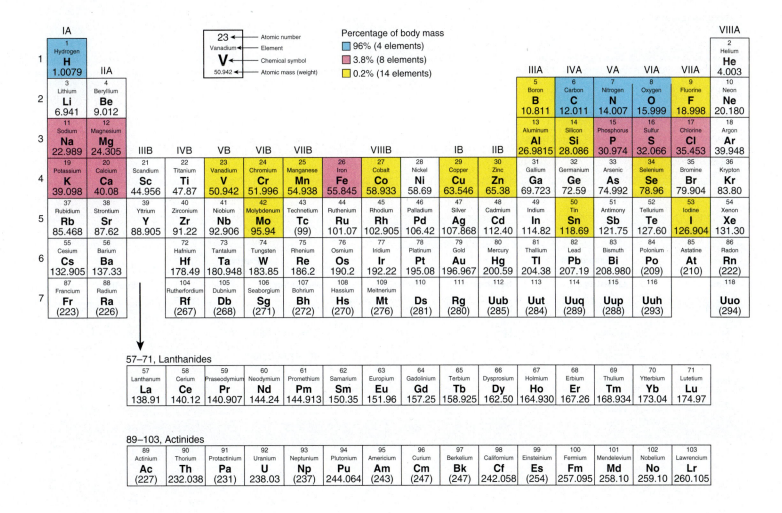

U.S. customary system

Parameter	Unit	Relation to other U.S. units	SI (metric) equivalent
Length	inch	1/12 foot	2.54 centimeters
foot	12 inches	0.305 meter	
yard	36 inches	0.914 meters	
mile	5,280 feet	1.609 kilometers	
Mass	grain	1/1000 pound	64.799 milligrams
dram	1/16 ounce	1.772 grams	
ounce	16 drams	28.350 grams	
pound	16 ounces	453.6 grams	
ton	2,000 pounds	907.18 kilograms	
Volume (Liquid)	ounce	1/16 pint	29.574 milliliters
pint	16 ounces	0.473 liter	
quart	2 pints	0.946 liter	
gallon	4 quarts	3.785 liters	
Volume (Dry)	pint	1/2 quart	0.551 liter
quart	2 pints	1.101 liters	
peck	8 quarts	8.810 liters	
bushel	4 pecks	35.239 liters	

International system (SI)

Base units			Prefixes		
Unit	Quantity	Symbol	Prefix	Multiplier	Symbol
meter	length	m	tera-	$10^{12} = 1{,}000{,}000{,}000{,}000$	T
kilogram	mass	kg	giga-	$10^9 = 1{,}000{,}000{,}000$	G
second	time	s	mega-	$10^6 = 1{,}000{,}000$	M
liter	volume	L	kilo-	$10^3 = 1{,}000$	k
mole	amount of matter	mol	hecto-	$10^2 = 100$	h
			deca-	$10^1 = 10$	da
			deci-	$10^{-1} = 0.1$	d
			centi-	$10^{-2} = 0.01$	c
			milli-	$10^{-3} = 0.001$	m
			micro-	$10^{-6} = 0.000{,}001$	μ
			nano-	$10^{-9} = 0.000{,}000{,}001$	n
			pico-	$10^{-12} = 0.000{,}000{,}000{,}001$	p

Temperature conversion
Fahrenheit (F) to Celsius (C)
$$°C = (°F - 32) \div 1.8$$
Celsius (C) to Fahrenheit (F)
$$°F = (°C \times 1.8) + 32$$

U.S. to SI (metric) conversion		
When you know	**Multiply by**	**To find**
inches	2.54	centimeters
feet	30.48	centimeters
yards	0.91	meters
miles	1.61	kilometers
ounces	28.35	grams
pounds	0.45	kilograms
tons	0.91	metric tons
fluid ounces	29.57	milliliters
pints	0.47	liters
quarts	0.95	liters
gallons	3.79	liters

SI (metric) to U.S. conversion		
When you know	**Multiply by**	**To find**
millimeters	0.04	inches
centimeters	0.39	inches
meters	3.28	feet
kilometers	0.62	miles
liters	1.06	quarts
cubic meters	35.32	cubic feet
grams	0.035	ounces
kilograms	2.21	pounds

Chapter 1
1. a; 2. d; 3. c; 4. b; 5. d; 6. c; 7. c; 8. b; 9. d; 10. c; 11. d; 12. a; 13. b; 14. d; 15. c

Chapter 2
1. d; 2. d; 3. d; 4. c; 5. c; 6. b; 7. a; 8. a; 9. c; 10. d; 11. c; 12. c; 13. b; 14. c; 15. b

Chapter 3
1. c; 2. d; 3. a; 4. b; 5. d; 6. c; 7. b; 8. b; 9. d; 10. d; 11. b; 12. a; 13. b; 14. d; 15. d

Chapter 4
1. b; 2. d; 3. a; 4. c; 5. c; 6. b; 7. d; 8. b; 9. b; 10. b; 11. d; 12. a; 13. c; 14. b; 15. d

Chapter 5
1. c; 2. b; 3. b; 4. a; 5. d; 6.d ; 7. c; 8. d; 9. a; 10. a; 11. c; 12. b; 13. d; 14. d; 15. b

Chapter 6
1. c; 2. b; 3. d; 4. a; 5. c; 6. b; 7. b; 8. b; 9. c; 10. d; 11. a; 12. d; 13. d; 14. c; 15. c

Chapter 7
1. b; 2. a; 3. c; 4. d; 5. d; 6. c; 7. a; 8. d; 9. c; 10. a; 11. b; 12. b; 13. d; 14. b; 15. a

Chapter 8
1. d; 2. c; 3. a; 4. b; 5. c; 6. b; 7. c; 8. c; 9. d; 10. d; 11. b; 12. a; 13. b; 14. c; 15. d; 16. a

Chapter 9
1. d; 2. b; 3. b; 4. d; 5. c; 6. a; 7. b; 8. d; 9.c; 10 b; 11. d; 12.c; 13. b; 14. a; 15. c

Chapter 10
1. b; 2. b; 3. c; 4. a; 5. d; 6. d; 7. b; 8. c; 9. b; 10. a; 11. c; 12. b; 13. b; 14. c; 15. c

Chapter 11
1. c; 2. b; 3. c; 4. a; 5. d; 6. d; 7. c; 8. b; 9. b; 10. a; 11. c; 12. b; 13. d; 14. c; 15. b

Chapter 12
1. b; 2. d; 3. a; 4. c; 5. d; 6. b; 7. d; 8. b; 9. b; 10. a; 11. d; 12. c; 13. a; 14. c; 15. c; 16. d

Chapter 13
1. d; 2. c; 3. b; 4. a; 5. d; 6. c; 7. c; 8. d; 9. a; 10. a; 11. c; 12. d; 13. a; 14. b; 15. b; 16. d; 17. c; 18. c

Chapter 14
1. c; 2. d; 3. a; 4. b; 5. c; 6. a; 7. d; 8. c; 9. a; 10. d; 11. c; 12. d; 13. c; 14. a; 15.b; 16. a; 17. c

Chapter 15
1. a; 2. c; 3. c; 4. a; 5. c; 6. b; 7. a; 8. d; 9. c; 10. b; 11. d; 12. c; 13. c; 14. a; 15. a; 16. b; 17. d; 18. c

Chapter 16
1. b; 2. c; 3. d; 4. c; 5. a; 6. c; 7. c; 8. c; 9. b; 10. d; 11. d; 12. b; 13. d; 14. c; 15. b

Glossary

Pronunciation Key

1. The most strongly accented syllable appears in capital letters, for example, bilateral (bī-LAT-er-al) and diagnosis (dī-ag-NŌ-sis).
2. If there is a secondary accent, it is noted by a prime ('), for example, constitution (kon'-sti-TOO-shun) and physiology (fiz'-ē-OL-ō-jē). Any additional secondary accents are also noted by a prime, for example, decarboxylation (dē'-kar-bok'-si-LĀ-shun).
3. Vowels marked by a line above the letter are pronounced with the long sound, as in the following common words:
   ā as in māke   ō as in pōle
   ē as in bē   ū as in cute
   ī as in īvy
4. Vowels not marked by a line above the letter are pronounced with the short sound, as in the following words:
   a as in above or at   o as in not
   e as in bet   u as in bud
   i as in sip

5. Other vowel sounds are indicated as follows:
   oy as in oil
   oo as in root
6. Consonant sounds are pronounced as in the following words:
   b as in bat   m as in mother
   ch as in chair   n as in no
   d as in dog   p as in pick
   f as in father   r as in rib
   g as in get   s as in so
   h as in hat   t as in tea
   j as in jump   v as in very
   k as in can   w as in welcome
   ks as in tax   z as in zero
   kw as in quit   zh as in lesion
   l as in let

A

Abdomen (ab-DŌ-men or AB-dō-men) The area between the diaphragm and pelvis.

Abdominal cavity (ab-DŌM-i-nal) Superior portion of the abdominopelvic cavity that contains the stomach, spleen, liver, gallbladder, most of the small intestine, and part of the large intestine.

Abdominal thrust maneuver A first-aid procedure for choking. Employs a quick, upward thrust against the diaphragm that forces air out of the lungs with sufficient force to eject any lodged material. Also called the **Heimlich maneuver** (HĪM-lik).

Abdominopelvic cavity (ab-dom'-i-nō-PEL-vic) A body cavity that is subdivided into a superior abdominal cavity and an inferior pelvic cavity.

Abduction (ab-DUK-shun) Movement away from the midline of the body.

Abortion (a-BOR-shun) The premature loss (spontaneous) or removal (induced) of the embryo or nonviable fetus; miscarriage due to a failure in the normal process of developing or maturing.

Abscess (AB-ses) A localized collection of pus and liquefied tissue in a cavity.

Absorption (ab-SORP-shun) Intake of fluids or other substances by cells of the skin or mucous membranes; the passage of digested foods from the gastrointestinal tract into blood or lymph.

Accommodation (a-kom-ō-DĀ-shun) An increase in the curvature of the lens of the eye to adjust for near vision.

Acetabulum (as'-e-TAB-ū-lum) The rounded cavity on the external surface of the hip bone that receives the head of the femur.

Acetylcholine (ACh) (as'-e-til-KŌ-lēn) A neurotransmitter liberated by many peripheral nervous system neurons and some central nervous system neurons. It is excitatory at neuromuscular junctions but inhibitory at some other synapses (for example, it slows heart rate).

Achalasia (ak'-a-LĀ-zē-a) A condition, in which the lower esophageal sphincter fails to relax normally as food approaches. A whole meal may become lodged in the esophagus and enter the stom-

ach very slowly. Distension of the esophagus results in chest pain that is often confused with pain originating from the heart.

Acid (AS-id) A proton donor, or a substance that dissociates into hydrogen ions (H⁺); characterized by an excess of hydrogen ions and a pH less than 7.

Acidosis (as-i-DŌ-sis) A condition in which blood pH is below 7.35. Also known as **acidemia**.

Acini (AS-i-nē) Groups of cells in the pancreas that secrete digestive enzymes.

Acoustic (a-KOOS-tik) Pertaining to sound or the sense of hearing.

Acquired immunodeficiency syndrome (AIDS) A disease caused by the human immunodeficiency virus (HIV). Characterized by a positive HIV-antibody test, low helper T cell count, and certain indicator diseases (for example Kaposi's sarcoma, Pneumocystis carinii pneumonia, tuberculosis, fungal diseases). Other symptoms include fever or night sweats, coughing, sore throat, fatigue, body aches, weight loss, and enlarged lymph nodes.

Acrosome (AK-rō-sōm) A lysosomelike organelle in the head of a sperm cell containing enzymes that facilitate the penetration of a sperm cell into a secondary oocyte.

Actin (AK-tin) A contractile protein that is part of the thin filaments in muscle fibers.

Action potential An electrical signal that propagates along the membrane of a neuron or muscle fiber (cell); a rapid change in membrane potential that involves a depolarization followed by a repolarization. Also called a **nerve action potential** or **nerve impulse** as it relates to a neuron, and a **muscle action potential** as it relates to a muscle fiber.

Active transport The movement of substances across cell membranes against a concentration gradient, requiring the expenditure of cellular energy (ATP).

Acute (a-KŪT) Having rapid onset, severe symptoms, and a short course; not chronic.

Glossary 515

Adaptation (ad'-ap-TĀ-shun) The adjustment of the pupil of the eye to changes in light intensity. The property by which a sensory neuron relays a decreased frequency of action potentials from a receptor, even though the strength of the stimulus remains constant; the decrease in perception of a sensation over time while the stimulus is still present.

Adduction (ad-DUK-shun) Movement toward the midline of the body.

Adenoid (AD-e-noyd) The pharyngeal tonsil.

Adenosine triphosphate (ATP) (a-DEN-ō-sēn trī-FOS-fāt) The main energy currency in living cells; used to transfer the chemical energy needed for metabolic reactions. ATP consists of the purine base *adenine* and the five-carbon sugar *ribose*, to which are added, in linear array, three *phosphate* groups.

Adenylate cyclase (a-DEN-i-lāt SĪ-klās) An enzyme that is activated when certain neurotransmitters or hormones bind to their receptors; the enzyme that converts ATP into cyclic AMP, an important second messenger.

Adipocyte (AD-i-pō-sīt) Fat cell, derived from a fibroblast.

Adipose tissue (AD-i-pōz) Tissue composed of adipocytes specialized for triglyceride storage and present in the form of soft pads between various organs for support, protection, and insulation.

Adrenal cortex (a-DRĒ-nal KOR-teks) The outer portion of an adrenal gland, divided into three zones; the zona glomerulosa secretes mineralocorticoids, the zona fasciculata secretes glucocorticoids, and the zona reticularis secretes androgens.

Adrenal glands Two glands located superior to each kidney. Also called the **suprarenal glands** (soo'-pra-RĒ-nal).

Adrenal medulla (me-DŪL-a) The inner part of an adrenal gland, consisting of cells that secrete epinephrine, norepinephrine, and a small amount of dopamine in response to stimulation by sympathetic preganglionic neurons.

Adrenergic neuron (ad'-ren-ER-jik) A neuron that releases epinephrine (adrenaline) or norepinephrine (noradrenaline) as its neurotransmitter.

Adrenocorticotropic hormone (ACTH) (ad-rē'-nō-kor-ti-kō-TRŌP-ik) A hormone produced by the anterior pituitary that influences the production and secretion of certain hormones of the adrenal cortex.

Adventitia (ad-ven-TISH-a) The outermost connective tissue covering of a structure or organ not covered by a serous coat.

Aerobic (air-Ō-bik) Requiring molecular oxygen.

Afferent arteriole (AF-er-ent ar-TĒ-rē-ōl) A blood vessel of a kidney that divides into the capillary network called a glomerulus; there is one afferent arteriole for each glomerulus.

Agglutination (a-gloo'-ti-NĀ-shun) Clumping of microorganisms or blood cells, typically due to an antigen–antibody reaction.

Aggregated lymphatic follicles (AG-re-gā-ted lim-FAT-its FOL-i-kalz) Clusters of lymph nodules that are most numerous in the ileum. Also called **Peyer's patches** (PĪ-erz).

Albinism (AL-bin-izm) Abnormal, nonpathological, partial, or total absence of pigment in skin, hair, and eyes.

Albumin (al-BŪ-min) The most abundant (60%) and smallest of the plasma proteins; it is the main contributor to blood colloid osmotic pressure (BCOP).

Aldosterone (al-DOS-ter-ōn) A mineralocorticoid produced by the adrenal cortex that promotes sodium and water reabsorption by the kidneys and potassium excretion in urine.

Alkaline (AL-ka-līn) Containing more hydroxide ions (OH^-) than hydrogen ions (H^+); a pH higher than 7.

Alkalosis (al-ka-LŌ-sis) A condition in which blood pH is higher than 7.45. Also known as **alkalemia**.

Allantois (a-LAN-tō-is) A small, vascularized outpouching of the yolk sac that serves as an early site for blood formation and development of the urinary bladder.

Alleles (a-LĒLZ) Alternate forms of a single gene that control the same inherited trait (such as type A blood) and are located at the same position on homologous chromosomes.

Allergen (AL-er-jen) An antigen that evokes a hypersensitivity reaction.

Alpha cell (AL-fa) A type of cell in the pancreatic islets (islets of Langerhans) in the pancreas that secretes the hormone glucagon.

Alveolar duct (al-VĒ-ō-lar) Branch of a respiratory bronchiole around which alveoli and alveolar sacs are arranged.

Alveolar macrophage (MAK-rō-fāj) Highly phagocytic cell found in the alveolar walls of the lungs. Also called a **dust cell**.

Alveolar pressure Air pressure within the lungs.

Alveolar sac A cluster of alveoli that share a common opening.

Alveolus (al-VĒ-ō-lus) A small hollow or cavity; an air sac in the lungs; milk-secreting portion of a mammary gland. *Plural is* **alveoli** (al-VĒ-ol-ī).

Alzheimer disease (AD) (ALTZ-hī-mer) Disabling neurological disorder characterized by dysfunction and death of specific cerebral neurons, resulting in widespread intellectual impairment, personality changes, and fluctuations in alertness.

Amnesia (am-NE-zē-a) A lack or loss of memory.

Amenorrhea (ā-men-ō-RĒ-a) Absence of menstruation.

Amino acid (a-MĒ-nō) An organic acid, containing an acidic carboxyl group (–COOH) and a basic amino group ($–NH_2$); the monomer used to synthesize polypeptides and proteins.

Amnion (AM-nē-on) A thin, protective fetal membrane that develops from the epiblast; holds the fetus suspended in amniotic fluid. Also called the "**bag of waters.**"

Amniotic fluid (am'-nē-OT-ik) Fluid in the amniotic cavity, the space between the developing embryo (or fetus) and amnion; the fluid is initially produced as a filtrate from maternal blood and later includes fetal urine. It functions as a shock absorber, helps regulate fetal body temperature, and helps prevent desiccation.

Amphiarthrosis (am'-fē-ar-THRŌ-sis) A slightly movable joint, in which the articulating bony surfaces are separated by fibrous connective tissue or fibrocartilage to which both are attached; types are syndesmosis and symphysis.

Ampulla (am-PUL-la) A saclike dilation of a canal or duct.

Anabolism (a-NAB-ō-lizm) Synthetic, energy-requiring reactions whereby small molecules are built up into larger ones.

Anaerobic (an-ar-Ō-bik) Not requiring oxygen.

Anal canal (Ā-nal) The last 2 or 3 cm (1 in.) of the rectum; opens to the exterior through the anus.

Anal column A longitudinal fold in the mucous membrane of the anal canal that contains a network of arteries and veins.

Anal triangle The subdivision of the female or male perineum that contains the anus.

Analgesia (an-al-JĒ-zē-a) Pain relief; absence of the sensation of pain.

Anaphase (AN-a-fāz) The third stage of mitosis in which the chromatids that have separated at the centromeres move to opposite poles of the cell.

Anaphylaxis (an'-a-fi-LAK-sis) A hypersensitivity (allergic) reaction in which IgE antibodies attach to mast cells and basophils, causing them to produce mediators of anaphylaxis (histamine, leukotrienes, kinins, and prostaglandins) that bring about increased blood permeability, increased smooth muscle contraction, and increased mucus production. Examples are hay fever, hives, and anaphylactic shock.

Anastomosis (a-nas-tō-MŌ-sis) An anatomical connection between tubular structure, especially arteries.

Anatomical position (an'-a-TOM-i-kal) A position of the body universally used in anatomical descriptions in which the body is erect, the head is level, the eyes face forward, the upper limbs are at the sides, the palms face forward, and the feet are flat on the floor.

Anatomic dead space Spaces of the nose, pharynx, larynx, trachea, bronchi, and bronchioles totaling about 150 mL; air in the anatomic dead space does not reach the alveoli to participate in gas exchange.

Anatomy (a-NAT-ō-mē) The structure or study of structure of the body and the relation of its parts to each other.

Androgens (AN-drō-jenz) Masculinizing sex hormones produced by the testes in males and the adrenal cortex in both genders; also responsible for libido (sexual desire); the two main androgens are testosterone and dihydrotestosterone.

Anemia (a-NĒ-mē-a) Condition of the blood in which the number of functional red blood cells or their hemoglobin content is below normal.

Anesthesia (an'-es-THĒ-zē-a) A total or partial loss of feeling or sensation; may be general or local.

Aneurysm (AN-ū-rizm) A saclike enlargement of a blood vessel caused by a weakening of its wall.

Angina pectoris (an-JI-na *or* AN-ji-na PEK-tō-ris) A pain in the chest, jaw, shoulder, or upper limb related to reduced coronary circulation due to coronary artery disease (CAD) or spasms of vascular smooth muscle in coronary arteries.

Angiotensin (an-jē-ō-TEN-sin) Either of two forms of a protein associated with regulation of blood pressure. Angiotensin I is produced by the action of renin on angiotensinogen and is converted by the action of ACE (angiotensin-converting enzyme) into angiotensin II, which stimulates aldosterone secretion by the adrenal cortex, stimulates the sensation of thirst, and causes vasoconstriction with resulting increase in systemic vascular resistance.

Anion (AN-ī-on) A negatively charged ion. Examples are the chloride ion (Cl^-) and bicarbonate ion (HCO_3^-).

Anoxia (an-OK-sē-a) Deficiency of oxygen.

Antagonist (an-TAG-ō-nist) A muscle that has an action opposite that of the prime mover (agonist) and yields to the movement of the prime mover.

Anterior (an-TĒR-ē-or) Nearer to or at the front of the body. Equivalent to **ventral** in bipeds.

Anterior pituitary (pi-TOO-i-tār-ē) Anterior lobe of the pituitary gland. Also called the **adenohypophysis** (ad'-e-nō-hī-POF-i-sis).

Anterior root The structure composed of axons of motor (efferent) neurons that emerges from the anterior aspect of the spinal cord and extends laterally to join a posterior root, forming a spinal nerve. Also called a **ventral root**.

Anterolateral pathway (an'-ter-ō-LAT-er-al) Sensory pathway that conveys information related to pain, temperature, tickle, and itch.

Antibody (AN-ti-bod'-ē) A protein produced by plasma cells in response to a specific antigen; the antibody combines with that antigen to neutralize, inhibit, or destroy it. Also called an **immunoglobulin** (im-ū-nō-GLOB-ū-lin) or **Ig**.

Antibody-mediated immunity That component of immunity in which B lymphocytes (B cells) develop into plasma cells that produce antibodies that destroy antigens. Also called **humoral immunity** (HŪ-mor-al).

Anticoagulant (an-tī-cō-AG-ū-lant) A substance that can delay, suppress, or prevent the clotting of blood.

Antidiuretic (an'-ti-dī-ū-RET-ik) Substance that inhibits urine formation.

Antidiuretic hormone (ADH) Hormone produced by neurosecretory cells in the hypothalamus that stimulates water reabsorption from kidney tubule cells into the blood and vasoconstriction of arterioles. Also called **vasopressin** (vāz-ō-PRES-in).

Antigen (AN-ti-jen) A substance that has the ability to provoke an immune response and the ability to react with the antibodies or cells that result from the immune response; contraction of *anti*body *gen*erator.

Antigen-presenting cell (APC) Special class of migratory cell that processes and presents antigens to T cells during an immune response; APCs include macrophages, B cells, and dendritic cells, which are present in the skin, mucous membranes, and lymph nodes.

Anuria (an-Ū-rē-a) Daily urine output of less than 50 mL.

Anus (Ā-nus) The distal end and outlet of the rectum.

Aorta (ā-OR-ta) The main systemic trunk of the arterial system of the body that emerges from the left ventricle.

Aortic body (ā-OR-tik) Cluster of chemoreceptors on or near the arch of the aorta that respond to changes in blood levels of oxygen, carbon dioxide, and hydrogen ions (H^+).

Aortic reflex A reflex that helps maintain normal systemic blood pressure; initiated by baroreceptors in the wall of the ascending aorta and arch of the aorta.

Apex (Ā-peks) The pointed end of a conical structure, such as the apex of the heart.

Aphasia (a-FĀ-zē-a) Loss of ability to express oneself properly through speech or loss of verbal comprehension.

Apnea (AP-nē-a) Temporary cessation of breathing.

Apneustic area (ap-NOO-stik) A part of the respiratory center in the pons that sends stimulatory nerve impulses to the inspiratory area that activate and prolong inhalation and inhibit exhalation.

Aponeurosis (ap'-ō-noo-RŌ-sis) A sheetlike tendon joining one muscle with another or with bone.

Apoptosis (ap'-ō-TŌ-sis *or* ap'-ōp-TŌ-sis) Programmed cell death; a normal type of cell death that removes unneeded cells during embryological development, regulates the number of cells in tissues, and eliminates many potentially dangerous cells such as cancer cells. During apoptosis, the DNA fragments, the nucleus condenses, mitochondria cease to function, and the cytoplasm shrinks, but the plasma membrane remains intact. Phagocytes engulf and digest the apoptotic cells, and an inflammatory response does not occur.

Aqueous humor (AK-wē-us HŪ-mer) The watery fluid, similar in composition to cerebrospinal fluid, that fills the anterior cavity of the eye.

Arachnoid mater (a-RAK-noyd MĀ-ter) The middle of the three meninges (coverings) of the brain and spinal cord. Also termed the **arachnoid**.

Arachnoid villus (VIL-us) Berrylike tuft of the arachnoid mater that protrudes into the superior sagittal sinus and through which cerebrospinal fluid is reabsorbed into the bloodstream.

Areola (a-RĒ-ō-la) Any tiny space in a tissue. The pigmented ring around the nipple of the breast.

Arm The part of the upper limb from the shoulder to the elbow.

Arrhythmia (a-RITH-mē-a) An irregular heart rhythm. Also called a **dysrhythmia**.

Arteriole (ar-TĒ-rē-ōl) A small, almost microscopic, artery that delivers blood to a capillary.

Arteriosclerosis (ar-tē-rē-ō-skle-RŌ-sis) Group of diseases characterized by thickening of the walls of arteries and loss of elasticity.

Artery (AR-ter-ē) A blood vessel that carries blood away from the heart.

Arthritis (ar-THRĪ-tis) Inflammation of a joint.

Arthrology (ar-THROL-ō-jē) The study or description of joints.

Arthroscopy (ar-THROS-co-pē) A procedure for examining the interior of a joint, usually the knee, by inserting an arthroscope into a small incision; used to determine extent of damage, remove torn cartilage, repair cruciate ligaments, and obtain samples for analysis.

Arthrosis (ar-THRŌ-sis) A joint or articulation.

Articular capsule (ar-TIK-ū-lar) Sleevelike structure around a synovial joint composed of a fibrous capsule and a synovial membrane.

Articular cartilage (KAR-ti-lij) Hyaline cartilage attached to articular bone surfaces.

Articular disc Fibrocartilage pad between articular surfaces of bones of some synovial joints. Also called a **meniscus** (men-IS-kus).

Articulation (ar-tik'-ū-Lā-shun) A joint; a point of contact between bones, cartilage and bones, or teeth and bones.

Ascending colon (KŌ-lon) The part of the large intestine that passes superiorly from the cecum to the inferior border of the liver, where it bends at the right colic (hepatic) flexure to become the transverse colon.

Ascites (as-SĪ-tēz) Abnormal accumulation of serous fluid in the peritoneal cavity.

Association areas Large cortical regions on the lateral surfaces of the occipital, parietal, and temporal lobes and on the frontal lobes anterior to the motor areas connected by many motor and sensory axons to other parts of the cortex. The association areas are concerned with motor patterns, memory, concepts of word-hearing and word-seeing, reasoning, will, judgment, and personality traits.

Asthma (AZ-ma) Usually allergic reaction characterized by smooth muscle spasms in bronchi resulting in wheezing and difficult breathing. Also called **bronchial asthma**.

Astigmatism (a-STIG-ma-tizm) An irregularity of the lens or cornea of the eye causing the image to be out of focus and producing faulty vision.

Astrocyte (AS-trō-sīt) A neuroglial cell having a star shape that participates in brain development and the metabolism of neurotransmitters, helps form the blood–brain barrier, helps maintain the proper balance of K⁺ for generation of nerve impulses, and provides a link between neurons and blood vessels.

Ataxia (a-TAK-sē-a) A lack of muscular coordination, lack of precision.

Atherosclerotic plaque (ath'-er-ō-skle-RO-tic PLAK) A lesion that results from accumulated cholesterol and smooth muscle fibers (cells) of the tunica media of an artery; may become obstructive.

Atom Unit of matter that makes up a chemical element; consists of a nucleus (containing positively charged protons and uncharged neutrons) and negatively charged electrons that orbit the nucleus.

Atomic mass (weight) Average mass of all stable atoms of an element, reflecting the relative proportion of atoms with different mass numbers.

Atomic number Number of protons in an atom.

Atrial fibrillation (Ā-trē-al fib-ri-LĀ-shun) Asynchronous contraction of cardiac muscle fibers in the atria that results in the cessation of atrial pumping.

Atrial natriuretic pentide (ANP) (na'-trē-ū-RET-ik) Peptide hormone, produced by the atria of the heart in response to stretching, that inhibits aldosterone production and thus lowers blood pressure; causes natriuresis, increased urinary excretion of sodium.

Atrioventricular (AV) bundle (ā'-trē-ō-ven-TRIK-ū-lar) The part of the conduction system of the heart that begins at the atrioventricular (AV) node, passes through the cardiac skeleton separating the atria and the ventricles, then extends a short distance down the interventricular septum before splitting into right and left bundle branches. Also called the **bundle of His** (HISS).

Atrioventricular (AV) node The part of the conduction system of the heart made up of a compact mass of conducting cells located in the septum between the two atria.

Atrioventricular (AV) valve A heart valve made up of membranous flaps or cusps that allows blood to flow in one direction only, from an atrium into a ventricle.

Atrium (Ā-trē-um) A superior chamber of the heart.

Atrophy (AT-rō-fē) Wasting away or decrease in size of a part, due to a failure, abnormality of nutrition, or lack of use.

Auditory ossicle (AW-di-tō-rē OS-si-kul) One of the three small bones of the middle ear called the **malleus, incus,** and **stapes**.

Auditory tube The tube that connects the middle ear with the nose and nasopharynx region of the throat. Also called the **eustachian tube** (ū-STĀ-shun *or* ū-STĀ-kē-an) or **pharyngotympanic tube**.

Auscultation (aws-kul-TĀ-shun) Examination by listening to sounds in the body.

Autoimmunity An immunological response against a person's own tissues.

Autolysis (aw-TOL-i-sis) Self-destruction of cells by their own lysosomal digestive enzymes after death or in a pathological process.

Autonomic ganglion (aw'-tō-NOM-ik GANG-lē-on) A cluster of cell bodies of sympathetic or parasympathetic neurons located outside the central nervous system.

Autonomic nervous system (ANS) Visceral sensory (afferent) and visceral motor (efferent) neurons. Autonomic motor neurons, both sympathetic and parasympathetic, conduct nerve impulses from the central nervous system to smooth muscle, cardiac muscle, and glands. So named because this part of the nervous system was thought to be self-governing or spontaneous.

Autophagy (aw-TOF-a-jē) Process by which worn-out organelles are digested within lysosomes.

Autopsy (AW-top-sē) The examination of the body after death.

Autosome (AW-tō-sōm) Any chromosome other than the X and Y chromosomes (sex chromosomes).

Axilla (ak-SIL-a) The small hollow beneath the arm where it joins the body at the shoulders. Also called the **armpit**.

Axon (AK-son) The usually single, long process of a nerve cell that propagates a nerve impulse toward the axon terminals.

Axon terminal Terminal branch of an axon where synaptic vesicles undergo exocytosis to release neurotransmitter molecules.

B

B cell A lymphocyte that can develop into a clone of antibody-producing plasma cells or memory cells when properly stimulated by a specific antigen.

Babinski sign (ba-BIN-skē) Extension of the great toe, with or without fanning of the other toes, in response to stimulation of the outer margin of the sole; normal up to 18 months of age and indicative of damage to descending motor pathways such as the corticospinal tracts after that.

Back The posterior part of the body; the dorsum.

Ball-and-socket joint A synovial joint in which the rounded surface of one bone moves within a cup-shaped depression or socket of another bone, as in the shoulder or hip joint.

Baroreceptor (bar'-ō-re-SEP-tor) Neuron capable of responding to changes in blood, air, or fluid pressure. Also called a **pressoreceptor**.

Basal ganglia (GANG-glē-a) Paired clusters of gray matter deep in each cerebral hemisphere including the globus pallidus, putamen, and caudate nucleus. Nearby structures that are functionally linked to the basal ganglia are the substantia nigra of the midbrain and the subthalamic nuclei of the diencephalon.

Basal metabolic rate (BMR) (BĀ-sal met'-a-BOL-ik) The rate of metabolism measured understandard or basal conditions (awake, at rest, fasting).

Base (BĀS) A nonacid or a proton acceptor, characterized by excess of hydroxide ions (OH^-) and a pH greater than 7. A ring-shaped, nitrogen-containing organic molecule that is one of the components of a nucleotide, namely, adenine, guanine, cytosine, thymine, and uracil; also known as a **nitrogenous base**.

Basement membrane Thin, extracellular layer between epithelium and connective tissue consisting of a basal lamina and a reticular lamina.

Basilar membrane (BĀS-i-lar) A membrane in the cochlea of the internal ear that separates the cochlear duct from the scala tympani and on which the spiral organ (organ of Corti) rests.

Basophil (BĀ-sō-fil) A type of white blood cell characterized by a pale nucleus and large granules that stain blue-purple with basic dyes.

Belly The abdomen. The gaster or prominent, fleshy part of a skeletal muscle.

Beta cell (BĀ-ta) A type of cell in the pancreatic islets (islets of Langerhans) in the pancreas that secretes the hormone insulin.

Bicuspid valve (bī-KUS-pid) Atrioventricular (AV) valve on the left side of the heart. Also called the **mitral valve**.

Bilateral (bī-LAT-er-al) Pertaining to two sides of the body.

Bile (BĪL) A secretion of the liver consisting of water, bile salts, bile pigments, cholesterol, lecithin, and several ions; it emulsifies lipids prior to their digestion.

Bilirubin (bil-ē-ROO-bin) An orange pigment that is one of the end products of hemoglobin breakdown in the hepatocytes and is excreted as a waste material in bile.

Blastocyst (BLAS-tō-sist) In the development of an embryo, a hollow ball of cells that consists of a blastocyst cavity (the internal cavity), trophoblast (outer cells), and inner cell mass.

Blastomere (BLAS-tō-mēr) One of the cells resulting from the cleavage of a fertilized ovum.

Blind spot Area in the retina at the end of the optic (II) nerve in which there are no photoreceptors.

Blood The fluid that circulates through the heart, arteries, capillaries, and veins and that constitutes the chief means of transport within the body.

Blood–brain barrier (BBB) A barrier consisting of specialized brain capillaries and astrocytes that prevents the passage of materials from the blood to the cerebrospinal fluid and brain.

Blood pressure (BP) Force exerted by blood against the walls of blood vessels due to contraction of the heart and influenced by the elasticity of the vessel walls; clinically, a measure of the pressure in arteries during ventricular systole and ventricular diastole. *See also* **mean arterial blood pressure**.

Blood–testis barrier (BTB) A barrier formed by Sertoli cells that prevents an immune response against antigens produced by spermatogenic cells by isolating the cells from the blood.

Body cavity A space within the body that contains various internal organs.

Body fluid Body water and its dissolved substances; constitutes about 60% of total body mass.

Bolus (BŌ-lus) A soft, rounded mass, usually food, that is swallowed.

Brachial plexus (BRĀ-kē-al PLEK-sus) A network of nerve axons of the ventral rami of spinal nerves C5, C6, C7, C8, and T1. The nerves that emerge from the brachial plexus supply the upper limb.

Bradycardia (brăd'-i-KAR-dē-a) A slow resting heart or pulse rate (under 50 beats per minute).

Brain The part of the central nervous system contained within the cranial cavity.

Brain stem The portion of the brain immediately superior to the spinal cord, made up of the medulla oblongata, pons, and midbrain.

Brain waves Electrical signals that can be recorded from the skin of the head due to electrical activity of brain neurons.

Broca's area (BRŌ-kaz) Motor area of the brain in the frontal lobe that translates thoughts into speech. Also called the **motor speech area**.

Bronchi (BRONG-kī) Branches of the respiratory passageway including primary bronchi (the two divisions of the trachea), secondary or lobar bronchi (divisions of the primary bronchi that are distributed to the lobes of the lungs), and tertiary or segmental bronchi (divisions of the secondary bronchi that are distributed to bronchopulmonary segments of the lungs). *Singular* is **bronchus**.

Bronchial tree The trachea, bronchi, and their branching structures up to and including the terminal bronchioles.

Bronchiole (BRONG-kē-ōl) Branch of a tertiary bronchus further dividing into terminal bronchioles (distributed to lobules of the lungs), which divide into respiratory bronchioles (distributed to alveolar sacs).

Bronchitis (brong-KĪ-tis) Inflammation of the mucous membrane of the bronchial tree; characterized by hypertrophy and hyperplasia of seromucous glands and goblet cells that line the bronchi and which results in a productive cough.

Buccal (BUK-al) Pertaining to the cheek or mouth.

Buffer system (BUF-er) A pair of chemicals—one a weak acid and the other the salt of the weak acid, which functions as a weak base—that resists changes in pH.

Bulbourethral gland (bul'-bō-ū-RĒ-thral) One of a pair of glands located inferior to the prostate on either side of the urethra that secretes an alkaline fluid into the urethra. Also called a **Cowper's gland** (KOW-perz).

Bulimia (boo-LIM-ē-a *or* boo-LĒ-mē-a) A disorder characterized by overeating at least twice a week followed by purging by self-induced vomiting, strict dieting or fasting, vigorous exercise, or use of laxatives or diuretics. Also called **binge–purge syndrome**.

Bulk flow The movement of large numbers of ions, molecules, or particles in the same direction due to pressure differences (osmotic, hydrostatic, or air pressure).

Bulk-phase endocytosis (pi'-nō-sī-TŌ-sis) A process by which most body cells can ingest membrane-surrounded droplets of interstitial fluid.

Bundle branch One of the two branches of the atrioventricular (AV) bundle made up of specialized muscle fibers (cells) that transmit electrical impulses to the ventricles.

Bursa (BUR-sa) A sac or pouch of synovial fluid located at friction points, especially about joints.

Buttocks (BUT-oks) The two fleshy masses on the posterior aspect of the inferior trunk, formed by the gluteal muscles.

C

Calcaneal tendon (kal-KĀ-nē-al) The tendon of the soleus, gastrocnemius, and plantaris muscles at the back of the heel. Also called the **Achilles tendon** (a-KIL-ēz).

Calcification (kal-si-fi-KĀ-shun) Deposition of mineral salts, primarily hydroxyapatite, in a framework formed by collagen fibers in which the tissue hardens. Also called **mineralization** (min'-e-ral-i-ZĀ-shun).

Calcitonin (CT) (kal-si-TŌ-nin) A hormone produced by the parafollicular cells of the thyroid gland that can lower the amount of blood calcium and phosphates by inhibiting bone resorption (breakdown of bone extracellular matrix) and by accelerating uptake of calcium and phosphates into bone extracellular matrix.

Calculus (KAL-kū-lus) A stone, or insoluble mass of crystallized salts or other material, formed within the body, as in the gallbladder, kidney, or urinary bladder.

Callus (KAL-lus) A growth of new bone tissue in and around a fractured area, ultimately replaced by mature bone. An acquired, localized thickening.

Calorie (KAL-ō-rē) A unit of heat. A calorie (cal) is the standard unit and is the amount of heat needed to raise the temperature of 1 g of water from 14 °C to 15 °C. The **kilocalorie (kcal)** or **Calorie** (spelled with an uppercase C), used to express the caloric value of foods and to measure metabolic rate, is equal to 1000 cal.

Calyx (KĀL-iks) Any cuplike division of the kidney pelvis. *Plural is* **calyces** (KĀ-li-sēz).

Canal (ka-NAL) A narrow tube, channel, or passageway.

Canaliculus (kan'-a-LIK-ū-lus) A small channel or canal, as in bones, where they connect lacunae. *Plural is* **canaliculi** (kan'-a-LIK-ū-lī).

Cancellous (KAN-sel-us) Having a reticular or latticework structure, as in spongy bone tissue.

Capacitation (ka'-pas-i-TĀ-shun) The functional changes that sperm undergo in the female reproductive tract that allow them to fertilize a secondary oocyte.

Capillary (KAP-i-lar'-ē) A microscopic blood vessel located between an arteriole and venule through which materials are exchanged between blood and interstitial fluid.

Carbohydrate (kar'-bō-HĪ-drāt) An organic compound containing carbon, hydrogen, and oxygen in a particular amount and arrangement and composed of monosaccharide subunits; usually has the general formula $(CH_2O)_n$.

Carcinogen (kar-SIN-ō-jen) A chemical substance or radiation that causes cancer.

Cardiac arrest (KAR-dē-ak) Cessation of an effective heartbeat in which the heart is completely stopped or in ventricular fibrillation.

Cardiac cycle A complete heartbeat consisting of systole (contraction) and diastole (relaxation) of both atria plus systole and diastole of both ventricles.

Cardiac muscle Striated muscle fibers (cells) that form the wall of the heart; stimulated by an intrinsic conduction system and regulated by autonomic motor neurons.

Cardiac notch An angular notch in the anterior border of the left lung into which part of the heart fits.

Cardiac output (CO) The volume of blood pumped from one ventricle of the heart (usually measured from the left ventricle) in 1 min; normally about 5.2 liters/min in an adult at rest.

Cardiology (kar-dē-OL-ō-jē) The study of the heart and diseases associated with it.

Cardiovascular center (kar-dē-ō-VAS-kū-lar) Groups of neurons scattered within the medulla oblongata that regulate heart rate, force of contraction, and blood vessel diameter.

Carotene (KAR-o-tēn) Antioxidant precursor of vitamin A, which is needed for synthesis of photopigments; yellow-orange pigment present in the stratum corneum of the epidermis. Accounts for the yellowish coloration of skin. Also termed **beta-carotene**.

Carotid body (ka-ROT-id) Cluster of chemoreceptors on or near the carotid sinus that respond to changes in blood levels of oxygen, carbon dioxide, and hydrogen ions.

Carotid sinus A dilated region of the internal carotid artery just above the point where it branches from the common carotid artery; it contains baroreceptors that monitor blood pressure.

Carotid sinus reflex A reflex that helps maintain normal blood pressure in the brain. Nerve impulses propagate from the carotid sinus baroreceptors over sensory axons in the glossopharyngeal (IX) nerves to the cardiovascular center in the medulla oblongata.

Carpal bones The eight bones of the wrist. Also called **carpals**.

Carpus (KAR-pus) A collective term for the eight bones of the wrist.

Cartilage (KAR-ti-lij) A type of connective tissue consisting of chondrocytes in lacunae embedded in a dense network of collagen and elastic fibers and an extracellular matrix of chondroitin sulfate.

Cartilaginous joint (kar′-ti-LAJ-i-nus) A joint without a synovial (joint) cavity where the articulating bones are held tightly together by cartilage, allowing little or no movement.

Cast A small mass of hardened material formed within a cavity in the body and then discharged from the body; can originate in different areas and can be composed of various materials.

Catabolism (ka-TAB-ō-lizm) Chemical reactions that break down complex organic compounds into simple ones, with the net release of energy.

Cataract (KAT-a-rakt) Loss of transparency of the lens of the eye or its capsule or both.

Cation (KAT-Ī-on) A positively charged ion. An example is a sodium ion (Na$^+$).

Cauda equina (KAW-da ē-KWĪ-na) A tail-like array of roots of spinal nerves at the inferior end of the spinal cord.

Cecum (SĒ-kum) A blind pouch at the proximal end of the large intestine that attaches to the ileum.

Cell The basic structural and functional unit of all organisms; the smallest structure capable of performing all the activities vital to life.

Cell cycle Growth and division of a single cell into two identical cells; consists of interphase and cell division.

Cell division Process by which a cell reproduces itself that consists of a nuclear division (mitosis) and a cytoplasmic division (cytokinesis); types include somatic and reproductive cell division.

Cell-mediated immunity That component of immunity in which specially sensitized T lymphocytes (T cells) attach to antigens to destroy them. Also called **cellular immunity**.

Cementum (se-MEN-tum) Calcified tissue covering the root of a tooth.

Center of ossification (os′-i-fi-KĀ-shun) An area in the cartilage model of a future bone where the cartilage cells hypertrophy and then secrete enzymes that result in the calcification of their matrix, resulting in the death of the cartilage cells, followed by the invasion of the area by osteoblasts that then lay down bone.

Central canal A microscopic tube running the length of the spinal cord in the gray commissure. A circular channel running longitudinally in the center of an osteon (haversian system) of mature compact bone, containing blood and lymphatic vessels and nerves. Also called a **haversian canal** (ha-VER-shan).

Central fovea (FŌ-vē-a) A depression in the center of the macula lutea of the retina, containing cones only and lacking blood vessels; the area of highest visual acuity (sharpness of vision).

Central nervous system (CNS) That portion of the nervous system that consists of the brain and spinal cord.

Centrioles (SEN-trē-ōlz) Paired, cylindrical structures of a centrosome, each consisting of a ring of microtubules and arranged at right angles to each other.

Centromere (SEN-trō-mēr) The constricted portion of a chromosome where the two chromatids are joined; serves as the point of attachment for the microtubules that pull chromatids during anaphase of cell division.

Centrosome (SEN-trō-sōm) A dense network of small protein fibers near the nucleus of a cell, containing a pair of centrioles and pericentriolar material.

Cephalic (se-FAL-ik) Pertaining to the head; superior in position.

Cerebellum (ser-e-BEL-um) The part of the brain lying posterior to the medulla oblongata and pons; governs balance and coordinates skilled movements.

Cerebral aqueduct (SER-ē-bral AK-we-dukt) A channel through the midbrain connecting the third and fourth ventricles and containing cerebrospinal fluid. Also termed the **aqueduct of Sylvius**.

Cerebral arterial circle A ring of arteries forming an anastomosis at the base of the brain between the internal carotid and basilar arteries and arteries supplying the cerebral cortex. Also called the **circle of Willis**.

Cerebral cortex The surface of the cerebral hemispheres, 2–4 mm thick, consisting of gray matter; arranged in six layers of neuronal cell bodies in most areas.

Cerebrospinal fluid (CSF) (se-rē′-brō-SPĪ-nal) A fluid produced by ependymal cells that cover choroid plexuses in the ventricles of the brain; the fluid circulates in the ventricles, the central canal, and the subarachnoid space around the brain and spinal cord.

Cerebrovascular accident (CVA) (se-rē′-brō-VAS-kū-lar) Destruction of brain tissue (infarction) resulting from obstruction or rupture of blood vessels that supply the brain. Also called a **stroke** or **brain attack**.

Cerebrum (SER-e-brum or se-RĒ-brum) The two hemispheres of the forebrain (derived from the telencephalon), making up the largest part of the brain.

Cerumen (se-ROO-men) Waxlike secretion produced by ceruminous glands in the external auditory meatus (ear canal). Also termed **ear wax**.

Ceruminous gland (se-ROO-mi-nus) A modified sudoriferous (sweat) gland in the external auditory meatus that secretes cerumen (ear wax).

Cervical ganglion (SER-vi-kul GANG-glē-on) A cluster of cell bodies of postganglionic sympathetic neurons located in the neck, near the vertebral column.

Cervical plexus (PLEK-sus) A network formed by nerve axons from the ventral rami of the first four cervical nerves and receiving gray rami communicates from the superior cervical ganglion.

Cervix (SER-viks) Neck; any constricted portion of an organ, such as the inferior cylindrical part of the uterus.

Chemical bond Force of attraction in a molecule or compound that holds its atoms together. Examples include ionic and covalent bonds.

Chemical element Unit of matter that cannot be broken apart into a simpler substance by ordinary chemical reactions. Examples include hydrogen (H), carbon (C), and oxygen (O).

Chemical reaction The combination or separation of atoms in which chemical bonds are formed or broken and new products with different properties are produced.

Chemoreceptor (kē′-mō-rē-SEP-tor) Sensory receptor that detects the presence of a specific chemical.

Chemotaxis (kē-mō-TAK-sis) Attraction of phagocytes to microbes by a chemical stimulus.

Chiasm (KĪ-azm) A crossing; especially the crossing of axons in the optic (II) nerve as they extend toward the opposite optic tract.

Chief cell The secreting cell of a gastric gland that produces pepsinogen, the precursor of the enzyme pepsin, and the enzyme gastric lipase. Also called a **zymogenic cell** (zī′-mō- JEN-ik). Cell in the parathyroid glands that secretes parathyroid hormone (PTH). Also called a **principal cell**.

Chiropractic (kī-rō-PRAK-tik) A system of treating disease by using one's hands to manipulate body parts, mostly the vertebral column.

Cholecystectomy (kō′-lē-sis-TEK-tō-mē) Surgical removal of the gallbladder.

Cholecystitis (kō′-lē-sis-TĪ-tis) Inflammation of the gallbladder.

Cholesterol (kō-LES-te-rol) Classified as a lipid, the most abundant steroid in animal tissues; located in cell membranes and used for the synthesis of steroid hormones and bile salts.

Cholinergic neuron (kō′-lin-ER-jik) A neuron that liberates acetylcholine as its neurotransmitter.

Chondrocyte (KON-drō-sīt) Cell of mature cartilage.

Chondroitin sulfate (kon-DROY-tin) An amorphous extracellular matrix material found outside connective tissue cells.

Chordae tendineae (KOR-dē TEN-di-nē-ē) Tendonlike, fibrous cords that connect atrioventricular valves of the heart with papillary muscles.

Chorion (KŌ-rē-on) The most superficial fetal membrane that becomes the principal embryonic portion of the placenta; serves a protective and nutritive function.

Chorionic villi (kō-rē-ON-ik VIL-lī) Fingerlike projections of the chorion that grow into the endometrium and contain fetal blood vessels.

Choroid (KŌ-royd) One of the vascular coats of the eyeball.

Choroid plexus (PLEK-sus) A network of capillaries located in the roof of each of the four ventricles of the brain; ependymal cells around choroid plexuses produce cerebrospinal fluid.

Chromatid (KRŌ-ma-tid) One of a pair of identical connected nucleoprotein strands that are joined at the centromere and separate during cell division, each becoming a chromosome of one of the two resulting cells.

Chromatin (KRŌ-ma-tin) The threadlike mass of genetic material, consisting of DNA and histone proteins, that is present in the nucleus of a nondividing or interphase cell.

Chromatolysis (krō-ma-TOL-i-sis) The breakdown of Nissl bodies into finely granular masses in the cell body of a neuron whose axon has been damaged.

Chromosome (KRŌ-mō-sōm) One of the small, threadlike structures in the nucleus of a cell, normally 46 in a human diploid cell, that bears the genetic material; composed of DNA and proteins (histones) that form a delicate chromatin thread during interphase; becomes packaged into compact rodlike structures that are visible under the light microscope during cell division.

Chronic (KRON-ik) Long term or frequently recurring; applied to a disease that is not acute.

Chronic obstructive pulmonary disease (COPD) A disease, such as bronchitis or emphysema, in which there is some degree of obstruction of airways and consequent increase in airway resistance.

Chyle (KĪL) The milky-appearing fluid found in the lacteals of the small intestine after absorption of lipids in food.

Chylomicron (kī-lō-MĪ-kron) Protein-coated sphericalstructure that contains triglycerides, phospholipids, and cholesterol and is absorbed into the lacteal of a villus in the small intestine.

Chyme (KĪM) The semifluid mixture of partly digested food and digestive secretions found in the stomach and small intestine during digestion of a meal.

Ciliary body (SIL-ē-ar′-ē) One of the three parts of the vascular tunic of the eyeball, the others being the choroid and the iris; includes the ciliary muscle and the ciliary processes.

Cilium (SIL-ē-um) A hair or hairlike process projecting from a cell that may be used to move the entire cell or to move substances along the surface of the cell. *Plural is* **cilia.**

Circadian rhythm (ser-KĀ-dē-an) A cycle of active and nonactive periods in organisms determined by internal mechanisms and repeating about every 24 hours.

Circular folds Permanent, deep, transverse folds in the mucosa and submucosa of the small intestine that increase the surface area for absorption. Also called **plicae circulares** (PLĪ-kē SER-kū-lar-ēs).

Circulation time Time required for blood to pass from the right atrium, through pulmonary circulation, back to the left ventricle, through systemic circulation to the foot, and back again to the right atrium; normally about 1 min.

Circumduction (ser′-kum-DUK-shun) A movement at a synovial joint in which the distal end of a bone moves in a circle while the proximal end remains relatively stable.

Cirrhosis (si-RŌ-sis) A liver disorder in which the parenchymal cells are destroyed and replaced by connective tissue.

Cisterna chyli (sis-TER-na KĪ-lē) The origin of the thoracic duct.

Cleavage The rapid mitotic divisions following the fertilization of a secondary oocyte, resulting in an increased number of progressively smaller cells, called blastomeres.

Climacteric (klī-mak-TER-ik) Cessation of the reproductive function in the female or decreased testicular activity in the male.

Climax The peak period or moments of greatest intensity during sexual excitement.

Clitoris (KLI-to-ris) An erectile organ of the female, located at the anterior junction of the labia minora, that is homologous to the male penis.

Clone (KLŌN) A population of identical cells.

Clot The end result of a series of biochemical reactions that changes liquid plasma into a gelatinous mass; specifically, the conversion of fibrinogen into a tangle of polymerized fibrin molecules.

Clot retraction (rē-TRAK-shun) The consolidation of a fibrin clot to pull a damaged tissue together.

Clotting Process by which a blood clot is formed. Also known as **coagulation** (cō-ag-ū-LĀ-shun).

Coccyx (KOK-six) The fused bones at the inferior end of the vertebral column.

Cochlea (KŌK-lē-a) A winding, cone-shaped tube forming a portion of the inner ear and containing the spiral organ (organ of Corti).

Cochlear duct The membranous cochlea consisting of a spirally arranged tube enclosed in the bony cochlea and lying along its outer wall. Also called the **scala media** (SCA-la MĒ-dē-a).

Coenzyme A nonprotein organic molecule that is associated with and activates an enzyme; many are derived from vitamins. An example is nico tinamide adenine dinucleotide (NAD), derived from the B vitamin niacin.

Coitus (KŌ-i-tus) Sexual intercourse.

Collagen (KOL-a-jen) A protein that is the main organic constituent of connective tissue.

Collateral circulation The alternate route taken by blood through an anastomosis.

Colliculus (ko-LIK-ū-lus) A small elevation.

Colloid (KOL-loyd) The material that accumulates in the center of thyroid follicles, consisting of thyroglobulin and stored thyroid hormones.

Colon The portion of the large intestine consisting of ascending, transverse, descending, and sigmoid portions.

Colostrum (kō-LOS-trum) A thin, cloudy fluid secreted by the mammary glands a few days prior to or after delivery before true milk is produced.

Column (KOL-um) Group of white matter tracts in the spinal cord.

Common bile duct A tube formed by the union of the common hepatic duct and the cystic duct that empties bile into the duodenum at the hepatopancreatic ampulla (ampulla of Vater).

Compact (dense) bone tissue Bone tissue that contains few spaces between osteons (haversian systems); forms the external portion of all bones and the bulk of the diaphysis (shaft) of long bones; is found immediately deep to the periosteum and external to spongy bone.

Complement (KOM-ple-ment) A group of at least 20 normally inactive proteins found in plasma that forms a component of innate and adaptive nonspecific resistance and immunity by bringing about cytolysis, inflammation, and opsonization.

Compound A substance that can be broken down into two or more other substances by chemical means.

Concha (KONG-ka) A scroll-like bone found in the nose. *Plural is* **conchae** (KONG-kē).

Concussion (kon-KUSH-un) Traumatic injury to the brain that produces no visible bruising but may result in abrupt, temporary loss of consciousness.

Conduction system A group of autorhythmic cardiac muscle fibers that generates and distributes electrical impulses to stimulate coordinated contraction of the heart chambers; includes the sinoatrial (SA) node, the atrioventricular (AV) node, the atrioventricular (AV) bundle, the right and left bundle branches, and the Purkinje fibers.

Conductivity (kon′-duk-TIV-i-tē) The ability of a cell to propagate (conduct) action potentials along its plasma membrane; characteristic of neurons and muscle fibers (cells).

Condyloid joint (KON-di-loyd) A synovial joint structured so that an oval-shaped condyle of one bone fits into an elliptical cavity of another bone, permitting side-to-side and back-and-forth movements, such as the joint at the wrist between the radius and carpals.

Cone (KŌN) The type of photoreceptor in the retina that is specialized for highly acute color vision in bright light.

Congenital (kon-JEN-i-tal) Present at the time of birth.

Conjunctiva (kon′-junk-TĪ-va) The delicate membrane covering the eyeball and lining the eyes.

Connective tissue One of the most abundant of the four basic tissue types in the body, performing the functions of binding and supporting; consists of relatively few cells in a generous extracellular matrix (the ground substance and fibers between the cells).

Consciousness (KON-shus-nes) A state of wakefulness in which an individual is fully alert, aware, and oriented, partly as a result of feedback between the cerebral cortex and reticular activating system.

Continuous conduction (kon-DUK-shun) Propagation of an action potential (nerve impulse) in a step-by-step depolarization of each adjacent area of an axon membrane.

Contraception (kon′-tra-SEP-shun) The prevention of fertilization or impregnation without destroying fertility.

Contractility (kon′-trak-TIL-i-tē) The ability of cells or parts of cells to actively generate force to undergo shortening for movements. Muscle fibers (cells) exhibit a high degree of contractility.

Contralateral (kon′-tra-LAT-er-al) On the opposite side; affecting the opposite side of the body.

Control center The component of a feedback system, such as the brain, that determines the point at which a controlled condition, such as body temperature, is maintained.

Conus medullaris (KŌ-nus med-ū-LAR-is) The tapered portion of the spinal cord inferior to the lumbar enlargement.

Convergence (con-VER-jens) A synaptic arrangement in which the synaptic end bulbs of several presynaptic neurons terminate on one postsynaptic neuron. The medial movement of the two eyeballs so that both are directed toward a near object being viewed in order to produce a single image.

Convulsion (con-VUL-shun) Violent, involuntary contractions or spasms of an entire group of muscles.

Cornea (KOR-nē-a) The nonvascular, transparent fibrous coat through which the iris of the eye can be seen.

Corona radiata The innermost layer of granulosa cells that is firmly attached to the zona pellucida around a secondary oocyte.

Coronary artery disease (CAD) A condition such as atherosclerosis that causes narrowing of coronary arteries so that blood flow to the heart is reduced. The result is **coronary heart disease (CHD)**, in which the heart muscle receives inadequate blood flow due to an interruption of its blood supply.

Coronary circulation The pathway followed by the blood from the ascending aorta through the blood vessels supplying the heart and returning to the right atrium. Also called **cardiac circulation**.

Coronary sinus (SĪ-nus) A wide venous channel on the posterior surface of the heart that collects the blood from the coronary circulation and returns it to the right atrium.

Corpus albicans (KOR-pus AL-bi-kanz) A white fibrous patch in the ovary that forms after the corpus luteum regresses.

Corpus callosum (kal-LŌ-sum) The great commissure of the brain between the cerebral hemispheres.

Corpus luteum (LOO-tē-um) A yellowish body in the ovary formed when a follicle has discharged its secondary oocyte; secretes estrogens, progesterone, relaxin, and inhibin.

Cortex (KOR-teks) An outer layer of an organ. The convoluted layer of gray matter covering each cerebral hemisphere.

Costal (KOS-tal) Pertaining to a rib.

Costal cartilage (KAR-ti-lij) Hyaline cartilage that attaches a rib to the sternum.

Cramp A spasmodic, usually painful contraction of a muscle.

Cranial cavity (KRĀ-nē-al) A body cavity formed by the cranial bones and containing the brain.

Cranial nerve One of 12 pairs of nerves that leave the brain; pass through foramina in the skull; and supply sensory and motor neurons to the head, neck, part of the trunk, and viscera of the thorax and abdomen. Each is designated by a Roman numeral and a name.

Cranium (KRĀ-nē-um) The skeleton of the skull that protects the brain and the organs of sight, hearing, and balance; includes the frontal, parietal, temporal, occipital, sphenoid, and ethmoid bones.

Creatine phosphate (KRĒ-a-tin FOS-fāt) Molecule in striated muscle fibers that contains high-energy phosphate bonds; used to gener-

ate ATP rapidly from ADP by transfer of a phosphate group. Also called **phosphocreatine** (fos'-fō-KRĒ-a-tin).

Crenation (krē-NĀ-shun) The shrinkage of red blood cells into knobbed, starry forms when they are placed in a hypertonic solution.

Crista (KRIS-ta) A crest or ridged structure. A small elevation in the ampulla of each semicircular duct that contains receptors for dynamic equilibrium.

Crossing-over The exchange of a portion of one chromatid with another during meiosis. It permits an exchange of genes among chromatids and is one factor that results in genetic variation of progeny.

Cryptorchidism (krip-TOR-ki-dizm) The condition of undescended testes.

Cupula (KUP-ū-la) A mass of gelatinous material covering the hair cells of a crista; a sensory receptor in the ampulla of a semicircular canal stimulated when the head moves.

Cushing syndrome (KUSH-ing) Condition caused by a hypersecretion of glucocorticoids characterized by spindly legs, "moon face," "buffalo hump," pendulous abdomen, flushed facial skin, poor wound healing, hyperglycemia, osteoporosis, hypertension, and increased susceptibility to disease.

Cutaneous (kū-TĀ-nē-us) Pertaining to the skin.

Cyanosis (sī-a-NŌ-sis) A blue or dark purple discoloration, most easily seen in nail beds and mucous membranes, that results from an increased concentration of deoxygenated (reduced) hemoglobin (more than 5 gm/dL).

Cyclic AMP (cyclic adenosine-3′,5′-monophosphate) Molecule formed from ATP by the action of the enzyme adenylate cyclase; serves as second messenger for some hormones and neurotransmitters.

Cyst (SIST) A sac with a distinct connective tissue wall, containing a fluid or other material.

Cystic duct (SIS-tik) The duct that carries bile between the gallbladder and the common bile duct.

Cystitis (sis-TĪ-tis) Inflammation of the urinary bladder.

Cytolysis (sī-TOL-i-sis) The rupture of living cells in which the contents leak out.

Cytokinesis (sī-tō-ki-NĒ-sis) Distribution of the cytoplasm into two separate cells during cell division; coordinated with nuclear division (mitosis).

Cytoplasm (SĪ-tō-plazm) Cytosol plus all organelles except the nucleus.

Cytoskeleton (sī′-tō-SKEL-e-ton) Complex internal structure of cytoplasm consisting of microfilaments, microtubules, and intermediate filaments.

Cytosol (SĪ-tō-sol) Fluid located within cells. Also called **intracellular fluid (ICF)** (in′-tra-SEL-ū-lar).

D

Deciduous (dē-SID-ū-us) Falling off or being shed seasonally or at a particular stage of development. In the body, referring to the first set of teeth.

Deep Away from the surface of the body or an organ.

Deep fascia (FASH-ē-a) A sheet of connective tissue wrapped around a muscle to hold it in place.

Defecation (def-e-KĀ-shun) The discharge of feces from the rectum.

Deglutition (dē-gloo-TISH-un) The act of swallowing.

Dehydration (dē-hī-DRĀ-shun) Excessive loss of water from the body or its parts.

Demineralization (de-min′-er-al-i-ZĀ-shun) Loss of calcium and phosphorus from bones.

Denaturation (de-nā-chur-Ā-shun) Disruption of the tertiary structure of a protein by heat, changes in pH, or other physical or chemical methods, in which the protein loses its physical properties and biological activity.

Dendrite (DEN-drīt) A neuronal process that carries electrical signals toward the cell body.

Dendritic cell (den-DRIT-ik) One type of antigen-presenting cell with long branchlike projections that commonly is present in mucosal linings such as the vagina, in the skin (Langerhans cells in the epidermis), and in lymph nodes.

Dental caries (KA-rēz) Gradual demineralization of the enamel and dentin of a tooth that may invade the pulp and alveolar bone. Also called **tooth decay**.

Dentin (DEN-tin) The bony tissues of a tooth enclosing the pulp cavity.

Dentition (den-TI-shun) The eruption of teeth. The number, shape, and arrangement of teeth.

Deoxyribonucleic acid (DNA) (dē-ok′-sē-rī′-bō-noo-KLĒ-ik) A nucleic acid constructed of nucleotides consisting of one of four bases (adenine, cytosine, guanine, or thymine), deoxyribose, and a phosphate group; encoded in the nucleotides is genetic information.

Depolarization (dē-pō-lar-i-ZĀ-shun) A reduction of voltage across a plasma meembrane; expressed as a change toward less negative (more positive) voltages on the interior surface of the plasma membrane.

Depression (de-PRESH-un) Movement in which a part of the body moves inferiorly.

Dermal papilla (pa-PILL-a) Fingerlike projection of the papillary region of the dermis that may contain blood capillaries or corpuscles of touch (Meissner corpuscles).

Dermatology (der-ma-TOL-ō-jē) The medical specialty dealing with diseases of the skin.

Dermatome (DER-ma-tōm) The cutaneous area developed from one embryonic spinal cord segment and receiving most of its sensory innervation from one spinal nerve. An instrument for incising the skin or cutting thin transplants of skin.

Dermis (DER-mis) A layer of dense irregular connective tissue lying deep to the epidermis.

Descending colon (KŌ-lon) The part of the large intestine descending from the left colic (splenic) flexure to the level of the left iliac crest.

Detritus (de-TRĪ-tus) Particulate matter produced by or remaining after the wearing away or disintegration of a substance or tissue; scales, crusts, or loosened skin.

Detrusor muscle (de-TROO-ser) Smooth muscle that forms the wall of the urinary bladder.

Developmental biology The study of development from the fertilized egg to the adult form.

Diagnosis (dī-ag-NŌ-sis) Distinguishing one disease from another or determining the nature of a disease from signs and symptoms by inspection, palpation, laboratory tests, and other means.

Dialysis (dī-AL-i-sis) The removal of waste products from blood by diffusion through a selectively permeable membrane.

Diaphragm (DĪ-a-fram) Any partition that separates one area from another, especially the dome-shaped skeletal muscle between the thoracic and abdominal cavities. Also a dome-shaped device that is placed over the cervix, usually with a spermicide, to prevent conception.

Diaphysis (dī-AF-i-sis) The shaft of a long bone.

Diarrhea (dī-a-RĒ-a) Frequent defecation of liquid feces caused by increased motility of the intestines.

Diarthrosis (dī-ar-THRŌ-sis) A freely movable joint; types are gliding, hinge, pivot, condyloid, saddle, and ball-and-socket.

Diastole (dī-AS-tō-lē) In the cardiac cycle, the phase of relaxation or dilation of the heart muscle, especially of the ventricles.

Diastolic blood pressure (dī-as-TOL-ik) The force exerted by blood on arterial walls during ventricular relaxation; the lowest blood pressure measured in the large arteries, normally about 70 mmHg in a young adult.

Diencephalon (dī′-en-SEF-a-lon) A part of the brain consisting of the thalamus, hypothalamus, epithalamus, and subthalamus.

Diffusion (dif-Ū-zhun) A passive process in which there is a net or greater movement of molecules or ions from a region of high concentration to a region of low concentration until equilibrium is reached.

Digestion (dī-JES-chun) The mechanical and chemical breakdown of food to simple molecules that can be absorbed and used by body cells.

Dilate (DĪ-lāt) To expand or swell.

Diploid (DIP-loyd) Having the number of chromosomes characteristically found in the somatic cells of an organism; having two haploid sets of chromosomes, one each from the mother and father. Symbolized 2n.

Direct motor pathways Collections of upper motor neurons with cell bodies in the motor cortex that project axons into the spinal cord, where they synapse with lower motor neurons or interneurons in the anterior horns. Also called the **pyramidal pathways** (pi-RAM-i-dal).

Disease Any change from a state of health.

Dislocation (dis-lō-KĀ-shun) Displacement of a bone from a joint with tearing of ligaments, tendons, and articular capsules. Also called **luxation** (luks-Ā-shun).

Dissect (di-SEKT) To separate tissues and parts of a cadaver or an organ for anatomical study.

Distal (DIS-tal) Farther from the attachment of a limb to the trunk; farther from the point of origin or attachment.

Diuretic (dī-ū-RET-ik) A chemical that increases urine volume by decreasing reabsorption of water, usually by inhibiting sodium reabsorption.

Diverticulum (dī-ver-TIK-ū-lum) A sac or pouch in the wall of a canal or organ, especially in the colon.

Dominant allele An allele that overrides the influence of an alternate allele on the homologous chromosome; the allele that is expressed.

Dorsiflexion (dor′-si-FLEK-shun) Bending the foot in the direction of the dorsum (upper surface).

Ductus arteriosus (DUK-tus ar-tē-rē-Ō-sus) A small vessel connecting the pulmonary trunk with the aorta; found only in the fetus.

Ductus (vas) deferens (DEF-er-ens) The duct that carries sperm from the epididymis to the ejaculatory duct. Also called the **seminal duct**.

Ductus epididymis (ep′-i-DID-i-mis) A tightly coiled tube inside the epididymis, distinguished into a head, body, and tail, in which sperm undergo maturation.

Ductus venosus (ve-NŌ-sus) A small vessel in the fetus that helps the circulation bypass the liver.

Duodenum (doo′-ō-DĒ-num or doo-OD-e-num) The first 25 cm (10 in.) of the small intestine, which connects the stomach and the jejunum.

Dura mater (DOO-ra MĀ-ter) The outermost of the three meninges (coverings) of the brain and spinal cord.

Dynamic equilibrium (ē-kwi-LIB-rē-um) The maintenance of body position, mainly the head, in response to sudden movements such as rotation.

Dysfunction (dis-FUNK-shun) Absence of completely normal function.

Dysmenorrhea (dis′-men-ō-RĒ-a) Painful menstruation.

Dyspnea (DISP-nē-a) Shortness of breath.

E

Eardrum A thin, semitransparent partition of fibrous connective tissue between the external auditory meatus and the middle ear. Also called the **tympanic membrane**.

Ectoderm The primary germ layer that gives rise to the nervous system and the epidermis of skin and its derivatives.

Ectopic (ek-TOP-ik) Out of the normal location, as in ectopic pregnancy.

Edema (e-DĒ-ma) An abnormal accumulation of interstitial fluid.

Effector (e-FEK-tor) An organ of the body, either a muscle or a gland, that is innervated by somatic or autonomic motor neurons.

Efferent arteriole (EF-er-ent ar-TĒ-rē-ōl) A vessel of the renal vascular system that carries blood from a glomerulus to a peritubular capillary.

Efferent ducts (EF-er-ent) A series of coiled tubes that transport sperm from the rete testis to the epididymis.

Ejaculation (e-jak-ū-LĀ-shun) The reflex ejection or expulsion of semen from the penis.

Ejaculatory duct (e-JAK-ū-la-tō-rē) A tube that transports sperm from the ductus (vas) deferens to the prostatic urethra.

Elasticity (e-las-TIS-i-tē) The ability of a tissue to return to its original shape after contraction or extension.

Electrocardiogram (**ECG** or **EKG**) (e-lek′-trō-KAR-dē-ō-gram) A recording of the electrical changes that accompany the cardiac cycle

that can be detected at the surface of the body; may be resting, stress, or ambulatory.

Electroencephalogram (EEG) (e-lek′-trō-en-SEF-a-lō-gram) A recording of the electrical activity of the brain from the scalp surface; used to diagnose certain diseases (such as epilepsy), furnish information regarding sleep and wakefulness, and confirm brain death.

Electrolyte (ē-LEK-trō-līt) Any compound that separates into ions when dissolved in water and that conducts electricity.

Electromyography (e-lek′-trō-mī-OG-ra-fē) Evaluation of the electrical activity of resting and contracting muscle to ascertain causes of muscular weakness, paralysis, involuntary twitching, and abnormal levels of muscle enzymes; also used as part of biofeedback studies.

Electron transport chain A sequence of electron carrier molecules on the inner mitochondrial membrane that undergo oxidation and reduction as they synthesize ATP.

Elevation (el-e-VĀ-shun) Movement in which a part of the body moves superiorly.

Embolism (EM-bō-lizm) Obstruction or closure of a vessel by an embolus.

Embolus (EM-bō-lus) A blood clot, bubble of air or fat from broken bones, mass of bacteria, or other debris or foreign material transported by the blood.

Embryo (EM-brē-ō) The young of any organism in an early stage of development; in humans, the developing organism from fertilization to the end of the eighth week of development.

Embryology (em′-brē-OL-ō-jē) The study of development from the fertilized egg to the end of the eighth week of development.

Emesis (EM-e-sis) Vomiting.

Emigration (em′-e-GRĀ-shun) Process whereby white blood cells (WBCs) leave the bloodstream by rolling along the endothelium, sticking to it, and squeezing between the endothelial cells. Adhesion molecules help WBCs stick to the endothelium. Also known as **migration** or **extravasation.**

Emission (ē-MISH-un) Propulsion of sperm into the urethra due to peristaltic contractions of the ducts of the testes, epididymides, and ductus (vas) deferens as a result of sympathetic stimulation.

Emmetropia (em′-e-TRŌ-pē-a) Normal vision in which light rays are focused exactly on the retina.

Emphysema (em′-fi′-SĒ-ma) A lung disorder in which alveolar walls disintegrate, producing abnormally large air spaces and loss of elasticity in the lungs; typically caused by exposure to cigarette smoke.

Emulsification (ē-mul′-si-fi-KĀ-shun) The dispersion of large lipid globules into smaller, uniformly distributed particles in the presence of bile.

Enamel (e-NAM-el) The hard, white substance covering the crown of a tooth.

Endocardium (en-dō-KAR-dē-um) The layer of the heart wall, composed of endothelium and connective tissue, that lines the inside of the heart and covers the valves and tendons that hold the valves open.

Endochondral ossification (en′-dō-KON-dral os′-i-fi-KĀ-shun) The replacement of cartilage by bone. Also called **intracartilaginous ossification** (in′-tra-kar′-ti-LAJ-i-nus).

Endocrine gland (EN-dō-krin) A gland that secretes hormones into interstitial fluid and then the blood; a ductless gland.

Endocrinology (en'-dō-kri-NOL-ō-jē) The science concerned with the structure and functions of endocrine glands and the diagnosis and treatment of disorders of the endocrine system.

Endocytosis (en'-dō-sī-TŌ-sis) The uptake into a cell of large molecules and particles in which a segment of plasma membrane surrounds the substance, encloses it, and brings it in; includes phagocytosis, bulk phase endocytosis, and receptor-mediated endocytosis.

Endoderm (EN-dō-derm) A primary germ layer of the developing embryo; gives rise to the gastrointestinal tract, urinary bladder, urethra, and respiratory tract.

Endodontics (en'-dō-DON-tiks) The branch of dentistry concerned with the prevention, diagnosis, and treatment of diseases that affect the pulp, root, periodontal ligament, and alveolar bone.

Endogenous (en-DOJ-e-nus) Growing from or beginning within the organism.

Endolymph (EN-dō-limf') The fluid within the membranous labyrinth of the internal ear.

Endometrium (en'-dō-MĪ-trē-um) The mucous membrane lining the uterus.

Endoplasmic reticulum (ER) (en'-do-PLAZ-mik re-TIK-ū-lum) A network of channels running through the cytoplasm of a cell that serves in intracellular transportation, support, storage, synthesis, and packaging of molecules. Portions of ER where ribosomes are attached to the outer surface are called **rough ER**; portions that have no ribosomes are called **smooth ER**.

Endorphin (en-DOR-fin) A neuropeptide in the central nervous system that acts as a painkiller.

Endosteum (en-DOS-tē-um) The membrane that lines the medullary (marrow) cavity of bones, consisting of osteogenic cells and scattered osteoclasts.

Endothelium (en'-dō-THĪ-lē-um) The layer of simple squamous epithelium that lines the cavities of the heart, blood vessels, and lymphatic vessels.

Energy The capacity to do work.

Enkephalin (en-KEF-a-lin) A peptide found in the central nervous system that acts as a painkiller.

Enteric nervous system (EN-ter-ik) The part of the nervous system that is embedded in the submucosa and muscularis of the gastrointestinal (GI) tract; governs motility and secretions of the GI tract.

Enzyme (EN-zīm) A substance that accelerates chemical reactions, usually a protein.

Eosinophil (ē'-ō-SIN-ō-fil) A type of white blood cell characterized by granules that stain red or pink with acid dyes.

Ependymal cells (e-PEN-de-mal) Neuroglial cell that cover choroid plexuses and produce cerebrospinal fluid (CSF); they also line the ventricles of the brain and probably assist in the circulation of CSF.

Epicardium (ep'-i-KAR-dē-um) The thin outer layer of the heart wall, composed of serous tissue and mesothelium. Also called the **visceral pericardium**.

Epidemiology (ep'-i-dē-mē-OL-ō-jē) Study of the occurrence and distribution of diseases and disorders in human populations.

Epidermis (ep-i-DERM-is) The superficial, thinner layer of skin, composed of keratinized stratified squamous epithelium.

Epididymis (ep'-i-DID-i-mis) A comma-shaped organ that lies along the posterior border of the testis and contains the ductus epididymis, in which sperm undergo maturation. *Plural is* **epididymides** (ep'-i-DID-i-mi-dēz).

Epidural space (ep'-i-DOO-ral) A space between the spinal dura mater and the vertebral canal, containing areolar connective tissue and a plexus of veins.

Epiglottis (ep'-i-GLOT-is) A large, leaf-shaped piece of elastic cartilage lying on top of the larynx, attached to the thyroid cartilage and its unattached portion is free to move up and down to cover the glottis (vocal folds and rima glottidis) during swallowing.

Epinephrine (ep-ē-NEF-rin) Hormone secreted by the adrenal medulla that produces actions similar to those that result from sympathetic stimulation. Also called **adrenaline** (a-DREN-a-lin).

Epineurium (ep'-i-NOO-rē-um) The superficial connective tissue covering around an entire nerve.

Epiphyseal line (ep'-i-FIZ-ē-al) The remnant of the epiphyseal plate in the metaphysis of a long bone.

Epiphyseal plate (ep'-i-FIZ-ē-al) The hyaline cartilage plate in the metaphysis of a long bone; site of lengthwise growth of long bones.

Epiphysis (ē-PIF-i-sis) The end of a long bone, usually larger in diameter than the shaft (diaphysis).

Episiotomy (e-piz'-ē-OT-ō-mē) A cut made with surgical scissors to avoid tearing of the perineum at the end of the second stage of labor.

Epistaxis (ep'-i-STAK-sis) Loss of blood from the nose due to trauma, infection, allergy, neoplasm, and bleeding disorders. Also called **nosebleed**.

Epithelial tissue (ep'-i-THĒ-lē-al) The tissue that forms innermost and outermost surfaces of body structures and forms the secreting portion of glands.

Erectile dysfunction Failure to maintain an erection long enough for sexual intercourse. Also known as **impotence** (IM-pō-tens).

Erection (ē-REK-shun) The enlarged and stiff state of the penis or clitoris resulting from the engorgement of the spongy erectile tissue with blood.

Erythema (er'-i-THĒ-ma) Skin redness usually caused by dilation of the capillaries.

Erythrocyte (e-RITH-rō-sīt) A mature red blood cell.

Erythropoiesis (e-rith'-rō-poy-Ē-sis) The process by which red blood cells are formed.

Erythropoietin (e-rith'-rō-POY-e-tin) A hormone released by the kidneys that stimulates red blood cell production.

Esophagus (e-SOF-a-gus) The hollow muscular tube that connects the pharynx and the stomach.

Essential amino acids Those 10 amino acids that cannot be synthesized by the human body at an adequate rate to meet its needs and therefore must be obtained from the diet.

Estrogens (ES-tro-jenz) Feminizing sex hormones produced by the ovaries; govern development of oocytes, maintenance of female reproductive structures, and appearance of secondary sex characteristics; also affect fluid and electrolyte balance, and protein anabolism. Examples are β-estradiol, estrone, and estriol.

Etiology (ē′-tē-OL-ō-jē) The study of the causes of disease, including theories of the origin and organisms (if any) involved.

Eupnea (ŪP-nē-a) Normal quiet breathing.

Eversion (ē-VER-zhun) The movement of the sole laterally at the ankle joint or of an atrioventricular valve into an atrium during ventricular contraction.

Excitability (ek-sīt′-a-BIL-i-tē) The ability of muscle fibers to receive and respond to stimuli; the ability of neurons to respond to stimuli and generate nerve impulses.

Excrement (EKS-kre-ment) Material eliminated from the body as waste, especially fecal matter.

Excretion (eks-KRĒ-shun) The process of eliminating waste products from the body; also the products excreted.

Exhalation (eks-ha-LĀ-shun) Breathing out; expelling air from the lungs into the atmosphere. Also called **expiration**.

Exocrine gland (EK-sō-krin) A gland that secretes its products into ducts that carry the secretions into body cavities, into the lumen of an organ, or to the outer surface of the body.

Exocytosis (ex′-ō-sī-TŌ-sis) A process in which membrane-enclosed secretory vesicles form inside the cell, fuse with the plasma membrane, and release their contents into the interstitial fluid; achieves secretion of materials from a cell.

Exogenous (ex-SOJ-e-nus) Originating outside an organ or part.

Expiratory reserve volume (eks-PĪ-ra-tō-rē) The volume of air in excess of tidal volume that can be exhaled forcibly; about 1200 mL in males and 700 mL in females.

Extensibility (ek-sten′-si-BIL-i-tē) The ability of muscle tissue to stretch when it is pulled.

Extension (ek-STEN-shun) An increase in the angle between two bones; restoring a body part to its anatomical position after flexion.

External Located on or near the surface.

External auditory canal (AW-di-tōr-ē) or **meatus** (mē-Ā-tus) A curved tube in the temporal bone that leads to the middle ear.

External ear The outer ear, consisting of the pinna, external auditory canal, and tympanic membrane (eardrum).

External nares (NĀ-rez) The external nostrils, or the openings into the nasal cavity on the exterior of the body.

External respiration The exchange of respiratory gases between the lungs and blood. Also called **pulmonary respiration**.

Exteroceptor (eks′-ter-ō-SEP-tor) A sensory receptor adapted for the reception of stimuli from outside the body.

Extracellular fluid (ECF) Fluid outside body cells, such as interstitial fluid and plasma.

Extracellular matrix (MĀ-triks) The ground substance and fibers between cells in a connective tissue.

Extrinsic (ek-STRIN-sik) Of external origin.

Extrinsic pathway (of blood clotting) Sequence of reactions leading to blood clotting that is initiated by the release of tissue factor (TF), also known as thromboplastin, that leaks into the blood from damaged cells outside the blood vessels.

Exudate (EKS-oo-dāt) Escaping fluid or semifluid material that oozes from a space and that may contain serum, pus, and cellular debris.

Eyebrow The hairy ridge superior to the eye.

F

Face The anterior aspect of the head.

Facilitated diffusion (fa-SIL-i-tā-ted dif-Ū-zhun) Diffusion in which a substance not soluble by itself in lipids diffuses across a selectively permeable membrane with the help of a transporter protein.

Fascia (FASH-ē-a) A fibrous membrane covering, supporting, and separating muscles.

Fascicle (FAS-i-kul) A small bundle or cluster, especially of nerve or muscle fibers (cells). Also called a **fasciculus** (fa-SIK-ū-lus). *Plural is* **fasciculi** (fa-SIK-yū-lī).

Fasciculation (fa-sik′-ū-LĀ-shun) Abnormal, spontaneous twitch of all skeletal muscle fibers in one motor unit that is visible at the skin surface; not associated with movement of the affected muscle; present in progressive diseases of motor neurons, for example, poliomyelitis.

Fauces (FAW-sēz) The opening from the mouth into the pharynx.

Feces (FĒ-sēz) Material discharged from the rectum and made up of bacteria, excretions, and food residue. Also called **stool**.

Feedback system A sequence of events in which information about the status of a situation is continually reported (fed back) to a control center.

Female reproductive cycle General term for the ovarian and uterine cycles, the hormonal changes that accompany them, and cyclic changes in the breasts and cervix; includes changes in the endometrium of a nonpregnant female that prepares the lining of the uterus to receive a fertilized ovum.

Fertilization (fer′-ti-li-ZĀ-shun) Penetration of a secondary oocyte by a sperm cell, meiotic division of secondary oocyte to form an ovum, and subsequent union of the nuclei of the gametes.

Fetal circulation The cardiovascular system of the fetus, including the placenta and special blood vessels involved in the exchange of materials between fetus and mother.

Fetus (FĒ-tus) In humans, the developing organism *in utero* from the beginning of the ninth week to birth.

Fever An elevation in body temperature above the normal temperature of 37 °C (98.6 °F) due to a resetting of the hypothalamic thermostat.

Fibrillation (fi-bri-LĀ-shun) Abnormal, spontaneous twitch of a single skeletal muscle fiber (cell) that can be detected with electromyography but is not visible at the skin surface; not associated with movement of the affected muscle; present in certain disorders of motor neurons, for example, amyotrophic lateral sclerosis (ALS). With reference to cardiac muscle, *see* **Atrial fibrillation** and **Ventricular fibrillation**.

Fibrin (FĪ-brin) An insoluble protein that is essential to blood clotting; formed from fibrinogen by the action of thrombin.

Fibrinogen (fī-BRIN-ō-jen) A clotting factor in blood plasma that by the action of thrombin is converted to fibrin.

Fibrinolysis (fī-bri-NOL-i-sis) Dissolution of a blood clot by the action of a proteolytic enzyme, such as plasmin (fibrinolysin), that dissolves fibrin threads and inactivates fibrinogen and other blood-clotting factors.

Fibroblast (FĪ-brō-blast) A large, flat cell that secretes most of the extracellular matrix material of areolar and dense connective tissues.

Fibrous joint (FĪ-brus) A joint that allows little or no movement, such as a suture or a syndesmosis.

Fibrous tunic (TOO-nik) The superficial coat of the eyeball, made up of the posterior sclera and the anterior cornea.

Fight-or-flight response The effects produced upon stimulation of the sympathetic division of the autonomic nervous system.

Filtration (fil-TRĀ-shun) The flow of a liquid through a filter (or membrane that acts like a filter) due to a hydrostatic pressure; occurs in capillaries due to blood pressure.

Filtration membrane Site of blood filtration in nephrons of the kidneys, consisting of the endothelium and basement membrane of the glomerulus and the epithelium of the visceral layer of the glomerular (Bowman's) capsule.

Fissure (FISH-ur) A groove, fold, or slit that may be normal or abnormal.

Fixed macrophage (MAK-rō-fāj) Stationary phagocytic cell found in the liver, lungs, brain, spleen, lymph nodes, subcutaneous tissue, and red bone marrow. Also called a **histiocyte** (HIS-tē-ō-sīt).

Flaccid (FLAS-sid) Relaxed, flabby, or soft; lacking muscle tone.

Flagellum (fla-JEL-um) A hairlike, motile process on the extremity of a bacterium, protozoan, or sperm cell. *Plural is* **flagella** (fla-JEL-a).

Flatus (FLĀ-tus) Gas in the stomach or intestines, commonly used to denote expulsion of gas through the anus.

Flexion (FLEK-shun) Movement in which there is a decrease in the angle between two bones.

Flexor reflex A protective reflex in which flexor muscles are stimulated while extensor muscles are inhibited.

Follicle (FOL-i-kul) A small secretory sac or cavity; the group of cells that contains a developing oocyte in the ovaries.

Follicle-stimulating hormone (FSH) Hormone secreted by the anterior pituitary that initiates development of ova and stimulates the ovaries to secrete estrogens in females, and initiates sperm production in males.

Fontanel (fon'-ta-NEL) A space filled with mesenchyme where bone formation is not yet complete, especially between the cranial bones of an infant's skull.

Foot The terminal part of the lower limb, from the ankle to the toes.

Foramen (fō-RĀ-men) A passage or opening; a communication between two cavities of an organ, or a hole in a bone for passage of vessels or nerves. *Plural is* **foramina** (fō-RĀM-i-na).

Foramen ovale (fō-RĀ-men ō-VAL-ē) An opening in the fetal heart in the septum between the right and left atria. A hole in the greater wing of the sphenoid bone that transmits the mandibular branch of the trigeminal (V) nerve.

Forearm (FOR-arm) The part of the upper limb between the elbow and the wrist.

Fossa (FOS-a) A furrow or shallow depression.

Fourth ventricle (VEN-tri-kul) A cavity filled with cerebrospinal fluid within the brain lying between the cerebellum and the medulla oblongata and pons.

Fracture (FRAK-choor) Any break in a bone.

Frenulum (FREN-ū-lum) A small fold of mucous membrane that connects two parts and limits movement.

Frontal plane A plane at a right angle to a midsagittal plane that divides the body or organs into anterior and posterior portions. Also called a **coronal plane** (kō-RŌ-nal).

Functional residual capacity (re-ZID-ū-al) The sum of residual volume plus expiratory reserve volume; about 2400 mL in males and 1800 mL in females.

Fundus (FUN-dus) The part of a hollow organ farthest from the opening.

Fungiform papilla (FUN-ji-form pa-PIL-a) A mushroomlike elevation on the upper surface of the tongue appearing as a red dot; most contain taste buds.

G

Gallbladder A small pouch, located inferior to the liver, that stores bile and empties by means of the cystic duct.

Gallstone A solid mass, usually containing cholesterol, in the gallbladder or a bile-containing duct; formed anywhere between bile canaliculi in the liver and the hepatopancreatic ampulla (ampulla of Vater), where bile enters the duodenum. Also called a **biliary calculus.**

Gamete (GAM-ēt) A male or female reproductive cell; a sperm cell or secondary oocyte.

Ganglion (GANG-glē-on) Usually, a group of neuronal cell bodies lying outside the central nervous system (CNS). *Plural is* **ganglia** (GANG-glē-a).

Gastric glands (GAS-trik) Glands in the mucosa of the stomach composed of cells that empty their secretions into narrow channels called gastric pits. Types of cells are chief cells (secrete pepsinogen), parietal cells (secrete hydrochloric acid and intrinsic factor), surface mucous and mucous neck cells (secrete mucus), and G cells (secrete gastrin).

Gastroenterology (gas'-trō-en'-ter-OL-ō-jē) The medical specialty that deals with the structure, function, diagnosis, and treatment of diseases of the stomach and intestines.

Gastrointestinal (gas-trō-in-TES-ti-nal) **(GI) tract** A continuous tube extending from the mouth to the anus. Also called the **alimentary canal** (al'-i-MEN-tar-ē).

Gastrulation (gas'-troo-LĀ-shun) The migration of groups of cells from the epiblast that transform a bilaminar embryonic disc into a trilaminar embryonic disc that consists of the three primary germ layers; transformation of the blastula into the gastrula.

Gene (JĒN) Biological unit of heredity; a segment of DNA located in a definite position on a particular chromosome; a sequence of DNA that codes for a particular mRNA, rRNA, or tRNA.

Generator potential The graded depolarization that results in a change in the resting membrane potential in a receptor (specialized neuronal ending); may trigger a nerve action potential (nerve impulse) if depolarization reaches threshold.

Genetic engineering The manufacture and manipulation of genetic material.

Genetics The study of genes and heredity.

Genitalia (jen'-i-TĀ-lē-a) Reproductive organs.

Genome (JĒ-nōm) The complete set of genes of an organism.

Genotype (JĒ-nō-tīp) The genetic makeup of an individual; the combination of alleles present at one or more chromosomal locations, as distinguished from the appearance, or phenotype, that results from those alleles.

Geriatrics (jer'-ē-AT-riks) The branch of medicine devoted to the medical problems and care of elderly persons.

Gestation (jes-TĀ-shun) The period of development from fertilization to birth.

Gingivae (jin-JI-vē) Gums. They cover the alveolar processes of the mandible and maxilla and extend slightly into each socket.

Gland Specialized epithelial cell or cells that secrete substances; may be exocrine or endocrine.

Glans penis (GLANZ PĒ-nis) The slightly enlarged region at the distal end of the penis.

Glaucoma (glaw-KŌ-ma) An eye disorder in which there is increased intraocular pressure due to an excess of aqueous humor.

Gliding joint A synovial joint having articulating surfaces that are usually flat, permitting only side-to-side and back-and-forth movements, as between carpal bones, tarsal bones, and the scapula and clavicle.

Glomerular capsule (glō-MER-ū-lar) A double-walled globe at the proximal end of a nephron that encloses the glomerular capillaries. Also called **Bowman's capsule** (BŌ-manz).

Glomerular filtrate (glō-MER-ū-lar FIL-trāt) The fluid produced when blood is filtered by the filtration membrane in the glomeruli of the kidneys.

Glomerular filtration The first step in urine formation in which substances in blood pass through the filtration membrane and the filtrate enters the proximal convoluted tubule of a nephron.

Glomerular filtration rate (GFR) The total volume of fluid that enters all the glomerular (Bowman's) capsules of the kidneys in 1 min; about 100–125 mL/min.

Glomerulus (glō-MER-ū-lus) A rounded mass of nerves or blood vessels, especially the microscopic tuft of capillaries that is surrounded by the glomerular (Bowman's) capsule of each kidney tubule. *Plural* is **glomeruli**.

Glottis (GLOT-is) The vocal folds (true vocal cords) in the larynx plus the space between them (rima glottidis).

Glucagon (GLOO-ka-gon) A hormone produced by the alpha cells of the pancreatic islets (islets of Langerhans) that increases blood glucose level.

Glucocorticoids (gloo-kō-KOR-ti-koyds) Hormones secreted by the cortex of the adrenal gland, especially cortisol, that influence glucose metabolism.

Gluconeogenesis (gloo'-kō-nē-ō-JEN-e-sis) The synthesis of glucose from certain amino acids or lactic acid.

Glucose (GLOO-kōs) A six-carbon sugar, $C_6H_{12}O_6$, that is a major energy source for the production of ATP by body cells.

Glucosuria (gloo'-kō-SOO-rē-a) The presence of glucose in the urine; may be temporary or pathological.

Glycogen (GLĪ-kō-jen) A highly branched polymer of glucose containing thousands of subunits; functions as a compact store of glucose molecules in liver and muscle fibers (cells).

Glycogenesis (glī'-kō-JEN-e-sis) The chemical reactions by which many molecules of glucose are used to synthesize glycogen.

Glycogenolysis (glī-kō-je-NOL-i-sis) The breakdown of glycogen into glucose.

Glycolysis (glī-KOL-i-sis) Series of chemical reactions in the cytosol of a cell in which a molecule of glucose is split into two molecules of pyruvic acid with the net production of two ATPs.

Goblet cell A goblet-shaped unicellular gland that secretes mucus; present in epithelium of the airways and intestines.

Goiter (GOY-ter) An enlarged thyroid gland.

Golgi complex (GOL-jē) An organelle in the cytoplasm of cells consisting of three to twenty flattened sacs (cisternae), stacked on one another, with expanded areas at their ends; functions in processing, sorting, packaging, and delivering proteins and lipids to the plasma membrane, lysosomes, and secretory vesicles.

Gomphosis (gom-FŌ-sis) A fibrous joint in which a cone-shaped peg fits into a socket.

Gonad (GŌ-nad) A gland that produces gametes and hormones; the ovary in the female and the testis in the male.

Gonadotropic hormone Anterior pituitary hormone that affects the gonads.

Gray matter Area in the central nervous system and ganglia containing neuronal cell bodies, dendrites, unmyelinated axons, axon terminals, and neuroglia; Nissl bodies impart a gray color and there is little or no myelin in gray matter.

Greater omentum (ō-MEN-tum) A large fold in the serosa of the stomach that hangs down like an apron anterior to the intestines.

Greater vestibular glands (ves-TIB-ū-lar) A pair of glands on either side of the vaginal orifice that open by a duct into the space between the hymen and the labia minora. Also called **Bartholin's glands** (BAR-tō-linz).

Groin (GROYN) The depression between the thigh and the trunk; the inguinal region.

Gross anatomy The branch of anatomy that deals with structures that can be studied without using a microscope. Also called **macroscopic anatomy**.

Growth An increase in size due to an increase in (1) the number of cells, (2) the size of existing cells as internal components increase in size, or (3) the size of intercellular substances.

Gustatory (GUS-ta-tō'-rē) Pertaining to taste.

Gynecology (gī'-ne-KOL-ō-jē) The branch of medicine dealing with the study and treatment of disorders of the female reproductive system.

Gynecomastia (gīn′-e-kō-MAS-tē-a) Excessive growth (benign) of the male mammary glands due to secretion of estrogens by an adrenal gland tumor (feminizing adenoma).

Gyrus (JĪ-rus) One of the folds of the cerebral cortex of the brain. *Plural* is **gyri** (JĪ-rī). Also called a **convolution**.

H

Hair A threadlike structure composed of dead, keratinized cells produced by hair follicles that develops in the dermis. Also called a **pilus** (PĪ-lus).

Hair follicle (FOL-li-kul) Structure, composed of epithelium and surrounding the root of a hair, from which hair develops.

Hair root plexus (PLEK-sus) A network of dendrites arranged around the root of a hair as free or naked nerve endings that are stimulated when a hair shaft is moved.

Hand The terminal portion of an upper limb, including the carpus, metacarpus, and phalanges.

Haploid (HAP-loyd) Having half the number of chromosomes characteristically found in the somatic cells of an organism; characteristic of mature gametes. Symbolized *n*.

Hard palate (PAL-at) The anterior portion of the roof of the mouth, formed by the maxillae and palatine bones and lined by mucous membrane.

Head The superior part of a human, cephalic to the neck. The superior or proximal part of a structure.

Heart A hollow muscular organ lying slightly to the left of the midline of the chest that pumps the blood through the cardiovascular system.

Heart block An arrhythmia (dysrhythmia) of the heart in which the atria and ventricles contract independently because of a blocking of electrical impulses through the heart at some point in the conduction system.

Heart murmur (MER-mer) An abnormal sound that consists of a flow noise that is heard before, between, or after the normal heart sounds, or that may mask normal heart sounds.

Heat exhaustion Condition characterized by cool, clammy skin, profuse perspiration, and fluid and electrolyte (especially sodium and chloride) loss that results in muscle cramps, dizziness, vomiting, and fainting. Also called **heat prostration**.

Heat stroke Condition produced when the body cannot easily lose heat and characterized by reduced perspiration and elevated body temperature. Also called **sunstroke**.

Hematocrit (Hct) (hē-MAT-ō-krit) The percentage of blood made up of red blood cells. Usually measured by centrifuging a blood sample in a graduated tube and then reading the volume of red blood cells and dividing it by the total volume of blood in the sample.

Hematology (hē′-ma-TOL-ō-jē) The study of blood.

Hemiplegia (hem-i-PLĒ-jē-a) Paralysis of the upper limb, trunk, and lower limb on one side of the body.

Hemoglobin (Hb) (hē′-mō-GLŌ-bin) A substance in red blood cells consisting of the protein globin and the iron-containing red pigment heme that transports most of the oxygen and some carbon dioxide in blood.

Hemolysis (hē-MOL-i-sis) The escape of hemoglobin from the interior of a red blood cell into the surrounding medium; results from disruption of the cell membrane by toxins or drugs, freezing or thawing, or hypotonic solutions.

Hemolytic disease of the newborn A hemolytic anemia of a newborn child that results from the destruction of the infant's erythrocytes (red blood cells) by antibodies produced by the mother; usually the antibodies are due to an Rh blood type incompatibility. Also called **erythroblastosis fetalis** (e-rith′-rō-blas-TŌ-sis fe-TAL-is).

Hemophilia (hē′-mō-FIL-ē-a) A hereditary blood disorder where there is a deficient production of certain factors involved in blood clotting, resulting in excessive bleeding into joints, deep tissues, and elsewhere.

Hemopoiesis (hē-mō-poy-Ē-sis) Blood cell production, which occurs in red bone marrow after birth. Also called **hematopoiesis** (hem′-a-tō-poy-Ē-sis).

Hemorrhage (HEM-or-rij) Bleeding; the escape of blood from blood vessels, especially when the loss is profuse.

Hemorrhoids (HEM-ō-royds) Dilated or varicosed blood vessels (usually veins) in the anal region. Also called **piles**.

Hemostasis (hē-MŌ-stā-sis) The stoppage of bleeding.

Heparin (HEP-a-rin) An anticoagulant given to slow the conversion of prothrombin to thrombin, thus reducing the risk of blood clot formation; found in basophils, mast cells, and various other tissues, especially the liver and lungs.

Hepatic (he-PAT-ik) Refers to the liver.

Hepatic duct A duct that receives bile from the bile capillaries. Small hepatic ducts merge to form the larger right and left hepatic ducts that unite to leave the liver as the common hepatic duct.

Hepatic portal circulation The flow of blood from the gastrointestinal organs to the liver before returning to the heart.

Hepatocyte (he-PAT-ō-cyte) A liver cell.

Hernia (HER-nē-a) The protrusion or projection of an organ or part of an organ through a membrane or cavity wall, usually the abdominal cavity.

Herniated disc (HER-nē-ā′-ted) A rupture of an intervertebral disc so that the nucleus pulposus protrudes into the vertebral cavity. Also called a **slipped disc**.

Heterozygous (he-ter-ō-ZĪ-gus) Possessing different alleles on homologous chromosomes for a particular hereditary trait.

Hiatus (hī-Ā-tus) An opening; a foramen.

Hinge joint A synovial joint in which a convex surface of one bone fits into a concave surface of another bone, such as the elbow, knee, ankle, and interphalangeal joints. Also called a **ginglymus joint** (JIN-gli-mus).

Hirsutism (HER-soot-izm) An excessive growth of hair in females and children, with a distribution similar to that in adult males, due to the conversion of vellus hairs into large terminal hairs in response to higher-than-normal levels of androgens.

Histamine (HISS-ta-mēn) Substance found in many cells, especially mast cells, basophils, and platelets, released when the cells are injured; results in vasodilation, increased permeability of blood vessels, and constriction of bronchioles.

Histology (hiss-TOL-ō-jē) Microscopic study of the structure of tissues.

Homeostasis (hō′-mē-ō-STĀ-sis) The condition in which the body's internal environment remains relatively constant, within physiological limits.

Hyperthermia (hī′-per-THERM-ē-a) An elevated body temperature.

Hypertonia (hī′-per-TŌ-nē-a) Increased muscle tone that is expressed as spasticity or rigidity.

Hypertonic (hī′-per-TON-ik) Solution that causes cells to shrink due to loss of water by osmosis.

Hypertrophy (hī-PER-trō-fē) An excessive enlargement or overgrowth of tissue without cell division.

Hyperventilation (hī′-per-ven-ti-LĀ-shun) A rate of respiration higher than that required to maintain a normal partial pressure of carbon dioxide in the blood.

Hypoglycemia (hī′-pō-glī-SĒ-mē-a) An abnormally low concentration of glucose in the blood; can result from excess insulin (injected or secreted).

Hypokalemia (hī′-pō-ka-LĒ-mē-a) Deficiency of potassium ions in the blood.

Hyponatremia (hī′-pō-na-TRĒ-mē-a) Deficiency of sodium ions in the blood.

Hypophyseal fossa (hī′-pō-FIZ-ē-al FOS-a) A depression on the superior surface of the sphenoid bone that houses the pituitary gland.

Hypophysis (hī-POF-i-sis) Pituitary gland.

Hyposecretion (hī′-pō-se-KRĒ-shun) Underactivity of glands resulting in diminished secretion.

Hypothalamus (hī′-pō-THAL-a-mus) A portion of the diencephalon, lying beneath the thalamus and forming the floor and part of the wall of the third ventricle.

Hypothermia (hī′-pō-THER-mē-a) Lowering of body temperature below 35 °C (95 °F); in surgical procedures, it refers to deliberate cooling of the body to slow down metabolism and reduce oxygen needs of tissues.

Hypotonia (hī′-pō-TŌ-nē-a) Decreased or lost muscle tone in which muscles appear flaccid.

Hypotonic (hī′-pō-TON-ik) Solution that causes cells to swell and perhaps rupture due to gain of water by osmosis.

Hypoventilation (hī-pō-ven-ti-LĀ-shun) A rate of respiration lower than that required to maintain a normal partial pressure of carbon dioxide in plasma.

Hypovolemic shock (hī-pō-vō-LĒ-mik) A type of shock characterized by decreased blood volume; may be caused by acute hemorrhage or excessive loss of other body fluids, for example, by vomiting, diarrhea, or excessive sweating.

Hypoxia (hī-POKS-ē-a) Lack of adequate oxygen at the tissue level.

Hysterectomy (hiss-te-REK-tō-mē) The surgical removal of the uterus.

Homologous chromosomes (hō-MOL-ō-gus) Two chromosomes that belong to a pair.

Homozygous (hō-mō-ZĪ-gus) Possessing the same alleles on homologous chromosomes for a particular hereditary trait.

Hormone (HOR-mōn) A secretion of endocrine cells that alters the physiological activity of target cells of the body.

Horn An area of gray matter (anterior, lateral, or posterior) in the spinal cord.

Human chorionic gonadotropin (hCG) (kō-rē-ON-ik gō-nad-ō-TRŌ-pin) A hormone produced by the developing placenta that maintains the corpus luteum.

Human growth hormone (hGH) Hormone secreted by the anterior pituitary that stimulates growth of body tissues, especially skeletal and muscular tissues. Also known as **somatotropin** and **somatotropic hormone (STH)**.

Hyaluronic acid (hī′a-loo-RON-ik) A viscous, amorphous extracellular material that binds cells together, lubricates joints, and maintains the shape of the eyeballs.

Hymen (HĪ-men) A thin fold of vascularized mucous membrane at the vaginal orifice.

Hypercalcemia (hī′-per-kal-SĒ-mē-a) An excess of calcium in the blood.

Hypercapnia (hī′-per-KAP-nē-a) An abnormal increase in the amount of carbon dioxide in the blood.

Hyperextension (hī′-per-ek-STEN-shun) Continuation of extension beyond the anatomical position, as in bending the head backward.

Hyperglycemia (hī′-per-glī-SĒ-mē-a) An elevated blood glucose level.

Hyperkalemia (hī′-per-kā-LĒ-mē-a) An excess of potassium ions in the blood.

Hypermetropia (hī′-per-mē-TRŌ-pē-a) A condition in which visual images are focused behind the retina, with resulting defective vision of near objects; farsightedness.

Hyperplasia (hī′-per-PLĀ-zē-a) An abnormal increase in the number of normal cells in a tissue or organ, increasing its size.

Hyperpolarization (hī′-per-PŌL-a-ri-zā′-shun) Increase in the internal negativity across a cell membrane, thus increasing the voltage and moving it farther away from the threshold value.

Hypersecretion (hī′-per-se-KRĒ-shun) Overactivity of glands resulting in excessive secretion.

Hypersensitivity (hī′-per-sen-si-TI-vi-tē) Overreaction to an allergen that results in pathological changes in tissues. Also called **allergy**.

Hypertension (hī′-per-TEN-shun) High blood pressure.

I

Ileocecal sphincter (il-ē-ō-SĒ-kal) A fold of mucous membrane that guards the opening from the ileum into the large intestine. Also called the **ileocecal valve**.

Ileum (IL-ē-um) The terminal part of the small intestine.

Immunity (im-Ū-ni-tē) The state of being resistant to injury, particularly by poisons, foreign proteins, and invading pathogens.

Immunoglobulin (Ig) (im-ū-nō-GLOB-ū-lin) An antibody synthesized by plasma cells derived from B lymphocytes in response to the introduction of an antigen. Immunoglobulins are divided into five kinds (IgG, IgM, IgA, IgD, IgE).

Immunology (im′-ū-NOL-ō-jē) The study of the responses of the body when challenged by antigens.

Implantation (im-plan-TĀ-shun) The insertion of a tissue or a part into the body. The attachment of the blastocyst to the stratum basalis of the endometrium about 6 days after fertilization.

Incontinence (in-KON-ti-nens) Inability to retain urine, semen, or feces through loss of sphincter control.

Indirect motor pathways Motor tracts that convey information from the brain down the spinal cord for automatic movements, coordination of body movements with visual stimuli, skeletal muscle tone and posture, and balance. Also known as **extrapyramidal pathways**.

Infarction (in-FARK-shun) A localized area of necrotic tissue, produced by inadequate oxygenation of the tissue.

Infection (in-FEK-shun) Invasion and multiplication of microorganisms in body tissues, which may be inapparent or characterized by cellular injury.

Inferior (in-FĒR-ē-or) Away from the head or toward the lower part of a structure. Also called **caudad** (KAW-dad).

Inferior vena cava (IVC) (VĒ-na CĀ-va) Large vein that collects deoxygenated blood from parts of the body inferior to the heart and returns it to the right atrium.

Infertility Inability to conceive or to cause conception. Also called **sterility**.

Inflammation (in′-fla-MĀ-shun) Localized, protective response to tissue injury designed to destroy, dilute, or wall off the infecting agent or injured tissue; characterized by redness, pain, heat, swelling, and sometimes loss of function.

Inflation reflex Reflex that prevents overinflation of the lungs. Also called the **Hering–Breuer reflex**.

Ingestion (in-JES-chun) The taking in of food, liquids, or drugs, by mouth.

Inguinal (IN-gwi-nal) Pertaining to the groin.

Inguinal canal An oblique passageway in the anterior abdominal wall just superior and parallel to the medial half of the inguinal ligament that transmits the spermatic cord and ilioinguinal nerve in the male and round ligament of the uterus and ilioinguinal nerve in the female.

Inhalation (in-ha-LĀ-shun) The act of drawing air into the lungs. Also termed **inspiration**.

Inheritance The acquisition of body traits by transmission of genetic information from parents to offspring.

Inhibin (in-HIB-in) A hormone secreted by the gonads that inhibits release of follicle-stimulating hormone (FSH) by the anterior pituitary.

Inhibiting hormone Hormone secreted by the hypothalamus that can suppress secretion of hormones by the anterior pituitary.

Inner cell mass A region of cells of a blastocyst that differentiates into the three primary germ layers—ectoderm, mesoderm, and endoderm—from which all tissues and organs develop.

Inorganic compound (in′-or-GAN-ik) Compound that usually lacks carbon, usually is small, and often contains ionic bonds. Examples include water and many acids, bases, and salts.

Insertion (in-SER-shun) The attachment of a muscle tendon to a movable bone or the end opposite the origin.

Inspiratory capacity (in-SPĪ-ra-tor-ē) Total inspiratory capacity of the lungs; the total of tidal volume plus inspiratory reserve volume; averages 3600 mL in males.

Inspiratory reserve volume (in-SPĪ-ra-tor-ē) Additional inspired air over and above tidal volume; averages 3100 mL.

Insulin (IN-soo-lin) A hormone produced by the beta cells of a pancreatic islet (islet of Langerhans) that decreases the blood glucose level.

Insulinlike growth factor (IGF) Small protein, produced by the liver and other tissues in response to stimulation by human growth hormone (hGH), that mediates most of the effects of human growth hormone. Previously called **somatomedin** (sō′-ma-tō-MĒ-din).

Integumentary (in-teg′-ū-MEN-tar-ē) Relating to the skin.

Intercalated disc (in-TER-ka-lāt-ed) An irregular transverse thickening of sarcolemma that contains desmosomes, which hold cardiac muscle fibers (cells) together, and gap junctions, which aid in conduction of muscle action potentials from one fiber to the next.

Intercostal nerve (in′-ter-KOS-tal) A nerve supplying a muscle located between the ribs.

Interferons (IFNs) (in′-ter-FĒR-ons) Antiviral proteins produced by virus-infected host cells; induce uninfected host cells to synthesize proteins that inhibit viral replication and enhance phagocytic activity of macrophages; types include alpha interferon, beta interferon, and gamma interferon.

Internal Away from the surface of the body.

Internal capsule A large tract of projection fibers lateral to the thalamus that is the major connection between the cerebral cortex and the brain stem and spinal cord; contains axons of sensory neurons carrying auditory, visual, and somatic sensory signals to the cerebral cortex plus axons of motor neurons descending from the cerebral cortex to the thalamus, subthalamus, brain stem, and spinal cord.

Internal ear The inner ear or labyrinth, lying inside the temporal bone, containing the organs of hearing and balance.

Internal nares (NĀ-rez) The two openings posterior to the nasal cavities opening into the nasopharynx. Also called the **choanae** (kō-Ā-nē).

Internal respiration The exchange of respiratory gases between blood and body cells. Also called **tissue respiration**.

Interneurons (in′-ter-NOO-ronz) Neurons whose axons extend only for a short distance and contact nearby neurons in the brain, spinal cord, or a ganglion; the vast majority of neurons in the body are interneurons.

Interoceptor (in′-ter-ō-SEP-tor) Sensory receptor located in blood vessels and viscera that provides information about the body's internal environment.

Interphase (IN-ter-fāz) The period of the cell cycle between cell divisions, consisting of the G_1-(gap or growth) phase, when the cell is engaged in growth, metabolism, and production of substances required for division; S-(synthesis) phase, during which chromosomes are replicated; and G_2-phase.

Interstitial fluid (in′-ter-STISH-al) The portion of extracellular fluid that fills the microscopic spaces between the cells of tissues; the internal environment of the body. Also called **inter cellular** or **tissue fluid**.

Interstitial growth Growth from within, as in the growth of cartilage. Also called **endogenous growth** (en-DOJ-e-nus).

Intervertebral disc (in'-ter-VER-te-bral) A pad of fibrocartilage located between the bodies of two vertebrae.

Intestinal gland A gland that opens onto the surface of the intestinal mucosa and secretes digestive enzymes. Also called a **crypt of Lieberkühn** (LĒ-ber-kun).

Intracellular fluid (ICF) (in'-tra-SEL-ū-lar) Fluid located within cells. Also called **cytosol** (SĪ-tō-sol).

Intramembranous ossification (in'-tra-MEM-bra-nus os'-i'-fi-KĀ-shun) The method of bone formation in which the bone is formed directly within mesenchyme arranged in sheetlike layers that resemble membranes.

Intraocular pressure (IOP) (in'-tra-OK-ū-lar) Pressure in the eyeball, produced mainly by aqueous humor.

Intrapleural pressure Air pressure between the two pleurae of the lungs, usually subatmospheric. Also called **intrathoracic pressure**.

Intrinsic (in-TRIN-sik) Of internal origin.

Intrinsic pathway (of blood clotting) Sequence of reactions leading to blood clotting that is initiated by damage to blood vessel endothelium or platelets; activators of this pathway are contained within blood itself or are in direct contact with blood.

Intrinsic factor (IF) A glycoprotein, synthesized and secreted by the parietal cells of the gastric mucosa, that facilitates vitamin B$_{12}$ absorption in the small intestine.

In utero (Ū-ter-ō) Within the uterus.

Invagination (in-vaj'-i-NĀ-shun) The pushing of the wall of a cavity into the cavity itself.

Inversion (in-VER-zhun) The movement of the sole medially at the ankle joint.

Ion (Ī-on) Any charged particle or group of particles; usually formed when a substance, such as a salt, dissolves and dissociates.

Ionization (ī'-on-i-ZĀ-shun) Separation of inorganic acids, bases, and salts into ions when dissolved in water. Also called **dissociation**.

Iris The colored portion of the vascular tunic of the eyeball seen through the cornea that contains circular and radial smooth muscle; the hole in the center of the iris is the pupil.

Irritable bowel syndrome (IBS) Disease of the entire gastrointestinal tract in which a person reacts to stress by developing symptoms (such as cramping and abdominal pain) associated with alternating patterns of diarrhea and constipation. Excessive amounts of mucus may appear in feces, and other symptoms include flatulence, nausea, and loss of appetite. Also known as **irritable colon** or **spastic colitis**.

Ischemia (is-KĒ-mē-a) A lack of sufficient blood to a body part due to obstruction or constriction of a blood vessel.

Isometric contraction (ī'-sō-MET-rik) A muscle contraction in which tension on the muscle increases, but there is only minimal muscle shortening so that no visible movement is produced.

Isotonic (ī'-sō-TON-ik) Having equal tension or tone. A solution having the same concentration of impermeable solutes as cytosol.

Isotonic contraction Contraction in which the tension remains the same; occurs when a constant load is moved through the range of motions possible at a joint.

Isotopes (Ī-sō-tōps') Chemical elements that have the same number of protons but different numbers of neutrons. Radioactive isotopes change into other elements with the emission of alpha or beta particles or gamma rays.

Isthmus (IS-mus) A narrow strip of tissue or narrow passage connecting two larger parts.

J

Jaundice (JAWN-dis) A condition characterized by yellowness of the skin, the white of the eyes, mucous membranes, and body fluids because of a buildup of bilirubin.

Jejunum (je-JOO-num) The middle part of the small intestine.

Joint kinesthetic receptor (kin'-es-THET-ik) A proprioceptive receptor located in a joint, stimulated by joint movement.

K

Keratin (KER-a-tin) An insoluble protein found in the hair, nails, and other keratinized tissues of the epidermis.

Keratinocyte (ke-RAT-in'-ō-sīt) The most numerous of the epidermal cells; produces keratin.

Ketone bodies (KĒ-tōn) Substances produced primarily during excessive triglyceride catabolism, such as acetone, acetoacetic acid, and b-hydroxybutyric acid.

Ketosis (kē-TŌ-sis) Abnormal condition marked by excessive production of ketone bodies.

Kidney (KID-nē) One of the paired reddish organs located in the lumbar region that regulates the composition, volume, and pressure of blood and produces urine.

Kidney stone A solid mass, usually consisting of calcium oxalate, uric acid, or calcium phosphate crystals, that may form in any portion of the urinary tract. Also called a **renal calculus** (KAL-kū-lus).

Kinesiology (ki-nē'-sē-OL-ō-jē) The study of the movement of body parts.

Kinesthesia (kin-es-THĒ-zē-a) The perception of the extent and direction of movement of body parts; this sense is possible due to nerve impulses generated by proprioceptors.

Korotkoff sounds (kō-ROT-kof) The various sounds that are heard while taking blood pressure.

Krebs cycle A series of biochemical reactions that occurs in the matrix of mitochondria in which electrons are transferred to coenzymes and carbon dioxide is formed. The electrons carried by the coenzymes then enter the electron transport chain, which generates a large quantity of ATP. Also called the **citric acid cycle** or **tricarboxylic acid (TCA) cycle**.

Kyphosis (kī-FŌ-sis) An exaggeration of the thoracic curve of the vertebral column, resulting in a "round-shouldered" appearance. Also called **hunchback**.

L

Labia majora (LĀ-bē-a ma-JŌ-ra) Two longitudinal folds of skin extending downward and backward from the mons pubis of the female.

Labia minora (min-OR-a) Two small folds of mucous membrane lying medial to the labia majora of the female.

Labial frenulum (LĀ-bē-al FREN-ū-lum) A medial fold of mucous membrane between the inner surface of the lip and the gums.

Labor The process of giving birth in which a fetus is expelled from the uterus through the vagina.

Lacrimal canal (LAK-ri-mal) A duct, one on each eyelid, beginning at the punctum at the medial margin of an eyelid and conveying tears medially into the nasolacrimal sac.

Lacrimal gland Secretory cells, located at the superior anterolateral portion of each orbit, that secrete tears into excretory ducts that open onto the surface of the conjunctiva.

Lacrimal sac The superior expanded portion of the nasolacrimal duct that receives the tears from a lacrimal canal.

Lactation (lak-TĀ-shun) The secretion and ejection of milk by the mammary glands.

Lacteal (LAK-tē-al) One of many lymphatic vessels in villi of the intestines that absorb triglycerides and other lipids from digested food.

Lacuna (la-KOO-na) A small, hollow space, such as that found in bones in which the osteocytes lie. *Plural* is **lacunae** (la-KOO-nē).

Lamellae (la-MEL-ē) Concentric rings of hard, calcified extracellular matrix found in compact bone.

Lamina propria (PRŌ-prē-a) The connective tissue layer of a mucosa.

Langerhans cell (LANG-er-hans) Epidermal dendritic cell that functions as an antigen-presenting cell (APC) during an immune response.

Lanugo (la-NOO-gō) Fine downy hairs that cover the fetus.

Large intestine The portion of the gastrointestinal tract extending from the ileum of the small intestine to the anus, divided structurally into the cecum, colon, rectum, and anal canal.

Laryngopharynx (la-rin′-gō-FAR-inks) The inferior portion of the pharynx, extending downward from the level of the hyoid bone that divides posteriorly into the esophagus and anteriorly into the larynx. Also called the **hypopharynx**.

Larynx (LAR-inks) The voice box, a short passageway that connects the pharynx with the trachea.

Lateral (LAT-er-al) Farther from the midline of the body or a structure.

Lateral ventricle (VEN-tri-kul) A cavity within a cerebral hemisphere that communicates with the lateral ventricle in the other cerebral hemisphere and with the third ventricle by way of the interventricular foramen.

Leg The part of the lower limb between the knee and the ankle.

Lens A transparent organ constructed of proteins (crystallins) lying posterior to the pupil and iris of the eyeball and anterior to the vitreous body.

Lesion (LĒ-zhun) Any localized, abnormal change in a body tissue.

Leukemia (loo-KĒ-mē-a) A malignant disease of the blood-forming tissues characterized by either uncontrolled production and accumulation of immature leukocytes in which many cells fail to reach maturity (acute) or an accumulation of mature leukocytes in the blood because they do not die at the end of their normal life span (chronic).

Leukocyte (LOO-kō-sīt) A white blood cell.

Leukocytosis (loo′-kō-sī-TŌ-sis) An increase in the number of white blood cells, above 10,000 per μL, characteristic of many infections and other disorders.

Leukopenia (loo-kō-PĒ-nē-a) A decrease in the number of white blood cells below 5000 cells per μL.

Leydig cell (LĪ-dig) A type of cell that secretes testosterone; located in the connective tissue between seminiferous tubules in a mature testis. Also known as **interstitial cell of Leydig**.

Libido (li-BĒ-dō) Sexual desire.

Ligament (LIG-a-ment) Dense regular connective tissue that attaches bone to bone.

Ligand (LĪ-gand) A chemical substance that binds to a specific receptor.

Limbic system A part of the forebrain, sometimes termed the visceral brain, concerned with various aspects of emotion and behavior; includes the limbic lobe, dentate gyrus, amygdala, septal nuclei, mammillary bodies, anterior thalamic nucleus, olfactory bulbs, and bundles of myelinated axons.

Lipase An enzyme that splits fatty acids from triglycerides and phospholipids.

Lipid (LIP-id) An organic compound composed of carbon, hydrogen, and oxygen that is usually insoluble in water, but soluble in alcohol, ether, and chloroform; examples include triglycerides (fats and oils), phospholipids, steroids, and eicosanoids.

Lipid bilayer Arrangement of phospholipid, glycolipid, and cholesterol molecules in two parallel layers in which the hydrophilic "heads" face outward and the hydrophobic "tails" face inward; found in cellular membranes.

Lipogenesis (li-pō-GEN-e-sis) The synthesis of triglycerides.

Lipolysis (lip-OL-i-sis) The splitting of fatty acids from a triglyceride or phospholipid.

Lipoprotein (lip′-ō-PRŌ-tēn) One of several types of particles containing lipids (cholesterol and triglycerides) and proteins that make it water soluble for transport in the blood; high levels of **low-density lipoproteins (LDLs)** are associated with increased risk of atherosclerosis, whereas high levels of **high-density lipoproteins (HDLs)** are associated with decreased risk of atherosclerosis.

Liver Large organ under the diaphragm that occupies most of the right hypochondriac region and part of the epigastric region. Functionally, it produces bile and synthesizes most plasma proteins; interconverts nutrients; detoxifies substances; stores glycogen, iron, and vitamins; carries on phagocytosis of worn-out blood cells and bacteria; and helps synthesize the active form of vitamin D.

Lordosis (lor-DŌ-sis) An exaggeration of the lumbar curve of the vertebral column. Also called **hollow back**.

Lower limb The appendage attached at the pelvic (hip) girdle, consisting of the thigh, knee, leg, ankle, foot, and toes. Also called **lower extremity**.

Lumbar (LUM-bar) Region of the back and side between the ribs and pelvis; loin.

Lumbar plexus (PLEK-sus) A network formed by the anterior (ventral) branches of spinal nerves L1 through L4.

Lumen (LOO-men) The space within an artery, vein, intestine, renal tubule, or other tubular structure.

Lungs Main organs of respiration that lie on either side of the heart in the thoracic cavity.

Lunula (LOO-noo-la) The moon-shaped white area at the base of a nail.

Luteinizing hormone (LH) (LOO-tē-in'-īz-ing) A hormone secreted by the anterior pituitary that stimulates ovulation, stimulates progesterone secretion by the corpus luteum, and readies the mammary glands for milk secretion in females; stimulates testosterone secretion by the testes in males.

Lymph (LIMF) Fluid confined in lymphatic vessels and flowing through the lymphatic system until it is returned to the blood.

Lymph node An oval or bean-shaped structure located along lymphatic vessels.

Lymphatic capillary (lim-FAT-ik) Closed-ended microscopic lymphatic vessel that begins in spaces between cells and converges with other lymphatic capillaries to form lymphatic vessels.

Lymphatic tissue A specialized form of reticular tissue that contains large numbers of lymphocytes.

Lymphatic vessel A large vessel that collects lymph from lymphatic capillaries and converges with other lymphatic vessels to form the thoracic or right lymphatic ducts.

Lymphocyte (LIM-fō-sīt) A type of white blood cell that helps carry out cell-mediated and antibody-mediated immune responses; found in blood and in lymphatic tissues.

Lysosome (LĪ-sō-sōm) An organelle in the cytoplasm of a cell, enclosed by a single membrane and containing powerful digestive enzymes.

Lysozyme (LĪ-sō-zīm) A bactericidal enzyme found in tears, saliva, perspiration, nasal secretions, and tissue fluids.

M

Macrophage (MAK-rō-fāj) Phagocytic cell derived from a monocyte; may be fixed or wandering.

Macula (MAK-ū-la) A discolored spot or a colored area. A small, thickened region on the wall of the utricle and saccule that contains receptors for static equilibrium.

Macula lutea (MAK-ū-la LOO-tē-a) The yellow spot in the center of the retina.

Major histocompatibility (MHC) antigens Surface proteins on white blood cells and other nucleated cells that are unique for each person (except for identical siblings); used to type tissues and help prevent rejection of transplanted tissues. Also known as **human leukocyte antigens (HLAs)**.

Malignant (ma-LIG-nant) Referring to diseases that tend to become worse and cause death, especially the invasion and spreading of cancer.

Mammary gland (MAM-ar-ē) Modified sudoriferous (sweat) gland of females that produces milk for the nourishment of the young.

Marrow (MAR-ō) Soft, spongelike material in the cavities of bones. Red bone marrow produces blood cells; yellow bone marrow contains adipose tissue that stores triglycerides.

Mast cell A cell found in areolar connective tissue that releases histamine, a dilator of small blood vessels, during inflammation.

Mastication (mas'-ti-KĀ-shun) Chewing.

Matter Anything that occupies space and has mass.

Mature follicle (FOL-i-kul) A large, fluid-filled follicle containing a secondary oocyte and surrounding granulosa cells that secrete estrogens. Also called a **Graafian follicle** (GRAF-ē-an).

Meatus (mē-Ā-tus) A passage or opening, especially the external portion of a canal.

Mechanoreceptor (me-KAN-ō-rē-sep-tor) Sensory receptor that detects mechanical deformation of the receptor itself or adjacent cells; stimuli so detected include those related to touch, pressure, vibration, proprioception, hearing, equilibrium, and blood pressure.

Medial (MĒ-dē-al) Nearer the midline of the body or a structure.

Mediastinum (mē'-dē-as-TĪ-num) The broad, median partition between the pleurae of the lungs, that extends from the sternum to the vertebral column in the thoracic cavity.

Medulla (me-DUL-la) An inner portion of an organ, such as the medulla of the kidneys.

Medulla oblongata (me-DUL-la ob'-long-GA-ta) The most inferior part of the brain stem. Also termed the **medulla**.

Medullary cavity (MED-ū-lar'-ē) The space within the diaphysis of a bone that contains yellow bone marrow. Also called the **marrow cavity**.

Medullary rhythmicity area (rith-MIS-i-tē) The neurons of the respiratory center in the medulla oblongata that control the basic rhythm of respiration.

Meiosis (mē-Ō-sis) A type of cell division that occurs during production of gametes, involving two successive nuclear divisions that result in daughter cells with the haploid (n) number of chromosomes.

Meissner corpuscle (MĪZ-ner) The sensory receptor for the sensation of touch; found in dermal papillae, especially in the palms and soles. Also called a **corpuscle of touch**.

Melanin (MEL-a-nin) A dark black, brown, or yellow pigment found in some parts of the body such as the skin, hair, and pigmented layer of the retina.

Melanocyte (MEL-a-nō-sīt') A pigmented cell, located between or beneath cells of the deepest layer of the epidermis, that synthesizes melanin.

Melanocyte-stimulating hormone (MSH) A hormone secreted by the anterior pituitary that stimulates the dispersion of melanin granules in melanocytes in amphibians; continued administration produces darkening of skin in humans.

Melatonin (mel-a-TŌN-in) A hormone secreted by the pineal gland that helps set the timing of the body's biological clock.

Membrane (MEM-brān) A thin, flexible sheet of tissue composed of an epithelial layer and an underlying connective tissue layer, as in an epithelial membrane, or of areolar connective tissue only, as in a synovial membrane.

Membranous labyrinth (mem-BRA-nus LAB-i-rinth) The part of the labyrinth of the internal ear that is located inside the bony

labyrinth and separated from it by the perilymph; made up of the semicircular ducts, the saccule and utricle, and the cochlear duct.

Menarche (me-NAR-kē) The first menses (menstrual flow) and beginning of ovarian and uterine cycles.

Meninges (me-NIN-jēz) Three membranes covering the brain and spinal cord, called the dura mater, arachnoid mater, and pia mater. *Singular* is **meninx** (MEN-inks).

Menopause (MEN-ō-pawz) The termination of the menstrual cycles.

Menstruation (men'-stroo-Ā-shun) Periodic discharge of blood, tissue fluid, mucus, and epithelial cells that usually lasts for 5 days; caused by a sudden reduction in estrogens and progesterone. Also called the **menstrual phase** or **menses**.

Merkel disc (MER-kel) Modified epidermal cell in the stratum basale of hairless skin that functions as a cutaneous receptor for touch. Also called a **tactile disc**.

Mesenchyme (MEZ-en-kīm) An embryonic connective tissue from which almost all other connective tissues arise.

Mesentery (MEZ-en-ter'-ē) A fold of peritoneum that attaches the small intestine to the posterior abdominal wall.

Mesoderm (MEZ-ō-derm) The middle primary germ layer that gives rise to connective tissues, blood and blood vessels, and muscles.

Metabolism (me-TAB-ō-lizm) All the biochemical reactions that occur within an organism, including the synthetic (anabolic) reactions and decomposition (catabolic) reactions.

Metacarpus (met'-a-KAR-pus) A collective term for the five bones that make up the palm.

Metaphase (MET-a-phāz) The second stage of mitosis, in which chromatid pairs line up on the metaphase plate of the cell.

Metaphysis (me-TAF-i-sis) Region of a long bone between the diaphysis and epiphysis that contains the epiphyseal plate in a growing bone.

Metastasis (me-TAS-ta-sis) The spread of cancer to surrounding tissues (local) or to other body sites (distant).

Metatarsus (met'-a-TAR-sus) A collective term for the five bones located in the foot between the tarsals and the phalanges.

Micelle (mī-SEL) A spherical aggregate of bile salts that dissolves fatty acids and monoglycerides so that they can be absorbed into small intestinal epithelial cells.

Microglia (mī-krō-GLĒ-a) Neuroglial cells that carry on phagocytosis.

Microvilli (mī'-krō-VIL-ē) Microscopic, fingerlike projections of the plasma membranes of cells that increase surface area for absorption, especially in the small intestine and proximal convoluted tubules of the kidneys.

Micturition (mik'-choo-RISH-un) The act of expelling urine from the urinary bladder. Also called **urination** (ū-ri-NĀ-shun).

Midbrain The part of the brain between the pons and the diencephalon. Also called the **mesencephalon** (mes'-en-SEF-a-lon).

Middle ear A small, epithelial-lined cavity hollowed out of the temporal bone, separated from the external ear by the eardrum and from the internal ear by a thin bony partition containing the oval and round windows; extending across the middle ear are the three auditory ossicles. Also called the **tympanic cavity** (tim-PAN-ik).

Midline An imaginary vertical line that divides the body into equal left and right sides.

Midsagittal plane A vertical plane through the midline of the body that divides the body or organs into *equal* right and left sides. Also called a **median plane**.

Milk ejection reflex Contraction of alveolar cells to force milk into ducts of mammary glands, stimulated by oxytocin (TO), which is released from the posterior pituitary in response to suckling action. Also called the **milk letdown reflex**.

Mineral Inorganic, homogeneous solid substance that may perform a function vital to life; examples include calcium and phosphorus.

Mineralocorticoids (min'-er-al-ō-KOR-ti-koyds) A group of hormones of the adrenal cortex that help regulate sodium and potassium balance.

Minute ventilation (MV) Total volume of air inhaled and exhaled per minute; about 6000 mL at rest.

Mitochondrion (mī'-tō-KON-drē-on) A double-membraned organelle that plays a central role in the production of ATP; known as the "powerhouse" of the cell.

Mitosis (mī-TŌ-sis) The orderly division of the nucleus of a cell that ensures that each new nucleus has the same number and kind of chromosomes as the original parent nucleus. The process includes the replication of chromosomes and the distribution of the two sets of chromosomes into two separate and equal nuclei.

Mitotic spindle (mī-TOT-ik) Collective term for a football-shaped assembly of microtubules that is responsible for the movement of chromosomes during cell division.

Modality (mō-DAL-i-tē) Any of the specific sensory entities, such as vision, smell, taste, or touch.

Molecule (MOL-e-kūl) The chemical combination of two or more atoms covalently bonded together.

Monocyte (MON-ō-sīt') The largest type of white blood cell, characterized by agranular cytoplasm.

Monounsaturated fat A fatty acid that contains one double covalent bond between its carbon atoms; it is not completely saturated with hydrogen atoms. Plentiful in triglycerides of olive and peanut oils.

Mons pubis (MONZ PŪ-bis) The rounded, fatty prominence over the pubic symphysis, covered by coarse pubic hair.

Morula (MOR-ū-la) A solid sphere of cells produced by successive cleavages of a fertilized ovum about four days after fertilization.

Motor end plate Region of the sarcolemma of a muscle fiber (cell) that includes acetylcholine (ACh) receptors, which bind ACh released by synaptic end bulbs of somatic motor neurons.

Motor neurons (NOO-ronz) Neurons that conduct impulses from the brain toward the spinal cord or out of the brain and spinal cord into cranial or spinal nerves to effectors that may be either muscles or glands. Also called **efferent neurons** (EF-er-ent).

Motor unit A motor neuron together with the muscle fibers (cells) it stimulates.

Mucin (MŪ-sin) A protein found in mucus.

Mucous cell (MŪ-kus) A unicellular gland that secretes mucus. Two types are mucous neck cells and surface mucous cells in the stomach.

Mucous membrane A membrane that lines a body cavity that opens to the exterior. Also called the **mucosa** (mū-KŌ-sa).

Mucus The thick fluid secretion of goblet cells, mucous cells, mucous glands, and mucous membranes.

Muscle An organ composed of one of three types of muscular tissue (skeletal, cardiac, or smooth), specialized for contraction to produce voluntary or involuntary movement of parts of the body.

Muscle action potential A stimulating impulse that propagates along the sarcolemma and transverse tubules; in skeletal muscle, it is generated by acetylcholine, which increases the permeability of the sarcolemma to sodium ions (Na$^+$).

Muscle fatigue (fa-TĒG) Inability of a muscle to maintain its strength of contraction or tension; may be related to insufficient oxygen, depletion of glycogen, and/or lactic acid buildup.

Muscular tissue A tissue specialized to produce motion in response to muscle action potentials by its qualities of contractility, extensibility, elasticity, and excitability; types include skeletal, cardiac, and smooth.

Muscle tone A sustained, partial contraction of portions of a skeletal or smooth muscle in response to activation of stretch receptors or a baseline level of action potentials in the innervating motor neurons.

Muscular dystrophies (DIS-trō-fēz′) Inherited muscle-destroying diseases, characterized by degeneration of muscle fibers (cells), which causes progressive atrophy of the skeletal muscle.

Muscularis (MUS-kū-la′-ris) A muscular layer (coat or tunic) of an organ.

Muscularis mucosae (mū-KŌ-sē) A thin layer of smooth muscle fibers that underlie the lamina propria of the mucosa of the gastrointestinal tract.

Mutation (mū-TĀ-shun) Any change in the sequence of bases in a DNA molecule resulting in a permanent alteration in some inheritable trait.

Myasthenia gravis (mī-as-THĒ-nē-a) Weakness and fatigue of skeletal muscles caused by antibodies directed against acetylcholine receptors.

Myelin sheath (MĪ-e-lin) Multilayered lipid and protein covering, formed by Schwann cells and oligodendrocytes, around axons of many peripheral and central nervous system neurons.

Myocardial infarction (MI) (mī′-ō-KAR-dē-al in-FARK-shun) Gross necrosis of myocardial tissue due to interrupted blood supply. Also called a **heart attack**.

Myocardium (mī′-ō-KAR-dē-um) The middle layer of the heart wall, made up of cardiac muscle tissue, lying between the epicardium and the endocardium and constituting the bulk of the heart.

Myofibril (mī-ō-FĪ-bril) A threadlike structure, extending longitudinally through a muscle fiber (cell) consisting mainly of thick filaments (myosin) and thin filaments (actin, troponin, and tropomyosin).

Myoglobin (mī-ō-GLŌ-bin) The oxygen-binding, iron-containing protein present in the sarcoplasm of muscle fibers (cells); contributes the red color to muscle.

Myogram (MĪ-ō-gram) The record or tracing produced by a myograph, an apparatus that measures and records the force of muscular contractions.

Myology (mī-OL-ō-jē) The study of muscles.

Myometrium (mī′-ō-MĒ-trē-um) The smooth muscle layer of the uterus.

Myopathy (mī-OP-a-thē) Any abnormal condition or disease of muscle tissue.

Myopia (mī-Ō-pē-a) Defect in vision in which objects can be seen distinctly only when close to the eyes; nearsightedness.

Myosin (MĪ-ō-sin) The contractile protein that makes up the thick filaments of muscle fibers.

N

Nail A hard plate, composed largely of keratin, that develops from the epidermis of the skin to form a protective covering on the dorsal surface of the distal phalanges of the fingers and toes.

Nail matrix (MĀ-triks) The part of the nail beneath the body and root from which the nail is produced.

Nasal cavity (NĀ-zal) A mucosa-lined cavity on either side of the nasal septum that opens onto the face at the external nares and into the nasopharynx at the internal nares.

Nasal septum (SEP-tum) A vertical partition composed of bone (perpendicular plate of ethmoid and vomer) and cartilage, covered with a mucous membrane, separating the nasal cavity into left and right sides.

Nasolacrimal duct (nā′-zō-LAK-ri-mal) A canal that transports the lacrimal secretion (tears) from the nasolacrimal sac into the nose.

Nasopharynx (nā′-zō-FAR-inks) The superior portion of the pharynx, lying posterior to the nose and extending inferiorly to the soft palate.

Neck The part of the body connecting the head and the trunk. A constricted portion of an organ such as the neck of the femur or uterus.

Necrosis (ne-KRŌ-sis) A pathological type of cell death that results from disease, injury, or lack of blood supply in which many adjacent cells swell, burst, and spill their contents into the interstitial fluid, triggering an inflammatory response.

Negative feedback The principle governing most control systems; a mechanism of response in which a stimulus initiates actions that reverse or reduce the stimulus.

Neonatal (nē-ō-NĀ-tal) Pertaining to the first four weeks after birth.

Neoplasm (NĒ-ō-plazm) A new growth that may be benign or malignant.

Nephron (NEF-ron) The functional unit of the kidney.

Nerve A cordlike bundle of neuronal axons and/or dendrites and associated connective tissue coursing together outside the central nervous system.

Nerve fiber General term for any process (axon or dendrite) projecting from the cell body of a neuron.

Nerve impulse A wave of depolarization and repolarization that self-propagates along the plasma membrane of a neuron; also called a **nerve action potential**.

Nervous tissue Tissue containing neurons that initiate and conduct nerve impulses to coordinate homeostasis, and neuroglia that provide support and nourishment to neurons.

Net filtration pressure (NFP) Net pressure that promotes fluid outflow at the arterial end of a capillary, and fluid inflow at the venous end of a capillary; net pressure that promotes glomerular filtration in the kidneys.

Neural plate (Noo-ral) A thickening of ectoderm, induced by the notochord, that forms early in the third week of development and represents the beginning of the development of the nervous system.

Neuralgia (noo-RAL-jē-a) Attacks of pain along the entire course or branch of a peripheral sensory nerve.

Neuritis (noo-RĪ-tis) Inflammation of one or more nerves.

Neuroglia (noo-RŌG-lē-a) Cells of the nervous system that perform various supportive functions. The neuroglia of the central nervous system are the astrocytes, oligodendrocytes, microglia, and ependymal cells; neuroglia of the peripheral nervous system include Schwann cells and satellite cells. Also called **glial cells** (GLĒ-al).

Neurolemma (noo-rō-LEM-ma) The peripheral, nucleated cytoplasmic layer of the Schwann cell. Also called **sheath of Schwann** (SCHVON).

Neurology (noo-ROL-ō-jē) The study of the normal functioning and disorders of the nervous system.

Neuromuscular junction (noo-rō-MUS-kū-lar) A synapse between the axon terminals of a somatic motor neuron and the sarcolemma of a muscle fiber (cell).

Neuron (NOO-ron) A nerve cell, consisting of a cell body, dendrites, and an axon.

Neurosecretory cell (noo-rō-SEC-re-tō-rē) A neuron that secretes a hypothalamic releasing hormone or inhibiting hormone into blood capillaries of the hypothalmus; a neuron that secretes oxytocin or antidiuretic hormone into blood capillaries of the posterior pituitary.

Neurotransmitter (noo'-rō-TRANS-mit-er) One of a variety of molecules within axon terminals that are released into the synaptic cleft in response to a nerve impulse, and that change the membrane potential of the postsynaptic neuron.

Neurulation (noo-roo-LĀ-shun) The process by which the neural plate, neural folds, and neural tube form.

Neutrophil (NOO-trō-fil) A type of white blood cell characterized by granules that stain pale lilac with a combination of acidic and basic dyes.

Nipple A pigmented, wrinkled projection on the surface of the breast that is the location of the openings of the lactiferous ducts for milk release.

Nociceptor (nō'-sē-SEP-tor) A free (naked) nerve ending that detects painful stimuli.

Node of Ranvier (ron-vē-Ā) A space, along a myelinated axon, between the individual Schwann cells that form the myelin sheath and the neurolemma. Also called a **neurofibral node**.

Norepinephrine (NE) (nor'-ep-ē-NEF-rin) A hormone secreted by the adrenal medulla that produces actions similar to those that result from sympathetic stimulation. Also called **noradrenaline** (nor-a-DREN-a-lin).

Notochord (NŌ-tō-cord) A flexible rod of mesodermal tissue that helps form part of the backbone and intervertebral discs.

Nuclear medicine The branch of medicine concerned with the use of radioisotopes in the diagnosis and therapy of disease.

Nucleic acid (noo-KLĒ-ic) An organic compound that is a long polymer of nucleotides, with each nucleotide containing a pentose sugar, a phosphate group, and one of four possible nitrogenous bases (adenine, cytosine, guanine, and thymine or uracil).

Nucleolus (noo-KLĒ-ō-lus) Spherical body within a cell nucleus composed of protein, DNA, and RNA that is the site of the assembly of small and large ribosomal subunits.

Nucleus (NOO-klē-us) A spherical or oval organelle of a cell that contains the hereditary factors of the cell, called genes. A cluster of unmyelinated nerve cell bodies in the central nervous system. The central part of an atom made up of protons and neutrons.

Nucleus pulposus (pul-PŌ-sus) A soft, pulpy, highly elastic substance in the center of an intervertebral disc; a remnant of the notochord.

Nutrient (NOO-trē-ent) A chemical substance in food that provides energy, forms new body components, or assists in various body functions.

O

Obesity (ō-BĒS-i-tē) Body weight more than 20% above a desirable standard due to excessive accumulation of fat.

Oblique plane (ō-BLĒK) A plane that passes through the body or an organ at an angle between the transverse plane and either the midsagittal, parasagittal, or frontal plane.

Obstetrics (ob-STET-riks) The specialized branch of medicine that deals with pregnancy, labor, and the period of time immediately after delivery (about 6 weeks).

Olfactory (ōl-FAK-tō-rē) Pertaining to smell.

Olfactory bulb A mass of gray matter containing cell bodies of neurons that form synapses with neurons of the olfactory (I) nerve, lying inferior to the frontal lobe of the cerebrum on either side of the crista galli of the ethmoid bone.

Olfactory receptor A bipolar neuron with its cell body lying between supporting cells located in the mucous membrane lining the superior portion of each nasal cavity; transduces odors into neural signals.

Olfactory tract A bundle of axons that extends from the olfactory bulb posteriorly to olfactory regions of the cerebral cortex.

Oligodendrocyte (ol'-i-gō-DEN-drō-sīt) A neuroglial cell that supports neurons and produces a myelin sheath around axons of neurons of the central nervous system.

Oligospermia (ol'-i-gō-SPER-mē-a) A deficiency of sperm cells in the semen.

Oncogenes (ONG-kō-jēnz) Cancer-causing genes; they derive from normal genes, termed proto-oncogenes, that encode proteins involved in cell growth or cell regulation but have the ability to transform a normal cell into a cancerous cell when they are mutated or inappropriately activated. One example is *p53*.

Oncology (ong-KOL-ō-jē) The study of tumors.

Oogenesis (ō'-ō-JEN-e-sis) Formation and development of female gametes (oocytes).

Oophorectomy (ō'-of-ō-REK-tō-me) Surgical removal of the ovaries.

Ophthalmic (of-THAL-mik) Pertaining to the eye.

Ophthalmologist (of'-thal-MOL-ō-jist) A physician who specializes in the diagnosis and treatment of eye disorders using drugs, surgery, and corrective lenses.

Ophthalmology (of'-thal-MOL-ō-jē) The study of the structure, function, and diseases of the eye.

Opsin (OP-sin) The glycoprotein portion of a photopigment.

Opsonization (op-sō-ni-ZĀ-shun) The action of some antibodies that renders bacteria and other foreign cells more susceptible to phagocytosis.

Optic (OP-tik) Refers to the eye, vision, or properties of light.

Optic chiasm (KĪ-azm) A crossing point of the optic (II) nerves, anterior to the pituitary gland. Also called the **optic chiasma** (kī-AZ-ma).

Optic disc A small area of the retina containing openings through which the axons of the ganglion cells emerge as the optic nerve (cranial nerve II). Also called the **blind spot**.

Optician (op-TISH-an) A technician who fits, adjusts, and dispenses corrective lenses on prescription of an ophthalmologist or optometrist.

Optic tract A bundle of axons that carry nerve impulses from the retina of the eye between the optic chiasm and the thalamus.

Optometrist (op-TOM-e-trist) Specialist with a doctorate degree in optometry who is licensed to examine and test the eyes and treat visual defects by prescribing corrective lenses.

Orbit (OR-bit) The bony, pyramidal-shaped cavity of the skull that holds the eyeball.

Organ A structure composed of two or more different kinds of tissues with a specific function and usually a recognizable shape.

Organelle (or-gan-EL) A permanent structure within a cell with characteristic shape that is specialized to serve a specific function in cellular activities.

Organic compound (or-GAN-ik) Compound that always contains carbon in which the atoms are held together by covalent bonds. Examples include carbohydrates, lipids, proteins, and nucleic acids (DNA and RNA).

Organism (OR-ga-nizm) A total living form; one individual.

Orgasm (OR-gazm) Sensory and motor events involved in ejaculation for the male and involuntary contraction of the perineal muscles in the female at the climax of sexual intercourse.

Orifice (OR-i-fis) Any aperture or opening.

Origin (OR-i-jin) The attachment of a muscle tendon to a stationary bone or the end opposite the insertion.

Oropharynx (or'-ō-FAR-inks) The intermediate portion of the pharynx, lying posterior to the mouth and extending from the soft palate to the hyoid bone.

Orthopedics (or'-thō-PĒ-diks) The branch of medicine that deals with the preservation and restoration of the skeletal system, articulations, and associated structures.

Osmoreceptor (oz'-mō-re-SEP-tor) Receptor in the hypothalamus that is sensitive to changes in blood osmolarity and, in response to high osmolarity (low water concentration), stimulates synthesis and release of antidiuretic hormone (ADH).

Osmosis (os-MŌ-sis) The net movement of water molecules through a selectively permeable membrane from an area of higher water concentration to an area of lower water concentration until equilibrium is reached.

Osmotic pressure The pressure required to prevent the movement of pure water into a solution containing solutes when the solutions are separated by a selectively permeable membrane.

Osseous (OS-ē-us) Bony.

Ossicle (OS-si-kul) One of the small bones of the middle ear (malleus, incus, stapes).

Ossification (os'-i-fi-KĀ-shun) Formation of bone. Also called **osteogenesis**.

Osteoblast (OS-tē-ō-blast) Cell formed from an osteogenic cell that participates in bone formation by secreting some organic components and inorganic salts.

Osteoclast (OS-tē-ō-clast') A large, multinuclear cell that resorbs (destroys) bone extracellular matrix.

Osteocyte (OS-tē-ō-sīt') A mature bone cell that maintains the daily activities of bone tissue.

Osteogenic layer (os'-tē-ō-JEN-ik) The inner layer of the periosteum that contains cells responsible for forming new bone during growth and repair.

Osteology (os'-tē-OL-ō-jē) The study of bones.

Osteon (OS-tē-on) The basic unit of structure in adult compact bone, consisting of a central (haversian) canal with its concentrically arranged lamellae, lacunae, osteocytes, and canaliculi. Also called an **haversian system** (ha-VER-shan).

Osteoporosis (os'-tē-ō-pō-RO-sis) Age-related disorder characterized by decreased bone mass and increased susceptibility to fractures, often as a result of decreased levels of estrogens.

Otic (Ō-tik) Pertaining to the ear.

Otolith (Ō-tō-lith) A particle of calcium carbonate embedded in the otolithic membrane that functions in maintaining static equilibrium.

Otolithic membrane (ō-tō-LITH-ik) Thick, gelatinous, glycoprotein layer located directly over hair cells of the macula in the saccule and utricle of the internal ear.

Otorhinolaryngology (ō'-tō-rī-nō-lar'-in-GOL-ō-jē) The branch of medicine that deals with the diagnosis and treatment of diseases of the ears, nose, and throat.

Oval window A small, membrane-covered opening between the middle ear and inner ear into which the footplate of the stapes fits.

Ovarian cycle (ō-VAR-ē-an) A monthly series of events in the ovary associated with the maturation of a secondary oocyte.

Ovarian follicle (FOL-i-kul) A general name for oocytes (immature ova) in any stage of development, along with their surrounding epithelial cells.

Ovary (Ō-var-ē) Female gonad that produces oocytes and hormones (estrogens, progesterone, inhibin, and relaxin).

Ovulation (ov-ū-LĀ-shun) The rupture of a mature ovarian (graafian) follicle with discharge of a secondary oocyte into the pelvic cavity.

Ovum (Ō-vum) The female reproductive or germ cell; an egg cell; arises through completion of meiosis in a secondary oocyte after penetration by a sperm.

Oxidation (ok-si-DĀ-shun) The removal of electrons from a molecule or, less commonly, the addition of oxygen to a molecule that results in a decrease in the energy content of the molecule. The oxidation of glucose in the body is called **cellular respiration**.

Oxyhemoglobin (Hb–O₂) (ok′-sē-HĒ-mō-glō-bin) Hemoglobin combined with oxygen.

Oxytocin (OT) (ok′-sē-TŌ-sin) A hormone secreted by neurosecretory cells in the hypothalamus that stimulates contraction of smooth muscle in the pregnant uterus and myoepithelial cells around the ducts of mammary glands.

P

P wave The deflection wave of an electrocardiogram that signifies atrial depolarization.

Pacinian corpuscle (pa-SIN-ē-an) Oval-shaped pressure receptor located in the dermis or subcutaneous tissue and consisting of concentric layers of connective tissue wrapped around the dendrites of a sensory neuron. Also called a **lamellated corpuscle**.

Palate (PAL-at) The horizontal structure separating the oral and the nasal cavities; the roof of the mouth.

Palpate (PAL-pāt) To examine by touch; to feel.

Pancreas (PAN-krē-as) A soft, oblong organ lying along the greater curvature of the stomach and connected by a duct to the duodenum. It is both an exocrine gland (secreting pancreatic juice) and an endocrine gland (secreting insulin and glucagon).

Pancreatic duct (pan′-krē-AT-ik) A single large tube that unites with the common bile duct from the liver and gallbladder and drains pancreatic juice into the duodenum.

Pancreatic islet (Ī-let) A cluster of endocrine gland cells in the pancreas that secretes insulin, glucagon, somatostatin, and pancreatic polypeptide. Also called an **islet of Langerhans** (LANG-er-hanz).

Papanicolaou test (pap′-a-NIK-ō-la-oo) A cytological staining test for the detection and diagnosis of premalignant and malignant conditions of the female genital tract. Cells scraped from the epithelium of the cervix of the uterus are examined microscopically. Also called a **Pap test** or **Pap smear**.

Papilla (pa-PIL-a) A small nipple-shaped projection or elevation.

Paralysis (pa-RAL-a-sis) Loss or impairment of motor function due to a lesion of nervous or muscular origin.

Paranasal sinus (par′-a-NĀ-zal SĪ-nus) A mucus-lined air cavity in a skull bone that communicates with the nasal cavity. Paranasal sinuses are located in the frontal, maxillary, ethmoid, and sphenoid bones.

Paraplegia (par-a-PLĒ-jē-a) Paralysis of both lower limbs.

Parasagittal plane (par-a-SAJ-i-tal) A vertical plane that does not pass through the midline and that divides the body or organs into *unequal* left and right portions.

Parasympathetic division (par′-a-sim-pa-THET-ik) One of the two subdivisions of the autonomic nervous system, having cell bodies of preganglionic neurons in nuclei in the brain stem and in the lateral gray horn of the sacral portion of the spinal cord; primarily concerned with activities that conserve and restore body energy.

Parathyroid gland (par′-a-THĪ-royd) One of usually four small endocrine glands embedded in the posterior surfaces of the lateral lobes of the thyroid gland.

Parathyroid hormone (PTH) A hormone secreted by the chief (principal) cells of the parathyroid glands that increases blood calcium level and decreases blood phosphate level.

Parenchyma (par-EN-ki-ma) The functional parts of any organ, as opposed to tissue that forms its stroma or framework.

Parietal (pa-RĪ-e-tal) Pertaining to or forming the outer wall of a body cavity.

Parietal cell A type of secretory cell in gastric glands that produces hydrochloric acid and intrinsic factor. Also called an **oxyntic cell**.

Parkinson disease (PD) Progressive degeneration of the basal ganglia and substantia nigra of the cerebrum resulting in decreased production of dopamine (DA) that leads to tremor, slowing of voluntary movements, and muscle weakness.

Parotid gland (pa-ROT-id) One of the paired salivary glands located inferior and anterior to the ears and connected to the oral cavity via a duct that opens into the inside of the cheek opposite the maxillary (upper) second molar tooth.

Parturition (par′-too-RISH-un) Act of giving birth to young; childbirth, delivery.

Patellar reflex (pa-TELL-ar) Extension of the leg by contraction of the quadriceps femoris muscle in response to tapping the patellar ligament. Also called the **knee jerk reflex**.

Patent ductus arteriosus (PĀ-tent DUK-tus ar-tēr-ē-Ō-sus) Congenital anatomical heart defect in which the fetal connection between the aorta and pulmonary trunk remains open instead of closing completely after birth.

Pathogen (PATH-ō-jen) A disease-producing microbe.

Pathological anatomy (path′-ō-LOJ-i-kal) The study of structural changes caused by disease.

Pectoral (PEK-tō-ral) Pertaining to the chest or breast.

Pediatrician (pē′-dē-a-TRISH-un) A physician who specializes in the care and treatment of children.

Pedicel (PED-i-sel) Footlike structure, as on podocytes of a glomerulus.

Pelvic cavity (PEL-vik) Inferior portion of the abdominopelvic cavity that contains the urinary bladder, sigmoid colon, rectum, and internal female and male reproductive structures.

Pelvis The basinlike structure formed by the two hip bones, the sacrum, and the coccyx. The expanded, proximal portion of the ureter, lying within the kidney and into which the major calyces open.

Penis (PĒ-nis) The organ of urination and copulation in males; used to deposit semen into the female vagina.

Pepsin Protein-digesting enzyme secreted by chief cells of the stomach in the inactive form pepsinogen, which is converted to active pepsin by hydrochloric acid.

Peptic ulcer An ulcer that develops in areas of the gastrointestinal tract exposed to hydrochloric acid; classified as a gastric ulcer if in the lesser curvature of the stomach and as a duodenal ulcer if in the first part of the duodenum.

Percussion (per-KUSH-un) The act of striking (percussing) an underlying part of the body with short, sharp blows as an aid in diagnosing the part by the quality of the sound produced.

Perforating canal (PER-fō-rā'-ting) A minute passageway by means of which blood vessels and nerves from the periosteum penetrate into compact bone. Also called **volkmann's canal** (FŌLK-manz).

Pericardial cavity (per'-i-KAR-dē-al) Small potential space between the visceral and parietal layers of the serous pericardium that contains pericardial fluid.

Pericardium (per'-i-KAR-dē-um) A loose-fitting membrane that encloses the heart, consisting of a superficial fibrous layer and a deep serous layer.

Perichondrium (per'-i-KON-drē-um) The membrane that covers cartilage.

Perilymph (PER-i-limf) The fluid contained between the bony and membranous labyrinths of the inner ear.

Perineum (per'-i-NĒ-um) The pelvic floor; the space between the anus and the scrotum in the male and between the anus and the vulva in the female.

Periodontal disease (per-ē-ō-DON-tal) A collective term for conditions characterized by degeneration of gingivae, alveolar bone, periodontal ligament, and cementum.

Periodontal ligament The periosteum lining the alveoli (sockets) for the teeth in the alveolar processes of the mandible and maxillae.

Periosteum (per'-ē-OS-tē-um) The membrane that covers bone and consists of connective tissue, osteogenic cells, and osteoblasts; is essential for bone growth, repair, and nutrition.

Peripheral (pe-RIF-er-al) Located on the outer part or a surface of the body.

Peripheral nervous system (PNS) The part of the nervous system that lies outside the central nervous system, consisting of nerves and ganglia.

Peristalsis (per'-i-STAL-sis) Successive muscular contractions along the wall of a hollow muscular structure.

Peritoneum (per'-i-tō-NĒ-um) The largest serous membrane of the body that lines the abdominal cavity and covers the viscera within the cavity.

Peritonitis (per'-i-tō-NĪ-tis) Inflammation of the peritoneum.

Peroxisome (per-OK-si-sōm) Organelle similar in structure to a lysosome that contains enzymes that use molecular oxygen to oxidize various organic compounds; such reactions produce hydrogen peroxide; abundant in liver cells.

Perspiration Sweat; produced by sudoriferous (sweat) glands and containing water, salts, urea, uric acid, amino acids, ammonia, sugar, lactic acid, and ascorbic acid. Helps maintain body temperature and eliminate wastes.

pH A measure of the concentration of hydrogen ions (H^+) in a solution. The pH scale extends from 0 to 14, with a value of 7 expressing neutrality, values lower than 7 expressing increasing acidity, and values higher than 7 expressing increasing alkalinity.

Phagocytosis (fag'-ō-sī-TŌ-sis) The process by which phagocytes ingest particulate matter; the ingestion and destruction of microbes, cell debris, and other foreign matter.

Phalanx (FĀ-lanks) The bone of a finger or toe. *Plural* is **phalanges** (fa-LAN-jēz).

Pharmacology (far'-ma-KOL-ō-jē) The science that deals with the effects and uses of drugs in the treatment of disease.

Pharynx (FAR-inks) The throat; a tube that starts at the internal nares and runs partway down the neck, where it opens into the esophagus posteriorly and the larynx anteriorly.

Phenotype (FĒ-nō-tīp) The observable expression of genotype; physical characteristics of an organism determined by genetic makeup and influenced by interaction between genes and internal and external environmental factors.

Phlebitis (fle-BĪ-tis) Inflammation of a vein, usually in a lower limb.

Photopigment A substance that can absorb light and undergo structural changes that can lead to the development of a receptor potential. An example is rhodopsin. In the eye, also called **visual pigment**.

Photoreceptor Receptor that detects light shining on the retina of the eye.

Physiology (fiz'-ē-OL-ō-jē) Science that deals with the functions of an organism or its parts.

Pia mater (PĪ-a MĀ-ter *or* PĒ-a MĀ-ter) The innermost of the three meninges (coverings) of the brain and spinal cord.

Pineal gland (PĪN-ē-al) A cone-shaped gland located in the roof of the third ventricle that secretes melatonin.

Pinna (PIN-na) The projecting part of the external ear composed of elastic cartilage and covered by skin and shaped like the flared end of a trumpet. Also called the **auricle** (AW-ri-kul).

Pituitary gland (pi-TOO-i-tār-ē) A small endocrine gland occupying the hypophyseal fossa of the sphenoid bone and attached to the hypothalamus by the infundibulum. Also called the **hypophysis** (hī-POF-i-sis).

Pivot joint A synovial joint in which a rounded, pointed, or conical surface of one bone articulates with a ring formed partly by another bone and partly by a ligament, as in the joint between the atlas and axis and between the proximal ends of the radius and ulna.

Placenta (pla-SEN-ta) The special structure through which the exchange of materials between fetal and maternal circulations occurs. Also called the **afterbirth**.

Plantar flexion (PLAN-tar FLEK-shun) Bending the foot in the direction of the plantar surface (sole).

Plaque (PLAK) A layer of dense proteins on the inside of a plasma membrane in adherens junctions and desmosomes. A mass of bacterial cells, dextran (polysaccharide), and other debris that adheres to teeth (dental plaque). See also atherosclerotic plaque.

Plasma (PLAZ-ma) The extracellular fluid found in blood vessels; blood minus the formed elements.

Plasma cell Cell that develops from a B cell (lymphocyte) and produces antibodies.

Plasma (cell) membrane Outer, limiting membrane that separates the cell's internal parts from extracellular fluid or the external environment.

Platelet (PLĀT-let) A fragment of cytoplasm enclosed in a cell membrane and lacking a nucleus; found in the circulating blood; plays a role in hemostasis. Also called a **thrombocyte** (THROM-bō-sīt).

Platelet plug Aggregation of platelets at a site where a blood vessel is damaged that helps stop or slow blood loss.

Pleura (PLOOR-a) The serous membrane that covers the lungs and lines the walls of the chest and the diaphragm.

Pleural cavity Small potential space between the visceral and parietal pleurae.

Plexus (PLEK-sus) A network of nerves, veins, or lymphatic vessels.

Pluripotent (plu-RIP-ō-tent) **stem cell** Immature stem cell in red bone marrow that gives rise to precursors of all the different mature blood cells.

Pneumotaxic area (noo-mō-TAK-sik) A part of the respiratory center in the pons that continually sends inhibitory nerve impulses to the inspiratory area, limiting inhalation and facilitating exhalation.

Podiatry (pō-DĪ-a-trē) The diagnosis and treatment of foot disorders.

Polar body The smaller cell resulting from the unequal division of primary and secondary oocytes during meiosis. The polar body has no function and degenerates.

Polycythemia (pol′-ē-sī-THĒ-mē-a) Disorder characterized by an above-normal hematocrit (above 55%) in which hypertension, thrombosis, and hemorrhage can occur.

Polysaccharide (pol′-ē-SAK-a-rīd) A carbohydrate in which three or more monosaccharides are joined chemically.

Polyunsaturated fat (pol′-ē-un-SACH-ū-rā′-ted) A fatty acid that contains more than one double covalent bond between its carbon atoms; abundant in triglycerides of corn oil, safflower oil, and cottonseed oil.

Polyuria (pol′-ē-U-rē-a) An excessive production of urine.

Pons (PONZ) The part of the brain stem that forms a "bridge" between the medulla oblongata and the midbrain, anterior to the cerebellum.

Positive feedback A feedback mechanism in which the response enhances the original stimulus.

Postcentral gyrus A gyrus of the cerebral cortex located immediately posterior to the central sulcus; contains the primary somatosensory area.

Posterior (pos-TĒR-ē-or) Nearer to or at the back of the body. Equivalent to **dorsal** in bipeds.

Posterior column–medial lemniscus pathways (lem-NIS-kus) Sensory pathways that carry information related to proprioception, touch pressure, and vibration. First-order neurons project from the spinal cord to the ipsilateral (same side) medulla in the posterior columns. Second-order neurons project from the medulla to the contralateral (opposite side) thalamus in the medial lemniscus. Third-order neurons project from the thalamus to the somatosensory cortex (postcentral gyrus) on the same side.

Posterior pituitary (pi-TOO-i-tār-ē) Posterior lobe of the pituitary gland. Also called the **neurohypophysis** (noo-rō-hī-POF-i-sis).

Posterior root The structure composed of sensory axons lying between a spinal nerve and the dorsolateral aspect of the spinal cord. Also called the **dorsal (sensory) root**.

Posterior root ganglion (GANG-glē-on) A group of cell bodies of sensory neurons and their supporting cells located along the posterior root of a spinal nerve. Also called a **dorsal (sensory) root ganglion**.

Postganglionic neuron (pōst′-gang-lē-ON-ik NOO-ron) The second autonomic motor neuron in an autonomic pathway, having its cell body and dendrites located in an autonomic ganglion and its unmyelinated axon ending at cardiac muscle, smooth muscle, or a gland.

Postsynaptic neuron (pōst-sin-AP-tik) The nerve cell that is activated by the release of a neurotransmitter from another neuron and carries nerve impulses away from the synapse.

Precapillary sphincter (SFINGK-ter) A ring of smooth muscle fibers (cells) at the site of origin of true capillaries that regulate blood flow into true capillaries.

Precentral gyrus (JĪ-rus) A gyrus of the cerebral cortex located immediately anterior to the central sulcus; contains the primary motor area.

Preganglionic neuron (prē′-gang-lē-ON-ik) The first autonomic motor neuron in an autonomic pathway, with its cell body and dendrites in the brain or spinal cord and its myelinated axon ending at an autonomic ganglion, where it synapses with a postganglionic neuron.

Pregnancy Sequence of events that normally includes fertilization, implantation, embryonic growth, and fetal growth and terminates in birth.

Premenstrual syndrome (PMS) Moderate to severe physical and emotional stress ocurring late in the postovulatory phase of the menstrual cycle and sometimes overlapping with menstruation.

Prepuce (PRĒ-poos) The loose-fitting skin covering the glans of the penis and clitoris. Also called the **foreskin**.

Presbyopia (prez-bē-Ō-pē-a) A loss of elasticity of the lens of the eye due to advancing age with resulting inability to focus clearly on near objects.

Presynaptic neuron (prē-sin-AP-tik) A neuron that propagates nerve impulses toward a synapse.

Prevertebral ganglion (prē-VER-te-bral GANG-lē-on) A cluster of cell bodies of postganglionic sympathetic neurons anterior to the spinal column and close to large abdominal arteries. Also called a **collateral ganglion**.

Primary germ layer One of three layers of embryonic tissue, called ectoderm, mesoderm, and endoderm, that give rise to all tissues and organs of the body.

Primary motor area A region of the cerebral cortex in the precentral gyrus of the frontal lobe of the cerebrum that controls specific muscles or groups of muscles.

Primary somatosensory area (sō-ma-tō-SEN-sō-re) A region of the cerebral cortex posterior to the central sulcus in the postcentral gyrus of the parietal lobe of the cerebrum that localizes exactly the points of the body where somatic sensations originate.

Prime mover The muscle directly responsible for producing a desired motion. Also called an **agonist** (AG-ō-nist).

Primitive gut Embryonic structure formed from the dorsal part of the yolk sac that gives rise to most of the gastrointestinal tract.

Proctology (prok-TOL-ō-jē) The branch of medicine concerned with the rectum and its disorders.

Progeny (PROJ-e-nē) Offspring or descendants.

Progesterone (prō-JES-te-rōn) A female sex hormone produced by the ovaries that helps prepare the endometrium of the uterus for implantation of a fertilized ovum and the mammary glands for milk secretion.

Prognosis (prog-NŌ-sis) A forecast of the probable results of a disorder; the outlook for recovery.

Prolactin (PRL) (prō-LAK-tin) A hormone secreted by the anterior pituitary that initiates and maintains milk secretion by the mammary glands.

Prolapse (PRŌ-laps) A dropping or falling down of an organ, especially the uterus or rectum.

Proliferation (prō-lif'-er-Ā-shun) Rapid and repeated reproduction of new parts, especially cells.

Pronation (prō-NĀ-shun) A movement of the forearm in which the palm is turned posteriorly.

Prophase (PRŌ-fāz) The first stage of mitosis during which chromatid pairs are formed and aggregate around the metaphase plate of the cell.

Proprioception (prō-prē-ō-SEP-shun) The perception of the position of body parts, especially the limbs, independent of vision; this sense is possible due to nerve impulses generated by proprioceptors.

Proprioceptor (prō'-prē-ō-SEP-tor) A receptor located in muscles, tendons, joints, or the internal ear (muscle spindles, tendon organs, joint kinesthetic receptors, and hair cells of the vestibular apparatus) that provides information about body position and movements.

Prostaglandin (PG) (pros'-ta-GLAN-din) A membrane-associated lipid; released in small quantities and acts as a local hormone.

Prostate (PROS-tāt) A doughnut-shaped gland inferior to the urinary bladder that surrounds the superior portion of the male urethra and secretes a slightly acidic solution that contributes to sperm motility and viability.

Protein An organic compound consisting of carbon, hydrogen, oxygen, nitrogen, and sometimes sulfur and phosphorus; synthesized on ribosomes and made up of amino acids linked by peptide bonds.

Proteasome (PRŌ-tē-a-sōm) Tiny cellular organelle in the cytosol and nucleus containing proteases that destroy unneeded, damaged, or faulty proteins.

Prothrombin (prō-THROM-bin) An inactive blood-clotting factor synthesized by the liver, released into the blood, and converted to active thrombin in the process of blood clotting by the activated enzyme prothrombinase.

Protraction (prō-TRAK-shun) The movement of the mandible or clavicle forward on a plane parallel with the ground.

Proximal (PROK-si-mal) Nearer the attachment of a limb to the trunk; nearer to the point of origin or attachment.

Pseudopods (SOO-dō-pods) Temporary protrusions of the leading edge of a migrating cell; cellular projections that surround a particle undergoing phagocytosis.

Ptosis (TŌ-sis) Drooping, as of the eyelid or the kidney.

Puberty (PŪ-ber-tē) The time of life during which the secondary sex characteristics begin to appear and the capability for sexual reproduction is possible; usually occurs between the ages of 10 and 17.

Pubic symphysis (SIM-fi-sis) A slightly movable cartilaginous joint between the anterior surfaces of the hip bones.

Puerperium (pū'-er-PER-ē-um) The period immediately after childbirth, usually 4–6 weeks.

Pulmonary (PUL-mo-ner'-ē) Concerning or affected by the lungs.

Pulmonary circulation The flow of deoxygenated blood from the right ventricle to the lungs and the return of oxygenated blood from the lungs to the left atrium.

Pulmonary edema (e-DĒ-ma) An abnormal accumulation of interstitial fluid in the tissue spaces and alveoli of the lungs due to increased pulmonary capillary permeability or increased pulmonary capillary pressure.

Pulmonary embolism (PE) (EM-bō-lizm) The presence of a blood clot or a foreign substance in a pulmonary arterial blood vessel that obstructs circulation to lung tissue.

Pulmonary ventilation (ven-ti-LĀ-shun) The inflow (inhalation) and outflow (exhalation) of air between the atmosphere and the lungs. Also called **breathing**.

Pulp cavity A cavity within the crown and neck of a tooth, which is filled with pulp, a connective tissue containing blood vessels, nerves, and lymphatic vessels.

Pulse (PULS) The rhythmic expansion and elastic recoil of a systemic artery after each contraction of the left ventricle.

Pupil The hole in the center of the iris, the area through which light enters the posterior cavity of the eyeball.

Purkinje fiber (pur-KIN-jē) Muscle fiber (cell) in the ventricular tissue of the heart specialized for conducting an action potential to the myocardium; part of the conduction system of the heart.

Pus The liquid product of inflammation containing leukocytes or their remains and debris of dead cells.

Pyloric sphincter (pī-LOR-ik) A thickened ring of smooth muscle through which the pylorus of the stomach communicates with the duodenum. Also called the **pyloric valve**.

Pyramid (PIR-a-mid) A pointed or cone-shaped structure. One of two roughly triangular structures on the anterior aspect of the medulla oblongata composed of the largest motor tracts that run from the cerebral cortex to the spinal cord. A triangular structure in the renal medulla.

Pyramidal tracts (pathways) (pi-RAM-i-dal) *See* **Direct motor pathways**.

Q

QRS wave The deflection wave of an electrocardiogram that represents the onset of ventricular depolarization.

Quadriplegia (kwod'-ri-PLĒ-jē-a) Paralysis of four limbs: two upper and two lower.

R

Radiographic anatomy (rā'-dē-ō-GRAF-ic) Diagnostic branch of anatomy that includes the use of x rays.

Rapid eye movement (REM) sleep Stage of sleep in which dreaming occurs, lasting for 5 to 10 minutes several times during a sleep cycle; characterized by rapid movements of the eyes beneath the eyelids.

Receptor (rē-SEP-tor) A specialized cell or a distal portion of a neuron that responds to a specific sensory modality, such as touch, pressure, cold, light, or sound, and converts it to an electrical signal (generator or receptor potential). A specific molecule or cluster of molecules that recognizes and binds a particular ligand.

Receptor-mediated endocytosis (en'-dō-sī-TŌ-sis) A highly selective process whereby cells take up specific ligands, which usually are large molecules or particles, by enveloping them within a sac of plasma membrane. Ligands are eventually broken down by enzymes in lysosomes.

Recessive allele (rē-SESS-iv) An allele whose presence is masked in the presence of a dominant allele on the homologous chromosome.

Recombinant DNA Synthetic DNA, formed by joining a fragment of DNA from one source to a portion of DNA from another.

Recovery oxygen consumption Elevated oxygen use after exercise ends due to metabolic changes that start during exercise and continue after exercise. Previously called **oxygen debt**.

Recruitment (rē-KROOT-ment) The process of increasing the number of active motor units. Also called **motor unit summation**.

Rectum (REK-tum) The last 20 cm (8 in.) of the gastrointestinal tract, from the sigmoid colon to the anus.

Reduction (rē-DUK-shun) The addition of electrons to a molecule or, less commonly, the removal of oxygen from a molecule that results in an increase in the energy content of the molecule.

Referred pain Pain that is felt at a site remote from the place of origin.

Reflex Fast response to a change (stimulus) in the internal or external environment that attempts to restore homeostasis.

Reflex arc The most basic conduction pathway through the nervous system, connecting a receptor and an effector and consisting of a receptor, a sensory neuron, an integrating center in the central nervous system, a motor neuron, and an effector.

Refraction (rē-FRAK-shun) The bending of light as it passes from one medium to another.

Refractory period (re-FRAK-to-rē) A time period during which an excitable cell (neuron or muscle fiber) cannot respond to a stimulus that is usually adequate to evoke an action potential.

Regional anatomy The division of anatomy dealing with a specific region of the body, such as the head, neck, chest, or abdomen.

Regurgitation (rē-gur'-ji-TĀ-shun) Return of solids or fluids to the mouth from the stomach; backward flow of blood through incompletely closed heart valves.

Relaxin (RLX) A female hormone produced by the ovaries and placenta that increases flexibility of the pubic symphysis and helps dilate the uterine cervix to ease delivery of a baby.

Releasing hormone Hormone secreted by the hypothalamus that can stimulate secretion of hormones of the anterior pituitary.

Remodeling (rē-MOD-el-ing) Replacement of old bone by new bone tissue.

Renal (RĒ-nal) Pertaining to the kidneys.

Renal corpuscle (KOR-pus-l) A glomerular (Bowman's) capsule and its enclosed glomerulus.

Renal pelvis A cavity in the center of the kidney formed by the expanded, proximal portion of the ureter, lying within the kidney, and into which the major calyces open.

Renal pyramid (PIR-a-mid) A triangular structure in the renal medulla containing the straight segments of renal tubules and the vasa recta.

Renin (RĒ-nin) An enzyme released by the kidney into the plasma, where it converts angiotensinogen into angiotensin I.

Renin–angiotensin–aldosterone (RAA) pathway A mechanism for the control of blood pressure, initiated by the secretion of renin by the kidney in response to low blood pressure; renin catalyzes formation of angiotensin I, which is converted to angiotensin II by angiotensin-converting enzyme (ACE), and angiotensin II stimulates secretion of aldosterone.

Repolarization (rē-pō-lar-i-ZĀ-shun) Restoration of a resting membrane potential after depolarization.

Reproduction (rē-prō-DUK-shun) The formation of new cells for growth, repair, or replacement; the production of a new individual.

Reproductive cell division Type of cell division in which gametes (sperm and oocytes) are produced; consists of meiosis and cytokinesis.

Residual volume (re-ZID-ū-al) The volume of air still contained in the lungs after a maximal exhalation; about 1200 mL in males and 1100 mL in females.

Resistance (re-ZIS-tans) Hindrance (impedance) to blood flow as a result of higher viscosity, longer total blood vessel length, and smaller blood vessel radius. Ability to ward off disease. The hindrance encountered by electrical charges as they move from one point to another. The hindrance encountered by air as it moves through the respiratory passageways.

Respiration (res-pi-RĀ-shun) Overall exchange of gases between the atmosphere, blood, and body cells consisting of pulmonary ventilation, external respiration, and internal respiration.

Respiratory center Neurons in the pons and medulla oblongata of the brain stem that regulate the rate and depth of pulmonary ventilation.

Respiratory membrane Structure in the lungs consisting of the alveolar wall and its basement membrane and a capillary endothelium and its basement membrane through which the diffusion of respiratory gases occurs.

Resting membrane potential The voltage difference between the inside and outside of a cell membrane when the cell is not responding to a stimulus; in many neurons and muscle fibers it is −70 to −90 mV, with the inside of the cell negative relative to the outside.

Retention (rē-TEN-shun) A failure to void urine due to obstruction, nervous contraction of the urethra, or absence of sensation of desire to urinate.

Reticular activating system (RAS) (re-TIK-ū-lar) A portion of the reticular formation that has many ascending connections with the cerebral cortex; when this area of the brain stem is active, nerve impulses pass to the thalamus and widespread areas of the cerebral cortex, resulting in generalized alertness or arousal from sleep.

Reticular formation A network of small groups of neuronal cell bodies scattered among bundles of axons (mixed gray and white matter) beginning in the medulla oblongata and extending superiorly through the central part of the brain stem.

Reticulocyte (re-TIK-ū-lō-sīt) An immature red blood cell.

Reticulum (re-TIK-ū-lum) A network.

Retina (RET-i-na) The deep coat of the posterior portion of the eyeball consisting of nervous tissue (where the process of vision begins) and a pigmented layer of epithelial cells that contact the choroid.

Retinal (RE-ti-nal) A derivative of vitamin A that functions as the light-absorbing portion of the photopigment rhodopsin.

Retraction (rē-TRAK-shun) The movement of a protracted part of the body posteriorly on a plane parallel to the ground, as in pulling the lower jaw back in line with the upper jaw.

Retrograde degeneration (RE-trō-grād dē-jen-er-Ā-shun) Changes that occur in the proximal portion of a damaged axon only as far as the first node of Ranvier; similar to changes that occur during Wallerian degeneration.

Retroperitoneal (re′-trō-per-i-tō-NĒ-al) External to the peritoneal lining of the abdominal cavity.

Rh factor An inherited antigen on the surface of red blood cells in Rh⁺ individuals; not present in Rh⁻ individuals.

Rhinology (rī-NOL-ō-jē) The study of the nose and its disorders.

Rhodopsin (rō-DOP-sin) The photopigment in rods of the retina, consisting of a glycoprotein called opsin and a derivative of vitamin A called retinal.

Ribonucleic acid (RNA) (rī-bō-noo-KLĒ-ik) A single-stranded nucleic acid made up of nucleotides, each consisting of a nitrogenous base (adenine, cytosine, guanine, or uracil), ribose, and a phosphate group; three types are messenger RNA (mRNA), transfer RNA (tRNA), and ribosomal RNA (rRNA), each of which has a specific role during protein synthesis.

Ribosome (RĪ-bō-sōm) An organelle in the cytoplasm of cells, composed of a small subunit and a large subunit that contain ribosomal RNA and ribosomal proteins; the site of protein synthesis.

Rigidity (ri-JID-i-tē) Hypertonia characterized by increased muscle tone, but reflexes are not affected.

Rigor mortis State of partial contraction of muscles after death due to lack of ATP; myosin heads (crossbridges) remain attached to actin, thus preventing relaxation.

Rod One of two types of photoreceptor in the retina of the eye; specialized for vision in dim light.

Root canal A narrow extension of the pulp cavity lying within the root of a tooth.

Rotation (rō-TĀ-shun) Moving a bone around its own axis, with no other movement.

Round window A small opening between the middle and internal ear, directly inferior to the oval window, covered by the secondary tympanic membrane.

Ruffini corpuscle (roo-FĒ-nē) A sensory receptor embedded deeply in the dermis and deeper tissues that detects stretching of the skin. Also called a **type II cutaneous mechanoreceptor**.

Rugae (ROO-gē) Large folds in the mucosa of an empty hollow organ, such as the stomach and vagina.

S

Saccule (SAK-ūl) The inferior and smaller of the two chambers in the membranous labyrinth inside the vestibule of the internal ear containing a receptor organ for static equilibrium.

Sacral plexus (SĀ-kral PLEK-sus) A network formed by the ventral branches of spinal nerves L4 through S3.

Sacral promontory (PROM-on-tor′-ē) The superior surface of the body of the first sacral vertebra that projects anteriorly into the pelvic cavity; a line from the sacral promontory to the superior border of the pubic symphysis divides the abdominal and pelvic cavities.

Saddle joint A synovial joint in which the articular surface of one bone is saddle shaped and the articular surface of the other bone is shaped like the legs of the rider sitting in the saddle, as in the joint between the trapezium and the metacarpal of the thumb.

Sagittal plane (SAJ-i-tal) A plane that divides the body or organs into left and right portions. Such a plane may be **midsagittal (median)**, in which the divisions are equal, or **parasagittal**, in which the divisions are unequal.

Saliva (sa-LĪ-va) A clear, alkaline, somewhat viscous secretion produced mostly by the three pairs of salivary glands; contains various salts, mucin, lysozyme, salivary amylase, and lingual lipase (produced by glands in the tongue).

Salivary amylase (SAL-i-ver-ē AM-i-lās) An enzyme in saliva that initiates the chemical breakdown of starch.

Salivary gland One of three pairs of glands that lie external to the mouth and pour their secretory product (saliva) into ducts that empty into the oral cavity; the parotid, submandibular, and sublingual glands.

Salt A substance that, when dissolved in water, ionizes into cations and anions, neither of which are hydrogen ions (H⁺) nor hydroxide ions (OH⁻).

Saltatory conduction (sal-ta-TŌ-rē) The propagation of an action potential (nerve impulse) along the exposed parts of a myelinated axon. The action potential appears at successive nodes of Ranvier and therefore seems to leap from node to node.

Sarcolemma (sar′-kō-LEM-ma) The cell membrane of a muscle fiber (cell), especially of a skeletal muscle fiber.

Sarcomere (SAR-kō-mēr) A contractile unit in a striated muscle fiber (cell) extending from one Z disc to the next Z disc.

Sarcoplasm (SAR-kō-plazm) The cytoplasm of a muscle fiber (cell).

Sarcoplasmic reticulum (sar′-kō-PLAZ-mik re-TIK-ū-lum) A network of saccules and tubes surrounding myofibrils of a muscle fiber (cell), comparable to endoplasmic reticulum; functions to reabsorb calcium ions during relaxation and to release them to cause contraction.

Saturated fat A fatty acid that contains only single bonds (no double bonds) between its carbon atoms; all carbon atoms are bonded to the maximum number of hydrogen atoms; prevalent in triglycerides of animal products such as meat, milk, milk products, and eggs.

Scala tympani (SKA-la TIM-pan-ē) The inferior spiral-shaped channel of the bony cochlea, filled with perilymph.

Scala vestibuli (ves-TIB-ū-lē) The superior spiral-shaped channel of the bony cochlea, filled with perilymph.

Schwann cell (SCHVON) A neuroglial cell of the peripheral nervous system that forms the myelin sheath and neurolemma around a nerve axon by wrapping around the axon in a jelly-roll fashion.

Sciatica (sī-AT-i-ka) Inflammation and pain along the sciatic nerve; felt along the posterior aspect of the thigh extending down the inside of the leg.

Sclera (SKLE-ra) The white coat of fibrous tissue that forms the superficial protective covering over the eyeball except in the most anterior portion; the posterior portion of the fibrous tunic.

Scleral venous sinus A circular venous sinus located at the junction of the sclera and the cornea through which aqueous humor drains from the anterior chamber of the eyeball into the blood. Also called the **canal of Schlemm** (SHLEM).

Sclerosis (skle-RŌ-sis) A hardening with loss of elasticity of tissues.

Scoliosis (skō'-lē-Ō-sis) An abnormal lateral curvature from the normal vertical line of the backbone.

Scrotum (SKRŌ-tum) A skin-covered pouch that contains the testes and their accessory structures.

Sebaceous gland (se-BĀ-shus) An exocrine gland in the dermis of the skin, almost always associated with a hair follicle, that secretes sebum. Also called an **oil gland.**

Sebum (SĒ-bum) Secretion of sebaceous (oil) glands.

Secondary response Accelerated, more intense cell-mediated or antibody-mediated immune response upon a subsequent exposure to an antigen after the primary response.

Secondary sex characteristic A characteristic of the male or female body that develops at puberty under the influence of sex hormones but is not directly involved in sexual reproduction; examples are the distribution of body hair, voice pitch, body shape, and muscle development.

Second messenger An intracellular mediator molecule that is produced in response to a first messenger (hormone or neurotransmitter) binding to its receptor in the plasma membrane of a target cell. Initiates a cascade of chemical reactions that produce characteristic effects for that particular target cell.

Secretion (se-KRĒ-shun) Production and release from a cell or a gland of a physiologically active substance.

Selective permeability (per'-mē-a-BIL-i-tē) The property of a membrane by which it permits the passage of certain substances but restricts the passage of others.

Semen (SĒ-men) A fluid discharged at ejaculation by a male that consists of a mixture of sperm and the secretions of the seminiferous tubules, seminal vesicles, prostate, and bulbourethral (Cowper's) glands.

Semicircular canals (sem-ī-SER-kū-lar) Three bony channels (anterior, posterior, lateral), filled with perilymph, in which lie the membranous semicircular canals filled with endolymph. They contain receptors for equilibrium.

Semicircular ducts The membranous semicircular canals filled with endolymph and floating in the perilymph of the bony semicircular canals; they contain cristae that are concerned with dynamic equilibrium.

Semilunar valve (sem'-ē-LOO-nar) A valve between the aorta or the pulmonary trunk and a ventricle of the heart.

Seminal vesicle (SEM-i-nal VES-i-kul) One of a pair of convoluted, pouchlike structures, lying posterior and inferior to the urinary bladder and anterior to the rectum, that secrete a component of semen into the ejaculatory ducts. Also termed a **seminal gland**.

Seminiferous tubule (sem'-ī-NI-fer-us TOO-būl) A tightly coiled duct, located in the testis, where sperm are produced.

Sensation A state of awareness of external or internal conditions of the body.

Sensory neurons (NOO-ronz) Neurons that carry sensory information from cranial and spinal nerves into the brain and spinal cord or from a lower to a higher level in the spinal cord and brain. Also called **afferent neurons** (AF-er-ent).

Septal defect An opening in the septum (interatrial or interventricular) between the left and right sides of the heart.

Septum (SEP-tum) A wall dividing two cavities.

Serous membrane (SIR-us) A membrane that lines a body cavity that does not open to the exterior. The external layer of an organ formed by a serous membrane. The membrane that lines the pleural, pericardial, and peritoneal cavities. Also called a **serosa** (se-RŌ-sa).

Sertoli cell (ser-TŌ-lē) A supporting cell in the seminiferous tubules that secretes fluid for supplying nutrients to sperm and the hormone inhibin, removes excess cytoplasm from spermatogenic cells, and mediates the effects of FSH and testosterone on spermatogenesis. Also called a **sustentacular cell** (sus'-ten-TAK-ū-lar).

Serum Blood plasma minus its clotting proteins.

Sesamoid bones (SES-a-moyd) Small bones usually found in tendons.

Sex chromosomes The twenty-third pair of chromosomes, designated X and Y, which determine the genetic sex of an individual; in males, the pair is XY; in females, XX.

Sexual intercourse The insertion of the erect penis of a male into the vagina of a female. Also called **coitus** (KŌ-i-tus).

Shivering Involuntary contraction of skeletal muscles that generates heat.

Shock Failure of the cardiovascular system to deliver adequate amounts of oxygen and nutrients to meet the metabolic needs of the body due to inadequate cardiac output. It is characterized by hypotension; clammy, cool, and pale skin; sweating; reduced urine formation; altered mental state; acidosis; tachycardia; weak, rapid pulse; and thirst. Types include hypovolemic, cardiogenic, vascular, and obstructive.

Shoulder joint A synovial joint where the humerus articulates with the scapula.

Sigmoid colon (SIG-moyd KŌ-lon) The S-shaped part of the large intestine that begins at the level of the left iliac crest, projects medially, and terminates at the rectum at about the level of the third sacral vertebra.

Sign Any objective evidence of disease that can be observed or measured such as a lesion, swelling, or fever.

Sinoatrial (SA) node (si-nō-Ā-trē-al) A small mass of cardiac muscle fibers (cells) located in the right atrium inferior to the opening of the superior vena cava that spontaneously depolarize and generate a cardiac action potential about 100 times per minute. Also called the **pacemaker**.

Sinus (SĪ-nus) A hollow in a bone (paranasal sinus) or other tissue; a channel for blood (vascular sinus); any cavity having a narrow opening.

Sinusoid (SĪ-nū-soyd) A large, thin-walled, and leaky type of capillary, having large intercellular clefts that may allow proteins and blood cells to pass from a tissue into the bloodstream; present in the liver, spleen, anterior pituitary, parathyroid glands, and red bone marrow.

Skeletal muscle An organ specialized for contraction, composed of striated muscle fibers (cells), supported by connective tissue, attached to a bone by a tendon or an aponeurosis, and stimulated by somatic motor neurons.

Skin The external covering of the body that consists of a superficial, thinner epidermis (epithelial tissue) and a deep, thicker dermis (connective tissue) that is anchored to the subcutaneous layer.

Skull The skeleton of the head consisting of the cranial and facial bones.

Sleep A state of partial unconsciousness from which a person can be aroused; associated with a low level of activity in the reticular activating system.

Sliding-filament mechanism The explanation of how thick and thin filaments slide relative to one another during striated muscle contraction to decrease sarcomere length.

Small intestine A long tube of the gastrointestinal tract that begins at the pyloric sphincter of the stomach, coils through the central and inferior part of the abdominal cavity, and ends at the large intestine; divided into three segments: duodenum, jejunum, and ileum.

Smooth muscle A tissue specialized for contraction, composed of smooth muscle fibers (cells), located in the walls of hollow internal organs, except for the heart, and innervated by autonomic motor neurons.

Sodium-potassium pump An active transport pump located in the plasma membrane that transports sodium ions out of the cell and potassium ions into the cell at the expense of cellular ATP. It functions to keep the ionic concentrations of these ions at physiological levels.

Soft palate (PAL-at) The posterior portion of the roof of the mouth, extending from the palatine bones to the uvula. It is a muscular partition lined with mucous membrane.

Solution A homogeneous molecular or ionic dispersion of one or more substances (solutes) in a dissolving medium (solvent) that is usually liquid.

Somatic cell division (sō-MAT-ik) Type of cell division in which a single starting parent cell duplicates itself to produce two identical cells; consists of mitosis and cytokinesis.

Somatic nervous system (SNS) The portion of the peripheral nervous system consisting of somatic sensory (afferent) neurons and somatic motor (efferent) neurons.

Spasm (SPAZM) A sudden, involuntary contraction of skeletal muscles.

Spasticity (spas-TIS-i-tē) Hypertonia characterized by increased muscle tone, increased tendon reflexes, and pathological reflexes (Babinski sign).

Spermatic cord (sper-MAT-ik) A supporting structure of the male reproductive system, extending from a testis to the deep inguinal ring, that includes the ductus (vas) deferens, arteries, veins, lymphatic vessels, nerves, cremaster muscle, and connective tissue.

Spermatogenesis (sper'-ma-tō-JEN-e-sis) The formation and development of sperm in the seminiferous tubules of the testes.

Sperm cell A mature male gamete. Also termed a **spermatozoon** (sper'-ma-tō-ZŌ-on).

Spermiogenesis (sper'-mē-ō-JEN-e-sis) The maturation of spermatids into sperm.

Sphincter (SFINGK-ter) A circular muscle that constricts an opening.

Sphygmomanometer (sfig'-mō-ma-NOM-e-ter) An instrument for measuring arterial blood pressure.

Spinal cord (SPĪ-nal) A mass of nerve tissue located in the vertebral cavity from which 31 pairs of spinal nerves originate.

Spinal nerve One of the 31 pairs of nerves that originate on the spinal cord from posterior and anterior roots.

Spinal shock A period from several days to several weeks following transection of the spinal cord and characterized by the abolition of all reflex activity.

Spinothalamic tracts (spī-nō-tha-LAM-ik) Sensory (ascending) tracts that convey information up the spinal cord to the thalamus for sensations of pain, temperature, itch, and tickle.

Spiral organ The organ of hearing, consisting of supporting cells and hair cells that rest on the basilar membrane and extend into the endolymph of the cochlear duct. Also called the **organ of Corti** (KOR-tē).

Spirometer (spī-ROM-e-ter) An apparatus used to measure lung volumes and capacities.

Spleen (SPLĒN) Large mass of lymphatic tissue between the fundus of the stomach and the diaphragm that functions in formation of blood cells during early fetal development, phagocytosis of ruptured blood cells, and proliferation of B cells during immune responses.

Spongy (cancellous) bone tissue Bone tissue that consists of an irregular latticework of thin plates of bone called trabeculae; spaces between trabeculae of some bones are filled with red bone marrow; found inside short, flat, and irregular bones and in the epiphyses (ends) of long bones.

Sprain Forcible wrenching or twisting of a joint with partial rupture or other injury to its attachments without dislocation.

Squamous (SKWĀ-mus) Flat or scalelike.

Starvation (star-VĀ-shun) The loss of energy stores in the form of glycogen, triglycerides, and proteins due to inadequate intake of nutrients or inability to digest, absorb, or metabolize ingested nutrients.

Stasis (STĀ-sis) Stagnation or halt of normal flow of fluids, as blood or urine, or of the intestinal contents.

Static equilibrium (ē-kwi-LIB-rē-um) The maintenance of posture in response to changes in the orientation of the body, mainly the head, relative to the ground.

Stellate reticuloendothelial cell (STEL-āt re-tik′-ū-lō-en′-dō-THĒ-lē-al) Phagocytic cell within a sinusoid of the liver. Also called a **Kupffer cell** (KOOP-fer).

Stem cell Unspecialized cell that has the ability to divide for indefinite periods and give rise to specialized cells.

Stenosis (sten-Ō-sis) An abnormal narrowing or constriction of a duct or opening.

Sterile (STE-ril) Free from any living microorganisms. Unable to conceive or produce offspring.

Sterilization (ster′-i-li-ZĀ-shun) Elimination of all living microorganisms. Any procedure that renders an individual incapable of reproduction (for example, castration, vasectomy, hysterectomy, or oophorectomy).

Stimulus Any stress that changes a controlled condition; any change in the internal or external environment that excites a sensory receptor, a neuron, or a muscle fiber.

Stomach The J-shaped enlargement of the gastrointestinal tract directly inferior to the diaphragm in the epigastric, umbilical, and left hypochondriac regions of the abdomen, between the esophagus and small intestine.

Stratum (STRĀ-tum) A layer.

Stressor A stress that is extreme, unusual, or long-lasting and triggers the stress response.

Stress response Wide-ranging set of bodily changes, triggered by a stressor, that gears the body to meet an emergency. Also known as **general adaptation syndrome (GAS)**.

Stretch receptor Receptor in the walls of blood vessels, airways, or organs that monitors the amount of stretching. Also termed a **baroreceptor.**

Stretch reflex A monosynaptic reflex triggered by sudden stretching of muscle spindles within a muscle that elicits contraction of that same muscle. Also called a **tendon jerk**.

Stroke volume The volume of blood ejected by either ventricle during one systole; about 70 mL in an adult at rest.

Subarachnoid space (sub′-a-RAK-noyd) A space between the arachnoid mater and the pia mater that surrounds the brain and spinal cord and through which cerebrospinal fluid circulates.

Subcutaneous (sub′-kū-TĀ-nē-us) Beneath the skin. Also called **hypodermic** (hi-pō-DER-mik).

Subcutaneous layer A continuous sheet of areolar connective tissue and adipose tissue between the dermis of the skin and the deep fascia of the muscles.

Subdural space (sub-DOO-ral) A space between the dura mater and the arachnoid mater of the brain and spinal cord that contains a small amount of fluid.

Sublingual gland (sub-LING-gwal) One of a pair of salivary glands situated in the floor of the mouth deep to the mucous membrane and to the side of the lingual frenulum, with a duct that opens into the floor of the mouth.

Submandibular gland (sub′-man-DIB-ū-lar) One of a pair of salivary glands found inferior to the base of the tongue deep to the mucous membrane in the posterior part of the floor of the mouth, posterior to the sublingual glands, with a duct situated to the side of the lingual frenulum. Also called the **submaxillary gland** (sub′-MAK-si-ler-ē).

Submucosa (sub-mū-KŌ-sa) A layer of connective tissue located deep to a mucous membrane, as in the gastrointestinal tract or the urinary bladder; the submucosa connects the mucosa to the muscularis layer.

Substrate A molecule upon which an enzyme acts.

Sudoriferous gland (soo′-dor-IF-er-us) An apocrine or eccrine exocrine gland in the dermis or subcutaneous layer that produces perspiration. Also called a **sweat gland**.

Sulcus (SUL-kus) A groove or depression between parts, especially between the convolutions of the brain. *Plural* is **sulci** (SUL-sī).

Summation (sum-MĀ-shun) The addition of the excitatory and inhibitory effects of many stimuli applied to a neuron. The increased strength of muscle contraction that results when stimuli follow one another in rapid succession.

Superficial (soo′-per-FISH-al) Located on or near the surface of the body or an organ.

Superior (soo-PĒR-ē-or) Toward the head or upper part of a structure.

Superior vena cava (SVC) (VĒ-na CĀ-va) Large vein that collects blood from parts of the body superior to the heart and returns it to the right atrium.

Supination (soo-pi-NĀ-shun) A movement of the forearm in which the palm is turned anteriorly.

Surface anatomy The study of the structures that can be identified from the outside of the body.

Surfactant (sur-FAK-tant) Complex mixture of phospholipids and lipoproteins, produced by type II alveolar (septal) cells in the lungs, that decreases surface tension.

Susceptibility (sus-sep′-ti-BIL-i-tē) Lack of resistance to the damaging effects of an agent such as a pathogen.

Suspensory ligament (sus-PEN-so-rē LIG-a-ment) A fold of peritoneum extending laterally from the surface of the ovary to the pelvic wall.

Sutural bone (SOO-cher-al) A small bone located within a suture between certain cranial bones.

Suture (SOO-cher) An immovable or slightly movable fibrous joint that joins skull bones.

Sympathetic division (sim′-pa-THET-ik) One of the two subdivisions of the autonomic nervous system, having cell bodies of preganglionic neurons in the lateral gray columns of the thoracic segment and the first two or three lumbar segments of the spinal cord; primarily concerned with processes involving the expenditure of energy.

Sympathetic trunk ganglion (GANG-glē-on) A cluster of cell bodies of sympathetic postganglionic neurons lateral to the vertebral column, close to the body of a vertebra. These ganglia extend inferiorly through the neck, thorax, and abdomen to the coccyx on both sides of the vertebral column and are connected to one another to form a chain on each side of the vertebral column. Also called **sympathetic chain** or **vertebral chain ganglia**.

Sympathomimetic (sim'-pa-thō-mi-MET-ik) Producing effects that mimic those brought about by the sympathetic division of the autonomic nervous system.

Symphysis (SIM-fi-sis) A line of union. A slightly movable fibrocartilaginous joint such as the pubic symphysis.

Symptom (SIMP-tum) A subjective change in body function not apparent to an observer, such as pain or nausea, that indicates the presence of a disease or disorder of the body.

Synapse (SYN-aps) The functional junction between two neurons or between a neuron and an effector, such as a muscle or gland; may be electrical or chemical.

Synaptic cleft (sin-AP-tik) The narrow gap at a chemical synapse that separates the axon terminal of one neuron from another neuron or muscle fiber (cell) and across which a neurotransmitter diffuses to affect the postsynaptic cell.

Synaptic end bulb Expanded distal end of an axon terminal that contains synaptic vesicles. Also called a **synaptic knob**.

Synaptic vesicle Membrane-enclosed sac in a synaptic end bulb that stores neurotransmitters.

Synarthrosis (sin'-ar-THRŌ-sis) An immovable or slightly movable joint such as a suture, gomphosis, and synchondrosis.

Synchondrosis (sin'-kon-DRŌ-sis) A cartilaginous joint in which the connecting material is hyaline cartilage.

Syndesmosis (sin'-dez-MŌ-sis) A slightly movable joint in which articulating bones are united by fibrous connective tissue.

Syndrome (SIN-drōm) A group of signs and symptoms that occur together in a pattern that is characteristic of a particular disease or abnormal condition.

Synergist (SIN-er-jist) A muscle that assists the prime mover by reducing undesired action or unnecessary movement.

Synostosis (sin'-os-TŌ-sis) A joint in which the dense fibrous connective tissue that unites bones at a suture has been replaced by bone, resulting in a complete fusion across the suture line.

Synovial cavity (si-NŌ-vē-al) The space between the articulating bones of a synovial joint, filled with synovial fluid. Also called a **joint cavity**.

Synovial fluid Secretion of synovial membranes that lubricates joints and nourishes articular cartilage.

Synovial joint A fully movable or diarthrotic joint in which a synovial (joint) cavity is present between the two articulating bones.

Synovial membrane The deeper of the two layers of the articular capsule of a synovial joint, composed of areolar connective tissue that secretes synovial fluid into the synovial (joint) cavity.

System An association of organs that have a common function.

Systemic (sis-TEM-ik) Affecting the whole body; generalized.

Systemic anatomy The anatomic study of particular systems of the body, such as the skeletal, muscular, nervous, cardiovascular, or urinary systems.

Systemic circulation The routes through which oxygenated blood flows from the left ventricle through the aorta to all the organs of the body except the lungs and deoxygenated blood returns to the right atrium.

Systemic vascular resistance (SVR) All the vascular resistance offered by systemic blood vessels. Also called **total peripheral resistance**.

Systole (SIS-tō-lē) In the cardiac cycle, the phase of contraction of the heart muscle, especially of the ventricles.

Systolic blood pressure (sis-TOL-ik) The force exerted by blood on arterial walls during ventricular contraction; the highest pressure measured in the large arteries, about 110 mmHg under normal conditions for a young adult.

T

T cell A lymphocyte that becomes immunocompetent in the thymus and can differentiate into a helper T cell or a cytotoxic T cell, both of which function in cell-mediated immunity.

T wave The deflection wave of an electrocardiogram that represents ventricular repolarization.

Tachycardia (tak'-i-KAR-dē-a) An abnormally rapid resting heartbeat or pulse rate (over 100 beats per minute).

Tactile (TAK-tīl) Pertaining to the sense of touch.

Target cell A cell whose activity is affected by a particular hormone.

Tarsal bones The seven bones of the ankle. Also called **tarsals**.

Tarsal gland Sebaceous (oil) gland that opens on the edge of each eyelid. Also called a **Meibomian gland** (mī-BŌ-mē-an).

Tarsal plate A thin, elongated sheet of connective tissue, one in each eyelid, giving the eyelid form and support. The aponeurosis of the levator palpebrae superioris is attached to the tarsal plate of the superior eyelid.

Tarsus (TAR-sus) A collective term for the seven bones of the ankle.

Tectorial membrane (tek-TŌ-rē-al) A gelatinous membrane projecting over and in contact with the hair cells of the spiral organ (organ of Corti) in the cochlear duct.

Teeth (TĒ-TH) Accessory structures of digestion, composed of calcified connective tissue and embedded in bony sockets of the mandible and maxilla, that cut, shred, crush, and grind food. Also called **dentes** (DEN-tēz).

Telophase (TEL-ō-fāz) The final stage of mitosis in which two nuclei become established.

Tendon (TEN-don) A white fibrous cord of dense regular connective tissue that attaches muscle to bone.

Tendon organ A proprioceptive receptor, sensitive to changes in muscle tension and force of contraction, found chiefly near the junctions of tendons and muscles. Also called a **Golgi tendon organ** (GOL-jē).

Tendon reflex A polysynaptic, ipsilateral reflex that protects tendons and their associated muscles from damage that might be brought about by excessive tension. The receptors involved are called tendon organs (Golgi tendon organs).

Teratogen (TER-a-tō-jen) Any agent or factor that causes physical defects in a developing embryo.

Testis (TES-tis) Male gonad that produces sperm and the hormones testosterone and inhibin. Also called a **testicle**.

Testosterone (tes-TOS-te-rōn) A male sex hormone (androgen) secreted by interstitial endocrinocytes (Leydig cells) of a mature testis; needed for development of sperm; together with a second androgen termed dihydrotestosterone (DHT), controls the growth and development of male reproductive organs, secondary sex characteristics, and body growth.

Tetany (TET-a-nē) Hyperexcitability of neurons and muscle fibers (cells) caused by hypocalcemia and characterized by intermittent or continuous tonic muscular contractions; may be due to hypoparathyroidism.

Thalamus (THAL-a-mus) A large, oval structure located bilaterally on either side of the third ventricle, consisting of two masses of gray matter organized into nuclei; main relay center for sensory impulses ascending to the cerebral cortex.

Thermoreceptor (THER-mō-rē-sep-tor) Sensory receptor that detects changes in temperature.

Thigh The portion of the lower limb between the hip and the knee.

Third ventricle (VEN-tri-kul) A slitlike cavity between the right and left halves of the thalamus and between the lateral ventricles of the brain.

Thirst center A cluster of neurons in the hypothalamus that is sensitive to the osmotic pressure of extracellular fluid and brings about the sensation of thirst.

Thoracic cavity (thō-RAS-ik) A cavity that contains two pleural cavities, the mediastinum, and the pericardial cavity.

Thoracic duct A lymphatic vessel that begins as a dilation called the cisterna chyli, receives lymph from the left side of the head, neck, and chest, the left arm, and the entire body below the ribs, and empties into the left subclavian vein. Also called the **left lymphatic duct** (lim-FAT-ik).

Thoracolumbar outflow (thō′-ra-kō-LUM-bar) The axons of sympathetic preganglionic neurons, which have their cell bodies in the lateral gray columns of the thoracic segments and first two or three lumbar segments of the spinal cord.

Thorax (THŌ-raks) The chest.

Threshold potential The membrane voltage that must be reached to trigger an action potential.

Threshold stimulus Any stimulus strong enough to initiate an action potential or activate a sensory receptor.

Thrombin (THROM-bin) The active enzyme formed from prothrombin that converts fibrinogen to fibrin during the formation of a blood clot.

Thrombolytic agent (throm-bō-LIT-ik) Chemical substance injected into the body to dissolve blood clots and restore circulation; mechanism of action is direct or indirect activation of plasminogen; examples include tissue plasminogen activator (t-PA), streptokinase, and urokinase.

Thrombosis (throm-BŌ-sis) The formation of a clot in an unbroken blood vessel, usually a vein.

Thrombus (THROM-bus) A stationary clot formed in an unbroken blood vessel, usually a vein.

Thymus (THĪ-mus) A bilobed organ, located in the superior mediastinum posterior to the sternum and between the lungs, in which T cells develop immunocompetence.

Thyroglobulin (TGB) (thī-rō-GLŌ-bū-lin) A large glycoprotein molecule produced by follicular cells of the thyroid gland in which some tyrosines are iodinated and coupled to form thyroid hormones.

Thyroid cartilage (THĪ-royd KAR-ti-lij) The largest single cartilage of the larynx, consisting of two fused plates that form the anterior wall of the larynx.

Thyroid follicle (FOL-i-kul) Spherical sac that forms the parenchyma of the thyroid gland and consists of follicular cells that produce thyroxine (T_4) and triiodothyronine (T_3).

Thyroid gland An endocrine gland with right and left lateral lobes on either side of the trachea connected by an isthmus; located anterior to the trachea just inferior to the cricoid cartilage; secretes thyroxine (T_4), triiodothyronine (T_3), and calcitonin (CT).

Thyroid-stimulating hormone (TSH) A hormone secreted by the anterior pituitary that stimulates the synthesis and secretion of thyroxine (T_4) and triiodothyronine (T_3).

Thyroxine (T_4) (thī-ROK-sēn) A hormone secreted by the thyroid gland that regulates metabolism, growth and development, and the activity of the nervous system.

Tic Spasmodic, involuntary twitching of muscles that are normally under voluntary control.

Tidal volume The volume of air breathed in and out in any one breath; about 500 mL in quiet, resting conditions.

Tissue A group of similar cells and their intercellular substance joined together to perform a specific function.

Tissue factor (TF) A factor, or collection of factors, whose appearance initiates the blood clotting process. Also called **thromboplastin** (throm-bō-PLAS-tin).

Tissue plasminogen activator (t-PA) (plaz-MIN-ō-gen) An enzyme that dissolves small blood clots by initiating a process that converts plasminogen to plasmin, which degrades the fibrin of a clot.

Tongue A large skeletal muscle covered by a mucous membrane located on the floor of the oral cavity.

Tonicity (tō-NIS-i-tē) A measure of the concentration of impermeable solute particles in a solution relative to cytosol. When cells are bathed in an isotonic solution, they neither shrink nor swell.

Tonsil (TON-sil) An aggregation of large lymphatic nodules embedded in the mucous membrane of the throat.

Torn cartilage A tearing of an articular disc (meniscus) in the knee.

Total lung capacity The sum of tidal volume, inspiratory reserve volume, expiratory reserve volume, and residual volume; about 6000 mL in males.

Trabecula (tra-BEK-ū-la) Irregular latticework of thin plates of spongy bone. Fibrous cord of connective tissue serving as supporting fiber by forming a septum extending into an organ from its wall or capsule. *Plural* is **trabeculae** (tra-BEK-ū-lē).

Trachea (TRĀ-kē-a) Tubular air passageway extending from the larynx to the fifth thoracic vertebra. Also called the **windpipe**.

Tract A bundle of nerve axons in the central nervous system.

Transcription (trans-KRIP-shun) The first step in the expression of genetic information in which a single strand of DNA serves as a template for the formation of an RNA molecule.

Translation (trans-LĀ-shun) The synthesis of a new protein on a ribosome as dictated by the sequence of codons in messenger RNA.

Transverse colon (trans-VERS KŌ-lon) The portion of the large intestine extending across the abdomen from the right colic (hepatic) flexure to the left colic (splenic) flexure.

Transverse fissure (FISH-er) The deep cleft that separates the cerebrum from the cerebellum.

Transverse plane A plane that divides the body or organs into superior and inferior portions. Also called a **horizontal plane**.

Transverse tubules (T tubules) (TOO-būls) Small, cylindrical invaginations of the sarcolemma of striated muscle fibers (cells) that conduct muscle action potentials toward the center of the muscle fiber.

Trauma (TRAW-ma) An injury, either a physical wound or psychic disorder, caused by an external agent or force, such as a physical blow or emotional shock; the agent or force that causes the injury.

Tremor (TREM-or) Rhythmic, involuntary, purposeless contraction of opposing muscle groups.

Tricuspid valve (trī-KUS-pid) Atrioventricular (AV) valve on the right side of the heart.

Triglyceride (trī-GLI-cer-īd) A lipid formed from one molecule of glycerol and three molecules of fatty acids that may be either solid (fats) or liquid (oils) at room temperature; the body's most highly concentrated source of chemical potential energy. Found mainly within adipocytes. Also called a **neutral fat** or a **triacylglycerol**.

Triiodothyronine (T₃) (trī-ī-ō-dō-THĪ-rō-nēn) A hormone produced by the thyroid gland that regulates metabolism, growth and development, and the activity of the nervous system.

Trophoblast (TRŌF-ō-blast) The superficial covering of cells of the blastocyst.

Tropic hormone (TRŌ-pik) A hormone whose target is another endocrine gland.

Trunk The part of the body to which the head and upper and lower limbs are attached.

Tubal ligation (lī-GĀ-shun) A sterilization procedure in which the uterine (fallopian) tubes are tied and cut.

Tubular reabsorption (TOO-byū-lar rē-ab-SORP-shun) The movement of filtrate from renal tubules back into blood in response to the body's specific needs.

Tubular secretion The movement of substances in blood into renal tubular fluid in response to the body's specific needs.

Tumor suppressor gene A gene coding for a protein that normally inhibits cell division; loss or alteration of a tumor suppressor gene called *p53* is the most common genetic change in a wide variety of cancer cells.

Tunica externa (TOO-nik-a eks-TER-na) The superficial coat of an artery or vein, composed mostly of elastic and collagen fibers. Also called the **adventitia** (ad-ven-TISH-a).

Tunica interna (in-TER-na) The deep coat of an artery or vein, consisting of a lining of endothelium, basement membrane, and internal elastic lamina (in an artery). Also called the **tunica intima** (IN-ti-ma).

Tunica media (MĒ-dē-a) The intermediate coat of an artery or vein, composed of smooth muscle and elastic fibers.

Twitch contraction Brief contraction of all muscle fibers (cells) in a motor unit triggered by a single action potential in its motor neuron.

U

Umbilical cord (um-BIL-i-kal) The long, ropelike structure containing the umbilical arteries and vein that connect the fetus to the placenta.

Umbilicus (um-BIL-i-kus *or* um-bil-Ī-kus) A small scar on the abdomen that marks the former attachment of the umbilical cord to the fetus. Also called the **navel**.

Upper limb The appendage attached at the shoulder girdle, consisting of the arm, forearm, wrist, hand, and digits. Also called **upper extremity**.

Uremia (ū-RĒ-mē-a) Accumulation of toxic levels of urea and other nitrogenous waste products in the blood, usually resulting from severe kidney malfunction.

Ureter (Ū-rē-ter) One of two tubes that connect the kidney with the urinary bladder.

Urethra (ū-RĒ-thra) The duct from the urinary bladder to the exterior of the body that conveys urine in females and urine and semen in males.

Urinary bladder (Ū-ri-ner-ē) A hollow, muscular organ situated in the pelvic cavity posterior to the pubic symphysis; receives urine via two ureters and stores urine until it is excreted through the urethra.

Urine The fluid produced by the kidneys that contains wastes and excess materials; excreted from the body through the urethra.

Urology (ū-ROL-ō-jē) The specialized branch of medicine that deals with the structure, function, and diseases of the male and female urinary systems and the male reproductive system.

Uterine tube (Ū-ter-in) Duct that transports ova from the ovary to the uterus. Also called the **fallopian tube** (fal-LŌ-pē-an) or **oviduct**.

Uterus (Ū-te-rus) The hollow, muscular organ in females that is the site of menstruation, implantation, development of the fetus, and labor. Also called the **womb**.

Utricle (Ū-tri-kul) The larger of the two divisions of the membranous labyrinth located inside the vestibule of the inner ear, containing a receptor organ for static equilibrium.

Uvula (Ū-vū-la) A soft, fleshy mass, especially the V-shaped pendant part, descending from the soft palate.

V

Vagina (va-JĪna) A muscular, tubular organ that leads from the uterus to the vestibule, situated between the urinary bladder and the rectum of the female.

Vallate papilla (VAL-āt pa-PIL-a) One of the circular projections that is arranged in an inverted V-shaped row at the back of the tongue; the largest of the elevations on the upper surface of the tongue that contains taste buds. Also called **circumvallate papilla**.

Valence (VĀ-lens) The combining capacity of an atom; the number of deficit or extra electrons in the outermost electron shell of an atom.

Varicose (VAR-i-kōs) Pertaining to an unnatural swelling, as in the case of a varicose vein.

Vasa recta (VĀ-sa REK-ta) Extensions of the efferent arteriole of a juxtamedullary nephron that run alongside the loop of the nephron (Henle) in the medullary region of the kidney.

Vasa vasorum (va-SŌ-rum) Blood vessels that supply nutrients to the larger arteries and veins.

Vascular (VAS-kū-lar) Pertaining to or containing many blood vessels.

Vascular spasm Contraction of the smooth muscle in the wall of a damaged blood vessel to prevent blood loss.

Vascular (venous) sinus A vein with a thin endothelial wall that lacks a tunica media and externa and is supported by surrounding tissue.

Vascular tunic (TOO-nik) The middle layer of the eyeball, composed of the choroid, ciliary body, and iris. Also called the **uvea** (Ū-ve-a).

Vasectomy (va-SEK-tō-mē) A means of sterilization of males in which a portion of each ductus (vas) deferens is removed.

Vasoconstriction (vāz-ō-kon-STRIK-shun) A decrease in the size of the lumen of a blood vessel caused by contraction of the smooth muscle in the wall of the vessel.

Vasodilation (vāz′-ō-DĪ-lā-shun) An increase in the size of the lumen of a blood vessel caused by relaxation of the smooth muscle in the wall of the vessel.

Vein (VĀN) A blood vessel that conveys blood from tissues back to the heart.

Vena cava (VĒ-na KĀ-va) One of two large veins that open into the right atrium, returning to the heart all of the deoxygenated blood from the systemic circulation except from the coronary circulation.

Ventral (VEN-tral) Pertaining to the anterior or front side of the body; opposite of dorsal.

Ventricle (VEN-tri-kul) A cavity in the brain filled with cerebrospinal fluid. An inferior chamber of the heart.

Ventricular fibrillation (ven-TRIK-ū-lar fib-ri-LĀ-shun) Asynchronous ventricular contractions; unless reversed by defibrillation, results in heart failure.

Venule (VEN-ūl) A small vein that collects blood from capillaries and delivers it to a vein.

Vermiform appendix (VER-mi-form a-PEN-diks) A twisted, coiled tube attached to the cecum.

Vertebral cavity (VER-te-bral) A space within the vertebral column formed by the vertebral foramina of all the vertebrae and containing the spinal cord.

Vertebral column The 26 vertebrae of an adult and 33 vertebrae of a child; encloses and protects the spinal cord and serves as a point of attachment for the ribs and back muscles. Also called the **backbone, spine,** or **spinal column**.

Vesicle (VES-i-kul) A small bladder or sac containing liquid.

Vestibular apparatus (ves-TIB-ū-lar) Collective term for the organs of equilibrium, which includes the saccule, utricle, and semicircular ducts.

Vestibular membrane The membrane that separates the cochlear duct from the scala vestibuli.

Vestibule (VES-ti-būl) A small space or cavity at the beginning of a canal, especially the inner ear, larynx, mouth, nose, and vagina.

Villus (VIL-lus) A projection of the intestinal mucosal cells containing connective tissue, blood vessels, and a lymphatic vessel; functions in the absorption of the end products of digestion. *Plural* is **villi** (VIL-ī).

Viscera (VIS-er-a) The organs inside the thoracic and abdominopelvic cavities. *Singular* is **viscus** (VIS-kus).

Visceral (VIS-er-al) Pertaining to the organs or to the covering of an organ.

Visceral effectors (e-FEK-torz) Organs of the thoracic and abdominopelvic cavities that respond to neural stimulation, including cardiac muscle, smooth muscle, and glands.

Vital capacity The sum of inspiratory reserve volume, tidal volume, and expiratory reserve volume; about 4800 mL in males.

Vital signs Signs necessary to life that include temperature (T), pulse (P), respiratory rate (RR), and blood pressure (BP).

Vitamin An organic molecule necessary in trace amounts that acts as a catalyst in normal metabolic processes in the body.

Vitreous body (VIT-rē-us) A soft, jellylike substance that fills the vitreous chamber of the eyeball, lying between the lens and the retina.

Vocal folds Pair of mucous membrane folds below the ventricular folds that function in voice production. Also called **true vocal cords**.

Voltage-gated channel An ion channel in a plasma membrane composed of integral proteins that functions like a gate to permit or restrict the movement of ions across the membrane in response to changes in the voltage.

Vulva (VUL-va) Collective designation for the external genitals of the female. Also called the **pudendum** (poo-DEN-dum).

W

Wallerian degeneration (wal-LE-rē-an) Degeneration of the portion of the axon and myelin sheath of a neuron distal to the site of injury.

Wandering macrophage (MAK-rō-fāj) Phagocytic cell that develops from a monocyte, leaves the blood, and migrates to infected tissues.

Wave summation (sum-MĀ-shun) The increased strength of muscle contraction that results when muscle action potentials occur one after another in rapid succession.

White matter Aggregations or bundles of myelinated and unmyelinated axons located in the brain and spinal cord.

X

Xiphoid (ZĪ-foyd) Sword-shaped. The inferior portion of the sternum is the **xiphoid process**.

Y

Yolk sac An extraembryonic membrane composed of the exocoelomic membrane and hypoblast. It transfers nutrients to the embryo, is a source of blood cells, contains primordial germ cells that migrate into the gonads to form primitive germ cells, forms part of the gut, and helps prevent desiccation of the embryo.

Z

Zona pellucida (pe-LOO-si-da) Clear glycoprotein layer between a secondary oocyte and the surrounding granulosa cells of the corona radiata.

Zygote (ZĪ-g-ot) The single cell resulting from the union of male and female gametes; the fertilized ovum.

Line Art Credits

Chapter 1
Figures 1.1, 1.8: Kevin Somerville. Figure 1.2: DNA Illustrations. Figure 1.4: Precision Graphics. Figure 1.5: Molly Borman. Figure 1.6: DNA Illustrations/Imagineering. Figure 1.7: Imagineering.

Chapter 2
Figures 2.1, 2.2: Precision Graphics. Figures 2.3–2.6, 2.8–2.10–2.12, 2.14–2.22: Imagineering. Figures 2.13: Precision Graphics/Imagineering.

Chapter 3
Figures 3.1–3.10: Tomo Narashima/Imagineering. Figures 3.12–3.21: Imagineering. Figures 3.22–3.26, 3.28: Kevin Somerville/Imagineering. Figure 3.27: Kevin Somerville. Figure 3.29: Precision Graphics.

Chapter 4
Figures 4.1–4.8: Kevin Somerville. Figure 4.9: Kevin Somerville/Imagineering.

Chapter 5
Figures 5.1, 5.3, 5.5, 5.7, 5.9, 5.11–5.13, 5.16, Table 5.1: John Gibb. Figures 5.2, 5.6: Kevin Somerville. Figures 5.4, 5.18: Imagineering. Figures 5.8, 5.10, 5.11, 5.15: John Gibb/Imagineering.

Chapter 6
Figure 6.1: Imagineering. Figure 6.2: Kevin Somerville /Imagineering. Figures 6.3–6.10: Imagineering. Figure 6.11: Kevin Somerville. Figures 6.12–6.24: John Gibb.

Chapter 7
Figures 7.1, 7.4, 7.8–7.10, 7.18 : Kevin Somerville. Figures 7.5–7.7, 7.11, 7.15, 7.16: Imagineering. Figures 7.3, 7.12–7.14, 7.17: Imagineering/Kevin Somerville. Figure 7.19: John Gibb/Kevin Somerville. Figure 7.20: Leonaard Dank/Imagineering.

Chapter 8
Figure 8.1: DNA Illustrations/Kevin Somerville. Figure 8.2: Kevin Somerville. Figures 8.3, 8.9, 8.10: Imagineering. Figures 8.4, 8.8, 8.11–8.14: Tomo Narashima. Figure 8.5: Molly Borman. Figures 8.6, 8.7: Sharon Ellis.

Chapter 9
Figures 9.1, 9.19: Kevin Somerville. Figures 9.2, 9.3, 9.6, 9.8, 9.11, 9.13, 9.15, 9.16, 9.18, 9.20: Imagineering. Figures 9.4, 9.7, 9.10, 9.12, 9.14: Lynn O'Kelley/Imagineering. Figure 9.21: Kevin Somerville/Lynn O'Kelley/Steve Oh.

Chapter 10
Figures 10.1, 10.2, 10.4–10.8: Imagineering.

Chapter 11
Figures 11.1: Imagineering. Figures 11.4, 11.8, 11.11, 11.15, 11.16: Kevin Somerville. Figures 11.2, 11.3, 11.5, 11.6, 11.10, 11.14, 11.17: Kevin Somerville/Imagineering. Figures 11.7, 11.13, 11.18, 11.19: Imagineering. Figure 11.12: Kevin Somerville/John Gibb. Figure 11.20: Kevin Somerville/John Gibb/Imagineering.

Chapter 12
Figures 12.1, 12.2: John Gibb/Molly Borman/Kevin Somerville/Imagineering. Figures 12.3, 12.7, 12.8–12.12: Imagineering. Figure 12.4: Kevin Somerville/Imagineering. Figure 12.5: Steve Oh. Figure 12.6: Steve Oh/Kevin Somerville/Sharon Ellis. Figure 12.13: Kevin Somerville.

Chapter 13
Figure 13.1: Molly Borman/Kevin Somerville. Figure 13.2: Kevin Somerville. Figure 13.3: Kevin Somerville/John Gibb. Figures 13.4–13.8: Imagineering. Figure 13.9: Imagineering/Kevin Somerville.

Chapter 14
Figure 14.1: Nadine Sokol, Kevin Somerville/Steve Oh. Figure 14.2: Imagineering/Kevin Somerville. Figure 14.3: DNA Illustrations/Nadine Sokol. Figures 14.4, 14.8, 14.10–14.19: Imagineering. Figure 14.5: Steve Oh/Kevin Somerville. Figures 14.6, 14.9: Kevin Somerville. Figure 14.7: Steve Oh.

Chapter 15
Figure 15.1: Kevin Somerville. Figure 15.2: Steve Oh/Imagineering. Figures 15.3, 15.5–15.7, 15.13: Kevin Somerville/Imagineering. Figures 15.4, 15.8–15.12, 15.14, 15.15: Imagineering.

Chapter 16
Figures 16.1–16.3, 16.6, 16.8, 16.9, 16.17: Kevin Somerville/Imagineering. Figures 16.4, 16.7, 16.13, 16.14, 16.16: Kevin Somerville. Figures 16.5, 16.10–16.12, 16.18: Imagineering.

Chapter 1

Page 1: Robert Clark/NG Image Collection; page 3: Rubberball Productions/Getty Images; page 7 (center left): Joel Sartore/NG Image Collection; page 7 (top left): Masterfile; page 7 (top right): Bruce Dale/NG Image Collection; page 7 (bottom left): Katrina Wittkamp/Getty Images, Inc.; page 7 (bottom right): Juergen Berger/Photo Researchers, Inc.; page 7 (center right): Richard Lord/The Image Works; page 12 (top): Dissection Shawn Miller; Photograph Mark Nielsen; page 12 (center): Dissection Shawn Miller; Photograph Mark Nielsen; page 12 (bottom): Dissection Shawn Miller; Photograph Mark Nielsen; page 15 (top): Biophoto Associates/Photo Researchers; page 15 (center): Simon Fraser/Photo Researchers, Inc.; page 15 (bottom): ©Camal/Phototake; page 15 (just below top): Scott Camazine/Photo Researchers, Inc. page 15 (just above bottom photo): Courtesy Andrew Joseph Tortora and Damaris Soler; page 17: Salisbury/Photo Researchers, Inc..

Chapter 2

Pages 20–21: James P. Blair/NG Image Collection; page 25 (center right): Heather Perry/NG Image Collection; page 25 (bottom): SIU/Visuals Unlimited; page 34: Stacy Gold/NG Image Collection; page 37 (top): Becky Hale/NG Image Collection; page 37 (bottom left): Don Farrall/Photodisc/Getty Images; page 37 (bottom right): A.T. Willett/Alamy; page 40 (top right): D.W. Fawcett/Photo Researchers, Inc.; page 40 (center left): Courtesy Michael Ross, University of Florida; page 40 (center left): Courtesy Michael Ross, University of Florida; page 46: Catherine Karnow/NG Image Collection.

Chapter 3

Pages 48–49: Jonathan Blair/NG Image Collection; page 58: Andy Washnik; page 61: David Phillips/Photo Researchers, Inc.; pages 68–69: Courtesy Michael Ross, University of Florida; page 74 (top): Mark Nielsen; page 74 (center): Mark Nielsen; page 74 (bottom): Mark Nielsen; page 75 (top): Mark Nielsen; page 75 (center): Mark Nielsen; page 75 (bottom): Mark Nielsen; page 77 (top): Mark Nielsen; page 77 (center): Mark Nielsen; page 77 (bottom): Mark Nielsen; page 79 (top): Mark Nielsen; page 79 (center): Mark Nielsen; page 79 (bottom): Mark Nielsen; page 80 (top): Courtesy Michael Ross, University of Florida; page 80 (center): Mark Nielsen; page 80 (center inset): Mark Nielsen; page 80 (bottom): Mark Nielsen; page 83: Jim Richardson/NG Image Collection; page 87: Roy Toft/NG Image Collection; page 88: Courtesy Michael Ross, University of Florida.

Chapter 4

Pages 90–91: Dawn Kish/NG Image Collection; page 103 (top): Sheila Terry/Science Photo Library/Photo Researchers, Inc.; page 103 (center): St. Stephen s Hospital/Science Photo Library/Photo Researchers, Inc.; page 103 (bottom): St. Stephen s Hospital/Science Photo Library/Photo Researchers, Inc.; page 104 (top right): Jean Claude Revy/Phototake; page 104 (center): Jean Claude Revy/Phototake; page 106 (left): Alain Dex/Photo Researchers, Inc.; page 106 (right): Biophoto Associates/Photo Researchers, Inc.; page 108: Biophoto Associates/Photo Researchers, Inc.; page 109: Digital Vision/Getty Images; page 111 (top): St. Stephen s Hospital/Science Photo Library/Photo Researchers, Inc.; page 111 (bottom): Alain Dex/Photo Researchers, Inc..

Chapter 5

Pages 112–113: NASA/NG Image Collection; page 116: John Burbidge/Photo Researchers; page 120 (top left): Courtesy Dr. Brent Layton; page 120 (top right): Courtesy Dr. Brent Layton; page 120 (bottom left): Courtesy Per Amundson, M.D.; page 120 (bottom right): Courtesy Dr. Brent Layton; pages 138–139: John Wilson White; page 143 (left): Grant Halverson/Getty Images, Inc.; page 143 (right): CNRI/Photo Researchers, Inc; page 144 (left): P. Motta/Photo Researchers, Inc.; page 144 (right): P. Motta/Photo Researchers, Inc.; page 145: CNRI/Photo Researchers, Inc; page 146: John Burbidge/Photo Researchers; page 149: Visuals Unlimited; page 150: John Wilson White.

Chapter 6

Pages 152–153: Jodi Cobb/NG Image Collection; page 154 (center left): John Wiley & Sons; page 154 (center right): Courtesy Michael Ross, University of Florida; page 154 (bottom right inset): Mark Nielsen; page 155 (bottom left): Mark Nielsen; page 155 (center right): Mark Nielsen; page 158: Courtesy Hiroyouki Sasaki, Yale E. Goldman and Clara Franzini–Armstrong; page 169 (center): PhotoDisc, Inc./Getty Images; page 169 (right): Derek E. Rothchild/Getty Images, Inc.; page 185 (center): Mark Nielsen; page 188: Photodisc Blue/Getty Images.

Chapter 7

Pages 190–191: ©AP/Wide World Photos; page 204: Joel Sartore/NG Image Collection; page 223: Bill Varie/Alamy.

Chapter 8

Pages 226–227: Masterfile; page 228 (center left): Plush Studios/Blend Images/Getty Images; page 228 (bottom left): Stephen Alvarez/NG Image Collection; page 228 (center right): Masterfile; page 228 (bottom right): Stacy Gold/NG Image Collection; page 229 (center left): Vladimir Pcholkin/Getty Images; page 229 (center right): Antonia Reeve/Photo Researchers, Inc; page 229 (bottom left): Somos/Veer/Getty Images; page 229 (bottom right): Blend Images/Getty Images; page 231 (top right), page 231 (center right), page 231 (center left), page 235: Masterfile; page 238: Paul Parker/Photo Researchers, Inc; page 249: Keen Press/NG Image Collection.

Chapter 9

Pages 252–253: Mike Powell/Getty Images, Inc.; page 260 (top left): From New England Journal of Medicine, February 18, 1999, vol. 340, No. 7, page 524. Photo provided courtesy of Robert Gagel, Department of Internal Medicine, University of Texas M.D. Anderson Cancer Center, Houston Texas; page 260 (bottom): Lester Bergman/The Bergman Collection; page 260 (top right): DIBYANGSHU SARKAR/AFP/Getty Images, Inc.; page 262: Courtesy Michael Ross, University of Florida; page 264 (left): Lester V. Bergman/©Corbis; page 264 (right): Lester Bergman/

The Bergman Collection; page 266: Courtesy Michael Ross, University of Florida; page 268: Tim Laman/NG Image Collection; page 269: Mark Nielsen; page 272: Biophoto Associates/Photo Researchers, Inc.; page 275: Courtesy Michael Ross, University of Florida; page 283: Barcroft Images/Fame Pictures; page 285 (bottom left): Lester Bergman/The Bergman Collection; page 285 (top right): Lester Bergman/The Bergman Collection.

Chapter 10

Pages 286–287: Look at Sciences/Phototake; page 289: Timothey Kosachev/iStockphoto; page 290 (top) Mark Nielsen; page 290 (bottom): Courtesy Michael Ross, University of Florida; page 290 (bottom): Courtesy Michael Ross, University of Florida; page 290 (bottom): Courtesy Michael Ross, University of Florida; page 290 (bottom): Courtesy Michael Ross, University of Florida; page 290 (bottom): Courtesy Michael Ross, University of Florida; page 295 (bottom): Dennis Kunkel/Phototake; page 296 (top): Jean Claude Revy/Phototake; page 296 (bottom): Jean Claude Revy/Phototake; page 298: Tim Boyle/Getty Images, Inc.; page 302 (center): Mark Nielsen; page 303 (center): Jean Claude Revy/Phototake; page 303 (bottom): Jean Claude Revy/Phototake; page 305: Yoav Levy/Phototake.

Chapter 11

Pages 308–309: UPI Photo/Jeremy L. Grisham/Landov LLC; page 315: Antonia Reeve/Photo Researchers, Inc.; page 318 (center): Image Source/Getty Images, Inc.; page 318 (top): Image Source/Getty Images, Inc.; page 319 (left): Vu/Cabisco/Visuals Unlimited; page 319 (right): W. Ober/Visuals Unlimited; page 321: Dennis Strete; page 325: Courtesy Michael Ross, University of Florida; page 335: U.S.Marine Corps photo by Lance Cpl. Ben J. Flores; page 336: Jim Cummins/Taxi/Getty Images; page 340: Charles Thatcher/Getty Images.

Chapter 12

Pages 342–343: Taylor S. Kennedy/NG Image Collection; page 349: Mark Nielsen; page 353: Lynn Johnson/NG Image Collection; page 361 (center): Chris Whittle/Splash News/NewsCom; page 361 (right): ISM/Phototake; page 363: Karen Kasmauski/NG Image Collection; page 367: National Cancer Insitute.

Chapter 13

Pages 370–371: Barry Bishop/NG Image Collection; page 381: Steve Raymer/NG Image Collection; page 390: Michael Schmitt/Getty Images, Inc.; page 391 (top left): James L. Stanfield/NG Image Collection; page 391 (top right): David De Lossy/Getty Images; page 395: Jay Dickman/NG Image Collection.

Chapter 14

Pages 398–399: Richard Nowitz/NG Image Collection; page 408: Dissection Shawn Miller, Photograph Mark Nielsen; page 423 (top): Neustockimages/iStockphoto; page 423 (center): Judith Glick/Phototake; page 426: Susan Van Etten/PhotoEdit; page 427: Masterfile; page 429: Masterfile; page 432: Masterfile; page 434: Donna Day/Getty Images, Inc..

Chapter 15

Pages 436–437: Tobias Titz/Getty Images, Inc.; page 439: Dissection Shawn Miller, Photograph Mark Nielsen; page 442: Dennis Strete; page 462: BSIP/Phototake; page 465: AJ Photo//Photo Researchers, Inc..

Chapter 16

Pages 468–469: Pablo Corral Vega/NG Image Collection; page page 486 (diaphragm): Photodisc Blue/Getty Images; page 486 (cervical cap): Gary Parker/Photo Researchers, Inc; page 486 (condoms): BSIP/Photo Researchers, Inc. ; page 486 (spermicides): Aaron Haupt/Photo Researchers, Inc.; page 486 (intrauterine devices): Saturn Stills//Photo Researchers, Inc.; page 486 (contraceptive patch): Gusto/Photo Researchers, Inc.; page 486 (birth control pills): Saturn Stills/Photo Researchers, Inc.; page 486 (vaginal ring): Phanie/Photo Researchers, Inc.; page 487 (top right): Paul Figura/Stone/Getty Images, Inc.; page 487 (center): Veronique Burger/Photo Researchers, Inc; page 494 (top left): Photo provided courtesy of Kohei Shiota, Congenital Anomaly Research Center, Kyoto University, Graduate School of Medicine; page 494 (top center): Courtesy National Museum of Health and Medicine, Armed Forces Institute of Pathology; page 494 (top right): Courtesy National Museum of Health and Medicine, Armed Forces Institute of Pathology; page 494 (bottom left): Courtesy National Museum of Health and Medicine, Armed Forces Institute of Pathology; page 494 (bottom right): Courtesy National Museum of Health and Medicine, Armed Forces Institute of Pathology; page 495 (top left): Last Refuge, Ltd/Phototake; page 495 (top center): Photo provided courtesy of Kohei Shiota, Congenital Anomaly Research Center, Kyoto University, Graduate School of Medicine; page 495 (top right): Photo provided courtesy of Kohei Shiota, Congenital Anomaly Research Center, Kyoto University, Graduate School of Medicine; page 495 (bottom): Mauro Fermariello/Photo Researchers, Inc. ; page 501: Jason Edwards/NG Image Collection; page 502 (far left): li chaoshu/iStockphoto; page 502 (left): Vikram Raghuvanshi Photography/iStockphoto; page 502 (center): Ana Abejon/iStockphoto; page 502 (right): Joey Boylan/iStockphoto; page 503 (left): Jeffrey Shanes/iStockphoto; page 503 (center): Anne Clark/iStockphoto; page 503 (right): Yin Yang/iStockphoto; page 505: BSIP/Photo Researchers, Inc.; page 507: Joey Boylan/iStockphoto; page 508: David M. Phillips/Photo Researchers, Inc..

Index

Autonomic nervous system
  enteric nervous system and, 403
  heart rate and, 319
  involuntary functions and, 193
  parasympathetic division of, 212–213, 212f, 213t
  somatic nervous system compared, 208, 208f, 210t
  sympathetic division of, 210, 211f
  two-neuron pathway and, 208
Autonomic sensory receptors, 193
Autoregulation, 324
Autorhythmic cells, 167
Axial skeleton, 122, 122f
Axillary hair, 97
Axillary veins, 328
Axon terminal, 199
Axons, 195

B

B cells, 352, 353f, 354
B lymphocytes, 354
Ball-and-socket joints, 141, 141f
Baroreceptor reflex, 334
Baroreceptors, 334
Barrier methods, contraception, 485
Basal cell carcinoma, 105
Basal cells, 232
Basal metabolic rate (BMR), 427–430, 427f
Basal surface, 72
Basement membrane, 72, 326f
Bases, 32, 32f
Basilar membrane, 243
Basilic veins, 328
Basophils, 289f, 290
Behavior therapy, obesity and, 425
Belly, skeletal muscles and, 168
Biceps brachii, 180t
Biceps femoris, 183t
Biconcave shape, 288
Bile pigment, 410
Bile salts, 410
Bilirubin, 293f, 449t
Biliverdin, 293f
Biochemical functions, 3
Biofeedback, 231
Bipolar neurons, 195
Birth, 498
Birth control pills, 485, 487
Blackheads, 98
Blastocyst, 488, 489f
Blastocyst cavity, 488
Blastomeres, 488
Blood
  artificial blood, 298
  capillaries, 95
  chemical composition testing and, 300
  clotting and, 294–295
  common medical blood tests, 300, 301f
  compatibility, 296–298

components of, 289f
disease diagnosis and, 300
flow through heart, 314–315, 314f
formed elements, 288–289
groups and types, 296–298, 297f
hemolytic disease and, 298
major functions of, 288
obtaining for testing, 299
oxygen and carbon dioxide transport in, 383–384, 384f–385f
plasma, 288
Rh incompatibility and, 298
transfusions and, 296–299
Blood-brain barrier, 196
Blood cells
  origin and development of, 291–293, 291f
  recycling of, 292
  red bone marrow stem cells and, 291–292
Blood colloidal osmotic pressure, 446f
Blood composition, kidneys and, 439
Blood pressure
  blood flow and, 332, 333f
  exercise and, 336–337, 336f
  factors influencing, 332–334, 334f
  hormones and, 336
  kidneys and, 438
  regulation of, 334–337
Blood smears, 290f
Blood vessels
  arteries and arterioles, 321–324
  capillaries, 324–325
  of heart, 313f
  hepatic portal circulations, 330–331
  types of, 320–321, 320f
  veins and venules, 326–330
Blood volume
  blood pressure and, 333
  increased, 457f
  kidneys and, 438
Body cavities, 12–14, 13f
Body fluids, 31–34
  electrolyte balance and, 454–455, 454f
  renin-angiotensin-aldosterone system and, 456–458, 457f
  water balance, 455–456, 455f
  water distribution, 453, 453f
Body hair, 97
Body mass index (BMI), 424, 425f
Body systems
  structural organization and, 2f–3f
  vital functions and, 3–6, 4f–5f
Body temperature, 319, 427–430, 427f
Bolus, 404
Bone
  bone mass, 144, 144f
  bone tissue, 78t
  cranial bones, 123, 124f–125f
  facial bones, 123–126
  fractures and, 120–121, 120f, 121f
  healing process and, 121f

microscopic structure of, 116–117, 116f
ossification and, 117, 118f
remodeling and, 119, 121f
structure and types of, 114–117
Bony callus formation, 121f
Brachial veins, 328
Brachialis, 180t
Brachiocephalic trunk, 322
Brachiocephalic veins, 328
Brachioradialis, 180t
Brain
  ascending and descending pathways through, 207, 207f
  blood pressure and, 334
  breathing regulation and, 386–389, 387f
  composition of, 202–205, 202f–203f
  cranial nerves and, 214–215
  maps of body within, 204–205, 205f
Brainstem, 202, 386–387, 387f
Breastfeeding, 500, 501f
Breathing
  control centers, 388, 388f
  muscles and, 175, 175f, 175t, 377–378
  patterns of, 380, 380t
  process of, 374–380, 376f–377f
  regulation of, 386–387, 387f
Breech position, 498
Bronchi, 373
Bronchial artery, 322
Bronchioles, 373, 375f
Buccinator muscle, 173t
Buffers, 33
Bulbourethral gland, 470
Bulbs, 96
Burns, skin, 102–103, 103f
Bursae, 136

C

Calcaneus, 133
Calcification, 227
Calcitonin, 262, 265f
Calcitriol, 119, 264
Calcium homeostasis, 102, 119, 119f
Calculi, urinary, 458
Calories, 416
Calyces, 440
Canaliculi, 117
Cancer, skin, 105–106, 106f
Capacitation, 485
Capillaries, 320f, 324–325, 325f
Capillary membranes, 382
Capsular hydrostatic pressure, 446f
Carbohydrates, 34–36, 35f, 418
Carbon dioxide conversion, 384
Carbon dioxide transportation, 384, 384f–385f
Carbonic acid-bicarbonate buffer system, 33, 458–460, 459f, 460f
Carbonic anhydrase, 384, 447
Carboxypeptidase, 410

H

Hair, 96–98, 97f
Hair cells, 243
Hair color, 97
Hair follicles, 96
Hair matrix, 96
Hair root plexuses, 96, 229
Hamstrings, 183t
Hands, 181, 181f, 181t
Haploid cells, 71
Haversian canal, 117
Head and neck muscles, 173–174, 173f, 173t, 174f
Healing process, bones and, 121f
Hearing, 241–245, 243f. *See also* Ear
Heart
    blood flow through, 314–315, 314f
    cardiac cycle and, 316–317, 317f
    cardiac output and, 318–319, 318f
    electrical signals and, 315–316, 315f
    endocrine cells and, 274f
    heart block, 316
    heart rate, 318–319
    location of, 311, 311f
    major blood vessels of, 313f
    major parts of, 313f
    microscopic structure of, 312f
    normal and obstructed arteries, 319, 319f
    valves and, 312
Heat loss, 427–430, 427f
Heat production, 427–430, 427f
Hematocrit, 288
Hematocrit tests, 301t
Hematoma, fracture, 121f
Hematopoiesis, 291
Hemidesmosomes, 72
Hemodialysis, 462
Hemoglobin, 95, 290, 290f, 382–384
Hemoglobin-based oxygen carriers (HBOCs), 298
Hemolysis, 296
Hemolytic anemia, 300
Hemolytic disease of newborn, 298
Hemopoiesis, 291
Hemorrhage, 294
Hemorrhagic anemia, 300
Hemostasis, 294–295, 295f
Hepatic portal circulations, 330–331, 330f–331f
Hepatic portal vein, 330–331
Hiccuping, 380t, 388
High-density lipoproteins, 421
Highly active antiretroviral therapy (HAART), 362
Hinge joints, 140, 140f
Hirsutism, 97
Histamine, 353f, 360
Histology, 71
Homeostasis

aging and disease and, 10
calcium, 119, 119f
defined, 8
Homeostatic balance, 8–10
Homogeneous liquid, 288
Homologous chromosomes, 70
Homunculi, 204
Hormones
    blood pressure and, 336
    digestion control and, 412, 412t
    endocrine system and, 8, 255
    heart rate and, 319
    hormonal contraception, 485
    hormonal control in man, 475, 475f
    hormone receptors, 255
    kidneys and, 439
    metabolic activities and, 420–421
    pregnancy and, 496
    receptor-mediated endocytosis and, 63
Human chorionic gonadotropin (hCG), 275f, 490, 496
Human chorionic somatomammotropin (hCS), 496
Human immunodeficiency virus (HIV), 362–364, 363f
Human placental lactogen (hPL), 496
Humerus, 132
Hydrochloric acid, 407f
Hydrogen bonds, 29, 29f
Hydrogen ions, 32
Hydrolysis, 32
Hydrophilic substances, 31
Hydrophobic molecules, 31
Hydrostatic pressure, 324
Hydroxide ions, 32
Hymen, 476, 478f
Hyoid bone, 126f, 127
Hyperglycemia, 267
Hypertension, 337
Hyperthyroidism, 264
Hypertonic solutions, 61, 61f
Hypoglycemia, 267
Hypothalamic hormones, 259–261, 259t
Hypothalamus, 263–264, 264f, 271, 271f, 272–273
Hypothermia, 430
Hypothyroidism, 264
Hypotonic solutions, 61, 61f
Hypovolemic shock, 335

I

Ibuprofen, 430
Ileum, 408
Immune response
    adaptive immunity, 350, 354–359, 354f–355f, 360, 360t
    allergies and, 360
    antibody-mediated immunity, 354, 358–359, 358f
    antigens and, 356, 356f

autoimmune diseases, 361
cell-mediated immunity, 356, 357f
first line of, 350–352, 350f–351f
immunoglobulins and, 359, 359t
immunological memory, 359, 359f
innate immunity, 350, 351–353
second line of, 352–353, 352f–353f
Immunoglobulins, 359, 359t
Immunological memory, 359, 359f
Impacted fractures, 120, 120f
Implantation, 488, 489f
Impulse transmission, neuromuscular junction and, 160, 160f
Incontinence, 452
Inferior mesenteric artery, 322
Inferior nasal conchae, 126
Inferior oblique muscle, 174t
Inferior phrenic arteries, 322
Inferior vena cava, 328
Inflammation, 353, 353f
Inflation reflex, breathing and, 388
Infraspinatus muscle, 179t
Infundibulum, 476, 477f
Ingestion, food, 404, 404f
Inhalation, 377–378, 377f
Inhibin, 273, 475, 477f
Innate immunity, 350, 351–353
Inner cell mass, 488
Inner ear, 242f
Inorganic compounds, 32
Inspiratory area, 386, 387f
Inspiratory reserve volume, 378
Insulin, 267, 267f
Integral proteins, 52
Integrative function, 193
Integumentary system
    age-related changes and, 105–106, 105f
    components of, 92–93, 92f–93f
    functions of, 4f
    glands of the skin, 98, 98f
    hair, 96–98, 97f
    nails, 99, 99f
    skin, 100–105
Intercalated discs, 167, 167f
Intercostal nerves, 217
Interferons, 352, 352f
Intermediate filaments, 53, 165
Internal intercostal muscles, 377f
Internal intercostals, 175t
Internal jugular veins, 328
Internal oblique muscle, 177t
Internal respiration, 375, 382–385, 383f
Internal urethral sphincter, 450f
Interneurons, 195
Interosseous membranes, 136, 137f
Interphase, 67, 68f
Interstitial fluid, 57
Intracellular fluid, 453
Intramembranous ossification, 117
Intrauterine devices, 485

Ionic bonds, 27, 27f
Ions, 27, 27f, 319
Iris, 236
Iron-binding proteins, 352, 352f
Iron-deficiency anemia, 300
Irregular bones, 114
Islets of Langerhans, 266–267
Isometric contractions, 164
Isotonic contractions, 164
Isotonic solutions, 61, 61f
Isotopes, 25, 25f

J

Jejunum, 408
Joint movements, 138–142, 138f–139f
Joints, 136, 137f
Juxtaglomerular cells, 457f
Juxtamedullary nephrons, 442

K

Keratin, 94, 100
Keratinization, 95
Keratinocytes, 100
Ketone bodies in urine, 449t
Ketones, 420
Kidney stones, 458
Kidneys
    blood composition and pH and, 439
    blood volume regulation and, 438
    as complex filter, 440–443
    electrolyte balance and, 454–455, 454f
    functions of, 438–439
    hormone production and, 439
    internal structure of, 440f–441f
    pH buffer systems and, 458–459, 459f
    pH change compensation and, 460,
        460f–461f
    renal failure and dialysis, 462
    renin-angiotensin-aldosterone system
        and, 456–458, 457f
    stone formation and, 458
    waste excretion and, 439
    water balance and, 455–456, 455f
    water distribution and, 453, 453f
Kinases, 39
Kinetic energy, 22
Knee
    anatomy of, 142, 142f
    injuries of, 143, 143f
    replacement of, 145, 145f
Krebs cycle, 419, 419f
Kyphosis, 127

L

Labia, 477f
Labia majora, 478
Labia minora, 478
Labor, 498–500, 499f
Lacrimal bones, 126

Lacrimal fluid, 235
Lacrimal glands, 235
Lactation, 500, 501f
Lacteals, 411
Lactic acid, 162f, 163
Lactose, 35f
Lacunae, 117
Lambdoid suture, 126
Laminae, 127
Large intestine, 401f, 412–414, 413f, 415t
Larynx, 373
Latent period, muscle twitch and, 163, 163f
Lateral rectus muscle, 174t
Lateral surface, 72
Laughing, 380t
Left common carotid artery, 322
Leg muscles, 184, 184f, 184t
Leptin, 275f, 424
Leukemia, 300
Leukocytes, 289f, 290, 449t
Leukocytosis, 292
Leukotrienes, 275f
Levator palpebrae superioris muscle, 174t
Levator scapulae, 176t
Leydig cells, 472, 472f
Life processes, 6, 7f
Ligaments, 136
Light, vision and, 238–239, 238f–239f
Limbic system, breathing and, 388
Lipid bilayer, 52
Lipids, 36, 37f, 420–421, 421f
Lipogenesis, 420
Lipolysis, 420
Lipoproteins, 420
Liquids, 22
Liver, 401f, 410, 410f, 415t
Lobes, 202
Lobules, 472f, 478
Long bones, 114–116, 115f
Longitudinal arch, 133
Loose connective tissue, 78t
Lordosis, 127
Low-density lipoproteins, 421
Lower body, 132–133, 134f–135f
Lower respiratory tract, 373–374, 373f
Lumbar vertebrae, 127
Lumen, 321
Lungs
    anatomy of, 373–374
    capacity of, 378, 391
    pH change compensation and, 460,
        460f–461f
    smoking and, 391–392, 391f
    ventilation and, 377–378
    volumes and, 378, 379f
Lunula, 99
Lupus, 361
Luteinizing hormone (LH), 273, 475
Lymph, 57, 344, 347f
Lymph circulation, 347, 347f

Lymph nodes, 344, 347f, 348f
Lymphatic capillaries, 344, 347f, 348f
Lymphatic ducts, 347f, 348f
Lymphatic nodules, 344
Lymphatic system. *See also* Immune response
    circulation of lymph, 347, 347f
    functions of, 5f, 344
    immune reactions and, 348–349
    primary organs and tissues, 346f
    secondary organs and tissues, 346f
    structures and organs, 344–347, 345f
Lymphatic vessels, 344, 347f, 348f
Lymphocytes, 289f, 290, 348
Lymphoid stem cells, 290f, 291
Lysosomes, 55, 55f
Lysozyme, 235, 409f

M

Magnetic resonance imaging (MRI), 15f
Major histocompatibility complex (MHC),
    356
Male birth control pills, 487
Male-pattern baldness, 97
Male reproductive system
    aging and, 502–503, 502f–503f
    hormonal control, 475, 475f
    organs of, 470–472, 471f
    path of sperm, 474, 474f
    sperm production and, 472–475
Male sexual characteristics, 475
Malignant melanoma, 106, 106f
Maltose, 35f
Mammary glands, 478, 479f
Mandible, 123
Manubrium, 131
Mass number, 24
Masseter muscle, 173t
Mastoid process, 123
Matter
    chemical elements and, 23, 23f
    difference from energy, 22–23, 22f
Mature (graafian) follicle, 480
Maxillae, 123
Maxillary sinus, 123
Mean arterial pressure, 332
Mechanical energy, 23
Medial rectus muscle, 174t
Median antebrachial veins, 328
Median cubital veins, 328
Mediastinum, 311
Medical imaging, 15f
Medulla, 269
Medulla oblongata, 206
Megakaryocytes, 290
Meiosis, 67–71, 70f, 71t, 473
Meissner corpuscles, 95, 102, 229
Melanin, 95–96
Melanocytes, 95
Melanoma, malignant, 106, 106f
Melatonin, 274

Membranes, 82–83
Menarche, 502
Meninges, 200, 200f
Menopause, 144, 503
Menstrual phase, reproductive cycle, 482, 482f
Menstruation, 482, 482f
Mental foramina, 126
Merkel (tactile) discs, 229
Mesenchyme, 117
Mesoderm, 490–491
Mesothelium, 83
Messenger RNA (mRNA), 65
Metabolic activities, hormones and chemical levels and, 420–421
Metabolism, 6
Metabolization, nutrients and, 418–422
Metacarpals, 132
Metaphase, 67, 69f
Metastasization, 105
Metatarsals, 133
Micelles, 410
Microbes in urine, 449t
Microbial infection, 102
Microfilaments, 53
Microglia, 196
Microtubules, 53
Microvilli, 53
Micturition reflex, 452, 452f
Milk ducts, 479f
Mineralocorticoids, 270–271
Minerals, 416
Minoxidil (Rogaine), 97
Minute ventilation (MV), 379
Mitochondria, 56, 56f
Mitochondrial crista, 56
Mitochondrial matrix, 56
Mitosis, 67–71, 68f–69f, 71t
Modified respiratory movements, 380, 380t
Molecular formula, 28
Molecules, 3, 28–29, 28f
Moles, 95, 106, 106f
Monocytes, 289f, 290
Monosaccharides, 34, 35f
Monounsaturated fats, 36
Mons pubis, 477f
Morula, 488, 489f
Motor function, 193
Motor neurons, 195
Motor unit, 164
Mouth
    anatomy of, 400f
    digestion and, 404, 404f, 415t
    innate immunity and, 350f
Movement, 6, 7f
Mucosa, 82, 402, 403f
Mucous membrane, 82
Multi-unit smooth muscle tissue, 166, 166f
Multiple sclerosis, 361
Multipolar neurons, 195

Muscle contraction, sliding filament theory and, 158–159, 158f
Muscle tone, 164
Muscular system
    cardiac muscle tissue, 155, 167, 167f
    functions of, 4f
    movement and, 156–165
    muscle fibers, 81, 157–158, 165
    muscle twitch, 163–165, 163f, 164f
    muscular tissue, 72, 80f, 81, 154–155, 154f
    nerve signals, 160–163
    skeletal muscle tissue, 154
    skeletal muscles, 168–184
    sliding filament theory and, 158–159, 158f
    smooth muscle tissue, 155, 165–166
Muscularis, 402, 403f
Mutation, 42
Myelin, 195
Myeloid stem cells, 290f, 291
Myocardial infarction, 319
Myocardial ischemia, 319
Myocardium, 312
Myofibrils, 156f–157f
Myofilaments, 158
Myoglobin, 162f, 163
Myometrium, 476, 477f
Myosin, 157f, 169
MyPyramid nutritional guidelines, 416, 417f
Myxedema, 264

N

Nails, 99, 99f
Nasal bones, 123
Nasal epithelium, 373
Naturally acquired active immunity, 360
Naturally acquired passive immunity, 360
Neck and back muscles, 178, 178f, 178t
Negative feedback system, 8–9, 9f
Nephritis, 462
Nephron loop, 442, 448–449, 448f
Nephrons, 442, 442f–443f, 444–445, 445f
Nerve cells, 81, 81f
Nerve endings, 95
Nerve fibers, 195
Nerve signals, skeletal muscle contraction and, 160–163
Nervous system
    action potentials and, 197, 197f
    autonomic nervous system, 208–213
    central nervous system, 200–208
    functions of, 4f, 192–193, 193f
    neuroglia and, 196, 196f
    neuron communication, 100f
    neurons and, 192–195, 194f
    peripheral nervous system, 214–219
    somatic nervous system, 208, 208f, 210t
    structural overview of, 192f
    synapses, 198–199, 199f
Nervous tissue, 72, 81
Net filtration pressure, 446, 446f

Neural tunic (retina), 236
Neurogenic shock, 335
Neuroglia, 81, 81f, 192, 196, 196f
Neuromuscular junction, impulse transmission and, 160, 160f
Neurons, 81, 81f, 192–195, 194f
Neurosecretory cells, 259
Neurotransmitters, 199
Neurulation, 491
Neutrons, 24
Neutrophils, 289f, 290
Nevus, 95
Nicotinamide adenine dinucleotide, 418
Nipples, 479f
Nitrogenous bases, 41
Nociceptors, 102, 230
Non-endocrine glands, 274, 274f
Non-polar covalent bonds, 29
Non-polar molecules, 29
Nonsteroid hormones, 255–257, 256f
Norepinephrine, 272
Nose, 350f, 373
Notochord, 491
Nuclear envelope, 57
Nuclear pores, 57
Nuclei, 195
Nucleic acids, 40–42, 40f–41f
Nucleolus, 57
Nucleotides, 40, 41f
Nucleus, 24, 50, 56–57, 56f
Nutrients
    carbohydrate conversion and, 418
    metabolization and, 418–422
    nutritional guidelines, 416, 417f
    small intestine absorption of, 410–411, 411f
Nutritional guidelines, 416, 417f

O

Obesity, 424–425
Oblique plane, 12, 12f
Obstructive pulmonary diseases, 379f, 391
Obturator foramina, 132
Occipital bone, 123
Occipital condyles, 123
Occipitofrontalis muscle, 173t
Octet rule, 26
Odorants, 232
Olfaction, 232
Olfactory epithelium, 232–233, 233f
Olfactory receptors, 232–233
Oligodendrocytes, 196
Oocytes, 476
Oogenesis, 476, 480–482, 481f
Oogonia, 480
Open fractures, 120, 120f
Opioid analgesics, 231
Optic chiasm, 240
Optic foramen, 123
Oral contraceptives, 485

Sympathetic division of autonomic nervous system, 210, 211f
Symphysis, 136, 137f
Symporters, 447
Synapses, 198–199, 199f
Synarthrosis, 136
Synchondrosis, 136, 137f
Syndesmosis, 136, 137f
Synergists, 169
Synovial fluid, 83, 136
Synovial joints, 136, 137f, 138f–139f, 140f–141f
Synovial membranes, 82, 83, 136
Synthesis reactions, 30f, 31
Syphilis, 486
Systemic lupus erythematosus (SLE), 361
Systole, 316

T

T cells, 352, 353f, 354–356
T lymphocytes, 354
T-tubule, 158
Talus, 133
Target cells, 255, 255f
Tarsals, 133
Tastants, 234
Taste pores, 234
Tears, 235
Teeth, 415t
Telomeres, 84, 84f
Telophase, 67, 68f
Temporal bones, 123
Temporal summation, 164
Temporalis muscle, 173t
Temporomandibular joints (TMJ), 123
Tendons, 136, 168
Tensor fasciae latae, 183t
Teres major muscle, 179t
Teres minor muscle, 179t
Terminal ganglia, 213
Terminator, 65
Tertiary structures, 38f
Testes, 470, 471f
Testosterone, 36, 272, 470, 472f, 475
Tetanus, 164, 164f
Tetrads, 70
Thalassemia, 300
Thermic effect of food, 424
Thick filaments, 157f, 158
Thick skin, 94
Thigh muscles, 183t
Thin filaments, 157f, 158
Thin skin, 94
Third-degree burns, 103f
Thoracic aorta, 321
Thoracic duct, 344, 347f
Thoracic vertebrae, 127
Thorax muscles, 175, 175f, 175t, 176, 176f, 176t

Threshold, 197
Thrombus, 294
Thymosin, 274f
Thymus, 274f, 344, 347f
Thyroid gland, 262–264, 262f
Thyroid hormone secretion, 263–264, 264f
Thyrotropin-releasing hormone (TRH), 259, 259t, 263
Thyroxine, 262
Tibia, 133
Tibial tuberosity, 133
Tibialis anterior, 184t
Tibialis posterior, 184t
Tidal volume, 378, 379f
Tight cell junctions, 72
Tissues
    aging and, 84
    connective tissue, 76–80
    cultures and, 83
    engineering and, 83
    epithelial tissue, 72–76
    hypoxia and, 292
    muscular tissue, 81
    nervous tissue, 81
    tissue level, 2f
    types of, 71–72
Toes, muscles that move, 184, 184f, 184t
Tongue, 234, 234f, 415t
Tonsils, 347f
Total lung capacity, 378
Trabeculae, 117
Trachea, 373
Trans fatty acids, 36
Transcellular route, reabsorption, 447
Transcription, 65, 65f
Transcutaneous electric nerve stimulator (TENS), 231
Transdermal patches, 102
Transduction, 236
Transfer RNA (tRNA), 67
Transferrin, 293f
Transitional cells, 73
Translation, 66f, 67
Transport proteins, 59
Transportation, blood and, 288
Transverse abdominis muscle, 177t
Transverse arch, 133
Transverse plane, 12, 12f
Transverse tubule, 158
Trapezius, 176t
Triceps brachii, 180t
Tricyclic antidepressants, 231
Triglycerides, 36, 37f
Triiodothyronine, 262
Tripeptides, 38
Trophoblast, 488
Tropomyosin, 157f, 169
Troponin, 157f, 169
True ribs, 131

Trypsin, 410
Tubal ligation, 485
Tubular reabsorption, 444–445
Tubular secretion, 444–445
Tunica externa, 326f
Tunica intima, 326f
Tunica media, 326f
Twitch, muscle, 163–165, 163f, 164f
Type 1 diabetes, 361, 422–424
Type 2 diabetes, 422–424
Type II alveolar cells, 381

U

Ulna, 132, 180, 180f, 180t
Ulnar veins, 328
Ultrasound, 15f
Umbilical cord, 491, 491f
Unipolar neurons, 195
Universal solvent, 31
Upper body, 130–131, 130f–131f
Upper limb muscles, 179, 179f, 179t
Upper respiratory tract, 372–374, 372f
Ureters, 450f
Urethra, 351f, 450f
Urinary bladder
    micturition reflex and, 452, 452f
    structures of, 450, 450f–451f
    temporary storage and, 450
    urethral structures and, 451, 451f
Urinary system. See also Kidneys
    abnormal constituents in urine, 449t
    body fluid composition and, 453–458
    components of, 438, 439f
    functions of, 5f
    glomerular filtration and, 446–447, 446f
    nephron loop, 448–449, 448f
    reabsorption and secretion, 447–448
    urine characteristics, 449t
    urine formation and, 444–445, 445f
Urinary tract infections, 451
Urination, 452
Urobilinogen, 293f, 449t
Uterine cavity, 476
Uterine tubes, 476, 477f
Uterus, 476, 477f
UV radiation, 105–106

V

Vaccination, 360
Vagina, 351f, 476, 477f
Vaginal pouch, 485
Valence shell, 25
Valves, heart, 312
Vas deferens, 470
Vascular resistance, blood pressure and, 333
Vascular spasm, 294, 295f
Vascular tunic, 236
Vasectomy, 485